COLLEGE PHYSICS

VOLUME II

[Waves, Vibrations & Sound ♦ Optics (Geometrical & Physical)]

Dr. A. B. Gupta

CELEBRATING 55 YEARS 1960-2015

Since 1960

[*For the Undergraduate (B.Sc./B.Tech.) students of Physics and Engineering*]

COLLEGE PHYSICS

VOLUME II

[Waves, Vibrations & Sound ◆ Optics (Geometrical & Physical)]

A. B. GUPTA, Ph. D.

Professor and Head (Retired)
Physics Unit and Dean,
Indian Statistical Institute, Kolkata.

BOOKS AND ALLIED (P) Ltd.

No. 1-E(1), 'Shubham Plaza' (1st Floor)
83/1 Beliaghata Main Road, Kolkata 700010 (India)
Tel : (033) 6535-3844 • (033) 6548-5530
e-mail : booksandallied1960@gmail.com • booksmktng@gmail.com

COLLEGE PHYSICS
(VOLUME II)

First Published : September, 2003
Second Edition (thoroughly revised) : 2007
Reprinted : 2010, 2013, 2014
Reprinted : 2017

Publisher : Arunabha Sen
BOOKS & ALLIED (P) LTD.
8/1 Chintamoni Das Lane, Kolkata 700009

Printer : EUREKA
91/A, Baithakkhana Road, Kolkata 700009

Typesetter : BOOKS & ALLIED (P) LTD.
8/1 Chintamoni Das Lane, Kolkata 700009

Project Team : Salil Das and Dinesh Bhunya

ISBN 81-87134-03-8

***Price* : ₹ 425.00**

Preface to the Second Edition

The new edition of a book is a kind of rebirth of the original baby in the sense that the baby grows up, the bulk increases, the physiognomy changes and new features develop. In presenting the new edition of 'COLLEGE PHYSICS II', I strongly feel the readers will find in it a completely revised, enlarged and upgraded version of the same baby with the basic nature and character retained unimpaired. With a view to modernisation, the book has been brought in step with the rapidly changing syllabus of Physics of different Indian Universities and Institutes. New formalism, newer type of problems, reorganisation in a number of areas, let alone the cosmetic brush up here and there, deserve special mention in ths context.

I ventured, in the previous edition, to draw attention of the students to some conceptual aspects and important information dressed under the head '*Notes*' strewn almost throughout the book. A small but interesting survey convinced me that most of the readers seldom take note of the '*Notes*', and skip over them presuming they are not of much importance. Undaunted, however, I have retained them in this edition as well, but changed the '*Notes*' into '*dark • balls*' in a final desperate attempt to attract the eye balls of the readers.

I shall be failing in my duty, if I do not recall, with a deep sense of gratitude, the response I received regarding the earlier edition of the work from an wider circle of the academic community of the country. The exhaustion of the previous edition in a rather short span of time is entirely an outcome of their interest in the work. Hope, the present edition would be able to satisfy them all the more. What else could be more rewarding to an author? I frankly do not know.

At our end, I have much pleasure in recording my indebtedness to Mr. Arunabha Sen, Supreme of Books & Allied (P) Ltd. and Mr. Salil Das of D.T.P. Unit for their zealous support and unfailing cooperation at all stages of production. To others of Books & Allied (P) Ltd., I am equally grateful for the homely atmosphere in which they used to receive me whenever I visited the Firm.

Good Friday, 2007

A.B. Gupta

Preface to the First Edition

The present work constitutes the second volume of the 'College Physics' series, written essentially on the basis of COSIP-syllabus as suggested by UGC. The volume includes *'Waves, Vibrations and Sound'* and *'Optics (Geometrical and Physical)'* and would cater to the need of degree students of Indian universities.

As in my other works on Physics, the endeavour has been to lay emphasis on the fundamentals and more depth of treatment. The subject matter has been developed with clarity and conciseness, and no compromise is made with rigour. The physical concepts and principles have been explained simply but adequately. The unnecessary mass of experimental materials has been curtailed to the minimum, keeping in view the time-frame for completion of the syllabus. To combat the ordeal of examination, special care has been taken to make the presentation in such a stepwise balanced fashion that a serious student would be able to face with confidence the questions of University examinations. To enrich the conceptual aspect, they should carefully go through the materials under *'Notes'*. The carrier of contents of higher physics is essentially mathematics and no pains have been spared to apply whatever mathematics is whenever needed.

The work is well-illustrated with suitable diagrams to enable the textual material make an imprint in the minds of young leaners. Problem-solving in Physics is of vital importance and is the touch-stone to test if the physical principles have been well-grasped and assimilated. Numerous varied problems, drawn mostly from university examination papers, have been incorporated at the end of each chapter. A large number of them has been fully worked out under *'Illustrated examples'* and many more under *'Problems'* for practice. The rich and exhaustive collection of questions under *'Review questions'* would expose the students to diverse aspects of the subject matter learnt. The modern style is to set short answer and objective questions and we have therefore incorporated a good number of such questions as well under *'Short answer and Objective questions'* at the end of each chapter. We have preferred, barring some exceptions, to the SI-system of units throughout the book.

I acknowledge with a deep sense of gratitude my indebtedness to authors of standard works I freely consulted. My thanks are also due to all concerned of Messers Books & Allied (P) Ltd., especially to its director Mr. Arunabha Sen for his kind personal interest at all stages of production.

I shall indeed be happy if this volume, like its companion volume, is also warmly received by the academic community. Any comment or criticism on the book will be thankfully acknowledged.

Rathajatra, 2003 **A.B. Gupta**

Contents

Part I
Waves, Vibrations and Sound

Part II
Optics (Geometrical and Physical)

Part I

Waves, Vibrations and Sound

Chapter 1

Simple Harmonic Motion

1.1 Introduction

A body is said to execute *periodic motion* if its motion is repeated at regular intervals (periodicity) of time. In such motions, the body moves on the *same path* over and over again and the time it takes to go once round the path is called the (*time*) *period* of the motion. The number of times the path is described in unit time is called the *frequency* of the motion. The time period T thus bears a reciprocal relation with frequency, $n : n = 1/T$ or $T = 1/n$.

A periodic motion may be *linear* when it is restricted to a straight line, e.g., the motion of a mass suspended by a spring or (approximately) the motion of a simple pendulum with small amplitude. Periodic motion may as well be *circular* as in the motion of the hands of the clock, or *elliptical* as is the motion of planets round the sun.

In periodic motion, if the body moves to and fro over the same path at regular intervals, it is termed **vibratory** or **oscillatory** motion, e.g., the motion of a pendulum. A uniform circular motion is *periodic* but *not oscillatory*. So also is the planetary motion in elliptic paths round the sun.

The simplest type of oscillatory motion is the **simple harmonic motion** or SHM—a class of very important motion in physics and will be discussed in this chapter in detail.

1.2 Simple harmonic motion : Characteristics

Definition—*When a body moves to and fro in a straight line such that the restoring force acting on it (and hence its acceleration) is always directed towards a certain fixed point on its path and is proportional to its distance from that point, the body is said to execute* **simple harmonic motion.**

An analysis of the definition of SHM indicates the following **characteristics** of it.

1. The motion is *periodic* and *linear.*
2. The motion is *always directed towards some fixed point* in its path.

3. The *force* acting on it (and hence also the *acceleration*) is *proportional to the displacement* i.e., the distance from the fixed point.

The function of the force is to bring back the body to its equilibrium position and hence the name *restoring force.*

Examples of SHMs are : The oscillation of a liquid column in a U-tube, the motion of a simple pendulum for small displacements, the vertical oscillations (small) of a mass carried by an elastic string, a cantilever loaded at the free end etc.

A particle in simple harmonic motion is called a *simple* or *linear harmonic oscillator.*

1.3 Motion of a linear harmonic oscillator

Let x be the displacement at any time t of a particle of mass m executing simple harmonic motion. For small displacements, the restoring force F acting on it to bring it back to the equilibrium position is proportional to $x \Rightarrow F = -sx$ where s, the *force constant* of proportionality, is called *spring constant* or *stiffness factor.* The negative sign indicates that F and s are oppositely directed. Equating the restoring force to the force of inertia,

$$m\frac{d^2x}{dt^2} = -sx \;\Rightarrow\; m\frac{d^2x}{dt^2} + sx = 0$$

$$\therefore \quad \boxed{\frac{d^2x}{dt^2} + \omega^2 x = 0} \qquad (1.3.1)$$

where $\omega^2 = s/m$.

This equation (1.3.1) is the *general* **differential equation** of motion of a *linear harmonic oscillator.*

Solutions of the equation—Let $x = e^{nt}$ be a trial solution. So, $dx/dt = ne^{nt}$ and $d^2x/dt^2 = n^2e^{nt} = n^2x$.

$\therefore$ From equation (1.3.1)

$$(n^2 + \omega^2)x = 0 \;\Rightarrow\; n^2 + \omega^2 = 0 \quad (\because\; x \neq 0)$$

$$\therefore \quad n = \pm j\omega, \quad \text{where } j = \sqrt{-1}.$$

$\therefore$ General solution of (1.3.1) is

$$\boxed{x = Ce^{j\omega t} + De^{-j\omega t}} \qquad (1.3.2)$$

where C and D are two complex constants.

Writing $C = C_1 + jC_2$, $D = D_1 + jD_2$ and using $e^{\pm j\theta} = \cos\theta \pm j\sin\theta$, we get, from the equation (1.3.2),

$$\begin{aligned} x = &(C_1 + D_1)\cos\omega t + (D_2 - C_2)\sin\omega t \\ &+ j\left[(C_2 + D_2)\cos\omega t + (C_1 - D_1)\sin\omega t\right] \end{aligned} \qquad (1.3.3)$$

Since x is *real*, the acceptable *physical solution* is either the real or the imaginary part of (1.3.3)

$$\therefore \quad x = a\cos\omega t + b\sin\omega t \tag{1.3.4}$$

where a, b are two *real* arbitrary constants.

Eq. (1.3.4) gives the *general physical solution* of the differential equation (1.3.1).

Let $a = A\sin\phi$ and $b = A\cos\phi \;\Rightarrow\; A = \sqrt{a^2+b^2}$ and $\tan\phi = a/b$. So, the equation (1.3.4) becomes

$$x = A\sin(\omega t + \phi) \tag{1.3.5}$$

Substituting $\phi = (\pi/2 + \theta)$, equation (1.3.5) changes into

$$x = A\cos(\omega t + \theta) \tag{1.3.6}$$

Each of (1.3.2), (1.3.4), (1.3.5) and (1.3.6) gives the general solution of the differential equation (1.3.1) : **exponential, sine, cosine** etc. forms. For any particular problem, one is to select the appropriate one from among them. In general, the solution taken is : $x = A\sin(\omega t + \phi)$.

The quantity A in (1.3.5) and (1.3.6) is the *maximum displacement* of the particle from the mean position and is known as the **amplitude** of the oscillation. If T be the *period*, then the displacement x at time t and $t+T$ must be the same. So, $\sin(\omega t+\phi) = \sin(2\pi + \omega t + \phi) = \sin[\omega(t+T)+\phi]$. Thus $\omega T = 2\pi$.

$$\therefore \quad \textit{Time period, } T = \frac{2\pi}{\omega}.$$

$$\textit{Frequency} \text{ of oscillation, } n = \frac{1}{T} = \frac{\omega}{2\pi}$$

The quantity $\omega(=2\pi n)$ is known as the **angular frequency**. The angle $(\omega t + \phi)$ in (1.3.5) or $(\omega t + \theta)$ in (1.3.6) is called the **phase angle** or **phase**. At $t = 0$, the phase angle is ϕ or θ. This *initial phase* is called the **epoch**.

• Simple harmonic motion is also called **sinusoidal** or **co-sinusoidal motion,** since an SHM is represented by the relations

$$x = A\sin(\omega t + \phi)$$
$$\text{or,} \quad x = A\cos(\omega t + \phi)$$

with the usual meanings of the symbols.

1.3.1 Evaluation of the constants : Different cases

Let us take the general solution of the differential equation (1.3.1)

$$x = a\cos\omega t + b\sin\omega t \tag{1.3.7}$$

where the two arbitrary constants a, b can be evaluated from the *initial conditions*.

1. Let the particle be displaced by x_0 from the mean position and released. So, the initial conditions are : (i) At $t = 0$, $x = x_0$; (ii) At $t = 0$, $\dot{x} = dx/dt = 0$, since the velocity of the particle is zero at the instant of release.

Substituting condition (i) in (1.3.7), we immediately get $x_0 = a$.

$$\therefore \quad x = x_0 \cos \omega t + b \sin \omega t$$

$$\therefore \quad \dot{x} = dx/dt = -x_0 \omega \sin \omega t + b\omega \cos \omega t$$

At $t = 0$, $\dot{x} = 0$ by condition (ii). This examples $0 = b\omega \Rightarrow b = 0$.

$\therefore$ Eq. (1.3.7) reduces to

$$\boxed{x = x_0 \cos \omega t.} \tag{1.3.8}$$

2. Let the particle be imparted an impulse at $t = 0$, when it was at the mean position. So, the initial conditions are : (i) At $t = 0$, $x = 0$; (ii) At $t = 0$, $\dot{x} = \dot{x}_0$.

Imposing the condition (i) in (1.3.7), $a = 0$.

$$\therefore \quad x = b \sin \omega t$$

$$\therefore \quad \dot{x} = b\omega \cos \omega t$$

$\therefore$ By the condition (ii) above, we have

$$\dot{x}_0 = b\omega \quad \Rightarrow \quad b = \dot{x}_0/\omega$$

$\therefore$ The solution becomes :

$$\boxed{x = \frac{\dot{x}_0}{\omega} \sin \omega t} \tag{1.3.9}$$

3. The particle is given both an initial displacement x_0 and an initial velocity $\dot{x}_0$. So, the initial conditions are : (i) At $t = 0$, $x = x_0$; (ii) At $t = 0$, $\dot{x} = \dot{x}_0$.

Imposing condition (i) in (1.3.7), we get $x_0 = a$. Therefore,

$$x = x_0 \cos \omega t + b \sin \omega t$$

$$\therefore \quad \dot{x} = -x_0 \omega \sin \omega t + b\omega \cos \omega t.$$

$\therefore$ Using the condition (ii), $\quad \dot{x}_0 = b\omega \Rightarrow b = \dot{x}_0/\omega$.

$\therefore$ The solution becomes :

$$\boxed{x = x_0 \cos \omega t + \frac{\dot{x}_0}{\omega} \sin \omega t} \tag{1.3.10}$$

1.4 Velocity and acceleration under SHM

The displacement x of the particle at any time t is given by

$$x = a \sin(\omega t + \phi), \quad \text{where } a = \text{amplitude} \tag{1.4.1}$$

$$\therefore \text{ **Velocity**, } v = \dot{x} = \frac{dx}{dt} = a\omega \cos(\omega t + \phi) = a\omega \sqrt{1 - \frac{x^2}{a^2}} = \boldsymbol{\omega \sqrt{a^2 - x^2}}. \tag{1.4.2}$$

$$\textbf{Acceleration,} \quad \ddot{x} = -a\omega^2 \sin(\omega t + \phi) = -\boldsymbol{\omega^2 x} \quad \text{using (1.4.1)} \qquad (1.4.3)$$

$\therefore$ **Maximum velocity** $= (v)_{\max} = \boldsymbol{a\omega}$, when $x = 0$, i.e., *the velocity is maximum when the particle is at the mean position*, during oscillation.

The velocity expression (1.4.2) also shows that it varies harmonically with the same frequency ω.

Since $x = a\sin(\omega t + \phi)$ and $v = a\omega\cos(\omega t + \phi) = a\omega\sin(\pi/2 + \omega t + \phi)$, the velocity of linear harmonic oscillator at any time t *leads* the displacement by a *phase angle* $\pi/2$ radian, i.e., the two are in *quadrature*.

$$\textbf{Maximum acceleration} = (\ddot{x})_{\max} = (\omega^2 x)_{\max} = \boldsymbol{a\omega^2}$$

and occurs when $x = x_{\max} = a$, i.e., *the acceleration is maximum when the particle is at the two extreme positions* during oscillation.

The acceleration expression (1.4.3) shows that it also varies harmonically with the same frequency ω.

Since the velocity, $v = a\omega\cos(\omega t + \phi)$ and the acceleration $= -a\omega^2 \sin(\omega t + \phi) = -a\omega^2\cos(\pi/2 + \omega t + \phi)$, the acceleration of a linear harmonic oscillator *leads* the velocity by a *phase angle* $\pi/2$ radian. Also, by comparing the expressions for displacement and acceleration, it is seen that the acceleration is *ahead* of displacement *in phase* by π i.e., they are in *antiphase*.

1.5 Energy of a linear harmonic oscillator

The total energy E of a linear harmonic oscillator at any time t is the sum of its kinetic energy, E_k and potential energy E_p at that instant.

$$\therefore \quad E = E_p + E_k.$$

Potential energy—The potential energy is equal to the work done against the restoring force F for a displacement x from the mean position.

$$\begin{aligned}\therefore \quad E_p &= \int_0^x sx\,dx = \tfrac{1}{2}sx^2 \\ &= \tfrac{1}{2}sa^2\cos^2(\omega t + \phi) \\ &= \tfrac{1}{2}m\omega^2 a^2 \cos^2(\omega t + \phi).\end{aligned}$$

Kinetic energy—The kinetic energy of the linear harmonic oscillator is

$$E_k = \tfrac{1}{2}m\,(dx/dt)^2 = \tfrac{1}{2}m\dot{x}^2 = \tfrac{1}{2}m\omega^2 a^2 \sin^2(\omega t + \phi)$$

$$\begin{aligned}\therefore \textbf{ Total energy,} \quad \boldsymbol{E} &= \boldsymbol{E_p + E_k} \\ &= \tfrac{1}{2}m\omega^2 a^2\cos^2(\omega t + \phi) + \tfrac{1}{2}m\omega^2 a^2 \sin^2(\omega t + \phi) \\ &= \boldsymbol{\tfrac{1}{2}m\omega^2 a^2} = \textbf{constant}.\end{aligned}$$

Clearly, the maximum potential energy, E_{pm} is given by

$$E_{pm} = \tfrac{1}{2}m\omega^2a^2 = E$$

and the maximum kinetic energy E_{km} is given by

$$E_{km} = \frac{1}{2}m\omega^2a^2 = E.$$

Thus the maximum kinetic and potential energies are equal, being equal to the total energy.

$$\text{i.e., } E_{pm} = E_{km} = E$$

Average kinetic energy over a time period T, is

$$\langle E_k \rangle = \frac{1}{T}\int_0^T \frac{1}{2}m\left(\frac{dx}{dt}\right)^2 dt = \frac{1}{T}\cdot\frac{1}{2}m\int_0^T \left(\frac{dx}{dt}\right)^2 dt$$

$$= \frac{m}{2T}\omega^2a^2\int_0^T \sin^2(\omega t + \phi)dt$$

$$= \frac{m}{2T}\omega^2a^2\cdot\frac{1}{2}T = \frac{1}{4}m\omega^2a^2 = \frac{1}{2}E_{km} = \frac{1}{2}E \quad (1.5.1)$$

Average potential energy over a time period T, is

$$\langle E_p \rangle = \frac{1}{T}\left[\frac{1}{2}s\int_0^T x^2dt\right] = \frac{s}{2T}a^2\int_0^T \cos^2(\omega t + \phi)dt$$

$$= \frac{s}{2T}a^2\cdot\frac{1}{2}T = \frac{1}{4}sa^2 = \frac{1}{4}m\omega^2a^2 = \frac{1}{2}E_{pm} = \frac{1}{2}E. \quad (1.5.2)$$

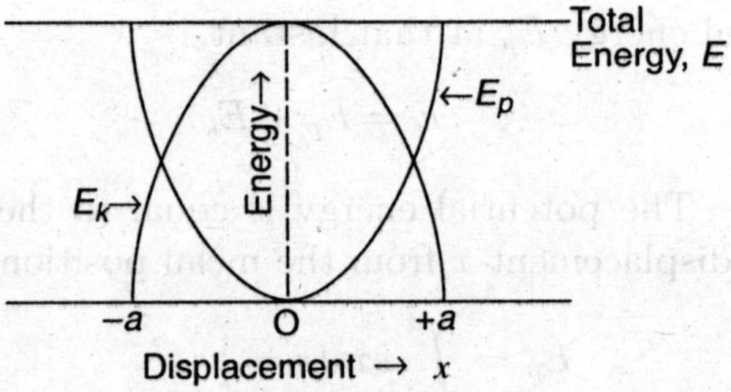

Fig 1.1

Equations (1.5.1) and (1.5.2) show that *the average kinetic and potential energies are equal and each is equal to half of the corresponding maximum energies.*

We have already obtained that the total energy

$$E = E_p + E_k = \tfrac{1}{2}m\omega^2a^2 = \text{constant} = E_{pm} = E_{km}.$$

So, as the particle oscillates, its *energy continuously changes between kinetic and potential forms;* one increases at the cost of the other. The variation of the total energy, kinetic energy and potential energy of a linear harmonic oscillator with displacement is shown in Fig. 1.1.

1.6 Equation of SHM from energy consideration

We shall now find the equation for SHM from the consideration of energy.

The potential energy E_p of a linear oscillator for a displacement x is the work done against the restoring force sx.

$$\therefore \quad E_p = \int_0^2 sx\,dx = \tfrac{1}{2}sx^2.$$

The instantaneous kinetic energy of the oscillator E_k is given by

$$E_k = \tfrac{1}{2}m\left(\frac{dx}{dt}\right)^2.$$

$$\therefore \quad \text{Total energy,} \quad E = E_p + E_k$$

$$= \tfrac{1}{2}m\left(\frac{dx}{dt}\right)^2 + \tfrac{1}{2}sx^2.$$

The total energy remains constant by the conservation of energy principle and this implies that $dE/dt = 0$.

$$\therefore \quad \frac{d}{dt}\left[\frac{1}{2}m\left(\frac{dx}{dt}\right)^2 + \frac{1}{2}sx^2\right] = 0$$

$$\Rightarrow \quad m\left(\frac{dx}{dt}\right)\frac{d^2x}{dt^2} + sx\left(\frac{dx}{dt}\right) = 0$$

$$\Rightarrow \quad \boxed{m\frac{d^2x}{dt^2} + sx = 0} \qquad (\because\ dx/dt \neq 0) \qquad (1.6.1)$$

This is the *equation of motion* of a linear harmonic oscillator *from energy consideration*, being the same as (1.3.1).

1.7 Electric analogue of harmonic oscillator

Consider a circuit, as illustrated in Fig. 1.2, consisting of an inductor of inductance L (and negligible resistance) connected to a capacitor of capacitance C.

The capacitor is first charged by a high tension battery; the electrostatic energy resides in its dielectric medium. The capacitor next begins to discharge through inductor and a current I flows through it. This sets up a magnetic field around the coil, the electrostatic energy being converted into magnetic energy. Next the magnetic field starts dying and induces charges on the capacitor. There is thus a *continuous exchange* of energy between the capacitor and the inductor.

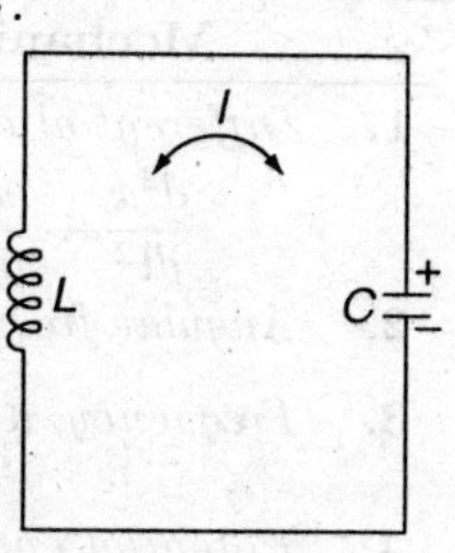

Fig 1.2

If L be the self-inductance of the coil and dI/dt the rate of growth of current in it,

$$\text{voltage difference across } L = -L\frac{dI}{dt}; \text{ voltage difference across } C = \frac{q}{C},$$

where q is the instantaneous charge.

Since there is no source of emf in the circuit,

$$L\frac{dI}{dt} + \frac{q}{C} = 0 \Rightarrow L\frac{d^2q}{dt^2} + \frac{q}{C} = 0, \quad \text{since } I = \frac{dq}{dt}$$

$$\therefore \quad \boxed{\frac{d^2q}{dt^2} + \frac{1}{LC}q = 0} \qquad (1.7.1)$$

which looks similar to the *standard differential equation* of an SHM:

$$\frac{d^2x}{dt^2} + \omega^2 x = 0, \quad \text{where } \omega^2 = 1/LC.$$

Eq. (1.7.1) is thus the *differential equation* of SHM for an *electrical circuit*, having the solution

$$q = q_0 \sin(\omega t + \phi)$$

where q_0 is the maximum charge, called the *amplitude of charge.*

• What *oscillates* in a simple harmonic electric circuit is the *charge on the capacitor.* The charge varies harmonically with time and gives rise to *electrical oscillations.*

• An LC-circuit *cannot* be realized in practice to obtain simple harmonic oscillations. This is because an inductance L has always some ohmic resistance, however small. This causes a continuous loss of energy due to joule heating and the amplitude of SHM slowly dies down to zero. If however the energy loss is compensated by supply of energy to the circuit from an external source by using a valve or a transistor circuit, simple harmonic electrical oscillations can be maintained.

1.7.1 Points of similarity between mechanical and electrical oscillations

The left column corresponds to the mechanical oscillator while the right column to the electrical oscillator.

Mechanical Oscillation	Electrical Oscillation
1. *Differential equation:* $\frac{d^2x}{dt^2} + \frac{s}{m}x = 0$	1. *Differential equation:* $\frac{d^2q}{dt^2} + \frac{1}{LC}q = 0$
2. *Angular frequency,* $\omega = \sqrt{s/m}$	2. *Angular frequency,* $\omega = \sqrt{1/LC}$
3. *Frequency,* $n = \frac{1}{2\pi}\sqrt{s/m}$	3. *Frequency,* $\nu = \frac{1}{2\pi\sqrt{LC}}$
4. *Potential energy,* $E_p = \frac{1}{2}sx^2$	4. *Electrostatic energy,* $E_{\text{elec}} = \frac{1}{2}\frac{q^2}{C}$
5. *Kinetic energy,* $E_k = \frac{1}{2}m\dot{x}^2$	5. *Magnetic energy,* $E_{\text{mag}} = \frac{1}{2}L\dot{q}^2$

Mechanical Oscillation	Electrical Oscillation
6. *Total energy*, $E = \frac{1}{2}m\dot{x}^2 + \frac{1}{2}sx^2$	6. *Total energy*, $E = \frac{1}{2}L\dot{q}^2 + \frac{1}{2}\frac{q^2}{C}$
7. *Stiffness*, s	7. *Stiffness*, $1/C$, reciprocal of capacitance
8. *Pliability*, $1/s$	8. *Pliability*, C, capacitance
9. *Displacement*, $x = a\sin(\omega t + \phi)$	9. *Charge equation*, $q = q_0\sin(\omega t + \phi)$
10. *Amplitude*, a	10. *Amplitude*, q_0
11. *Initial phase angle*, ϕ	11. *Initial phase*, ϕ
12. *Mass*, m	12. *Inductance*, L
13. *Velocity*, dx/dt	13. *Current*, dq/dt

• The students should verify that they fully understand the above points of similarity between the mechanical and the electrical oscillators.

1.8 Superposition of simple harmonic motions

We shall discuss here the resultant motion obtained by *superposition* of two simple harmonic motions under *different situations.*

1. *Two SHMs of the same frequency but different phases and amplitudes, acting along the same straight line.*

Let $x_1 = a_1\cos(\omega t + \phi_1)$ and $x_2 = a_2\cos(\omega t + \phi_2)$ be the two SHMs of amplitudes a_1, a_2 and initial phases ϕ_1, ϕ_2 acting along x-axis.

∴ Instantaneous resultant displacement x is given by

$$\begin{aligned} x = x_1 + x_2 &= a_1\cos(\omega t + \phi_1) + a_2\cos(\omega t + \phi_2) \\ &= (a_1\cos\phi_1 + a_2\cos\phi_2)\cos\omega t - (a_1\sin\phi_1 + a_2\sin\phi_2)\sin\omega t \\ &= A\cos\phi\cos\omega t - A\sin\phi\sin\omega t \\ &= A\cos(\omega t + \phi) \end{aligned} \tag{1.8.1}$$

where
$$A\cos\phi = a_1\cos\phi_1 + a_2\cos\phi_2$$
$$A\sin\phi = a_1\sin\phi_1 + a_2\sin\phi_2$$

$$\therefore \quad A^2 = a_1^2 + a_2^2 + 2a_1a_2\cos(\phi_1 - \phi_2) \tag{1.8.2}$$

$$\tan\phi = \frac{a_1\sin\phi_1 + a_2\sin\phi_2}{a_1\cos\phi_1 + a_2\cos\phi_2} \tag{1.8.3}$$

Eq. (1.8.1) shows that the resultant motion *is also an SHM* of the *same* frequency ω of the components.

• The above analysis can be extended to n number of SHMs when we shall get structurally the same equation (1.8.1) for the resultant motion but in this case

$$A\sin\phi = \sum_i a_i\sin\phi_i; \quad A\cos\phi = \sum_i a_i\cos\phi_i$$

$$A^2 = \left[\sum_i a_i \sin\phi_i\right]^2 + \left[\sum_i a_i \cos\phi_i\right]^2,$$

$$\text{and,}\quad \tan\phi = \sum_i a_i \sin\phi_i \Big/ \sum_i a_i \cos\phi_i$$

• The individual motions being cosine functions, the resultant motion is given by $x = A\cos(\omega t + \phi)$. If however the individual motions are sine functions, the resultant is $x = A\sin(\omega t + \phi)$.

2. *Two simple harmonic motions of the same frequency but different amplitudes and phases at right angles to each other.*

Let the two SHMs be given by the following equations.

$$x = a_1 \sin(\omega t + \phi_1)$$

$$y = a_2 \sin(\omega t + \phi_2)$$

where a_1, a_2 are the amplitudes and ϕ_1, ϕ_2 the initial phases of the x-motion and the y-motion respectively.

We may rewrite

$$\begin{aligned} y &= a_2 \sin\left[(\omega t + \phi_1) - (\phi_1 - \phi_2)\right] \\ &= a_2 \sin(\omega t + \phi_1)\cos(\phi_1 - \phi_2) - a_2 \cos(\omega t + \phi_1)\sin(\phi_1 - \phi_2) \\ &= a_2 \frac{x}{a_1}\cos(\phi_1 - \phi_2) - a_2\sqrt{1 - x^2/a_1^2}\,\sin(\phi_1 - \phi_2) \\ &= \frac{a_2}{a_1} x \cos(\phi_1 - \phi_2) - \frac{a_2}{a_1}\sqrt{a_1^2 - x^2}\,\sin(\phi_1 - \phi_2) \end{aligned}$$

$$\therefore\quad \sqrt{a_1^2 - x^2}\,\sin(\phi_1 - \phi_2) = -\frac{a_1}{a_2} y + x\cos(\phi_1 - \phi_2)$$

Squaring and rearranging the above exprression, we obtain,

$$a_1^2 \sin^2(\phi_1 - \phi_2) = x^2 + \frac{a_1^2}{a_2^2} y^2 - 2\frac{a_1}{a_2} xy \cos(\phi_1 - \phi_2)$$

$$\text{or,}\quad \frac{x^2}{a_1^2} + \frac{y^2}{a_2^2} - 2\frac{xy}{a_1 a_2}\cos(\phi_1 - \phi_2) = \sin^2(\phi_1 - \phi_2) \qquad (1.8.4)$$

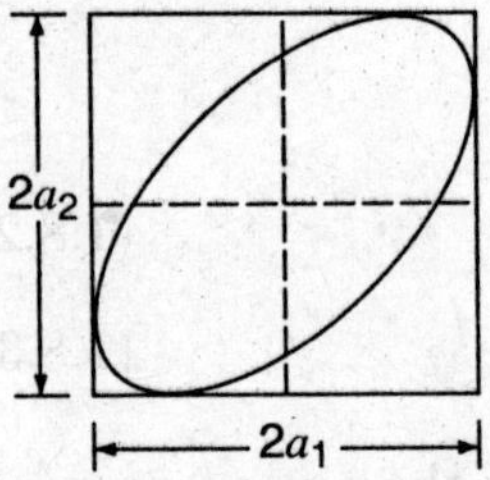

Fig 1.3

Eq. (1.8.4) gives the resultant motion and represents an *ellipse* with the major and the minor axes inclined to the co-ordinate axes. The ellipse can be inscribed in a rectangle of sides $2a_1$ and $2a_2$ (Fig. 1.3). The inclination θ of the major axis with x-axis is given by

$$\tan 2\theta = \frac{2a_1 a_2}{a_1^2 - a_2^2}\cos(\phi_1 - \phi_2) \qquad (1.8.5)$$

Important special cases : Case 1. $\phi_1 = \phi_2$; the two SHMs are *in phase.*

$$\therefore \quad \cos(\phi_1 - \phi_2) = \cos 0 = 1 \text{ and } \sin(\phi_1 - \phi_2) = 0$$

$\therefore$ From (1.8.4) we obtain,

$$\frac{x^2}{a_1^2} + \frac{y^2}{a_2^2} - 2\frac{xy}{a_1 a_2} = 0 \Rightarrow \left(\frac{x}{a_1} - \frac{y}{a_2}\right)^2 = 0$$

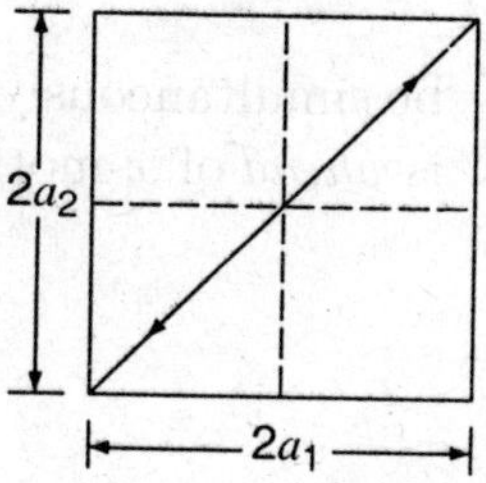

Fig 1.4

which represents a pair of coincident **straight lines** through the origin in the first and third quadrants (Fig. 1.4) with a *slope* of magnitude a_2/a_1.

Case 2. $\phi_1 - \phi_2 = \pi$, i.e., the two motions are in *opposite phase.* So, $\cos(\phi_1 - \phi_2) = \cos\pi = -1$; $\sin(\phi_1 - \phi_2) = \sin\pi = 0$.

$\therefore$ Eq. (1.8.4) reduces to:

$$\frac{x^2}{a_1^2} + \frac{y^2}{a_2^2} + \frac{2xy}{a_1 a_2} = 0 \Rightarrow \left(\frac{x}{a_1} + \frac{y}{a_2}\right)^2 = 0$$

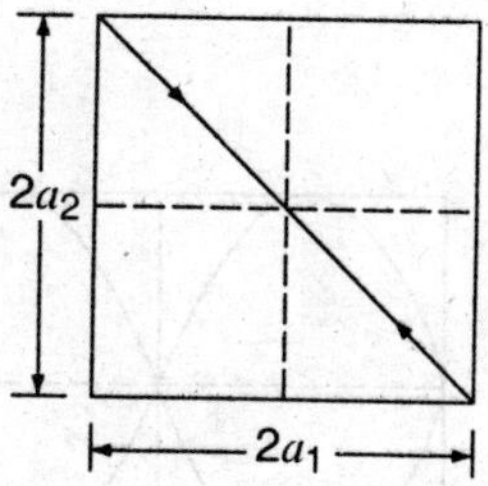

Fig 1.5

which also represents a pair of coincident **straight lines**, through the origin, in the second and fourth quadrants with a *slope* $-a_2/a_1$ as illustrated in Fig. 1.5.

Case 3. $\phi_1 - \phi_2 = \pi/2$, i.e., the two motions are *in quadrature.* So, $\cos(\phi_1 - \phi_2) = 0$ and $\sin(\phi_1 - \phi_2) = 1$.

$\therefore$ Eq. (1.8.4) reduces to

$$\frac{x^2}{a_1^2} + \frac{y^2}{a_2^2} = 1$$

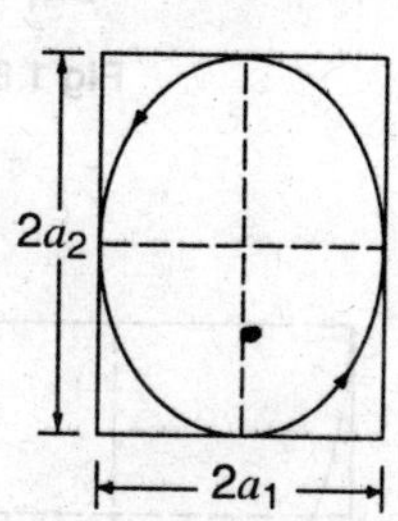

Fig 1.6

which represents a **symmetrical ellipse** of semi-axes a_1 and a_2 coinciding with the co-ordinate axes as shown in Fig. 1.6.

Case 4. $a_1 = a_2 = a$ and $\phi_1 - \phi_2 = \pi$. So, (1.8.4) reduces to

$$x^2 + y^2 = a^2$$

which is a **circle** of radius a with centre at the origin of the coordinate axes. The resultant motion is an *uniform circular motion* of angular frequency ω (Fig. 1.7).

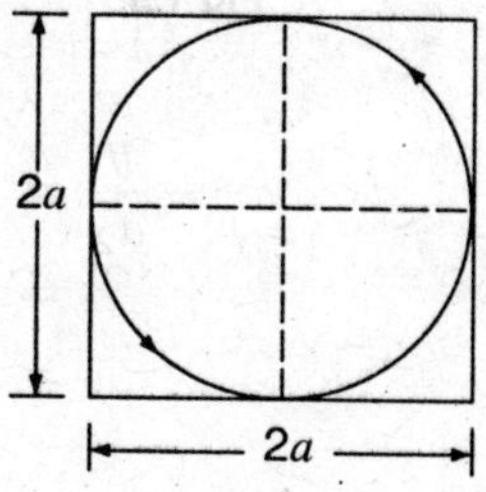

Fig 1.7

3. *Two simple harmonic motions of different amplitudes and commensurate frequencies at right angles to each other.*

(a) Frequencies are in the ratio 1 : 2 (i.e., the periods in the ratio 2 : 1)

Let the two simple harmonic motions be represented by

$$x = a_1 \sin \omega t$$
$$y = a_2 \sin(2\omega t + \phi)$$

be simultaneously impressed on a particle along x and y-axes respectively; the y-motion is *ahead* of x-motion *in phase* by ϕ.

$$\therefore \quad \frac{y}{a_2} = \sin 2\omega t \cos\phi + \cos 2\omega t \sin\phi$$

$$= 2\sin\omega t \cos\omega t \cos\phi + (1 - 2\sin^2\omega t)\sin\phi$$

$$= 2\frac{x}{a_1}\sqrt{1 - \frac{x^2}{a_1^2}}\cos\phi + \left(1 - 2\frac{x^2}{a_1^2}\right)\sin\phi$$

$$\text{or,} \quad \frac{y}{a_2} + \left(2\frac{x^2}{a_1^2} - 1\right)\sin\phi = 2\frac{x}{a_1}\sqrt{1 - \frac{x^2}{a_1^2}}\cos\phi \qquad (1.8.6)$$

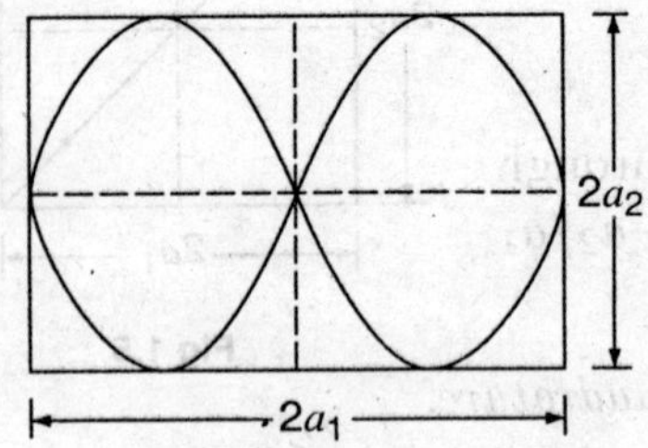

Fig 1.8

(i) If $\phi = 0$, then the relation (1.8.6) reduces to

$$\frac{y}{a_2} = \frac{2x}{a_1}\sqrt{1 - \frac{x^2}{a_1^2}} \quad \Rightarrow \quad \frac{y^2}{a_2^2} + \frac{4x^2}{a_1^2}\left(\frac{x^2}{a_1^2} - 1\right) = 0 \qquad (1.8.7)$$

Eq. (1.8.7) is one of fourth degree in x and represents a curve consisting of **two loops** (Fig. 1.8) resembling the arithmetical figure 8. If the phase difference ϕ is changed, the shape of the loops changes.

(ii) If $\phi = \pi/2$, then (1.8.6) reduces to

$$\frac{y}{a_2} + \left(2\frac{x^2}{a_1^2} - 1\right) = 0 \quad \text{or,} \quad x^2 = -\frac{a_1^2}{2a_2}(y - a_2) \qquad (1.8.8)$$

which represents a **parabola** (Fig. 1.9), **symmetrical** about the y-axis with vertex at $(0, a_2)$.

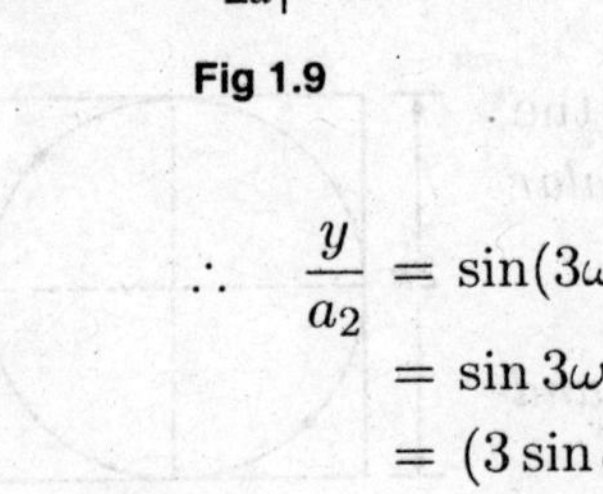

Fig 1.9

(b) Frequencies are in the ratio 1 : 3.

Let the two SHMs be represented by the equations

$$x = a_1 \sin\omega t;$$
$$y = a_2 \sin(3\omega t + \phi)$$

$$\therefore \quad \frac{y}{a_2} = \sin(3\omega t + \phi)$$

$$= \sin 3\omega t \cos\phi + \cos 3\omega t \sin\phi$$

$$= (3\sin\omega t - 4\sin^3\omega t)\cos\phi + (4\cos^3\omega t - 3\cos\omega t)\sin\phi$$

$$= \left(3\frac{x}{a_1} - 4\frac{x^3}{a_1^3}\right)\cos\phi + \left[4\left(1 - \frac{x^2}{a_1^2}\right)^{3/2} - 3\left(1 - \frac{x^2}{a_1^2}\right)^{1/2}\right]\sin\phi \qquad (1.8.9)$$

If $\phi = 0$, then (1.8.9) reduces to the following form :

$$\frac{y}{a_2} - \frac{3x}{a_1} + 4\frac{x^3}{a_1^3} = 0$$

which gives the following **cubic curve** as illustrated in Fig. 1.10.

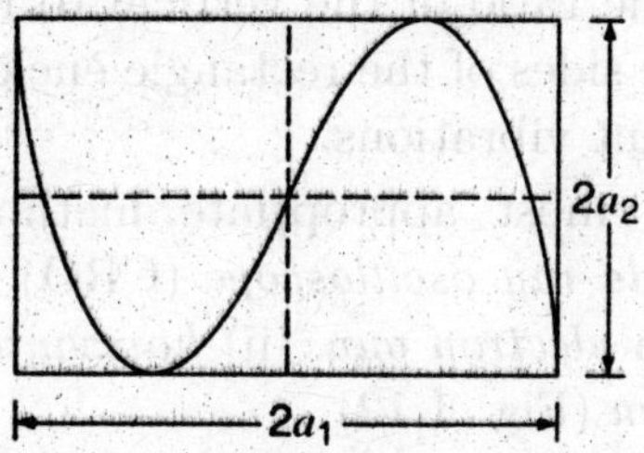

Fig 1.10

If however, $\phi = \pi/2$, then the equation (1.8.9) reduces to

$$\frac{y}{a_2} = 4\left(1 - \frac{x^2}{a_1^2}\right)^{3/2} - 3\left(1 - \frac{x^2}{a_1^2}\right)^{1/2} = \left(1 - \frac{x^2}{a_1^2}\right)^{1/2}\left[1 - \frac{4x^2}{a_1^2}\right]$$

$$\therefore \quad \frac{y^2}{a_2^2} = \left(1 - \frac{x^2}{a_1^2}\right)\left(1 - \frac{4x^2}{a_1^2}\right)^2$$

which is a sixth degree equation in x and represents a curve with *three loops* as shown in Fig. 1.11. If the phase difference ϕ changes, the shape of the loop also changes.

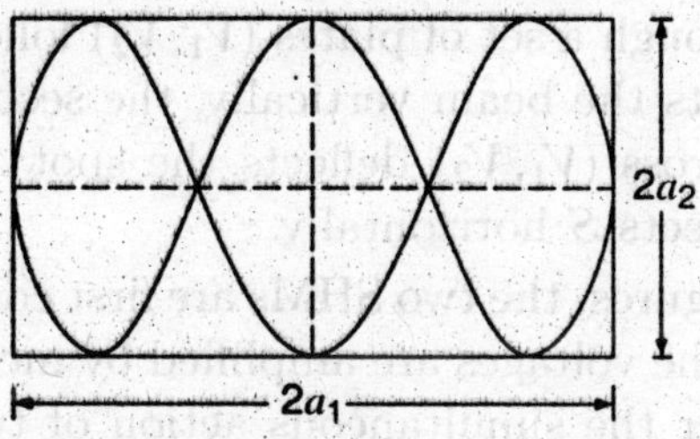

Fig 1.11

• In general, it may be said that the *number of loops* in the path will be *equal to the ratio of the frequencies.*

1.9 Lissajous figures : Demonstration

If two simple harmonic motions, at right angles to each other, are superposed then the resultant curves formed by the superposition are called **Lissajous figures**.

The shape of the curves traced out is determined, as seen already by the following factors : (i) the *relative amplitudes*, (ii) the *periods* (or frequencies), and (iii) the *initial phases* of the component SHMs.

The curves shown in Fig. 1.3 to Fig. 1.11 are all examples of Lissajous figures. From Lissajous figure, one can readily find out the ratio of the frequencies of the component vibrations. The method consists in drawing a horizontal and a vertical line to intersect the Lissajous curve and determine the number of cuts each line makes with it. The ratio of the number of cuts made by horizontal line to the number of cuts made by the vertical one gives the ratio of the vertical to horizontal frequency. As regards amplitudes, the ratio of the sides of the rectangle enclosing the curve is the ratio of the amplitudes of the component vibrations.

Demonstration—The most appropriate method for demonstrating Lissajous figures is to use the *cathode ray oscilloscope* (CRO)—essentially a cathode ray tube (CRT)—consisting of (i) an *electron gun*, (ii) *horizontal* and *vertical deflecting systems* and (iii) a *fluorescent screen* (Fig. 1.12).

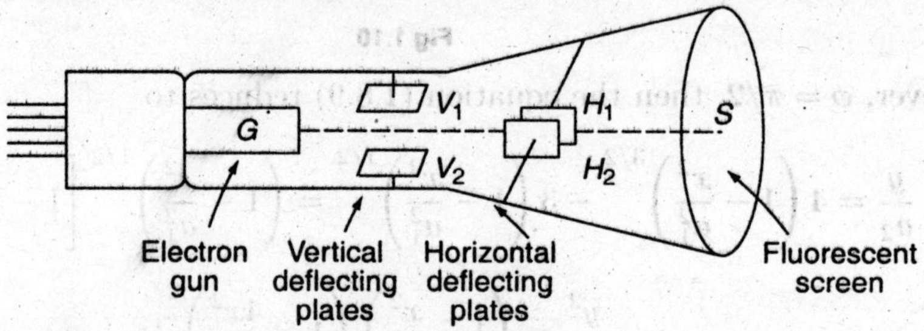

Fig 1.12

The electron gun G produces a pencil of high-speed electrons to be focussed on the screen to form a visible spot S on it. On its way from the gun to screen, the electron beam passes through a set of plates (V_1, V_2) followed by a second set (H_1, H_2). While the first set deflects the beam vertically, the second set introduces a horizontal deflection. A voltage across (V_1, V_2) deflects the spot S vertically on the screen and that across (H_1, H_2) deflects S horizontally.

To demonstrate the figures, the two SHMs are first converted into sinusoidal voltages using two microphones, the voltages are amplified by means of amplifiers and fed to the two sets of plates. Under the simultaneous action of two sinusoidal voltages at right angles to each other, the spot S traces out and visually displays, the *Lissajous figures*.

1.10 Uniform circular motion from two SHMs at right angles

Let two SHMs acting at *right angles* to each other be impressed simultaneously on a particle and be represented by

$$x = a\sin\omega t; \quad y = a\cos\omega t \tag{1.10.1}$$

the y-equation *leading* in phase by $\pi/2$.

$$\therefore \quad x^2 + y^2 = a^2(\sin^2\omega t + \cos^2\omega t). \quad \text{or,} \quad x^2 + y^2 = a^2 \tag{1.10.2}$$

This is the equation of a *circle* of radius a with centre at origin. The particle thus executes a *uniform circular motion* with angular frequency ω.

At $t = 0$, $x = 0$ and $y = a$ implying that the particle is at A (Fig. 1.13). At $t = T/4$ (T = time period), $\omega t = \omega T/4 = \pi/2$ i.e., $x = a$, $y = 0$. The particle is at B. So the particle traces the circle *clockwise*. So the resultant motion represented by (1.10.2) is a **clockwise** *circular motion*.

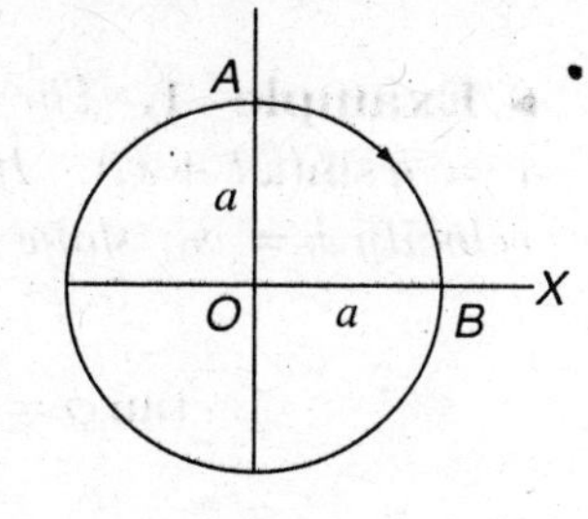

Fig 1.13

• Similarly, it can be easily shown that the resultant of the two SHMs at right angles to each other :

$$x = a \sin \omega t; \quad y = -a \cos \omega t \tag{1.10.3}$$

is also a *uniform circular motion* but traced in the **anticlockwise** direction. Here y-motion *lags* behind x-motion by $\pi/2$, since $-\cos \omega t = \sin(\omega t - \pi/2)$.

Thus *two perpendicular SHMs of the same amplitude and frequency* (or period) *and differing in phase by* $\pi/2$ *give rise to a uniform circular motion*, the *sense* depending on which of the two SHMs *leads* or *lags* the other.

1.11 Combination of two opposite circular motions

We shall prove that if a uniform circular motion described *clockwise* is superposed on another uniform circular motion of the same radius and angular frequency, but described *anticlockwise,* the resultant motion is an SHM along x-axis having the same frequency but with an amplitude equal to the diameter of the either motion.

Proof. The *clockwise* circular motion of radius a and angular frequency ω, according to the Art. 1.10, can be resolved into two SHMs at right angles :

$$x = a \sin \omega t;$$
$$y = a \cos \omega t$$

The *anticlockwise* circular motion of the same radius and angular frequency is similarly resolvable into the following two SHMs at right angles:

$$x = a \sin \omega t;$$
$$y = -a \cos \omega t$$

On their superposition, therefore, we have

$$x = a \sin \omega t + a \sin \omega t = 2a \sin \omega t \tag{1.11.1}$$
$$y = a \cos \omega t - a \cos \omega t = 0 \tag{1.11.2}$$

Eq. (1.11.1) represents an SHM along x-axis of angular frequency ω and amplitude $2a$. There is however, *no* motion along the y-axis. Hence is the proposition.

1.12 Illustrated Examples

► **Example 1.** *The displacement of a simple harmonic oscillator is given by $x = a\sin(\omega t + \phi)$. If the oscillations started at time $t = 0$ from a position x_0 with velocity $\dot{x} = v_0$, show that*

$$\tan\phi = \omega x_0/v_0 \quad \textit{and} \quad a = \left(x_0^2 + \frac{v_0^2}{\omega^2}\right)^{1/2}$$ **(Punjab Univ.)**

Solution. We have, by the problem, the displacement of the oscillator

$$x = a\sin(\omega t + \phi)$$
$$\therefore \quad \dot{x} = a\omega\cos(\omega t + \phi)$$

At $t = 0$, $x = x_0$. $\therefore\ x_0 = a\sin\phi$; Also at $t = 0$, $\dot{x} = v_0$. $\therefore\ v_0 = a\omega\cos\phi$.

$$\therefore \quad \tan\phi = \frac{\sin\phi}{\cos\phi} = \frac{x_0/a}{v_0/a\omega} = \frac{\omega x_0}{v_0}$$

Since $\sin\phi = x_0/a$ and $\cos\phi = v_0/a\omega$, we have, by squaring and adding,

$$\frac{x_0^2}{a^2} + \frac{v_0^2}{a^2\omega^2} = 1 \Rightarrow \left(x_0^2 + \frac{v_0^2}{\omega^2}\right) = a^2$$

$$\therefore \quad a = \left(x_0^2 + \frac{v_0^2}{\omega^2}\right)^{1/2}.$$

► **Example 2.** *A body executes SHM and has velocities* 80 cm/s *and* 60 cm/s *when the displacements are* 3 cm *and* 4 cm *respectively. Calculate the amplitude of vibration and the time taken to travel* 2.5 cm *from the positive extremity of oscillation.*

(Punjab Univ.)

Solution. Velocity, $v = \omega\sqrt{a^2 - x^2}$. $\therefore\quad 80 = \omega\sqrt{a^2 - 3^2}$ and $60 = \omega\sqrt{a^2 - 4^2}$

$$\therefore \quad \frac{80}{60} = \frac{\sqrt{a^2 - 9}}{\sqrt{a^2 - 16}} \Rightarrow (a^2 - 16)80^2 = (a^2 - 9)60^2$$

$$\Rightarrow 16(a^2 - 16) = 9(a^2 - 9) \Rightarrow \mathbf{a = 5\ cm}.$$

$$\therefore\ 80 = \omega\sqrt{5^2 - 9} \Rightarrow \omega = 20\ \text{rad/s}.$$

$\therefore$ Time t to reach the positive extremity is given by :

$$5 = 5\sin\omega t \Rightarrow \omega t = \pi/2$$

$$\therefore \quad t = \frac{\pi}{2\omega} = \frac{\pi}{2 \times 20} = \frac{\pi}{40}\ \text{sec}.$$

Distance of the point 2.5 cm from positive extremity is $5 - 2.5 = 2.5$ cm from mean position.

$\therefore$ Time τ to reach a point 2.5 cm from the mean position is given by

$$2.5 = 5 \sin \omega\tau \Rightarrow \sin \omega\tau = \tfrac{1}{2}.$$

$$\therefore \quad \omega\tau = \frac{\pi}{6} \Rightarrow \tau = \frac{\pi}{6\omega} = \frac{\pi}{6 \times 20} = \frac{\pi}{120} \text{ sec}$$

$$\therefore \text{ Time to reach the point in reference} = \frac{\pi}{40} - \frac{\pi}{120} = \frac{2\pi}{120} = \frac{\boldsymbol{\pi}}{\mathbf{60}} \text{ sec.}$$

▶ **Example 3.** *When the displacement is one-half of the maximum amplitude, what fraction of the total energy is kinetic and what fraction is potential in SHM?*

(Lucknow Univ.)

Solution. The expression for the total energy E of SHM is given by

$$E = \tfrac{1}{2} m a^2 \omega^2.$$

But potential energy, $\quad E_p = \tfrac{1}{2} m\omega^2 x^2$. Here $x = a/2$.

$$\therefore \quad E_p = \tfrac{1}{2} m\omega^2 (a/2)^2 = \tfrac{1}{8} m a^2 \omega^2 = \tfrac{1}{4} E.$$

Since $E = E_k$ (kinetic energy) $+ E_p$, we have

$$\begin{aligned} E_k &= \tfrac{1}{2} m a^2 \omega^2 - \tfrac{1}{8} m a^2 \omega^2 \\ &= \tfrac{3}{8} m a^2 \omega^2 = \tfrac{3}{4} \times \tfrac{1}{2} m a^2 \omega^2 = \tfrac{3}{4} E \end{aligned}$$

$\therefore$ When the displacement is one-half of the maximum amplitude, $\mathbf{\tfrac{1}{4}}$th of total energy is potential and $\mathbf{\tfrac{3}{4}}$ is kinetic.

▶ **Example 4.** *Calculate the ratio of the displacement to amplitude for a simple harmonic motion, when the kinetic energy is 90% of the total energy.* **(Punjab Univ.)**

Solution. Total energy, $E = \tfrac{1}{2} m a^2 \omega^2$. Let x be the displacement when E_k (kinetic energy) $= 90\%$ of E.

Since $E_k = 90\% E$, then the potential energy, $E_p = 10\% E$. But $E_p = \tfrac{1}{2} m\omega^2 x^2$.

$$\therefore \tfrac{1}{2} m\omega^2 x^2 / \tfrac{1}{2} m\omega^2 a^2 = x^2/a^2 = 10/100 = 0.1$$

$$\therefore \quad \frac{\text{displacement}}{\text{amplitude}} = \frac{x}{a} = \sqrt{0.1} \simeq \mathbf{0.32}.$$

► **Example 5.** *At any instant of motion, the kinetic energy is suddenly increased by a blow in the ratio* $(1+\beta) : 1$. *Show that the amplitude is increased in the ratio* $\sqrt{1+\beta v^2/a^2\omega^2} : 1$.

Solution. The total energy of the vibration of mass m *before the blow* is

$$E = E_k + E_p = \tfrac{1}{2}mv^2 + \tfrac{1}{2}m\omega^2 x^2 = \tfrac{1}{2}m\omega^2 A^2$$

where A is the initial amplitude.

After the blow, $E_k = \frac{1}{2}mv^2(1+\beta)$. Total energy after the blow, $E' = \frac{1}{2}m\omega^2 a^2$, where a is the amplitude after blow.

$$\begin{aligned}\therefore\quad \tfrac{1}{2}m\omega^2 a^2 &= \tfrac{1}{2}mv^2(1+\beta) + \tfrac{1}{2}m\omega^2 x^2\\ &= \tfrac{1}{2}mv^2 + \tfrac{1}{2}m\omega^2 x^2 + \tfrac{1}{2}mv^2\beta\\ &= \tfrac{1}{2}m\omega^2 A^2 + \tfrac{1}{2}mv^2\beta\end{aligned}$$

$$\therefore\quad \frac{a^2}{A^2} = 1 + \frac{\beta v^2}{A^2\omega^2} \quad\Rightarrow\quad \boldsymbol{a : A = \sqrt{1 + \frac{\beta v^2}{A^2\omega^2}} : 1}$$

► **Example 6.** *A simple harmonic oscillator of mass* 0.2 g *vibrates with an amplitude* 4 cm. *When the displacement is zero, the velocity is* 1 m/s. *Calculate the frequency and the energy of the oscillator.*

Solution. The displacement equation gives, with usual symbols :

$$\begin{aligned} x &= a\sin(\omega t + \phi)\\ &= 4\sin(\omega t+\phi), \quad \text{since } a = 4 \text{ cm}\\ \therefore\quad \dot{x} &= 4\omega\cos(\omega t+\phi) \end{aligned} \tag{1.12.1}$$

Now at $x = 0$ (zero displacement), $\sin(\omega t+\phi) = 0 \Rightarrow (\omega t+\phi) = 0$.
$\therefore$ $\cos(\omega t+\phi) = 1$. Therefore, from (1.12.1), we obtain

$$100 = 4\omega \quad\Rightarrow\quad \omega = 25 \text{ rad/s}$$

$$\therefore \text{ Frequency, } n = 25/2\pi = \mathbf{3.98\ Hz.}$$

Energy of the oscillator $= \frac{1}{2}ma^2\omega^2 = \frac{1}{2}\times 0.2\times 4^2\times 25^2 = \mathbf{10^3}$ **erg.**

► **Example 7.** *The oscillatory motion of a body is represented by* $y = ae^{j\omega t}$, *where the symbols have usual meanings. Show that the motion is simple harmonic.*

(Punjab Univ.)

Solution. By the problem, we have $y = ae^{j\omega t}$.

$$\therefore\quad \frac{dy}{dt} = aj\omega e^{j\omega t} \quad\Rightarrow\quad \frac{d^2y}{dt^2} = aj^2\omega^2 e^{j\omega t} = -a\omega^2 e^{j\omega t} = -\omega^2 y$$

$$\therefore \quad \frac{d^2y}{dt^2} + \omega^2 y = 0.$$

This is the *standard differential equation* of SHM. So, the motion represented by $y = ae^{j\omega t}$ is **simple harmonic.**

▶ **Example 8.** *The displacement of a moving particle at any time t is given by :*

$$y = a\cos\omega t + b\sin\omega t.$$

Show that the motion is simple harmonic. **(Lucknow Univ.)**

Solution. We have, the displacement equation

$$y = a\cos\omega t + b\sin\omega t$$

$$\begin{aligned} \therefore \quad dy/dt &= -a\omega\sin\omega t + b\omega\cos\omega t \\ \therefore \quad d^2y/dt^2 &= -a\omega^2\cos\omega t - b\omega^2\sin\omega t \\ &= -\omega^2(a\cos\omega t + b\sin\omega t) \\ &= -\omega^2 y \end{aligned}$$

$$\therefore \quad \frac{d^2y}{dt^2} + \omega^2 y = 0$$

This is the *standard differential equation* for simple harmonic motion. Hence the equation $y = a\cos\omega t + b\sin\omega t$ represents a **simple harmonic motion.**

▶ **Example 9.** *A particle executing SHM has displacements x_1 and x_2 when its velocities are v_1 and v_2 respectively. Show that the angular frequency ω, amplitude a and period T are*

$$\text{(i) } \omega = \sqrt{\frac{v_1^2 - v_2^2}{x_2^2 - x_1^2}}, \quad \text{(ii) } a = \sqrt{\frac{x_2^2 v_1^2 - x_1^2 v_2^2}{v_1^2 - v_2^2}}, \quad \text{(iii) } T = 2\pi\sqrt{\frac{x_2^2 - x_1^2}{v_1^2 - v_2^2}}.$$

Solution. Let $x = a\sin(\omega t - \theta)$ be the equation of the motion of the particle under simple harmonic motion.

$$\therefore \quad v = \dot{x} = a\omega\cos(\omega t - \theta)$$

$$\therefore \quad \sin(\omega t - \theta) = x/a; \quad \cos(\omega t - \theta) = v/a\omega. \tag{1.12.2}$$

Squaring and adding the equations in (1.12.2), we have

$$\frac{x^2}{a^2} + \frac{v^2}{a^2\omega^2} = 1.$$

Imposing the given conditions, we readily obtain

$$\frac{x_1^2}{a^2} + \frac{v_1^2}{a^2\omega^2} = 1 = \frac{x_2^2}{a^2} + \frac{v_2^2}{a^2\omega^2} \tag{1.12.3}$$

$$\therefore \quad \frac{x_2^2 - x_1^2}{a^2} = \frac{v_1^2 - v_2^2}{a^2\omega^2} \Rightarrow \omega^2 = \frac{v_1^2 - v_2^2}{x_2^2 - x_1^2} \Rightarrow \omega = \sqrt{\frac{v_1^2 - v_2^2}{x_2^2 - x_1^2}} \tag{1.12.4}$$

Substituting for ω^2 in the first equation of (1.12.3)

$$\frac{x_1^2}{a^2} + \frac{v_1^2(x_2^2 - x_1^2)}{a^2(v_1^2 - v_2^2)} = 1 \Rightarrow \frac{x_2^2 v_1^2 - x_1^2 v_2^2}{a^2(v_1^2 - v_2^2)} = 1$$

$$\therefore \ a^2 = \frac{x_2^2 v_1^2 - x_1^2 v_2^2}{v_1^2 - v_2^2} \Rightarrow a = \sqrt{\frac{x_2^2 v_1^2 - x_1^2 v_2^2}{v_1^2 - v_2^2}}.$$

$$\text{Time period,} \quad T = \frac{2\pi}{\omega} = 2\pi\sqrt{\frac{x_2^2 - x_1^2}{v_1^2 - v_2^2}}$$

► **Example 10.** *A sphere of mass M capable of producing simple harmonic motion along the x-axis is at rest. The stiffness constant is s. Another sphere of mass m moving with velocity u_0 hits the first sphere elastically such that the line joining the centres is along x-axis. Find the displacement of the first sphere at any subsequent instant in terms of $M, m, , u_0$ and s.* **(Calcutta Hons.)**

Solution. If v_1 and v_2 be the velocities of mass m and M respectively after collision along x-axis, then the conservation of linear momentum dictates

$$mu_0 = mv_1 + Mv_2 \tag{1.12.5}$$

Again, the collision being elastic, the relative velocity between the masses before collision will be equal and opposite to that after collision

$$\therefore \quad u_0 - 0 = v_2 - v_1 \Rightarrow v_1 = v_2 - u_0$$

$\therefore$ From equation (1.12.5), we obtain

$$mu_0 = m(v_2 - u_0) + Mv_2$$

$$\Rightarrow \quad v_2 = \frac{2mu_0}{m + M} \tag{1.12.6}$$

The equation of motion of the mass M having displacement x is given by

$$M\frac{d^2x}{dt^2} + sx = 0$$

$$\Rightarrow \quad \frac{d^2x}{dt^2} + \omega^2 x = 0, \quad \text{where } \omega^2 = s/M$$

The solution of the above standard differential equation of SHM is

$$x = A\cos\omega t + B\sin\omega t$$

But, at $t = 0$, $x = 0 \Rightarrow A = 0 \Rightarrow x = B\sin\omega t \Rightarrow \dot{x} = B\omega\cos\omega t$

Again, at $t = 0$, $\dot{x} = v_2$. So, $v_2 = B\omega \Rightarrow B = v_2/\omega$.

$\therefore$ Displacement of first sphere, $x = \dfrac{v_2}{\omega}\sin\omega t = \dfrac{2mu_0}{m+M}\sqrt{\dfrac{M}{s}}\sin\sqrt{\dfrac{s}{M}}\,t$

▶ **Example 11.** *Find the resultant of a large number (n) of simple harmonic vibrations of equal amplitude a and equal successive phase difference δ.*

Solution. We represent the successive SHMs by exponential functions :

$$ae^{j\omega t},\quad ae^{j(\omega t+\delta)},\quad ae^{j(\omega t+2\delta)},\quad \cdots,\quad ae^{j[\omega t+(n-1)\delta]}$$

Let the resultant SHM be represented by $Re^{j(\omega t+\phi)}$ where R is the amplitude of the resultant vibration and ϕ its phase difference with respect to the first component $ae^{j\omega t}$.

$$\therefore\quad Re^{j(\omega t+\phi)} = ae^{j\omega t} + ae^{j(\omega t+\delta)} + ae^{j(\omega t+2\delta)} + \cdots + ae^{j[\omega t+(n-1)\delta]}$$

$$= ae^{j\omega t}\left[1 + e^{j\delta} + e^{j2\delta} + \cdots + e^{j(n-1)\delta}\right].$$

The expression within the bracket is a G.P. series with common ratio $e^{j\delta}$.

$$\therefore\quad Re^{j(\omega t+\phi)} = ae^{j\omega t}\frac{1-e^{jn\delta}}{1-e^{j\delta}}$$

$$= ae^{j\omega t}\frac{e^{jn\delta/2}}{e^{j\delta/2}}\left[\frac{e^{-jn\delta/2}-e^{jn\delta/2}}{e^{-j\delta/2}-e^{j\delta/2}}\right]$$

$$= ae^{j[\omega t+(n-1)\delta/2]}\,\frac{\sin n\delta/2}{\sin\delta/2} \qquad (1.12.7)$$

Equating the imaginary part from both sides of (1.12.7) we obtain,

$$R\sin(\omega t+\phi) = a\frac{\sin n\delta/2}{\sin\delta/2}\sin\left[\omega t + (n-1)\delta/2\right]$$

$\therefore$ **Resultant amplitude,** $R = a\dfrac{\sin n\delta/2}{\sin\delta/2}$ and **phase difference** $\phi = (n-1)\delta/2$.

• When n is very large, δ is very small, and also $\phi \simeq n\delta/2$ or $(n-1)\delta/2$ and $\sin\delta/2 \to \delta/2 = \phi/n$. Therefore, the amplitude .

$$\boxed{R = a\frac{\sin\phi}{\phi/n} = na\frac{\sin\phi}{\phi}} \qquad (1.12.8)$$

which is rather an important relation.

• When $\phi = \pi/2$, $\sin\phi = 1$, and $R = 2na/\pi$. When $\phi = \pi$, $\sin\phi = 0$ and $R = 0$. When $\phi = 3\pi/2$, $\sin\phi = -1$ and $R = -2na/3\pi$.

Review Questions

1. Define simple harmonic motion. Derive the differential equation of motion of a simple harmonic oscillator and its various solutions.
(Calcutta, Burdwan and Kanpur Univ)

2. Set up the differential equation for SHM and give the different forms of the physical solution. Show that the solution is periodic in nature.

3. Give the forms of SHM for the following initial conditions : (a) at $t = 0$, $x = a$ and $\dot{x} = 0$; (b) at $t = 0$, $x = 0$ and $\dot{x} = u_0$.

4. Find an expression for the velocity of a simple harmonic oscillator. The velocity of simple harmonic oscillator at any time t leads the displacement by a phase angle $\pi/2$ radian. Explain why. **(Guru Nanak Univ.)**

5. Find an expression for the acceleration of a simple harmonic oscillator. Show that for a body executing SHM, the acceleration leads the velocity by $\pi/2$ and displacement by π. **(Punjab Univ.)**

6. Derive an expression for the total energy of a harmonic oscillator and show that it is a constant and is proportional to the square of the amplitude.
(Punjab Univ.)

7. (a) Derive the differential equation for the simple harmonic motion from energy considerations.

   (b) Show that for a particle executing SHM, the average value of kinetic energy and potential energy is the same and each is equal to half the total energy.
(Himachal and Guru Nanak Univ.)

8. Calculate the resultant of two SHMs of the same frequency acting along the same line but differing in phase. What is the amplitude when the phase difference is $\pi/2$? **(Calcutta Univ.)**

9. A particle is subjected simultaneously to two SHMs of the same period but of different amplitudes and phases in perpendicular directions. Find the expression for the resultant motion. For what conditions the path may be a straight line, ellipse or circle? Discuss the different important cases?
(Vidyasagar, Burdwan and Lucknow Univ.)

10. What are Lissajous figures? Find the Lissajous figures formed by the superposition of two simple harmonic vibrations at right angles when their periods are in the ratio 2 : 1 and there is a phase difference 0 or $\pi/2$.
(Kurukshetra and Lucknow)

11. Find the resultant of n number of SHMs of equal amplitude and successive phase angle δ. What happens when the quantity n is very large?

12. Show that the resultant motion produced by the superposition of two SHM's $x = a \sin \omega t$ and $y = b \sin(\omega t + \phi)$ is an ellipse, in general. When can it be linear, if at all?

13. Show that the composition of two equal and opposite circular motions leads to a linear harmonic motion of amplitude twice that of any circular motion.

Short Answer and Objective Questions

1. What is the distance covered by a particle, executing SHM of amplitude A in one time period.

2. Write down the characteristics of simple harmonic motion and the differential equation of a linear harmonic oscillator.

3. Simple harmonic motion is also called sinusoidal or co-sinusoidal motion. Justify. **(Punjab Univ.)**

4. Write down the dimensions of the force constant of a vibrating spring?

5. Name a periodic motion which is not oscillatory. What will be the period of a simple pendulum in a satellite?

6. Derive a relation between restoring force and potential energy. Show $\langle K.E.\rangle = \langle P.E.\rangle = \frac{1}{2}E_{\text{total}}$.

7. What does oscillate in a simple harmonic electric oscillator? Can this be realized in practice? Explain.

8. What are Lissajous figures? How will you demonstrate them experimentally? Write down two practical applications of Lissajous figures.

9. The SI-unit of force constant or spring constant is

   (a) Nm, (b) Nm^{-1}, (c) Nm^{-2}, (d) none of these.

   (Himachal Univ.)

10. Which one of the following is *not* essential for simple harmonic motion?

    (a) inertia, (b) restoring force, (c) elasticity, (d) gravity.

    (Himachal Univ.)

11. The velocity of a particle executing SHM is maximum at a point where displacement is

    (a) zero, (b) minimum, (c) maximum, (d) none of these.

12. For a particle executing SHM, the phase difference between the displacement and the velocity is

(a) $\pi/2$, (b) π, (c) 0, (d) $-\pi/2$.

(Himachal Univ.)

13. Two SHMs of the same amplitude, period and phase act at right angles to each other. The resultant vibration will be

(a) circular, (b) elliptical, (c) straight line, (d) none.

14. Which one of the following statements is true?

(a) All periodic motions are simple harmonic.

(b) All simple harmonic motions are periodic.

(c) Potential energy of an S.H. oscillator $= \frac{1}{2}\times$ restoring force $\times$ displacement.

(d) All the above statements are correct.

15. Which one of the following relations is *not* true for SHM?

(a) $\langle K.E.\rangle = \langle P.E.\rangle = \frac{1}{2}E_{\text{total}}$. (b) $T = 2\pi\sqrt{\frac{\text{Displacement}}{\text{Acceleration}}}$.

(c) $m\frac{d^2x}{dt^2} + \frac{s}{m}x = 0$. (d) None of these.

Problems

1. A particle executing SHM along a straight line has a velocity 4 cm/s at a distance of 3 cm from the mean position and 3 cm/s at a distance of 4 cm from it. Find the time it takes to travel 2.5 cm from positive extremity of oscillation. [2.19s]

2. A mass m suspended from a spring of stiffness s executes SHM. Calculate its time period. If we use a stiffer spring of stiffness s', without changing the mass, how will the period of oscillation change?

(Himachal Univ.) [$T = 2\pi\sqrt{m/s}$; $T' = 2m\sqrt{m/s'}$]

3. A lift is ascending at an acceleration of 3 cm/s^2. What is the period of oscillation of a simple pendulum of length 1 m suspended in the lift?

(Punjab Univ.) [1.756s]

4. A mass stands on a platform that vibrates simple harmonically in a vertical direction at a frequency of 5 Hz. Show that the mass will lose contact with the platform if the displacement exceeds 10^{-2} m.

5. A particle executing SHM is at its mean position at time $t = 0$. At time, $t = 0.1$s and $t = 0.2$s, the displacements are 1 cm and 1.5 cm respectively. Determine the frequency and the amplitude. [1.15 Hz, 1.51 cm]

6. A tunnel is dug across the earth passing through its centre. If a body of mass 1 kg is dropped into it, will it come out at the other end? If not, what type of motion does it execute? If the radius of the earth is 6400 km and g on the surface of the earth is 10 ms^{-2}, calculate the possible time period of the body. **(Vidyasagar Univ.)** [83.2 minutes]

7. What is the ratio of kinetic energy at displacement one fourth to one third of the amplitude in case of SHM? **(Punjab Univ.)** [1.055]

8. A linear harmonic oscillator is characterized by $y = a\cos\omega t$. Calculate the displacement at which the kinetic energy is equal to its potential energy. **(Punjab Univ.)** [$\pm a/\sqrt{2}$]

9. A particle of mass m is attached to the mid-point of a weightless string of length l stretched vertically. The ends of the string are fixed to rigid supports. Show that the period of small horizontal oscillations of the mass is :

$$T = \pi\sqrt{\frac{2ml}{(T_1 + T_2)}},$$

where T_1, T_2 are the tensions in the two halves of the string.

10. A weightless spring is suspended vertically from a rigid support. Two equal masses are attached to the free end of the spring so that the spring elongates by x_0. If one mass drops off, find the motion of the remaining mass. [$x = \frac{1}{2}x_0 \cos\frac{2g}{x_0}t$]

11. A cylinder of mass M and cross-sectional area A floats vertically in a liquid of density ρ. If it is depressed slightly and then released, find the frequency of vertical oscillations. **(Calcutta Univ.)** [$\frac{1}{2\pi}\sqrt{A\rho g/M}$]

12. The equation of motion of a particle in the x-direction is : $x = 4\sin^2(\omega t - \pi/6)$ cm. Show that the particle executes SHM and find (a) the mean position, (b) the frequency and (c) the amplitude of oscillation. [2 cm; ω/π Hz; 2 cm]

13. Two vibrations, at right angles to each other, are described by the following equations : $x = 3\cos 4\pi t$ and $y = 3\cos(8\pi t + \pi/3)$, where x and y are expressed in centimetres and t in seconds. Construct the Lissajous figure of the combined motion. **(Burdwan Hons.)** [$x^2(1 - x^2/9) = \{y^2 - (x^2 - 9/2)^2\}$]

14. The equation of motion of a particle is $x = 4\sin(10\pi t - \pi/6)$ cm along the x-direction. Show that the particle is executing SHM and find the frequency and amplitude of oscillation. [5 Hz, 4 cm]

15. A particle is under SHM with amplitude a. Show that the velocity of the particle at a distance $\sqrt{3}a/2$ is half its maximum.

16. $3f + 12x = 0$ represents the relation between the acceleration f and displacement x of a particle executing SHM along x-axis. Find the time period of motion. $[\pi]$

    Find the percentage increase in the time period of a simple pendulum if its length be increased by 69%.

17. Two simple harmonic motions of the same frequency act on a particle simultaneously in the same direction. Their amplitudes are 1.0 cm. and 2.0 cm. respectively. The phase angle of the second with respect to the first is 60° and that of the third with respect to the second is 30°. Obtain the amplitude and the phase angle.

18. A particle is executing SHM in a straight line. When its displacement from the mean position is half its amplitude, what percentage of its total energy is kinetic and what percentage potential? At what displacement will each kind of energy be 50%?

19. A particle of mass 0.5 kg executes an SHM under a force $F = -50x$ where F is in Newton and x in metre. If it crosses the centre of oscillation with a speed of 10 m/s, find the amplitude of the motion.

Chapter 2

Damped Simple Harmonic Motion

2.1 Introduction : Free vibration

Whenever an elastic body is displaced slightly from its position of stable equilibrium and left to itself, it tends to come back to its original position due to the restoring force or force of restitution. In doing so, the body usually overshoots the equilibrium position and vibrates to and fro about it with its own characteristic frequency. Such vibrations are known as *free* or *non-resisted vibration* of the body. The frequency of these vibrations depends on the mass and the elastic properties of the body and is termed *natural frequency.* The reciprocal of this frequency is called the *natural period* of vibration.

Such vibrations may be transverse as in pendulums or longitudinal as in a stretched spiral spring or torsional as in a twisted wire with a load at its free end. All such vibrations will be simple harmonic if the restoring force is proportional to displacement, *provided* it is *small.*

Ideally, the body should vibrate with natural frequency for an indefinite period with constant amplitude. But in reality, the free vibrations are seen to diminish gradually in amplitude (Fig. 2.1) with the passage of time and finally stops altogether, the body coming to rest at its mean position.

The decay of vibration is due to the fact that it is *resisted by frictional forces*—external or internal—such as the viscosity of the medium, the internal friction due to relative sliding of different layers of the vibrator in motion etc. For instance, in a simple pendulum energy is lost due to friction at the supports and resistance of air.

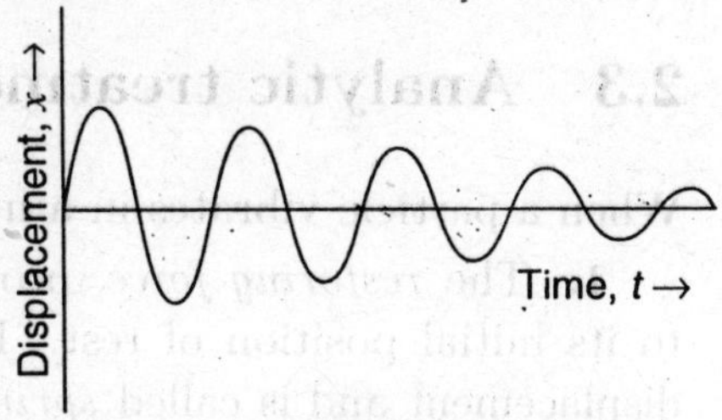

Fig 2.1

This phenomenon of the decay of amplitude with time is called **damping**. The vibration is said to be *damped* or *resisted vibration.* As we shall see, the frequency of damped oscillation is *slightly less* than that of the undamped natural oscillation.

2.2 Damping force is only velocity-dependent

Let a body execute SHM being represented by the equation

$$x = a \sin \omega t \tag{2.2.1}$$

and be acted upon by a *small* damping force $\vec{F}$.

Also let, as a general case, $\vec{F}$ be a function of displacement, velocity and acceleration.

$$\therefore \quad \vec{F} = A + B\vec{x} + C(d\vec{x}/dt) + D(d^2\vec{x}/dt^2) \tag{2.2.2}$$

$\therefore$ Work done by $\vec{F}$ in displacing the body by a small amount $d\vec{x}$ is given by

$$dW = \vec{F}\cdot d\vec{x} = \vec{F}\cdot \frac{d\vec{x}}{dt}dt = \vec{F}\cdot \vec{v}\,dt$$

$\therefore$ Work done per cycle, $W = \int_0^T dW$, where T is the period of vibration.

$$\therefore \quad W = \int_0^T \left[A + B\vec{x} + C(d\vec{x}/dt) + D(d^2\vec{x}/dt^2)\right](d\vec{x}/dt)dt, \quad \text{using (2.2.2).}$$

But $\quad x = a\sin\omega t \Rightarrow dx/dt = a\omega\cos\omega t$ and $d^2x/dt^2 = -a\omega^2 \sin\omega t$.

$$\therefore \quad W = \int_0^T Aa\omega\cos\omega t\,dt + \int_0^T (B - D\omega^2)\,a^2\omega\sin\omega t\cos\omega t\,dt + \int_0^T Ca^2\omega^2\cos^2\omega t\,dt$$

$$= 0 + 0 + \tfrac{1}{2}TCa^2 = \pi a^2\omega \quad (\because\ T = 2\pi/\omega)$$

on evaluating the integrals separately.

Thus, work is done *only* by that term of damping force $\vec{F}$ which is *proportional to the velocity* of the vibrating body; all other terms do not contribute anything. So, ***the damping force depends only upon velocity and is independent of displacement and acceleration.***

2.3 Analytic treatment of damped vibrations

When a particle vibrates in a medium, there are *two* forces acting on it.

1. The *restoring force*, proportional to displacement x, tending to bring it back to its initial position of rest. It is thus $-sx$ where s is the restoring force per unit displacement and is called *spring constant*, or *stiffness factor*.

2. The *retarding* or *resistive force* is proportional to the instantaneous velocity (assumed small) of the particle. It is thus $-r(dx/dt)$ where r, the resistive force per unit velocity is called the *damping constant.*

The $-$ sign indicates that both the forces act opposite to the direction of motion of the particle.

The net force on the vibrator is thus $-sx - r(dx/dt)$ so that its equation of motion is

$$m\frac{d^2x}{dt^2} = -sx - r\frac{dx}{dt}$$

where m is the mass of the mechanical vibrator (particle).

$$\text{or,} \quad \boxed{\frac{d^2x}{dt^2} + 2k\frac{dx}{dt} + \omega_0^2 x = 0} \tag{2.3.1}$$

where $2k = r/m$ and $\omega_0^2 = s/m$. The quantity $2k$ gives the resistive force per unit mass per unit velocity and k is called *damping coefficient;* ω_0^2 is the resistive force per unit mass per unit displacement.

Eq. (2.3.1) is known as the *differential equation of damped vibration* or *damped* simple harmonic motion.

Solution of equation—Let $x = e^{\lambda t}$ be a trial solution of the equation.

$$\therefore \quad \frac{dx}{dt} = \lambda e^{\lambda t} = \lambda x; \quad \frac{d^2x}{dt^2} = \lambda^2 e^{\lambda t} = \lambda^2 x$$

From (2.3.1), therefore, we have on substitution,

$$(\lambda^2 + 2k\lambda + \omega_0^2)x = 0$$

$$\Rightarrow \quad \lambda^2 + 2k\lambda + \omega_0^2 = 0 \quad (\because\ x \neq 0)$$

$$\therefore \quad \lambda = -k \pm \sqrt{k^2 - \omega_0^2}$$

Thus, the general solution of (2.3.1) is given by

$$x = A_1 e^{(-k+\sqrt{k^2-\omega_0^2})t} + A_2 e^{(-k-\sqrt{k^2-\omega_0^2})t}$$

$$= e^{-kt}\left(A_1 e^{\sqrt{k^2-\omega_0^2}t} + A_2 e^{-\sqrt{k^2-\omega_0^2}t}\right) \tag{2.3.2}$$

where A_1 and A_2 are two arbitrary constants.

Three different cases may now arise. These are

Case 1. Heavy damping : $k > \omega_0$. So $\sqrt{k^2 - \omega_0^2}$ is real. Let $\sqrt{k^2 - \omega_0^2} = p$.

$$\therefore \quad x = e^{-kt}(A_1 e^{pt} + A_2 e^{-pt}) \quad \text{from (2.3.2)}$$

Put $A = A_1 + A_2$ and $B = A_1 - A_2$. $\therefore$ $A_1 = \frac{1}{2}(A + B)$; $A_2 = \frac{1}{2}(A - B)$

$$\therefore \quad x = e^{-kt}\left[\frac{A+B}{2}e^{pt} + \frac{A-B}{2}e^{-pt}\right]$$

$$= e^{-kt}\left[A\frac{e^{pt} + e^{-pt}}{2} + B\frac{e^{pt} - e^{-pt}}{2}\right]$$

$$= e^{-kt}(A\cosh pt + B\sinh pt)$$

The displacement thus consists of *two* terms both of which decay exponentially to zero with increasing time.

The values of the constants A and B depend on the initial conditions. If $x = 0$ at $t = 0$, $A = 0$ and

$$x = Be^{-kt} \sinh pt = Be^{-kt} \sinh \sqrt{k^2 - \omega_0^2}\, t.$$

The above is *not* a periodic function of time. Such a motion is thus called *aperiodic* or *overdamped.* There is no vibration at all; the particle passes its equilibrium position at most once before coming to rest asymptotically. Such a motion is exhibited by a pendulum vibrating in a highly viscous liquid or by an aperiodic moving coil galvanometer.

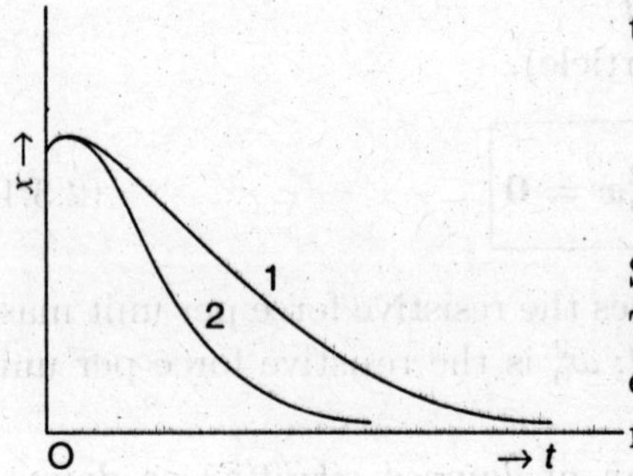

Fig 2.2

The variation of x with t is shown by Curve 1 in Fig. 2.2. Because of the factor e^{-kt}, x decreases exponentially with t before being rested eventually.

Case 2. Critical damping : $k = \omega_0$. The effect of damping term is balanced by that of the stiffness term. To get the solution, in this case, let $k \to \omega_0$, so that $\sqrt{k^2 - \omega_0^2} = \delta$, say where $\delta \to 0$.

$\therefore$ From equation (2.3.2), we obtain

$$x = e^{-kt}\left(A_1 e^{\delta t} + A_2 e^{-\delta t}\right)$$
$$= e^{-kt}\left[A_1(1 + \delta t + \cdots) + A_2(1 - \delta t + \cdots)\right],$$

since $\delta \to 0$, terms higher than linear in δt are neglected.

$$\therefore \quad x = e^{-kt}\left[(A_1 + A_2) + \delta(A_1 - A_2)t\right]$$
$$= e^{-kt}(A + Bt), \tag{2.3.3}$$

where $A = A_1 + A_2$ and $B = \delta(A_1 - A_2)$. The constant A has the dimensions of length and B that of the velocity.

Here also the motion is *aperiodic*, but the decay is quickest, $x \to 0$ quicker than the overdamped case. The motion is said to be *critically damped.* The property of critical damping is used in instruments like ballistic galvanometers. For comparison with the overdamped case, the variation of x with t is shown in curve 2 of Fig. 2.2, the curve 1 representing the overdamped case.

Case 3. Light damping : $k < \omega_0$. So $\sqrt{k^2 - \omega_0^2}$ is imaginary. Let us write

$$\sqrt{k^2 - \omega_0^2} = j\omega \quad \Rightarrow \quad \omega = \sqrt{\omega_0^2 - k^2}.$$

From (2.3.2), therefore, we obtain

$$x = e^{-kt}\left[A_1 e^{j\omega t} + A_2 e^{-j\omega t}\right].$$

Putting $A_1 = ae^{j\phi}/2j$ and $A_2 = ae^{-j\phi}/2j$, we get. from the above equation,

$$x = ae^{-kt}\left[\frac{e^{j(\omega t+\phi)} - e^{-j(\omega t+\phi)}}{2j}\right] = A_0 e^{-kt}\sin(\omega t + \phi) \tag{2.3.4}$$

where A_0 and ϕ are constants, the values of which depend on the initial conditions.

Eq. (2.3.4) shows that the motion is *damped oscillatory* with amplitude $A_0 e^{-kt}$ which goes on decreasing with time and with a *new angular frequency*, $\omega = \sqrt{\omega_0^2 - k^2}$ which is less than ω_0. This is the *underdamped* case; the variation of x with t in this case is shown in the adjoining Fig. 2.3.

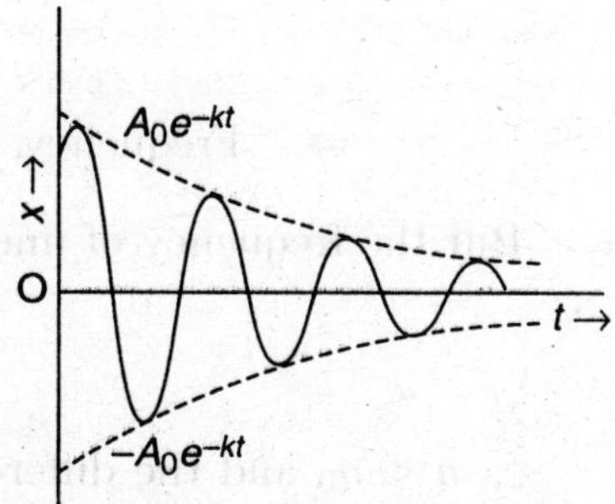

Fig 2.3

Values of A_0 and ϕ — Let the displacement $x = x_0$ (maximum) at $t = 0$. From (2.3.4), therefore, we have

$$x_0 = A_0 \sin\phi. \tag{2.3.5}$$

Also, differentiating (2.3.4), we obtain

$$dx/dt = -A_0 k e^{-kt}\sin(\omega t + \phi) + A_0\omega e^{-kt}\cos(\omega t + \phi).$$

But at $t = 0$, $dx/dt = 0$, so that from above, we obtain

$$-A_0 k \sin\phi + A_0\omega\cos\phi = 0$$

$$\Rightarrow \quad \tan\phi = \frac{A_0\omega}{A_0 k} = \frac{\omega}{k} = \frac{\sqrt{\omega_0^2 - k^2}}{k}$$

$$\therefore \quad \sin\phi = \frac{\omega}{\sqrt{\omega^2 + k^2}} = \frac{\omega}{\omega_0} \quad (\because\ \omega^2 = \omega_0^2 - k^2)$$

Substituting the value of $\sin\phi$ in (2.3.5), we have

$$x_0 = A_0\omega/\omega_0 \ \Rightarrow\ A_0 = \frac{x_0\omega_0}{\omega} = \frac{x_0\omega_0}{\sqrt{\omega_0^2 - k^2}}$$

$$x = \frac{x_0\omega_0}{\omega}e^{-kt}\sin(\omega t + \phi), \quad \text{from (2.3.4)}$$

$$= \frac{x_0\omega_0}{\sqrt{\omega_0^2 - k^2}}e^{-kt}\sin\left(\sqrt{\omega_0^2 - k^2}\,t + \phi\right)$$

$$= A_0 e^{-kt}\sin\left(\sqrt{\omega_0^2 - k^2}\,t + \phi\right) \tag{2.3.6}$$

where $A_0 = x_0\omega_0/\sqrt{\omega_0^2 - k^2}$.

The graph showing the displacement x with time t is given in Fig. 2.3. The time period

$$T = \frac{2\pi}{\omega} = \frac{2\pi}{\sqrt{\omega_0^2 - k^2}}$$

$$\Rightarrow \quad \text{Frequency, } n = \frac{1}{T} = \frac{1}{2\pi}\sqrt{\omega_0^2 - k^2} = \frac{1}{2\pi}\sqrt{s/m - r^2/4m^2}.$$

But the frequency of undamped vibration is given by

$$n_0 = \frac{1}{2\pi}\sqrt{\frac{s}{m}}.$$

$\therefore$ $n < n_0$ and the difference $n_0 - n$ depends on the damping factor r. Larger the value of r, more is the difference.

2.4 Electric analogue : Discharge of a capacitor

Consider the circuit as illustrated in Fig. 2.4 where R is a resistance, L an inductance and C a capacitance. They are connected in series, as shown, to a battery B and a key K. By pressing K downward, the capacitor is charged to a value q_0, say. If K is next released, the capacitor will slowly discharge through R and L giving a current I at any instant t. Then, if q be the charge on capacitor at t, the potential difference across capacitor $= q/C$; the potential difference across resistor $= RI$; the back emf in the inductor $= -L(dI/dt)$.

$$\therefore \quad \frac{q}{C} + RI = -L\frac{dI}{dt}$$

$$\Rightarrow \quad L\frac{dI}{dt} + RI + \frac{q}{C} = 0. \tag{2.4.1}$$

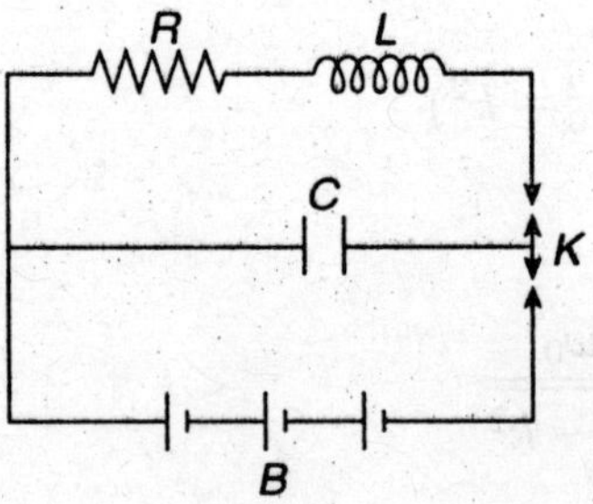

Fig 2.4

But $I = dq/dt$ and $dI/dt = d^2q/dt^2$.

$\therefore$ From (2.4.1), we have

$$L\frac{d^2q}{dt^2} + R\frac{dq}{dt} + \frac{q}{C} = 0$$

$$\text{or,} \quad \frac{d^2q}{dt^2} + 2k\frac{dq}{dt} + \omega_0^2 q = 0 \tag{2.4.2}$$

where $2k = R/L$ and $\omega_0^2 = 1/LC$.

Eq. (2.4.2) has identical structure with that of (2.3.1). Proceeding similarly, therefore, the general solution of (2.4.2) is

$$q = e^{-kt}\left[A_1 e^{\sqrt{k^2-\omega_0^2}t} + A_2 e^{-\sqrt{k^2-\omega_0^2}t}\right] \tag{2.4.3}$$

where A_1, A_2 are arbitrary constants having the dimensions of charge and are to be determined from initial conditions.

The motion represented by (2.4.3) depends on the relative values of k and ω_0, and as before, *three* different *cases* may arise. These are:

Case 1. Heavy damping : $k^2 > \omega_0^2 \Rightarrow R^2/4L^2 > 1/LC$. So, $\sqrt{k^2 - \omega_0^2}$ is real. The charge consists of two terms both of which decay exponentially. The charge on the capacitor decreases rapidly with time to zero. The motion is *aperiodic*, i.e., *non-oscillatory* and *overdamped.*

Case 2. Critical damping : $k^2 = \omega_0^2 \Rightarrow R^2/4L^2 = 1/LC$.

The charge, in this case, at any instant is given, as before, by

$$q = e^{-kt}(A + Bt) \tag{2.4.4}$$

the constant A having the dimensions of charge and B those of current.

Here also the motion is *aperiodic* or *non-oscillatory*. But the decay is quickest. The discharge is said to be *critically damped.*

Case 3. Light damping : $k^2 < \omega_0^2$, i.e., $R^2/4L^2 < 1/LC$. So $\sqrt{k^2 - \omega_0^2}$ is imaginary and, proceeding as before, the solution becomes

$$q = A_0 e^{-kt}\sin(\omega t + \phi) \tag{2.4.5}$$

where $\omega = \sqrt{\omega_0^2 - k^2}$ and A_0 and ϕ are constants. Calculations give us

$$\tan\phi = \frac{\omega}{k} = \sqrt{\frac{\omega_0 - k^2}{k}} = \sqrt{\frac{1}{LC} - \frac{R^2}{4L^2}} \bigg/ \frac{R}{2L}$$

$$A_0 = \frac{q_0\omega_0}{\sqrt{\omega_0^2 - k^2}} = \frac{q_0\sqrt{1/LC}}{\sqrt{1/LC - R^2/4L^2}}$$

$$\therefore \quad q = A_0 e^{-Rt/2L}\sin\left[\sqrt{\frac{1}{LC} - \frac{R^2}{4l^2}}t + \phi\right] \tag{2.4.6}$$

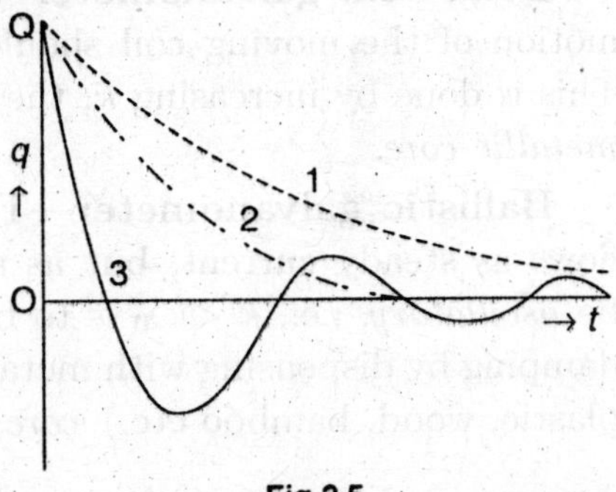

Fig 2.5

Eq. (2.4.6) clearly shows that the discharge of the capacitor in this case is **damped oscillatory.**

The discharge in the above three cases are illustrated in Fig. 2.5 by curves 1, 2 and 3 respectively.

Time period and frequency of oscillation—The time period of oscillation T is given by

$$T = \frac{2\pi}{\omega} = \frac{2\pi}{\sqrt{\omega_0^2 - k^2}} = \frac{2\pi}{\sqrt{1/LC - R^2/4L^2}} \tag{2.4.7}$$

$$\therefore \quad \text{Frequency, } n = \frac{1}{T} = \frac{1}{2\pi}\sqrt{\frac{1}{LC} - \frac{R^2}{4L^2}} \tag{2.4.8}$$

• If R is *negligible*, then however we have,

$$T = 2\pi\sqrt{LC},$$

$$n = \frac{1}{2\pi\sqrt{LC}}.$$

Effect of resistance on amplitude and frequency of vibration—The amplitude of damped vibration is given at any time t by

$$A = A_0 e^{-Rt/2L}$$

$\therefore$ *Higher* the value of R, *smaller* is $e^{-Rt/2L}$ and it follows that *as R increases, the amplitude A* for a given t *decreases* and conversely.

The frequency n is given by (2.4.8). So, *with increasing R,* the n-value or *the frequency decreases.*

2.4.1 Critical damping and design of electric instruments

In damped vibration, if $k > \omega_0$, the motion is *overdamped* and *non-oscillatory.* If $k < \omega_0$, the motion is *damped oscillatory.* Critical damping corresponds to $k = \omega_0$, when the motion is *neither overdamped nor oscillatory* and is thus a *limiting case* of behaviour. In designing electric instruments like *dead-beat* and *ballistic galvanometers*, the property of critical damping is used.

Dead beat galvanometer—This is used to measure *steady current.* So, here, the motion of the moving coil should be *non-oscillatory,* i.e., $k > \omega_0$ is to be satisfied. This is done by increasing k, the electromagnetic damping, by winding the coil over a *metallic core.*

Ballistic galvanometer—This is used to measure the *charge, not* the one that flows as steady current, but as a *sudden discharge.* So the motion of the coil must be *oscillatory,* i.e., $k < \omega_0$ is to be ensured. This is done by reducing electromagnetic damping by dispensing with metallic core and winding the coil over *non-metallic* (ivory, plastic, wood, bamboo etc.) *core,* instead.

2.5 Logarithmic decrement

An important quantity in the steady state of damped oscillations is the *logarithmic decrement.*

When the motion is damped oscillatory, the amplitude falls off with time. The amplitude at any time t is given by

$$A = A_0 e^{-kt}$$

Let the oscillation start from the mean position. Then after a time $t = T/4$, where T is the period, the particle goes to the extreme position. If A_1 be the first amplitude then we may write

$$A_1 = A_0 e^{-kT/4}.$$

The particle will again be at the extreme position on the *same side* after a time T, i.e., after $(T + T/4)$ from start. If A_2 be the second amplitude

$$A_2 = A_0 e^{-k(T+T/4)}$$

Similarly, if A_3, A_4, A_5, etc. be the 3rd, 4th, 5th, etc. amplitudes on the *same side*,

$$\begin{aligned} A_3 &= A_0 e^{-k(2T+T/4)} \\ A_4 &= A_0 e^{-k(3T+T/4)} \\ &\cdots\cdots\cdots\cdots\cdots\cdots \\ A_n &= A_0 e^{-k[(n-1)T+T/4]} \end{aligned}$$

$$\therefore \quad \frac{A_1}{A_2} = \frac{A_2}{A_3} = \frac{A_3}{A_4} = \cdots = \frac{A_{n-1}}{A_n} = e^{kT} = x, \text{ say.}$$

The quantity x is called the *decrement*. The *logarithmic decrement* is defined by

$$\lambda = \ln x = kT. \tag{2.5.1}$$

Definition—*The logarithmic decrement is defined as the natural logarithm of the ratio of two successive amplitudes on the same side of equilibrium position of a damped oscillation.*

The value of A_0 can be evaluated by determining λ experimentally.

Value of A_0—We have, $A_1 = A_0 e^{-kT/4}$. This implies

$$A_0 = A_1 e^{kT/4} = A_1 e^{\lambda/4} = A_1 \left(1 + \frac{\lambda}{4}\right) \tag{2.5.2}$$

since $\lambda \ll 1$, the higher powers of λ have been neglected.

λ for a mechanical oscillator—For a mechanical oscillator, the time period $T = 2\pi/\omega = 2\pi/\sqrt{\omega_0^2 - k^2}$.

$\therefore$ Logarithmic decrement, λ is given by

$$\begin{aligned} \lambda = kT &= \frac{r}{2m} T = \frac{r}{2m} \cdot \frac{2\pi}{\omega} = \frac{\pi r}{m\omega} \\ &= \frac{\pi r}{m\sqrt{\omega_0^2 - k^2}} = \frac{\pi r}{m\sqrt{s/m - r^2/4m^2}}. \end{aligned} \tag{2.5.3}$$

As $\lambda = kT$, *the logarithmic decrement gives a method of evaluating k, the damping coefficient* and hence r ($\because\ k = r/2m$), the *damping factor*.

λ for an electrical oscillator—For an electrical oscillator, we have $k = R/2L$ and $\omega_0^2 = 1/LC$.

∴ Logarithmic decrement,

$$\lambda = \frac{R}{2L}T = \frac{\pi R}{L\omega} = \frac{\pi R}{L\sqrt{\omega_0^2 - k^2}} \quad (\because\ T = 2\pi/\omega)$$

$$= \frac{\pi R}{L\sqrt{1/LC - R^2/4L^2}}. \tag{2.5.4}$$

Experimental determination of λ—We have the relation

$$\frac{A_1}{A_2} = \frac{A_2}{A_3} = \cdots = \frac{A_{n-1}}{A_n} = e^{kT} \Rightarrow \ln\frac{A_{n-1}}{A_n} = kT = \lambda$$

where λ is the *logarithmic decrement*, determined by comparing two amplitudes, n periods apart.

The values of amplitudes are determined at intervals of each time period and a graph is plotted between the intervals T, $2T$, $3T$, $\cdots$ and the corresponding value of $\ln(A_0/A_1)$, $\ln(A_0/A_2)$, $\ln(A_0/A_3)$, $\cdots$. The graph (Fig. 2.6) is a straight line, the slope of which is

$$\tan\theta = \frac{BC}{AC} = \frac{\ln(A_0/A_n)}{nT} = \lambda.$$

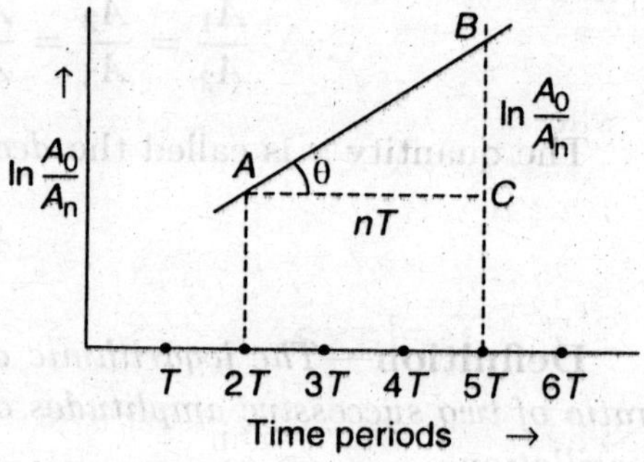

Fig 2.6

• The *physical significance of* λ is that it gives the measure of the rate at which the amplitude of vibration of a damped oscillatory system decays.

2.6 Energy relations : Equation of damped SHM

Let x be the instantaneous displacement of the particle of mass m, executing damped simple harmonic motion.

∴ Its instantaneous potential energy, $E_p = \int_0^x sx\,dx = \frac{1}{2}sx^2$.

Instantaneous kinetic energy, $E_k = \frac{1}{2}m(dx/dt)^2$.

$$\therefore \quad \text{Total energy,} \quad E = E_k + E_p$$

$$= \frac{1}{2}m\left(\frac{dx}{dt}\right)^2 + \frac{1}{2}sx^2 \tag{2.6.1}$$

The instantaneous rate of dissipation of energy is due to the resistive force $k(dx/dt)$ and is the product of the force and velocity.

$$\therefore \quad -\frac{dE}{dt} = k\left(\frac{dx}{dt}\right)\cdot\left(\frac{dx}{dt}\right) = k\left(\frac{dx}{dt}\right)^2 \tag{2.6.2}$$

But, from (2.6.1) above, we obtain

$$\frac{dE}{dt} = m\frac{dx}{dt}\cdot\frac{d^2x}{dt^2} + sx\frac{dx}{dt} \tag{2.6.3}$$

$\therefore$ From equation (2.6.2) and (2.6.3), we obtain

$$m\frac{d^2x}{dt^2} + k\frac{dx}{dt} + sx = 0 \tag{2.6.4}$$

which is the *differential equation* for damped vibration obtained from *energy principle.*

If x_1, x_2 be the two *successive* maximum displacements on the *same side*, the time-interval between them is T, the period of vibration. At x_1, x_2 the particle is momentarily at rest so that its energy is wholly potential. The energy loss in time T, the period, is therefore

$$\Delta E = \tfrac{1}{2}s(x_1^2 - x_2^2) = \tfrac{1}{2}sx_1^2(1 - x_2^2/x_1^2) = E(1 - e^{-2\lambda})$$

where $E\left(=\frac{1}{2}sx_1^2\right)$ is the energy at the beginning and λ the logarithmic decrement.

$$\therefore \quad \Delta E/E = 1 - e^{-2\lambda} \simeq 2\lambda, \quad \lambda \text{ being small.}$$

2.7 Relaxation time of damped oscillatory system

The amplitude of damped oscillatory motion decreases with time. In this context, a quantity called *relaxation time* or *modulus of decay* has been defined.

Definition—*The* ***relaxation time*** τ, *also called the modulus of decay, is defined as the time in which the amplitude of damped oscillatory motion falls to* $1/e$ *of its original value.*

The amplitude of the damped oscillator at any time t is given by

$$A = A_0 e^{-kt}$$

where k is the damping coefficient.

If τ be the relaxation time, then by definition,

$$(1/e)\cdot A_0 = A_0 e^{-k\tau} \quad \Rightarrow \quad e^{-k\tau} = 1/e = e^{-1}$$

$$\therefore \quad k\tau = 1$$

$$\therefore \quad \boxed{\textbf{relaxation time,} \quad \tau = 1/k} \tag{2.7.1}$$

The *relaxation time* thus *varies inversely as the damping coefficient.*

• Since $1/e \simeq 0.37$, the amplitude of a damped oscillator falls to about 37% of its initial value in time τ. *Higher the relaxation time, slower is the rate of fall and so also the rate of energy dissipation.*

For a **mechanical oscillator**, $2k = r/m$, where r is the damping constant.

$$\therefore \quad \tau = \frac{1}{k} = \frac{1}{r/2m} = \frac{2m}{r} \tag{2.7.2}$$

The *relaxation time* is thus *inversely proportional* to the *damping factor*.

For an **electrical oscillator**, $2k = R/L$, where R is resistance and L the inductance.

$$\therefore \quad \tau = \frac{1}{k} = \frac{1}{R/2L} = \frac{2L}{R} \tag{2.7.3}$$

The *relaxation time thus varies inversely as the resistance* of the oscillator circuit.

2.8 Quality factor or Q-factor

The amplitude of a damped oscillatory motion decays with time because of the dissipation of energy. In this context, a term *quality factor*, or *Q-factor*, has been defined.

Definition—*The quality factor is defined as the angle in radian through which the damped oscillator oscillates as its energy reduces to E_0/e from E_0, the maximum value of energy.*

It measures the rate of decay of energy of the damped vibrator.

For an *undamped vibrator*, the total energy (kinetic + potential) E at any time t is given by

$$E = \tfrac{1}{2} m\omega_0^2 A^2$$

But for a *damped oscillator*, $A = A_0 e^{-kt}$ and in place of ω_0, we have $\omega = \sqrt{\omega_0^2 - k^2}$.

$\therefore$ *Energy of damped oscillator*, $E = \frac{1}{2} m\omega^2 A_0^2 e^{-2kt} = E_0 e^{-2kt}$

$$\text{where} \quad E_0 = \tfrac{1}{2} m\omega^2 A_0^2 \tag{2.8.1}$$

If $t = -1/2k$, then $E = E_0/e$. So, in time $t = 1/2k = m/r$, the energy of the oscillator decays to $(1/e)E_0$ from E_0. In this time, the oscillator vibrates through an angle θ radian, given by

$$\theta = \omega t = \omega \times \frac{m}{r}.$$

So, by the very definition, the **quality factor**, Q is given by the relation

$$\boxed{Q = \frac{m\omega}{r}} \tag{2.8.2}$$

$\therefore$ For a *mechanical* oscillator, $Q = \omega m/r = \omega/2k \quad (\because\ 2k = r/m)$. (2.8.2a)

For an *electrical* oscillator, $Q = \omega/2k = L\omega/R \quad (\because\ 2k = R/L)$. (2.8.2b)

Quality factor is thus inversely proportional to the damping coefficient k. Lower the damping, higher is the *Q-value* and conversely. *Q-factor* thus measures the *efficiency* of the oscillator.

• If τ be the time for the energy to decay to $1/e$ of its initial value, then $\tau = 1/2k = m/r$. But $Q = m\omega/r = \omega\tau \Rightarrow r = Q/\omega$. This τ is called *energy decay time.*

Effect of large Q-value on frequency—We have, the relation

$$\boxed{Q = \frac{m\omega}{r} = \frac{m\sqrt{\omega_0^2 - k^2}}{r}.} \tag{2.8.3}$$

When Q is *very large*, necessarily *damping is much less* and $\omega = \sqrt{\omega_0^2 - k^2} \simeq \omega_0$.

$$\therefore \quad Q = \frac{m\omega_0}{r} = \frac{m}{r}\left(\frac{s}{m}\right)^{1/2} = \text{constant.}$$

As $\omega \simeq \omega_0$ = constant, a *large quality factor has little or no effect on the frequency.*

Alternative definition of Q—The energy of a damped oscillator is given by the relation $E = E_0 e^{-2kt}$.

$\therefore$ Power dissipation is given by

$$P = -\frac{dE}{dt} = 2kE_0 e^{-2kt} = 2kE$$

$$\text{or,} \quad -dE = 2kE\,dt = 2\frac{r}{2m}E\,dt$$

$$= \frac{r}{m}E\,dt$$

If $dt = T$ = the time period $= 2\pi/\omega$, then we can compute the loss of energy in one period, i.e., *loss of energy per cycle* as

$$-dE_r = \frac{r}{m}ET = \frac{r}{m}E\frac{2\pi}{\omega}$$

$$\text{or,} \quad \frac{\omega m}{r} = -\frac{2\pi E}{dE_r}$$

$$\Rightarrow \quad \boxed{\boldsymbol{Q = -2\pi \times \frac{E}{dE_r}}} \tag{2.8.4}$$

This relation gives an alternative definition for the quality factor Q.

$$\therefore \ \textbf{Quality factor, } Q = 2\pi \times \frac{\textbf{Energy of the oscillator}}{\textbf{Energy lost per cycle}}$$

2.8.1 Amplitude, energy, logarithmic decrement and relaxation time in terms of Q-factor

The **amplitude**, A at any time t, of the motion is given by

$A = A_0 e^{-kt}$ and $Q = \omega/2k$ from equation (2.8.2a). $\therefore$ $A = A_0 e^{-\omega t/2Q}$.

$\therefore$ The **energy**, $E = \frac{1}{2}m\omega^2 A_0^2 e^{-2kt} = E_0 e^{-2kt}$, from (2.8.1)

$$= E_0 e^{-\omega t/Q}.$$

The **logarithmic decrement**, $\lambda = kT = k\dfrac{2\pi}{\omega} = \dfrac{\pi}{Q}$

The **relaxation time**, $\tau = \dfrac{1}{k} = \dfrac{2Q}{\omega}$

2.9 Illustrated Examples

► **Example 1.** *A mass of* 1 kg *is suspended from a spring of stiffness constant* 25 Nm^{-1}. *If the undamped frequency is* $2/\sqrt{3}$ *times the damped frequency, calculate the damping factor.* **(Himachal Univ.)**

Solution. Undamped frequency, $n_0 = \dfrac{1}{2\pi}\sqrt{s/m}$; and the damped frequency,

$$n_1 = \frac{1}{2\pi}\sqrt{s/m - r^2/4m^2}.$$

$$\therefore \quad \frac{n_0}{n_1} = \frac{\sqrt{s/m}}{\sqrt{s/m - r^2/4m^2}} = \frac{2}{\sqrt{3}}, \text{ by the problem.}$$

$$\therefore \quad 3\frac{s}{m} = 4\left(\frac{s}{m} - \frac{r^2}{4m^2}\right) \Rightarrow \frac{s}{m} = \frac{r^2}{m^2} \Rightarrow r^2 = sm.$$

By the problem, $s = 25\ \text{Nm}^{-1}$, $m = 1$ kg, and thus we have

$$r = \sqrt{25 \times 1} = \mathbf{5\ kg.s^{-1}} \quad (\because\ \text{N} = \text{kg.ms}^{-2})$$

► **Example 2.** *In a one dimensional motion of a mass* 10 g, *it is acted on by a restoring force* 10 dyne/cm *and a resisting force* 2 dyne sec/cm. *Find* (i) *if the motion is aperiodic or oscillatory,* (ii) *the resisting force per unit velocity that will make the motion critically damped and* (iii) *the mass for which the given forces will make the motion critically damped.* **(Calcutta Hons.)**

Solution. By the problem, $m = 10$ g, $s = 10$ dyne/cm and $r = 2$ dyne sec/cm.

$$\therefore \quad 2k = r/m = 2/10 = 0.2 \text{ neper/s}; \quad \omega_0 = \sqrt{s/m} = \sqrt{10/10} = 1 \text{ rad/s}$$

(i) Since $k\ (= 0.1) < \omega_0 (= 1)$, the motion is **oscillatory**.

(ii) For critical damping, $k = \omega_0 \Rightarrow r^2/4m^2 = s/m \Rightarrow r = 2\sqrt{sm} = 2\sqrt{10 \times 10}$

$$= \mathbf{20\ dyne\ sec/cm}$$

(iii) For critical damping, $r^2/4m^2 = s/m \Rightarrow m = r^2/4s = 4/(4 \times 10) = \mathbf{0.1\ g}$

▶ **Example 3.** *In an oscillatory circuit,* $L = 0.5$ H, $C = 1.8\ \mu$F. *What is the maximum value of resistance to be connected so that the circuit may produce oscillations.* **(Himachal Univ.)**

Solution. For the circuit to produce oscillation, the maximum resistance that can be allowed is provided by the condition of critical damping.

$$\frac{R^2}{4L^2} = \frac{1}{LC} \quad \Rightarrow \quad R = 2\sqrt{\frac{L}{C}}$$

$$\therefore \quad R = 2\sqrt{0.5/1.8 \times 10^{-6}} = \mathbf{1054\ \Omega}$$

▶ **Example 4.** *A damped vibrating system, starting from rest, reaches a first amplitude of* 500 mm *which reduces to* 50 mm *in that direction after* 100 *oscillations, each of period* 2.3 sec. *Find the damping constant and the correction for the first displacement for damping.* **(Rajasthani Univ.)**

Solution. If $A_1, A_2, \cdots$ etc. be the successive amplitudes, the damping constant is

$$k = \ln\frac{A_1}{A_2} = \ln\frac{A_2}{A_3} = \cdots.$$

The amplitude of n oscillations after the first is A_{n+1}.

$$\therefore \quad \ln\frac{A_1}{A_{n+1}} = \ln\frac{A_1}{A_2} \times \ln\frac{A_2}{A_3} \times \cdots \times \ln\frac{A_n}{A_{n+1}}$$

$$= n\lambda = nkT, \text{ where } T \text{ is the time period.}$$

By the problem, $A_1 = 500$ and $A_{n+1} = 50$; also, $n = 100$, $T = 2.3$ s

$$\therefore \quad \ln(500/50) = 100 \times k \times 2.3$$

$$\text{or,} \quad \ln 10 = 100 \times k \times 2.3$$

$$\therefore \quad 100k = 1 \quad (\because \ \ln 10 = 2.3)$$

$$\therefore \quad k = 1/100 = \mathbf{0.01}$$

When the system has its first displacement, it has executed one quarter vibration. So if A_0 be the displacement if there were no damping,

$$A_0 = A_1\left(1 + \frac{\lambda}{4}\right), \quad \text{where } \lambda = \text{logarithmic decrement}$$

$$= A_1\left(1 + \frac{kT}{4}\right) \quad \text{since } \lambda = kT$$

$$= 500\left(1 + \frac{1}{4} \times 0.01 \times 2.3\right) = 502.9 \text{ mm}$$

$$\therefore \quad \text{Necessary correction} = \mathbf{+2.9\ mm}$$

▶ **Example 5.** *A massless spring, whose upper end is fixed to a rigid support, carries a horizontal disc of mass* 100 g *at the lower end. The system oscillates with a frequency* 10 Hz *and the amplitude of damped oscillations reduces to half its undamped value in* 1 minute. *Find* (i) *the resistive force constant,* (ii) *the relaxation time of the system,* (iii) *the quality factor and* (iv) *the force constant of the spring.* **(Burdwan Hons.)**

Solution. If A_0 be the undamped amplitude, the damped amplitude at time t will be A_0e^{-kt}.

$$\therefore \quad \tfrac{1}{2}A_0 = A_0e^{-kt} = A_0e^{-60k} \quad (\because\ t = 1\,\text{min} = 60\,\text{s})$$

$$\therefore \quad 60k = \ln 2 \Rightarrow k = \ln 2/60$$

(i) Resistive force constant, r is given by

$$r = 2mk = 2 \times 100 \times \ln 2/60$$
$$= \mathbf{2.31\ dyne.sec/cm}$$

(ii) Relaxation time, $\tau = 1/k = 60/\ln 2 = \mathbf{86.57\ s}$

(iii) Quality factor,

$$Q = \pi/\lambda, \quad \lambda = \text{logarithmic decrement}$$
$$= \pi/kT, \quad T = \text{time period}$$
$$= \pi/k(2\pi/\omega) = \omega/2k$$

But, $\omega = 2\pi n = 2\pi \times 10 = 20\pi$

$$\therefore \quad Q = 20\pi/2(\ln 2/60) = 600\pi/\ln 2 = \mathbf{2720}.$$

(iv) Force constant of the spring, s is given by

$$s = m\omega_0^2 = m(\omega^2 + k^2)$$
$$= 100(400\pi^2 + k^2)$$
$$= 100\left[400 \times 9.87 + (\ln 2/60)^2\right]$$
$$= \mathbf{3.95 \times 10^5\ dyne/cm}$$

▶ **Example 6.** *The quality factor of an undamped tuning fork of frequency* 256 Hz *is* 10^3. *Find the time taken for its energy to get reduced to* $1/e$ *of its energy in the absence of damping. How many oscillations are executed during this time?*

Solution. The time taken for the energy to decay to $1/e$ of its initial value is

$$t = \frac{Q}{\omega} = \frac{Q}{\sqrt{\omega_0^2 - k^2}};$$

where Q = quality factor, k = damping coefficient, ω_0 = undamped angular frequency.

Here $\omega = \omega_0$. Therefore, we have

$$t = \frac{Q}{\omega_0} = \frac{10^3}{2\pi \times 256} = \mathbf{0.62\ s}$$

$\therefore$ No. of oscillations in this time is $N = 256 \times t = 256 \times 0.62 = \mathbf{159}$

▶ **Example 7.** *In an LCR circuit, $L = 1$ mH, $C = 10\ \mu$F, $R = 0.4\ \Omega$. Calculate the Q-factor.* **(Punjab Univ.)**

Solution. In an LCR circuit, the Q-factor is given by

$$Q = \frac{L\omega}{R} = \frac{L}{R}\sqrt{\frac{1}{LC} - \frac{R^2}{4L^2}}$$

$$= \frac{10^{-3}}{4 \times 10^{-1}}\sqrt{\frac{1}{10^{-3} \times 10^{-5}} - \frac{4^2 \times 10^{-2}}{4 \times 10^{-6}}}$$

$$= \frac{10^{-3}}{4 \times 10^{-1}}\sqrt{10^8 - 4 \times 10^4}$$

$$\simeq \frac{10^{-2}}{4} \times 10^4 \simeq \mathbf{25}$$

▶ **Example 8.** *Show that the fractional change in the natural frequency of a damped simple harmonic oscillator is $1/8Q^2$, where Q is the quality factor of the oscillator.* **(Guru Nanak Univ.)**

Solution. Let ω_0 and ω be the angular frequencies of the undamped and damped oscillator respectively. Then, we have

$$\omega = \sqrt{\omega_0^2 - k^2} = \sqrt{\omega_0^2 - r^2/4m^2}$$

$$= \omega_0(1 - r^2/4m^2\omega_0^2)^{1/2}$$

$$= \omega_0\left(1 - \frac{1}{2}\frac{r^2}{4m^2\omega_0^2}\right)$$

since $r^2/4m^2\omega_0^2$ is very small for small values of r.

$$\therefore \quad \frac{\omega}{\omega_0} = 1 - \frac{1}{8}\frac{r^2}{m^2\omega_0^2} \Rightarrow 1 - \frac{\omega}{\omega_0} = \frac{1}{8}\frac{r^2}{m^2\omega_0^2}$$

$$\text{or,} \quad \frac{\omega_0 - \omega}{\omega_0} = \frac{1}{8}\frac{r^2}{m^2\omega_0^2} = \frac{\mathbf{1}}{\mathbf{8Q^2}}$$

since, the quality factor of the oscillator, $Q = m\omega_0/r$.

► **Example 9.** *The frequency of an underdamped harmonic oscillator is adjusted to be equal to half the frequency experienced by the oscillator without damping. Calculate the logarithmic decrement of the system.* **(Punjabi Univ.)**

Solution. Let ω_0 and ω be the frequencies of the undamped and underdamped oscillator respectively. Then, by the problem,

$$\tfrac{1}{2}\omega_0 = \omega = \sqrt{\omega_0^2 - k^2}, \quad k = \text{damping coefficient.}$$

$$\therefore \quad \frac{\omega_0^2}{4} = \omega_0^2 - k^2 \;\Rightarrow\; k^2 = \frac{3}{4}\omega_0^2$$

$$\therefore \quad k = \frac{\sqrt{3}}{2}\omega_0 = \sqrt{3}\omega$$

$\therefore$ Logarithmic decrement λ is given by

$$\lambda = kT = k\cdot\frac{2\pi}{\omega} = \sqrt{3}\omega\cdot\frac{2\pi}{\omega} = \mathbf{2\pi\sqrt{3}}$$

► **Example 10.** *Show that the rate of dissipation of energy of a damped oscillator equals the rate of doing work against the resistive force.*

Solution. The instantaneous total energy E is given by

$$E = \frac{1}{2}m\left(\frac{dx}{dt}\right)^2 + \frac{1}{2}sx^2$$

$$\therefore \quad \frac{dE}{dt} = m\frac{dx}{dt}\cdot\frac{d^2x}{dt^2} + sx\frac{dx}{dt} = \frac{dx}{dt}\left(m\frac{d^2x}{dt^2} + sx\right) = \frac{dx}{dt}\left(-r\frac{dx}{dt}\right)$$

$$\left[\because \; m\frac{d^2x}{dt} + r\frac{dx}{dt} + sx = 0\right]$$

$$= -F\frac{dx}{dt} = -\frac{dW}{dt}$$

where F = resistive force = $r(dx/dt)$ and $-F\,dx = -dW$ = work done against F for a small displacement dx.

Review Questions

1. A particle executes a one-dimensional motion along x-axis under (a) a restoring force proportional to its displacement from the equilibrium position and (b) a damping force proportional to its velocity. Set up the equation of motion of the particle and solve it.

2. Show that the logarithm of the ratio of successive amplitudes on either side of the equilibrium position is a constant.

3. Set up the differential equation for damped vibration of a system from energy principle and solve it. Give the conditions for the (a) overdamped (b) critically damped and (c) underdamped motion. Write down the time-displacement relationship in each of the above cases.

4. Define the following terms used in studying damped simple harmonic vibration of a particle : (a) decay constant; (b) modulus of decay, and (c) logarithmic decrement. Explain relaxation time.

5. What are damped vibrations? Obtain an expression for the displacement in a damped oscillatory motion. Discuss the case of heavy damping.

(Kurukshetra and Kanpur Univ)

6. A particle of mass m can move on x-axis. It is subjected to (a) a restoring force $-sx$ and (b) a frictional force $-r(dx/dt)$. Write the differential equation of the motion. Solve it and discuss the solution. **(Lucknow Univ.)**

7. Using the general solution of the equation of damped simple harmonic motion, discuss the case of critical damping. **(Punjab and Kanpur Univ.)**

8. Write down the equation of damped simple harmonic oscillator. Find the expression for displacement and discuss when we get oscillatory damped SHM.

(Guru Nanak and Himachal Univ.)

9. A charged capacitor discharges through an inductance and a resistance in series. Discuss the possible solutions and represent these by graphs. Deduce the condition under which the discharge is oscillatory. Does the presence of resistance affect the amplitude and frequency of damped oscillations? **(Punjab Univ.)**

10. Define logarithmic decrement and derive a relation for it. How can you determine its value experimentally? **(Guru Nanak and Punjabi Univ.)**

11. Define relaxation time of damped oscillatory system. Show that it varies inversely as damping constant. Derive an expression for the relaxation period of (a) a mechanical oscillator and (b) an electrical oscillator. How is it useful in determining the logarithmic decrement of a system? **(Punjabi Univ.)**

12. (a) Define quality factor, Q, of a damped oscillator. Deduce an expression for it for a mechanical and an electrical oscillator.

(b) Show that (i) lower the damping, higher will be the quality factor; (ii) for large quality factor, damping has little or no effect on frequency.

(Himachal and Punjab Univ.)

13. Discuss the methods (logarithmic decrement, relaxation time and quality factor) for quantitative measurements of damping effect in a damped simple harmonic oscillator. **(Himachal and Guru Nanak Univ.)**

14. Identify the forces acting on a damped natural ocillator. Write down the differential equation of damped oscillation along x-axis and obtain a solution for it to find a general solution for displacement with two constants.

15. Show that the ratio of energy lost per cycle to the energy stored in a damped oscillator is $2\pi/Q$, where Q is the quality factor. **(Punjabi Univ.)**

16. Show that the Q-factor represents the change in phase during which the energy of a damped oscillator reduces to $1/e$ of its initial value. **(Himachal Univ.)**

17. Show that $x = (A + Bt)e^{-rt/2m}$ is the solution of the differential equation

$$\frac{d^2x}{dt^2} + \frac{r}{m}\frac{dx}{dt} + \frac{s}{m}x = 0$$

for the critically damped oscillation. **(Himachal Univ.)**

18. Prove that the damping force is independent of the displacement and acceleration and is proportional to velocity only. **(Punjabi Univ.)**

Short Answer and Objective Questions

1. What is free vibration? What do you mean by damping?

2. Explain the terms : damping coefficient and stiffness factor.

3. Enumerate the conditions for the overdamped, critically damped and the underdamped vibrations.

4. Set up, from energy consideration, the differential equation for damped vibration.

5. Discuss the role of resistance on amplitude and frequency in the oscillatory discharge of a charged capacitor through an inductance and resistance.

6. Explain how the conditions of critical damping are used in designing of electrical instruments. **(Kurukshetra Univ.)**

7. Define logarithmic decrement of a damped oscillatory system. What is its physical significance? **(Guru Nanak Univ.)**

8. Define relaxation time of a damped oscillator and relate it with damping constant and logarithmic decrement. **(Punjabi Univ.)**

9. Express the amplitude, energy, logarithmic decrement and the relaxation time in terms of quality factor, Q.

10. Define quality factor of a damped oscillator. Show that lower the damping, higher is the quality factor.

11. Show that the unit of damping term k is s^{-1}. **(Himachal Univ.)**

12. The unit of force constant or spring constant in SI system is
(a) $N^{-1}m$ (b) Nm^{-1} (c) Nm^{-2} (d) Nm^2.

13. The relaxation time τ for a mechanical oscillator is related to damping constant r as
(a) $\tau = r/2m$ (b) $\tau = 2m/r$ (c) $\tau = 2mr$ (d) $\tau = r/m$.

14. The quality factor Q of an electric oscillator is
(a) $\omega R/L$ (b) $L\omega/R$ (c) $L/\omega R$ (d) ω/LR.

15. Which one of the following relations is valid?
(a) $\lambda = \pi/Q$ (b) $\tau = 1/k$ (c) $Q = \pi/k$ (d) All of these.

16. The relaxation time is defined as the time during which the amplitude of the damped oscillator
(a) Grows to e times the initial value,
(b) Decays to $1/e$ times the initial value,
(c) Grows to e^2 times the initial value,
(d) Decays to $1/e^2$ times the initial value. **(Himachal Univ.)**

17. Which one of the following statements is correct?
(a) Lower the damping, higher will be the quality factor,
(b) Higher the damping, higher will be quality factor,
(c) Damping is in no way related to quality factor,
(d) Quality factor is directly proportional to logarithmic decrement.

Problems

1. A particle of mass 1 g is displaced from its position of rest and released. If it is acted on by a restoring force 5 dyne/cm and a damping force 1 dyne sec/cm, examine if the motion is aperiodic or oscillatory. If the initial displacement of the particle is 5 cm, what would be the displacement after 5 second?
[Oscillatory; −0.134 cm]

2. In Ex. 1, under "Illustrated examples", the mass is given an initial impulse 20 g.cm/s while at rest. Obtain an expression for the displacement x at time t.
(Calcutta Hons.) [$x = \frac{2}{\sqrt{0.99}} e^{-0.1t} \sin \sqrt{0.99}t$]

3. A capacitor of capacitance 1 μF, an inductance of 0.2 H and a resistance of 800 Ω are connected in series. Is the circuit oscillatory? If so, calculate the frequency of oscillation. **(Punjab Univ.)** [Oscillatory; 159.2]

4. A simple pendulum has a period of 1 s and an amplitude 10°. After 10 complete oscillations, its amplitude is reduced to 5°. What is the relaxation time of the pendulum and what is its quality factor? **(Punjab Univ.)** [$\simeq$ 14.43 s; 45.33]

5. Deduce the frequency and quality factor of an LCR circuit with L = 2 mH, C = 5 μF and R = 0.2 Ω. **(Himachal Univ.)** [10^4; 100]

6. Examine if the discharge of a capacitor is oscillatory, given that R = 500 Ω, L = 0.1 H and C = 1 μF. **(Guru Nanak Univ.)** [Oscillatory]

7. In an oscillatory circuit L = 0.2 H, C = 0.0012 μF. What is the maximum value of resistance to be connected so that the circuit may produce oscillations? **(Himachal Univ.)** [2.58 × $10^4 \Omega$]

8. If the energy of a note of frequency 100 Hz decreases to one half of its original value in 1 s, calculate the Q-value. [906.3]

9. The radiated power P of an electron (damped oscillator) of mass m and charge e is given by $P = e^2\omega^4 x_0^2/12\pi\varepsilon_0 c^3$ Watt at a wavelength of 0.6 micron. Show that Q-value is about 10^7 and its free radiation life time (= time for its energy to decay to $1/e$ th of its initial value) is about 10^{-8} s, x_0 being the amplitude of the oscillator. Use memory or physical table for the values of e, m, c and ε_0.

10. The motion of a particle of mass 0.1 kg, subjected to a restoring force 0.1 Nm^{-1}, is critically damped. While at rest, its motion is started by an initial velocity of 0.5 m/s. Find the maximum displacement of the particle. [0.184 m]

11. In a one-dimensional motion of mass 10 g, it is acted on by a restoring force 10 dynes/cm and a resisting force 2 dyne/cm. Find (a) if the motion is oscillatory, (b) the resisting force per unit velocity which will make the motion critically damped and (c) the mass for which the given forces will make the motion critically damped.

Chapter 3

Forced Vibration and Resonance

3.1 Forced Vibration and Resonance : Introduction

The amplitude of *free vibrations* of a body gradually dies out, as already discussed, on account of the dissipation of energy in each vibration due to frictional forces. If however, the loss of energy is continuously replenished, the vibration could be maintained. If a *strong* external *periodic force* constantly acts upon the body, it would eventually continue to vibrate under the influence of that periodic force. If the frequency of the applied periodic force be *not* the same as the natural frequency of the vibrating body, disturbances occur at the initial stage of vibration—the *unsteady state.* While the body tends to vibrate with its characteristic natural frequency, the applied force tends to enforce its own frequency on it—sometime helping and sometime hindering the vibration. But eventually the natural vibrations of the body subside due to the resisting forces and the body finally vibrates with the frequency of the driving force with a constant amplitude — the *steady state.*

Such vibration where a body or a system is made to vibrate under the action of a strong impressed periodic force with a frequency equal to that of the impressed force, which is different from the natural frequency of the system, is called **forced vibration**.

The amplitude of forced vibrations is usually small. But if the frequency (hence the period) of the applied periodic force be equal to the natural frequency of the body, the vibrations are built up quickly as the force is always in phase with the vibration of the body. It thus gains in amplitude and finally begins to vibrate with considerably large amplitude even when the applied force is small. But as the motion grows, so does the resistance to it so that a steady state of oscillation is attained when the energy supplied by the external source is completely utilised in overcoming the frictional forces.

The particular type of forced vibration in which the natural frequency of the vibrating system agrees with that of the externally applied strong periodic force is called **sympathetic vibration** *or* **resonance** and the system is said to *resonate* with the impressed periodic force. A resonant vibration has a *very large amplitude* and in the ideal case of undamped vibration, it would be *infinitely* large.

In forced vibration, the vibrating system is called the *forced system* or the *driven system*, and the externally applied periodic force is termed the *forcing system* or *the driving system.*

• In the initial stage of forced vibration, if the natural frequency of the body be close to that of the external force, *beats* might be heard.

• Resonance is not limited to sound alone. There are many instances of *electrical resistance.* For instance, the tuning of a radio set to any desired station is an instance of electrical resonance. Similarly, there are instances also of *mechanical resonance.*

• Stringed musical instruments, whether Eastern or Western, use hollow enclosures usually wooden. The vibrations of the strings cause forced vibrations of the enclosure with the air inside. The emitted sound becomes intense as desired.

• If the stem of a vibrating tuning fork is pressed, say on a wooden table, it is found that the sound becomes louder. The vibrating fork induces a forced vibration to the table which causes vibration of a large volume of air with a frequency equal to that of the fork. And this is why the sound becomes louder. But this loud sound lasts for a much shorter time because of the conservation of energy principle.

• Resonant or sympathetic vibration is of much importance to the structural engineers as well. To them, the development of large stresses in structural elements is a serious headache, for these might cross the elastic limit and thereby cause heavy damage to the structure. And if a small periodic force acts on some machine or structure having a frequency equal to that of its natural frequency, then vibrations with large amplitude would set in. Proper attention is therefore to be paid to the external periodic force that may act on a structure while making designs for it.

• While marching over suspension bridges, soldiers are usually ordered to break steps. But why? For otherwise, the bridge may vibrate dangerously and give way when resonance occurs between the frequency of steps of the marching soldiers and the natural frequency of the suspension bridge. There are many records of such accidents in the history of the Army.

• The forced vibration is of much importance in many branches of physics and many examples of forced vibration may be cited. The diaphragm of a microphone vibrates being linked (by sound waves) to the vibration of say, a musical instrument; the diaphragm of a loudspeaker vibrates being linked by current oscillations to the output circuit of an amplifier; the receiving circuit of a radio oscillates being linked by radio waves to the transmitting circuit of the broadcasting station etc. etc. All these are examples of forced vibrations, where the forced system picks up energy from the forcing system and vibrates. This pick up of energy from the forcing system is *without any appreciable feedback.* Therefore, in such cases the energy transfer is essentially a *one-way process.* This may be due to the very weak coupling between the systems, or the driving system has a large reservoir of energy.

3.2 Analytic treatment of forced vibration

The equation of motion of a damped simple harmonic oscillator of mass m and having x as its displacement from the mean position at any time t is

$$m\frac{d^2x}{dt^2} + r\frac{dx}{dt} + sx = 0 \tag{3.2.1}$$

where sx is the restoring force, s being the *stiffness factor*; $r(dx/dt)$ the resisting force, r being the resisting force per unit velocity.

If the oscillator is acted upon by an external periodic force $F = F_0 e^{j\omega t}$, then its equation of motion is, using (3.2.1),

$$\frac{d^2x}{dt^2} + \frac{r}{m}\frac{dx}{dt} + \frac{s}{m}x = \frac{F_0}{m}e^{j\omega t}$$

$$\text{or,}\quad \frac{d^2x}{dt^2} + 2k\frac{dx}{dt} + \omega_0^2 x = \frac{F_0}{m}e^{j\omega t} \tag{3.2.2}$$

where $r/m = 2k$ and $s/m = \omega_0^2$.

The general solution of the inhomogeneous equation (3.2.2) consists, as is well-known, of two parts : (i) a *complimentary function* and (ii) the *particular integral.*

Transient state—If the motion of the oscillator is oscillatory, the complimentary function, as shown in Chapter : *Damped simple harmonic motion,* is

$$x = A_0 e^{-kt}\sin\left(\sqrt{\omega_0^2 - k^2}\cdot t + \phi\right)$$

$$= A_0 e^{-kt}\sin(\omega' t + \phi) \tag{3.2.3}$$

where $\omega' = \sqrt{\omega_0^2 - k^2}$.

The amplitude $A_0 e^{-kt}$ decays with time due to the exponential factor e^{-kt}. The quantity ω_0 is the natural frequency of the oscillator in the absence of the dissipative and the driving force.

This part of the solution is called the *transient term* as it dies out with time, persisting only for a short while. During this time, the natural vibration of the system prevails and the system behaves as if there is no external force.

Steady state—In the steady state, let the particular solution of (3.2.2) be

$$x = Ae^{j\omega t} \Rightarrow \frac{dx}{dt} = j\omega Ae^{j\omega t} \Rightarrow \frac{d^2x}{dt^2} = -\omega^2 Ae^{j\omega t}$$

Substituting these values in (3.2.2) and simplifying, we obtain

$$A(-\omega^2 m + j\omega r + s) = F_0$$

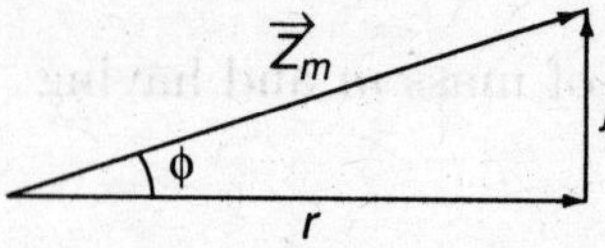

Fig 3.1

$$\therefore \quad A = \frac{F_0}{j\omega r + (s - \omega^2 m)} = \frac{-jF_0}{\omega r - j(s - \omega^2 m)}$$

$$= \frac{-jF_0}{\omega[r + j(\omega m - s/\omega)]} = \frac{-jF_0}{\omega \vec{Z}_m} \tag{3.2.4}$$

$$\text{where} \quad \vec{Z}_m = r + j(\omega m - s/\omega) \tag{3.2.5}$$

and is (Fig. 3.1) called the **mechanical impedance**, in analogy with the electrical impedance.

$$\vec{Z}_m = Z_m e^{j\phi}$$

$$\text{where} \quad Z_m = \sqrt{r^2 + (\omega m - s/\omega)^2}, \tag{3.2.6}$$

is the *magnitude* of $\vec{Z}_m$ and $\phi = \tan^{-1}[\omega m - s/\omega]/r$, the *phase difference* between the displacement x and the impressed force F.

$$\therefore \quad x = Ae^{j\omega t} = \frac{-jF_0}{\omega \vec{Z}_m} e^{j\omega t} = \frac{-jF_0}{\omega Z_m e^{j\phi}} e^{j\omega t} = \frac{-jF_0}{\omega Z_m} e^{j(\omega t - \varphi)} \tag{3.2.7}$$

Discussion : (i) Driving force $F = F_0 \cos \omega t$ — If the external periodic force be $F_0 \cos \omega t$, it is the real part of $F_0 e^{j\omega t}$.

Now, by (3.2.7),

$$x = \frac{-jF_0}{\omega Z_m} \cos(\omega t - \phi) + \frac{F_0}{\omega Z_m} \sin(\omega t - \phi) \tag{3.2.8}$$

So, the real part of this equation gives

$$x = \frac{F_0}{\omega Z_m} \sin(\omega t - \phi) = \frac{F_0}{\omega Z_m} \cos\left[\omega t - \left(\phi + \frac{\pi}{2}\right)\right] \tag{3.2.9}$$

Thus in the steady state, the *displacement amplitude* of the driven oscillator is *constant* and is $F_0/\omega Z_m$ and it always *lags behind* the driving force in phase by $(\phi + \pi/2)$.

(ii) Driving force $F = F_0 \sin \omega t$—If, however, the external periodic force be $F_0 \sin \omega t$, it is the imaginary part of $F_0 e^{j\omega t}$. From the imaginary part of (3.2.8),

$$x = \frac{-F_0}{\omega Z_m} \cos(\omega t - \phi) = \frac{F_0}{\omega Z_m} \sin\left[\omega t - \left(\phi + \frac{\pi}{2}\right)\right] \tag{3.2.10}$$

Here also, in the steady state, the *constant displacement amplitude* of the driven oscillator is $F_0/\omega Z_m$ and it *lags behind* the driving force in phase by $(\phi + \pi/2)$.

It is also noted that in the *steady state,* the system oscillates *simple harmonically* with the impressed frequency ω.

If the driving frequency ω and the natural damped frequency $\omega' (= \sqrt{\omega_0^2 - k^2})$ be nearly equal, their superposition will produce *beats* when transient persists. Such beats are aptly called *transient beats* which last longer, smaller the decay constant k.

• Transients are found to appear whenever external periodic force is either impressed or removed, e.g., they occur whenever a note starts or stops in musical instruments. They are also of much importance in connection with (i) the fidelity response of loudspeakers and microphones and (ii) the reverberration of sound in auditoriums which usually produces undesirable effects.

3.3 Variation of displacement amplitude with forcing frequency

The displacement amplitude of a forced oscillator, in the steady state, is given by

$$a = \frac{F_0}{\omega Z_m} = \frac{F_0}{\omega\sqrt{r^2 + (\omega m - s/\omega)^2}}, \quad \text{using (3.2.6) for } Z_m$$

$$= \frac{F_0}{\omega\sqrt{4k^2m^2 + (\omega m - \omega_0^2 m/\omega)^2}}, \quad (\because\ 2k = r/m;\ \omega_0^2 = s/m)$$

$$= \frac{F_0}{\sqrt{4k^2m^2\omega^2 + m^2(\omega^2 - \omega_0^2)^2}} = \frac{F_0/m}{\sqrt{4k^2\omega^2 + (\omega^2 - \omega_0^2)^2}}$$

$$= \frac{f_0}{\sqrt{4k^2\omega^2 + (\omega^2 - \omega_0^2)^2}}, \quad \text{where } f_0 = F_0/m. \tag{3.3.1}$$

We shall now discuss the variation of displacement amplitude with forcing frequency and consider the following *three* cases :

(i) Low forcing frequency ($\omega \ll \omega_0$)—In this case, ω^2 can be neglected compared to ω_0^2 and $4k^2\omega^2$ compared to $(\omega_0^2)^2$.

$$\therefore \quad a = \frac{f_0}{\omega_0^2} = \frac{F_0/m}{s/m} = \frac{F_0}{s} \tag{3.3.2}$$

So, *at very low forcing frequencies, the amplitude of forced vibration depends on the driving force F_0 and the stiffness constant s, but is independent of driving frequency.* Thus, the amplitude of the forced oscillator is ***stiffness controlled.***

(ii) High forcing frequency ($\omega \gg \omega_0$)—In this case, $\omega^2 \gg \omega_0^2$ and ω_0^2 may be neglected compared to ω^2. So, from (3.3.1),

$$\therefore \quad a = \frac{f_0}{\sqrt{4k^2\omega^2 + \omega^4}} \simeq \frac{f_0}{\omega^2} \tag{3.3.3}$$

provided the damping coefficient k is very small so that $4k^2\omega^2$ is negligible compared to ω^4.

As $\omega \to \infty$, $a \to 0$, i.e., *the amplitude approaches zero as the forcing frequency becomes very large.*

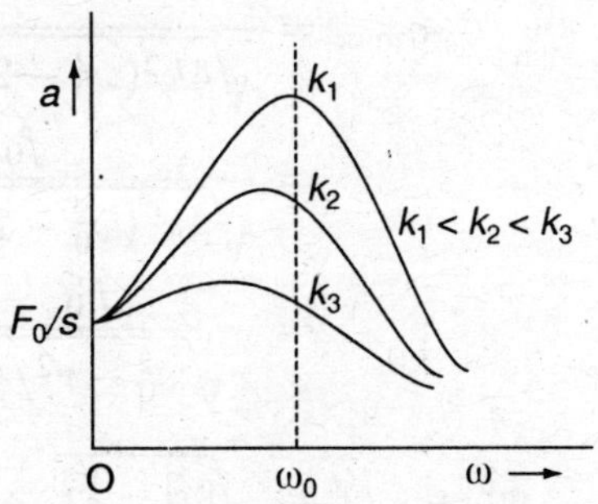

Fig 3.2

(iii) At frequency of amplitude resonance ($\omega = \omega_0$)—Here the driving frequency is equal to the frequency of the oscillator, i.e., it is a case of *amplitude resonance.*

$$\therefore \quad a = \frac{f_0}{\sqrt{4k^2\omega^2 + (\omega^2 - \omega_0^2)^2}} = \frac{f_0}{2k\omega}. \tag{3.3.4}$$

So, the *amplitude* of forced oscillation at resonance frequency *is inversely proportional to the damping coefficient k and the frequency ω*. **For small k, amplitude is large and conversely. If $k \to 0$, $a \to \infty$.**

The variation of displacement amplitude with frequency of the driver for *three* different k's is shown in Fig. 3.2. At $\omega = 0$, we have $a = F_0/s$ for all values of damping. As damping decreases, the maximum value of a increases and the frequency at which a_{max} occurs gets shifted towards undamped frequency ω_0. If $k = 0$ (no damping), a becomes infinite.

When however damping is present, the displacement resonance occurs at a frequency *slightly lower* than the resonance frequency, as discussed below.

For amplitude a to be maximum, $da/d\omega = 0$. This implies

$$\frac{d}{d\omega}\left[\frac{f_0}{\sqrt{4k^2\omega^2 + (\omega^2 - \omega_0^2)^2}}\right] = 0$$

$$\Rightarrow \quad \frac{-f_0[4\omega(\omega^2 - \omega_0^2) + 8k^2\omega]}{2[4k^2\omega^2 + (\omega^2 - \omega_0^2)^2]^{3/2}} = 0$$

$$\Rightarrow \quad 4\omega(\omega^2 - \omega_0^2) + 8k^2\omega = 0$$

$$\Rightarrow \quad \omega^2 - \omega_0^2 = -2k^2$$

$$\therefore \quad \omega = \sqrt{\omega_0^2 - 2k^2} = \omega_0\sqrt{1 - 2k^2/\omega_0^2} = \omega_0\sqrt{1 - r^2/2m^2\omega_0^2} \tag{3.3.5}$$

Thus the ***resonance frequency** for amplitude resonance* is **slightly less** *than the natural frequency* ω_0.

Maximum amplitude—By substituting (3.3.5) in the expression for a, we have,

$$a_{max} = \frac{f_0}{\sqrt{4k^2(\omega_0^2 - 2k^2) + (\omega_0^2 - 2k^2 - \omega_0^2)^2}}$$

$$= \frac{f_0}{\sqrt{4k^2(\omega_0^2 - 2k^2) + 4k^4}} = \frac{f_0}{2k\sqrt{\omega_0^2 - k^2}} = \frac{f_0}{(r/m)\sqrt{\omega_0^2 - r^2/4m^2}}$$

$$= \frac{F_0}{r\sqrt{\omega_0^2 - r^2/4m^2}} = \frac{F_0}{r\omega'} \tag{3.3.6}$$

$$\text{where, } \omega' = \sqrt{\omega_0^2 - k^2} = \sqrt{\omega_0^2 - r^2/4m^2} \tag{3.3.7}$$

= natural frequency of the damped oscillator.

3.4 Velocity of a forced oscillator

We shall now calculate the *velocity* of a forced oscillator when the driving force is (i) $F_0 \cos \omega t$ and (ii) $F_0 \sin \omega t$.

Case 1. When the driving force is $F = F_0 \cos \omega t$, the displacement of the forced oscillation is given by

$$x = \frac{F_0}{\omega Z_m} \sin(\omega t - \phi)$$

$$\therefore \quad \textbf{Velocity,} \; v = \frac{dx}{dt} = \frac{F_0}{\omega Z_m} \omega \cos(\omega t - \phi)$$

$$= \frac{F_0}{Z_m} \cos(\omega t - \phi) = v_0 \cos(\omega t - \phi) \tag{3.4.1}$$

$$= v_0 \sin \left[(\omega t - \phi) + \frac{\pi}{2} \right] \tag{3.4.1a}$$

where $v_0 = F_0/Z_m =$ **velocity amplitude**, and $Z_m = \sqrt{r^2 + (\omega m - s/\omega)^2}$.

Case 2. When the driving force is $F = F_0 \sin \omega t$, the displacement of the forced oscillation is given by

$$x = \frac{-F_0}{\omega Z_m} \cos(\omega t - \phi)$$

$$\therefore \quad \textbf{Velocity,} \; v = \frac{dx}{dt} = \frac{F_0}{Z_m} \sin(\omega t - \phi) \tag{3.4.2}$$

$$= v_0 \sin(\omega t - \phi)$$

$$= v_0 \cos \left[(\omega t - \phi) + \frac{\pi}{2} \right] \tag{3.4.2a}$$

where, as before, $v_0 = F_0/Z_m =$ **velocity amplitude,** and $Z_m = \sqrt{r^2 + (\omega m - s/\omega)^2}$.

Thus (i) *the velocity always leads the displacement by a phase angle* $\pi/2$ and (ii) *lags behind the driving force by a phase angle* ϕ. The *velocity amplitude* is $F_0/Z_m =$ amplitude of driving force/mechanical impedance.

3.4.1 Variation of velocity amplitude with driving force frequency

The *velocity amplitude* v_0, corresponding to driving force $F_0 \cos \omega t$ and $F_0 \sin \omega t$, is given by

$$v_0 = \frac{F_0}{Z_m} = \frac{F_0}{\sqrt{r^2 + (\omega m - s/\omega)^2}} \tag{3.4.3}$$

We shall now study the variation of velocity amplitude with driving frequency by considering its behaviour in *three* ranges of frequency : *low, high* and *resonant frequency.*

(i) Low frequency—For low frequencies, s/ω is a large quantity and Z_m essentially depends on s i.e., *stiffness controlled.*

If $\omega \to 0$, $m\omega \to 0$ and $s/\omega \to \infty$ and r^2 can also be neglected compared to s^2/ω^2

$$\therefore \quad v_0 = \frac{F_0}{Z_m} = \frac{F_0}{s/\omega} = F_0 \frac{\omega}{s} \Rightarrow \text{ when } \omega \to 0, \;\; v_0 \to 0$$

(ii) High frequency—At high frequencies, $m\omega$ is a large quantity and Z_m essentially depends on m, *i.e., mass* or *inertia controlled.*

If $\omega \to \infty$, $m\omega \to \infty$, but $s/\omega \to 0$ and r^2 can be neglected compared to $m^2\omega^2$.

$$\therefore \quad v_0 = \frac{F_0}{Z_m} = \frac{F_0}{m\omega} \Rightarrow \text{ when } \omega \to \infty, \;\; v_0 \to 0.$$

(iii) Resonant frequency—Here $\omega = \omega_0 = \sqrt{s/m}$, the natural frequency of the oscillator and this implies $\omega_0^2 = s/m \Rightarrow \omega_0 m - s/\omega_0 = 0$. So Z_m now has the *minimum* value given by $Z_m = r$.

$$\therefore \quad v_0 = \frac{F_0}{Z_m} = \frac{F_0}{r}$$

and this is the *maximum velocity amplitude.* This frequency ω_0 is called the *frequency of velocity resonance* and is equal to the natural frequency of the oscillator.

That for *velocity resonance,* $\omega = \omega_0 = \sqrt{s/m}$, can also be proved as under. Since $v_0 = F_0/Z_m$, it implies that v_0 = maximum, when Z_m = minimum, i.e., $dZ_m/d\omega = 0$.

$$\text{Now,} \quad \frac{dZ_m}{d\omega} = \frac{d}{d\omega}\left[r^2 + \left(\omega m - \frac{s}{\omega}\right)^2\right]^{1/2}$$

$$= \frac{1}{2}\left[r^2 + \left(\omega m - \frac{s}{\omega}\right)^2\right]^{-1/2} \cdot 2(\omega m - s/\omega)(m + s/\omega^2) = 0$$

$$\text{or,} \quad (\omega m - s/\omega)(m + s/\omega^2) = 0 \quad \Rightarrow \quad (\omega m - s/\omega)(\omega m + s/\omega)/\omega = 0$$

$$\text{or,} \quad \omega^2 m^2 - s^2/\omega^2 = 0 \Rightarrow \omega^4 = s^2/m^2$$

$$\therefore \quad \omega = \sqrt{s/m} = \omega_0.$$

The variation of velocity amplitude with the frequency of the driving force is represented graphically in Fig. 3.3 for the different values of r.

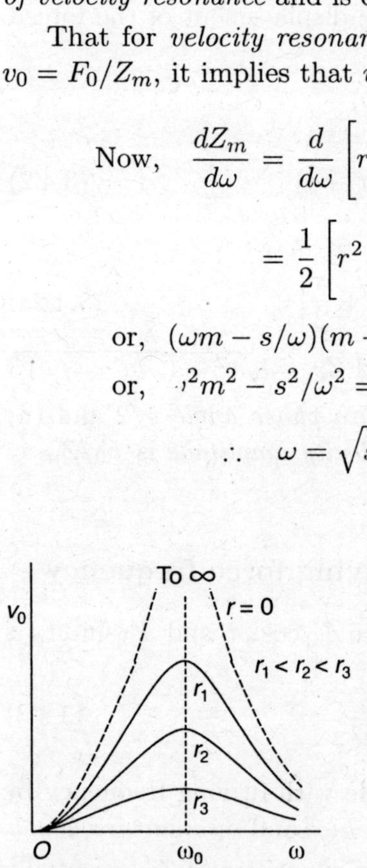

Fig 3.3

At resonance, $v_0 = F_0/r$. So the height and the sharpness of the peak is inversely proportional to r. For small r, the velocity amplitude at resonance is very large; this corresponds to curve r_1. Conversely, for large r, the velocity amplitude maximum has a smaller value. For low and high frequencies, however, the velocity amplitude tends to zero, as shown, in all the curves. At $\omega = 0$, $v = 0$. For zero damping, the velocity amplitude is infinite at $\omega = \omega_0$.

3.4.2 Variation of phase difference between velocity and driving frequency

With $F = F_0 \cos \omega t$ as the driving force, the velocity of the forced oscillator is, from equation (3.4.1),

$$v = \frac{F_0}{Z_m} \cos(\omega t - \phi)$$

$\therefore$ Phase difference between the velocity and driving force is $-\phi$, i.e., the velocity *lags begind* the driving force by an angle ϕ where ϕ is given by

$$\tan \phi = \frac{\omega m - s/\omega}{r}$$

(i) Low frequency—At low frequencies, $\omega \to 0$, $\Rightarrow m\omega \to 0$, but $s/\omega \to \infty$.

$$\therefore \quad \tan \phi = -\infty = \tan\left(-\frac{\pi}{2}\right)$$

$$\Rightarrow \quad \phi = -\frac{\pi}{2} \quad \Rightarrow \quad -\phi = \frac{\pi}{2}$$

Hence *the velocity leads the driving force by a phase angle $\pi/2$.*

(ii) High frequency—At high frequencies, $\omega \to \infty$, $\Rightarrow m\omega \to \infty$, but $s/\omega \to 0$.

$$\therefore \quad \tan \phi = +\infty = \tan\left(\frac{\pi}{2}\right) \quad \Rightarrow \quad \phi = \frac{\pi}{2} \quad \Rightarrow \quad -\phi = -\frac{\pi}{2}.$$

Hence *the velocity lags behind the driving force by a phase angle $\pi/2$.*

Thus, as the frequency of the driving force increases from $0 \to \omega_0$, the velocity *leads* the force by an angle $\pi/2$ which slowly decreases to zero when $\omega = \omega_0$ and the velocity is *in phase* with the force. If, however, the frequency is increased further, the velocity *lags* behind the force and as $\omega \to \infty$, the angle of lag approaches $\pi/2$. These characteristic features are represented graphically in the adjoining Fig. 3.4

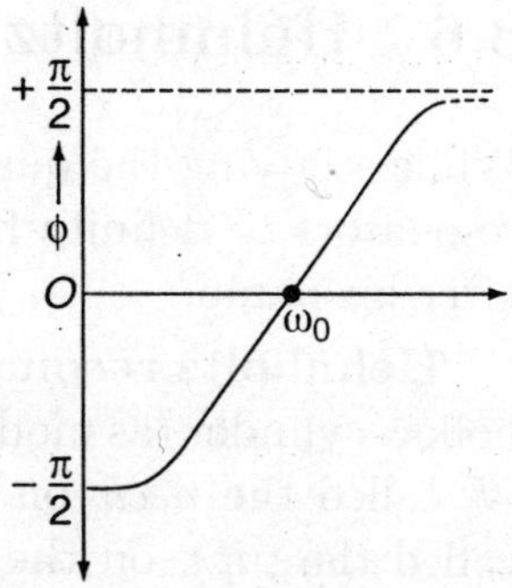

Fig 3.4

3.5 Acceleration amplitude of forced oscillator

We shall now discuss the behaviour of *acceleration* of a forced oscillator with the driving force frequency.

When the driving force is $F = F_0 \cos \omega t$, the displacement of forced oscillator in the steady state is obtained from (3.2.9).

$$x = \frac{F_0}{\omega Z_m} \sin(\omega t - \phi)$$

$$\therefore \textbf{ Acceleration, } \quad \frac{d^2x}{dt^2} = -\frac{F_0\omega^2}{\omega Z_m}\sin(\omega t - \phi)$$

$$= -\frac{F_0\omega}{Z_m}\sin(\omega t - \phi) \qquad (3.5.1)$$

$$\therefore \textbf{ Acceleration amplitude } = \frac{F_0\omega}{Z_m}, \text{ where}$$

$$Z_m = \text{mechanical impedance} = \sqrt{r^2 + (m\omega - s/\omega)^2}.$$

At **low frequencies** when $\omega \to 0$, $Z_m \to s/\omega$, in the line of argument given earlier.

$$\therefore \text{ Acceleration amplitude } = \frac{F_0\omega}{s/\omega} = \frac{F_0\omega^2}{s} \quad \textit{(stiffness controlled)},$$

being inversely proportional to s and directly to the square of the frequency.

For **large values of frequency**, s/ω is negligible, compared to ωm, and r^2 is negligible compared to $\omega^2 m^2$. So, $Z_m = \omega m$.

$$\therefore \text{ Acceleration amplitude} = \frac{F_0\omega}{\omega m} = \frac{F_0}{m} \quad (\textit{mass} \text{ or } \textit{inertia controlled}).$$

Thus, *for large frequencies, the acceleration amplitude depends on the driving force and the mass, but is independent of the frequency.*

3.6 Helmholtz resonator : Sound analysis

While studying the quality of musical notes, Helmholtz constructed a large number of resonators of definite frequencies, which are now known as **Helmholtz resonators**, after his name.

Helmholtz resonator—Helmholtz resonator consists of either a hollow sphere or a hollow cylinder (as modified by Koenig) of metal or glass with a wide cylindrical opening N, called the *neck*, on one side and a small aperture P at the end of a narrow stem, called the *pipe*, on the diametrically opposite side. The neck N receives the exciting waves, while the aperture P is applied to the ear through a tube. The resonator

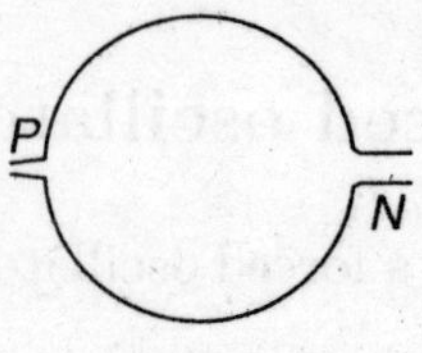

Fig 3.5

Fig 3.6

serves as an air cavity of dimensions small compared to the wavelength of sound to be detected. In the spherical type (Fig. 3.5) resonator, the volume of air cavity is fixed, while in the cylindrical type (Fig. 3.6), the volume is variable.

The internal capacity of the resonator is large compared to the neck or aperture that communicates with the atomosphere. This makes the condensation in the resonator practically constant and the to-and-fro motion of air in vibration is significant only at the neck, making the instrument more *selective* as resonator. The air in the vicinity of the neck thus *acts as a piston,* alternately compressing and rarefying the air in the resonator, *causing the air to suffer adiabatic pressure changes.*

Theory : Expression for natural frequency—Let V be the volume of the resonator, A the cross-sectional area of the neck of length l and ρ the density of air.

So, the mass of the air in the neck, that is, the equivalent *mass of the* 'piston' is $\rho l A$. If, at any instant, x be the displacement of the aerial piston outward, due to a slight fall of pressure, the increase in volume of air in the resonator is $dV = Ax$. As the pressure-change inside is *adiabatic,*

$$d(PV^\gamma) = 0 \;\Rightarrow\; dP = -\gamma P(dV/V) = -\gamma\frac{PAx}{V}$$

where γ is the usual ratio of the two specific heats of air.

The aerial piston is thus acted upon by a resultant force, directed inward, and of magnitude $AdP = -\gamma PA^2x/V$.

∴ Its equation of motion, neglecting damping, is given by

$$\rho l A\frac{d^2x}{dt^2} + \gamma\frac{PA^2x}{V} = 0$$

$$\Rightarrow \frac{d^2x}{dt^2} + \omega_0^2 x = 0 \tag{3.6.1}$$

where $\omega_0^2 = \gamma PA/(\rho Vl) = c^2A/Vl$, where $c = \sqrt{\gamma P/\rho}$, the velocity of sound in air.

So, the motion of air in the neck is *simple harmonic*; the *natural frequency* of vibration is

$$\boxed{n_0 = \frac{\omega_0}{2\pi} = \frac{c}{2\pi}\sqrt{\frac{A}{lV}}} \tag{3.6.2}$$

A resonator has thus a *single natural frequency* and when the frequency of incident sound coincides with n_0, a resonance occurs giving a loud sound. The resonator is then said to 'speak'. The resonator is *lightly damped* and *highly selective.* The other frequencies in the incident sound evoke no or little response.

• Due to open-end effect, the length l is to be replaced by $l + 0.6r$, where r is the radius of the neck, as found experimentally.

Analysis of a complex note—Helmholtz resonator is a very important analyser of sound. To find the components of a complex note, a series of resonators of different capacities and neck-areas are used. Depending on the freqencies of components, the appropriate resonator would 'speak'. It will thus pick up only the note of its own frequency out of a mixture of sounds. But the main difficulty in the process of analysis is that even for a short range of frequencies, an unusually large number of resonators are needed.

3.7 Mechanical and electrical analogues

An interesting and illuminating analogy may be drawn between the mechanical and electrical oscillations and the associated quantities.

The equation of forced vibration of a mechanical oscillator, in complex form, is

$$m\frac{d^2x}{dt^2}+r\frac{dx}{dt}+sx=F_0e^{j\omega t} \tag{3.7.1}$$

and that of an LCR-circuit, in series, is given by

$$L\frac{d^2q}{dt^2}+R\frac{dq}{dt}+\frac{q}{C}=V_0e^{j\omega t} \tag{3.7.2}$$

The two equations (3.7.1) and (3.7.2) look structurally alike and analogies between the mechanical and the electrical quantities are readily observed.

Mechanical		Electrical
1. mass, m	analogous to	self-inductance, L
2. resistive force per unit velocity, r	... ...	electric resistance, R
3. compliance, $1/s$	... ...	electrical capacitance, C
4. force, F_0	... ...	voltage, V_0
5. displacement, x	... ...	charge, q
6. velocity, $dx/dt(=v)$	... ...	electric current, $dq/dt\ (=I)$

The analogy forces us to term r as *mechanical resistance.* The steady-state current, in an a.c. circuit, is given by

$$I=\frac{V_0e^{j\omega t}}{\vec{Z}},\ \text{where}\ \vec{Z}=R+j\left(L\omega-\frac{1}{C\omega}\right)$$

and is termed the (complex) *impedance* of the circuit, having

$$\text{magnitude, } Z=\sqrt{R^2+(L\omega-1/C\omega)^2}$$
$$\text{phase angle, } \phi=\tan^{-1}(L\omega-1/C\omega)/R$$

The quantity $(L\omega-1/C\omega)$ is called the *reactance* X; $L\omega$ is the *inductive reactance* and $1/C\omega$, the *capacitive reactance.* The current (physical) *lags behind* the driving voltage by the angle ϕ.

We shall now take up the case of **mechanical oscillator**. The steady-state solution of (3.7.1) is given by

$$x=\frac{-jF_0e^{j\omega t}}{\omega\left[r+j(\omega m-s/\omega)\right]}=\frac{-jF_0e^{j\omega t}}{\omega\vec{Z}_m} \tag{3.7.3}$$

where $\vec{Z}_m=r+j(\omega m-s/\omega)=Z_me^{j\phi}$, having the magnitude $Z_m=\sqrt{r^2+(\omega m-s/\omega)^2}$ and phase angle $\phi=\tan^{-1}(\omega m-s/\omega)/r$.

In analogy with electric oscillator, $\vec{Z}_m$ is called the *mechanical impedance.* The quantity $\omega m - s/\omega$ is called the *mechanical reactance*, X_m and ωm the *inertial reactance* and s/ω the *stiffness reactance.* $1/s$ is called *compliance* by analogy.

The mechanical impedance is often expressed in *ohm* (mechanical) having the dimensions force/velocity i.e., mass/time with units kg/s in SI system.

3.8 Amplitude and velocity resonances

Although we have already discussed both these resonances, we shall summarise their *main features* once again hereunder.

A. Amplitude (or, **displacement**) **resonance**—It has been observed earlier that in forced vibration, the steady-state displacement and velocity amplitudes are both *dependent* on the *driving frequency.* If for a certain frequency of the driver, the displacement amplitude is a *maximum*, then we have the phenomenon of what is called **amplitude resonance** or **displacement resonance**.

It has been shown in Art. 3.4 that the value of the forcing frequency ω for which the *amplitude* is a *maximum* is given by

$$\omega^2 = \omega_0^2 - 2k^2 \quad \Rightarrow \quad \omega = \sqrt{\omega_0^2 - 2k^2}$$

which is *less* than both ω_0, the undamped natural angular frequency and $\sqrt{\omega_0^2 - k^2}$, the angular frequency of damped vibration of the forced oscillator.

And the maximum value of displacement amplitude is given by

$$a_{\max} = \frac{F/m}{2k\sqrt{\omega_0^2 - k^2}}$$

When the *amplitude* is *maximum,* the *potential energy,* $\frac{1}{2}sx^2$ of the forced oscillator is also a *maximum.* So, the amplitude resonance occurs along with the maximum potential energy of the forced oscillator.

The variation of displacement amplitude with driving frequency ω for *three* values of decay constant k or r has already been illustrated in Fig. 3.2 and discussed somewhat in detial. At $\omega = 0$, $a = F_0/s$ for all values of k. As k decreases, the frequency for which a is maximum moves towards ω_0, the undamped natural frequency. Also, the values of $a_{\max}$ increase with decreasing k, i.e., diminishing damping. If there is *no damping,* i.e., if $k = 0$, the amplitude a becomes ∞ at $\omega = \omega_0$.

B. Velocity resonance—We have seen earlier that with $F = F_0 \cos \omega t$ as the driving force, the velocity of the forced oscillator is

$$v = \frac{F_0 \cos(\omega t - \phi)}{\sqrt{r^2 + (\omega m - s/\omega)^2}} = v_0 \cos(\omega t - \phi), \text{ say,}$$

so that the velocity amplitude, $v_0 = F_0/\sqrt{r^2 + (\omega m - s/\omega)^2} = F_0/Z_m$.

We have also shown that as the frequency ω is varied, then v_0 becomes maximum when the mechanical reactance $\omega m - s/\omega = 0 \Rightarrow \omega = \sqrt{s/m} = \omega_0$, that is, at a frequency equal to the undamped natural frequency of the forced system.

This phenomenon of attainment of maximum for the velocity amplitude of forced oscillator at a certain frequency of the driving force (here $\omega = \omega_0$) *is known as the* **velocity resonance.**

The maximum value of the velocity amplitude v_0 is given by

$$v_{max} = F_0/r.$$

When the velocity amplitude is maximum, i.e., velocity resonance occurs, the kinetic energy $\frac{1}{2}mv^2$ of the forced system also simultaneously becomes maximum. Velocity resonance is therefore also called simply *resonance.*

The maximum kinetic energy : $E_{max} = \frac{1}{2}mv_{max}^2 = mF_0^2/2r^2$.

We have already illustrated in Fig. 3.3, the variation of velocity amplitude with frequency of the forcing system for different values of damping coefficient k or damping factor r. At $\omega = 0$, $v_0 = 0$ and v_{max} occurs at $\omega = \omega_0$ for all k's and v_{max} increases as r decreases i.e., as k decreases. For zero damping $v_{max} \to \infty$ at $\omega = \omega_0$.

Comparison with electrical case—In a series LCR circuit, resonance occurs when reactance vanishes. Analogically, *velocity resonance* occurs when mechanical reactance vanishes and velocity resonance is thus simply termed *resonance.* So, *resonance is said to occur at that frequency of forcing system for which the mechanical reactance is zero.* At resonance, therefore, the *mechanical impedance* is purely *resistive* and is equal to r.

Amplitude resonance is analogous electrically to capacitor charge resonance in a series LCR-circuit and is not very significant. When damping is small i.e., at small k-values, both amplitude resonance and velocity resonance occur practically at the same frequency (since $\omega = \sqrt{\omega_0^2 - 2k^2} \simeq \omega_0$ for very small k-value), being equal to the natural undamped frequency.

3.8.1 Sharpness of resonance : Effect of damping

The displacement amplitude of a forced oscillator is

$$a = \frac{F_0/m}{\sqrt{4k^2\omega^2 - (\omega^2 - \omega_0^2)^2}} = \frac{f_0}{\sqrt{4k^2\omega^2 - (\omega^2 - \omega_0^2)^2}}$$

At resonance, $\omega = \omega_0 \Rightarrow a = f_0/2k\omega$ where $f_0 = F_0/m$ = applied force per unit mass, k-the damping coefficient.

If the frequency ω of the applied force is increased or decreased from the resonant value ω_0, the amplitude always decreases, i.e., falls. If for a *slight* shift in the frequency ω of the applied force from its resonant value, the *fall* in amplitude is *rapid,* then *resonance* is said to be *sharp*; if however the *fall is gradual* and *slow* the *resonance* is said to be *flat.*

Effect of damping—The sharpness of resonance depends on damping, as is clear from Fig. 3.2 where amplitude vs. frequency curves for three different k-values have been drawn.

1. When $k = 0$, i.e., there is *no damping*, the amplitude a at resonance ($\omega = \omega_0$) is *infinite* and *falls rapidly* for a slight shift from resonant frequency, ω_0.

2. When k is small, i.e., *damping* is *small*, the amplitude is sufficiently large at resonance but *falls off rapidly* as the frequency of the applied force gets slightly shifted from the resonant frequency. The *resonance* is *sharp*.

3. When k is large, the amplitude at resonance is small and the amplitude falls off gradually such that even when the frequency of the applied force is little different from the resonant frequency, the amplitude is practically the same. The *resonance is flat.*

We may thus say that *when damping is small, the resonance is sharp and when damping is large, the resonance is flat.*

We shall again come back to the discussion of the sharpness of resonance in greater details in Art. 3.10 : *Sharpness of resonance revisted.*

3.9 Instantaneous and average powers

An oscillator when set to forced vibrations by an external driving force, some energy is dissipated in each cycle due to the damping resistance and the loss is to be replenished to sustain the steady state of the oscillator.

The power of the driver is the rate of its doing work. Now, in the steady state, the *instantaneous power* P of the driver is the product of the *instantaneous force*, $F(= F_0 \cos\omega t)$, and the *instantaneous velocity*, $v(= dx/dt)$ where x is the displacement.

∴ **Instantaneous power** supplied to the oscillator is

$$\begin{aligned} P &= \frac{dW}{dt} = (F_0 \cos\omega t)\frac{dx}{dt} \\ &= F_0 \cos\omega t \cdot \frac{F_0}{Z_m}\cos(\omega t - \phi), \quad \text{using (3.4.1)} \\ &= \frac{F_0^2}{Z_m}\cos\omega t \cos(\omega t - \phi) \\ &= \frac{F_0^2}{Z_m}(\cos^2\omega t\cos\phi + \sin\omega t\cos\omega t\sin\phi) \\ &= \frac{F_0^2}{Z_m}\left(\cos^2\omega t\cos\phi + \tfrac{1}{2}\sin 2\omega t\sin\phi\right) \end{aligned} \tag{3.9.1}$$

∴ **Average power** over a complete cycle is given by

$$P_{av} = \frac{1}{T}\int_0^T P\, dt,$$

where $T(= 2\pi/\omega)$ is the *period of the driving force.*

Now, the average value of $\cos^2 \omega t$, over a cycle, is $\frac{1}{2}$ and that of $\sin 2\omega t$ is zero.

$$\therefore \quad P_{av} = \frac{F_0^2}{2Z_m} \cos\phi \qquad (3.9.2)$$

the associated factor $\cos\phi$ is known, as in alternating current, the *power factor*.

$$\tan\phi = \frac{\omega m - s/\omega}{r}$$

$$\Rightarrow \cos\phi = \frac{r}{\sqrt{r^2 + (\omega m - s/\omega)^2}} = \frac{r}{Z_m}$$

$$\therefore \quad P_{av} = \frac{F_0^2}{2Z_m} \cos\phi = \frac{F_0^2}{2Z_m} \cdot \frac{r}{Z_m} = \frac{F_0^2 r}{2Z_m^2} \qquad (3.9.3)$$

• While the instantaneous power given by (3.9.1) can be, depending on time t, both positive and negative, the average power given by (3.9.3) however is always positive.

The *effective* or the *root mean square* (r.m.s) value of the simple harmonic driving force is $F_0/\sqrt{2}$ and the steady-state r.m.s value of velocity is $F_0/\sqrt{2}Z_m$.

$$\therefore \quad P_{av} = \frac{F_0^2}{2Z_m} \cos\phi = \frac{F_0}{\sqrt{2}} \cdot \frac{F_0}{\sqrt{2}Z_m} \cos\phi$$

$$= F_{rms} v_{rms} \cos\phi \qquad (3.9.4)$$

where F_{rms} and v_{rms} are the rms-values of the force and the velocity respectively. So, the average power is *less* than the product of F_{rms} and v_{rms} by the power factor $\cos\phi$.

At resonance, the mechanical reactance, $\omega m - s/\omega = 0 \Rightarrow \omega^2 = s/m = \omega_0^2$. Then $\cos\phi = r/r = 1$. So, the **average power in resonance** is

$$\boxed{(P_{av})_r = \frac{F_0^2}{2r}} \qquad (3.9.5)$$

and *resonance* and *unit power factor are synonymous*. Plainly (and also shown earlier), the force and the velocity at resonance are in phase.

The average power P_{av} depends on Z_m and in turn on frequency ω. The variation of P_{av} with frequency ω is illustrated graphically in Fig. 3.7.

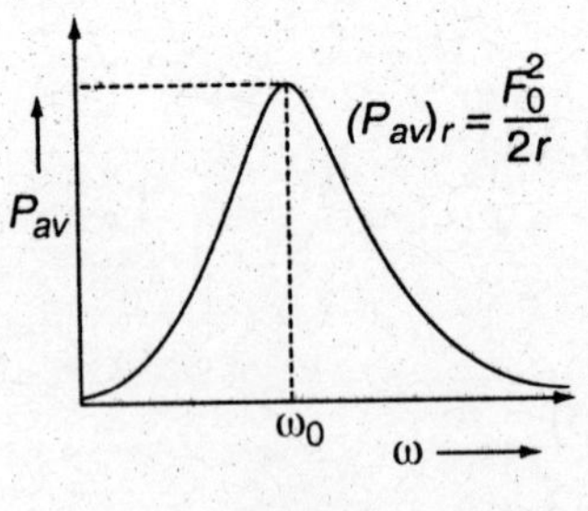

Fig 3.7

At **low frequency**, $\omega \to 0$, $m\omega \to 0$, $s/\omega \to \infty$. So, Z_m is very large $\Rightarrow$ P_{av} is very small and, for $\omega = 0$, $Z_m = \infty$, and $P_{av} = 0$.

At **high frequency**, $\omega \to \infty$, $m\omega \to \infty$, $s/\omega \to 0$. So, Z_m tends again to be very large and P_{av} is very small.

At **resonant frequency**, $\omega = \omega_0$. So, $Z_m = r$, $\cos\phi = 1$ and P_{av} is maximum = $F_0^2/2r$.

The graph shows that the peak of P_{av} corresponds to resonant frequency ω_0 and both the height of the peak and its sharpness is determined by damping resistance r. The variation of P_{av} with ω determines the *response of the oscillator to the driving force.* This graph of P_{av} vs. ω is called the **absorption curve** of the oscillator.

Average power supplied by driver = average power dissipated in overcoming resistive force — We shall now show that the the average power supplied by driver is consumed in overcoming the resistive force.

The resistive force on the damped oscillator is $r(dx/dt)$. So, the instantaneous rate of doing work against it is

$$\begin{aligned}\frac{dW}{dt} &= r\frac{dx}{dt}\left(\frac{dx}{dt}\right) = r\left(\frac{dx}{dt}\right)^2 \\ &= r\left[\frac{F_0}{Z_m}\cos(\omega t-\phi)\right]^2 \\ &= \frac{rF_0^2}{Z_m^2}\cos^2(\omega t-\phi)\end{aligned}$$

Over a complete cycle, the average value of $\cos^2(\omega t-\phi)$ is $\frac{1}{2}$. So, the average rate of doing work against the resistive force is

$$\left(\frac{dW}{dt}\right)_{av} = \frac{1}{2}\frac{rF_0^2}{Z_m^2} \tag{3.9.6}$$

Eq. (3.9.3) and (3.9.6) show that $P_{av} = (dW/dt)_{av} \Rightarrow$ the **average power supplied** *by the driving force to the oscillator is consumed in* **overcoming the resistive force.**

• It is interesting to compare the expression for the average power obtained here : $P_{av} = F_{rms}v_{rms}\cos\phi$ with the expression for power in a.c. circuits where it is given as : $P = (V_0/\sqrt{2})\cdot(I_0/\sqrt{2})\cos\phi$, where $\cos\phi$ is the *power factor* and $V_0/\sqrt{2}$ and $I_0/\sqrt{2}$ are respectively the rms voltage, V_{rms} and the rms current, I_{rms}.

The applied emf in a.c. circuits therefore corresponds to applied driving force and electric impedance Z_e to the mechanical impedance Z_m.

• Energy supplied to the forced oscillator is *not stored* in it. For, it is proved here that the energy supplied by the external driving force is equal to the energy dissipated against the resistive or frictional forces. The energy supplied only helps in maintenance of the oscillations of the forced oscillator.

3.10 Sharpness of resonance revisited

Earlier, in Art. 3.8.1, we explained in brief what is meant by sharpness of resonance. We shall go here a bit more deeply into it.

We have already seen that if P_{av} is plotted against the frequency ω of the driver to obtain the *absorption curve* of the oscillator, P_{av} atttains its peak value $F_0^2/2r$ at resonance frequency $\omega = \omega_0 = \sqrt{s/m}$ and falls off on either side of ω_0 (Fig. 3.7). When r is small, the peak value of P_{av} is high and conversely. The absorption curves also show that *smaller the damping, greater is the drop in* P_{av} when ω is shifted slightly from ω_0. This implies that resonance is *sharp* in case of *small damping* and is *flat* or *broad* when *damping* is *large.* A comparison of curves (a) and (b) readily convinces one of this fact.

Thus, *sharpness of resonance* may be *defined* as the rapidity with which average power P_{av} supplied by the driving force drops as its frequency ω differs from the resonance frequency ω_0.

We may look at higher sharpness of resonance for low damping and its converse from another slightly different kind of plotting, namely a plot $P_{av}/(P_{av})_r$ against ω. Two such curves for two r-values (r_1 and r_2) are shown in Fig. 3.8.

$$\text{Now,}\quad \frac{P_{av}}{(P_{av})_r} = \frac{F_0^2 r/2Z_m^2}{F_0^2/2r} = \frac{r^2}{Z_m^2} = \frac{r^2}{r^2 + (\omega m - s/\omega)^2} = \frac{4k^2m^2}{4k^2m^2 + m^2\left(\omega - \omega_0^2/\omega\right)^2}$$

$$= \frac{4k^2}{4k^2 + \omega_0^2\left(\dfrac{\omega}{\omega_0} - \dfrac{\omega_0}{\omega}\right)^2} = \frac{4k^2}{4k^2 + \omega_0^2\Delta^2} \tag{3.10.1}$$

where the quantity Δ is given by

$$\Delta = \frac{\omega}{\omega_0} - \frac{\omega_0}{\omega} \tag{3.10.2}$$

At resonance, $\omega = \omega_0$. So $\Delta = 0$ and $P_{av}/(P_{av})_r = 1$; but *below* and *above* the *resonance,* $\Delta \neq 0$ and $P_{av}/(P_{av})_r < 1$. *Quantitatively, the* **sharpness of resonance** S_r *is defined as the reciprocal of* $|\Delta|$ *at which* $P_{av}/(P_{av})_r = \frac{1}{2}$. From Fig. 3.8, we note that there are *two* values of ω, namely ω_1 and ω_2 for which P_{av} is half its resonance value. These two frequencies ω_1 and ω_2 are known as **half-power** frequencies — $\omega_1 (< \omega_0)$, the *lower half-power frequency* and $\omega_2 (> \omega_0)$ the *upper half-power frequency* respectively.

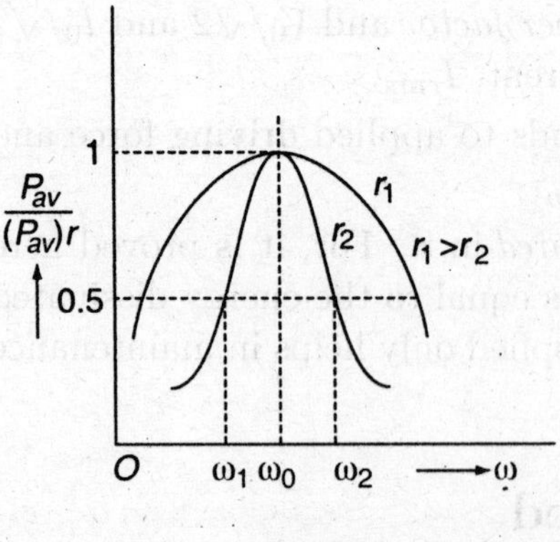

Fig 3.8

$\therefore$ From (3.10.1), we obtain by imposing the quantitative definition of sharpness

$$\frac{4k^2}{4k^2 + \omega_0^2\Delta^2} = \frac{1}{2} \Rightarrow \omega_0^2\Delta^2 = 4k^2$$

$$\text{or,}\quad \omega_0\Delta = \pm 2k \Rightarrow \omega - \omega_0^2/\omega = \pm 2k, \text{ using (3.10.2)}$$

$$\text{or,}\quad \omega^2 \mp 2k\omega - \omega_0^2 = 0 \tag{3.10.3}$$

This is a quadratic equation in ω having ω_1 and ω_2 (the two half-power frequencies) as its roots.

$$\therefore\quad \omega_1 = \sqrt{k^2 + \omega_0^2} - k; \quad \omega_2 = \sqrt{k^2 + \omega_0^2} + k \tag{3.10.4}$$

restricting ourselves to the positive roots only.

$$\therefore \quad S_r = \textit{shartpness of resonance} = \frac{1}{|\Delta|} = \frac{\omega_0}{2k} = \frac{\omega_0 m}{r} \tag{3.10.5}$$

But $2k = \omega_2 - \omega_1$, using (3.10.4). So from (3.10.5), we obtain

$$\boxed{S_r = \frac{\omega_0}{\omega_2 - \omega_1}} \tag{3.10.6}$$

and the quantity $\omega_0/(\omega_2-\omega_1)$ is called the **quality factor Q**, in analogy with electrical circuit resonance.

The frequency range $(\omega_2 - \omega_1)$ is called the **half-power band width** or simply *bandwidth.*

$$\therefore \quad \boxed{Q = S_r = \frac{\omega_0}{(\omega_2 - \omega_1)}} \tag{3.10.7}$$

Thus, *Q is a measure of the sharpness of resonance.* It is obvious that smaller the damping, smaller is the bandwith and higher is the quality factor. So, the response of the forced oscillator to the driving force is highly *selective*, being too *rich* at resonant frequency and too *poor* at off-resonant frequencies.

The maximum kinetic energy at resonance is

$$E_{\max} = \tfrac{1}{2} m (v_0)^2_{\max} = m F_0^2 / 2r^2 \tag{3.10.8}$$

The energy dissipated per cycle at resonance is

$$E_{dis} = F_0^2 T/2r = F_0^2 \pi/(\omega_0 r) \quad \text{[using (3.9.5)]} \tag{3.10.9}$$

$$\therefore \quad \frac{E_{\max}}{E_{dis}} = \frac{mF_0^2}{2r^2} \times \frac{\omega_0 r}{F_0^2 \pi} = \frac{\omega_0 m}{2\pi r} = \frac{Q}{2\pi} \quad (\because\ Q = \omega_0 m/r) \tag{3.10.10}$$

$$\therefore \quad Q = 2\pi \frac{E_{\max}}{E_{dis}} = 2\pi \times \frac{\text{maximum kinetic energy at resonance}}{\text{energy dissipated per cycle at resonance}}$$

which may be taken as an *alternative definition* of the quality factor Q.

• $Q = \omega_0 m/r$ is analogous to the quality factor of an electrical circuit given by $Q = \omega_0 L/R$ where $\omega_0 = 1/\sqrt{LC}$. This has been discussed in detail under Chapter : *Alternating Current in* **Electricity and magnetism** part. Q-value is also known as the **figure of merit.**

Quality factor in terms of wavelength—Let n_1, n_2 and n_0 be the frequencies, and λ_1, λ_2 and λ_0 the wavelengths corresponding to the angular frequencies ω_1, ω_2 and ω_0 so that

$$\begin{aligned} \omega_0 &= 2\pi n_0 = 2\pi c/\lambda_0; \\ \omega_1 &= 2\pi n_1 = 2\pi c/\lambda_1 \\ \omega_2 &= 2\pi n_2 = 2\pi c/\lambda_2, \quad \text{where } c = \text{velocity.} \end{aligned}$$

$$\therefore \quad Q = \frac{\omega_0}{\omega_2 - \omega_1} = \frac{2\pi c/\lambda_0}{2\pi c(1/\lambda_2 - 1/\lambda_1)} = \frac{1/\lambda_0}{(1/\lambda_2 - 1/\lambda_1)}$$

$$= \frac{1}{\lambda_0}\frac{\lambda_1\lambda_2}{\lambda_1 - \lambda_2} = \frac{\lambda_0}{\lambda_1 - \lambda_2} \tag{3.10.11}$$

since $\lambda_1\lambda_2 = \lambda_0^2$, approximately.

Quality factor as a measure of sharpness of resonance—The quality factor measures the *sharpness of resonance* or the *sharpness of tuning. Higher* the *quality factor* Q for a given resonant frequency ω_0, *smaller* is the *bandwidth* $(\omega_2 - \omega_1)$, i.e., *sharper is the tuning.* This feature is utilized in increasing the *selectivity* of radio sets. The sharpness of response of a given circuit reproduces the radio signals without interference from signals of frequencies close to it.

***Q*-value as amplification factor**—The displacement amplitude in forced oscillation is

$$a = \frac{f_0}{\sqrt{4k^2\omega^2 + (\omega^2 - \omega_0^2)^2}}$$

When $\omega \to 0$, the amplitude is:

$$a_0 = \frac{f_0}{\omega_0^2} = \frac{f_0}{s/m} = \frac{f_0 m}{s} = \frac{F_0}{s}$$

The displacement at resonance is given by

$$a_{\max} = \frac{F_0}{\omega' r}, \quad \text{where } \omega' = \sqrt{\omega_0^2 - k^2} = \sqrt{s/m - r^2/4m^2}$$

The **amplification factor** *is the ratio of the maximum displacement at resonance to the maximum displacement when the frequency of the applied force approaches zero.*

$$\text{Now,} \quad \left(\frac{a_{\max}}{a_0}\right)^2 = \frac{F_0^2}{\omega'^2 r^2}\cdot\frac{s^2}{F_0^2} = \frac{s^2}{\omega'^2 r^2} = \frac{m^2\omega_0^4}{r^2(\omega_0^2 - r^2/4m^2)}$$

$$= \frac{m^2\omega_0^4}{r^2\omega_0^2\left(1 - \dfrac{r^2}{4m^2\omega_0^2}\right)} = \frac{Q^2}{\left(1 - \dfrac{1}{4Q^2}\right)}$$

$$(\because s/m = \omega_0^2; \;\; Q = m\omega_0/r)$$

$$\therefore \quad \frac{a_{\max}}{a_0} = \frac{Q}{(1 - 1/4Q^2)^{1/2}} = \frac{Q}{1 - 1/8Q^2} \tag{3.10.12}$$

If Q is very large, $1/8Q^2$ is very small compared to 1 and can therefore be neglected.

$$\therefore \quad \boxed{\frac{a_{\max}}{a_0} = Q}$$

So, it may be concluded that the Q-value of an oscillatory system is equal to the *amplification factor* : the displacement at low frequencies is amplified by a factor Q at displacement resonance.

3.11 Illustrated Examples

▶ **Example 1.** *An object of mass* 2 kg *hangs from a massless spring that gets extended by* 2.5 cm *when the object is attached. The top end of the spring oscillates simple harmonically with an amplitude of* 2 mm. *If Q of the system be* 20 *and* $g = 10$ m/s^2, *find* (i) *the angular frequency ω_0 of free undamped oscillation,* (ii) *the amplitude of free oscillations at $\omega = \omega_0$ and* (iii) *the mean power input to maintain forced oscillations at an angular frequency* 1% *greater than ω_0.* **(Burdwan Hons.)**

Solution. Let s be the spring constant. So at equilibrium of the suspended mass,

$$s \times 2.5 \times 10^{-2} = 2 \times 10 \quad \Rightarrow \quad s = 8 \times 10^2 = 800 \text{ N/m}^{-1}.$$

(i) Free undamped frequency, $\omega_0 = \sqrt{s/m} = \sqrt{800/2} = \mathbf{20\ rad.\ s^{-1}}$

(ii) Since $Q(= 20)$ is large, k is small and the displacement amplitude becomes maximum at $\omega \simeq \omega_0$.

∴ Maximum displacement amplitude,

$$a_{\max} = \frac{F_0/m}{2k\sqrt{\omega_0^2 - k^2}} \simeq \frac{F_0}{2mk\omega_0} = \frac{F_0 Q}{m\omega_0^2}$$

$$\therefore \quad a_{\max} = \frac{800 \times 2 \times 10^{-3} \times 20}{2 \times 20^2} = 40 \times 10^{-3} \text{ m} = \mathbf{4\ cm}$$

$$\left(\because \ F_0 = s \times 2 \times 10^{-3} = 800 \times 2 \times 10^{-3}\right)$$

(iii) Mean power input, $P_{av} = F_0^2 r / 2Z_m^2$.

$$\text{Here } Z_m^2 = r^2 + (\omega m - s/\omega)^2. \quad \text{But } r = \frac{m\omega_0}{Q} = \frac{2 \times 20}{20} = 2 \text{ kg s}^{-1}.$$

$$\text{Also, } \omega = \frac{101}{100}\omega_0 = \frac{101}{100} \times 20 = 20.2 \text{ rad.s}^{-1}.$$

$$\therefore \quad Z_m^2 = 4 + (20.2 \times 2 - 800/20.2)^2 = 4.64 \text{ kg}^2\text{s}^{-2}$$

$$\therefore \quad P_{av} = \frac{800^2 \times 2^2 \times 10^{-6} \times 2}{2 \times 4.64} = \mathbf{0.55\ W}$$

▶ **Example 2.** *An object of mass* 0.1 kg *hangs from a spring of spring constant* 100 Nm^{-1}. *A resistive force rv acts on it where v is the velocity in* ms^{-1} *and* $r = 1$ kg.s^{-1}. *If the object be subjected to a driving force $F = F_0 \cos \omega t$ with $F_0 = 2$ N and $\omega = 50$ rad.s^{-1}, then calculate the amplitude of oscillation and the phase, relative to the force, in the steady state.*

Solution. Here, $\omega_0 = \sqrt{s/m} = \sqrt{100/0.1} = \sqrt{1000}$ rad.s^{-1}, $f_0 = F_0/m = 2/0.1 =$ 20 N.kg^{-1} and $k = r/2m = 1/2 \times 0.1 = 5$ s^{-1}.

In the steady state, the amplitude of oscillation is

$$a = \frac{f_0}{\sqrt{4k^2\omega^2 + (\omega^2 - \omega_0^2)^2}} = \frac{20}{\sqrt{4 \times 5^2 \times 50^2 + (50^2 - 1000)^2}}$$

$$= 0.0126 \text{ m} = \mathbf{1.26 \text{ cm}}$$

If ϕ be the phase angle by which the oscillations lag behind the force, then

$$\tan\phi = \frac{2k\omega}{\omega^2 - \omega_0^2} = \frac{2 \times 5 \times 50}{50^2 - 1000} = \tan 161.7°. \quad \therefore \quad \boldsymbol{\phi = 161.7°}$$

► **Example 3.** *If ω_1 and ω_2 are the two half-power frequencies and ω_0 the resonant frequency of a forced oscillator, prove that $\omega_0 = \sqrt{\omega_1\omega_2}$.*

Solution. The half-power frequencies ω_1 and ω_2 are the two positive roots of the quadratic equation

$$\omega^2 \mp 2k\omega - \omega_0^2 = 0$$

$$\therefore \quad \omega_1 = \sqrt{k^2 + \omega_0^2} - k; \quad \omega_2 = \sqrt{k^2 + \omega_0^2} + k$$

$$\therefore \quad \omega_1\omega_2 = (k^2 + \omega_0^2) - k^2 = \omega_0^2$$

$$\Rightarrow \quad \boldsymbol{\omega_0 = \sqrt{\omega_1\omega_2}}$$

► **Example 4.** *A damped harmonic oscillator, with small damping, is driven by a force $F = F_0 \cos\omega t$ where the amplitude F_0 is kept constant, but its angular frequency ω is varied. The steady state amplitude at very low ω is 0.1 mm and attains a maximum value of 10 cm when ω = 100 rad.s^{-1}. Find the Q-value of the system and the time taken for its energy being reduced to 1/e of its initial value. What is the half-width of the power resonance curve?*

Solution. (a) Since the damping is small, Q-value is very large.

$$\therefore \quad \frac{a_{\max}}{a_0} \simeq Q$$

where $a_{\max}$ is the maximum amplitude and a_0 the amplitude when $\omega \to 0$ (very low).

Here, $a_0 = 0.1$ mm $= 0.01$ cm and $a_{\max} = 10$ cm. Therefore,

$$Q = \frac{10}{0.01} = \mathbf{1000}$$

(b) The time in which energy falls to $1/e$ of its initial value is termed the *energy decay time*, τ.

$$\therefore \quad \tau = \frac{Q}{\omega_0} = \frac{1000}{100} = \mathbf{10 \text{ s}}$$

(c) The half-width of power resonance curve is given by

$$\frac{1}{2}(\omega_2 - \omega_1) = \frac{1}{2}\frac{\omega_0}{Q} = \frac{1}{2}\cdot\frac{100}{1000} = \frac{1}{2\times 10} = \mathbf{0.05\ rad/s}$$

▶ **Example 5.** *Show that for a forced vibration, the total energy of the vibrating system is not constant and that in such a case,*

$$\frac{\textit{average potential energy}}{\textit{average kinetic energy}} = \frac{\omega_0^2}{\omega^2}$$

where $\omega_0 =$ *natural angular frequency* $= \sqrt{s/m}$. **(Calcutta Hons.)**

Solution. The displacement x of the particle of mass m in forced vibration, is given by

$$x = a\cos(\omega t - \phi)$$

$$\therefore \quad \text{velocity,} \quad \dot{x} = -a\omega \sin(\omega t - \phi)$$

$\therefore$ Kinetic energy of the particle $E_K = \frac{1}{2}m\dot{x}^2 = \frac{1}{2}ma^2\omega^2\sin^2(\omega t - \phi)$

Potential energy of the particle, $E_P = \frac{1}{2}sx^2 = \frac{1}{2}sa^2\cos^2(\omega t - \phi)$

$\therefore$ Sum of the kinetic and the potential energy $E_K + E_P$ is plainly not constant.

$$\text{Now,} \quad \bar{E}_K = \text{average K.E.} = \tfrac{1}{2}ma^2\omega^2\cdot\tfrac{1}{2} = \tfrac{1}{4}ma^2\omega^2$$

$$\bar{E}_P = \text{average P.E.} = \tfrac{1}{2}sa^2\cdot\tfrac{1}{2} = \tfrac{1}{4}sa^2$$

since, over a full cycle, the average values of $\sin^2(\omega t - \phi)$ and $\cos^2(\omega t - \phi)$ are each equal to $\frac{1}{2}$.

$$\therefore \quad \frac{\bar{E}_P}{\bar{E}_K} = \frac{\frac{1}{4}sa^2}{\frac{1}{4}ma^2\omega^2} = \frac{s}{m\omega^2} = \boldsymbol{\frac{\omega_0^2}{\omega^2}}$$

where $\omega_0 =$ natural angular frequency $= \sqrt{s/m}$.

▶ **Example 6.** *A root mean square voltage of* 100 V *is applied to a series resonant circuit having* $R = 10\ \Omega$, $L = 10$ mH *and* $C = 1\ \mu$F. *Calculate* (i) *the natural frequency,* (ii) *the current in resonance,* (iii) *the Q-value of the circuit at resonance, and* (iv) *the bandwidth of the circuit.*

Solution. (i) The natural frequency, $n = \dfrac{1}{2\pi\sqrt{LC}}$, with usual meanings of symbols.

$$\therefore \quad n = \frac{1}{2\times 3.14\sqrt{10\times 10^{-3}\times 1\times 10^{-6}}} = \mathbf{1592\ Hz}$$

(ii) Current at resonance, $I_{\max} = \dfrac{V_{\max}}{R} = \dfrac{\sqrt{2}V_{rms}}{R}$

$$= \frac{\sqrt{2}\times 100}{10} = 10\sqrt{2} = \mathbf{14.14\ A}$$

(iii) Q-value of circuit $= \omega_0 \dfrac{L}{R} = \dfrac{1}{\sqrt{LC}} \cdot \dfrac{L}{R} = \dfrac{1}{R}\sqrt{\dfrac{L}{C}}$

$$\therefore \quad Q = \frac{1}{10}\sqrt{\frac{10 \times 10^{-3}}{1 \times 10^{-6}}} = \frac{1}{10} \times \frac{10^{-1}}{10^{-3}} = \mathbf{10}$$

(iv) Bandwidth of the circuit $= \omega_2 - \omega_1$ and $Q = \omega_0/(\omega_2 - \omega_1)$.

$$\therefore \quad \text{Bandwidth, } \omega_2 - \omega_1 = \frac{\omega_0}{Q} = \frac{1}{\sqrt{LC}} \cdot \frac{1}{Q} = \frac{1}{10} \times \frac{1}{\sqrt{10 \times 10^{-3} \times 1 \times 10^{-6}}}$$

$$= \frac{1}{10} \times \frac{1}{10^{-4}} = \frac{1}{10^{-3}} = \mathbf{1000\ rad/s.}$$

Review Questions

1. Explain free vibration, forced vibration and resonance. Give one example of each. **(Himachal Univ.)**

2. Define and explain what you mean by the impedance of an electrical circuit. Find an expression for it.

3. Find the value of the mechanical impedance of a physical system subjected to driving forces. What is its physical significance? **(Punjabi Univ.)**

4. Distinguish between transient and steady states in a forced oscillator. Explain the transient and steady state behaviour of a mechanical oscillator driven by a periodic force $F = F_0 e^{j\omega t}$. Show that the driven oscillator is always behind the driving force in phase. **(Kanpur and Punjab Univ.)**

5. Discuss the behaviour of displacement versus driving force frequency in case of a forced oscillator. Prove that the resultant frequency of the driving force is slightly less than the natural frequency of the displacement. **(Guru Nanak and Himachal Univ.)**

6. If the displacement y of a forced oscillator is given by $y = F_0 \sin(\omega t - \phi)/\omega Z_m$, with usual significance of the symbols, then show that the displacement resonance occurs at a frequency ω slightly less than $\omega_0 (= \sqrt{s/m})$, the frequency of free oscillations.

7. For a mechanical oscillator driven by a force $F = F_0 \cos \omega t$, discuss the variation of phase difference between displacement and driving force, with driving force frequency. **(Kanpur Univ.)**

8. Derive an expression for the velocity of a forced oscillator. Discuss the variation of velocity amplitude with the driving force frequency and show its behaviour

graphically. Discuss also the variation of phase difference between velocity and driving force frequency. **(Punjab Univ.)**

9. Describe, with the associated theory, a Helmholtz resonator. How is it used to analyse a complex note? **(Calcutta and Vidyasagar Univ.)**

10. Derive an expression for the average power supplied to a forced oscillator by an external driving force $F = F_0 \cos \omega t$. **(Himachal Univ.)**

11. Show graphically the variation of average power of a forced oscillator with driving force frequency. **(Punjab Univ.)**

12. Derive an expression for the power supplied to the forced oscillator and show that it is equal to the average power being dissipated.

13. Find the Q-value of an oscillator in terms of resonance absorption bandwidth. **(Punjab and Himachal Univ.)**

14. Show that the average power supplied by the driving periodic force to an oscillator under forced virbation is

$$P_{av} = \frac{(F_0^2 k/m)}{\omega_0^2 \Delta^2 + 4k^2},$$

with usual meanings of the symbols.

Draw curves of P_{av} vs. Δ and hence discuss the sharpness of resonance and its dependence on damping coefficient k.

15. Find the Q-value of an oscillator in terms of energy decay. What is figure of merit of a vibration and how is it defined in terms of bandwidth? 'Q is a measure of sharpness of resonance'.—Explain. **(Guru Nanak Univ.)**

16. If the displacement at resonance is given by

$$A_{\max} = F_0/\omega' r, \text{ where } \omega' = \sqrt{s/m - r^2/4m^2},$$

show that the displacement at low frequencies is amplified by a factor Q at displacement resonance. **(Guru Nanak and Himachal Univ.)**

17. Find the Q-value of an oscillatory system as an amplification factor. **(Punjab Univ.)**

18. Prove that the bandwidth of the resonance absorption curve defines the phase angle range $\tan \phi = \pm 1$. **(Guru Nanak and Punjab Univ.)**

19. Identify the forces acting on a particle under forced vibration. Formulate the differential equation of motion of the particle along x-axis. Solve the equation for velocity of the particle as a function of time. Hence fnd the velocity amplitude at velocity resonance.

20. Write short explanatory notes on :
    (a) Free vibration, forced vibration and resonance.
    (b) Electrical and mechanical impedance.
    (c) Mechanical reactance and its constituents.
    (d) Sharpness of resonance and its significance.
    (e) Helmholtz resonator and its applications.
    (f) Q-value and bandwidth, and figure of merit of a vibration.

Short Answer and Objective Questions

1. What is mechanical reactance? What are its important constituents? **(Guru Nanak Univ.)**
2. Of the natural frequency of damped oscillations and frequency of displacement resonance, which one is greater?
3. Why is there a large amplitude, when the frequency of external periodic force is the same as the natural frequency of the body? **(Punjab Univ.)**
4. Show that in a forced mechanical oscillator, the acceleration amplitude at high frequency is independent of the frequency.
5. Show that the displacement resonance occurs at a frequency slightly less than the frequency of velocity resonance.
6. What is sharpness of resonance? What are the effects of damping on sharpness of resonance? **(Kurukshetra and Lucknow Univ.)**
7. Why should two tuning forks be very accurately in unison to show resonance, while a tuning fork and air column require to be only approximately tuned? **(Kurukshetra Univ.)**
8. When is the power supplied to an oscillator by the driving force a maximum? **(Himachal Univ.)**
9. Show that, in the steady state, the amplitude and phase of a driven oscillator adjust themselves so that the average power supplied by the driving force just equals that being dissipated by the frictional force.
10. Show that the power dissipated per cycle by the frictional force $r\dot{y}$ is $\pi r \omega y_{\max}^2$ where $y_{\max} = F_0/\omega(Z_m)_{\min}$. **(Guru Nanak Univ.)**
11. Is the energy stored in a forced oscillator? Explain. **(Punjabi Univ.)**
12. What is the physical significance of the Q-value of a forced oscillator?

13. What is figure of merit of a vibration? How is it defined in terms of the bandwidth? **(Himachal Univ.)**

14. The quality factor Q is a measure of the sharpness of resonance in case of a forced oscillator. Explain. **(Guru Nanak Univ.)**

15. Complete the sentence "At resonance, the displacement...... the driving force by an angle......". **(Punjab Univ.)**

16. Which of the following is the phase relationship between the displacement x of the forced oscillator and the applied force F?

    (a) x lags behind F by ϕ, (b) x leads F by ϕ,

    (c) x lags behind F by $\phi + \pi/2$, (d) x leads F by $\phi + \pi/2$.

    (Himachal Univ.)

17. What is the value of the phase angle ϕ when the frequency of driving force is equal to the natural frequency of the undamped oscillator?

    (a) $-\pi/2$, (b) zero, (c) $\pi/2$, (d) π.

18. What is the phase difference between the driving force and velocity of forced oscillator?

    (a) ϕ, (b) $\pi/2 + \phi$, (c) $\pi - \phi/2$, (d) $\pi/2 - \phi$.

19. The quality factor of an electric oscillator is given by

    (a) $Q = L/\omega_0 R$, (b) $Q = R/\omega_0 L$, (c) $Q = \omega_0 L/R$, (d) $Q = \omega_0 LR$.

20. The quality factor is proportional

    (a) directly to the damping factor r,

    (b) inversely to the damping factor r,

    (c) directly to the square of the damping factor r,

    (d) inversely to the square of the damping factor r. **(Himachal Univ.)**

21. The resonant frequency of an electric oscillator is given by

    (a) $n = 2\pi/\sqrt{LC}$, (b) $n = 2\pi\sqrt{LC}$, (c) $n = 1/2\pi\sqrt{LC}$, (d) none of these.

Problems

1. A damped oscillator consists of a mass 200 g attached to a spring of constant 100 Nm^{-1} and damping constant 5 Nm^{-1}. It is driven by a force $F = 6\cos\omega t$ Newton, where $\omega = 30\ \text{s}^{-1}$. If the displacement in the steady state is $x = A\sin(\omega t - \phi)$ metre, find the values of A and ϕ. Also calculate the power supplied to the oscillator. **(Punjabi Univ.)** [3.52 mm; 28°6′; 2.8 W]

2. Light of wavelength 6×10^{-5} cm is emitted by an electron in an atom (a damped simple harmonic oscillator) with a quality factor 3×10^6. Find the width of the spectral line from such an atom from resonance bandwidth. [2×10^{-3} A]

3. The line width of orange line of Krypton 86 at $\lambda = 6058$ A is found to be 5.50×10^{-3} A. Calculate (a) line frequency and (b) the line width in Hertz.
(Guru Nanak Univ.) [4.95×10^{14} Hz; 5.45×10^{20} Hz]

4. The forced harmonic oscillations have displacement amplitudes at frequencies $\omega_1 = 400$ rad s^{-1} and $\omega_2 = 600$ rad s^{-1}. Find the resonant frequency at which the displacement amplitude is maximum. [510 rad s^{-1}]

5. A particle of mass 0.01 kg is subjected to a restoring force 0.5 N/m, a damping force 10^{-4} kg s^{-1} and an external sinusoidal force of constant amplitude. Find the frequency of the driving force for which there will be velocity resonance. Also calculate the Q-factor and the bandwidth of resonance.
[7.07 rad s^{-1}; 707, 10^{-2} rad s^{-1}]

6. To the lower end of a massless spring, of spring constant 10 N/m and hanging vertically from a rigid suppot, is connected to a particle of mass of 0.1 kg. A resistive force $-kv$ acts on the oscillating mass where v is the particle velocity and k a constant. Find (a) the value of k, (b) the Q-value of oscillator and (c) the percentage change in frequency of oscillation due to damping.
(Delhi Hons.) [8×10^{-3} kg/s, 125, 0.02]

7. A particle of mass 0.1 kg is suspended from a spring of spring constant 200 N/m. A resistive force $-kv$ acts on the particle, v being the particle velocity in m/s and $k = 1$ Ns/m. A periodic force 2.5 cos $50t$ in Newton is applied to the particle. Find the amplitude of oscillation of the particle and the phase difference between the driving force and the particle displacement in steady state. [3.54 cm, $-45°$]

Chapter 4

Coupled Oscillations

4.1 Introduction

In forced oscillations, discussed in the previous chapter, we assumed that the forcing or driving system remains essentially unaffected by the forced oscillations of the driven system. The forcing system acts only as a source of a periodic force with *no feed-back* of energy from the driven system. We shall now discuss the oscillations of coupled systems where we cannot disregard the *feed-back of energy* from the driven system to the driver.

Let us, for example, take two coupled oscillating systems as illustrated in Fig. 4.1. In (a), two simple pendulums A, B have their bobs connected by a spring S and, in (b) there are two coupled LC-circuits. These systems have *two degrees of freedom.* When the moving parts of such a system oscillate say, in the plane of the paper, there are only two variables by which the motion can be completely specified. In the two coupled pendulums, the two variables are the *displacements* of the two bobs, and in the two coupled LC-circuits, the two variables are the *charges* on the two capacitors or *currents* in the two circuits.

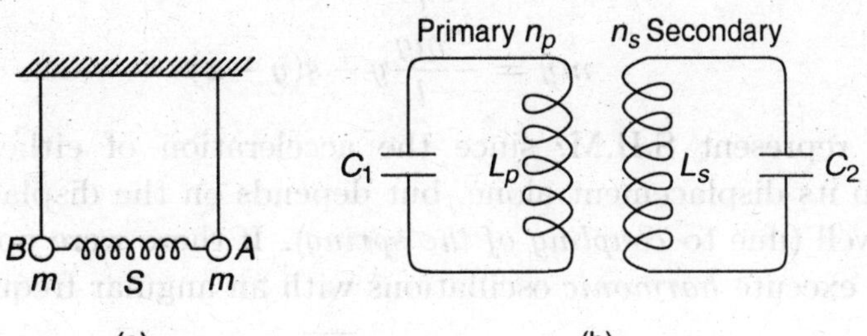

Fig 4.1

In the above examples, the two oscillators are coupled so tightly that one can no longer distinguish between the driver and the driven systems and both the oscillators are to be treated *at par.* In physics, there is no dearth of coupled oscillatory systems.

As a matter of fact, *oscillators rarely exist in isolation.* A string, for example, consists of a large number of particles coupled to one another. So are the atoms in a crystal lattice of a solid. The atoms are capable of oscillations in many different ways or modes, each mode having its own characteristic frequency.

4.2 Two coupled pendulums

In the light of the discussion of Art. 4.1., we may now *define a coupled oscillator.*

Coupled oscillator—*If two or more oscillators are linked together such that an exchange of energy occurs between them then they are called coupled oscillators.*

We shall consider a system of two identical pendulums A, B—each of mass m and length l—coupled through a *linear* spring S of force constant s. The separation between the bobs is such that the spring is relaxed in equilibrium position (Fig. 4.2). Let the pendulums be set into vibration with a *small* amplitude in the plane of the paper and let x and y be the displacements (in the same direction) at an instant t (Fig. 4.3). The spring is stretched or compressed depending on if $y > x$ or $x > y$. Assuming $y > x$, the tension $s(y - x)$ will act *against* the acceleration $\ddot{y}$ of pendulum B but *in favour* of the acceleration $\ddot{x}$ of pendulum A.

Fig 4.2

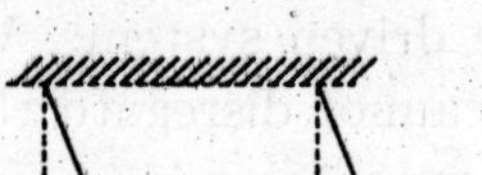

Fig 4.3

$\therefore$ The equations of motion of pendulums A and B, for small oscillations in a plane, are

$$m\ddot{x} = -\frac{mg}{l}x + s(y - x) \tag{4.2.1}$$

$$m\ddot{y} = -\frac{mg}{l}y - s(y - x) \tag{4.2.2}$$

which do *not* represent S.H.M. since the acceleration of either pendulum is not proportional to its displacement alone, but depends on the displacement of the other pendulum as well (due to *coupling of the spring*). If there were no spring, $s = 0$, each pendulum will execute *harmonic* oscillations with an angular frequency

$$\omega_0 = \sqrt{g/l}$$

In terms of ω_0, we have, from (4.2.1) and (4.2.2),

$$\ddot{x} = -\omega_0^2 x + \frac{s}{m}(y - x) \tag{4.2.3}$$

$$\ddot{y} = -\omega_0^2 y - \frac{s}{m}(y - x) \tag{4.2.4}$$

To find the *effect of coupling* on each pendulum, we choose *two new coordinates*

$$X = x + y; \quad Y = x - y$$

$$\therefore \; \ddot{X} = \ddot{x} + \ddot{y}; \quad \ddot{Y} = \ddot{x} - \ddot{y}$$

Adding and subtracting (4.2.3) and (4.2.4) respectively, we get, on using the new coordinates

$$\ddot{X} + \omega_0^2 X = 0 \tag{4.2.5}$$

$$\ddot{Y} + \left(\omega_0^2 + \frac{2s}{m}\right) Y = 0 \tag{4.2.6}$$

While (4.2.3) and (4.2.4) each have *two* variables, (4.2.5) and (4.2.6) have *only one* and they are the familiar equations of S.H.M.

The motion of the coupled system can thus be described in terms of two coordinates X and Y; each equation is a *linear differential equation of S.H.M. with constant coefficients with only one variable.* The two motions are completely *uncoupled* and *independent,* although the moving parts of the system are not. One motion can thus exist without the other.

If $Y = 0$, $x = y = a$, say for all times, so that the motion is completely described by (4.2.5) and the angular frequency of oscillation is given by

$$\omega_0 = \sqrt{g/l},$$

the same as that of either pendulum oscillating in isolation; the effect of stiffness of coupling is absent. The spring is at its natural length. This is because both the pendulums are swinging *in phase*—**in phase vibrations** (Fig. 4.4).

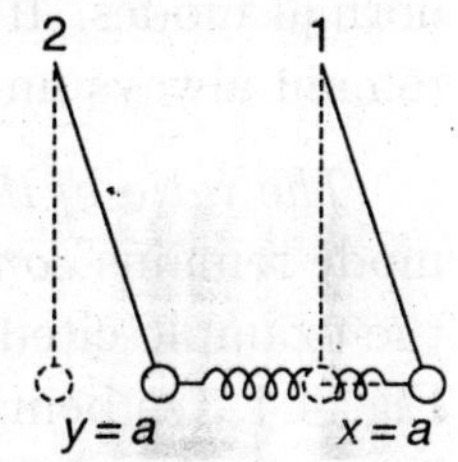

Fig 4.4

If $X = 0$, $y = -x$ for all times, so that the motion is completely described by (4.2.6) and the angular frequency is

$$\omega' = \sqrt{\omega_0^2 + \frac{2s}{m}}$$

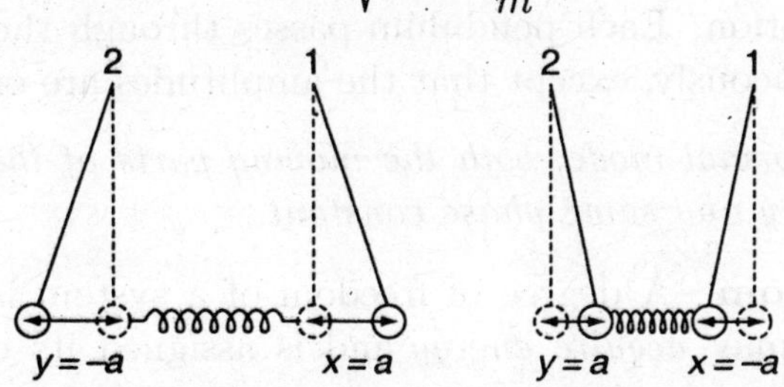

Fig 4.5

which is *greater than* ω_0, and the *effect of stiffness of coupling is present.* The spring is either *extended* or *compressed* during vibration and the two pendulums oscillating harmonically are always *out of phase*—**out of phase vibration** (Fig. 4.5)

4.3 Normal co-ordinates and normal modes

The significance of choosing X and Y in Art 4.2 to describe the motion of a coupled system is that these parameters illustrate what is called the *normal coordinates.*

The **normal coordinates** *of a coupled system are those coordinates that enable us to express the equations of motion of the system as a set of linear differential equations with constant coefficients and in which each equation contains only one dependent variable.*

In the example of Art 4.2, the normal coordinates are $X = x + y$ and $Y = x - y$ and equations (4.2.5) and (4.2.6) have only one dependent variable.

The S.H.M. associated with each normal co-ordinate is called a **normal mode** *of the coupled system. Each normal mode has its own characteristic frequency known as* **normal mode frequency** *or* **normal frequency**. In each normal mode, all the components of the system vibrate with the same normal frequency. The normal modes of vibration are *entirely independent of each other*, as the energy associated with a normal mode is never exchanged with that associated with another normal mode. The total energy of the oscillating system is equal to the sum of the energies of all the normal modes. If at any certain time, only one mode is excited, the other modes will remain always unexcited and they will have no energy of the vibrating mode.

The ratio of the displacements (y/x) *of the moving parts of the system* for a normal mode remains *constant and is known as the* **shape** *or* **configuration** *of that mode.* In the example cited in Art 4.2, $y/x = 1$ and the shape of the first mode is illustrated in Fig. 4.4. On being released, the two pendulums executes SHM with the same frequency $\omega_0 = \sqrt{g/l}$ and are always *in phase,* $y/x = 1$ throughout the motion. As already stated, this is the **in phase mode** of the system.

For the other normal mode shown in Fig. 4.5 where $y/x = -1$, the pendulums are displaced *equally* but in *opposite* directions. When released, each execute SHM with the same frequency $\omega' = \sqrt{g/l + 2s/m}$ but is *out of phase* by 180°, y/x being equal to -1 throughout the motion. Each pendulum passes through their respective positions of equilibrium simultaneously, except that the amplitudes are opposite in sign.

Thus *in a given normal mode, both the moving parts of the system execute SHM with the same frequency and same phase constant.*

Degrees of freedom—A degree of freedom of a system is the *independent way* by which the system may *acquire energy* and is assigned its own particular normal coordinates.

The *number* of degrees of freedom (and the number of normal coordinates) of a system is the number of possible ways of acquiring energy by it. Each has *two* degrees of freedom since it may have both *kinetic* and *potential energy.* In the previous example, the total energy of the oscillator represented by the normal coordinates X and Y

corresponding to the *two modes* are

$$E_X = a\dot{X}^2 + bX^2 \quad \text{and} \quad E_Y = c\dot{Y}^2 + dY^2$$

where a, b, c, d are constants, $a\dot{X}^2$ and $c\dot{Y}^2$ give the kinetic energy and bX^2 and dY^2 the potential energy.

• In general, the motion of a coupled system is *not* periodic. But if the system is started in *just* the *right ways*, the system will execute simple harmonic motion. In our example, the system had two degrees of freedom and exactly two possible ways of starting the system for its execution of simple harmonic motions—the *two* possible *normal modes.*

• When a system can be set into motion by an infinite number of ways, why are we interested in the so-called normal modes, particularly when the *just right ways* need be carefully sought for such vibrations? The answer is rather simple. This is because, once the normal modes are found, the problem of finding the general motion of the system, started in any arbitrary way, is completely solved.

4.4 Types of coupling

Two oscillators may be coupled in a number of different ways—the so-called *types of coupling*—to bring about energy-exchange between them.

The common coupling components for mechanical oscillators may be *stiffness* or *mass.* In case of electrical oscillators, the coupling components may be *capacitance* or *inductance.* Both the capacitance and inductance can store energy and thus coupling through them does not consume any power. One may as well use *resistance* as a coupling component, but the difficulty is that it involves the inevitable joule-loss of energy. So, with capacitive and inductive coupling, the energy-transfer occurs over a number of oscillations, but in resistive coupling the amplitude falls off rapidly.

4.5 General method for finding normal mode frequencies

Consider a coupled system of two identical simple pendulums of length l and bob-mass m, coupled by a spring of stiffness s. The equations of motion are given by (4.2.1) and (4.2.2) :

$$m\ddot{x} + \frac{mg}{l}x + s(x - y) = 0 \tag{4.5.1}$$

$$m\ddot{y} + \frac{mg}{l}x - s(x - y) = 0 \tag{4.5.2}$$

where x and y $(x > y)$ are the displacements of the pendulums from their mean positions.

While oscillating in a *normal mode*, each component of a coupled system oscillates with the frequency of that mode. Assuming that the coupled pendulums oscillate in one of its normal modes of frequency ω, the solutions of (4.5.1) and (4.5.2) are :

$$x = A\cos\omega t; \quad y = B\cos\omega t$$

where A and B are the displacement amplitudes at frequency ω.

$$\therefore \quad \ddot{x} = -A\omega^2\cos\omega t; \quad \ddot{y} = -B\omega^2\cos\omega t$$

Substituting these values in (4.5.1) and (4.5.2),

$$\left[-m\omega^2 A + \frac{mg}{l}A + s(A-B)\right]\cos\omega t = 0 \tag{4.5.3}$$

$$\text{and} \quad \left[-m\omega^2 B + \frac{mg}{l}B - s(A-B)\right]\cos\omega t = 0 \tag{4.5.4}$$

First normal mode frequency— Adding (4.5.3) and (4.5.4), we get

$$(A+B)(-m\omega^2 + mg/l) = 0 \tag{4.5.5}$$

$$\Rightarrow \omega^2 = g/l, \text{ since } A+B \neq 0, \text{ in general.}$$

$\therefore$ *First normal mode frequency,* ω_1 is given by

$$\boxed{\omega_1 = \sqrt{g/l}}$$

Second normal mode frequency—Subtracting (4.5.4) from (4.5.3), we obtain

$$(A-B)(-m\omega^2 + mg/l + 2s) = 0 \tag{4.5.6}$$

$$\Rightarrow \omega^2 = \frac{g}{l} + \frac{2s}{m}, \text{ since } (A-B) \neq 0, \text{ in general.}$$

$\therefore$ *Second normal mode frequency,* ω_2 is given by

$$\boxed{\omega_2 = \sqrt{\frac{g}{l} + \frac{2s}{m}}}$$

In phase condition—Putting $\omega^2 = g/l$ in (4.5.6), we have

$$(A-B)2s = 0 \Rightarrow A - B = 0 \Rightarrow A = B.$$

This gives us the *in phase condition of the vibration.*

Out of phase condition—Putting $\omega^2 = \frac{g}{l} + \frac{2s}{m}$ in equation (4.5.5),

$$-(A+B)2s = 0 \Rightarrow A + B = 0 \Rightarrow A = -B$$

which gives the *out of phase* condition.

4.6 Total energy of a stiffness-coupled system of two identical pendulums

The equations of motion of a stiffness-coupled system of two identical pendulums 1 and 2, each of mass m, in normal co-ordinates X and Y are given by (4.2.5) and (4.2.6) :

$$\ddot{X} + \omega_0^2 X = 0 \tag{4.6.1}$$

$$\ddot{Y} + (\omega_0^2 + 2s/m)Y = 0 \tag{4.6.2}$$

where $X = x + y$, $Y = x - y$, x and y are the displacements (in the same direction) of pendulums 1 and 2 respectively; $\omega_0^2 = g/l$ where l is the length of each pendulum and s the stiffness constant.

Solving (4.6.1) and (4.6.2), we obtain

$$X = x + y = X_0 \cos(\omega_1 t - \phi_1) \tag{4.6.3}$$

$$Y = x - y = Y_0 \cos(\omega_2 t - \phi_2) \tag{4.6.4}$$

where X_0, Y_0 are the *normal mode amplitudes*. The *normal mode frequencies* are however given by

$$\omega_1 = \omega_0 = \sqrt{\frac{g}{l}}$$

$$\omega_2 = \sqrt{\frac{g}{l} + \frac{2s}{m}}$$

Assume for simplification, $\phi_1 = \phi_2 = 0$ and $X_0 = Y_0 = 2a$.

$\therefore$ Eq. (4.6.3) and (4.6.4) reduce respectively to

$$X = x + y = 2a \cos \omega_1 t \tag{4.6.5}$$

$$Y = x - y = 2a \cos \omega_2 t \tag{4.6.6}$$

Adding the equations (4.6.5) and (4.6.6)

$$x = \tfrac{1}{2}(X + Y) = a \cos \omega_1 t + a \cos \omega_2 t$$

Similarly, by subtraction of (4.6.5) and (4.6.6),

$$y = \tfrac{1}{2}(X - Y) = a \cos \omega_1 t - a \cos \omega_2 t$$

Velocities are therefore given by:

$$\dot{x} = -a\omega_1 \sin \omega_1 t - a\omega_2 \sin \omega_2 t$$

$$\dot{y} = -a\omega_1 \sin \omega_1 t + a\omega_2 \sin \omega_2 t$$

Let the system be set to motion by displacing bob 1 by $x = 2a$, keeping $y = 0$ and $\dot{x} = \dot{y} = 0$ at $t = 0$ (i.e., both bobs are released from rest).

$\therefore$ Motions of pendulums 1 and 2 are given respectively by

$$x = a\cos\omega_1 t + a\cos\omega_2 t = 2a\cos\tfrac{1}{2}(\omega_2-\omega_1)t\cos\tfrac{1}{2}(\omega_1+\omega_2)t$$

$$y = a\cos\omega_1 t - a\cos\omega_2 t = 2a\sin\tfrac{1}{2}(\omega_2-\omega_1)t\sin\tfrac{1}{2}(\omega_1+\omega_2)t$$

Let $\frac{1}{2}(\omega_1+\omega_2) = \omega_a$ and $\frac{1}{2}(\omega_2-\omega_1) = \omega_m$, where ω_a and ω_m are respectively called the *average frequency* and *modulated* or **beat frequency**.

$\therefore$ Modulated amplitude of pendulum 1 is : $A = 2a\cos\omega_m t$

Modulated amplitude of pendulums 2 is : $B = 2a\sin\omega_m t$

Also, we have

$$x = A\cos\omega_a t;$$

$$y = -B\sin\omega_a t.$$

Let, for simplicity, the spring be *very weak* so that it does *not* store any energy. We may then take the modulated amplitudes $2a\cos\omega_m t$ and $2a\sin\omega_m t$ as constant over one cycle of average frequency.

$\therefore$ Energy of pendulum 1 :

$$E_1 = \tfrac{1}{2}m\dot{x}^2 = \tfrac{1}{2}m(A\omega_a)^2 = \tfrac{1}{2}m(2a\cos\omega_m t)^2\omega_a^2.$$

Energy of pendulum 2 :

$$E_2 = \tfrac{1}{2}m\dot{y}^2 = \tfrac{1}{2}m(B\omega_a)^2 = \tfrac{1}{2}m(2a\sin\omega_m t)^2\omega_a^2$$

$\therefore$ **Total energy**, $E = E_1 + E_2$

$$= 2ma^2\omega_a^2 = \text{constant.} \qquad (4.6.7)$$

Also, we have, $E_1 - E_2 = 2ma^2\omega_a^2(\cos^2\omega_m t - \sin^2\omega_m t) = E\cos 2\omega_m t$ (4.6.8)

Solving (4.6.7) and (4.6.8), we obtain

$$E_1 = \tfrac{1}{2}E(1+\cos 2\omega_m t) = \tfrac{1}{2}E\,[1+\cos(\omega_1-\omega_2)t]$$

$$E_2 = \tfrac{1}{2}E(1-\cos 2\omega_m t) = \tfrac{1}{2}E\,[1-\cos(\omega_1-\omega_2)t]$$

Thus, the *total energy is constant*, but it *flows back and forth between the two pendulums at the modulated or beat frequency*.

So, if bob 1 is displaced by $2a$ and released, it shows a cosine behaviour with frequency ω_a, the average of two normal mode frequencies. Its amplitude also varies as cosine law at a frequency ω_m, i.e., half the difference of two normal modes. Bob 2 however starts from zero and vibrates as sine law with the same frequency ω_a; its amplitude grows to $2a$ and then decays as sine law of frequency ω_m.

Fig. 4.6(a) illustrates the X and Y modes into which the initial configuration $x = 2a, y = 0$ can be decomposed. X mode implies $x = y = a$. So, $X_0 = x + y = 2a$ is the '*in phase*' mode. Y mode implies $x = a$, $y = -a$. So, $Y_0 = x - y = 2a$ is the '*out of phase*' mode.

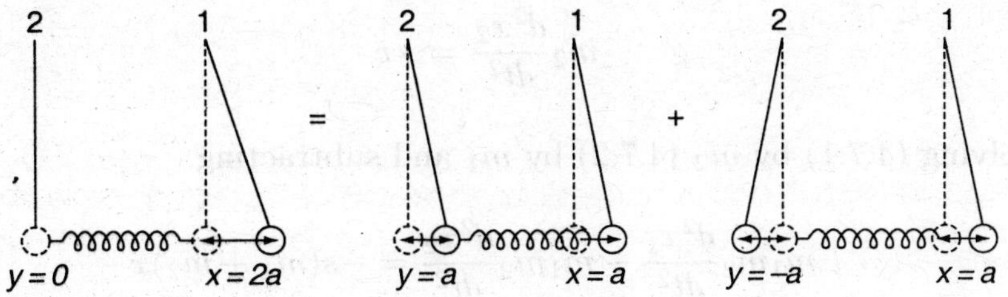

Fig 4.6(a)

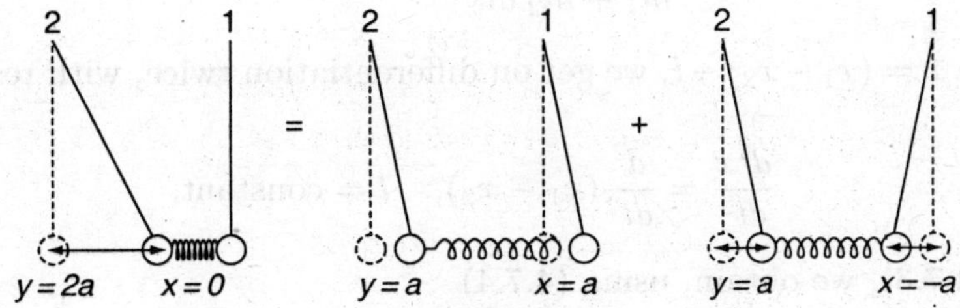

Fig 4.6(b)

But Y mode has a higher frequency. It will gain, after a number of vibrations, half a vibration (i.e., phase difference 180°) over X mode. Then, a combination of the two modes will make $x = 0$ and $y = 2a$ as shown in Fig. 4.6(b). Again, after some time, Y will gain another half a vibration to make $x = 2a$ and $y = 0$.

Thus, only *the pendulums exchange energy, the normal modes do not*, that is, '*in phase*' mode remains '*in phase*' and '*out of phase*' mode remains '*out of phase*'.

4.7 A two-body coupled oscillator

Let two masses m_1, m_2 be coupled by a spring of stiffness s. When displaced from equilibrium positions, the spring extends or contracts, depending on the displacement of the mass. A linear restoring force develops in the spring causing the masses to vibrate harmonically about their respective positions of equilibrium. Such a system is known as a *two-body oscillator* or *coupled oscillator*. The two atoms of a diatomic molecule, connected by a *bond*, is a familiar example of such a system.

Since there is *no* external force, the centre of mass of the system is either stationary or moves with a constant velocity. In the *centre-of-mass frame*, supposed stationary, the masses will move in opposite direction.

If l be the normal length of the spring and x_1, x_2 the coordinates of the ends of the spring, then the extension of the spring, $x = (x_1 - x_2) - l$; x will be *positive* if the spring is stretched, *zero* if of normal length and *negative* if compressed.

Let x be *positive*, when the forces exerted by the spring on the masses are equal and opposite and of magnitude sx. The equations of motion are thus

$$m_1 \frac{d^2 x_1}{dt^2} = -sx \tag{4.7.1}$$

$$m_2 \frac{d^2 x_2}{dt^2} = sx \tag{4.7.2}$$

Multiplying (4.7.1) by m_2 (4.7.2) by m_1 and subtracting,

$$m_1 m_2 \frac{d^2 x_1}{dt^2} - m_1 m_2 \frac{d^2 x_2}{dt^2} = -s(m_1 + m_2)x$$

$$\text{or,} \quad \frac{m_1 m_2}{m_1 + m_2} \frac{d^2}{dt^2}(x_1 - x_2) = -sx \tag{4.7.3}$$

Now, since $x = (x_1 - x_2) - l$, we get on differentiation twice, with respect to time t, we get

$$\frac{d^2 x}{dt^2} = \frac{d^2}{dt^2}(x_1 - x_2), \quad l = \text{constant.} \tag{4.7.4}$$

$\therefore$ From (4.7.3), we obtain, using (4.7.4)

$$\frac{m_1 m_2}{m_1 + m_2} \frac{d^2 x}{dt^2} = -sx$$

$$\Rightarrow \quad \boxed{\mu \frac{d^2 x}{dt^2} + sx = 0} \tag{4.7.5}$$

where $\mu = m_1 m_2/(m_1 + m_2)$, the **reduced mass** of the system.

Eq. (4.7.5) shows that *the motion is simple harmonic*, having the time period T and frequency n as

$$T = 2\pi\sqrt{\frac{\mu}{s}}; \quad n = \frac{1}{T} = \frac{1}{2\pi}\sqrt{\frac{s}{\mu}} \tag{4.7.6}$$

$$\therefore \text{ Angular frequency, } \omega = 2\pi n = \sqrt{s/\mu}$$

Both the masses m_1 and m_2 execute SHM along the same straight line, but the motions differ in phase by π radians.

4.8 Inductive coupling of two electric oscillators

Two resistance-free electric circuits—*LC* circuits—are shown in Fig. 4.7. A mutual inductance exists between the circuits whenever the magnetic flux due to current in one intercepts the other. When the current changes, so does the magnetic flux and an emf is induced in both the circuits. Whatever occurs in one circuit, the other is affected by it and thus the two circuits are *coupled.* This is what is called **inductive coupling**. The familiar example is that of a *transformer* that works on the principle of mutual induction between its primary and the secondary coils.

Coefficient of mutual induction—Let n_p and n_s be the number of turns in the primary and the secondary of self-inductance L_p and L_s respectively (Fig. 4.7), and ϕ be the magnetic flux set up in the primary when unit current flows *per turn* in it.

Assuming no leakage of flux, the flux linked with each primary turn is $n_p\phi$.

$\therefore$ Total magnetic flux linked with the primary is $L_p = n_p(n_p\phi)$.

Similarly, if unit current flows per turn through the secondary, the magnetic flux produced is also ϕ.

Total magnetic flux linked with the secondary is $L_s = n_s(n_s\phi)$

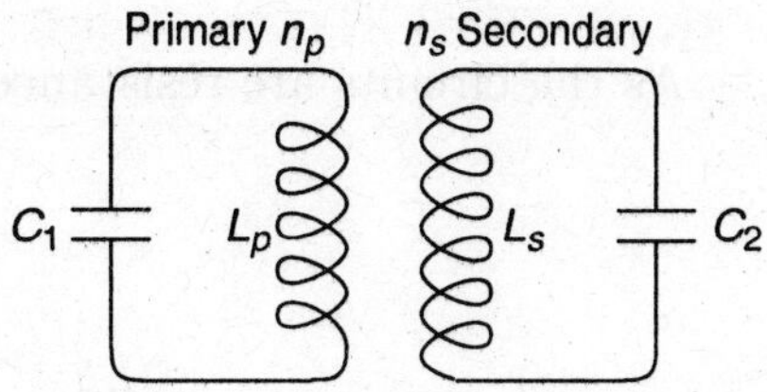

Fig 4.7

Assuming, again, that all the lines of magnetic flux of the primary are intercepted by the secondary, total flux linking the secondary $= n_s(n_p\phi)$.

$\therefore$ **Coefficient of mutual induction** between the circuits is

$$M = n_s n_p \phi = \sqrt{L_p L_s} \tag{4.8.1}$$

Coefficient of coupling—The above relation (4.8.1) holds only when there is no leakage of magnetic flux and it is an idealization. Some leakage however does take place in practice. So, $M < \sqrt{L_p L_s}$.

The *ratio* $M/\sqrt{L_pL_s} = k$ *is called the* **coefficient of coupling**. In the ideal case, $k = 1$ but, in practice, $k < 1$ always. If k is *small* ($\ll 1$), the circuits have *loose coupling.* If $k \to 1$, i.e., *very high,* the *coupling is tight.*

4.9 Energy transfer and normal mode frequencies of inductively coupled circuits

Let two resistance-free electric circuits (Fig. 4.7) be inductively coupled and made to oscillate with a frequency ω. Let the currents in the primary and the secondary be represented respectively by

$$I_p = I_1 e^{j\omega t} \quad \text{and} \quad I_s = I_2 e^{j\omega t}$$

$$\left.\begin{aligned} \therefore \text{ Induced emf in the primary} &= -L_p\frac{dI_p}{dt} = -j\omega L_p I_p \\ \text{Induced emf in the secondary} &= -L_s\frac{dI_s}{dt} = -j\omega L_s I_s \end{aligned}\right\} \tag{4.9.1}$$

$\therefore$ Potential difference across the capacitor C_1 in the primary is ($j = \sqrt{-1}$)

$$\frac{1}{C_1}\int I_p dt = \frac{1}{C_1}\int I_1 e^{j\omega t} dt = \frac{1}{C_1}\frac{I_1 e^{j\omega t}}{j\omega} = -\frac{jI_p}{\omega C_1} \tag{4.9.2}$$

Potential difference across the capacitor C_2 in the secondary is

$$\frac{1}{C_2}\int I_s dt = -\frac{jI_s}{\omega C_2}. \tag{4.9.3}$$

As the circuits are resistance-free, the emf equations of the circuits are

$$-L_p\frac{dI_p}{dt} + \frac{\int I_p dt}{C_1} - M\frac{dI_s}{dt} = 0 \tag{4.9.4}$$

$$\text{and}\quad -L_s\frac{dI_s}{dt} + \frac{\int I_s dt}{C_2} - M\frac{dI_p}{dt} = 0 \tag{4.9.5}$$

Now, using (4.9.1), (4.9.2) and (4.9.3), we may rewrite (4.9.4) and (4.9.5) as

$$j\omega L_p I_p - \frac{j}{\omega C_1} I_p + j\omega M I_s = 0 \tag{4.9.6}$$

$$j\omega L_s I_s - \frac{j}{\omega C_2} I_s + j\omega M I_p = 0 \tag{4.9.7}$$

Multiplying (4.9.6) by ω/jL_p and (4.9.7) by ω/jL_s,

$$\omega^2 I_p - \frac{I_p}{L_p C_1} + \frac{M}{L_p}\omega^2 I_s = 0 \tag{4.9.8}$$

$$\text{and}\quad \omega^2 I_s - \frac{I_s}{L_s C_2} + \frac{M}{L_s}\omega^2 I_p = 0 \tag{4.9.9}$$

But $1/L_pC_1 = \omega_1^2$, the square of the *natural frequency of primary* (LC-circuit) and $1/L_sC_2 = \omega_2^2$, the square of the *natural frequency of secondary.*

$\therefore$ Eq. (4.9.8) and (4.9.9) reduce to

$$I_p(\omega_1^2 - \omega^2) = (M/L_p)\omega^2 I_s$$

$$I_s(\omega_2^2 - \omega^2) = (M/L_s)\omega^2 I_p$$

Multiplying, we get $\quad I_pI_s(\omega_1^2 - \omega^2)(\omega_2^2 - \omega^2) = (M^2/L_pL_s)\omega^4 I_pI_s$

$$\Rightarrow\ (\omega_1^2 - \omega^2)(\omega_2^2 - \omega^2) = (M^2/L_pL_s)\omega^4 = k^2\omega^4$$

where $k = M/\sqrt{L_pL_s}$, the *coefficient of coupling.*

For simplification, let $\omega_1 = \omega_2 = \omega_0$. Then, we have

$$(\omega_0^2 - \omega^2)^2 = k^2\omega^4 \quad \Rightarrow \quad \omega_0^2 - \omega^2 = \pm k\omega^2$$

$$\therefore\quad \omega = \pm\frac{\omega_0}{\sqrt{1 \pm k}}$$

Since ω cannot be negative, we get two frequencies:

$$\omega_a = \frac{\omega_0}{\sqrt{1+k}}; \quad \omega_b = \frac{\omega_0}{\sqrt{1-k}}$$

which correspond to the *two normal mode frequencies.*

When $\omega_a = \omega_0/\sqrt{1+k}$, the secondary current I_s is *in phase* with the primary current I_p. So, ω_a is the frequency of the *phase mode.*

When $\omega_b = \omega_0/\sqrt{1-k}$, I_s is *out of phase* with I_p. So, ω_b is the frequency of the *out of phase mode.*

Loose and tight coupling—As already indicated that if $k \ll 1$, the *coupling* is said to be *loose*; then both the systems behave almost *independently* and $\omega_a \simeq \omega_b \simeq \omega_0$. *Coupling* is said to be *tight* if $k \to 1$ and then ω_a and ω_b *widely differ* from ω_0 and the peak values of current are displaced.

The variation of current amplitude with driving force frequency ω is illustrated in Fig. 4.8. For small k (curve 1), the peak values of current agree; for intermediate k-value (curve 2), the peak currents are displaced showing a clear dip; for large k (curve 3), the dip is more pronounced.

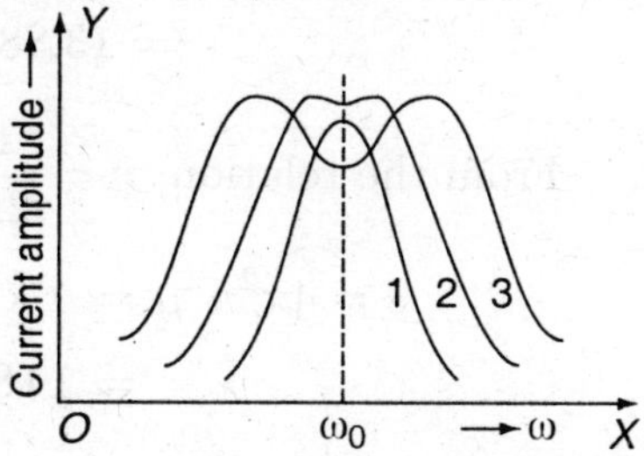

Fig 4.8

For small k (loose coupling), band width is negligible i.e., *resonance occurs at one frequency* i.e., ω. For large k (tight coupling), band width (i.e. $\omega_a \sim \omega_b$) is also large, that is, *resonance occurs at two widely different frequencies.*

4.10 Illustrated Examples

► **Example 1.** *The angular frequency of vibration of a carbon monoxide molecule* (CO) *is* 0.6×10^{15} s^{-1}. *Calculate the work required for stretching it by* 0.5 A *from its equilibrium position.* **(Punjab Univ.)**

Solution. The reduced mass of CO-molecule is given by

$$\mu = \frac{m_1 m_2}{m_1 + m_2} = \frac{12 \times 16}{12 + 16} \text{ u} = 6.85 \text{ u}$$
$$= 6.85 \times 1.67 \times 10^{-27} \text{ kg.}$$

Angular frequency, $\omega = 2\pi n = 0.6 \times 10^{15}$ s^{-1}. But, $\omega = \sqrt{s/\mu}$

$$\therefore \quad s = \omega^2 \mu = (0.6 \times 10^{15})^2 \times 6.85 \times 1.67 \times 10^{-27} \text{ Nm}^{-1}$$
$$= 4.118 \times 10^3 \text{ Nm}^{-1}.$$

Amount of stretching, $x = 0.5$ A $= 0.5 \times 10^{-10}$ m. $\therefore$ Required work, W is given by

$$W = \tfrac{1}{2}sx^2 = \tfrac{1}{2} \times 4.118 \times 10^3 \times (0.5 \times 10^{-10})^2$$
$$= \mathbf{5.15 \times 10^{-18}\ J}$$

► **Example 2.** *A sodium chloride molecule* (NaCl) *vibrates with the natural frequency* 1.14×10^{13} Hz. *Find the inter-atomic force constant for the molecule. Mass of* Na-*atom* = 23 u, *mass of* Cl-*atom* = 35 u. **(Punjab Univ.)**

Solution. The reduced mass μ of NaCl-molecule is given by

$$\mu = \frac{m_1 m_2}{m_1 + m_2} = \frac{23 \times 35}{23 + 35}\ \text{u} = 13.88\ \text{u}$$
$$= 13.88 \times 1.67 \times 10^{-27}\ \text{kg} = 23.18 \times 10^{-27}\ \text{kg}.$$

From the relation, $n = \dfrac{1}{2\pi}\sqrt{s/\mu}$, the interatomic force constant, s is given by

$$s = 4\pi^2 n^2 \mu = 4 \times (3.14)^2 \times (1.14 \times 10^{13})^2 \times 23.18 \times 10^{-27}\ \text{Nm}^{-1}$$
$$= \mathbf{118.9\ Nm^{-1}}$$

► **Example 3.** *Two identical particles, each of mass m are connected with two identical massless springs of spring constant k as shown below in* Fig. 4.9. *Show that the angular frequencies of the two normal modes of vertical oscillations are given by* $\omega^2 = (3 \pm \sqrt{5})k/2m$.

Solution. The restoring forces on mass A and B (Fig. 4.9) are respectively

$$-kx_a + k(x_b - x_a)$$
$$\text{and}\quad -k(x_b - x_a)$$

where x_a and x_b are the displacements of the mass A and B respectively at any time t.

$\therefore$ Equations of motion are:

$$m\ddot{x}_a = kx_a + k(x_b - x_a)$$
$$\text{and}\quad m\ddot{x}_b = -k(x_b - x_a)$$

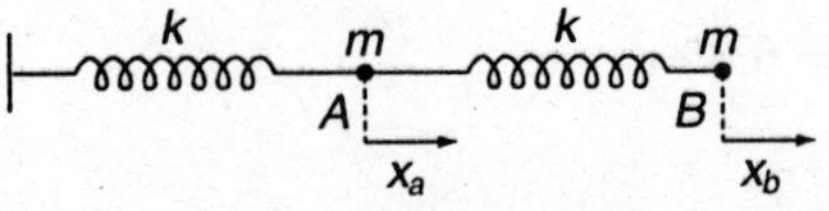

Fig 4.9

$$\text{or,}\quad \ddot{x}_a = -\frac{2k}{m}x_a + \frac{k}{m}x_b \qquad (4.10.1)$$

$$\text{and}\quad \ddot{x}_b = -\frac{k}{m}x_b + \frac{k}{m}x_a \qquad (4.10.2)$$

In a normal mode with frequency ω and phase constant ϕ,

$$x_a = A\cos(\omega t + \phi)$$
$$x_b = B\cos(\omega t + \phi)$$

Using these values in (4.10.1) and (4.10.2)

$$\frac{x_a}{x_b} = \frac{A}{B} = \frac{k/m}{2k/m - \omega^2} = \frac{k/m - \omega^2}{k/m} \tag{4.10.3}$$

$$\Rightarrow \left(\frac{2k}{m} - \omega^2\right)\left(\frac{k}{m} - \omega^2\right) = \frac{k^2}{m^2}$$

$$\Rightarrow \quad \omega^4 - \frac{3k}{m}\omega^2 + \frac{k^2}{m^2} = 0 \tag{4.10.4}$$

$$\therefore \quad \omega^2 = (3 \pm \sqrt{5})\frac{k}{2m}$$

which gives the angular frequencies of the two normal modes of oscillations.

▶ **Example 4.** *The equations of motion of two coupled oscillators are $\ddot{q}_1 + \sqrt{2}q_2 + 4q_1 = 0$ and $\ddot{q}_2 + \sqrt{2}q_1 + 5q_2 = 0$. Find the normal frequencies and the ratios of the amplitudes of the normal modes.* **(Burdwan Hons)**

Solution. Let $q_1 = Ae^{j\omega t}$ and $q_2 = Be^{j\omega t}$. Putting these values of q_1 and q_2 in the given equations of motion,

$$-(\omega^2 + 4)Ae^{j\omega t} + \sqrt{2}Be^{j\omega t} = 0 \quad \Rightarrow \quad (\omega^2 + 4)A + \sqrt{2}B = 0 \tag{4.10.5}$$

$$\text{and} \quad \sqrt{2}Ae^{j\omega t} + (\omega^2 + 5)Be^{j\omega t} = 0 \quad \Rightarrow \quad \sqrt{2}A + (\omega^2 + 5)B = 0 \tag{4.10.6}$$

Now, for non-zero solutions for A, B, we must have

$$\begin{vmatrix} \omega^2 + 4 & \sqrt{2} \\ \sqrt{2} & \omega^2 + 5 \end{vmatrix} = 0$$

$$\Rightarrow \quad (\omega^2 + 4)(\omega^2 + 5) - 2 = 0 \quad \Rightarrow \quad (\omega^2)^2 + 9\omega^2 + 18 = 0$$

$$\text{or,} \quad (\omega^2 + 3)(\omega^2 + 6) = 0 \qquad \therefore \quad \omega^2 = -3, -6$$

$\therefore$ *Normal frequencies* are $\boldsymbol{\omega_1 = \sqrt{3}}$ **rad/s;** $\boldsymbol{\omega_2 = \sqrt{6}}$ **rad/s.**

Substituting $\omega^2 = 3$ in (4.10.5), $A + \sqrt{2}B = 0$. So the *ratio of amplitudes* in this mode is $\boldsymbol{A/B = -\sqrt{2}}$.

Substituting $\omega^2 = 6$ in (4.10.5), $-2A + \sqrt{2}B = 0$. So, for this mode, $\boldsymbol{A/B = 1/\sqrt{2}}$

▶ **Example 5.** *In a transformer, the mutual inductance of two coils is* 0.3 H, *whereas the self-inductances of primary and secondary are* 0.28 H *and* 0.36 H *respectively. Is the transformer loosely coupled or tightly?* **(Himachal Univ.)**

Solution. The coefficient of coupling, $k = M/\sqrt{L_p L_s}$. Here $M = 0.3$ H, $L_p = 0.28$ H and $L_s = 0.36$ H.

$$\therefore \quad k = \frac{0.3}{\sqrt{0.28 \times 0.36}} = \mathbf{0.95}$$

As k-value is very nearly equal to 1, the transformer is rather **tightly** coupled.

Review Questions

1. Define and explain normal coordinates, degrees of freedom and normal modes of vibration of an oscillatory system. **(Guru Nanak Univ.)**

2. What is a coupled oscillator? Consider a system of two identical pendulums coupled through a spring of stiffness s. Find the differential equation of the system in terms of normal coordinates X and Y. Show that when normal coordinate $Y = 0$, the frequency is the same as that of either pendulum in isolation. **(Himachal Univ.)**

3. Derive an expression for the total energy of a stiffness-coupled system of two identical pendulums. **(Punjab Univ.)**

4. What are the different types of coupling used for coupling two oscillators? Show that in the 'in phase mode', the frequency of oscillations is the same as that of uncoupled oscillators, whereas in the 'out of phase mode', the frequency is raised. **(Punjabi Univ.)**

5. Obtain an expression for the normal frequencies of stiffness-coupled pendulums.

6. Give a general method of finding the normal mode of frequencies. Prove that in normal mode of vibration of two coupled pendulums, the vibration of either coupled pendulum is given by

$$\omega = \sqrt{\frac{g}{l} + \frac{2s}{m}}$$ **(Himachal Univ.)**

7. Two identical pendulums form a coupled system. Show that the frequency of oscillation is the same as that of either pendulum in isolation and the stiffness of coupling has no effect when normal coordinate $Y = 0$, that is, when the displacements of individual particles are equal.

8. Two masses m_1 and m_2 are coupled by a spring of stiffness s. The masses are pulled apart and released. Show that their motion is simple harmonic and the system oscillates with frequency $\omega = \sqrt{s/\mu}$, where μ is the reduced mass of the system and is given by

$$\mu = \frac{m_1 m_2}{m_1 + m_2}$$ **(Punjab Univ.)**

9. Explain the inductance coupling of two electric oscillators and define coefficient of coupling. **(Guru Nanak and Punjabi Univ.)**

10. Explain the transfer of energy between two resistance-free electric circuits coupled inductively.

Short Answer and Objective Questions

1. Define : normal coordinates, normal modes of vibration and degrees of freedom.

2. What do you mean by coupled oscillator? What are the different types of coupling two oscillators?

3. Do normal modes of a coupled system of two oscillators exchange energy with each other?

4. Show that the normal modes are independent of each other and there is no exchange of energy between two coupled pendulums. **(Punjab Univ.)**

5. Define coefficient of coupling of two inductively coupled oscillators. When is the coupling loose or tight? **(Himachal Univ.)**

6. Which one of the following is correct for tight coupling of two inductively coupled oscillators?

   (a) $k = 0$, (b) $L_1C_1 = L_2C_2$, (c) $\omega_1 = \omega_2$, (d) none of these.

   (Guru Nanak Univ.)

7. The stiffness of coupling has no effect when both the oscillators move

   (a) out of phase, (b) in phase, (c) both (a) and (b), (d) none of these.

8. The normal mode beat frequency of two identical coupled pendulum is 5 Hz. The time interval between two maximum displacements of the pendulum will be

   (a) 0.1 s, (b) 0.2 s, (c) 1 s, (d) 5 s.

9. The coefficient of coupling of coupled electric oscillators is

   (a) $k = ML_pL_s$, (b) $k = \dfrac{M}{L_pL_s}$, (c) $k = \dfrac{M}{\sqrt{L_pL_s}}$, (d) $k = \dfrac{\sqrt{L_pL_s}}{M}$.

   (Himachal Univ.)

10. For tight coupling of oscillators, resonance occurs at

    (a) two widely separated frequencies, (c) single low frequency,

    (b) two close frequencies, (d) single high frequency.

Problems

1. Two equal masses m are connected with two identical massless springs of spring constant k. Show that the angular frequencies of the two normal modes are given by $\omega^2 = (3 \pm \sqrt{5})k/2m$ and the ratios of the amplitudes of the masses are $\frac{1}{2}(\sqrt{5}-1)$ in the slower mode and $\frac{1}{2}(\sqrt{5+1})$ in the faster mode.

2. Two particles of the same mass m are attached to a massless spring of length l with fixed ends stretched by a tension T. Find the normal frequencies of the masses if they are attached at $l/3$ and $2l/3$ from one end. Assume T to be constant during the vibration. $[\sqrt{3T/ml}; 3\sqrt{T/ml}]$

3. Show that the ratio of the amplitudes of the coupled masses, in the above problem, is 1 for the normal frequency $\sqrt{3T/ml}$ and -1 for $3\sqrt{T/ml}$. Show the displacements of the masses in normal modes.

   [masses move in (i) same, (ii) opposite directions]

4. Two identical pendulums each of length 0.5 m have their bobs connected by a weak spring. With one pendulum clamped, the period of the other is 1.4 s. With neither pendulum clamped, what are the periods of two normal modes?

   [$\simeq$ 1.42 s; $\simeq$ 1.38 s]

5. Two identical LC-circuits, each having a natural frequency of 500 Hz, are coupled inductively such that the coupling coefficient is 0.5. Find the frequencies of the two normal modes. [408 Hz; 707 Hz]

6. Two masses 0.01 kg and 0.03 kg are connected by a massless spring of constant 10 Nm^{-1}. Find the frequency of oscillation if the masses are free to move along the line joining their centres. [5.8 Hz]

7. The interatomic force constant of an HCl molecule is 5.4×10^2 Nm^{-1}. Find the fundamental frequency of vibration of the molecule. Mass of H-atom $= 1.67 \times 10^{-27}$ kg; mass of Cl-atom $= 5.845 \times 10^{-26}$ kg. [9.18×10^{13} Hz]

Chapter 5

Transverse Waves

5.1 Wave motion : Types of waves

A **wave motion** is a *form of disturbance that propagates through a medium by the repeated periodic motion of the particles* in the medium *about their mean* (equilibrium) *positions.* The disturbance is handed over from one particle to its neighbouring one *without any bodily transference* of the medium.

In general, there are *two* types of wave motion—*transverse* and *longitudinal.*

Transverse waves—If the particles of the medium vibrate about their mean positions at right angles to the direction of propagation of the wave, the wave is said to be a *transverse wave.*

Longitudinal wave—If however the particles of the medium vibrate about their mean positions in the direction of propagation of the wave, the wave is termed a *longitudinal wave.*

Waves along a stretched string are transverse waves. Electromagnetic waves are transverse in nature. But sound waves in solids and fluids, compressional waves along a string are examples of longitudinal waves.

5.2 Progressive wave and its characteristics

A wave motion propagation from one point to another in a medium *without* being subjected to any *boundary condition* is known as a *progressive wave.* It is *characteristically* different, as we shall see in Chapter 7, from the so-called *stationary* or *standing wave.*

A progressive wave has a number of **characteristics :**

1. The waves are produced by the *periodic vibrations* of the particles of the medium.

2. Each particle vibrates about its mean position with the *same amplitude and period.* The simplest case is the simple harmonic vibrations.

3. The *medium does not move* with the wave; it is only the *energy* which *is carried* by the advancement of the waveform in the direction of wave propagation.

4. The *phase difference* between any two vibrating particles in the line of propagation is *proportional to the distance* between them.

5. The *wave velocity* in a given medium is a constant and is determined by its *density* and *elasticity.* However, the particle velocity of the vibrator goes on *changing,* from *maximum* in the mean position to *minimum* at the two extremities.

5.3 Equation of a progressive wave

Let a transverse wave in a string travel along $+x$ direction from left to right starting from the point A. A particle on the right of A will start vibrating a time later than one on the left (Fig. 5.1).

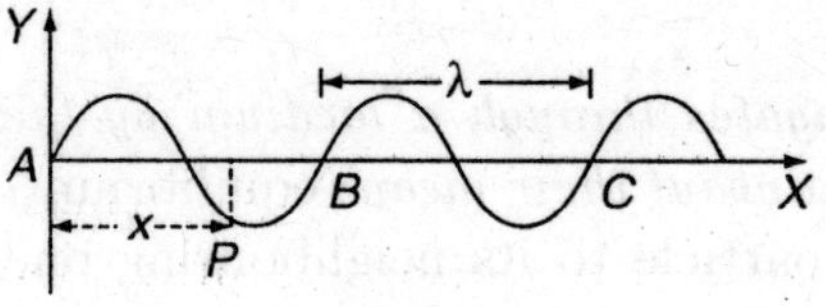

Fig 5.1

As each particle executes S.H.M., the equation of motion of any particle, say A, is given by

$$y = a \sin \omega t$$

where a is the amplitude, y the displacement at time t and ω the angular frequency $= 2\pi n$ where n is the frequency of vibration.

$$\therefore \quad y = a \sin 2\pi nt$$

When A is in its mean position of vibration, if the particles B, C etc. are also at their mean positions then A, B, C etc. are in the same state of vibration, i.e., in the *same phase.* The distance BC between two consecutive particles in the same phase is known as the *wavelength,* λ. The phase changes by 2π in going from A to B or from B to C etc.

$\therefore$ While going from A to any point P, distant x from A, the phase change will be given by

$$\phi = \frac{2\pi}{\lambda} x$$

$\therefore$ Displacement of P is given by :

$$\begin{aligned} y &= a \sin(\omega t - \phi) \\ &= a \sin\left(2\pi nt - \frac{2\pi}{\lambda} x\right) \\ &= a \sin\left(2\pi \frac{c}{\lambda} t - \frac{2\pi}{\lambda} x\right) \end{aligned}$$

which c is the velocity of the wave in the medium.

$$\therefore \quad \boxed{\boldsymbol{y = a \sin \frac{2\pi}{\lambda}(ct - x)}} \qquad (5.3.1)$$

This is quite a *general equation* giving the displacement of any particle distant x from the fixed point A at any time t.

So, this is the **general equation** *of a progressive harmonic wave.*

Eq. (5.3.1) may also be given the following *alternative form*

$$y = a\sin\left(\frac{2\pi c}{\lambda}t - \frac{2\pi}{\lambda}x\right)$$

$$\text{or,}\quad \boxed{\boldsymbol{y = a\sin(\omega t - kx)}} \tag{5.3.2}$$

since $2\pi c/\lambda = 2\pi n = \omega$ and $k = 2\pi/\lambda$, called the *propagation constant.*

In the exponential form, the displacement equation is given by

$$\boxed{\boldsymbol{y = ae^{j(\omega t - kx)}}} \tag{5.3.3}$$

where $j = \sqrt{-1}$.

• Note that the following equations different in form

$$\left.\begin{aligned} y &= a\sin\frac{2\pi}{\lambda}(ct - x) \\ y &= a\sin(\omega t - kx) \\ \text{and}\quad y &= ae^{j(\omega t - kx)} \end{aligned}\right\} \tag{5.3.4}$$

all denote a progressive wave motion moving along the $+x$ direction (i.e., to the right).

• If however the wave travels along $-x$ direction (i.e., to the left), the sign of ϕ is to be changed as the vibration at x begins earlier than that at $x = 0$. So, the corresponding equations will be given by the following set obtained by replacing x by $-x$ in the set of equations (5.3.4) :

$$\left.\begin{aligned} y &= a\sin\frac{2\pi}{\lambda}(ct + x), \\ y &= a\sin(\omega t + kx) \\ \text{and}\quad y &= ae^{j(\omega t + kx)}. \end{aligned}\right\} \tag{5.3.5}$$

5.4 Volumetric strain, particle velocity and wave velocity

Before we proceed further, we like to define some relevant important terms.

Volumetric strain—Let a wave be set up in a cylinder of cross-section unity, the axis of the cylinder coinciding with the direction of propagation ACE of the wave (Fig. 5.2). Let $ABCDE$ represent the displacement curve of the wave—the *waveform*.

Consider P, Q to be two close points dx apart. The displacement at $P = PR$ and that at $Q = QS$. So, the volume of the gas limited by P and Q is $PQ \times 1 = PQ$.

When the wave is in the position as shown in Fig. 5.2, the change in volume between the points P and Q is given by

$$QS - PR = ST = dy$$

$$\therefore \quad \textit{Volume strain} = \frac{\text{change in volume}}{\text{original volume}}$$

$$= \frac{ST}{PQ} = \frac{ST}{RT} = \frac{dy}{dx}$$

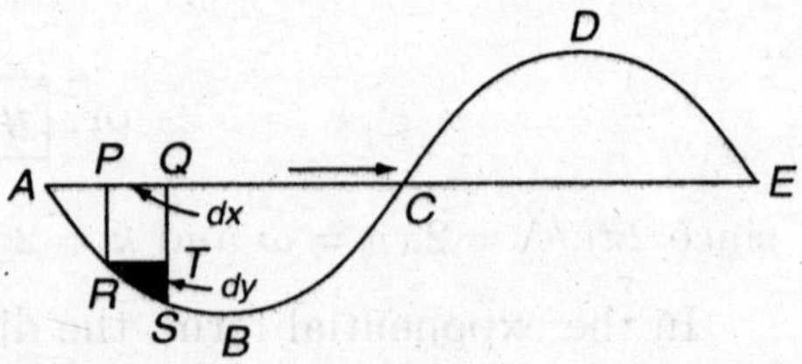

Fig 5.2

$\therefore$ **Volume strain = Slope of the displacement curve.**

Thus the *volumetric strain at a given point is measured by the slope of the displacement curve at that point.*

Particle velocity—Let the equation of the wave be given as

$$y = a \sin \frac{2\pi}{\lambda}(ct - x) \tag{5.4.1}$$

Differentiating (5.4.1) with respect to time t, we obtain

$$\textit{Particle velocity} = \frac{dy}{dt} = a\frac{2\pi c}{\lambda} \cos \frac{2\pi}{\lambda}(ct - x) \tag{5.4.2}$$

Differentiating (5.4.1) with respect to displacement x, we get

$$\text{Volume strain} = \frac{dy}{dx} = -a\frac{2\pi}{\lambda} \cos \frac{2\pi}{\lambda}(ct - x) \tag{5.4.3}$$

If dy/dx is *positive, a rarefaction* occurs; if however dy/dx is *negative, a compression* takes place.

Comparing (5.4.2) and (5.4.3), we obtain

$$\boxed{\frac{dy}{dt} = -c\frac{dy}{dx}} \tag{5.4.4}$$

$\therefore$ **Particle velocity = wave velocity × volume strain.**

This is an important relation existing between the volume strain, the particle velocity and the wave velocity of a prograssive wave.

5.5 Transverse waves in a string : Expression for velocity

A string is defined as a *perfectly uniform* and *flexible filament* of solid matter whose length is large compared to the radius.

When a string is stretched between two fixed points (Fig. 5.3) and a portion of it is plucked perpendicular to its length, *transverse vibrations* are set up in it. The particles vibrate at right angles to its length and the movement is handed over from particle to particle. This results in the propagation of a transverse wave along the string with a velocity determined by some constants of the string.

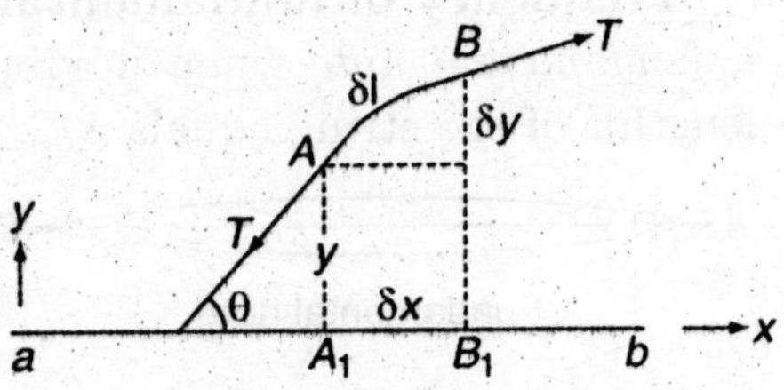

Fig 5.3

Expression for velocity—Let a string be stretched between two fixed points and let ab be its rest position along x-axis (Fig. 5.3). Let a portion of the string be now drawn along y-axis, i.e., at right angles to its length such that the *amplitude* of vibration is *small.* Since the string is perfectly flexible, we can take the *tension* T of the string to remain essentially *constant* along the string, acting tangentially at each point as the waves travel along it.

Consider a small segment $AB(=\delta l)$ of the string in the displaced position. Let θ be the angle made by the tangent at A with ab.

Transverse component of the tension at $A = T\sin\theta \simeq T\tan\theta = T(\partial y/\partial x)$, since the angle θ is very small.

Transverse component of tension at $B = T\dfrac{\partial}{\partial x}\left(y + \dfrac{\partial y}{\partial x}\delta x\right) = T\dfrac{\partial y}{\partial x} + T\dfrac{\partial^2 y}{\partial x^2}\delta x.$

$\therefore$ Resultant force along y-direction on the element $AB = T\dfrac{\partial^2 y}{\partial x^2}\delta x.$

If ρ be the mass per unit length of the string, $\rho\delta x$ is the mass of the element AB.

$$\therefore \quad \rho\delta x \times \text{acceleration of } AB = T\frac{\partial^2 y}{\partial x^2}\delta x$$

$$\text{or,} \quad \rho\delta x\frac{\partial^2 y}{\partial t^2} = T\frac{\partial^2 y}{\partial x^2}\delta x$$

$$\text{or,} \quad \frac{\partial^2 y}{\partial t^2} = \frac{T}{\rho}\frac{\partial^2 y}{\partial x^2}$$

$$\text{or,} \quad \boxed{\frac{\partial^2 y}{\partial t^2} = c^2\frac{\partial^2 y}{\partial x^2}} \tag{5.5.1}$$

where c has been substituted for $\sqrt{T/\rho}$.

This is the **differential wave equation** for transverse wave motion along the stretched string.

$$\therefore \quad \text{Wave velocity,} \quad c = \sqrt{T/\rho}. \tag{5.5.2}$$

Frequency of fundamental note—A string fixed at both ends is said to produce a *fundamental note* when it vibrates as a whole in *one segment* (Fig. 5.4). So the length l of the string equals $\lambda/2$, λ being the wavelength of the note.

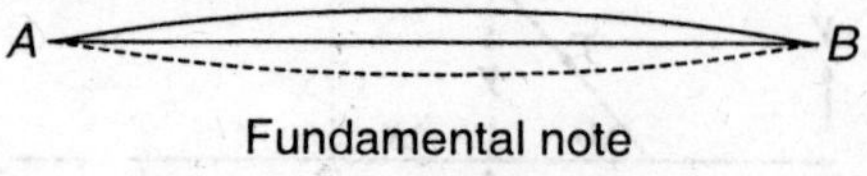

Fig 5.4

$$\therefore \lambda = 2l \Rightarrow n = \text{frequency} = c/\lambda$$

$$\therefore n = \frac{1}{2l}\sqrt{\frac{T}{\rho}}$$

Thus, the frequency of the fundamental note (string vibration) is given by

$$\boxed{n_1 = \frac{1}{2l}\sqrt{\frac{T}{\rho}}} \tag{5.5.3}$$

Frequency of first octave—A string fixed at both ends produces its *first octave*, when it vibrates in *two segments* (Fig. 5.5). So, in this case, $\lambda = l$.

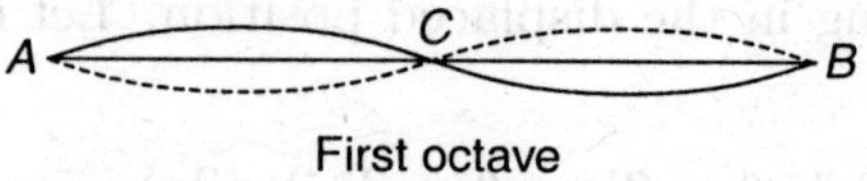

Fig 5.5

$\therefore$ Frequency of first octave:

$$n_2 = \frac{c}{\lambda} = \frac{1}{l}\sqrt{\frac{T}{\rho}} = \frac{2}{2l}\sqrt{\frac{T}{\rho}}$$

Generalising the situation, we find that, if the string vibrates in p number of segments, the corresponding frequency is given by

$$\boxed{n_p = \frac{p}{2l}\sqrt{\frac{T}{\rho}}}$$

• In case of the fundamental note, both the end points A and B are *nodes*. In case of first octave there is, in addition to the two ends A and B, a *node* at the mid-point C.

• The dimensions of $c\ (= \sqrt{T/\rho})$ are those of the velocity $[LT^{-1}]$.

• The wave equation (5.5.1) has been derived by considering the vibrations of the string as a prototype. Any partial differential equation of the form

$$\frac{\partial^2 \psi}{\partial t^2} = c^2 \frac{\partial^2 \psi}{\partial x^2}$$

represents a **general wave equation** where $\psi(x, y, z, t)$ is a function of space and time coordinates and c is the velocity with which the wave travels in the medium with its form unchanged.

The wave variable ψ represents the transverse displacement y in the case of string, pressure in the case of sound waves in a gas and electric field vector $\vec{E}$ in the case of electromagnetic waves etc.

• Read also the contents of Art 5.11 in this context.

5.6 Possible solutions of the wave equation

The form of the wave equation plainly indicates that its solution must be a linear function of the variables x and t. A function of the form

$$y = f_1(ct - x),$$
$$\text{or,}\quad y = f_2(ct + x)$$

is one of its solutions.

$\therefore$ General solution is : $y = f_1(ct - x) + f_2(ct + x)$

Verification—If $y = f_1(ct - x)$, then we have, on differentiation,

$$\frac{dy}{dt} = cf_1'(ct - x) \Rightarrow \frac{d^2y}{dt^2} = c^2 f_1''(ct - x)$$

$$\text{or,}\quad \frac{dy}{dx} = -f_1'(ct - x) \Rightarrow \frac{d^2y}{dx^2} = f_1''(ct - x)$$

$$\therefore \quad \boxed{\frac{d^2y}{dt^2} = c^2\frac{d^2y}{dx^2}}$$

Similarly, it can also be verified that $f_2(ct+x)$ is also a solution of the wave equation.

Harmonic solutions—In harmonic progressive wave, every particle of the medium executes S.H.M. that the can be represented by a sine or a cosine function.

For **sine function,** $y = f(ct - x) = a \sin k(ct - x)$, where a is the amplitude and k is a constant. The wave profile is regularly repeated at a distance $2\pi/k$, for $a \sin k(ct - x + 2\pi/k) = a \sin k(ct - x)$.

$\therefore 2\pi/k = \lambda$, the wavelength $\Rightarrow k = 2\pi/\lambda$, the *propagation constant.*

$$\therefore \quad y = a \sin k(ct - x) = a \sin \frac{2\pi}{\lambda}(ct - x)$$

$$= a \sin\left(\frac{2\pi ct}{\lambda} - \frac{2\pi}{\lambda}x\right) = a \sin 2\pi(nt - x/\lambda)$$

$$= a \sin(\omega t - kx), \quad \omega = 2\pi n \text{ where } n = \text{frequency.}$$

Similarly, for **cosine function,** we have

$$y = a \cos \frac{2\pi}{\lambda}(ct - x)$$
$$= a \cos 2\pi(nt - x/\lambda)$$
$$= a \cos(\omega t - kx).$$

Exponential form—The exponential form : $y = ae^{j(\omega t - kx)}$, $(j = \sqrt{-1})$, is an equally valid solution as it contains both the sine and cosine functions.

While the real part of the expression $y = ae^{j(\omega t - kx)}$ is the cosine function,

$$y = a\cos(\omega t - kx) \tag{5.6.1}$$

the imaginary part corresponds to the sine function

$$y = a\sin(\omega t - kx) \tag{5.6.2}$$

Each of the equations (5.6.1) and (5.6.2) represents a wave moving to the right, that is, the $+x$ direction.

Conversely, the expression $y = ae^{j(\omega t + kx)}$ gives

$$y = a\cos(\omega t + kx) \tag{5.6.3}$$

$$\text{and} \quad y = a\sin(\omega t + kx), \tag{5.6.4}$$

Each of the equations (5.6.3) and (5.6.4) represents similarly a wave moving to the left, i.e., the $-x$ direction.

• It can be readily verified that the following expression

$$y = x^2 + c^2t^2$$

is also a solution of one-dimensional wave equation and represents a wave travelling along x-axis with a velocity c.

5.7 Characteristic impedance of a string

A wave, be it a mechanical, electromagnetic or electric, while propagating through a medium experiences an *impedance.* If there is *no dissipative* or *resistive mechanism,* the impedance will be determined by : (a) *inertia* and *elasticity* in case of mechanical waves and (b) *inductance* and *capacitance* for electric waves—the two parameters that *store energy.* The impedance in such a case is a real quantity but becomes complex when there is a *loss mechanism.*

Transverse impedance (of a string)—The impedance offered (by a string) to transverse progressive waves is termed *transverse impedance.* It is defined *as the ratio of the transverse force to the transverse velocity.*

$$\therefore \quad \text{Impedance, } Z = \frac{\text{Transverse force}}{\text{Transverse velocity}} = \frac{F}{v}$$

Let a continuous string stretch from the left end $x = 0$ to the right and a progressive wave set up in it by an impressed oscillating force $F = F_0e^{j\omega t}$ acting at $x = 0$ end (Fig. 5.6). The force acts in the plane of the paper in a *transverse* direction, i.e., at right angles to the equilibrium position of the string.

If T be the constant tension of the string, then at $x = 0$, the applied force F is

$$F = F_0e^{j\omega t} = -T\sin\theta,$$

the component of T that opposes the force, θ being as shown in Fig. 5.6(b).

$$\therefore\ F = -T\tan\theta = -T\left(\frac{dy}{dx}\right)_{x=0}$$

since θ is a very small angle, we write $\sin\theta \simeq \tan\theta$.

The displacement of the progressive wave is given by

$$\vec{y} = \vec{a}e^{j(\omega t - kx)} \qquad (5.7.1)$$

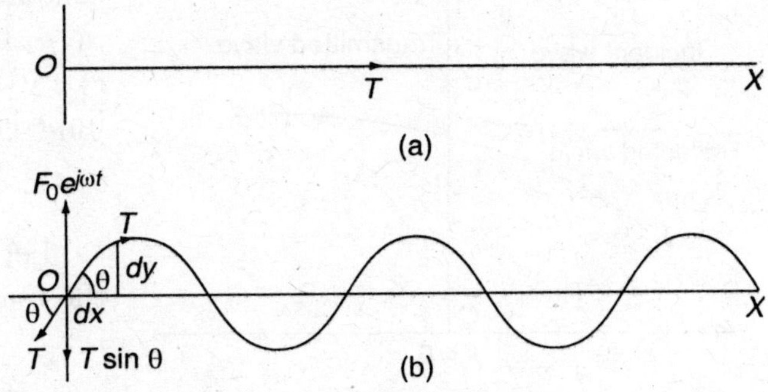

Fig 5.6

where $\vec{a}$, the amplitude may be *complex* due to its phase relation with F, and $k = 2\pi/\lambda = \omega/c$, the *propagation constant* and ω the *angular frequency* of the applied periodic force and c the wave velocity.

$$\therefore\ \text{At } x = 0,\quad F_0e^{j\omega t} = -T\left(\frac{dy}{dx}\right)_{x=0} = jkT\vec{a}e^{j\omega t}.$$

$$\therefore\quad \vec{a} = \frac{F_0}{jkT} \quad\Rightarrow\quad \vec{y} = \frac{F_0}{jT}\times\frac{c}{\omega}e^{j(\omega t - kx)},\quad \text{using (5.7.1)}$$

$$\therefore\quad \text{Transverse velocity, } \vec{v} = \frac{d\vec{y}}{dt} = \vec{a}j\omega e^{j(\omega t - kx)} = F_0\frac{\omega}{kT}e^{j(\omega t - kx)} = v_0e^{j(\omega t - kx)}$$

where $v_0 = F_0\omega/kT$.

$$\therefore\quad \textit{Velocity amplitude,}\quad v_0 = F_0c/T$$

$$\therefore\quad \text{Impedance, } \boxed{\boldsymbol{Z = \frac{F_0}{v_0} = \frac{T}{c} = \rho c}} \quad (\because\ c = \sqrt{T/\rho}) \qquad (5.7.2)$$

where ρ is the mass per unit length, i.e., the *linear density.*

• Note that the characteristic impedance *depends* upon (i) the *mass per unit length*, ρ and (ii) the *velocity of the wave*, c. These again depend on *inertia* and *elasticity* of the medium—the two *energy storing* parameters.

5.8 Reflection and transmission of a wave on a string due to a discontinuity

The characteristic impedance offered by a string to the waves travelling along it is just seen to be ρc. If there is a *sudden change* in impedance due to a *discontinuity* at the junction of two media, the wave suffers a *reflection* at the boundary. This is irrespective of the nature of the wave—sound waves along a string, voltage or current waves along a transmission line or an electromagnetic wave in a medium.

Let a string consist of two parts joined smoothly at $x = 0$ (the point P), having constant tension T throughout (Fig. 5.7). If ρ_1, ρ_2 be the linear densities of the parts and c_1, c_2 the wave velocities in them, then their impedance are

$$Z_1 = \rho_1 c_1 \quad \text{and} \quad Z_2 = \rho_2 c_2.$$

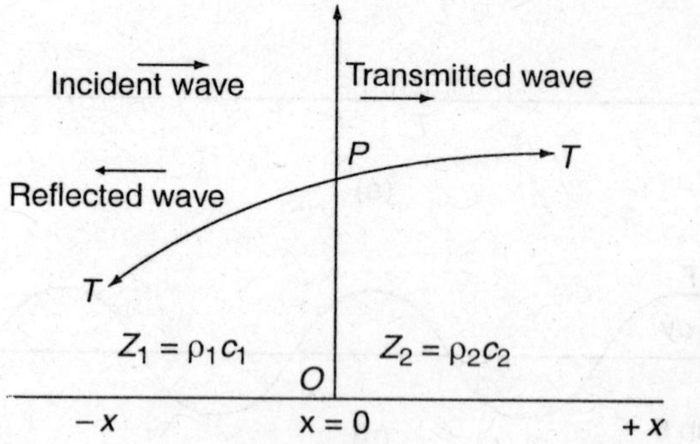

Fig 5.7

Let the incident wave be represented by the real part of the displacement equation

$$y_i = A_1 e^{j(\omega t - k_1 x)}$$

A_1 being the amplitude of the incident wave travelling along $+x$ direction with velocity c_1 and propagation constant k_1.

Then the reflected and transmitted waves will be given by

$$y_r = B_1 e^{j(\omega t + k_1 x)}$$
$$y_t = A_2 e^{j(\omega t - k_2 x)}$$

where B_1 = amplitude of reflected wave moving along $-x$ direction with velocity c_1 and $k_1 = \omega / c_1$; A_2 = amplitude of transmitted wave travelling along $+x$ direction with velocity c_2 and $k_2 = \omega / c_2$.

Now, the displacement y immediately to the left and to the right of $x = 0$ must be the same so that there is *no discontinuity of displacement* (**geometrical condition**) :

$$\therefore \quad y_i + y_r = y_t$$

$$\text{or,} \quad A_1 e^{j(\omega t - k_1 x)} + B_1 e^{j(\omega t + k_1 x)} = A_2 e^{j(\omega t - k_2 x)}$$

$$\text{At } x = 0, \quad \text{therefore,} \quad A_1 e^{j\omega t} + B_1 e^{j\omega t} = A_2 e^{j\omega t}$$

$$\Rightarrow \quad \boxed{A_1 + B_1 = A_2} \tag{5.8.1}$$

Also, the transverse force $T(\partial y / \partial x)_{x=0}$ must be *continuous across the boundary*, implying that the *gradient* (dy/dx) *must be the same on both sides*; for otherwise, a finite transverse force will act on an infinitesimal mass of the string and will result in an infinite acceleration which is absurd. This is the **dynamical condition** :

$$T\frac{dy_i}{dx} + T\frac{dy_r}{dx} = T\frac{dy_t}{dx}$$

$$\text{or,} \quad -Tk_1 A_1 e^{j(\omega t - k_1 x)} + Tk_1 B_1 e^{j(\omega t + k_1 x)} = -Tk_2 A_2 e^{j(\omega t - k_2 x)}$$

$$\text{At, } x = 0, \quad -Tk_1 A_1 e^{j\omega t} + Tk_1 B_1 e^{j\omega t} = -Tk_2 A_2 e^{j\omega t}$$

$$\Rightarrow \quad -Tk_1 A_1 + Tk_1 B_1 = -Tk_2 A_2$$

$$\text{or,} \quad -\omega \frac{T}{c_1} A_1 + \omega \frac{T}{c_1} B_1 = -\omega \frac{T}{c_2} A_2$$

$$\Rightarrow \quad -\rho_1 c_1 A_1 + \rho_1 c_1 B_1 = -\rho_2 c_2 A_2 \quad (\because T/c_1 = \rho_1 c_1 \text{ etc.})$$

$$\text{or,} \quad -Z_1 A_1 + Z_1 B_1 = -Z_2 A_2 \quad (\because \rho_1 c_1 = Z_1 \text{ etc.})$$

$$\Rightarrow \quad \boxed{(A_1 - B_1)Z_1 = A_2 Z_2} \tag{5.8.2}$$

Reflection coefficient of amplitude—The *reflection coefficient of amplitude* is defined as the ratio of amplitude of the reflected wave to that of the incident wave.

$$\therefore \quad \text{Reflection coefficient} = \frac{\text{Amplitude of reflected wave}}{\text{Amplitude of incident wave}} = \frac{B_1}{A_1}$$

But from (5.8.2), using (5.8.1), we obtain

$$\frac{A_2}{A_1 - B_1} = \frac{A_1 + B_1}{A_1 - B_1} = \frac{Z_1}{Z_2}$$

$$\therefore \quad \text{Reflection coefficient} = \boxed{\frac{B_1}{A_1} = \frac{Z_1 - Z_2}{Z_1 + Z_2}} \tag{5.8.3}$$

Transmission coefficient of amplitude—The *transmission coefficient of amplitude* is defined as the ratio of the amplitude of the transmitted wave to that of the incident wave.

$$\therefore \quad \text{Transmission coefficient} = \frac{\text{Amplitude of transmitted wave}}{\text{Amplitude of incident wave}} = \frac{A_2}{A_1}$$

But from (5.8.2), using (5.8.1), we obtain

$$(A_1 - A_2 + A_1)Z_1 = A_2 Z_2$$

$$\Rightarrow \quad 2A_1 Z_1 = A_2(Z_1 + Z_2)$$

$$\therefore \quad \text{Transmission coefficient} = \boxed{\frac{A_2}{A_1} = \frac{2Z_1}{Z_1 + Z_2}} \tag{5.8.4}$$

• The expression for both the reflection and the transmission coefficients are *independent of* ω, that is, they hold for waves of all frequencies. They are real quantities and are thus free from phase changes except however for a phase change of π on reflection from a *fixed end*.

Amplitude after reflection at a super dense medium ($Z_2 \simeq \infty$)—If $Z_2 \simeq \infty$, i.e., $x = 0$ is a *fixed end*, then from (5.8.3)

$$\frac{B_1}{A_1} = \frac{(Z_1/Z_2) - 1}{(Z_1/Z_2) + 1} = \frac{-1}{+1} = -1$$

The incident wave is thus *completely reflected with a phase change of* π. There is *no transmitted wave.* A group of waves having a number of component frequencies will retain its shape even after reflection.

Amplitude after reflection at a rarer medium ($Z_2 \simeq 0$)—If $Z_2 \simeq 0$, i.e., $x = 0$ is a *free end*, then from (5.8.3)

$$\frac{B_1}{A_1} = \frac{Z_1}{Z_1} = 1;$$

But, the transmission coefficient : $\dfrac{A_2}{A_1} = 2$, using (5.8.4)

The amplitude of transmitted wave A_2 is double that of the incident wave A_1. As $B_1 = A_1$, the incident wave is *completely reflected without any phase change.*

• Since the amplitude of the transmitted wave A_2 is twice the amplitude of the incident wave A_1, the intensity of the transmitted wave becomes four times that of the incident wave. This results in a very loud sound, called a **flick**. This explains why we hear a 'flick' at the end of a whip or a free end of a string.

5.9 Reflection and transmission of energy

The total energy per unit length of a string of linear density ρ, if treated as a simple harmonic oscillator of amplitude A and angular frequency ω, is

$$E = \tfrac{1}{2}\rho\omega^2 A^2.$$

As the wave moves along the string with a velocity c, each unit length of the string takes up its oscillation so that :

Rate of energy flow along the string, $\mathcal{R}$ = energy × velocity = $\frac{1}{2}\rho\omega^2 A^2 c$.

Let the reflection occur at $x = 0$. So the rate of energy arriving at the boundary $x = 0$ with the incident wave is

$$\mathcal{R}_a = \tfrac{1}{2}\rho_1\omega^2 A_1^2 c_1 = \tfrac{1}{2}\omega^2 A_1^2 Z_1 \quad (\because\ Z_1 = \rho_1 c_1) \tag{5.9.1}$$

where Z_1 is the impedance of the first medium.

The energy leaves the boundary through the reflected and transmitted waves.

∴ Rate at which energy leaves the boundary is

$$\begin{aligned}\mathcal{R}_l &= \tfrac{1}{2}\rho_1\omega^2 B_1^2 c_1 + \tfrac{1}{2}\rho_2\omega^2 A_2^2 c_2\\ &= \tfrac{1}{2}\omega^2 B_1^2 Z_1 + \tfrac{1}{2}\omega^2 A_2^2 Z_2\end{aligned}$$

where Z_2 is the impedance of the second medium.

Putting the values of B_1 and A_2, from (5.8.3) and (5.8.4),

$$\mathcal{R}_l = \frac{1}{2}\omega^2 A_1^2 \frac{Z_1(Z_1 - Z_2)^2 + 4Z_1^2 Z_2}{(Z_1 + Z_2)^2} = \frac{1}{2} Z_1\omega^2 A_1^2 \tag{5.9.2}$$

$$\therefore \quad \boxed{\mathcal{R}_a = \mathcal{R}_l,}$$

using the relations (5.9.1) and (5.9.2).

Thus, *all the energy arriving at a boundary in the incident wave leaves the boundary in the reflected and transmitted wave.*

Reflected and transmitted energy coefficients—The reflected energy coefficient is the ratio of reflected energy and the incident energy.

$$\therefore \quad \textit{Reflected energy coefficient} = \frac{\frac{1}{2}Z_1\omega_1^2 B_1^2}{\frac{1}{2}Z_1\omega^2 A_1^2} = \frac{B_1^2}{A_1^2} = \left(\frac{Z_1 - Z_2}{Z_1 + Z_2}\right)^2 \tag{5.9.3}$$

Similarly, the transmitted energy coefficient is the ratio of the transmitted energy and the incident energy.

$$\therefore \quad \textit{Transmitted energy coefficient} = \frac{\frac{1}{2}Z_2\omega^2 A_2^2}{\frac{1}{2}Z_1\omega^2 A_1^2} = \frac{Z_2 A_2^2}{Z_1 A_1^2} = \frac{4Z_1 Z_2}{(Z_1 + Z_2)^2} \tag{5.9.4}$$

5.10 Impedance matching and its uses

A wave meeting a boundary between two media with different impedances gets partly reflected and partly transmitted.

The reflected energy coefficient $= (Z_1 - Z_2)^2/(Z_1 + Z_2)^2$, from (5.9.3).

$\therefore$ When $Z_1 = Z_2$, the reflected energy coefficient $= 0$.

The transmitted energy coefficient $= 4Z_1Z_2/(Z_1 + Z_2)^2$, from (5.9.4)

$\therefore$ When $Z_1 = Z_2$, the transmitted energy coefficient $= 1$, so that : *Transmitted energy = Incident energy* and *no* energy is *reflected* at the boundary.

In such a case, *the impedances are said to be* **perfectly matched.**

• **Conditions for matching when $Z_1 \neq Z_2$**—If the impedances of the two media are *not* the same, then also *impedance matching* is possible by *coupling* a piece of string of length l and of impedance Z under the following proviso :

$$\text{(i) } Z = \sqrt{Z_1 Z_2} \quad \text{and} \quad \text{(ii) } l = \lambda/4.$$

So the required conditions are : (i) The coupling element should have an impedance equal to the geometric mean of the impedances of the two media and (ii) the length of the coupling element should have a length equal to the quarter wavelength of the incident wave in the medium of coupling element.

Uses—The above matching conditions are quite general and apply to all types of waves travelling through two media of different impedances. The impedance matching thus finds applications in many fields :

1. The transfer of power from a generator to a load is maximum if there is impedance matching between the generator and the load.

2. Two transmission lines having different impedances are matched by the insertion of $\lambda/4$-stubs of the line between them.

3. In electronic power amplifier or radio set, the impedance of output circuit is matched with that of the loud-speaker by a suitable coupling transformer.

4. Optical instruments have lenses coated with a dielectric layer of thickness $\lambda/4$ (quarter wave layer) and having an index $n = \sqrt{n_1 n_2}$ where n_1 is the index of air ($n_1 = 1$) and n_2 that of the material of the lens. Such coatings are called **anti-reflection films** and their application is known as *blooming*. Exact matching however is almost an impossible task.

5.11 Stationary or standing waves in a string

So long we discussed the progressive waves in a string. We shall now consider hereunder the *stationary* or *standing waves* in a string.

Let a string AB of length l be rigidly fixed at both ends (Fig. 5.8) i.e., at $x = 0$ and at $x = l$ and a monochromatic wave of angular frequency ω and amplitude a propagate along $+x$ direction. So the wave equation is

$$y_1 = ae^{j(\omega t - kx)}$$

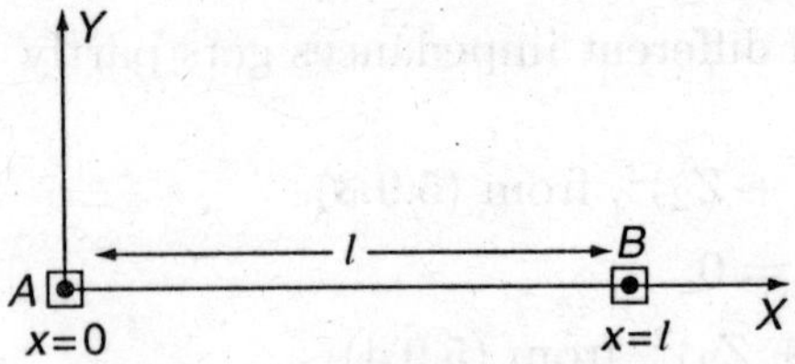

Fig 5.8

where $k(= \omega/c)$ is the propagation constant.

The wave gets reflected from the fixed end B and if b be the amplitude of the reflected wave travelling along $-x$ direction, the reflected wave equation is

$$y_2 = be^{j(\omega t + kx)}$$

If y be the displacement of a point in the string due to both these waves, then

$$y = ae^{j(\omega t - kx)} + be^{j(\omega t + kx)} \tag{5.11.1}$$

The **boundary conditions** to be satisfied at all times are the following : (i) At $x = 0$, $y = 0$; at $x = l$, $y = 0$.

Applying the first boundary condition to (5.11.1), we obtain

$$0 = ae^{j\omega t} + be^{j\omega t} \quad \Rightarrow \quad \boxed{a = -b} \tag{5.11.2}$$

This means that the wave is completely reflected at either end with a *reversal of phase*, i.e., a *phase change π occurs.*

From (5.11.1), using (5.11.2), we obtain

$$\begin{aligned} y &= ae^{j\omega t}\left[e^{-jkx} - e^{+jkx}\right] \\ &= -2jae^{j\omega t}\sin kx \end{aligned} \tag{5.11.3}$$

This satisfies the *time-independent standing wave-form* of the wave equation

$$\frac{d^2y}{dx^2} + k^2y = 0, \text{ for}$$

$$\frac{dy}{dt} = (-2j)a(j\omega)e^{j\omega t}\sin kx, \quad \text{using (5.11.3).}$$

$$\therefore \quad \frac{d^2y}{dt^2} = (-2j)a(-\omega^2)e^{j\omega t}\sin kx = -\omega^2 y$$

$$\therefore \quad \frac{1}{c^2}\frac{d^2y}{dt^2} = -\frac{\omega^2}{c^2}y = -k^2 y$$

Substituting in the *standard differential wave equation*, we obtain

$$\frac{d^2y}{dx^2} = \frac{1}{c^2}\frac{d^2y}{dt^2} = -k^2 y$$

$$\Rightarrow \quad \frac{d^2y}{dx^2} + k^2 y = 0.$$

Applying the second boundary condition, we get from the relation (5.11.3),

$$0 = -2jae^{j\omega t}\sin kl$$

$$\Rightarrow \quad \sin kl = \sin\frac{\omega}{c}l = 0$$

$$\Rightarrow \quad \boxed{\frac{\omega l}{c} = n\pi} \tag{5.11.4}$$

where n stands for an integer.

This imposes a *restriction* on the value of the *allowed* frequencies (*quantization*). We have

$$\frac{\omega_n l}{c} = n\pi \quad \Rightarrow \quad \frac{2\pi\nu_n}{c} = \frac{n\pi}{l}$$

$$\Rightarrow \quad \boxed{\nu_n = \frac{nc}{2l}} \tag{5.11.5}$$

where n is an integer.

If λ_n be the corresponding wavelength

$$\lambda_n = \frac{c}{\nu_n} = \frac{2l}{n} \quad \Rightarrow \quad \boxed{l = \frac{n\lambda_n}{2}} \tag{5.11.6}$$

The *allowed frequencies thus define the length of the string as an exact number of half wavelengths.*

Eigen frequencies—The frequencies as obtained from (5.11.5) for various integral values of n are called the *normal frequencies* or *normal modes of vibration.* They are often termed **eigen frequencies** or *proper frequencies.*

The lowest frequency corresponds to $n = 1$ and is called the *fundamental.* The frequencies corresponding to $n > 1$, i.e., $2, 3, 4, \cdots$ etc. are known as *harmonics.* For *first harmonic* $n = 2$, for *second harmonic* $n = 3$ etc. These are all shown in Fig. 5.9. In general, all the normal modes may be present simultaneously, the general displacement being the superposition of the displacements due to various eigen frequencies.

Nodes and antinodes—For $n > 1$, there are, along the string, a number of positions that are always at rest. These positions are given by

$$\sin \frac{n\pi x}{l} = 0 \Rightarrow \frac{n\pi x}{l} = r\pi$$
$$(r = 0, 1, 2, \cdots, n)$$

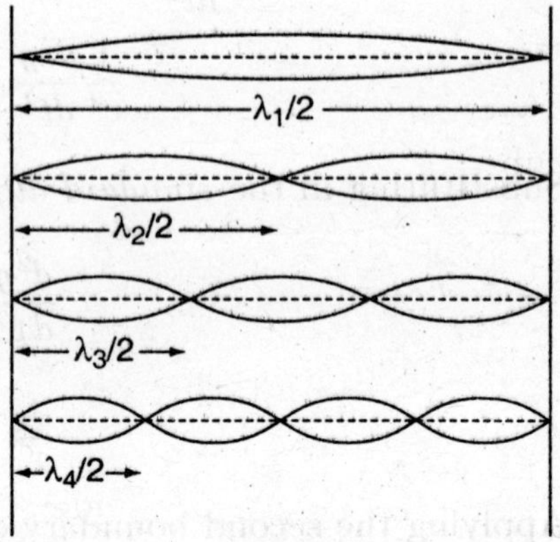

Fig 5.9

The positions of two fixed ends corresponds to $r = 0$ and $r = n$; $r = 0$ gives $x = 0$ and $r = n$ gives $x = l$. Between these two, there are $n - 1$ positions for the nth harmonic which are *equally spaced* along the string where the displacement $y = 0$ for all t. These positions are called **nodes** or *nodal points* of the standing waves. At the nodes, the reflection is complete so that the amplitude of incident wave and that of the reflected progressive wave are equal and opposite. Between two consecutive nodes, and *equally spaced*, are the positions where the *displacement* is *maximum*. These positions of maximum displacement are called **antinodes**.

Equation for standing waves—Eq. (5.11.3) for nth harmonic may be expressed as

$$y_n = -2aje^{j\omega_n t} \sin \frac{\omega_n}{c} x \quad (\because k_n = \omega_n / c)$$
$$= 2a(-j)\,[\cos \omega_n t + j \sin \omega_n t] \sin \frac{\omega_n}{c} x$$
$$= (A_n \cos \omega_n t + B_n \sin \omega_n t) \frac{\omega_n}{c} x \tag{5.11.7}$$

where $A_n = -2aj$, $B_n = 2a$.

Eq. (5.11.6) is the equation for *standing or stationary waves* in the string.

5.12 Energy of a vibrating string

We shall now find expression for the energy of a vibrating string is partly kinetic and partly potential.

Let dx be an elemental length of a vibrating stretched string of length l and let ρ be its mass per unit length.

$\therefore$ Kinetic energy of the element dx of the string,

$$dE_k = \tfrac{1}{2} \text{ mass} \times \text{velocity}^2 = \tfrac{1}{2} \rho dx\, \dot{y}^2$$

∴ Total kinetic energy of the string of length l is given by

$$E_k = \frac{1}{2}\int_0^l \rho \dot{y}^2\, dx. \tag{5.12.1}$$

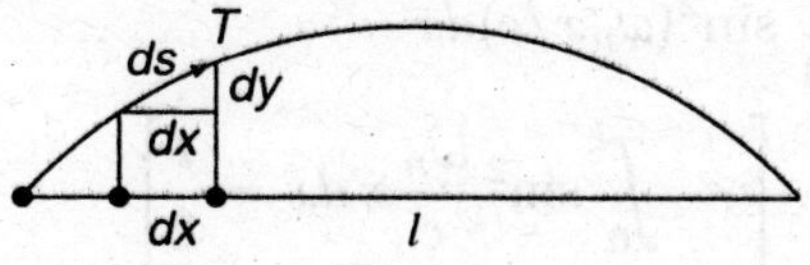

Fig 5.10

If the tension T of the string extends an elemental length dx in the normal position (Fig. 5.10) of the string to ds in the vibrating position, the work done by it is the potential energy of the string.

∴ Potential energy of the string-element dx is

$$\begin{aligned} dE_p &= T(ds - dx) \\ &= T\left[(dx^2 + dy^2)^{\frac{1}{2}} - dx\right] \\ &= T\left[dx\left\{1 + \left(\frac{dy}{dx}\right)^2\right\}^{\frac{1}{2}} - dx\right] \\ &= T\left[dx\left\{1 + \frac{1}{2}\left(\frac{dy}{dx}\right)^2 + \cdots\right\} - dx\right] \\ &= \frac{1}{2}T\left(\frac{dy}{dx}\right)^2 dx, \end{aligned}$$

neglecting the higher powers of (dy/dx) which are all negligibly small and taking tension T as constant.

∴ Total potential energy of the string is given by

$$E_p = \frac{1}{2}T\int_0^l \left(\frac{dy}{dx}\right)^2 dx \tag{5.12.2}$$

Energy in each normal mode—In normal mode, every point in the string vibrates with the same frequency and at any moment all parts vibrate either in the same or opposite phase.

∴ Kinetic energy in nth mode, $E_{nk} = \frac{1}{2}\int_0^l \rho \dot{y}_n^2 dx.$

Potential energy in the nth mode, E_{np} is given by

$$E_{np} = \frac{1}{2}T\int_0^l \left(\frac{dy_n}{dx}\right)^2 dx$$

But from the relation (5.11.7), we have

$$y_n = (A_n \cos\omega_n t + B_n \sin\omega_n t)\sin\frac{\omega_n}{c}x$$

$$\therefore \quad \dot{y}_n = (-A_n\omega_n \sin\omega_n t + B_n\omega_n \cos\omega_n t)\sin\frac{\omega_n}{c}x$$

$$\text{and} \quad \frac{dy_n}{dx} = \frac{\omega_n}{c}(A_n \cos\omega_n t + B_n \sin\omega_n t)\cos\frac{\omega_n}{c}x$$

$$\therefore \quad E_{nk} = \tfrac{1}{2}\rho\omega_n^2(-A_n \sin\omega_n t + B_n\cos\omega_n t)^2 \int_0^l \sin^2(\omega_n x/c)dx$$

$$= \tfrac{1}{4}l\rho\omega_n^2(-A_n\sin\omega_n t + B_n\cos\omega_n t)^2 \quad \left[\because \int_0^l \sin^2\frac{\omega_n}{c}x\,dx = \frac{l}{2}\right]$$

$$E_{np} = \tfrac{1}{2}T\frac{\omega_n^2}{c^2}(A_n\cos\omega_n t + B_n\sin\omega_n t)^2\int_0^l \cos^2\frac{\omega_n}{c}x\,dx$$

$$= \tfrac{1}{4}lT\frac{\omega_n^2}{c^2}(A_n\cos\omega_n t + B_n\sin\omega_n t)^2 \quad \left[\because \int_0^l \cos^2\frac{\omega_n}{c}x\,dx = \frac{l}{2}\right]$$

$$= \tfrac{1}{4}l\rho\omega_n^2(A_n\cos\omega_n t + B_n\sin\omega_n t)^2 \quad (\because c^2 = T/\rho)$$

$\therefore$ **Total energy** of vibration E_n of the string is given by

$$E_n = E_{nk} + E_{np}$$

$$= \tfrac{1}{4}l\rho\omega_n^2\left[(-A_n\sin\omega_n t + B_n\cos\omega_n t)^2 + (A_n\cos\omega_n t + B_n\sin\omega_n t)^2\right]$$

$$= \tfrac{1}{4}l\rho\omega_n^2(A_n^2 + B_n^2)$$

$$= \tfrac{1}{4}m\omega_n^2(A_n^2 + B_n^2) \tag{5.12.3}$$

where $\rho l = m =$ mass of the entire string.

• If in the resultant vibration of the string, there are more than one mode, then the total energy of the resultant vibration is obtained by adding the energy of each mode as if it existed *independently* of the other modes.

$$\therefore \quad E_{\text{Total}} = \sum E_n = \frac{1}{4}m\sum_n \omega_n^2(A_n^2 + B_n^2). \tag{5.12.4}$$

5.13 Illustrated Examples

► **Example 1.** *A simple harmonic wave is travelling along x-axis and is given by*

$$y = 5\sin 2\pi(0.2t - 0.5x)$$

where x is in metres and t is in seconds.

Calculate the amplitude, frequency, wavelength, wave velocity, particle velocity, velocity amplitude, particle acceleration and acceleration amplitude. **(Punjab Univ.)**

Solution. We have :

$$y = 5\sin 2\pi(0.2t - 0.5x)\text{ m}$$

Amplitude, $a =$ **5 m**; frequency $n =$ **0.2 s^{-1}**; $(x/\lambda) = 0.5x$

$\Rightarrow$ wavelength $\lambda = 1/0.5 = \mathbf{2\ m}$; wave velocity, $v = n\lambda = 0.2 \times 2 = \mathbf{0.4\ ms^{-1}}$.

Particle velocity, $dy/dt = 5\cos 2\pi(0.2t - 0.5x) \times 0.2 = \mathbf{\cos 2\pi(0.2t - 0.5x)}$.

$\therefore$ Particle velocity amplitude $= \mathbf{1\ ms^{-1}}$

Particle acceleration, $d^2y/dt^2 = -\sin 2\pi(0.2t - 0.5x) \times 0.2$

$$= \mathbf{-0.2 \sin 2\pi(0.2t - 0.5x)}.$$

$\therefore$ Particle acceleration amplitude $= \mathbf{-0.2\ ms^{-2}}$

► **Example 2.** *Write the equation of the wave travelling in the $-x$-direction and having an amplitude* 0.01 m, *a frequency* 550 Hz *and a speed* 330 m/s. **(Lucknow Univ.)**

Solution. The equation of a wave travelling along $-x$ direction is :

$$y = a \sin \frac{2\pi}{\lambda}(ct + x) \tag{5.13.1}$$

By the problem, $a = 0.01$ m; $v = 330$ m/s and $n = 550$ Hz. Therefore,

$$\text{wavelength, } \lambda = \frac{c}{n} = \frac{330}{550} = \frac{3}{5} \text{ m}$$

Substituting these values in (5.13.1), the required equation is given by

$$y = 0.01 \sin \left[2\pi \times \frac{5}{3}(330t + x)\right]$$

$$= \mathbf{0.01 \sin 2\pi\ (550t + 5x/3)}$$

► **Example 3.** *If a wave is represented by the equation $y = a \sin k(ct - x)$, prove that $d^2y/dt^2 = c^2(d^2y/dx^2)$.* **(Punjab Univ.)**

Solution. It is given that the wave equation is

$$y = a \sin k(ct - x)$$

$$\therefore \quad dy/dt = akc \cos k(ct - x);$$

$$d^2y/dt^2 = -k^2c^2 a \sin k(ct - x)$$

$$dy/dx = -ak \cos k(ct - x);$$

$$d^2y/dx^2 = -k^2 a \sin k(ct - x)$$

$$\therefore \quad \mathbf{\frac{d^2y}{dt^2} = c^2 \frac{d^2y}{dx^2}}$$

► **Example 4.** *Show that $y = x^2 + c^2t^2$ is a solution of one dimensional wave equation.*

Solution. It is given that

$$y = x^2 + c^2t^2.$$

$$\therefore \quad \frac{dy}{dt} = 2c^2t; \quad \frac{d^2y}{dt^2} = 2c^2.$$

$$\text{Also,} \ \frac{dy}{dx} = 2x; \quad \frac{d^2y}{dx^2} = 2.$$

$$\therefore \quad \frac{d^2y}{dt^2} = c^2\frac{d^2y}{dx^2}$$

which is the one-dimensional wave equation travelling along $+x$ direction with a velocity c. **Hence** the result.

▸ **Example 5.** *A string* 1.3 m *in length is divided into three segments such that their frequencies are in the ratio* 2 : 3 : 4. *Find the length of each segment.* **(Punjab Univ.)**

Solution. The frequency of vibrations of a string is given by

$$n = \frac{1}{2l}\sqrt{\frac{T}{\rho}}$$

$$\Rightarrow \ n \propto \frac{1}{l}, \quad \text{for constant } T \text{ and } \rho.$$

The frequencies here are in the ratio 2 : 3 : 4. So, the lengths of the segments l_1, l_2, l_3 must be in the ratio $\frac{1}{2} : \frac{1}{3} : \frac{1}{4}$ or 6 : 4 : 3.

Total length of the string, $l = l_1 + l_2 + l_3 = 1.3$ m.

$$\therefore \quad l_1 = \frac{1\cdot 3}{6+4+3} \times 6 = \mathbf{0.6\ m}$$

$$l_2 = \frac{1\cdot 3}{6+4+3} \times 4 = \mathbf{0.4\ m}$$

$$l_3 = \frac{1\cdot 3}{6+4+3} \times 3 = \mathbf{0.3\ m}$$

▸ **Example 6.** *A string AB of length* 100 cm *is under uniform tension. A bridge is kept at C such that BC* = 60 cm. *When the string is subjected to vibration, BC part vibrates with a frequency* 252 Hz. *Find the frequency of both the parts of the string when the bridge is moved through* 10 cm *towards A.* **(Vidyasagar Univ.)**

Solution. When BC is 60 cm., the frequency of BC part is :

$$n = \frac{1}{2l}\sqrt{\frac{T}{\rho}} \Rightarrow 252 = \frac{1}{2 \times 60}\sqrt{\frac{T}{\rho}} \Rightarrow \sqrt{\frac{T}{\rho}} = 252 \times 120.$$

When the bridge moves towards A by 10 cm to reach D, $BD = 60 + 10 = 70$ cm and naturally, $DA = 100 - 70 = 30$ cm.

$\therefore$ Frequency of BD-part, $n_1 = \dfrac{1}{2 \times 70}\sqrt{\dfrac{T}{\rho}}$

$$= \frac{1}{2 \times 70} \times 252 \times 120 = \mathbf{216\ Hz.}$$

Frequency of DA part,

$$n_2 = \frac{1}{2 \times 30} \times 252 \times 120 = \mathbf{504\ Hz.}$$

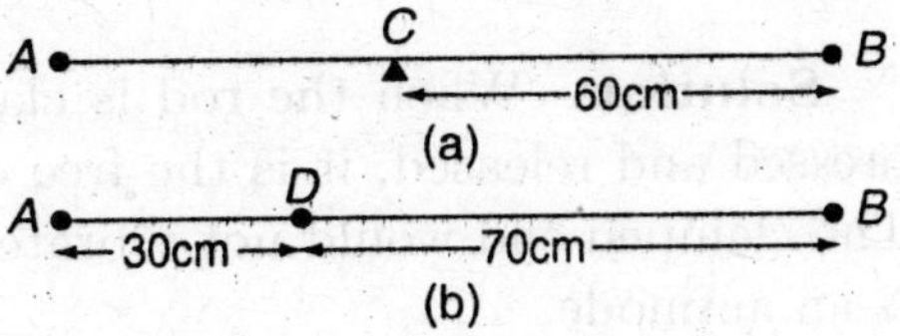

Fig 5.11

▶ **Example 7.** *A stretched string is observed to vibrate with a frequency* 30 Hz *in the fundamental note when the length of the string is* 60 cm. *The string has a mass of* 0.5 g/cm.. *Find the velocity of propagation of the transverse wave and compute the tension of the string.* **(Burdwan Univ.)**

Solution. Frequency of fundamental note, $n = 30$ Hz; length of the string, $l = 60$ cm; mass per unit length $\rho = 0.5$ g/cm. If T be the tension of the string, then

$$n = \frac{1}{2l}\sqrt{\frac{T}{\rho}} \Rightarrow T = 4n^2l^2\rho$$

$$= 4 \times 30^2 \times 60^2 \times 0.5$$

$$= \mathbf{6.48 \times 10^6\ dyne/cm}$$

$\therefore$ Velocity of propagation of wave, $c = \sqrt{T/\rho} = n \times 2l.$

$$= 30 \times 2 \times 60 = \mathbf{3600\ cm/s}$$

▶ **Example 8.** *Two strings of linear densities* 0.5 g/m *and* 1.5 g/m *are joined together and stretched with a certain force. Calculate* (a) *the ratio of wave speeds in two strings and* (b) *the reflection and transmission coefficients of energy.* **(Guru Nanak Univ.)**

Solution. Here $\rho_1 = 0.5 \times 10^{-3}$ kg/m; $\rho_2 = 1.5 \times 10^{-3}$ kg/m.

(a) $\dfrac{c_1}{c_2}$ = ratio of wave speeds $= \sqrt{\dfrac{T}{\rho_1}} \times \sqrt{\dfrac{\rho_2}{T}} = \sqrt{\dfrac{\rho_2}{\rho_1}} = \sqrt{\dfrac{1.5 \times 10^{-3}}{0.5 \times 10^{-3}}} = \sqrt{3} = \mathbf{1.73}$

(b) $\dfrac{Z_1}{Z_2} = \dfrac{\rho_1 c_1}{\rho_2 c_2} = \dfrac{\rho_1}{\rho_2} \times \dfrac{c_1}{c_2} = 1.73 \times \dfrac{0.5 \times 10^{-3}}{1.5 \times 10^{-3}} = 0.58. \quad \therefore Z_1 = 0.58 Z_2.$

$\therefore$ Reflected energy coefficient $= \left(\dfrac{Z_1 - Z_2}{Z_1 + Z_2}\right)^2 = \left(\dfrac{0.58Z_2 - Z_2}{0.58Z_2 + Z_2}\right)^2 = \left(\dfrac{0.42}{1.58}\right)^2 = \mathbf{0.07}$

Transmitted energy coefficient $= \dfrac{4Z_1Z_2}{(Z_1 + Z_2)^2} = \dfrac{4 \times 0.58 Z_2^2}{(1.58 Z_2)^2} = \mathbf{0.93}$

► **Example 9.** *When a rod is clamped at one end horizontally and the free end is pressed down and released, show that for the smallest frequency of vibration, the wavelength is four times the length of the rod.*

Solution. When the rod is clamped at one end horizontally and the free end is pressed and released, it is the free end that would vibrate with maximum amplitude. The clamped end would not vibrate at all. So the clamped end is a node, the free end is an antinode.

$\therefore$ Length of the rod, l = minimum distance between a node and an antinode = $\lambda/4$.

$$\therefore \quad l = \frac{\lambda}{4} \Rightarrow \lambda = \mathbf{4l}.$$

This is the maximum wavelength and thus corresponds to smallest frequency. **Hence.**

Review Questions

1. What is meant by wave motion? What are transverse and longitudinal waves?
2. State the characteristics of the progressive wave. Derive an equation for the progressive wave of wavelength λ and amplitude a moving with a velocity v in the $+x$ direction. **(Lucknow and Burdwan Univ.)**
3. Show that the slope of the displacement curve for a wave gives the volumetric strain in the medium in which it travels.
4. Establish the relation between wave velocity and particle velocity in a progressive wave. Hence deduce the differential equation of the plane progressive wave.
5. Derive a relation between the wave velocity v, particle velocity dy/dt and the volumetric strain dy/dx. **(Vidyasagar and Kanpur Univ.)**
6. Deduce an expression for the velocity of transverse waves in a string. Hence find the frequency of the fundamental note and the first octave for a vibrating string fixed at both the ends. **(Calcutta and Burdwan Univ)**
7. Prove that the wave equation for a transverse wave in a string is given by
$$\frac{\partial^2 y}{\partial x^2} = \frac{1}{c^2}\frac{\partial^2 y}{\partial t^2}$$
where $c = \sqrt{T/\rho}$, T being the tension and ρ the linear density of the string. What are its possible solutions? **(Guru Nanak and Himachal Univ.)**
8. Derive the relation for the characteristic impedance of a string. Explain the factors on which it depends. **(Punjabi Univ.)**

9. What happens if incident waves fall at a common boundary of two strings having different impedances but joined in such a way that the tension is the same throughout?

10. Define reflection coefficient and transmission coefficient of amplitude. What is the amplitude of transverse waves reflected at the dense medium (i.e., $Z_2 = \infty$)? **(Punjab Univ.)**

11. Show that all the energy in the incident wave arriving at the boundary leaves the boundary in the reflected and the transmitted wave. Define reflection and transmission coefficients of energy. **(Guru Nanak and Punjab Univ.)**

12. State the conditions for perfect impedance matching between two media. What are the uses of impedance matching? Give its application for anti-reflection films. **(Guru Nanak Univ.)**

13. Discuss analytically the formation of standing waves in a string of fixed length. What are its eigen frequencies? How does the position of nodal points change with eigen frequencies? **(Punjab Univ.)**

14. Derive an expression for the potential and kinetic energy of a vibrating string. Hence calculate the energy in each normal mode of a vibrating string. **(Guru Nanak Univ.)**

Short Answer and Objective Questions

1. What is a wave motion? What are transverse and longitudinal waves? Give examples.

2. Enumerate the characteristic features of a progressive wave.

3. Show that : particle velocity = wave velocity × slope of displacement curve (or strain).

4. Show that $y = x^2 + c^2t^2$ is a solution of one dimensional wave equation. **(Punjab Univ.)**

5. What are the factors on which the characteristic impedance of a string depends?

6. Explain why we hear a 'flick' at the end of a string. **(Guru Nanak Univ.)**

7. State the conditions of perfect impedance matching between two media. Give the uses of impedance matching.

8. Two electric transmission cables are joined at a point. What special care should be taken for proper transmission of power? **(Guru Nanak Univ.)**

9. The differential equation of wave motion is given by

(a) $\frac{d^2y}{dx^2} = \frac{1}{c^2}\frac{d^2y}{dt^2}$ (b) $\frac{d^2y}{dt^2} = \frac{1}{c^2}\frac{d^2y}{dx^2}$ (c) $\frac{d^2y}{dx^2} = c\frac{d^2y}{dx^2}$ (d) $\frac{d^2y}{dx^2} = \frac{1}{c}\frac{d^2y}{dt^2}$

10. The path difference x and the phase difference ϕ are related as

(a) $\phi = \frac{2\pi}{\lambda}x$ (b) $x = \frac{2\pi}{\lambda}\phi$ (c) $\phi = 2\pi\lambda x$ (d) $x = 2\pi\lambda\phi$

11. The velocity of transverse waves in a string is given by

(a) $c = \sqrt{T}/m$ (b) $c = T/\sqrt{m}$ (c) $c = T/m$ (d) none of these.

where T is the tension of the string and m its linear density.

12. The condition for impedance matching may be written as

(a) $Z_1 > Z_2$ (b) $Z_1 = Z_2$ (c) $Z_1 < Z_2$ (d) $Z_2 = \infty$.

13. A string 1.3 m long is divided into three segments such that their frequencies are in the ratio 2 : 3 : 4, then the lengths of the segments are in the ratio

(a) 6 : 4 : 3 (b) 2 : 3 : 4 (c) 4 : 3 : 2 (d) none of these

Problems

1. If a wave of frequency 500 Hz travels with a velocity of 200 ms^{-1}, then find the change in phase at a given point in space between a time interval of 10^{-3} sec. Also find the path difference between two points which differ in phase by $\pi/2$ radian. **(Punjab Univ.)** [π, 0.1 m]

2. The displacement equation for a transverse plane wave at any instant is

$$y(x,t) = 0.03\sin(3\pi t - 0.03\pi x)$$

where x and t are in metres and seconds. Calculate the wavelength, frequency and velocity of the wave and also the phase difference between two particles 0.05 m apart at the same instant. [66.67 m, 1.5 Hz, 100 ms^{-1}, 0.0015 π]

3. A wave of frequency 400 Hz travels with a velocity 800 m/s. How far are two points situated whose displacements differ by $\pi/4$ in phase? **(Punjab Univ.)** [0.25 m]

4. Prove that $\psi = Ae^{i(\omega t - kx)}$ is a solution of the wave equation moving along $+x$ direction. **(Guru Nanak Univ.)**

5. A string is stretched with a force of 1000 N. Its linear density is 0.1 kg m^{-1}. One end of the string ($x = 0$) oscillates with an amplitude of 0.05 m and frequency 400 Hz so that the travelling waves are set up in the $+x$ direction. Calculate the velocity and wavelength of the wave and write down the equation of the travelling wave. [100 ms^{-1}, 0.25 m; $y = 0.05 \sin 8\pi(100t - x)$]

6. Two wires of steel of the same length are stretched on a sonometer. The tension of the first and the second are 8 kg wt and 2 kg wt respectively. Find the ratio of the fundamental notes emitted by the two wires when the diameter of cross-section of the first is half that of the second. **(Calcutta Univ.)** [4 : 1]

7. Two strings of linear densities 0.25 g/cm and 1.0 g/cm are joined together and stretched with a force of 100 N. Calculate the coefficients of reflection and transmission of amplitude. **(Punjab Univ.)** [−0.333; 0.667]

8. A wire of linear density 0.01 g/cm is stretched with a tension of 10 N. It resonates at a frequency of 200 Hz. The next higher frequency at which it resonates is 240 Hz. Find the length of the wire.

9. Equation of a progressive wave is given by

$$y = \sin\left(500\pi t - \frac{\pi}{70}x\right) \text{ units.}$$

What are the amplitude, frequency and wavelength of the wave?
(Vidyasagar Univ.) [1 unit, 250 unit, 140 unit]

10. A wave travelling in a rod meets a metallic joint where the Young's modulus is twice its value in the rod but the density remains the same. Find the reflection and transmission coefficients of amplitude. **(Punjab Univ.)** [0.17, 0.83]

Chapter 6

Longitudinal Waves

6.1 Velocity of sound in a solid medium

Sound is a form of longitudinal wave motion and can be excited in a solid medium, such as a rod whose thickness is small compared to its length, either by rubbing it along its length with a resined cloth or by stroking at one end with a hammer. We shall first find an expression for the velocity of sound in a solid to be followed subsequently the same in a fluid medium.

Expression for velocity—Consider a rod of uniform cross-sectional area α, with Y and ρ as the Young's modulus and density respectively of its material (Fig. 6.1). Let longitudinal sound waves propagate through it, from left to right, along x-axis, the axis of the rod. Choose two planes A, B perpendicular to the axis of the rod, initially at distances x and $x + \delta x$ from the origin O. Let them get displaced by y and $y + \delta y$ to planes A' and B' respectively as the waves travel through the medium. The new distance between the planes is

$$(x + \delta x + y + \delta y) - (x + y) = \delta x + \delta y = \delta x + \frac{\partial y}{\partial x}\delta x$$

since $\partial y/\partial x$ is the rate of change of displacement y with distance x. So, the increase in distance between the planes, δx apart, is $(\partial y/\partial x)\delta x$.

$\therefore$ Extension of the rod per unit length $= \partial y/\partial x$.

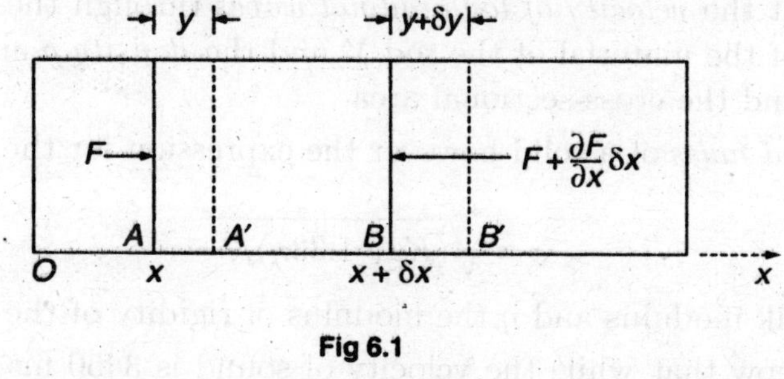

Fig 6.1

If F be the force acting on the first plane A, then the stress on it is F/α.

$$\therefore \text{ Young's modulus, } Y = \frac{F/\alpha}{\partial y/\partial x}$$

$$\Rightarrow F = Y\alpha\frac{\partial y}{\partial x}$$

Force acting on the second plane B however is

$$F + \frac{\partial F}{\partial x}\delta x = Y\alpha\frac{\partial y}{\partial x} + \frac{\partial}{\partial x}\left(Y\alpha\frac{\partial y}{\partial x}\right)\delta x$$

$$= Y\alpha\frac{\partial y}{\partial x} + Y\alpha\frac{\partial^2 y}{\partial x^2}\delta x$$

$$\therefore \text{ Resultant force on the layer } AB \text{ is } Y\alpha\frac{\partial^2 y}{\partial x^2}\delta x$$

But the mass of the layer $AB = \alpha\rho\delta x$ and its acceleration is $\partial^2 y/\partial t^2$.

$$\therefore \alpha\rho\delta x\frac{\partial^2 y}{\partial t^2} = Y\alpha\frac{\partial^2 y}{\partial x^2}\delta x$$

$$\Rightarrow \boxed{\frac{\partial^2 y}{\partial t^2} = \frac{Y}{\rho}\frac{\partial^2 y}{\partial x^2}} \qquad (6.1.1)$$

which is the *differential equation of longitudinal wave propagation* along the rod.

Comparing this relation with the standard differential equation for wave motion

$$\frac{d^2 y}{dt^2} = c^2\frac{d^2 y}{dx^2}$$

the velocity c of the longitudinal wave is given by $c^2 = Y/\rho$

$$\therefore \boxed{c = \sqrt{\frac{Y}{\rho}}} \qquad (6.1.2)$$

This shows that the *velocity of longitudinal waves* through the rod *depends* on the *Young's modulus* of the material of the rod Y and the *density* ρ and is *independent* of the applied force and the cross-sectional area.

- In an *extended mass* of a solid however the expression for the velocity is different and is given by

$$c = \sqrt{(K + 1.33\eta)/\rho} \qquad (6.1.3)$$

where K is the bulk modulus and η the modulus of rigidity of the solid.

Calculations show that while the velocity of sound is 3450 ms^{-1} in a brass bar, it is 4270 ms^{-1} in an extended mass of the same metal.

6.2 Velocity of sound waves (longitudinal) in a fluid

Consider a fluid cylinder of cross-section α and two right plane sections A and B in the tube (Fig. 6.2). Let longitudinal waves travel from *left to right.* The waves will produce *rarefactions* and *condensations* in the medium which will travel from left to right along the axis of the tube—the x-axis.

Let the undisturbed positions of planes A, B, measured from any arbitrary zero along x-axis be x and $x+\delta x$. At some later instant, when the medium gets disturbed, let the plane A be displaced to A' along the tube by an amount y. At the same instant, let B be displaced to B' by an amount $y+\delta y = y + \dfrac{\partial y}{\partial x}\delta x$.

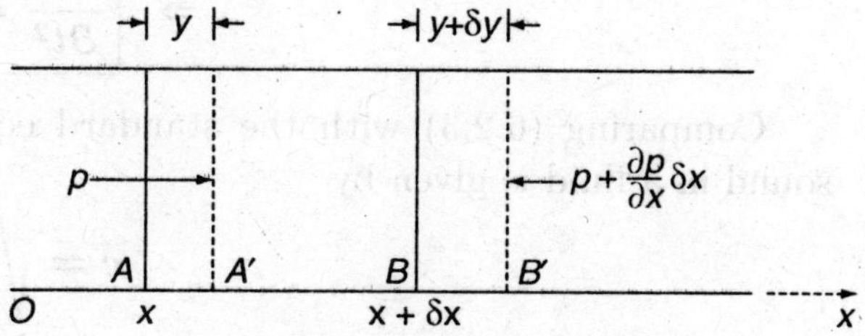

Fig 6.2

The disturbance has changed the *volume* of the fluid between the planes A, B with a consequent variation in *pressure* from point to point along the tube.

Now, the original volume of fluid between $A, B = \delta x \times \alpha = \alpha\delta x$.

The new volume of fluid between $A', B' = (\delta x + \delta y)\alpha = \alpha\delta x + \alpha(\partial y/\partial x)\delta x$.

$\therefore$ Increase in volume $= \alpha(\partial y/\partial x)\delta x$

$\therefore$ Volume strain = change in volume to original volume $= -(\partial y/\partial x)$.

The mass of the fluid between A, $B = \alpha\rho\delta x$, where ρ is the density of the fluid. Let the excess pressure i.e., *pressure above the normal* be p at plane A. We assume the excess pressure at plane A' to be also p. At B, and hence also at B', the excess pressure will be $p + (\partial p/\partial x)\delta x$. The two pressures act in *opposite* directions so that the resultant pressure on element $A'B' = (\partial p/\partial x)\delta x$ and changes its volume, and a resultant force $\alpha(\partial p/\partial x)\delta x$ acts in the direction B' to A' on the contained fluid.

But, we have the relation

stress = modulus of elasticity × strain.

$$\therefore \quad p = -K\left(\frac{\partial y}{\partial x}\right) \tag{6.2.1}$$

where K is the bulk modulus, for here a *change in volume* is involved.

And the resultant force from B' to A' = *mass × acceleration.*

$$\therefore \quad \alpha\frac{\partial p}{\partial x}\delta x = \alpha\rho\delta x\left(-\frac{\partial^2 y}{\partial t^2}\right) \tag{6.2.2}$$

the negative sign only indicates that the acceleration is opposite in direction to that of the displacement.

$\therefore$ From (6.2.2) using (6.2.1)

$$\frac{\partial}{\partial x}\left(-K\frac{\partial y}{\partial x}\right)\delta x = \rho\delta x\left(-\frac{\partial^2 y}{\partial t^2}\right)$$

$$\Rightarrow \quad K\frac{\partial^2 y}{\partial x^2} = \rho\frac{\partial^2 y}{\partial t^2}$$

$$\Rightarrow \quad \boxed{\frac{\partial^2 y}{\partial t^2} = \frac{K}{\rho}\frac{\partial^2 y}{\partial x^2}} \tag{6.2.3}$$

Comparing (6.2.3) with the standard equation for wave motion, the velocity of sound in a fluid is given by

$$c = \sqrt{\frac{K}{\rho}}$$

which shows that the velocity of sound waves in a fluid *depends* only on the *bulk modulus* and the density of the fluid.

- The bulk modulus K is larger for liquids than for gases. The speed of sound in liquids is thus generally greater than that in gases.
- Whether the K-value of liquids is to be taken under adiabatic or isothermal condition is somewhat controversial. In many cases, the difference between the two K-values is not significant. A measurement of velocity gives a value for K, referred to as *acoustic bulk modulus* of the liquid.
- The speed of sound in liquids depends on temperature and to some extent on the pressure. In sea water, the speed increases with salinity and depth below the surface.

6.3 Velocity of sound in gases

When a sound wave propagates in a gas, the pressure-changes, that is, compressions and rarefactions, occur so rapidly, as pointed out by Laplace, that the generated heat cannot leak away to the surroundings. The process is thus *adiabatic* and the adiabatic ideal gas law holds.

$$\therefore \quad PV^\gamma = \text{const.} \quad (\gamma = C_P/C_V)$$

Differentiating, the above expression, we obtain

$$\gamma PV^{\gamma-1}dV + V^\gamma dP = 0$$

$$\Rightarrow \quad \gamma P\,dV = -V\,dP$$

$$\Rightarrow \quad \gamma P = -V\frac{dP}{dV} = \text{bulk modulus, } K$$

$$\therefore \quad \textbf{Velocity of sound,} \quad c = \sqrt{\frac{K}{\rho}} \Rightarrow \boxed{c = \sqrt{\frac{\gamma P}{\rho}}} \tag{6.3.1}$$

which is known as the **Laplace's equation** for sound velocity in gases.

Taking $\gamma = 1.40$ for air, the velocity of air at S.T.P. is given by

$$c = \sqrt{\frac{1.40 \times 76 \times 13.6 \times 980}{0.001293}} = 33100 \text{ cm/s} = \mathbf{331\ m/s}.$$

This is in good agreement with the experimentally obtained value for the velocity of sound in air.

6.3.1 Speed of sound and r.m.s. speed of gas molecules

A relationship can be obtained between the speed of sound in gases and the root mean square value of the speed of gas molecules. According to the kinetic theory, the gas pressure

$$P = \tfrac{1}{3}\rho v^2$$

where ρ is the density of the gas and v the r.m.s. speed of gas molecules.

$$\therefore \quad v = \sqrt{\frac{3P}{\rho}} = \sqrt{\frac{3}{\gamma}\cdot\frac{\gamma P}{\rho}} = c\sqrt{\frac{3}{\gamma}}, \text{ using (6.3.1)} \qquad (6.3.2)$$

where c is the speed of sound in the gas.

In air or diatomic gases, $\gamma = 1.40$ so that we have

$$v = c\sqrt{3/1.40} = 1.46c$$

The *r.m.s. speed of air molecules is thus nearly one and a half times greater than the speed of sound in air.*

6.4 Factors affecting sound velocity

Various physical factors such as pressure, temperature, density, humidity etc. may affect the velocity of sound in gases. We shall discuss these effects sequentially.

Effect of pressure—From Laplace's equation, the velocity of sound in gases is

$$c = \sqrt{\frac{\gamma P}{\rho}}$$

At constant temperature, the ratio P/ρ is a constant, by Boyle's law. Thus *the velocity of sound in gases, at constant temperature, is independent of pressure*—a result confirmed by direct experiments. At very high pressure, say 50 atmos, a slight (2.5%) increase is however observed due to deviations from Boyle's law.

Effect of temperature—A change in temperature brings about a change in density, if the pressure remains unchanged. If c_t and c_0 be the velocities at t°C and 0°C respectively,

$$c_t = \sqrt{\frac{\gamma P}{\rho_t}}; \quad c_0 = \sqrt{\frac{\gamma P}{\rho_0}}$$

$$\therefore \quad \frac{c_t}{c_0} = \sqrt{\frac{\rho_0}{\rho_t}} = \sqrt{\frac{\rho_t(1+\alpha t)}{\rho_t}}$$

$$= \sqrt{1+\alpha t} = 1 + \tfrac{1}{2}\alpha t \tag{6.4.1}$$

expanding binomially and neglecting higher powers of α, the coefficient of volume expansion of the gas at constant pressure. The value of α is 1/273 so that

$$\boxed{\frac{c_t}{c_0} = \sqrt{\frac{273+t}{273}} = \sqrt{\frac{T}{T_0}}} \tag{6.4.2}$$

Also, using equation (6.4.1), $c_t = c_0(1 + t/546)$

Thus, *the velocity of sound in a gas is proportional directly to the square root of its absolute temperature* and increases by 1/546 times its velocity at 0°C for every degree rise in temperature.

Effect of density—If v_a and v_b be the velocities of sound in gases a and b respectively having the same value of γ and at the same pressure, then

$$v_a = \sqrt{\frac{\gamma P}{\rho_a}}; \quad v_b = \sqrt{\frac{\gamma P}{\rho_b}},$$

$$\therefore \quad \boxed{\frac{v_b}{v_a} = \sqrt{\frac{\rho_a}{\rho_b}}} \tag{6.4.3}$$

Thus, *the velocity of sound in two gases at the same pressure and having the same value of γ is inversely proportional to the square root of their densities.*

Effect of humidity—Moist air is less dense than dry air. So the velocity of sound would be *greater in moist air than in dry air*—a result confirmed by experiments.

Let c_m be the velocity of sound in moist air of density ρ_m and at a pressure P, c_d the velocity in dry air, at the same pressure, of density ρ_d. Let p be the pressure exerted by moisture at t°C.

$\therefore$ Proportion of dry air in each unit volume of moist air is $(P-p)/P$ and that of moisture is p/P. So, the mass of unit volume of moist air at a given temperature is

$$\rho_m = (P-p)\frac{\rho_d}{P} + \frac{p}{P} \times 0.62\rho_d \qquad \left[\because \ \frac{\rho_d}{\rho_m} = \frac{1}{0.62}\right]$$

$$= (P - 0.38p)\frac{\rho_d}{P}$$

Therefore, we get from the above relation

$$\frac{\rho_m}{\rho_d} = \frac{P - 0.38p}{P}. \quad \text{But } \frac{c_d}{c_m} = \sqrt{\frac{\rho_m}{\rho_d}}$$

$$\therefore \quad \boxed{\frac{c_d}{c_m} = \sqrt{\frac{P-0.38p}{P}}} \tag{6.4.4}$$

Effect of the nature of the gas—From Laplace's equation, the velocity of sound in a gas is directly proportional to the square root of γ and inversely to the square root of its density. Now, γ is 1.66 for monoatomic gases, 1.41 for diatomic ones and still lower for polyatomic gases. Also, density of a gas is proportional to its molecular weight. *The velocity of sound in a gas would therefore depend both on its molecular structure and molecular weight.*

Effect of wind—The velocity of sound depends on the velocity of the medium through which it propagates. The propagation is helped or hindered according as the velocity of the medium is in or opposite to the direction of propagation. The observed velocity of sound is the *vector sum* of the *velocity in still medium* and the *velocity of the medium*—the maximum being $c+\omega$ and the minimum $c-\omega$, where c is the velocity of sound in still medium and ω that of the medium.

• In investgating the effect of physical factors on the velocity of sound, the following procedure may also be taken up. From Laplace's equation, we have

$$c = \sqrt{\frac{\gamma P}{\rho}} = \sqrt{\frac{\gamma R_0 T}{M}} \tag{6.4.5}$$

where M = molecular weight, R_0 = molar gas constant and T the absolute temperature.

The equation (6.4.5) immediately shows that the velocity of sound in a gas

(a) is *independent* of the *pressure* of the gas,

(b) varies directly as the *square root* of the *temperature*,

(c) depends on the *molecular weight* and the *nature* of the gas.

Effect of frequency and intensity—For audible sound, the velocity is found to be *independent of frequency.* This is why the music consisting of notes of different frequencies does not appear to have different characters at different distances. The kinetic theory of gases however predicts that for ultrasonics, the velocity should slightly increase with increase in frequency.

The velocity of sound is also *practically independent of its intensity* for sounds of ordinary intensity. However, for intense sounds such as explosive waves, this is not true. The velocity near the source is found to be higher than that of a wave of normal intensity and falls off gradually to the normal value as the distance from the source is gradually increased. This is presumably due to the bulk modulus being not a constant, the relation $c = \sqrt{K/\rho}$ does not hold.

6.5 Energy density of a plane progressive wave

When sound waves propagate through a medium, the particles in it are set into vibrations about their mean positions. The medium thus acquires an energy which is partly kinetic and partly potential. The *average energy in unit volume* of the medium is referred to as the *energy density.*

It should be carefully noted that there is no bodily transfer of the medium in the direction of propagation of the progressive wave. There is only a transfer of energy in the direction of propagation.

Let a plane, harmonic, longitudinal wave travel along the $+x$ direction. Then, the kinetic energy of a layer of the medium, of thickness δx and a unit cross-sectional area, normal to the direction of wave propagation is

$$dE_k = \frac{1}{2}\,\text{mass} \times (\text{velocity})^2 = \frac{1}{2}\rho_0 \delta x \left(\frac{dy}{dt}\right)^2$$

where ρ_0 is the density of the medium, the thickness δx is too small to ensure the same velocity, dy/dt for all the particles in the layer.

Now, for a harmonic progressive wave,

$$y = a\sin(\omega t - kx)$$

$$\therefore \quad \frac{dy}{dt} = a\omega\cos(\omega t - kx) \tag{6.5.1}$$

$$\therefore \quad dE_k = \tfrac{1}{2}\rho_0 \delta x \cdot a^2\omega^2\cos^2(\omega t - kx) \tag{6.5.2}$$

The time average of the $\cos^2$-term over a complete period is $\frac{1}{2}$.

$$\therefore \quad \overline{dE_k} = \tfrac{1}{4}\rho_0 \delta x\, a^2\,\omega^2 = \text{average K.E. of the layer.}$$

$\therefore$ Time **average of kinetic energy density**

$$\bar{E}_k = \frac{\overline{dE_k}}{\delta x} = \tfrac{1}{4}\rho_0 a^2\omega^2 = \pi^2 a^2 n^2 \rho_0 \quad (\because \ \omega = 2\pi n)$$

Again, the increase in potential energy of the layer, dE_P, is the work done in compressing the layer by an increase of pressure from $P_0 \to P$; so the excess pressure is $p = P - P_0$.

$$\therefore \quad dE_P = -\int_{P_0}^{P} P\,dV$$

where dV is the change in volume. The negative sign indicates that an increase in V implies a decrease in P.

Now, the bulk modulus of elasticity

$$K = -V_0\frac{dP}{dV} \Rightarrow dV = -\frac{V_0}{K}dP$$

$$\begin{aligned}\therefore \quad dE_P &= \frac{V_0}{K}\int_{P_0}^{P} P\,dP = \frac{V_0}{2K}(P^2 - P_0^2)\\ &= \frac{V_0}{2K}(P + P_0)(P - P_0)\\ &= \frac{V_0}{2K}(2P_0 + p)p \quad (\because \ p = P - P_0)\\ &= \frac{V_0}{2K}(2P_0 p + p^2)\end{aligned} \tag{6.5.3}$$

The excess pressure p, from Hooke's law, is given by

$$p = K\left(-\frac{\partial y}{\partial x}\right) = akK\cos(\omega t - kx) \tag{6.5.4}$$

$$(\because \; y = a\sin(\omega t - kx))$$

As p varies sinusoidally, its average value over a cycle is *zero*. But the average value of p^2 is $\frac{1}{2}(akK)^2$ or $\frac{1}{2}(c^2\rho_0 ka)^2$, since $c^2 = K/\rho_0$.

$\therefore$ Average potential energy of the layer is, from (6.5.3),

$$\overline{dE_P} = \frac{V_0}{4K}(c^2\rho_0 ka)^2 = \frac{V_0}{4}c^2\rho_0 k^2 a^2 \quad (\because \; c^2\rho_0 = K)$$

$$= \frac{V_0}{4}\rho_0\omega^2 a^2 = \frac{1}{4}\rho_0\delta x\omega^2 a^2 \quad (\because \; c^2k^2 = \omega^2;\; V_0 = \delta x)$$

$$\therefore \quad \boxed{\overline{dE_P} = \overline{dE_K}}$$

So, the **average potential energy density**

$$\bar{E}_P = \frac{\overline{dE_P}}{\delta x} = \tfrac{1}{4}\rho_0 a^2\omega^2 = \pi^2 a^2 n^2 \rho_0 \quad (\because \; \omega = 2\pi n) \tag{6.5.5}$$

$\therefore$ **Total average energy density**

$$\bar{E} = \bar{E}_K + \bar{E}_p = \tfrac{1}{2}\rho_0 a^2\omega^2 = 2\pi^2 a^2 n^2 \rho_0 \tag{6.5.6}$$

$$\text{Also,} \quad \bar{E}_K = \bar{E}_p = \tfrac{1}{4}\rho_0 a^2\omega^2 = \tfrac{1}{2}\left(\tfrac{1}{2}\rho_0 a^2\omega^2\right) = \tfrac{1}{2}\bar{E} \tag{6.5.7}$$

$\therefore$ **Average kinetic energy density = Average potential energy density**

= $\frac{1}{2}$ average total energy density

Maximum kinetic energy — From (6.5.2), the kinetic energy is maximum when the $\cos^2$-term is maximum, i.e., unity. So, Maximum kinetic energy,

$$\boxed{E_{\text{max}} = \tfrac{1}{2}\rho_0\delta x a^2\omega^2 = \tfrac{1}{2}ma^2\omega^2,}$$

where $m = \rho_0\delta x$, the mass of the layer.

6.6 Intensity of sound or acoustic intensity

The *intensity of sound*, also called **acoustic intensity**, is defined as *the amount of sound energy flowing through the medium per unit area normally per unit time* and is symbolized by I.

For plane progressive waves, the amount of energy that crosses unit area normally at unit time along the line of propagation is the energy contained in a cylinder of cross-sectional area unity and length equal to the phase velocity c.

$$\therefore \quad I = \text{average energy density} \times \text{phase velocity}$$
$$= \tfrac{1}{2}\rho_0 a^2\omega^2 \times c = 2\pi^2 n^2 a^2 \rho_0 \cdot c \quad (\because\ \omega = 2\pi n) \tag{6.6.1}$$

The acoustic intensity is thus proportional to the following factors.

(a) density of the medium,

(b) phase velocity of the wave,

(c) square of the amplitude of the wave, and

(d) square of the frequency of the wave.

• Since the intensity of the wave is the product of energy density and velocity of the wave, it is also known as **energy current**.

• The expressions for intensity and also the energy density (average) clearly indicate that they are *independent* of space and time coordinates (x and t).

Alternative expressions for *I*—From equation (6.5.4), the maximum acoustic pressure is

$$p_{\max} = akK = c^2\rho_0 ak \quad (\because\ c^2\rho_0 = K)$$
$$= c\rho_0\omega a \quad (\because\ c = \omega/k)$$

Therefore, (6.6.1) may be rewritten as

$$\boxed{I = \frac{1}{2}\rho_0 a^2\omega^2 c = \frac{p^2_{\max}}{2\rho_0 c} = \frac{p^2_{\text{rms}}}{\rho_0 c}} \tag{6.6.2}$$

This equation represents the **intensity of a plane wave** in terms of acoustic pressure—maximum or r.m.s.

The root mean square particle velocity u_{rms} is obtained from (6.5.1).

$$u^2_{\text{rms}} = \frac{1}{T}\int_0^T u^2 dt = \frac{\omega^2 a^2}{T}\int_0^T \cos^2(\omega t - kx)dt$$
$$= \frac{\omega^2 a^2}{T} \times \frac{T}{2} = \frac{\omega^2 a^2}{2}$$
$$\therefore \quad u_{rms} = \frac{\omega a}{\sqrt{2}}$$

$$\therefore \quad p_{\text{rms}}u_{\text{rms}} = \frac{c\rho_0\omega a}{\sqrt{2}} \cdot \frac{\omega a}{\sqrt{2}} = \frac{1}{2}c\rho_0\omega^2 a^2 = I. \tag{6.6.3}$$

Thus, *the acoustic intensity for a plane progressive wave is the product of the rms acoustic pressure and rms particle velocity.*

The SI unit of intensity is Watt/metre^2 i.e., W/m^2.

Note. If p be the acoustic pressure over a unit area in the direction of propagation and u the corresponding particle velocity, then the *instantaneous acoustic power* is pu. But $pu = c^2\rho_0 k\omega a^2 \cos^2(\omega t - kx)$. Since the average value of $\cos^2$-term over a cycle is $\frac{1}{2}$, we get :

Average acoustic power or **intensity**

$$I = \overline{pu} = \tfrac{1}{2}c^2\rho_0 k\omega a^2 = \tfrac{1}{2}\rho_0\omega^2 a^2 c \tag{6.6.4}$$

which is the same as equation (6.6.1).

Characteristic or specific acoustic impedance—In an electric circuit, the *instantaneous power* = *instantaneous voltage* × *instantaneous current* = VI. The *instantaneous acoustic power* is analogous to this expression. The acoustic pressure p corresponds to voltage V and the particle velocity u to current I.

Now, the ratio V/I is termed the *impedance* Z; the analogous ratio p/u is called the *characteristic* or *specific acoustic impedance*, Z_s.

The **specific acoustic impedance** *is thus defined as the ratio of the excess pressure p to the particle velocity u in wave motion.*

$$Z_s = \frac{p}{u} = \frac{p}{(dy/dt)}$$

But, from (6.5.3), and (6.5.1), we obtain the excess pressure

$$p = akK\cos(\omega t - kx)$$

$$(dy/dt) = a\omega\cos(\omega t - kx)$$

$$\therefore \quad Z_s = \frac{kK}{\omega} = \frac{c^2\rho_0 k}{\omega} \quad (\because\ c^2\rho_0 = K)$$

$$= \rho_0 c = \sqrt{K\rho_0} \quad (\because\ \omega = ck) \tag{6.6.5}$$

Z_s thus depends only on K and ρ_0 of the medium i.e., it is *medium-specific* or a *characteristic of the medium.* Hence the name.

Eq. (6.6.2) may be used to express the intensity I in terms of Z_s.

$$\boxed{I = \frac{p_{\text{rms}}^2}{Z_s}.} \tag{6.6.6}$$

6.7 Wave velocity and group velocity

Wave motion is a kind of disturbance that travels through a medium by the repeated periodic motion of its particles. The particles execute S.H.M. about their mean positions only, without travelling through the medium. Every particle starts vibrating a little later than its preceding one so that a progressive change of phase is seen as one travels from one particle to the next.

It is the phase relationship of the particles that is observed as a wave.

Wave velocity or phase velocity—It is the velocity with which the plane of equal phase (e.g., crest or trough etc.) travels through the medium.

A plane progressive harmonic wave propagating in the direction of $+x$ axis is

$$y = a\sin(\omega t - kx)$$

where y is the particle displacement at the position co-ordinate x and time t, a is the amplitude, ω the angular frequency ($= 2\pi n$), and k the propagation constant of the wave ($= 2\pi/\lambda$).

$$\therefore \quad \frac{\omega}{k} = \frac{2\pi n}{2\pi/\lambda} = n\lambda = c$$

This c is known as the *phase velocity of the wave* or *wave velocity.*

Thus, *the wave velocity or phase velocity is the velocity with which a plane progressive wave-front travels forward and which has a constant phase given by* $(\omega t - kx)$.

If at time $t + dt$, the constant phase $(\omega t - kx)$ advances to the point $x + dx$,

$$\omega t - kx = \omega(t + dt) - k(x + dx)$$

$$\Rightarrow \omega\, dt = k\, dx$$

$$\text{or,} \quad \frac{dx}{dt} = \frac{\omega}{k} = c \text{ (phase velocity),}$$

Alternatively : We have

$$\omega t - kx = \text{constant.}$$

Differentiating the above expression with respect to t, we obtain

$$\omega - k\frac{dx}{dt} = 0 \;\Rightarrow\; \frac{dx}{dt} = \frac{\omega}{k} = c \text{ (phase velocity).}$$

Group velocity—If two or more plane simple harmonic waves of the same amplitude but of different frequencies superpose on each other, a *group* of waves is formed. The amplitude of the group changes with distance. *The velocity with which the maximum amplitude of the wave-group travels is known as the* **group velocity**. *The energy of the wave group is transmitted with group velocity.*

The wave velocity of each component in a group would be different in all media except in vacuum. Hence such a group of waves will *disperse* with time as it travels through a medium.

A monochromatic wave train of finite length however can be treated as a wave group since it is theoretically formed by superposition of an infinite number of plane harmonic waves of continuously varying frequencies. Practically, however, the frequencies are within a band, the band width being determined by the length of the wave train. The shorter the length, wider is the band.

For small amplitude acoustic waves travelling in an isotropic homogeneous elastic medium, the wave velocity does not depend on the frequency—the velocities of the component waves are equal. So, the group does not change its form as it advances. For *monochromatic wave*, the group velocity and wave velocity are identical.

6.7.1 Expression for group velocity and its relation with wave velocity

Let a wave group be formed by superposition of two waves of equal amplitude a and slightly different angular frequencies ω and $\omega + d\omega$ moving with propagation constants k and $k + dk$ respectively. The resultant displacement of a particle at (x, t) due to the simultaneous action of the two waves is

$$y = a\sin(\omega t - kx) + a\sin\left[(\omega + d\omega)t - (k + dk)x\right] \tag{6.7.1}$$

$$= 2a\sin\left[\left(\omega + \frac{d\omega}{2}\right)t - \left(k + \frac{dk}{2}\right)x\right]\cos\left(\frac{t\,d\omega - x dk}{2}\right)$$

$$= 2a\cos\left(\frac{t\,d\omega - x\,dk}{2}\right)\sin\left[\left(\omega + \frac{d\omega}{2}\right)t - \left(k + \frac{dk}{2}\right)x\right] \tag{6.7.2}$$

In (6.7.2) the slowly varying cosine term can be treated as the *amplitude* of the resultant wave represented by the sine term.

∴ Angular frequency of resultant wave $= \omega + d\omega/2$, and propagation constant $= k + dk/2$ which are clearly the mean values for the component waves.

∴ Phase velocity of resultant wave is

$$c = \frac{\omega + d\omega/2}{k + dk/2} \simeq \frac{\omega}{k}$$

i.e., *very nearly equal to the phase velocity of the component waves.*

The amplitude represented by the cosine term advances with the group velocity c_g. If its phase $(t\,d\omega - x\,dk)/2$ progresses to $(x + dx)$ at time $(t + dt)$, then

$$t\,d\omega - x\,dk = (t + dt)d\omega - (x + dx)dk$$

$$\therefore \text{ Group velocity, } \quad \boxed{c_g = \frac{dx}{dt} = \frac{d\omega}{dk}} \tag{6.7.3}$$

Interrelation—Now, $\omega = ck \;\Rightarrow\; d\omega = c\,dk + k\,dc.$

$$\therefore \quad c_g = \frac{d\omega}{dk} = c + k\frac{dc}{dk} \tag{6.7.4}$$

$$\text{But } k = 2\pi/\lambda \Rightarrow \lambda = 2\pi/k \Rightarrow d\lambda/dk = -2\pi/k^2. \tag{6.7.5}$$

∴ Eq. (6.7.4) reduces to

$$c_g = c + k\frac{dc}{d\lambda}\cdot\frac{d\lambda}{dk} = c - \frac{2\pi}{k}\frac{dc}{d\lambda}, \text{ using (6.7.5)}$$

$$\text{or, } \quad \boxed{c_g = c - \lambda\frac{dc}{d\lambda}} \tag{6.7.6}$$

Equation (6.7.4) and (6.7.6) *relate the group velocity with the phase velocity.*

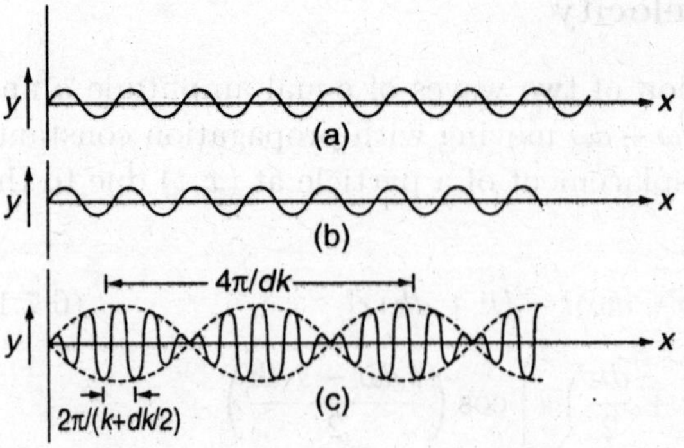

Fig 6.3

The two component waves and their resultant wave are shown in Fig. 6.3 (a), (b) and (c). The solid curve in (c) gives the sine term, and the envelope, as shown by dotted lines, gives the cosine term. The amplitude variation of the resultant wave given by the sine term is clearly shown in the figure. The amplitude varies simple harmonically with x and t. The waves are split into groups and the velocity of wave group or envelope is c_g.

• From (6.7.6), it is seen that *three* cases may arise : (i) $dc/d\lambda$ is a *positive* quantity, when $c_g < c$; (ii) $dc/d\lambda$ is *zero*, when $c_g = c$ and (iii) $dc/d\lambda$ is a *negative* quantity, when $c_g > c$. Hence, the *group velocity may be less than, equal to or greater than the wave velocity* according as $dc/d\lambda$ is *positive, zero* or a *negative* quantity.

6.8 Dispersion of waves : Dispersion relations

From relation (6.7.6), when in a medium $dc/d\lambda = 0$, i.e., the wave velocity does not vary with wavelength, the group velocity, c_g = wave velocity, c. All the waves in the group, although of different wavelengths or frequencies travel with the same velocity and remain together. Such a medium is called a *non-dispersive medium.* The familiar example is that of while light travelling through vacuum.

In a medium in which $dc/d\lambda \neq 0$, the group velocity will depend on wavelengths or frequencies. So, waves of different wavelengths will travel with different velocities and will thus get separated. *The process of separation of waves of different wavelengths (or frequencies) while travelling through a medium is termed* **dispersion** *of waves* and the medium is called a *dispersive medium.* The familiar example is the dispersion of white light into its constituent colours while passing through a glass prism—the *spectrum* of white light.

Dispersion relations—If $dc/d\lambda$ is *positive*, $c_g < c$, i.e., the group velocity is less than the wave velocity. The corresponding dispersion is called *normal dispersion.* In this case, the dispersion formula (6.7.4) becomes

$$\boxed{c_g = c - \lambda\frac{dc}{d\lambda}}$$

If $dc/d\lambda = 0$, $c_g = c$, i.e., the group velocity is the same as the wave velocity. There is therefore *no dispersion*; the medium is *non-dispersive* and the dispersion relation reduces to $c_g = c$.

If however $dc/d\lambda$ is *negative*, $c_g > c$, i.e., the group velocity is greater than the wave velocity. The corresponding dispersion is called *anomalous dispersion.*

In this case, the dispersion formula is given by

$$\boxed{c_g = c + \lambda \frac{dc}{d\lambda}}$$

Electromagnetic waves in vacuum are *non-dispersive.* In dielectrics, e.g., glass, there is *normal dispersion* of e.m. waves except at natural resonant frequency. In electric conductors, however, there is *anomalous dispersion.*

• When light travels through vacuum, all the waves of different wavelengths travel with the same velocity, i.e., there is no change of wave velocity with wavelength or $dc/d\lambda = 0, \Rightarrow \; c_g = c.$ *This explains why there is no difference in wave velocity and group velocity when light travels in vacuum.*

• One can easily *distinguish a dispersive medium* from *a non-dispersive* one. In a non-dispersive medium, the group velocity c_g is equal to its wave velocity c. For a dispersive medium however $c_g < c$, or, $c_g > c$.

• The phase or wave velocity should be carefully distinguished from group velocity. While the phase or wave velocity, $c = \omega/k$, i.e., the ratio of the angular frequency ω to the propagation constant k, the group velocity $c_g = d\omega/dk$, i.e., it is the rate of change of angular frequency ω with respect to the propagation constant k.

6.9 Doppler effect

Whenever there is a relative motion between the source of sound and the listener, the pitch (frequency) of the sound, as heard by the listener, would differ from the true pitch (frequency)—rising when they approach, falling as they recede. This phenomenon of the *apparent change in frequency of sound owing to the relative motion of the source and the listener is known as* **Doppler effect**, after the Austrain physicist C. H. Doppler.

The *cause of the phenomenon* however is not far to seek. As the stationary sounding body produces a note of frequency n, a listener at rest receives n waves per sound. But if the source be in motion and approaches the listener, the number of waves received by him per second would now be greater than n. The result is thus an apparent rise in frequency. When, however, the source recedes, the listener receives lesser number of waves per second and this causes an apparent fall in the pitch. Similar changes in frequency would also occur for a moving observer.

We shall now find analytical expressions for the apparent change in frequency i.e., the *Doppler shift* in different situations.

6.10 Expression for Doppler shifts : Different cases

In calculating the Dopper shift in the following different cases, we shall assume that the motions of the source and the observer are *in the same line* and that the medium is *at rest* (no wind).

1. Source moving, observer stationary : Suppose that a source S of sound, of frequency n and wavelength λ, **approaches** a stationary observer O with a velocity u (Fig. 6.4). Let c be the velocity of sound, so that $c = n\lambda$.

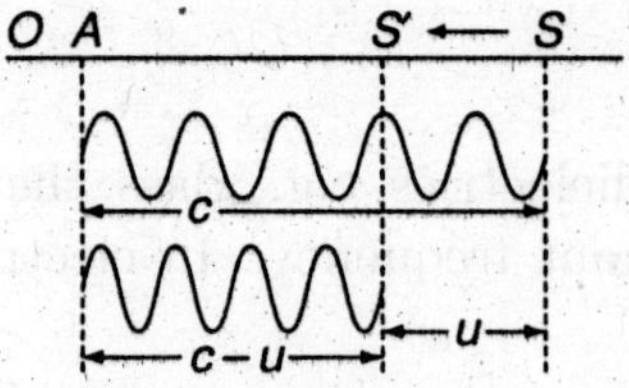

Fig 6.4

After one second, the source S advances a distance $SS' = u$, to occupy the position S'. During this period n-waves are emitted by the source; the first of them is now at A and the last one at S'. So n-waves produced per sec will now be crowded in the space $S'A = c - u$. So, by definition,

Reduced or *apparent wavelength,*

$$\lambda' = \frac{c-u}{n} = \frac{c-u}{c/\lambda} = \frac{\lambda(c-u)}{c}$$

$$\therefore \quad \lambda' < \lambda$$

$$\therefore \textit{ Apparent frequency, } n' = \frac{c}{\lambda'} = \frac{c}{\lambda(c-u)/c} = \frac{nc}{c-u}$$

$$\therefore \quad n' > n$$

$$\therefore \quad \textbf{Doppler shift, } dn = n' - n = \frac{nu}{c-u}$$

So, the **pitch** *of the sound* **rises** as the **source approaches** *the observer.*

For a **receding source**, the expression for Doppler shift can be readily obtained by simply changing u by $-u$.

$$\therefore \ \lambda' = \frac{\lambda(c+u)}{c} \Rightarrow \lambda' > \lambda$$

$$\text{and } \ n' = \frac{nc}{c+u} \Rightarrow n' < n$$

$$\therefore \quad \textbf{Doppler shift, } dn = n - n' = \frac{nu}{c+u}$$

This shows that, in this case the *wavelength rises* and the *frequency* or *pitch falls.*

• *Remember,* the motion of the source causes only an *apparent* change in wavelength.

2. Observer moving, source stationary : Let v be the velocity of **approach** of the observer O (Fig. 6.5) towards the stationary source S, emitting continuously waves of frequency n and wavelength λ. The waves that move past O in one second occupy the distance $OA = c$, the magnitude of the velocity of sound. At the end of one second, the observer is at O', where $OO' = v$. So the number of waves received per second by the observer is $(c+v)/\lambda = n(c+v)/c$, which is the apparent frequency n'.

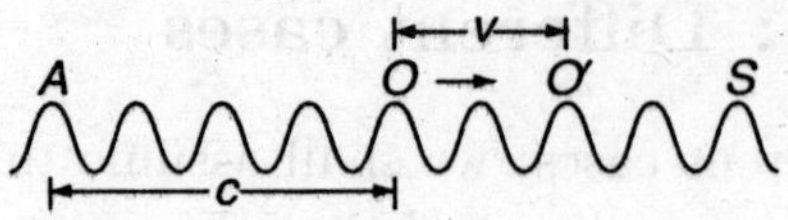

Fig 6.5

$$\therefore \quad \textit{Apparent frequency, } n' = \frac{n(c+v)}{c} \Rightarrow n' > n$$

$$\therefore \quad \textbf{Doppler shift, } dn = n' - n = \frac{nv}{c}$$

So, the *frequency* of sound waves *increases* when the observer *approaches* the stationary source of sound.

For a **receding observer**, we put , as before, $-v$ in place of v and thereby obtain:

$$\textit{Apparent frequency, } n' = \frac{n(c-v)}{c}$$

$$\Rightarrow n' < n$$

$$\therefore \quad \textbf{Doppler shift, } dn = n - n' = nv/c$$

The *pitch* now *falls* by the **same amount**, nv/c, as the observer *moves away* from the stationary source.

• In this case, the motion of the observer causes *a change in the number of waves received by him per second*, but there is *no change in the wavelength* of the note emitted by the source.

3. Source and observer both moving : This is a case that can be developed out of coupling of the previous two cases. So, in computing the apparent frequency, we would utilize the result of the last case, using the modified (i.e. *reduced*) wavelength derived in the first case.

Thus, for an observer and a source **approaching each other** with velocities v and u respectively, the apparent frequency n' is given by

$$n' = \frac{c+v}{\lambda'} = \frac{c+v}{c-u}n \quad (\because \text{ by Case 1, } \lambda' = (c-u)/n)$$

For the source and observer **receding from each other**, we are to change the sign of both u and v, so that

$$n' = \frac{c-v}{\lambda'} = \frac{c-v}{c+u}n$$

Effect of wind : In the cases considered so far, c is the velocity of sound in a stationary medium. If however a wind blows with a velocity ω, the sound velocity relative to the fixed reference is $c \pm \omega$, according as ω is in the direction of propagation of the waves or opposite to it.

Imposing the effect of wind in Case 3 above, we have for the apparent frequency

$$n' = n\frac{c+v\pm\omega}{c-u\pm\omega}, \quad \text{for } \textit{approach}$$

$$\text{and,} \quad n' = n\frac{c-v\pm\omega}{c+u\pm\omega}, \quad \text{for } \textit{recession.}$$

• When $\omega = 0$ and $v = 0$, $n' = nc/(c \pm u)$, which corresponds to Case 1.

• When $\omega = 0$ and $u = 0$, $n' = n(c \pm v)/c$, which corresponds to Case 2.

• When $\omega = 0$, it obviously gives Case 3. The above might therefore be considered as a *general* case.

4. Observer not in line with the motion of the source : Let the source move along PQ (Fig. 6.6) and the observer be stationed at O. During motion when the source is at S_1, $u\cos\theta$ is the component velocity towards O, so that the apparent frequency of the note as heard by the observer is

$$n' = n\frac{c}{c - u\cos\theta},$$

as in the case of *approaching source* in Case 1 and is greater than n.

When the source is at S_2, the component velocity is $u\cos\phi$ away from O. The apparent frequency is now lower and is given by

$$n'' = n\frac{c}{c + u\cos\phi},$$

as in the case of *receding source* discussed under Case 1.

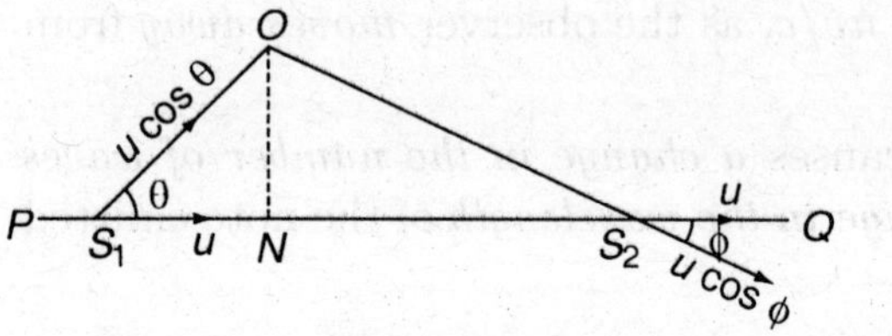

Fig 6.6

At N, the point of *closest approach* to observer, we have $\theta = \phi = 90°$, so that $n' = n'' = n$. The observer thus hears the same frequency.

• In the *laboratory*, Doppler effect can be **demonstrated** rather simply by mounting a reed at the end of a rod of about 0.5 m long, the other end being attached to a turntable rotating with high speed. The frequency of the reed whistling through air increases as the reed approaches, and it decreases when the reed recedes from the observer, confirming the Doppler phenomenon.

6.11 Illustrated Examples

► **Example 1.** *A sound note of frequency* 200 Hz *is produced in hydrogen. Find its wavelength, assuming velocity of sound in air is* 332 ms^{-1}, *and density of air is* 14.4 *times that of hydrogen.* **(Burdwan Univ.)**

Solution. If c_H and c_A be the velocities of sound in hydrogen and air respectively, and ρ_H and ρ_A their densities in order, then

$$\frac{c_H}{c_A} = \sqrt{\frac{\rho_A}{\rho_H}} = \sqrt{14.4} = 3.79$$

$$\therefore \quad c_H = c_A \times 3.79 = 332 \times 3.79 \text{ ms}^{-1} = 1258 \text{ ms}^{-1}$$

$\therefore$ Wavelength of sound in hydrogen = 1258/200 = **6.29 m**

► **Example 2.** *Calculate the increase in pressure when a sound wave produces an energy* 10^{-7} W/cm^2. *Velocity of sound* = 340 m/s *and density of air* = 0.001293 g/c.c. **(Kanpur Univ.)**

Solution. Intensity, $I = 10^{-7}\ \text{W/cm}^2 = 10^{-7}\ \text{Js}^{-1}\text{cm}^{-2} = 10^{-3}\ \text{Js}^{-1}\text{m}^{-2}$. Density of air, $\rho = 0.001293$ g/c.c. $= 0.001293 \times 10^3\ \text{kg/m}^3$. Velocity of sound, $c = 340$ m/s.

$$\therefore \quad \text{Excess pressure, } p_{\max} = \sqrt{2I\rho c}$$

$$= \sqrt{2 \times 10^{-3} \times 0.001293 \times 10^3 \times 340}$$

$$= \mathbf{0.94\ Nm^{-2}}$$

▶ **Example 3.** *A tuning fork has a frequency of* 256 *and on amplitude* 0.5 cm. *If the velocity of sound is* 40 m/s, *calculate the energy current. Density of air is* 0.001293 g/c.c. **(Kanpur Univ.)**

Solution. Given : $c = 340$ m/s; $\rho = 0.001293$ g/c.c. $= 0.001293 \times 10^3\ \text{kg/m}^3$, $n = 256$, $a = 0.5$ cm $= 0.5 \times 10^{-2}$ m.

$$\therefore \quad \text{Energy current} = \text{Intensity, } I = 2\pi^2 n^2 a^2 \rho c$$

$$= 2 \times 9.87 \times 256^2 \times (0.5 \times 10^{-2})^2 \times 0.001293 \times 10^3 \times 340$$

$$= 119\ \text{Js}^{-1}\text{m}^{-2} = \mathbf{119\ W/m^2}$$

▶ **Example 4.** *Calculate the group velocity when the following two waves*

$$y_1 = 10\sin(2\pi t - 5x)\text{m}$$

$$y_2 = 15\sin(5\pi t + 5x)\text{m}$$

are superposed on each other. **(Punjab Univ.)**

Solution. The wave equation is : $y = a\sin(\omega t - kx)$

$\therefore$ Angular frequency, $\omega_1 = 2\pi$, $\omega_2 = 5\pi$. Propagation constant, $k_1 = +5$, $k_2 = -5$.

$$\therefore \quad \text{Group velocity, } c_g = \frac{d\omega}{dk} = \frac{\omega_1 - \omega_2}{k_1 - k_2} = \frac{2\pi - 5\pi}{+5 - (-5)}$$

$$= -\frac{3}{10}\pi = \mathbf{-0.3\pi\ ms^{-1}}.$$

▶ **Example 5.** *Show that in a conducting medium, the group velocity*

$$c_g = c\left[1 + \frac{\lambda}{2\varepsilon_r}\frac{d\varepsilon_r}{d\lambda}\right]$$

where $\varepsilon_r = \varepsilon/\varepsilon_0 =$ *relative permittivity and* $\mu = 1$ *and comment on the type of dispersion.*

Solution. The wave velocity,

$$c = \frac{1}{\sqrt{\varepsilon_r \mu}} = \frac{1}{\sqrt{\varepsilon_r}} = \varepsilon_r^{-1/2} \quad (\because\ \mu = 1) \qquad (6.11.1)$$

$$\therefore \quad \frac{dc}{d\lambda} = -\frac{1}{2}\varepsilon_r^{-3/2}\frac{d\varepsilon_r}{d\lambda} = -\frac{1}{2\varepsilon_r}\varepsilon_r^{-1/2}\frac{d\varepsilon_r}{d\lambda} = -c\frac{1}{2\varepsilon_r}\frac{d\varepsilon_r}{d\lambda}, \quad \text{by (6.11.1)}$$

Now, we have, $c_g = c - \lambda\dfrac{dv}{d\lambda} = c - \lambda\left(-c\dfrac{1}{2\varepsilon_r}\dfrac{d\varepsilon_r}{d\lambda}\right)$

$$= c + \lambda c\frac{1}{2\varepsilon_r}\frac{d\varepsilon_r}{d\lambda} = c\left[1 + \frac{\lambda}{2\varepsilon_r}\frac{d\varepsilon_r}{d\lambda}\right]$$

Since the quantity in the bracket is greater than 1, $c_g > c$. So, it is a case of **anomalous dispersion**.

► **Example 6.** *An increase of pressure of* 100 kPa *causes a certain volume of water to decrease by* 5×10^{-3} *per cent of its original volume. Show that the speed of sound in water is about* 1414 m/s. **(Calcutta Univ.)**

Solution. The speed of sound in water is $v = \sqrt{K/\rho}$ where K = bulk modulus = $p/\frac{dV}{V} = 100 \times 10^3/5 \times 10^{-5} = 20 \times 10^8$ Pa and ρ = density = 1 g/c.c. = 10^3 kg/m^3.

$$\therefore \quad v = \sqrt{\frac{20 \times 10^8}{10^3}} = \sqrt{2} \times 10^3 = 1.414 \times 10^3$$

$$= \mathbf{1414\ m/s.}$$

► **Example 7.** *The phase velocity v of deep water waves of wavelength* λ *is given by* $v^2 = (g\lambda/2\pi) + (2\pi S/\rho\lambda)$, *where g is the acceleration due to gravity,* ρ *the density and S the surface tension of the liquid. Find the wavelength* λ_0 *of the waves which do not disperse in water. Show that for* $\lambda \ll \lambda_0$, *the group velocity* v_g *is* $\simeq 3v/2$ *and for* $\lambda \gg \lambda_0$, $v_g \simeq v/2$. **(Calcutta Hons.)**

Solution. For waves that do *not* disperse, $dv/d\lambda = 0$. Differentiating, the relation

$$v^2 = \frac{g\lambda}{2\pi} + \frac{2\pi S}{\rho\lambda}, \tag{6.11.2}$$

$$2v\frac{dv}{d\lambda} = \frac{g}{2\pi} - \frac{2\pi S}{\rho\lambda^2} = 0, \text{ since } \frac{dv}{d\lambda} = 0 \Rightarrow \lambda_0^2 = \frac{4\pi^2 S}{\rho g} \tag{6.11.3}$$

$$\therefore \quad \boldsymbol{\lambda_0 = 2\pi\sqrt{\frac{S}{\rho g}}}$$

Group velocity, $v_g = v - \lambda\dfrac{dv}{d\lambda} = v - \lambda\left(\dfrac{g}{4\pi v} - \dfrac{\pi S}{\rho v\lambda^2}\right)$, using $\dfrac{dv}{d\lambda}$ from (6.11.3)

$$= v - \lambda\left(\frac{\pi S}{\rho v\lambda_0^2} - \frac{\pi S}{\rho v\lambda^2}\right), \text{ by using } \lambda_0^2 \text{ from (6.11.4)}$$

$$= v - \frac{\pi S}{\rho v}\left(\frac{\lambda}{\lambda_0^2} - \frac{1}{\lambda}\right)$$

If $\lambda \ll \lambda_0$, $\quad v_g \simeq v + \dfrac{\pi S}{\rho v \lambda} = v + \dfrac{v}{2} \simeq \dfrac{\mathbf{3}}{\mathbf{2}}\boldsymbol{v}$, using (6.11.2)

If $\lambda \gg \lambda_0$, $\quad v_g \simeq v - \dfrac{\pi S \lambda}{\rho v \lambda_0^2} = v - \dfrac{v}{2} \simeq \dfrac{\boldsymbol{v}}{\mathbf{2}}$, using (6.11.2) and (6.11.3)

► **Example 8.** *Show that if the source moves away with the velocity of sound from an observer at rest, the frequency of vibration is halved.*

Solution. When the source is in motion and the observer is at rest, the apparent frequency

$$n' = \frac{nc}{c-u}$$

where c is the velocity of sound and u that of the source.

Here $u = -c$, as the source recedes away with the velocity of sound.

$$\therefore \quad n' = \frac{nc}{c-(-c)} = \frac{nc}{2c} = \frac{\boldsymbol{n}}{\mathbf{2}}$$

► **Example 9.** *An observer on a railway platform observed a train passing through the station at a speed u. Show that the frequency of the whistle changes by* $nc[2u/(c^2-u^2)]$.

Solution. Let n_1 be the apparent frequency when the train approaches the observer on the platform.

$$\therefore \quad n_1 = \frac{nc}{c-u}$$

If n_2 be the apparent frequency when the train moves away from the observer,

$$n_2 = \frac{nc}{c+u}$$

$$\therefore \quad \textbf{Change in frequency, } n_1 - n_2 = nc\left[\frac{1}{c-u} - \frac{1}{c+u}\right]$$

$$= \boldsymbol{nc}\frac{\mathbf{2}\boldsymbol{u}}{\boldsymbol{c}^{\mathbf{2}} - \boldsymbol{u}^{\mathbf{2}}}$$

► **Example 10.** *Calculate the percentage difference between the frequency of a note emitted by the whistle of a train approaching an observer with a velocity* 30 m/s *and that heard by the observer, the velocity of sound in air being* 350 m/s.

Solution. Apparent frequency in approach is given by

$$n' = \frac{c}{c-u}n = \frac{350}{350-30}n = \frac{350}{320}n = \frac{35}{32}n$$

$$\therefore \quad \text{Frequency difference} = n' - n = \frac{35}{32}n - n = \frac{3}{32}n$$

$$\therefore \quad \text{Percentage difference} = \mathbf{9.4\%}.$$

► **Example 11.** *A tuning fork of frequency* 440 *approaches a wall with a velocity* 4 m/s. *What will be the number of beats heard between the direct and the reflected sounds, if the velocity of sound is* 332 m/s*?*

Solution. Here n = frequency of the fork = 440; c = velocity of sound = 332 m/s.

Let the fork lie between the observer and the wall. So, as the fork approaches the wall, it moves away from the observer. The observer hears two sounds : (i) directly from the receding fork and (ii) that reflected from the walls.

The apparent frequency n_1 of the direct sound is given by

$$n_1 = \frac{nc}{c+u}, \quad \text{where } u = \text{velocity of fork.}$$

$$= \frac{440 \times 332}{332 + 4}$$

$$= 434.8 \text{ Hz.}$$

As the source moves towards the wall and the observer is at rest relative to wall, it is equivalent to the case of source moving towards a stationary observer.

∴ Apparent frequency n_2 of the reflected sound is given by

$$n_2 = \frac{nc}{c-u} = \frac{440 \times 332}{332 - 4}$$

$$= 445.4 \text{ Hz.}$$

∴ Number of beats heard = $n_1 \sim n_2 = 445.4 - 434.8 =$ **10.6/s.**

► **Example 12.** *A policeman on duty at a crossing challenges a motor driver for exceeding the speed limit of* 100 km/hour *by detecting a change of* 20 *vibrations in the horn note of frequency* 128 *as the car passes him. Is he correct? The velocity of sound in air* = 350 m/s.

Solution. As the car *approaches* the policeman, the apparent frequency n_1 is given by the relation

$$n_1 = \frac{nc}{c-u}$$

where n = original frequency, c = velocity of sound and u = velocity of source (car).

When the car *passes* the policeman, the apparent frequency is given by

$$n_2 = \frac{nc}{c+u}$$

$$\therefore \quad \text{Change in frequency} = n_1 - n_2 = nc\left[\frac{1}{c-u} - \frac{1}{c+u}\right]$$

$$= nc\frac{2u}{c^2 - u^2}$$

But, from the data, $n_1 - n_2 = 20$ Hz, $c = 350$ m/s and $n = 128$ Hz.

$$\therefore \quad 20 = 128 \times 350 \times \frac{2u}{350^2 - u^2} \Rightarrow u^2 + 4480u - 350^2 = 0.$$

$$\therefore \quad u = \frac{-4480 \pm \sqrt{4480^2 + 4 \times 350^2}}{2}$$

$$= 27.5 \text{ m/s} = \mathbf{99} \textbf{ km/hr}$$

∴ The motor driver has not crossed the speed limit of 100 km/hr. The policeman is **not correct**.

► **Example 13.** *Two aeroplanes are approaching each other with speeds* 540 km/h *each. The frequency of a note emitted by the first as heard by the passengers in the other is* 540 Hz. *Calculate the true frequency of the note heard by its passengers. Velocity of sound in air* = 350 m/s.

Solution. Let n be the true frequency. The velocity of sound $c = 350$ m/s and $u = 540$ km/h $= 150$ m/s $= v$, where u and v are the velocity of approach of the source and the observer.

∴ Apparent frequency due to the motion of the source (first aeroplane) and the observer (second plane) is

$$n_1 = \frac{c+v}{c-u} n \Rightarrow 540 = \frac{350+150}{350-150} n$$

$$\Rightarrow n = \frac{540 \times 200}{500} = \mathbf{216 \text{ Hz}}$$

► **Example 14.** *A siren is fitted in a car speeding towards a vertical wall with* 36 km/h. *A stationary observer, behind the car, listens to the siren sound directly from the source and that on reflection from the wall. Find the apparent frequency of the direct and the reflected waves. Speed of sound* = 340 m/s; *frequency of siren* = 500 Hz.

Solution. Speed of sound, $c = 340$ m/s; speed of the car, $u = 36$ km/h $= 10$ m/s. The observer is at rest w.r.t. the medium and the source is receding from the observer.

∴ Apparent frequency of direct waves, $n' = \dfrac{nc}{c+u} = \dfrac{500 \times 340}{340+10} = \mathbf{486 \text{ Hz}}$

Frequency received by the wall (approaching source)

$$n'' = \frac{nc}{c-u} = \frac{500 \times 340}{340 - 10} = \mathbf{515 \text{ Hz}}.$$

The wall reflects this sound without changing the frequency and as there is no relative motion between the observer and the wall, the frequency of the reflected wave, as heard by the observer, is **515 Hz.**

Review Questions

1. Show that the velocity of sound in a solid medium is $\sqrt{Y/\rho}$ where Y and ρ denote respectively the Young's modulus and density of the solid. **(Calcutta Univ.)**

2. Derive an expression for the velocity of longitudinal waves in a gaseous medium. **(Burdwan and Lucknow Univ.)**

3. What is meant by specific acoustic impedance? In what unit is it measured? **(Osmania Univ.)**

4. Starting from the relation $v = \sqrt{E/\rho}$ for the velocity of sound in a gas, derive Laplace's equation for the velocity of sound in a gaseous medium.

5. Discuss the variation of sound velocity with temperature, pressure, density, wind velocity and moisture. **(Burdwan and Vidyasagar Univ.)**

6. Obtain the expression for the total mechanical energy per unit volume of the medium when a wave travels through it. Prove that in a progressive wave half the energy is kinetic and half potential. Also show that the maximum kinetic energy is equal to the maximum potential energy. **(Himachal Univ.)**

7. Explain the term : intensity of an acoustic wave. Obtain expression for the intensity of the wave using expression for total energy possessed by a wave in terms of density of the medium, frequency, amplitude and velocity of the wave. **(Calcutta and Guru Nanak Univ.)**

8. Show that the intensity of the wave at a point is given by the relation

$$I = p_{\text{rms}}^2/\rho c$$

where p_{rms} is the root mean square value of the excess pressure.

9. Obtain an expression for the excess pressure at a point in a medium through which plane longitudinal waves are passing. Show that the pressure and displacement are in quadrature.

10. Find the expression for intensity of a plane progressive wave. How is the intensity related to the pressure amplitude in a longitudinal wave?

11. Define the terms : wave velocity and group velocity. Find a relation connecting the two. **(Punjabi and Himachal Univ.)**

12. What is dispersion of waves? Discuss and distinguish between normal dispersion, non-dispersion and anomalous dispersion. Give dispersion relations for them with examples. **(Guru Nanak Univ.)**

13. Show that the change in frequency due to Doppler effect is greater when the source approaches the observer than when the observer approaches the source with the same speed. **(Calcutta Univ.)**

14. Derive an expression for the change in frequency of a note due to the relative motion of the observer and the source. **(Shivaji Univ.)**

15. The pitch of a note is altered if the source or the observer moves relative to one another. Explain clearly the reason for the phenomenon. Deduce an expression showing the relation between the alteration of the pitch and the velocities of the source, observer and the medium. **(Calcutta Univ.)**

16. It was reported that a musical policeman could estimate the speed of a motor car by the drop in pitch of the horn sounded as the car passed him. Give a theoretical explanation to substantiate this report. **(Bombay Univ.)**

17. A train is approaching a vertical cliff with a velocity v km/hour. The driver sounds a whistle of pitch n and finds the frequency of the echo equal to n_1 where

$$n_1 = n\frac{c+v}{c-u}$$

Explain how the above result is obtained, if c is the velocity of sound in km/hour. **(Guru Nanak Univ.)**

Short Answer and Objective Questions

1. What is specific acoustic impedance? Give the name of its unit.

2. Starting from the relation $v = \sqrt{E/\rho}$, derive Laplace's equation $v = \sqrt{\gamma P/\rho}$ in a gaseous medium. The frequency of a fork is 260 Hz. How many times the fork has to vibrate for the sound to reach a distance of 66 m? Velocity of sound in air = 330 m/s.

3. Discuss the variation of sound velocity with (a) pressure, (b) temperature and (c) moisture. The frequency of a fork is 480 Hz. Sound travels 15 m in the time the fork vibrates 20 times. Find the velocity and wavelength of sound.

4. How does the intensity of an acoustic wave vary with amplitude, frequency and velocity of the wave, and the density of the medium?

5. Write down the expressions for the total energy and intensity of a progressive wave and show that they are independent of space and time coordinates. **(Himachal Univ.)**

6. What are wave velocity and group velocity? Write down the relation connecting the two.

7. Is group velocity less or greater than the wave velocity? Give your comments. **(Punjab and Punjabi Univ.)**
8. Discuss, in short, with examples the normal dispersion, anomalous dispersion and the non-dispersion.
9. Explain why there is no difference in wave velocity and group velocity when light travels in vacuum. **(Himachal Univ.)**
10. How dispersive medium is distinguished from a non-dispersive medium? **(Guru Nanak Univ.)**
11. Write down expressions for phase velocity and group velocity in terms of angular velocity and propagation constant (or number). **(Lucknow Univ.)**
12. If a fork of high pitch is rapidly moved to and fro between the ear and the surface of a flat wall, beats are heard. Explain why.
13. In a non-dispersive medium, the group velocity is

    (a) greater than phase velocity, (c) equal to phase velocity,

    (b) less than phase velocity, (d) independent of phase velocity.

    (Himachal Univ.)
14. The phase velocity c and the group velocity c_g are related in a normal dispersive medium as

    (a) $c_g > c$, (b) $c_g = c$, (c) $c_g = c^{-1}$, (d) $c_g < c$.

    (Himachal Univ.)
15. The phase velocity c and the group velocity c_g are related as

    (a) $c_g = c - \lambda \dfrac{dc}{d\lambda}$, (b) $c_g = c + \lambda \dfrac{d\lambda}{dc}$

    (c) $c = c_g + \dfrac{dc_g}{d\lambda}$, (d) $c = c_g - \lambda \dfrac{dc_g}{d\lambda}$.

    (Himachal Univ.)
16. The intensity of a progressive wave is

    (a) independent of space and time coordinates

    (b) dependent on space coordinate but independent of time-coordinate

    (c) independent of space coordinate but dependent on time coordinate

    (d) dependent on both space and time coordinates.
17. A tuning fork of frequency 90 is sounded and moved towards an observer with a velocity 1/10th of the velocity of sound. The observer will hear the note having a frequency.

    (a) 100 (b) 80 (c) 90 (d) none of these.

18. The Doppler effect is *not* observed if

(a) the source and the observer both move

(b) the source and the observer are both at rest

(c) the source is stationary, but the observer moves

(d) only the source moves.

19. When the source and the observer are moving away from each other, the apparent pitch (frequency)

(a) increases (b) decreases (c) remains the same (d) becomes infinite.

20. The change in frequency due to Doppler effect does not depend on

(a) the speed of the source

(b) the speed of the observer

(c) the frequency of the source

(d) separation between source and observer.

21. The engine of a train sounds a whistle at a frequency n. The frequency, as heard by a passenger in the train, is

(a) $> n$ (b) $< n$ (c) $= 1/n$ (d) $= n$.

Problems

1. A load of 20 kg is suspended by a steel wire. When rubbed with a resined cloth, its frequency is found to be 20 times its frequency when plucked. Find the cross-sectional area of the wire. Given, Y for steel = 19.6×10^{10} Nm^{-2}. [4×10^{-7} m^2]

2. The velocity of sound in air is 332.35 m/s and the density of air is 0.001293×10^3 kg/m^3. Calculate the adiabatic exponent γ for air. [1.41]

3. Find the temperature at which the velocity of sound in air becomes 1.5 times its velocity at 0°C. **(Calcutta and Vidyasagar Univ.)** [341.25°C]

4. The following two sinusoidal waves are superposed.

$$y_1 = 0.03\cos(7t - 10x)\text{m}$$
$$\text{and} \quad y_2 = 0.03\cos(5t - 8x)\text{m}$$

Calculate the group velocity. **(Guru Nanak Univ.)** [1 ms^{-1}]

5. A steel wire of diameter 0.5m is stretched by a load of 5 kg. wt. If $Y = 2 \times 10^{12}$ dyne/cm^2 for steel, find the ratio of the fundamental frequencies for longitudinal and transverse vibrations. [$\simeq 28.3$]

6. A source of sound emits energy equally in all directions at the rate of 0.5 joule per second. What is the intensity at a distance of 10 m? [4×10^{-6} W/cm^2]

7. A locomotive engine moves towards you with a ground speed of 18.8 km per hour blowing a whistle which has a frequency of 1000 Hz. At the same time, you are moving towards the engine in a car with an equal speed. What will be the apparent frequency of the whistle to you? Velocity of sound = 333 m/s.

(Calcutta Univ.) [1031]

8. A man standing on a railway platform notices the change in pitch of the note of the whistle of an engine as it passes by him. If the actual frequency of the whistle be 256 and the velocity of sound be 1/20th of that of the sound, what will be the frequencies of note heard by the man before and after the engine passes him?

(Gauhati Univ.) [2695; 243.5]

9. Two aeroplanes pass each other in opposite directions and one of them is blowing a whistle of frequency 540 Hz. Calculate the frequency of the note heard in the other plane (a) before and (b) after they have passed each other. Velocity of either plane is 540 km/hour and the velocity of sound = 350 m/s.

[1350 Hz; 216 Hz]

10. A policeman on duty challenges a motor driver for crossing the speed limit of 90 km/hour by detecting a change of 20 vibrations in the horn note of frequency 128 as the car passes him. Is he correct? Velocity of sound = 332 m/s.

(Punjab Univ.) [No]

11. The wavelength of a spectral line coming from a star is changed by the motion of the star from 6000 A to 6001 A. Find the velocity of the star with respect to the earth. Velocity of light = 3×10^{10} cm/s.

(Cambridge Univ.) [5×10^6 cm/s; receding]

12. A motor car approaches and passes a stationary observer with a speed of 32 km/h. If its horn emits a frequency of 300 Hz, what will be (a) the absolute change in the frequency and (b) the percentage change in frequency as the car passes the observer. The speed of sound = 1180 km/h.

[16.8; 5.6]

13. A source emitting a sound of frequency n is placed at a large distance from an observer. It starts moving towards the observer with a uniform acceleration a. The speed of sound in the medium is v. Show that the frequency heard by the observer just after the source starts is $2vn^2/(2vn - a)$.

Chapter 7

Superposition of waves: Interference, Beats and Stationary waves

7.1 Introduction

Waves are said to *interact* with each other whenever two or more wave-trains pass through the same medium *simultaneously.* The result of interaction can be obtained by the application of the **principle of superposition,** first enunciated by Huygen in connection with light waves. The principle states:

Whenever two or more wave trains are superposed, the resultant disturbance of a particle at any point in the medium is the vector sum of the individual displacements. The principle is valid for waves of very small amplitude and accounts for many important phenomena in acoustics. We shall discuss three such important cases of superposition, namely:

1. Interference—This is due to the superposition of two wave-trains of the *same frequency* and *equal* (or *nearly equal*) *amplitudes,* and having *a constant phase relation* between them.

2. Beats—This phenomenon is due to the superposition of two wave-trains of *nearly equal frequencies* and *amplitudes* moving in the *same direction.*

3. Stationary (standing) waves—This is due to the superposition of *two identical wave-trains* travelling in *opposite* directions.

7.2 Interference of sound

When two wave-trains of sound travel along the same line in a medium, each particle in the medium will be acted upon simultaneously by both. The resultant displacement of the particles will be the *algebraic sum* of the displacements due to individual wave-trains. If the two waves arrive at a point in the *same phase,* then the resultant amplitude will be the sum of two individual amplitudes, while at points where they arrive in *opposite phase*, the resultant amplitude will be equal to the difference of individual amplitudes. In the first case, the resultant amplitude at those points will be maximum and hence also the intensity of sound. In the second case, the resultant amplitude will be minimum or zero and so also the intensity.

This phenomenon of redistribution of energy due to superposition of two sound waves is known as **interference**.

It must be noted that at places of silence the *energy is not destroyed, nor energy is created* at points of increased sound. There is only a *redistribution*; the energy removed from points of silence gets concentrated at points of enhanced sound.

Example 1. In Fig. 7.1 is shown two waves, 1 and 2 of the *same frequency* and *nearly equal* amplitudes. They produce by superposition the resultant wave 3. Here crest of one falls over the crest of the other, trough over the trough. The resultant wave has an amplitude equal to the sum of the amplitudes of the components. It is a case **constructive interference**.

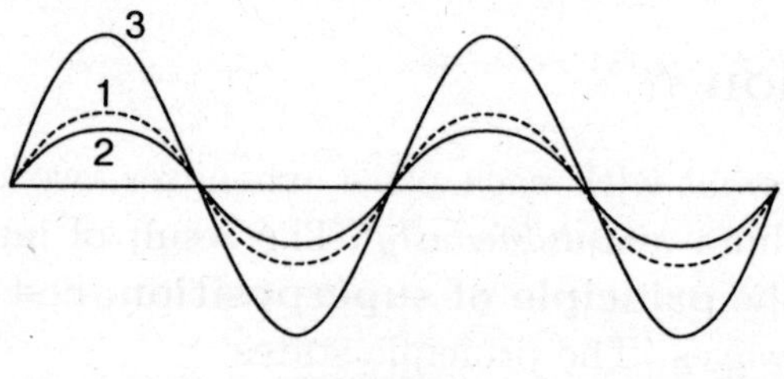

Fig 7.1

2. In Fig. 7.2 the two waves 1 and 2 of the *same frequency* and *equal* amplitudes reach in opposite phase, the crest of one falling on the trough of the other. The resultant disturbance is the undistured state 3. The resultant intensity is zero. It is a case of **destructive interfence**.

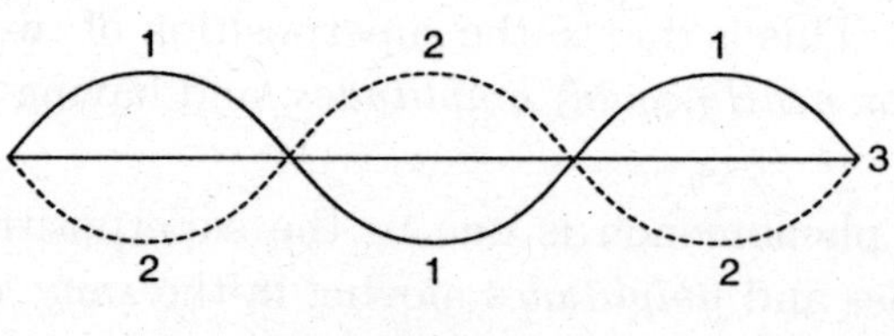

Fig 7.2

• Interference is a very *general* term used not only in acoustics, but also in optics where light waves interfere.

Conditions of interference—Interference can occur only if certain conditions are fulfilled. These are :

1. There must be a constant phase relation between the two wave-trains at a given point in the medium or, in other words, the two sources must be *coherent*. This is ensured if the two wave-trains are derived from the same source.

2. The two waves must produce displacements of the medium along the same straight line.

3. The frequency and the amplitude of the two waves should be equal, for then they can completely annual each other producing silence.

7.3 Experimental demonstration : Quincke's tube

The phenomenon of interference of sound can be demonstrated well by the following experiment due to Quincke.

Quincke's tube—The apparatus is a U-shaped tube BAE (Fig. 7.3) with two side tubes B and E. There is another U-tube CD that can slide in and out of BAE. A vibrating tuning fork T is held over the side tube B and the detector R (or ear) is placed in front of the side tube E. The sound waves start from the fork and travel along two different paths BAE and $BCDE$ and reunite at R. If the two paths are equal, the sound waves meet in phase at R and *interfere constructively* to produce an intense sound. The path BCD is then increased by drawing out CD when at a certain position no sound is heard. The path difference Δ between the two routes is now $\lambda/2$, where λ is the wavelength of sound. If CD is drawn out further, the intensity will again be maximum at a certain position when $\Delta = 2 \cdot \lambda/2 = \lambda$. On increasing the path-length, the intensity again becomes zero and so on.

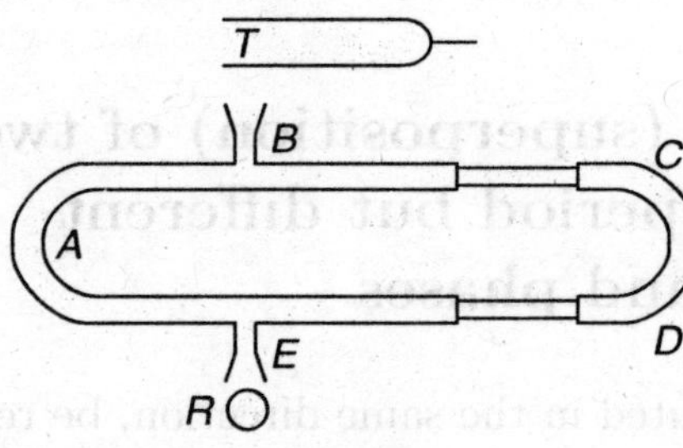

Fig 7.3

In general, if $\Delta = 2n \cdot \lambda/2$ ($n = 0, 1, 2, \cdots$ etc.), the sound is reinforced and if $\Delta = (2n+1)\lambda/2$, $n = 0, 1, 2, \cdots$ the sound is destroyed by mutual interference.

• It affords a *good method* for finding the velocity of sound. The tube CD is adjusted for maximum sound and it is next drawn out by a certain distance d for the next maximum sound. Then $\lambda = 2d$; but the velocity $c = n\lambda = 2nd$, n being the frequency of the fork.

7.4 Silence zones : Explanation

The existence of silence zones can be simply and readily explained by the phenomenon of interference.

When an intense source of sound is mounted at a distance from the observer, he may receive waves directly from the source and also waves reflected from some large surfaces, the intensity of which is practically the same as that of the direct beam. In such cases, the distances travelled by the direct beam and the reflected beam are not the same. If the path difference Δ between the two beams be an odd multiple of half a wavelength, i.e., $\Delta = (2n+1)\lambda/2$, practically *no sound is heard.* This explains the existence of **silence zones**.

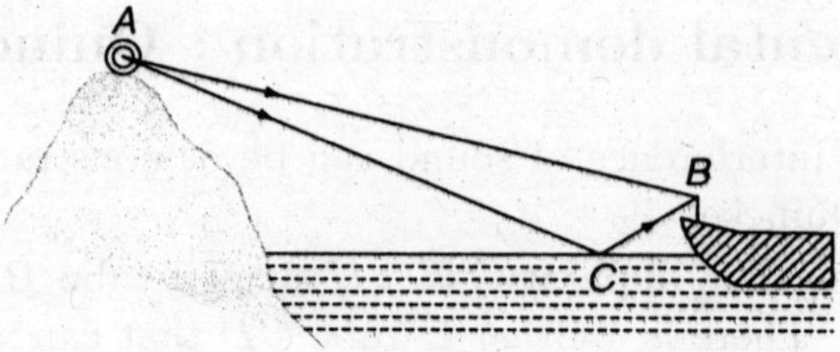

Fig 7.4

The silence zone is observed in case of ships approaching a fog siren. Let a ship B approach a fog siren A (Fig. 7.4) installed on a high rock. The sound from A reaches B by two paths : (i) *directly* along AB and (ii) along ACB *by* way of *reflection* at C from water sheet.

When the path difference Δ between ACB and AB is $\lambda/2$ or odd multiple of it, *no sound* will be heard at B. The sound however becomes audible if the ship moves inward or outward when Δ changes.

7.5 Interference (superposition) of two S.H.Ms of the same period but different amplitudes and phases

Let the two S.H.Ms, executed in the same direction, be represented as:

$$y_1 = a_1 \sin \frac{2\pi}{\lambda}(ct - x) \tag{7.5.1}$$

$$\text{and} \quad y_2 = a_2 \sin \left[\frac{2\pi}{\lambda}(ct - x) + \phi\right] \tag{7.5.2}$$

where a_1, a_2 are the amplitudes of the separate motions and ϕ the phase difference between them.

$\therefore$ The resultant displacement when the two S.H.Ms interfere is

$$\begin{aligned} y &= y_1 + y_2 \\ &= a_1 \sin \frac{2\pi}{\lambda}(ct - x) + a_2 \sin \left[\frac{2\pi}{\lambda}(ct - x) + \phi\right] \\ &= a_1 \sin \frac{2\pi}{\lambda}(ct - x) + a_2 \sin \frac{2\pi}{\lambda}(ct - x) \cos\phi + a_2 \cos \frac{2\pi}{\lambda}(ct - x) \sin\phi \end{aligned}$$

$$= [a_1 + a_2 \cos\phi] \sin\frac{2\pi}{\lambda}(ct - x) + [a_2 \sin\phi] \cos\frac{2\pi}{\lambda}(ct - x)$$

$$= a\cos\theta \sin\frac{2\pi}{\lambda}(ct - x) + a\sin\theta\cos\frac{2\pi}{\lambda}(ct - x)$$

$$= a\sin\left[\frac{2\pi}{\lambda}(ct - x) + \theta\right] \tag{7.5.3}$$

where we have made the following substitutions:

$$\left.\begin{aligned} a\cos\theta &= a_1 + a_2\cos\phi \\ a\sin\theta &= a_2\sin\phi \end{aligned}\right\} \tag{7.5.4}$$

$$\therefore\ a^2 = a_1^2 + a_2^2 + 2a_1a_2\cos\phi \tag{7.5.5}$$

$$\text{and,}\ \tan\theta = \frac{a_2\sin\phi}{a_1 + a_2\cos\phi} \tag{7.5.6}$$

The resultant wave train has thus the same frequency but a different amplitude and phase from those of the component wave-trains.

Observations: 1. When $\phi = 0$, we have from equation (7.5.5), $a^2 = a_1^2 + a_2^2 + 2a_1a_2 = (a_1 + a_2)^2 \Rightarrow a = a_1 + a_2$ and $\tan\theta = 0 \Rightarrow \theta = 0$, the resultant vibration is in phase with the components.

2. When however the phase difference $\phi = \pi$, we have $a = a_1 \sim a_2 = 0$, if $a_1 = a_2$.

So, if the phase difference between the two waves is zero, they reinforce each other leading to *constructive interfarence.* When however the phase difference is 180°, they destroy each other's effect—*destructive interference.*

7.6 Beats

The phenomenon of beats is an interesting example of interference of sound waves.

When two wave-trains of slightly different frequencies and amplitudes move along the same straight line in the same direction, the resultant displacement at a point alternately waxes and wanes in amplitude (and hence *in intensity*) *at regular intervals—as many times per second as is the difference in their frequencies.* This phenomenon of *periodic fluctuations*—waxing and waning—of sound due to *mutual interference* of two wave-trains is known as **beats**.

• If two tuning forks of frequencies, say 256 and 260 be sounded together, a clear fluctuation in intensity four times a second of sound would be discernible.

• The following ***conditions*** are ***to be satisfied*** to clearly hear the beats.

1. The difference in two frequencies should be restricted to 10; if the difference be greater than 10, the beats are too fast to be distinguished.

2. The difference in amplitude should also be small. Otherwise, the weaker sound would not appreciably affect the stronger one.

3. The two sound waves should be of the same type, that is, they should have the same waveform.

7.6.1 Graphical treatment of beats

The phenomenon of beats is illustrated graphically in Fig. 7.5. Let the curves C_1 and C_2, having the same amplitude and initial phase, represent the component vibrations. They have frequencies 5 and 6 respectively and by superposition they give the resultant vibration—the curve C.

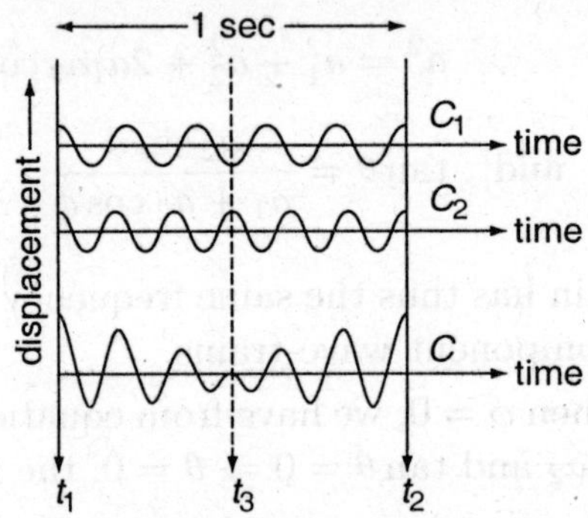

Fig 7.5

At times t_1 and t_2 the two waves agree in phase and the resultant displacements are *maximum* at t_1 and t_2. At t_3, they are in opposite phase and the resultant displacement is a *minimum*. As intensity depends on the amplitude of vibration, the sound heard for small intervals corresponding to the instants t_1 and t_2 is the most intense and is minimum at t_3. Such fluctuations of intensity appear as *beats*. Two consecutive maxima (or consecutive minima) give one beat. So Fig. 7.5 shows that one beat is produced in 1 sec.

Note that between the instants t_1 and t_2 the number of waves of C_2 is one more than that of C_1 and within a distance covered in one second, there is one pairs of maxima. The number of beats per second is thus 1, the difference of frequencies ($= 6 - 5$) of the component waves.

We shall now discuss the phenomenon of beats analytically.

7.6.2 Analytic treatment of beats

Let m and $m + n$ be the frequencies and a, b the amplitudes of the two interfering waves. Then the displacements of a particle due to the waves are

$$y_1 = a \sin 2\pi mt \tag{7.6.1}$$

$$y_2 = b \sin 2\pi(m + n)t \tag{7.6.2}$$

Hence the resultant displacement, when the two waves superpose, is given by

$$\begin{aligned} y = y_1 + y_2 &= a\sin 2\pi mt + b\sin 2\pi(m+n)t \\ &= a\sin 2\pi mt + b\,[\sin 2\pi mt\cos 2\pi nt + \cos 2\pi mt\sin 2\pi nt] \\ &= (a + b\cos 2\pi nt)\sin 2\pi mt + b\sin 2\pi nt\cos 2\pi mt \\ &= c\cos\phi\sin 2\pi mt + c\sin\phi\cos 2\pi mt = c\sin(2\pi mt + \phi) \end{aligned}$$

where the following substitutions have been made:

$$\left.\begin{aligned} c\cos\phi &= a + b\cos 2\pi nt \\ c\sin\phi &= b\sin 2\pi nt \end{aligned}\right\} \tag{7.6.3}$$

$$\therefore\ c^2 = a^2 + 2ab\cos 2\pi nt + b^2 \tag{7.6.4}$$

[Squaring and adding the equations (7.6.3)]

$$\text{and}\quad \tan\phi = \frac{b\sin 2\pi nt}{a + b\cos 2\pi nt} \tag{7.6.5}$$

Thus, the amplitude c of the resultant wave will vary with time, sometimes having maximum value and sometimes minimum.

c will be **maximum**, when $\cos 2\pi nt = 1$.

or, $2\pi nt = 0,\ 2\pi,\ 4\pi,\ 6\pi, \cdots$, etc.

or, $t = 0,\ 1/n,\ 2/n,\ 3/n, \cdots$, etc.

i.e., **integral multiples** *of* $1/n$ and the maximum value of $c = a + b$.

Similarly, c will be **minimum**, when $\cos 2\pi nt = -1$

or, $2\pi nt = \pi,\ 3\pi,\ 5\pi,\ 7\pi, \cdots$, etc.

or, $t = 1/2n,\ 3/2n,\ 5/2n, \cdots$, etc.,

i.e., **odd multiple** *of* $1/2n$ and the *minimum* value of $c = a - b$.

Thus, with the passage of time, the amplitude of the resultant wave at a given point varies between the maximum $a + b$ and the minimum $a - b$. If the *amplitudes* of the interacting waves be *equal*, (i.e., $a = b$), the variation is between $2a$ and 0.

The interval between two maxima (or minima) is $1/n$ sec. So the *number of beats per second* (two successive maxima or minima comprise one beat) *is n, the difference of the frequencyies* and is known as the **beat frequency**.

7.6.3 Applications of the phenomenon of beats

The phenomenon of beats finds the following practical applications in a number of different fields.

1. Determination of *unknown frequency* of a fork.
2. *Tuning* of the musical instruments.
3. *Detection* of harmful gases in mines (mine-safety).

Frequency determination—When two forks of nearly equal frequencies are sounded, they produce beats equal in number to the difference in their frequencies. This number is counted and say, it is x. If n be the known frequency of one of the forks, the frequency of the other one is $n' = n \pm x$.

To ascertain if $n' > n$, or, $n' < n$, the fork of the unknown frequency is slightly loaded with some wax. This causes a lowering in the frequency and hence the beat frequency would either increase or decrease. In case it increases, $n' < n$ and so $n' = n - x$. If, however, it decreases, $n' > n$ and so $n' = n + x$.

Tuning of musical instruments—Musical instruments like sitar, organ pipes, piano can be tuned by utilising the phenomenon of beats. The instrument is sounded along with another of known frequency with which it is to be tuned. If the frequencies are slightly out of unison, beats will be heard and appropriate adjustments are made till the beats disappear which ensures perfect unison.

Detection of fire-damp—Two identical organ pipes, one filled with pure air and the other by the mine air, are sounded together. If the mine air is pure, the pipes remain in tune and no beats are heard. But if it is impure and fire-damp is present, the velocity of sound changes and so the frequency of the note of the pipe. Thus beats are produced and the danger of explosion is forewarned.

7.7 Interference and beats compared

Although interference and beats both are the products of superposition of two wave-trains, there are some differences between the two as indicated in the following table.

	Interference		Beats
1.	Arises due to superposition of two wave-trains moving in the same or opposite directions.	**1.**	Arises due to superposition of two wave-trains moving in the same direction.
2.	The interacting wave-trains must have exactly the same frequency.	**2.**	The interacting wave-trains must have slightly different frequencies.
3.	The positions of maxima and minima are fixed so that the phase difference between the waves at any point remains constant.	**3.**	The phase difference between the waves at a point changes continuously from 0 to π and so the positions of maxima and minima are not fixed.
4.	The pattern is fixed in space, the amplitude at a given point remains constant. The amplitude however varies from point to point.	**4.**	The pattern moves in the direction of propagation and this progressive character causes amplitude at a point to vary from a maximum to a minimum.

7.8 Stationary waves : Analytic treatment

Stationary waves are the product of superpostion of two *identical* (amplitude, period, and hence frequency, are equal) *progressive waves moving along the same line* but in *opposite directions*.

The equation of a progressive wave moving along the positive x-direction is

$$y_1 = a \sin \frac{2\pi}{\lambda}(ct - x) \tag{7.8.1}$$

while that of an identical progressive wave moving the negative x-direction is

$$y_2 = a \sin \frac{2\pi}{\lambda}(ct + x) \tag{7.8.2}$$

where a is the amplitude, λ the wavelength, c the velocity of the wave and y the particle displacement at the point x in time t.

When the two such waves are superposed, the resultant particle displacement would be given by

$$\begin{aligned} y = y_1 + y_2 &= a\left\{\sin \frac{2\pi}{\lambda}(ct - x) + \sin \frac{2\pi}{\lambda}(ct + x)\right\} \\ &= 2a \sin \frac{2\pi}{\lambda} ct \cos \frac{2\pi}{\lambda} x \\ &= A \sin \frac{2\pi}{\lambda} ct, \end{aligned} \tag{7.8.3}$$

where $A = 2a \cos \frac{2\pi}{\lambda} x$, the amplitude of the stationary wave.

Eq. (7.8.3) is known as the *equation of* **stationary wave** and represents a simple harmonic motion of the same frequency (or period) of the individual waves but of amplitude $A = 2a \cos 2\pi x/\lambda$ which is *not* a constant but a periodic function of the position x of the particle. As (7.8.3) does not contain a term like $(ct - x)$, it does *not* represent a progressive wave; rather, at a particular instant the nature of the displacement is stationary. It represents a **stationary** or **standing wave**.

Amplitude variation—The amplitude A of the stationary wave is

$$A = 2a \cos 2\pi x/\lambda$$

$$\therefore \quad A = 0, \text{ when } \cos 2\pi x/\lambda = 0 \Rightarrow 2\pi x/\lambda = (2n+1)\pi/2$$

$$\Rightarrow x = (2n+1)\lambda/4, \quad n = 0, 1, 2, \cdots.$$

$\therefore$ At points $x = \lambda/4,\ 3\lambda/4,\ 5\lambda/4$ etc. the amplitude of the displacement is *zero*. Such points of zero displacement are called the *displacement nodes* or simply **nodes**. The distance between two consecutive nodes is plainly $2\lambda/4$ or $\lambda/2$.

But A is **maximum**, when $\cos 2\pi x/\lambda = \pm 1$ and then $A = 2a$. This occurs when,

$$2\pi x/\lambda = n\pi, \quad n = 0, 1, 2, \cdots,$$

$$\text{or,} \quad x = n\lambda/2$$

$\therefore$ At points $x = 0, \lambda/2, 2\lambda/2, 3\lambda/2$ etc., the amplitude of the displacement is maximum ($= 2a$). Such points of maximum displacement are called *displacement antinodes* or simply **antinodes**. The distance between two successive antinodes is also $\lambda/2$. Obviously antinodes are midway between two nodes.

Velocity and acceleration variation : The displacement of the particle at a given (x, t) is given by

$$\textbf{Displacement}: y = 2a\cos\frac{2\pi x}{\lambda}\sin\frac{2\pi ct}{\lambda}. \qquad (7.8.4)$$

$$\therefore \textbf{Particle velocity}: \frac{dy}{dt} = \frac{4\pi ac}{\lambda}\cos\frac{2\pi x}{\lambda}\cos\frac{2\pi ct}{\lambda}. \qquad (7.8.5)$$

$$\therefore \textbf{Particle acceleration}: \frac{d^2y}{dt^2} = -8\frac{\pi^2c^2}{\lambda^2}a\cos\frac{2\pi x}{\lambda}\sin\frac{2\pi ct}{\lambda} \qquad (7.8.6)$$

The equations show that the displacement, the velocity and the acceleration all vary with position and time. We shall emphasise, here, more on the variation with respect to position.

It appears from above that the velocity and acceleration both vanish at points given by $\cos 2\pi x/\lambda = 0$, i.e., exactly at **nodal points**. Similarly, they are seen to be maximum at **antinodal points**.

• Instead of studying amplitude variation, the existence of nodes and antinodes also follows from the displacement equation above.

Pressure variation (strain)—The *excess pressure* (the variation of pressure) p in a sound wave is given by the relation

$$p = -K\frac{dy}{dx} \qquad (7.8.7)$$

where K is the bulk modulus of elasticity and (dy/dx) is called the *strain* or *compression*.

For a stationary wave,

$$y = 2a\cos\frac{2\pi x}{\lambda}\sin\frac{2\pi ct}{\lambda}$$

$$\therefore \quad \frac{dy}{dx} = -\frac{4\pi a}{\lambda}\sin\frac{2\pi x}{\lambda}\sin\frac{2\pi ct}{\lambda}$$

$$\therefore \quad p = +\frac{4\pi aK}{\lambda}\sin\frac{2\pi x}{\lambda}\sin\frac{2\pi ct}{\lambda} \qquad (7.8.8)$$

$$= A'\sin\frac{2\pi ct}{\lambda} \qquad (7.8.9)$$

$$\text{where } A' = \frac{4\pi aK}{\lambda}\sin\frac{2\pi x}{\lambda}$$

is the amplitude of pressure variation.

Now, $A' = 0$, when $\sin 2\pi x/\lambda = 0 \Rightarrow x = n\lambda/2$.

Such points are called the **pressure nodes** and they correspond to *displacement antinodes.* Thus at the points of *maximum displacement,* there is *no pressure variation,* the pressure being normal.

At the displacement nodes, $x = (2n + 1)\lambda/4$, we have $\sin 2\pi x/\lambda = 1$ and $A' = 4\pi aK/\lambda$ = maximum, so that the maximum pressure variation—**pressure antinodes**—occurs at points of no displacement.

• Since pressure is proportional to strain, the *strain is maximum at the nodes and minimum at the antinodes.*

• It is left as an exercise to the readers to verify that the displacement is maximum at $t = (2m+1)T/4$, $m = 0, 1, 2, \cdots$, being alternately positive or negative, and minimum at $t = 0, T/2, 2T/2, \cdots, mT/2$, $m = 0, 1, 2, \cdots$, etc. T being the period of vibration of the particle.

• The strain is *maximum* at $t = (2m + 1)T/4$, $m = 0, 1, 2, \cdots$ and *minimum* at $t = mT/2$, $m = 0, 1, 2, \cdots$.

7.8.1 Chief characteristics of stationary waves

The chief *characteristics* of stationary or standing waves are thus:

1. The stationary waves are *not progressive* i.e., *do not travel* forward or backward. There is therefore *no transfer of energy from one particle to the other.* The displacement and velocity at any point continuously change from maximum positive to maximum negative and conversely.

2. All particles, barring those at nodes, execute S.H.M. with the same period as that of individual waves. Particles *at nodes* have *zero displacement, zero velocity* and *maximum change in strain* (pressure). *At antinodes, the strain is zero, but the displacement and velocity are maximum.*

3. The *distance between two consecutive nodes* or *consecutive antinodes is* $\lambda/2$, where λ is the wavelength of the stationary wave (and also of individual waves)

4. Twice in one vibration, the particles have maximum displacement—one positive and the other negative—at an interval of $T/2$, passing momentarily through position of rest.

5. Twice in one vibration, the particles pass through their mean positions, at an interval of $T/2$, in opposite directions with maximum velocity.

The medium is split into segments. For $T/2$, all the particles in a segment between two successive nodes move in one direction and in the next $T/2$ oppositely. At any instant, particles in two consecutive segments move oppositely.

7.8.2 Stationary waves by reflection : Phase change

If a progressive wave meets in its path a boundary surface, it gets reflected from that surface. The reflected wave necessarily travels backward and by way of superposition with the forward wave produces stationary waves.

We shall show that on account of reflection from the boundary surface, there is phase change and it depends on whether the boundary is perfectly rigid or yielding.

Let the incident wave, moving along positive x-direction, be

$$y_1 = a\sin\frac{2\pi}{\lambda}(ct - x) \tag{7.8.10}$$

The reflected wave moves in the negative x-direction. So it may be represented as

$$y_2 = ra\sin\frac{2\pi}{\lambda}(ct + x), \tag{7.8.11}$$

where r (≤ 1) is the ratio of reflected amplitude to the incident one and is called *displacement reflection coefficient.*

At $x = 0$, the resultant displacement of the particle is given by

$$\begin{aligned} y &= y_1 + y_2 \\ &= (1 + r)a\sin\frac{2\pi ct}{\lambda} \end{aligned} \tag{7.8.12}$$

Now, for the incident wave, we have

$$\frac{dy}{dx} = -\frac{2\pi a}{\lambda}\cos\frac{2\pi}{\lambda}(ct - x), \quad \text{using (7.8.10)}$$

$$\therefore \text{ Excess pressure } p_1 = \frac{2\pi aK}{\lambda}\cos\frac{2\pi}{\lambda}(ct - x) \tag{7.8.13}$$

since the excess pressure $p = -K(dy/dx)$.

Similarly, we shall have for the reflected wave

$$\begin{aligned} p_2 &= \text{excess pressure for reflected wave} \\ &= -\frac{2\pi aK}{\lambda}r\cos\frac{2\pi}{\lambda}(ct + x) \end{aligned} \tag{7.8.14}$$

At $x = 0$, the resultant excess pressure is

$$\begin{aligned} p &= p_1 + p_2, \text{ at } x = 0 \\ &= (1 - r)\frac{2\pi aK}{\lambda}\cos\frac{2\pi ct}{\lambda} \end{aligned} \tag{7.8.15}$$

Now, at a **rigid boundary**, $y = 0$. So $r = -1$ from (7.8.12). Therefore,

$$y_2 = -a\sin\frac{2\pi}{\lambda}(ct + x), \quad \text{using (7.8.11)}$$

$$p_2 = \frac{2\pi aK}{\lambda}\cos\frac{2\pi}{\lambda}(ct + x), \quad \text{using (7.8.14)}$$

Thus, *on reflection from a rigid surface there occurs a phase change by π of the displacement, but not of the pressure.*

At an **yielding surface**, $p = 0$, so that $r = 1$ from (7.8.15).

$$\therefore \quad y_2 = a \sin \frac{2\pi}{\lambda}(ct - x)$$

$$p_2 = -\frac{2\pi a K}{\lambda} \cos \frac{2\pi}{\lambda}(ct + x)$$

So, *reflection at an yielding surface introduces a phase change of π in the pressure, but not in displacement.*

• For a perfectly rigid reflector, $r = -1$ and for a reflector that can move (yield) freely $r = 1$. These two cases represent the two extreme situations. If the reflector is not perfectly rigid, r necessarily lies between 0 and -1.

7.8.3 Stationary waves in organ pipes

Stationary vibrations can be set up in the air-column enclosed in an organ pipe by a sudden disturbance communicated to it. The compressions and rarefactions generated by the excitation constitute the incident wave-train that interferes with the reflected wave-train moving in opposite direction and produces the standing waves in it. The vibration is always longitudinal.

An organ pipe may be of two kinds : **(a) closed organ pipe** and **(b) open organ pipe**. A *closed pipe* is one in which one end of the pipe is closed and the other end open, while in the *open pipe* both the ends of the pipe are open. It is always assumed that the diameter of the tube (closed or open) is small compared to its length and also the wavelength of sound.

Modes of vibration in closed organ pipes—Let a vibrating fork be held over the open end A. A longitudinal wave-train passes along the pipe and is reflected at the closed end that acts as a rigid wall. The two oppositely moving wave-trains superpose to give rise to standing waves in the air column.

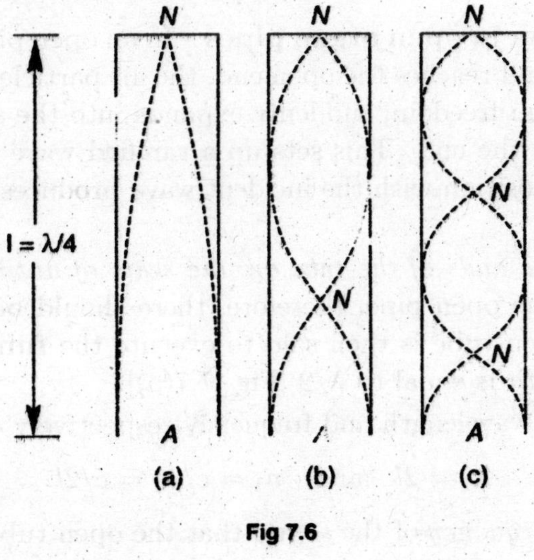

Fig 7.6

The air particles at open end enjoy the maximum freedom of movement and is the seat of an *antinode.* At the closed end, however, air particles can hardly vibrate and is thus always a *node.* This is the simplest mode of vibration in a closed pipe. Since the distance between a node and the next antinode is $\lambda/4$, the wavelength is *four times* the tube length. Under this condition, the closed pipe is said to execute its **fundamental** note or the **first harmonic** [Fig. 7.6(a)]

$$\therefore \quad \frac{\lambda_1}{4} = l, \quad \text{or,} \quad \lambda_1 = 4l, \quad l = \text{length of the tube}$$

$$\therefore \text{ Frequency, } n_1 = \frac{c}{\lambda_1} = \frac{c}{4l}, \; c = \text{velocity of sound.}$$

This is the *lowest frequency* the closed organ pipe can emit.

In the closed pipe, the open end is always an antinode and the closed end a node. The mode of vibration, next to fundamental, is thus *one* intermediate node and *one* antinode [Fig. 7.6(b)].

If n_2 and λ_2 be the frequency and wavelength respectively of the above mode, called the **first overtone,**

$$\lambda_2 = \frac{4}{3}l \quad \text{and so,} \quad n_2 = \frac{c}{\lambda_2} = 3\frac{c}{4l} = 3n_1.$$

It thus represents the **third harmonic** of the vibration.

For the **second overtone**, there will be *two* intermediate nodes and *two* antinodes [Fig. 7.6(c)] and its wavelength λ_3 and frequency n_3 are given by

$$\lambda_3 = \frac{4}{5}l \quad \text{and} \quad n_3 = \frac{c}{\lambda_3} = 5\frac{c}{4l} = 5n_1.$$

It thus represents the **fifth harmonic** of the vibration and so on.

Thus, the modes of vibration *in a closed organ pipe are such that only the odd harmonics are obtained.*

Modes of vibration in open organ pipes—In an open pipe, when a compression travels along the tube and reaches the open end, the air particles there being capable of vibration with maximum freedom, suddenly expands into the surrounding air causing a sudden rarefaction at the end. This sets up a rarefied wave that travels back along the tube and by superposition with the incident wave produces stationary vibration in the open tube.

Being open, *both the ends of the tube are the seats of antinodes.* For the simplest mode of vibration in the open pipe, therefore, there should be a node between these two antinodes. The open tube is then said to execute the **fundamental** or the **first harmonic** and its length is equal to $\lambda/2$ [Fig. 7.7(a)].

If λ_1 and n_1 be the wavelength and frequency respectively of the fundamental,

$$\lambda_1 = 2l \quad \text{and} \quad n_1 = c/\lambda_1 = c/2l.$$

This is the *lowest frequency* of the sound that the open tube can emit.

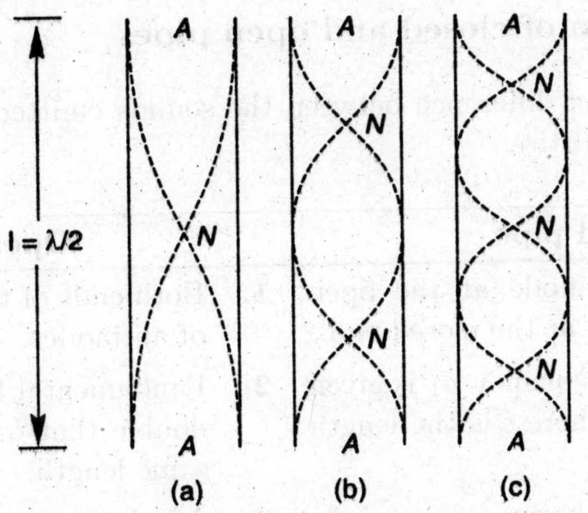

Fig 7.7

The next mode of vibration is the **first overtone**. It will have *two* intermediate nodes and *one* intermediate antinode [Fig. 7.7(b)], apart from two antinodes at the two open ends. In this condition, the pipe-length is equal to the wavelength. So, its wavelength is given by

$$\lambda_2 = l, \text{ and its frequency, } n_2 = c/\lambda_2 = c/l = 2n_1.$$

So, it is the **second harmonic** of the vibration.

In the next mode, the **second overtone**, there will be *three* intermediate nodes and *two* intermediate antinodes [Fig. 7.7(c)].

$$\therefore \quad \lambda_3 = \frac{2}{3}l$$

$$\text{and} \quad n_3 = \frac{c}{\lambda_3} = 3\frac{c}{2l} = 3n_1.$$

So, it is the **third harmonic**, and so on.

Thus, unlike closed pipes, *in open pipes, all the harmonics – odd and even – are obtained.*

• Whatever be the method of excitation of organ pipes, it is rather difficult to excite any single mode of vibration. More often than not, a number of modes are simultaneously excited, and their number and relative intensities determine the *quality* of the emitted sound.

• Sound of an open pipe is **more musical** than that of a closed pipe. In the sound from an open pipe, all the harmonics, unlike in the closed pipe where only odd harmonics are present, are excited. Since it is richer in harmonics, its sound is more musical and melodious.

• A number of factors may affect the frequency of the notes emitted by pipes. Some of them are *temperature, humidity, nature of the gas* and all of these affect the velocity of sound and hence also the emitted frequency.

7.8.4 Comparison of closed and open pipes

The following points of difference between the sounds emitted from closed and open organ pipes are noteworthy.

Closed pipe	Open pipe
1. Always an antinode at the open end and a node at the closed end.	1. Both ends of the pipe are the seats of antinodes.
2. Fundamental frequency n_1 is given by $n_1 = c/4l$ where l is the length of the pipe.	2. Fundamental frequency $n_1 = c/2l$, double that of a closed one of the same length.
3. Only the odd harmonics are obtained. So the quality of note is less sweet.	3. All the harmonics (odd and even) are obtained. So the notė-quality is sweeter.

7.9 Progressive vs. stationary waves

There is a number of distinction between the progressive and the stationary or standing waves. A critical study of this is both important and interesting. The different points of distinction are therefore shown below in a tabular form.

Progressive wave	Stationary wave
1. A progressive wave is due to the continuous periodic vibration of a part of the medium.	1. A product of superposition of two identical progressive waves moving oppositely in the same line.
2. The distubrance is transmitted from one particle to another in the direction of progpagation of the wave and the wave travels onward with a fixed velocity.	2. No bodily transference of disturbance occurs. The waveform consisting of condensations and rarefactions does not move but only appears and disappears.
3. Each particle of the medium undergoes the same cyclical changes as regards the amplitude, velocity, acceleration, variation in pressure etc., with the same frequency.	3. Amplitude, velocity, pressure-variations etc., change from point to point but constant at a point. At nodes, the displacement, velocity, acceleration are zero, but pressure variation is maximum. At antinodes, condition is opposite. The frequency is the same for all particles.

Progressive wave	Stationary wave
4. The vibration of one particle occurs a little earlier than its successor. So phases of particles continuously change; phase difference between two particles is proportional to their separation. Amplitude of vibration is the same.	**4.** All particles in the same segment (between two successive nodes or antinodes) vibrate in the same phase but with different amplitudes. Particles in two consecutive segments vibrate in opposite phases.
5. In a complete cycle, there is no instant when all the particles remain at rest.	**5.** Particles at nodes are permanently at rest; others are momentarily at rest at the same instant twice in a cycle.

7.10 Illustrated Examples

▶ **Example 1.** *Two tuning forks A and B, the frequency of B being* 512, *produce* 5 *beats per sec. A is filed and the beats are found to occur at shorter intervals. Find the frequency of A.* **(Punjab Univ.)**

Solution. Let n be the frequency of A. Since 5 beats are produced per sec, $n = 512 \pm 5$, i.e., either 517 or 507. Since filing of A increases its frequency and also since, by the problem, beat-frequency increases, the frequency of A must be greater than that of B. So, $n = \mathbf{517}$.

▶ **Example 2.** *Calculate the velocity of sound in a gas in which two waves of wavelengths* 50 cm *and* 50.5 cm *produce* 6 *beats per second.* **(Vidyasagar Univ.)**

Solution. Let n_1 = frequency of first wave; λ_1 = wavelength of first wave = 50 cm; n_2 = frequency of second wave; λ_2 = wavelength of second wave = 50.5 cm.

Now, $n_1\lambda_1 = c = n_2\lambda_2$, where c = velocity of sound in the gas.

$$\therefore \quad n_1 = c/50, \quad n_2 = c/50.5, \quad \therefore \; n_1 > n_2. \quad \text{But } n_1 - n_2 = 6,$$

$$\therefore \quad c\left[\frac{1}{50} - \frac{1}{50.5}\right] = 6.$$

$$\therefore \quad c = 30300 \text{ cm/s} = \mathbf{303 \; m/s}.$$

▶ **Example 3.** *A set of* 24 *tuning forks is arranged in a series of increasing frequency. If each fork produces* 4 *beats per sec with the preceding one and the last fork sounds the octave of the first, calculate the frequencies of the first and the last fork.* **(Bombay Univ.)**

Solution. Let n be the frequency of the first fork of the set. So, $2n$ is the frequency of the last one. Hence their difference is $2n - n = n$.

Since the difference in frequency between any two consecutive forks is 4, the difference between the first and the last fork is 4×23.

$$\therefore \quad n = 4 \times 23 = 92; \quad 2n = 184.$$

Thus the frequencies of the first and the last fork are **92** and **184** respectively.

► **Example 4.** *A cog wheel containing* 15 *cogs, revolves at the rate of* 1020 rev/min. *It now produces* 1 *beat* per sec *with a tuning fork. On increasing the speed of the wheel to* 1032 rev/min, 2 *beats* per sec *are heard. What is the frequency of the fork?*

Solution. Let n_1, n_2 be the frequencies of the note of the cog wheel in the first and the second case respectively and n be the frequency of the fork.

$$\therefore \quad n_1 = \frac{1020}{60} \times 15 = 255;$$

$$n_2 = \frac{1032}{60} \times 15 = 258.$$

By the problem, $n = n_1 \pm 1 = 255 \pm 1 = 256$ or 254

Also, $n = n_2 \pm 4 = 258 \pm 2 = 260$ or 256

$\therefore$ Frequency of the fork, $n =$ **256**.

► **Example 5.** *A closed pipe* 1 m *long and containing a gas resounds in its fundamental note to a tuning fork of frequency* 100 Hz. *What is the velocity of sound in the gas.* **(Ujjain Univ.)**

Solution. Length of closed pipe, $l = 1$ m. Wavelength of note, $\lambda = 4l = 4$ m. Frequency of the note, $n = 100$.

$\therefore$ Velocity of sound, $c = n\lambda = 100 \times 4 =$ **400 m/s.**

► **Example 6.** *Two tuning forks A and B of approximately the same frequency are used to obtain Lissajous figures. It is observed that the figures go through a cycle of change in* 20 sec. *On slightly loading B, the figures go through one cycle in* 15 sec. *If A has a frequency* 400 Hz, *find the frequency of B before and after loading.*

(Calcutta Hons.)

Solution. *Before loading,* the frequency of the pattern-change $= \frac{1}{20} = 0.05$ Hz.

After loading, the corresponding frequency $= \frac{1}{15} = 0.067$ Hz. Now loading decreases the frequency of B. But the freq-difference between A and B, when B is loaded, has increased. This implies the original frequency of B is less than that of A.

$\therefore$ Frequency of B (before loading) $= 400 - 0.05 =$ **399.95 Hz.**

Frequency of B (after loading) $= 400 - 0.067 =$ **399.933 Hz.**

Review Questions

1. What is interference of sound? Under what conditions can two sound waves produce interference? How will you demonstrate the phenomenon? **(Utkal Univ.)**

2. What are silence zones? How do you explain their existence in the neighbourhood of a powerful source of sound?

3. Find the resultant of two simple harmonic motions of the same period executed in the same direction but differing in phase and amplitude. **(Calcutta and Osmania Univ.)**

4. What are beats? How are they produced? How will you show that the number of beats produced by two tuning forks is equal to the difference of their frequencies? **(Calcutta and Burdwan Univ.)**

5. Investigate theoretically the formation of stationary waves and discuss their characteristic properties. **(Lucknow Univ.)**

6. Discuss analytically the modes of vibration in case of closed pipes. **(Kanpur Univ.)**

7. Show that the frequency of the fundamental note of an open organ pipe is twice that from a closed pipe of the same length. Discuss the mode of vibration of the closed and open pipes and hence explain why the sound emitted from the open pipe is more musical than for a closed one. **(Burdwan and Vidyasagar Univ.)**

8. Distinguish between the progressive and the stationary waves. **(Punjabi Univ.)**

9. Write short explantory notes on :

   (a) Stationary waves (b) Silence zones (c) Beats (d) Interference of sound.

Short Answer and Objective Questions

1. Explain what is meant by interference of sound.

2. What are beats? Distinguish between interference and beats.

3. What are silence zones? Explain in terms of interference of sound.

4. What are nodes and antiodes? What is their minimum separation in terms of wavelength?

5. The sound emitted by an open organ pipe is more musical compared to that of a closed organ pipe.—Explain.

6. Describe an experiment to demonstrate the interference of sound waves.

7. Compare acoustically an open organ pipe with a closed organ pipe.

8. Two wave-trains interfere constructively if the phase difference between them is

(a) $\pi/2$ (b) π (c) zero (d) none of these.

9. The distance between two consecutive antinodes is

(a) λ (b) $\lambda/2$ (c) $\lambda/4$ (d) none of these.

10. The amplitude of a stationary wave is zero at

(a) node (b) midway between node and antinode (c) antinode (d) no where.

11. The length of a closed organ pipe is 20 cm. The wavelength of the fundamental note is

(a) 80 cm (b) 60 cm (c) 40 cm (d) 20 cm.

Problems

1. Two tuning forks produce 4 beats when sounded together. One of them is in unison with 1.20 m length and the other one with 1.25 m of it. Calculate the frequencies of the forks. **(Kanpur Univ.)** [100, 96]

2. Two organ pipes when sounded together produce 4 beats per sec. Find their frequencies if one of them is 75 cm long and the other is 80 cm long. **(Kanpur Univ.)** [64, 60]

3. 64 tuning forks are arranged in order of increasing frequency and any two successive forks give 4 beats per sec when sounded together. If the last fork gives the octave of the first, calculate the frequency of the latter. **(Calcutta Univ.)** [252]

4. A fork of unknown frequency gives 4 beats per sec. when sounded with another fork of frequency 256 Hz. The fork is now loaded with a piece of wax and again 4 beats per sec is heard. What is the unknown frequency of the fork? [260 Hz]

5. What is the velocity of sound in a gas in which two waves of wavelengths 150 cm and 151 cm respectively produce 6 beats in 5 seconds. [317.8 m/s]

6. The note of an organ pipe beats 33 times in 10 sec with a tuning fork of frequency 256 and 48 times in 10 sec with one of frequency 264. What is the frequency of the organ pipe? [259.25]

7. Sound waves from a vibrating body reach a point by two paths. There is silence at the point when the paths differ by 12 cm or 37 cm. If the frequency of the fork be 1375, find the velocity of sound. [330 m/s]

8. A closed organ pipe 24.5 cm long is in unison with a tuning fork of frequency 256. If its length is increased by 2 mm, find the number of beats produced when the fork and pipe are sounded together. **(Andhra Univ.)** [2 per sec]

9. Two tuning forks of slightly different frequencies produce Lissajous figures. A circular figure is obtained that appears in every 10 sec. If one tuning fork has the frequency 300 Hz, calculate the frequency of the other. [300.1 Hz or 299.9 Hz]

10. A sonometer of 100 cm wire is divided into two parts by a knife edge. When the two parts with a difference in length of 10 cm are vibrated, four beats are produced per second. Find the frequencies of the two parts of sonometer wire. [22 Hz, 18 Hz]

11. A tuning fork is in unison with a stretched string of length 75 cm. If the length is reduced by 3 cm, the string produces 6 beats per second with the fork. Find the frequency of the fork. [144 Hz]

Chapter 8

Electromagnetic Waves

Maxwell's field equations provide us with all the information that can be derived from the classical theory of electric and magnetic fields. What in particular is of paramount importance is that the fields generated by the moving charges can leave the source to propagate through space in the form of waves, called *electromagnetic waves.* The situation finally culminated in the identification of light waves with electromagnetic waves and the consequent development of **electromagnetic theory** of light.

8.1 Pre-Maxwell Electrodynamics

Before the appearance of Maxwell in the scene, the basic laws of electricity and magnetism were expressed by the following equations.

$$\textbf{1.}\quad \vec{\nabla}\cdot\vec{D}=\rho \quad\Rightarrow\quad \vec{\nabla}\cdot\vec{E}=\rho/\varepsilon \tag{8.1.1}$$

$$\textbf{2.}\quad \vec{\nabla}\cdot\vec{B}=0 \quad\Rightarrow\quad \vec{\nabla}\cdot(\mu\vec{H})=0 \Rightarrow \vec{\nabla}\cdot\vec{H}=0 \tag{8.1.2}$$

$$\textbf{3.}\quad \vec{\nabla}\times\vec{E}=-\partial\vec{B}/\partial t \quad\Rightarrow\quad \vec{\nabla}\times\vec{E}=-\mu(\partial\vec{H}/\partial t) \tag{8.1.3}$$

$$\textbf{4.}\quad \vec{\nabla}\times\vec{H}=\vec{J} \tag{8.1.4}$$

where $\vec{D}$ is the electric displacement vector $=\varepsilon\vec{E}$, ε being the permittivity of the medium; ρ the charge density; $\vec{B}$ the magnetic induction vector $=\mu\vec{H}$, μ being the permeability of the medium; $\vec{E}$ the electric field vector; $\vec{H}$ the magnetic field vector and $\vec{J}$ the conduction current density $=\sigma\vec{E}$, σ being the conductivity of the medium.

• Of the above four equations, the first three are general equations valid for both static and dynamic fields. The fourth equation was derived from steady state observations—the *Ampere's law.*

• A critical examination shows that the ***Ampere's law is bound to fail in time-varying fields.*** Taking the divergence of both sides of (8.1.4), we get

$$\vec{\nabla}\cdot(\vec{\nabla}\times\vec{H})=\vec{\nabla}\cdot\vec{J}=0 \tag{8.1.5}$$

since the divergence of a curl always vanishes.

This is true for steady-state phenomenon. It is however incompatible, when current changes with time, with the *equation of continuity* that reflects the principle of conservation of charge :

$$\boxed{\vec{\nabla}\cdot\vec{J}+\frac{\partial\rho}{\partial t}=0.} \tag{8.1.6}$$

8.1.1 Maxwell's contribution

Maxwell could understand the situation well and suggested that the way out of the impasse lies in modifying Ampere's law which suffers from an incomplete definition of the total current density.

By using Gauss' law, the relation (8.1.6) may be written as

$$\vec{\nabla}\cdot\vec{J}=-\frac{\partial\rho}{\partial t}=-\frac{\partial}{\partial t}(\vec{\nabla}\cdot\vec{D})=\vec{\nabla}\cdot\left(-\frac{\partial\vec{D}}{\partial t}\right)$$

$$\Rightarrow\quad \vec{\nabla}\cdot\left(\vec{J}+\frac{\partial\vec{D}}{\partial t}\right)=0 \tag{8.1.7}$$

Maxwell, therefore, replaced $\vec{J}$ of Ampere's law by $\vec{J}+\frac{\partial\vec{D}}{\partial t}$ i.e., $\vec{J}+\varepsilon\frac{\partial\vec{E}}{\partial t}$ and modified the Ampere's law as :

$$\vec{\nabla}\times\vec{H}=\vec{J}+\frac{\partial\vec{D}}{\partial t}=\vec{J}+\varepsilon\frac{\partial\vec{E}}{\partial t} \tag{8.1.8}$$

which then becomes valid for steady-state phenomena and is also compatible with the equation of continuity.

The *modified* Ampere's law has now *two* terms—the term $\vec{J}$ is the usual '*conduction current density*'; the second term $\partial\vec{D}/\partial t$, arising from time-variation of electric displacement, is called '*displacement current density*' by Maxwell. The introduction of this '*displacement current*' *is the chief contribution of* Maxwell to *electromagnetic theory.*

Displacement current—But what exactly is meant by *displacement current?* In fact, displacement current, is *not* current in the sense of the motion of charges.

Consider the simple circuit as illustrated in Fig.8.1 consisting of a charged capacitor C with its plates connected by a conducting wire. The current that flows in the wire is equal to the time rate of change of charge on the plates, i.e.,

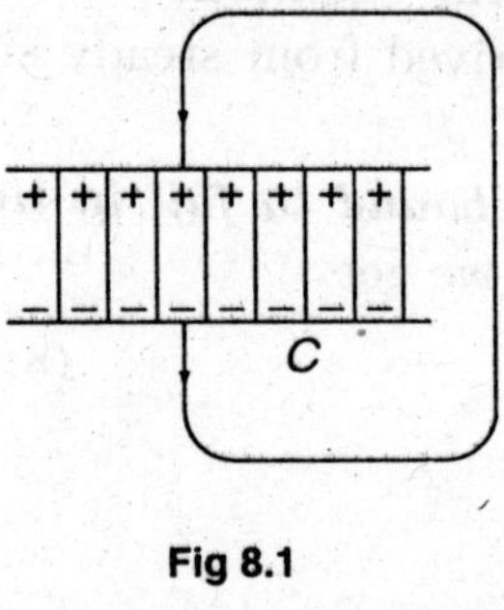

Fig 8.1

$$I=\frac{dQ}{dt} \tag{8.1.9}$$

where Q is the charge on the positive plate of the capacitor and is related to the field E in the capacitor by

$$E=\frac{\sigma}{\varepsilon_0}=\frac{Q}{\varepsilon_0 A} \tag{8.1.10}$$

where σ is the charge density and A is the area of the plate.

$$\therefore \quad I = \frac{d}{dt}(Q) = \frac{d}{dt}(\varepsilon_0 AE), \text{ using (8.1.10)}$$

$$= \varepsilon_0 A\frac{\partial E}{\partial t} = A\frac{\partial D}{\partial t}$$

$$\Rightarrow \quad \frac{\partial D}{\partial t} = \frac{I}{A}, \text{ the } \textit{current density}. \tag{8.1.11}$$

The quantity $\partial D/\partial t$ can thus be *interpreted* as the *density of some current*. This corresponds to the "current" that must flow in the space, including vacuum, between the charged plates of the capacitor when joined by a conducting wire, thus completing the conduction current.

8.1.2 Dimensions of displacement current density

To find the dimensions of displacement current density $\partial \vec{D}/\partial t$, we start as

$$\vec{D} = \varepsilon\vec{E} = \text{permittivity} \times \text{electric field}$$

$$= (\text{Farad/metre}) \times (\text{volt/metre})$$

$$= \text{Coulomb/(metre)}^2$$

$$\therefore \quad \text{Dimensions of } \frac{\partial \vec{D}}{\partial t} = \frac{1}{\text{sec.}} \cdot \frac{\text{Coulomb}}{(\text{metre})^2} = \frac{\textbf{Ampere}}{(\textbf{metre})^2},$$

the same as of current per unit area or the current density.

Relative size of two types of currents—Consider a metallic wire in which there is an electric field given by

$$\vec{E} = \vec{E}_0 e^{-j\omega t}$$

$$\therefore \quad \vec{J} = \sigma\vec{E} = \sigma\vec{E}_0 e^{-j\omega t} \Rightarrow |\vec{J}|^2 = \sigma^2|\vec{E}_0|^2$$

$$\vec{D} = \varepsilon_0\vec{E}_0 e^{-j\omega t} \Rightarrow \frac{\partial \vec{D}}{\partial t} = -j\omega\varepsilon_0\vec{E}_0 e^{-j\omega t}$$

$$\Rightarrow \left|\frac{\partial \vec{D}}{\partial t}\right|^2 = \omega^2\varepsilon_0^2|\vec{E}_0|^2 \tag{8.1.12}$$

$$\therefore \quad \frac{|\vec{J}|}{|\partial\vec{D}/\partial t|} = \frac{\sigma}{\omega\varepsilon_0}$$

If, for instance, the metal be copper, $\sigma = 5.9 \times 10^7$, then

$$\frac{\sigma}{\omega\varepsilon_0} \sim \frac{10^{19}}{\omega}$$

This shows that the ratio is very large for all frequencies and it implies that $|\vec{J}| \gg |\partial\vec{D}/\partial t|$. So, the displacement current is not as significant as the conduction current. Nevertheless, the very idea and introduction of displacement current has been of far-reaching consequences in the development of physics and a rich variety of new phenomena was unravelled by the use of displacement current.

8.2 Maxwell's field equations : Physical significance

The four equations which the field vectors $\vec{E}, \vec{D}, \vec{B}$ and $\vec{H}$ satisfy everywhere are the following.

1. $\vec{\nabla}\cdot\vec{D}=\rho$ or, $\vec{\nabla}\cdot\vec{E}=\rho/\varepsilon$. **(Gauss' law of electrostatics)**
2. $\vec{\nabla}\cdot\vec{B}=0$ or, $\vec{\nabla}\cdot\vec{H}=0$ **(Gauss' law of magnetostatics)**
3. $\vec{\nabla}\times\vec{E}=-\dfrac{\partial\vec{B}}{\partial t}$ or, $\vec{\nabla}\times\vec{E}=-\mu\dfrac{\partial\vec{H}}{\partial t}$ **(Faraday's law of induction)**
4. $\vec{\nabla}\times\vec{H}=\vec{J}+\dfrac{\partial\vec{D}}{\partial t}$ or, $\vec{\nabla}\times\vec{H}=\vec{J}+\varepsilon\dfrac{\partial\vec{E}}{\partial t}$ **(Modified Ampere's law)**

These equations constitute the fundamental equations of electromagnetic field and are known as **Maxwell's field equations**. Of these, note that the first two are *steady state equations* and the last two are *time-varying equations*.

Physical significance—The equation 1 is a restatement of *Gauss' law in electrostatics*, in the differential form.

The equation 2 is a restatement of *Gauss' law in magnetism, along with the fact* that magnetic poles always exist in pairs, i.e., *monopole does not exist*.

The equation 3, the first time-varying equation, is a restatement of *Faraday's law of electromagnetic induction*.

And the equation 4, the second time-varying equation, is a restatement of *Ampere's law modified by the introduction of displacement current* along with the conduction current. It tells us that *a changing electric field produces a changing magnetic field*, just as a changing magnetic field produces a changing electric field. And *an alternate production of electric and magnetic field* gives rise to the propagation of *electromagnetic waves*.

• Integrating equation 1, over a volume V, we have

$$\frac{1}{\varepsilon}\int_V \rho dV=\int_V \vec{\nabla}\cdot\vec{E}\,dV=\oint_S \vec{E}\cdot d\vec{S}, \text{ by divergence theorem.}$$

But, $\oint_S \vec{E}\cdot d\vec{S}$ is the electric flux through a closed surface and $\int_V \rho dV=q$, the total charge within the closed surface. So, $\oint_S \vec{E}\cdot d\vec{S}=q/\varepsilon$, which is the *Gauss' law*.

• Integrating equation 2, over a volume V,

$$0=\int_V \vec{\nabla}\cdot\vec{B}dV=\oint_S \vec{B}\cdot d\vec{S}, \text{ by divergence theorem.}$$

But, $\oint_S \vec{B}\cdot d\vec{S}$ is the magnetic flux through a closed surface and is zero, implying that the flux entering into the volume equals the flux leaving it. So, there is no *source* or *sink* in the volume i.e., *monopole cannot exist*.

• Integrating equation 3, over any closed surface bounding a path C

$$-\oint_S \frac{\partial \vec{B}}{\partial t}\cdot d\vec{S} = \oint_S (\vec{\nabla}\times\vec{E})\cdot d\vec{S} = \oint_C \vec{E}\cdot d\vec{l}, \text{ by Stokes' theorem.}$$

But $\oint_C \vec{E}\cdot d\vec{l}$ is the induced emf in a closed circuit and $\oint_S -\frac{\partial \vec{B}}{\partial t}\cdot d\vec{S}$ is the *negative* rate of change of magnetic flux through the circuit and they are equal. This is *Faraday's law* of electromagnetic induction.

• The basic laws of electricity and magnetism are thus contained succinctly in the four equations of Maxwell, along with an important of Ampere's law.

8.3 Component forms of Maxwell's equations

The Maxwell's four equations may also be given the component forms in terms of the Cartesian coordinates and are shown below.

1.
$$\vec{\nabla}\cdot\vec{E} = \left(\hat{i}\frac{\partial}{\partial x} + \hat{j}\frac{\partial}{\partial y} + \hat{k}\frac{\partial}{\partial z}\right)\cdot\left(\hat{i}E_x + \hat{j}E_y + \hat{k}E_z\right)$$
$$= \frac{\partial E_x}{\partial x} + \frac{\partial E_y}{\partial y} + \frac{\partial E_z}{\partial z}$$
$$\therefore \quad \boxed{\frac{\partial \boldsymbol{E_x}}{\partial \boldsymbol{x}} + \frac{\partial \boldsymbol{E_y}}{\partial \boldsymbol{y}} + \frac{\partial \boldsymbol{E_x}}{\partial \boldsymbol{z}} = \frac{\boldsymbol{\rho}}{\boldsymbol{\varepsilon}}} \tag{8.3.1}$$

2.
$$\vec{\nabla}\cdot\vec{H} = \left(\hat{i}\frac{\partial}{\partial x} + \hat{j}\frac{\partial}{\partial y} + \hat{k}\frac{\partial}{\partial z}\right)\cdot\left(\hat{i}H_x + \hat{j}H_y + \hat{k}H_z\right)$$
$$= \frac{\partial H_x}{\partial x} + \frac{\partial H_y}{\partial y} + \frac{\partial H_z}{\partial z}$$
$$\therefore \quad \boxed{\frac{\partial \boldsymbol{H_x}}{\partial \boldsymbol{x}} + \frac{\partial \boldsymbol{H_y}}{\partial \boldsymbol{y}} + \frac{\partial \boldsymbol{H_z}}{\partial \boldsymbol{z}} = \boldsymbol{0}} \tag{8.3.2}$$

3.
$$\vec{\nabla}\times\vec{E} = \begin{vmatrix} \hat{i} & \hat{j} & \hat{k} \\ \frac{\partial}{\partial x} & \frac{\partial}{\partial y} & \frac{\partial}{\partial z} \\ E_x & E_y & E_z \end{vmatrix}$$
$$= \hat{i}\left(\frac{\partial E_z}{\partial y} - \frac{\partial E_y}{\partial z}\right) + \hat{j}\left(\frac{\partial E_x}{\partial z} - \frac{\partial E_z}{\partial x}\right) + \hat{k}\left(\frac{\partial E_y}{\partial x} - \frac{\partial E_x}{\partial y}\right)$$

Also,
$$-\mu\frac{\partial \hat{H}}{\partial t} = -\mu\left(\hat{i}\frac{\partial H_x}{\partial t} + \hat{j}\frac{\partial H_y}{\partial t} + \hat{k}\frac{\partial H_z}{\partial t}\right)$$

Equating the components along the corresponding co-ordinates, we get

$$\left.\begin{aligned} -\mu\frac{\partial H_x}{\partial t} &= \frac{\partial E_z}{\partial y} - \frac{\partial E_y}{\partial z} \\ -\mu\frac{\partial H_y}{\partial t} &= \frac{\partial E_x}{\partial z} - \frac{\partial E_z}{\partial x} \\ -\mu\frac{\partial H_z}{\partial t} &= \frac{\partial E_y}{\partial x} - \frac{\partial E_x}{\partial y} \end{aligned}\right\} \tag{8.3.3}$$

4. $$\vec{\nabla} \times \vec{H} = \begin{vmatrix} i & j & k \\ \frac{\partial}{\partial x} & \frac{\partial}{\partial y} & \frac{\partial}{\partial z} \\ H_x & H_y & H_z \end{vmatrix}$$

$$= \hat{i}\left(\frac{\partial H_z}{\partial y} - \frac{\partial H_y}{\partial z}\right) + \hat{j}\left(\frac{\partial H_x}{\partial z} - \frac{\partial H_z}{\partial x}\right) + \hat{k}\left(\frac{\partial H_y}{\partial x} - \frac{\partial H_x}{\partial y}\right)$$

Also, $$\varepsilon\frac{\partial \vec{E}}{\partial t} = \varepsilon\left(\hat{i}\frac{\partial E_x}{\partial t} + \hat{j}\frac{\partial E_y}{\partial t} + \hat{k}\frac{\partial E_z}{\partial t}\right)$$

For a non-conducting medium, however, we have

$$\vec{\nabla} \times \vec{H} = \varepsilon\frac{\partial \vec{E}}{\partial t} \quad (\because\ J = 0)$$

$\therefore$ Equating the components along the corresponding co-ordinate axes,

$$\left.\begin{aligned} \varepsilon\frac{\partial E_x}{\partial t} &= \frac{\partial H_z}{\partial y} - \frac{\partial H_y}{\partial z} \\ \varepsilon\frac{\partial E_y}{\partial t} &= \frac{\partial H_x}{\partial z} - \frac{\partial H_z}{\partial x} \\ \varepsilon\frac{\partial E_z}{\partial t} &= \frac{\partial H_y}{\partial x} - \frac{\partial H_x}{\partial y} \end{aligned}\right\} \tag{8.3.4}$$

8.4 Decay of free charge : Relaxation time

An important deduction that follows immediately from Maxwell's equation is the decay of free charges.

From the fourth equation of Maxwell, we have

$$\vec{\nabla} \times \vec{H} = \vec{J} + \frac{\partial \vec{D}}{\partial t} = \sigma\vec{E} + \varepsilon\frac{\partial \vec{E}}{\partial t}$$

$$\therefore\ \vec{\nabla}\cdot(\vec{\nabla} \times \vec{H}) = \sigma(\vec{\nabla}\cdot\vec{E}) + \varepsilon\frac{\partial}{\partial t}(\vec{\nabla}\cdot\vec{E}),$$ assuming σ and ε as constants.

or, $$0 = \sigma\frac{\rho}{\varepsilon} + \frac{\partial \rho}{\partial t} \quad (\because\ \vec{\nabla}\cdot\vec{E} = \rho/\varepsilon)$$

Integrating, we obtain

$$\boxed{\rho = \rho_0 e^{-t\sigma/\varepsilon} = \rho_0 e^{-t/\tau}} \tag{8.4.1}$$

where $\tau = \varepsilon/\sigma$ and is called the **relaxation time**.

Eq. (8.4.1) shows that *free charge decays exponentially* at a rate independent of any other electromagnetic disturbance.

8.5 Wave equation in a dielectric (non-conducting) medium

We shall now see that as a result of Maxwell's set of equations, the electric and magnetic fields can travel through space in the form of waves, called *electromagnetic waves* (e.m. waves). We shall derive hereunder the relevant wave equation in a dielectric (i.e., non-conducting) medium.

Let μ be the permeability, σ the conductivity, ε the permittivity of a *dielectric* medium, assumed linear, i.e., ε, μ and σ are constants. As it is non-conducting, the conductivity $\sigma = 0$. Let it also be free from any charge source implying that charge density $\rho = 0$. Thus, in such a medium, Maxwell's equations for the electric field vector $\vec{E}$ and magnetic field vector $\vec{H}$ are :

$$\vec{\nabla}\cdot\vec{E} = 0 \tag{8.5.1}$$

$$\vec{\nabla}\cdot\vec{H} = 0 \tag{8.5.2}$$

$$\vec{\nabla}\times\vec{E} = -\mu\frac{\partial\vec{H}}{\partial t} \tag{8.5.3}$$

$$\vec{\nabla}\times\vec{H} = \varepsilon\frac{\partial\vec{E}}{\partial t} \tag{8.5.4}$$

since in a dielectric medium the current density $\vec{J} = \sigma\vec{E} = 0$.

Wave equation for $\vec{E}$: From (8.5.3), taking curl, we have,

$$\vec{\nabla}\times(\vec{\nabla}\times\vec{E}) = \vec{\nabla}\times\left(-\mu\frac{\partial\vec{H}}{\partial t}\right) = -\mu\left(\vec{\nabla}\times\frac{\partial\vec{H}}{\partial t}\right) = -\mu\frac{\partial}{\partial t}(\vec{\nabla}\times\vec{H})$$

$$\text{or,}\quad \vec{\nabla}(\vec{\nabla}\cdot\vec{E}) - \vec{E}(\vec{\nabla}\cdot\vec{\nabla}) = -\mu\frac{\partial}{\partial t}\left(\varepsilon\frac{\partial\vec{E}}{\partial t}\right),\quad \text{using (8.5.4)}$$

$$\text{or,}\quad -\nabla^2\vec{E} = -\mu\varepsilon\frac{\partial^2\vec{E}}{\partial t^2}\quad (\because\ \vec{\nabla}\cdot\vec{E} = 0\ \text{ by (8.5.1)})$$

$$\therefore\quad \boxed{\nabla^2\vec{E} = \mu\varepsilon\frac{\partial^2 E}{\partial t^2}} \tag{8.5.5}$$

Wave equation for $\vec{H}$: From (8.5.4), taking curl, we have,

$$\vec{\nabla}\times(\vec{\nabla}\times\vec{H})=\vec{\nabla}\times\left(\varepsilon\frac{\partial\vec{E}}{\partial t}\right)=\varepsilon\left(\vec{\nabla}\times\frac{\partial\vec{E}}{\partial t}\right)=\varepsilon\frac{\partial}{\partial t}(\vec{\nabla}\times\vec{E})$$

$$\text{or,}\quad \vec{\nabla}(\vec{\nabla}\cdot\vec{H})-\vec{H}(\vec{\nabla}\cdot\vec{\nabla})=\varepsilon\frac{\partial}{\partial t}\left(-\mu\frac{\partial\vec{H}}{\partial t}\right),\quad \text{using (8.5.3)}$$

$$\text{or,}\quad -\nabla^2\vec{H}=-\mu\varepsilon\frac{\partial^2\vec{H}}{\partial t^2},\quad (\because\ \vec{\nabla}\cdot\vec{H}=0\ \text{ by (8.5.2)})$$

$$\therefore\quad \boxed{\nabla^2\vec{H}=\mu\varepsilon\frac{\partial^2 H}{\partial t^2}} \tag{8.5.6}$$

Eq. (8.5.5) and (8.5.6) are respectively the *general wave equations* for the electric field $\vec{E}$ and the magnetic field $\vec{H}$ in a dielectric medium.

• Note that $\vec{E}$ and $\vec{H}$ satisfy the same wave equation and we can readily identify $v=\sqrt{1/\mu\varepsilon}$ as the speed of the wave.

8.6 Wave equation in a conducting medium

Let μ be the permeability, ε the permittivity and σ the conductivity of the medium which is free from any charge source so that the charge density $\rho=0$.

For such a medium the Maxwell's equations take the form :

$$\vec{\nabla}\cdot\vec{E}=0 \tag{8.6.1}$$

$$\vec{\nabla}\cdot\vec{H}=0 \tag{8.6.2}$$

$$\vec{\nabla}\times\vec{E}=-\mu\frac{\partial\vec{H}}{\partial t} \tag{8.6.3}$$

$$\vec{\nabla}\times\vec{H}=\vec{J}+\varepsilon\frac{\partial\vec{E}}{\partial t}=\sigma\vec{E}+\varepsilon\frac{\partial\vec{E}}{\partial t} \tag{8.6.4}$$

since the current density $\vec{J}=\sigma\vec{E}$, $\vec{E}$ and $\vec{H}$ being the electric and magnetic field vectors.

Wave equation for $\vec{E}$: From (8.6.3), taking curl, we obtain

$$\vec{\nabla}\times(\vec{\nabla}\times\vec{E})=\vec{\nabla}\times\left(-\mu\frac{\partial\vec{H}}{\partial t}\right)=-\mu\left(\vec{\nabla}\times\frac{\partial\vec{H}}{\partial t}\right)=-\mu\frac{\partial}{\partial t}(\vec{\nabla}\times\vec{H})$$

$$\text{or,}\quad \vec{\nabla}(\vec{\nabla}\cdot\vec{E})-\vec{E}(\vec{\nabla}\cdot\vec{\nabla})=-\mu\frac{\partial}{\partial t}\left(\sigma\vec{E}+\varepsilon\frac{\partial\vec{E}}{\partial t}\right),\quad \text{using (8.6.4).}$$

$$\text{or,}\quad -\nabla^2\vec{E}=-\mu\sigma\frac{\partial\vec{E}}{\partial t}-\mu\varepsilon\frac{\partial^2\vec{E}}{\partial t^2},\quad \text{using (8.6.1).}$$

$$\therefore\quad \boxed{\nabla^2\vec{E}=\mu\sigma\frac{\partial\vec{E}}{\partial t}+\mu\varepsilon\frac{\partial^2\vec{E}}{\partial t^2}} \tag{8.6.5}$$

Wave equation for $\vec{H}$: From (8.6.4), taking curl we have

$$\vec{\nabla} \times (\vec{\nabla} \times \vec{H}) = \vec{\nabla} \times \left(\sigma \vec{E} + \varepsilon \frac{\partial \vec{E}}{\partial t}\right) = \sigma(\vec{\nabla} \times \vec{E}) + \varepsilon \frac{\partial}{\partial t}(\vec{\nabla} \times \vec{E})$$

$$\text{or,} \quad \vec{\nabla}(\vec{\nabla} \cdot \vec{H}) - \vec{H}(\vec{\nabla} \cdot \vec{\nabla}) = -\mu\sigma \frac{\partial \vec{H}}{\partial t} - \mu\varepsilon \frac{\partial^2 \vec{H}}{\partial t^2}, \quad \text{using (8.6.3).}$$

$$\text{or,} \quad -\nabla^2 \vec{H} = -\mu\sigma \frac{\partial \vec{H}}{\partial t} - \mu\varepsilon \frac{\partial^2 H}{\partial t^2} \quad (\because \vec{\nabla} \cdot \vec{H} = 0, \text{ by (8.6.2)})$$

$$\therefore \quad \boxed{\nabla^2 \vec{H} = \mu\sigma \frac{\partial \vec{H}}{\partial t} + \mu\varepsilon \frac{\partial^2 \vec{H}}{\partial t^2}} \tag{8.6.6}$$

• Note that here also $\vec{E}$ and $\vec{H}$ satisfy the same wave equation.

• If the medium is *non-conducting*, i.e., a *dielectric*, $\sigma = 0$ and (8.6.5) and (8.6.6) reduce respectively to the following form

$$\nabla^2 \vec{E} = \mu\varepsilon \frac{\partial^2 \vec{E}}{\partial t^2} \tag{8.6.7}$$

$$\nabla^2 \vec{H} = \mu\varepsilon \frac{\partial^2 \vec{H}}{\partial t^2} \tag{8.6.8}$$

in agreement with what we obtained earlier in (8.5.5) and (8.5.6).

Dissipative term—Comparing (8.6.5) and (8.6.7), the term $\mu\sigma(\partial\vec{E}/\partial t)$ is the *dissipative term* that allows current to flow. Similarly, by comparing (8.6.6) and (8.6.8), we see that the *dissipative term* is $\mu\sigma(\partial\vec{H}/\partial t)$.

8.7 Velocity (or speed) of electromagnetic waves

We have for a *dielectric medium*, the following wave equations :

$$\nabla^2 \vec{E} = \mu\varepsilon \frac{\partial^2 \vec{E}}{\partial t^2} \tag{8.7.1}$$

$$\nabla^2 \vec{H} = \mu\varepsilon \frac{\partial^2 \vec{H}}{\partial t^2} \tag{8.7.2}$$

for the electric field $\vec{E}$ and the magnetic field $\vec{H}$ respectively.

They are of the form of the familiar wave equation :

$$\nabla^2 \phi = \frac{1}{v^2} \frac{\partial^2 \phi}{\partial t^2} \tag{8.7.3}$$

where v is the velocity (or speed) of propagation of the wave.

Comparing (8.7.1) and (8.7.2) with (8.7.3), it follows that *any time variation in electric and magnetic fields propagates with the same velocity*

$$\boxed{v = \frac{1}{\sqrt{\mu\varepsilon}}} \tag{8.7.4}$$

• One should bear in mind that while equation (8.7.3) is a *scalar* equation, each of (8.7.1) and (8.7.2) is a *vector* equation. The implication is straight-forward. Each component of $\vec{E}$ and $\vec{H}$ satisfies the scalar equation (8.7.3).

Speed of e.m. waves in vacuo — According to Maxwell's equations, therefore, the empty space supports the propagation of electromagnetic waves with a speed

$$c = \frac{1}{\sqrt{\mu_0\varepsilon_0}} = \frac{1}{\sqrt{4\pi \times 10^{-7} \times 8.8547 \times 10^{-12}}}$$
$$= \mathbf{2.99784 \times 10^8\ m/s.}$$

where $\mu_0 = 4\pi \times 10^{-7}$ and $\varepsilon_0 = 8.8547 \times 10^{-12}$ in SI units.

This is precisely the *speed of light* c in vacuum! Maxwell's equations thus vindicate that *light is an electromagnetic wave.* The *role played* by Maxwell's term $\mu\varepsilon(\partial\vec{E}/\partial t)$, in modified Ampere's law, in this regard is *crucial.* For without it, the wave equation would *not* have emerged at all.

Refractive index—We have seen that the electromagnetic waves propagate through a homogeneous (linear) medium with a speed v given by

$$v = \frac{1}{\sqrt{\mu\varepsilon}} \tag{8.7.5}$$

If the *medium is transparent,* μ is very nearly 1 (i.e., $\mu \simeq 1$) and ε greater than 1 ($\varepsilon > 1$). This implies that light should travel slower through matter, as is well-known from optics :

$$v = \frac{c}{n} \tag{8.7.6}$$

where c is the velocity of light in vacuo and n the refractive index of the material. So from (8.7.5) and (8.7.6) the index n is related to electric and magnetic properties of the material

$$\therefore \quad n = \frac{c}{v} = \frac{1/\sqrt{\mu_0\varepsilon_0}}{1/\sqrt{\mu\varepsilon}} = \sqrt{\frac{\mu\varepsilon}{\mu_0\varepsilon_0}} = \sqrt{\frac{\varepsilon}{\varepsilon_0}} \tag{8.7.7}$$

since in a non-magnetic dielectric, $\mu = \mu_0$.

$$\therefore \quad \boxed{n = \sqrt{\frac{\varepsilon}{\varepsilon_0}} = \sqrt{K}} \tag{8.7.8}$$

where the relative permittivity or $\varepsilon/\varepsilon_0 = K$, the dielectric constant of the medium.

• We know, again, that n varies with the wavelength λ or the *colour* of the incident light and this fact accounts for the **dispersion** of light. The electromagnetic explanation for the dispersion is also forthcoming, since the electric polarizability of a molecule, and also therefore the permittivity ε, depends on the rapidity of the oscillation of electric field vector $\vec{E}$.

• Eq. (8.7.8), i.e., $n^2 = K$ has been verified experimentally in case of non-polar molecules such as air, H_2, C_6H_6 etc.

8.8 Solution of wave equation : Dielectric medium

Consider the wave equation in a homogeneous, isotropic, linear dielectric medium.

The simplest type of wave which would be a solution of (8.7.1) or (8.7.2) is a **plane wave** — a good approximation to actual waves in many situations, and many elementary but important characteristics of e.m. waves can be obtained from it. We shall therefore consider plane wave solutions of (8.7.1) and (8.7.2).

In a **plane wave**, the wave amplitudes — the field vector components — are constants at a given time over all points in a plane perpendicular to the direction of propagation. This plane is called the *wavefront* that advances with a velocity v (Fig. 8.2) in a direction normal to it.

Plane wave solution — For the propagation of wave in any direction, we take the plane wave fields to be of the form

$$\vec{E} = \vec{E}_0 e^{j(\vec{k}\cdot\vec{r}-\omega t)} \tag{8.8.1}$$

$$\vec{H} = \vec{H}_0 e^{j(\vec{k}\cdot\vec{r}-\omega t)} \tag{8.8.2}$$

where $\vec{E}_0$ and $\vec{H}_0$ ae complex amplitudes, fixed in space and time and $\vec{k}$ is the propagation vector determining the direction of wave propagation.

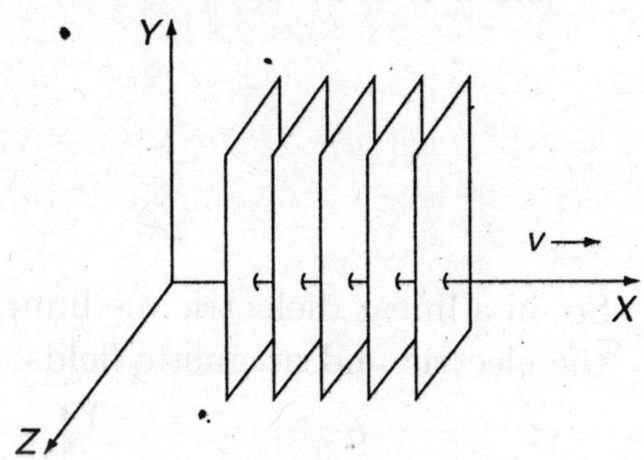

Fig 8.2

Substituting (8.8.1) in Maxwell's equations, $\vec{\nabla}\cdot\vec{E} = 0$ and $\vec{\nabla}\cdot\vec{H} = 0$, we obtain

$$\vec{k}\cdot\vec{E} = 0 \tag{8.8.3}$$

$$\text{and} \quad \vec{k}\cdot\vec{H} = 0 \tag{8.8.4}$$

It thus follows that both $\vec{E}$ and $\vec{H}$ are perpendicular to the direction of propagation vector $\vec{k}$. So, **an electromagnetic wave is a transverse wave.**

On substitution of (8.8.1) in Maxwell's equations

$$\vec{\nabla}\times\vec{E} = -\mu\frac{\partial\vec{H}}{\partial t}$$

$$\text{and} \quad \vec{\nabla}\times\vec{H} = \varepsilon\frac{\partial\vec{E}}{\partial t}$$

$$j\vec{k} \times \vec{E} = -\mu(-j\omega\vec{H}) \Rightarrow \vec{k} \times \vec{E} = \mu\omega\vec{H} \quad (8.8.5)$$

$$j\vec{k} \times \vec{H} = \varepsilon(-j\omega\vec{E}) \quad \Rightarrow \vec{k} \times \vec{H} = -\varepsilon\omega\vec{E} \quad (8.8.6)$$

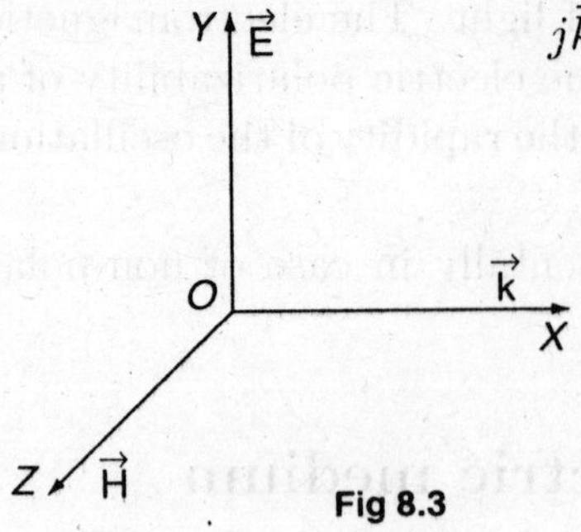

Fig 8.3

Eq. (8.8.5) shows that $\vec{H}$ is normal to both $\vec{k}$ and $\vec{E}$, while (8.8.6) implies that $\vec{E}$ is normal to both $\vec{k}$ and $\vec{H}$. It therefore follows that *the field vectors $\vec{E}$ and $\vec{H}$ are mutually perpendicular and they are also perpendicular to the direction of propagation $\vec{k}$. So $\vec{E}, \vec{H}$ and $\vec{k}$ constute a right handed orthogonal system* (Fig. 8.3).

• In taking the plane wave solutions (8.8.1) and (8.8.2) we assumed $\vec{E}$ and $\vec{H}$ to be in phase. *But are they really in phase?* The answer follows.

If we substitute (8.8.1) in the wave equations (8.5.5) and (8.5.6)

$$k^2 = \varepsilon\mu\omega^2 \quad (8.8.7)$$

Thus, for e.m. waves in a linear dielectric medium, k is real and hence (8.8.5) implies that *$\vec{E}$ and $\vec{H}$ are in phase.*

Since $\vec{E}$ is normal to $\vec{k}$, we have from (8.8.7), in terms of magnitudes,

$$kE = \mu\omega H$$

$$\text{or,} \quad \sqrt{\varepsilon}E = \sqrt{\mu}H \quad (8.8.8)$$

$$\Rightarrow \quad \tfrac{1}{2}\varepsilon E^2 = \tfrac{1}{2}\mu H^2$$

So, in a linear dielectric medium, the electromagnetic energy is equally shared between the electric and magnetic fields.

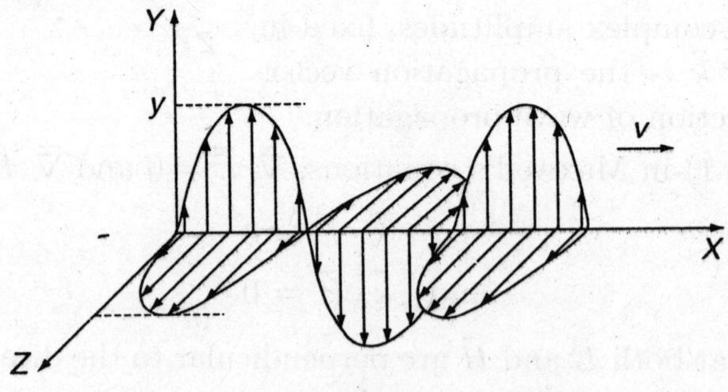

Fig 8.4

Fig. 8.4 shows the diagram of a *TEM* (transverse electro-magnetic) wave propagating along $+x$ direction at any time t. $\vec{E}$ at different x-values are indicated by arrowheads, parallel to y-axis, in xy-plane. $\vec{H}$ are also shown at those x-values by arrowheads, but parallel to z-axis, and lyng in zx-plane.

• The above analysis is also applicable to free space. Only ε and μ are to be replaced by ε_0 and μ_0 rrespectively.

• The ratio of the amplitudes of $\vec{E}$ and $\vec{H}$ at any point and at any instant, is

$$\frac{E_0}{H_0} = \sqrt{\frac{\mu}{\varepsilon}} = Z_0, \text{ say} \tag{8.8.9}$$

using the equation (8.8.8).

Z_0 has the dimensions of impedence and is called the *intrinsic impedance* of the medium concerned.

In free space, $Z_0 = \sqrt{\mu_0/\varepsilon_0} = \mathbf{377\,\Omega}$, putting the values of μ_0 and ε_0. We shall return to Z once again later.

8.9 Poynting vector and Poynting theorem

Poynting vector—Poynting vector is *defined as the cross-product of the electric field vector $\vec{E}$ and the magnetic field vector $\vec{H}$* and is symbolised by $\vec{S}$.

$$\therefore \quad \textbf{Poynting vector,} \quad \boxed{\vec{S} = \vec{E} \times \vec{H}}$$

For an electromagnetic wave, the direction of propagation of the wave is always at right angles to both $\vec{E}$ and $\vec{H}$; so the direction of Poynting vector is *always in the direction of propagation of the wave.*

But *what does this Poynting vector signify or represent?* It measures the electric power (or electric energy per unit time) per unit area for,

$$\vec{E} \times \vec{H} = \frac{\text{volt}}{\text{length}} \times \frac{\text{current}}{\text{length}} = \frac{\text{power}}{\text{area}}$$

The SI unit of Poynting vector is Watt/sq. metre (Wm^{-2}).

Thus, the **significance** of the *Poynting vector* is that it *measures the flow of electromagnetic energy per unit time per unit area placed normally to the direction of propagation of the electromagnetic wave.*

Poynting theorem—As the electromagnetic wave propagates from one point to another, there occurs a transit of electromagnetic energy. It is observed that in the process, the *law of conservation of energy* holds good and is referred to as **Poynting's theorem**.

From Maxwell's field equations, we have

$$\vec{\nabla} \times \vec{E} = -\mu \frac{\partial \vec{H}}{\partial t}; \quad \vec{\nabla} \times \vec{H} = \vec{J} + \varepsilon \frac{\partial \vec{E}}{\partial t} \tag{8.9.1}$$

$$\therefore \quad \vec{H} \cdot (\vec{\nabla} \times \vec{E}) = -\mu \frac{\partial \vec{H}}{\partial t} \cdot \vec{H} = -\tfrac{1}{2}\mu \frac{\partial H^2}{\partial t} \tag{8.9.2}$$

$$\text{and} \quad \vec{E} \cdot (\vec{\nabla} \times \vec{H}) = \vec{E} \cdot \vec{J} + \vec{E} \cdot \varepsilon \frac{\partial \vec{E}}{\partial t} = \vec{E} \cdot \vec{J} + \tfrac{1}{2}\varepsilon \frac{\partial E^2}{\partial t} \tag{8.9.3}$$

Subtracting (8.9.2) from (8.9.3), we obtain

$$\vec{E}\cdot(\vec{\nabla}\times\vec{H})-\vec{H}\cdot(\vec{\nabla}\times\vec{E})=\vec{E}\cdot\vec{J}+\tfrac{1}{2}\varepsilon\frac{\partial E^2}{\partial t}+\tfrac{1}{2}\mu\frac{\partial H^2}{\partial t}$$

$$=\vec{E}\cdot\vec{J}+\tfrac{1}{2}\frac{\partial}{\partial t}(\varepsilon E^2+\mu H^2) \qquad (8.9.4)$$

Using on L.H.S. of (8.9.4) the vector identity

$$\vec{A}\cdot(\vec{B}\times\vec{C})-\vec{C}\cdot(\vec{B}\times\vec{A})=\vec{B}\cdot(\vec{C}\times\vec{A})=-\vec{B}\cdot(\vec{A}\times\vec{C}),$$

we obain : $$-\vec{\nabla}\cdot(\vec{E}\times\vec{H})=\vec{E}\cdot\vec{J}+\frac{\partial}{\partial t}\left(\tfrac{1}{2}\varepsilon E^2+\tfrac{1}{2}\mu H^2\right)$$

$$\Rightarrow \quad \vec{\nabla}\cdot\vec{S}+\frac{\partial}{\partial t}\left(\tfrac{1}{2}\varepsilon E^2+\tfrac{1}{2}\mu H^2\right)=-\vec{E}\cdot\vec{J} \qquad (8.9.5)$$

Integrating (8.9.5) over a volume V bounded by a closed surface A,

$$\int_V \vec{\nabla}\cdot\vec{S}\,dV+\frac{\partial}{\partial t}\int_V\left(\tfrac{1}{2}\varepsilon E^2+\tfrac{1}{2}\mu H^2\right)dV=-\int_V \vec{E}\cdot\vec{J}\,dV$$

or, $$\int_A \vec{S}\cdot d\vec{A}+\frac{\partial}{\partial t}\int_V\left(\tfrac{1}{2}\varepsilon E^2+\tfrac{1}{2}\mu H^2\right)dV=-\int_V \vec{E}\cdot\vec{J}\,dV \qquad (8.9.6)$$

applying Gauss's divergence theorem, $d\vec{A}$ being an elemental area of the surface A.

Eq. (8.9.6) is the *statement of the well-known* **Poynting's theorem.**

Let us now explore what (8.9.6) does **signify.**

The **first term** $\int_A \vec{S}\cdot d\vec{A}$ is the *amount of energy flowing out per unit time*, i.e., rate of energy-flow out of the bounding surface A.

The **second term** $\frac{\partial}{\partial t}\int_V\left(\tfrac{1}{2}\varepsilon E^2+\tfrac{1}{2}\mu H^2\right)dV$ is the *time-rate of change of the total energy* (electric + magnetic) of the electromagnetic field stored in the volume V.

The **third term** $-\int_V \vec{E}\cdot\vec{J}\,dV$ is the *work done by the field* on the source with the energy dissipated in the form of *Joule heat* in the volume V.

Eq. (8.9.6) thus gives none else than the **law of conservation of energy.**

- In free-space, $\vec{J}=0$ and (8.9.6) then reduces to

$$\int_A \vec{S}\cdot d\vec{A}=-\frac{\partial}{\partial t}\int_V\left(\tfrac{1}{2}\varepsilon E^2+\tfrac{1}{2}\mu H^2\right)dV$$

which states that *the flow of energy per unit time across the boundary of a closed surface is equal to the time-rate of change of total energy of the electromagnetic field in the volume enclosed by the surface.*

8.10 Average values of Poynting vector

1. For sinusoidally time-varying fields—Let the electric and magnetic field vectors be represented as

$$\vec{E} = \vec{E}_0 \sin \omega t\ ; \quad \vec{H} = \vec{H}_0 \sin \omega t$$

The **average Poynting vector** is *defined as its average value over one complete period T of the electro-magnetic wave.*

$$\begin{aligned}
\therefore \quad \langle \vec{S} \rangle &= \frac{1}{T}\int_0^T (\vec{E} \times \vec{H})dt \\
&= \frac{1}{T}(\vec{E}_0 \times \vec{H}_0)\int_0^T \sin^2 \omega t\, dt \\
&= \frac{\omega}{2\pi}(\vec{E}_0 \times \vec{H}_0)\int_0^{2\pi/\omega} \frac{1-\cos 2\omega t}{2} dt \\
&= \frac{\omega}{4\pi}(\vec{E}_0 \times \vec{H}_0)\left[t - \frac{\sin 2\omega t}{2\omega}\right]_0^{2\pi/\omega} \\
&= \frac{\omega}{4\pi}(\vec{E}_0 \times \vec{H}_0)\frac{2\pi}{\omega} = \tfrac{1}{2}(\vec{E}_0 \times \vec{H}_0) = \frac{\vec{E}_0}{\sqrt{2}} \times \frac{\vec{H}_0}{\sqrt{2}} \\
&= \vec{E}_{\text{rms}} \times \vec{H}_{\text{rms}} \qquad (8.10.1)
\end{aligned}$$

2. Time-varying fields : varying as $e^{-j\omega t}$ — Let the physical fields be given by the real parts of their complex representation, i.e.,

$$\text{Re}\,\vec{E} = \text{Re}\,\vec{E}_0 e^{-j\omega t}; \quad \text{Re}\,\vec{H} = \text{Re}\,\vec{H}_0 e^{-j\omega t}$$

$$\therefore \quad \text{Re}\,\vec{E} = \tfrac{1}{2}(\vec{E} + \vec{E}^*); \quad \text{Re}\,\vec{H} = \tfrac{1}{2}(\vec{H} + \vec{H}^*)$$

$$\begin{aligned}
\therefore \quad \vec{S} &= \text{Re}\,\vec{E} \times \text{Re}\,\vec{H} \\
&= \tfrac{1}{2}\left[\vec{E} \times \vec{H} + \vec{E}^* \times \vec{H}^* + \vec{E} \times \vec{H}^* + \vec{E}^* \times \vec{H}\right] \qquad (8.10.2)
\end{aligned}$$

Since $\int_0^T e^{\pm 2j\omega t} dt = 0$, the time-averages of the first two terms on r.h.s of (8.10.2) vanish over a complete period $T = 2\pi/\omega$.

$$\langle \vec{S} \rangle = \tfrac{1}{4}(\vec{E} \times \vec{H}^* + \vec{E}^* \times \vec{H}) = \tfrac{1}{2}\,\textbf{Re}\,(\vec{E} \times \vec{H}^*) \qquad (8.10.3)$$

• The magnitude of $\langle \vec{S} \rangle$, i.e., $|\langle \vec{S} \rangle|$ is called the ***intensity of radiation,*** I. Then, the ***radiation pressure*** p is given by

$$\boxed{p = \frac{I}{c} = \frac{|\langle \vec{S} \rangle|}{c}} \qquad (8.10.4)$$

where c is the velocity of electromagnetic waves in vacuo.

8.11 Characteristics of e.m. waves in vacuo

The eletromagnetic waves in free space show a number of characteristics, as observed by our analysis so far.

1. The e.m. waves are *tansverse in nature.* The electric field $\vec{E}$, the magnetic field $\vec{H}$ and the direction of propagation of the wave are at right angles to each other.

2. The wave velocity c of e.m. waves in free space is given by $c = \sqrt{1/\varepsilon_0\mu_0}$, where ε_0 = permittivity, and μ_0 = permeability of free space.

3. The rate of flow of electromagnetic energy per unit area in the direction of wave propagation is given by *Poynting vector,* $\vec{S} = \vec{E} \times \vec{H}$ or $(\vec{E} \times \vec{B})/\mu$.

4. The field vector $\vec{E}$ and $\vec{H}$ are *in phase* implying that when $\vec{E}$ attains the maximum value $\vec{E}_0$, $\vec{H}$ also attains its maximum value $\vec{H}_0$.

5. The electromagnetic energy is *equally shared* between the electric and the magnetic fields : $\frac{1}{2}\varepsilon_0 E^2 = \frac{1}{2}\mu_0 H^2$ so that the *total energy density* U of e. m. waves is

$$U = \tfrac{1}{2}\varepsilon_0 E^2 + \tfrac{1}{2}\mu_0 H^2 = \varepsilon_0 E^2.$$

• In dielectric, the *energy density* of e.m. waves is given by

$$U = U_e + U_m = \tfrac{1}{2}\varepsilon E^2 + \tfrac{1}{2}\mu H^2 = \tfrac{1}{2}\varepsilon E^2 + \tfrac{1}{2}\varepsilon E^2 = \varepsilon E^2$$
$$= \varepsilon_r(\varepsilon_0 E^2) = \varepsilon_r \times \textit{energy density in free space.}$$

where ε_r = relative permittivity = $\varepsilon/\varepsilon_0$.

8.12 Characterization of light wave by electric vector

A light wave is an electromagnetic wave. The force exerted by its electric field vector $\vec{E}$ on a particle of charge q is $q\vec{E}$, i.e., of magnitude qE. If however the particle moves with a velocity $\vec{v}$, the force exerted by the magnetic vector $\vec{B}$ of e.m. wave is $q(\vec{v} \times \vec{B})$, i.e., having a maximum value qvB.

Now, for an electromagnetic wave, we have

$$\left|\frac{\vec{E}}{\vec{H}}\right| = \sqrt{\frac{\mu}{\varepsilon}} = \mu\sqrt{\frac{1}{\mu\varepsilon}} = \mu c \quad \therefore \quad \frac{|\vec{E}|}{|\vec{B}|} = c \;\Rightarrow\; B = \frac{E}{c}$$

$$\therefore \quad qvB = \frac{v}{c}qE \tag{8.12.1}$$

which shows that *magnetic force is (v/c) times the electric force.* Since $v \ll c$, this implies that electric force $\gg$ magnetc force. And that is why *a light wave is characterised by its electric vector* $\vec{E}$ which is also called the **light vector**. Every optical phenomenon—from the sensation of vision to photographic action — is due to the electric vector.

• To form an idea about the relative values of electric and magnetic forces, let us consider the light intensity of sunlight on the earth's surface. It is about 1 kW/m^2, giving 1 kV/m as the amplitude of electric vector $\vec{E}$, which can be easily measured. The corresponding amplitude of magnetic vector $\vec{H}$ is about 3×10^{-6} T, too weak to be accurately measured.

8.13 Electromagnetic waves in a conducting medium

Consider a linear, homogeneous, isotropic conducting medium of conductivity σ, having no free charge.

For such a medium, the wave equation, as already obtained, are

$$\nabla^2 \vec{E} - \sigma\mu \frac{\partial \vec{E}}{\partial t} - \varepsilon\mu \frac{\partial^2 \vec{E}}{\partial t^2} = 0 \tag{8.13.1}$$

$$\nabla^2 H - \sigma\mu \frac{\partial \vec{H}}{\partial t} - \varepsilon\mu \frac{\partial^2 \vec{H}}{\partial t^2} = 0 \tag{8.13.2}$$

Both the equations contain damping terms proportional to σ, the conductivity of the medium.

Plane wave solutions — The above equations admit plane wave solutions. Assuming the field vectors to vary harmonically with time,

$$\vec{E}(\vec{r}, t) = \vec{E}_0 e^{j(\vec{k}\cdot\vec{r} - \omega t)} \tag{8.13.3}$$

$$\vec{H}(\vec{r}, t) = \vec{H}_0 e^{j(\vec{k}\cdot\vec{r} - \omega t)} \tag{8.13.4}$$

where $\vec{E}_0$, $\vec{H}_0$ are complex wave amplitudes and $\vec{k} = k\hat{n}$, the propagation vector.

Substituting (8.13.3) in (8.13.1), we obtain

$$(-k^2 + \varepsilon\mu\omega^2 + j\sigma\mu\omega)\vec{E} = 0$$

$$\Rightarrow \quad k^2 = \varepsilon\mu\omega^2 + j\sigma\mu\omega \tag{8.13.5}$$

In *a conducting medium*, $\vec{k}$ is thus complex and let us take

$$k = \alpha + j\beta,$$

where the quantities α and β are both real.

$$\therefore \quad k^2 = \alpha^2 - \beta^2 + j2\alpha\beta \tag{8.13.6}$$

Comparing (8.13.5) and (8.13.6), we have

$$\alpha^2 - \beta^2 = \varepsilon\mu\omega^2,$$

$$\text{and} \qquad 2\alpha\beta = \sigma\mu\omega$$

Solving these two simultaneous equations, we obtain

$$\alpha = \omega\sqrt{\frac{\varepsilon\mu}{2}}\left[\sqrt{1+\left(\frac{\sigma}{\varepsilon\omega}\right)^2}+1\right]^{\frac{1}{2}} \tag{8.13.7}$$

$$\beta = \omega\sqrt{\frac{\varepsilon\mu}{2}}\left[\sqrt{1+\left(\frac{\sigma}{\varepsilon\omega}\right)^2}-1\right]^{\frac{1}{2}} \tag{8.13.8}$$

We take the positive square root only, as negative sign would make α, β complex. In terms of α, β the field vectors becomes, using (8.13.3) and (8.13.4)

$$\vec{E}(\vec{r},t) = \vec{E}_0 e^{-\beta r} \cdot e^{j(\alpha r - \omega t)} \tag{8.13.9}$$

$$\vec{H}(\vec{r},t) = \vec{H}_0 e^{-\beta r} e^{j(\alpha r - \omega t)} \tag{8.13.10}$$

The equations show an important fact that *a plane wave cannot propagate in a conducting medium without atienuation of its amplitude* (Fig. 8.5). When a plane wave propagates in a conductor, the oscillating $\vec{E}$ in the wave sets up current in it; and to drive the current, work must be done and some energy is dissipated as heat (joule-loss). *This results in the attenuation* of the amplitude of the wave. The quantity β *is a measure of attenuation* and is called the **absorption coefficient** of the medium.

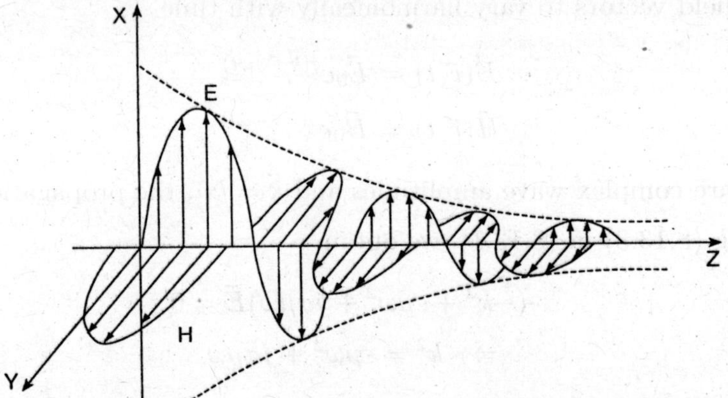

Fig. 8.5

• For a *non-conduting medium*, $\sigma = 0 \Rightarrow \beta = 0$ and there is *no attenuation.* For a **good conductor**, $\sigma/\varepsilon\omega \gg 1$ and we obtain

$$\boxed{\alpha = \beta = \sqrt{\frac{\mu\sigma\omega}{2}}} \tag{8.13.11}$$

using the relation $2\alpha\beta = \sigma\mu\omega$.

• **Skin depth** — We have just seen the attenuation of amplitude of e.m. waves with distance inside a conductor.

Skin depth δ is defined as that distance which the e.m. waves require in a conductor to reduce its amplitude by a factor $1/e$.

This occurs, from (8.13.9), when $r = 1/\beta = \sqrt{2/\mu\sigma\omega}$

$$\therefore \quad \textbf{Skin depth,} \quad \boxed{\delta = \sqrt{\frac{2}{\mu\sigma\omega}}} \tag{8.13.12}$$

Note that the skin depth is *inversely* proportional to the *square root of the frequency*. So, for waves of high frequency, δ is low. Thus, *the skin depth decreases as the frequency increases and vice versa.*

For instance, at microwave frequencies, δ for silver is extremely small $\sim 10^{-3}$ mm. The electric field of high frequency e.m. waves is thus confined to a small region on the surface of the conductor. This is why the phenomenon is called **skin effect**. Thus, *by using a thin coating of silver or copper, a poor conductor can be made a good conductor,* a technique used to reduce the material cost of good conductors.

• The skin depth of sea water at a frequency of 30 kHz is 10^{-1} m and explains why the *radio-communication with submarines becomes difficult at a depth of several metres.*

Substituting the solutions (8.13.3) and (8.13.4) in Maxwell's equations,

$$\left.\begin{aligned} \vec{\nabla}\cdot\vec{E} = 0 \quad &\text{and} \quad \vec{\nabla}\cdot\vec{H} = 0, \\ \vec{k}\cdot\vec{E} = 0 \quad &\text{and} \quad \vec{k}\cdot H = 0. \end{aligned}\right\} \tag{8.13.13}$$

This implies that $\vec{E}$ and $\vec{H}$ *are normal to the direction of propagation.*

Substituting the solutions, again, in Maxwell's equations

$$\left.\begin{aligned} \vec{\nabla}\times\vec{E} &= -\mu\frac{\partial\vec{H}}{\partial t} \\ \text{and} \quad \vec{\nabla}\times\vec{H} &= \sigma\vec{E} + \varepsilon\frac{\partial\vec{E}}{\partial t} \end{aligned}\right\}$$

$$\left.\begin{aligned} j\vec{k}\times\vec{E} = j\omega\mu\vec{H} \quad &\Rightarrow \quad \vec{k}\times\vec{E} = \mu\omega\vec{H} \\ \text{and} \quad j\vec{k}\times\vec{H} = \sigma\vec{E} - j\omega\varepsilon\vec{E} \quad &\Rightarrow \quad \vec{k}\times\vec{H} = -(\varepsilon\omega + j\sigma)\vec{E} \end{aligned}\right\} \tag{8.13.14}$$

implying that $\vec{E}$ and $\vec{H}$ *are mutually perpendicular, and they are also perpendcular to the direction of propagation.*

The attenuation of the amplitude for a progressive e.m. wave is shown graphically in Fig. 8.5.

Phase lag — From (8.13.14), we have

$$\begin{aligned} \vec{H} &= \frac{1}{\mu\omega}(\alpha + j\beta)(\hat{n}\times\vec{E}) \quad (\because \; \vec{k} = k\hat{n}; \; k = \alpha + j\beta) \\ &= \frac{1}{\mu\omega}\sqrt{\alpha^2 + \beta^2}\cdot e^{j\tan^{-1}(\beta/\alpha)}(\hat{n}\times\vec{E}). \end{aligned} \tag{8.13.15}$$

This indicates that, unlike in a dielectric, *$\vec{E}$ and $\vec{H}$ are not in phase in a conductor but $\vec{H}$ lags behind $\vec{E}$ by an angle* $\phi = \tan^{-1}(\beta/\alpha)$.

For a *good conductor,* $\alpha = \beta \Rightarrow \phi = \tan^{-1} 1 = \mathbf{45°}$.

Sharing of field energy — For a good conductor, the relative magnitudes of the field vectos $\vec{E}$ and $\vec{H}$ are given by

$$\frac{|\vec{H}|}{|\vec{E}|} = \frac{1}{\mu\omega}\sqrt{\alpha^2+\beta^2} = \frac{\sqrt{2}\alpha}{\mu\omega} \quad (\because \alpha = \beta)$$

$$= \frac{\sqrt{2}}{\mu\omega}\sqrt{\frac{\mu\sigma\omega}{2}} = \sqrt{\frac{\sigma}{\mu\omega}}$$

Thus, $|\vec{H}| \gg |\vec{E}|$ and therefore *the field energy is* **not equally shared** *by the field* $\vec{E}$ and the field $\vec{H}$, but is almost entirely *magntic* in character.

8.14 Reflection and refraction of plane waves at a plane interface

Let two linear dielectric ($\sigma = 0$) media 1 and 2, characterised by permittivities ε_1 and ε_2, and permeabilities μ_1 and μ_2 be separated by a plane boundary — the plane $x = 0$ (Fig. 8.6)

A plane e.m. wave is incident obliquely on the interface at O. In general, there will be both reflected and transmitted waves.

Let $\vec{E}, \vec{E}_1$ and $\vec{E}_2$ be the electric fields for the incident, reflected and transmitted waves respectively and be represented as

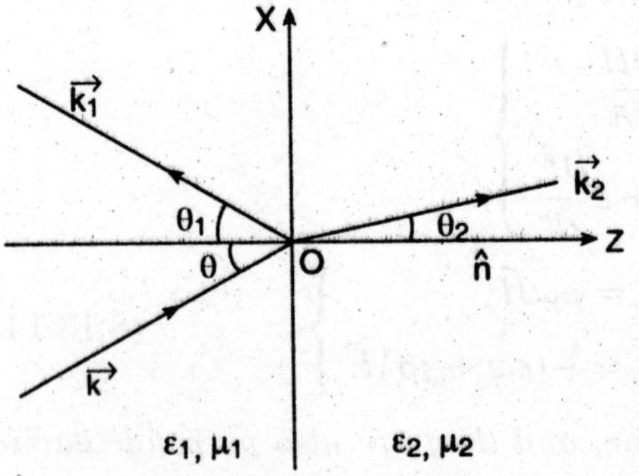

Fig 8.6

$$\left.\begin{aligned} \vec{E} &= \vec{E}_0 \exp\left[j(\vec{k}\cdot\vec{r} - \omega t)\right] \\ \vec{E}_1 &= \vec{E}_{10} \exp\left[j(\vec{k}_1\cdot\vec{r} - \omega_1 t)\right] \\ \vec{E}_2 &= \vec{E}_{20} \exp\left[j(\vec{k}_2\cdot\vec{r} - \omega_2 t)\right] \end{aligned}\right\} \quad (8.14.1)$$

where $\vec{E}_0, \vec{E}_{10}, \vec{E}_{20}$ are the time and space independent amplitudes which may be complex, $\vec{k}$'s the propagation vectors and ω's the frequencies of the incident, reflected or transmitted or refracted waves.

Now, the boundary condition to be satisfied by the fields in the plane $x = 0$ is that the tangential component of the electric field is continuous across the boundary.

$$\therefore \; (E_0)_t \exp\left[j(\vec{k}\cdot\vec{r} - \omega t)\right] + (E_{10})_t \exp\left[j(\vec{k}_1\cdot\vec{r} - \omega_1 t)\right]$$

$$= (E_{20})_t \exp\left[j(\vec{k}_2\cdot\vec{r} - \omega_2 t)\right] \quad (8.14.2)$$

where t denotes the tangential component.

Since (8.14.2) must be valid for *all times*, the exponential must be the same at the boundary for the three fields.

$$\therefore \quad \omega = \omega_1 = \omega_2 \tag{8.14.3}$$

i.e., *the frequency remains unchanged in the reflected and the refracted waves.*

Since (8.14.2) is also valid for *all space points* on the interface

$$\vec{k}\cdot\vec{r} = \vec{k}_1\cdot\vec{r} = \vec{k}_2\cdot\vec{r} \tag{8.14.4}$$

This shows that *all the propagation vectors are coplanar.* If $\vec{r}$ is chosen to lie on the boundary plane (i.e., $\hat{e}_n\cdot\vec{r} = 0$, $\hat{e}_n$ being unit vector normal) and in the plane of propagation vector, it follows that

$$k\sin\theta = k_1\sin\theta_1 = k_2\sin\theta_2 \tag{8.14.5}$$

Now, $\vec{k}$ and $\vec{k}_1$ are in the same medium and hence are equal in magnitude

$$\therefore \quad \boxed{\boldsymbol{\theta = \theta_1}} \tag{8.14.6}$$

i.e., **angle of incidence = angle of reflection**

Since, again, $\quad k_1\sin\theta_1 = k_2\sin\theta_2$, and $k = \omega\sqrt{\varepsilon\mu}$

$$\frac{\sin\theta_1}{\sin\theta_2} = \frac{k_2}{k_1} = \sqrt{\frac{\varepsilon_2\mu_2}{\varepsilon_1\mu_1}} \quad (\because\ \theta = \theta_1)$$

For non-magnetic materials, $\mu_2 = \mu_1$ so that from above, we obtain

$$\frac{\sin\theta_1}{\sin\theta_2} = \sqrt{\frac{\varepsilon_2}{\varepsilon_1}} = \frac{n_2}{n_1}$$

n_1, n_2 being the refractive indices of media 1 and 2, respectively.

$$\therefore \quad \boxed{\boldsymbol{n_1\sin\theta_1 = n_2\sin\theta_2}} \quad \cdots \quad \textbf{Snell's law} \tag{8.14.7}$$

Eq. (8.14.6) and (8.14.7) are the familiar simple laws of reflection and refraction of geometrical optics.

8.15 Impedance of a dielectric to electromagnetic waves

In electric circuits, the impedance Z is defined by the ratio of the applied voltage to the current I flowing through it.

$$\therefore \quad Z = V/I.$$

Impedance of a dielectric—For electromagnetic waves, *the impedance of a dielectric medium is the ratio of the instantaneous electric field intensity to the instantaneous magnetic field intensity.*

$\therefore$ For electromagnetic waves, $Z = E/H$. It can easily be seen that this Z also has the dimensions of ohm

Dimensions of E = volt/metre, those of H = amp/metre so that we obtain

$$\frac{E}{H} = \frac{\text{volt}}{\text{metre}} \times \frac{\text{metre}}{\text{ampere}} = \frac{\text{volt}}{\text{amp}} = \text{ohm}$$

The wave equation for the electric and magnetic field vectors for a plane polarized wave moving along the $+z$ axis and polarized such that $E_y = 0$ and $H_x = 0$, are

$$E_x = E_0 \sin \frac{2\pi}{\lambda}(vt - z) \tag{8.15.1}$$

$$H_y = H_0 \sin \frac{2\pi}{\lambda}(vt - z) \tag{8.15.2}$$

$$\therefore \quad \frac{\partial E_x}{\partial z} = -E_0 \frac{2\pi}{\lambda} \cos \frac{2\pi}{\lambda}(vt - z)$$

$$\frac{\partial H_y}{\partial t} = H_0 \frac{2\pi v}{\lambda} \cos \frac{2\pi}{\lambda}(vt - z)$$

Using the relation, $\dfrac{\partial E_x}{\partial z} = -\mu \dfrac{\partial H_y}{\partial t}$,

$$E_0 = \mu H_0 v \Rightarrow \frac{E_0}{H_0} = \mu v = \mu \frac{1}{\sqrt{\mu\varepsilon}} = \sqrt{\frac{\mu}{\varepsilon}} \tag{8.15.3}$$

$$\therefore \quad Z = \frac{E_0}{H_0} = \frac{E_x}{H_y} = \sqrt{\frac{\mu}{\varepsilon}} \quad \text{or,} \quad \left|\frac{\vec{E}}{\vec{H}}\right| = \sqrt{\frac{\mu}{\varepsilon}}$$

where μ and ε are respectively the permeability and permittivity of the dielectric medium.

Since the ratio $|\vec{E}/\vec{H}|$ is real and positive, the **electric and magnetic fields are in phase** with each other.

$\therefore$ *Characteristic impedance of a dieletric* medium to electromagnetic waves is

$$Z = \sqrt{\mu/\varepsilon}.$$

Computation of characteristic impedance of free space—For free space,

$$\mu = \mu_0 = 4\pi \times 10^{-7} \text{ kg m C}^2$$

$$\varepsilon = \varepsilon_0 = 1/(36\pi \times 10^9) \text{ C}^2\text{s}^2/\text{kg m}^3$$

$$\therefore \quad Z\,(\text{free space}) = Z_0 = \sqrt{\frac{\mu_0}{\varepsilon_0}} = \sqrt{4\pi \times 10^{-7} \times 36\pi \times 10^9} = \mathbf{377\ ohm.}$$

Refractive index and impedence—We have

$$\frac{Z\,(\text{free space})}{Z\,(\text{dielectric})} = \frac{\sqrt{\mu_0/\varepsilon_0}}{\sqrt{\mu/\varepsilon}} = \sqrt{\frac{\mu_0 \varepsilon}{\mu \varepsilon_0}} = \sqrt{\frac{\varepsilon_r}{\mu_r}} = \sqrt{\varepsilon_r} \quad (\because \mu_r \simeq 1)$$

$\therefore$ For a dielectric, $n = \sqrt{\varepsilon_r} = Z$ (free space)/Z (dielectric)

When the electromagnetic waves travel from a medium of index n_1 and impedance Z_1 to a medium of index n_2 and impedance Z_2, then the refractive index of the second medium with respect to the first is

$$n = \frac{n_2}{n_1} = \frac{Z\,(\text{free space})/Z_2}{Z\,(\text{free space})/Z_1} = \frac{Z_1}{Z_2} \tag{8.15.4}$$

8.16 Total energy density in dielectric

In Art. 8.15, we have obtained the following relation.

$$\frac{E_x}{H_y} = \sqrt{\frac{\mu}{\varepsilon}} \Rightarrow \varepsilon E_x^2 = \mu H_y^2$$

$$\therefore \quad \tfrac{1}{2}\varepsilon E_x^2 = \tfrac{1}{2}\mu H_y^2 \tag{8.16.1}$$

The quantities εE_x^2 and μH_y^2 have the dimensions of energy per unit volume or *energy density*. The quantity $\frac{1}{2}\varepsilon E_x^2$ is the *electrostatic energy per unit volume*, while $\frac{1}{2}\mu H_y^2$ is the *magnetic energy per unit volume*.

Total energy density, U_T is therefore given by

$$\begin{aligned} U_T &= \tfrac{1}{2}\varepsilon E_x^2 + \tfrac{1}{2}\mu H_y^2 \\ &= \varepsilon E_x^2, \quad \text{using (8.16.1)} \\ &= \varepsilon_0 \varepsilon_r E_x^2 \end{aligned}$$

For a plane polarized electromagnetic wave moving along positive z axis, $E_x = E$. Therefore, the total energy density is given by

$$U_T = \varepsilon_0 \varepsilon_r E^2 = \varepsilon_r(\varepsilon_0 E^2) \tag{8.16.2}$$

where $\varepsilon_0 E^2$ is the *total energy density in free space.*

$\therefore$ **Total energy density in dielectric**

$= \varepsilon_r \times$ **total energy density in free space.**

This is rather an important relation.

8.17 Reflection and transmission of electromagnetic waves : Normal incidence

Let an electromagnetic wave be incident normally on an infinite plane boundary that separates two media of impedances Z_1 and Z_2 (Fig. 8.7). While Z's are real for dielectrics, they are complex for conducting media. Also, let the electric field vectors be E_i, E_r and E_t for the incident, reflected and transmitted waves respectively and the corresponding magnetic field vectors be H_i, H_r and H_t.

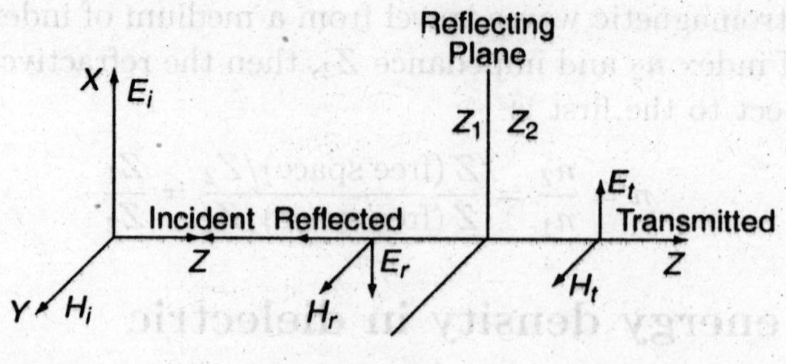

Fig 8.7

The energy flow condition of Poynting vector dictates that $\vec{E}_r \times \vec{H}_r$ is oppositely directed to $\vec{E}_i \times \vec{H}_i$. Let the incident wave travel along z-axis. The reflected wave will then travel along $-z$ axis (backward normal reflection). If $Z_2 < Z_1$, the electric vector gets reversed; if however $Z_2 > Z_1$, the magnetic vector is reversed. We shall restrict ourselves to $Z_2 < Z_1$.

By the boundary conditions, the components of $\vec{E}$ and $\vec{H}$ tangential to the boundary are continuous across the boundary. We may therefore write,

$$E_i + E_r = E_t \tag{8.17.1}$$

$$H_i + H_r = H_t \tag{8.17.2}$$

$$\text{But,} \quad Z_1 = \frac{E_i}{H_i}; \quad -Z_1 = \frac{E_r}{H_r} \quad \text{and} \quad Z_2 = \frac{E_t}{H_t}.$$

$$\therefore \; H_i = \frac{E_i}{Z_1}; \quad H_r = -\frac{E_r}{Z_1}; \quad H_t = \frac{E_t}{Z_2}.$$

$\therefore$ From the relation (8.17.2), we obtain

$$\frac{E_i}{Z_1} - \frac{E_r}{Z_1} = \frac{E_t}{Z_2} \Rightarrow E_i - E_r = E_t \frac{Z_1}{Z_2} \tag{8.17.3}$$

$$\therefore \quad \frac{E_i + E_r}{E_i - E_r} = \frac{Z_2}{Z_1}, \text{ using (8.17.1)}. \tag{8.17.4}$$

By componendo and dividendo, we obtain

$$\frac{E_r}{E_i} = \frac{Z_2 - Z_1}{Z_2 + Z_1} \tag{8.17.5}$$

$$\therefore \quad \frac{E_r}{E_i} + 1 = \frac{Z_2 - Z_1}{Z_2 + Z_1} + 1 \Rightarrow \frac{E_r + E_i}{E_i} = \frac{2Z_2}{Z_2 + Z_1}$$

$$\Rightarrow \frac{E_t}{E_i} = \frac{2Z_2}{Z_2 + Z_1}, \text{ using (8.17.1)}. \tag{8.17.6}$$

Reflection coefficient—The *reflection coefficient* R is defined as the ratio E_r/E_i.

$\therefore$ From (8.17.5),
$$R = \frac{E_r}{E_i} = \frac{Z_2 - Z_1}{Z_2 + Z_1} \tag{8.17.7}$$

Transmission coefficient—The *transmission coefficient* T is defined as the ratio E_t/E_i. $\therefore$ From (8.17.6),
$$T = \frac{E_t}{E_i} = \frac{2Z_2}{Z_2 + Z_1} \tag{8.17.8}$$

The *reflection and transmission coefficients* may also be expressed in terms of *refractive indices* n_1 and n_2 of the media, as shown.

From (8.17.4),
$$\frac{E_i + E_r}{E_i - E_r} = \frac{Z_2}{Z_1} = \frac{n_1}{n_2}, \quad \text{using (8.15.4)}$$

$$\therefore \quad R = \frac{E_r}{E_i} = \frac{n_1 - n_2}{n_1 + n_2} = \frac{1 - n_2/n_1}{1 + n_2/n_1} = \frac{1-n}{1+n} \tag{8.17.9}$$

$$\text{or,} \quad \boxed{\boldsymbol{R = \frac{1-n}{1+n}}} \tag{8.17.10}$$

where n is the index of second medium with respect to the first.

Adding 1 to both sides of (8.17.9),
$$\frac{E_r + E_i}{E_i} = \frac{E_t}{E_i} = \frac{2}{1+n}$$

$$\therefore \quad \boxed{\boldsymbol{T = \frac{E_t}{E_i} = \frac{2}{1+n}}} \tag{8.17.11}$$

Reflected energy coefficient—The *reflected energy coefficient* R_E is defined by $R_E = E_rH_r/E_iH_i$.

Let the electromagnetic wave travel in air (or free space) and strike a dielectric normally.
$$\therefore \quad \frac{E_i}{H_i} = \frac{E_r}{H_r} = Z_1 = \sqrt{\frac{\mu_0}{\varepsilon_0}}; \quad \frac{E_t}{H_t} = Z_2 = \sqrt{\frac{\mu}{\varepsilon}} \tag{8.17.12}$$

where μ's represent permeabilities, ε's the permittivities.
$$\therefore \quad H_i = \sqrt{\frac{\varepsilon_0}{\mu_0}}E_i; \quad H_r = \sqrt{\frac{\varepsilon_0}{\mu_0}}E_r; \quad H_t = \sqrt{\frac{\varepsilon}{\mu}}E_t, \quad \text{using (8.17.12).}$$

$$\therefore \quad R_E = \frac{E_rH_r}{E_iH_i} = \frac{\sqrt{\varepsilon_0/\mu_0}E_r^2}{\sqrt{\varepsilon_0/\mu_0}E_i^2} = \frac{E_r^2}{E_i^2} = \left(\frac{1-n}{1+n}\right)^2, \quad \text{using (8.17.9).}$$

For, instance, in air-glass interface, $n = 1.5$ and $R_E = (0.5/2.5)^2 = 4\%$.

Transmitted energy coefficient—The *transmitted energy coefficient* T_E is defined by

$$T_E = \frac{E_t H_t}{E_i H_i} = \frac{\sqrt{\varepsilon/\mu} E_t^2}{\sqrt{\varepsilon_0/\mu_0} E_i^2} = \sqrt{\frac{\varepsilon_r}{\mu_r}} \frac{E_t^2}{E_i^2}, \quad \text{using (8.17.12).}$$

For a non-magnetic substance $\mu_r = 1$ and $n = \sqrt{\varepsilon_r}$. Therefore, we have

$$T_E = n\frac{E_t^2}{E_i^2} = n\left(\frac{2}{1+n}\right)^2 = \frac{4n}{(1+n)^2}$$

using equation (8.17.11).

For instance, in air-glass interface, $n = 1.5$ so that we obtain

$$T_E = \frac{4 \times 1.5}{(1+1.5)^2} = \frac{6}{6.25} = 0.96 = 96\%.$$

Good conductors are good reflectors—Let the electromagnetic wave travel in air to strike *a perfect conductor* normally. A perfect conductor implies $Z_2 = 0$.

$$\therefore \quad \text{From (8.17.7),} \quad R = \frac{E_r}{E_i} = \frac{Z_2 - Z_1}{Z_2 + Z_1} = -1 \Rightarrow E_r = -E_i.$$

The incident electromagnetic wave is thus *completely reflected* by the conductor.

$$\text{From (8.17.8),} \quad T = \frac{E_t}{E_i} = \frac{2Z_2}{Z_2 + Z_1} = 0$$

There is thus *no transmitted wave.* This shows that *a perfect conductor is a perfect reflector* or good conductors are good reflectors of electromagnetic waves. That is why we find that light waves are reflected very well from polished metallic surfaces.

8.18 Dispersion of electromagnetic waves in a conductor

Let us now discuss the phenomenon of *dispersion* of electromagnetic waves in a conducting medium.

The phase velocity of electromagnetic waves in free space (i.e., vacuum) is given by

$$c = \sqrt{\frac{1}{\mu_0 \varepsilon_0}} \tag{8.18.1}$$

where μ_0 is the permeability and ε_0 the permittivity of free space.

The phase velocity of electromagnetic waves in a medium is given by

$$v = \sqrt{\frac{1}{\mu \varepsilon}} \tag{8.18.2}$$

where μ is permeability and ε the permittivity of the medium.

$$\therefore \quad \frac{c^2}{v^2} = \frac{\mu\varepsilon}{\mu_0\varepsilon_0} = \mu_r\varepsilon_r \Rightarrow c^2 = \varepsilon_r v^2 \quad (\because \mu_r = 1 \text{ for a conductor}) \tag{8.18.3}$$

where μ_r and ε_r are respectively the relative permeability and relative permittivity.

While for a conductor ε_r and v are functions of λ, the wavelength, c is a constant. So, by differentiating (8.18.3), with respect to λ, we obtain

$$\varepsilon_r 2v\frac{dv}{d\lambda} + v^2\frac{d\varepsilon_r}{d\lambda} = 0$$

$$\Rightarrow \quad \frac{dv}{d\lambda} = -\frac{v}{2\varepsilon_r}\frac{d\varepsilon_r}{d\lambda} \tag{8.18.4}$$

Now, the phase velocity v and group velocity v_g are related by the following equation :

$$v_g = v - \lambda\frac{dv}{d\lambda} \tag{8.18.5}$$

Substituting the value of $dv/d\lambda$ from (8.18.4) in relation (8.18.5), we get

$$v_g = v + \frac{\lambda v}{2\varepsilon_r}\frac{d\varepsilon_r}{d\lambda}$$

$$= v\left(1 + \frac{\lambda}{2\varepsilon_r}\frac{d\varepsilon_r}{d\lambda}\right)$$

The group velocity v_g is thus greater than the phase velocity v. When $v_g > v$, it is a case of **anomalous dispersion.**

The electromagnetic waves thus show *anomalous dispersion* in conductors.

8.19 Illustrated Examples

▶ **Example 1.** *An electromagnetic wave in empty space has a maximum electric field of* 1000 V/m. *What is the maximum value of magnetic field?* **(Himachal Univ.)**

Solution. We have, $B_0 = \mu_0 H_0$. But from (8.15.3), $E_0/B_0 = c$. Multiplying these two equations,

$$E_0 = \mu_0 H_0 c \Rightarrow H_0 = E_0/\mu_0 c.$$

$\therefore$ Maximum magnetic field, $H_0 = \dfrac{1000}{4\pi \times 10^{-7} \times 3 \times 10^8} =$ **2.65 unit.**

▶ **Example 2.** *If a* 500 W *laser beam is concentrated by a lens into a cross-sectional area of* 10^{-10} m^2*, find the value of Poynting vector and the amplitude of electric field. Given:* $\varepsilon_0 = 9 \times 10^{-12}$ SI unit. **(Himachal Univ.)**

Solution. Poynting vector, $\vec{S} = \frac{\text{power}}{\text{area}} = \frac{500}{10^{-10}} = \mathbf{5 \times 10^{12}\ W/m^2}$.

But average Poynting vector, $S_{av} = \frac{1}{2}c\varepsilon_0 E_0^2 \Rightarrow E_0 = \sqrt{\frac{2S_{av}}{c\varepsilon_0}}$

$$\therefore \quad E_0 = \left(\frac{2 \times 5 \times 10^{12}}{3 \times 10^8 \times 9 \times 10^{-12}}\right)^{1/2} = \mathbf{0.61 \times 10^8\ V/m.}$$

► **Example 3.** *Calculate the Poynting vector for a* 60 W *lamp at a distance of* 0.5 m *from it.* **(Guru Nanak Univ.)**

Solution. The light emitted from the lamp will spread out into a sphere of radius $r = 0.5$ m. The average power of the lamp, $P = 60$ W.

$$\therefore \text{ Average Poynting vector, } S_{av} = \frac{P}{4\pi r^2} = \frac{60}{4 \times 3.14 \times (0.5)^2} = \mathbf{19.1\ W/m^2}.$$

► **Example 4.** *Calculate the frequency at which the skin depth in sea water is* 1 m. *Given :* $\sigma = 4.3$ mho/m *and* $\mu = 4\pi \times 10^{-7}$ H/m. **(Himachal Univ.)**

Solution. The skin depth δ is given by

$$\delta^2 = \frac{2}{\omega\mu\sigma} = \frac{2}{(2\pi\nu)\mu\sigma} \quad \Rightarrow \nu = \frac{2}{2\pi\mu\sigma\delta^2}$$

Here, $\sigma = 4.3$ mho/m, $\mu = 4\pi \times 10^{-7}$ H/m and $\delta = 1$ m. Therefore, the frequency,

$$\nu = \frac{2}{2\pi \times 4\pi \times 10^{-7} \times 4.3 \times 1^2} = 589 \times 10^4 \text{ Hz} = \mathbf{58.9\ kHz}$$

► **Example 5.** *Calculate the skin depth for a frequency of* 10^{10} Hz *for silver. Given, for silver,* $\sigma = 2 \times 10^7$ mho/m, *and* $\mu = 4\pi \times 10^{-7}$ H/m. **(Punjabi Univ.)**

Solution. Here $\nu = 10^{10}$ Hz $\Rightarrow \omega = 2\pi\nu = 2\pi \times 10^{10}$, $\sigma = 2 \times 10^7$ mho/m, $\mu = 4\pi \times 10^{-7}$ H/m.

$$\therefore \quad \text{Skin depth, } \delta = \sqrt{2/\omega\mu\sigma} = \sqrt{2/(2\pi \times 10^{10} \times 4\pi \times 10^{-7} \times 2 \times 10^7)}$$

$$= \sqrt{1/(8\pi^2 \times 10^{10})} = \mathbf{1.12 \times 10^{-6}\ m}$$

► **Example 6.** *Explain the behaviour of pure water to electromagnetic waves of frequency* 200 Hz *and* 100 MHz, *when the dielectric constant of water is* 80 *and* $\sigma = 10^{-3}$ mho/m. **(Guru Nanak Univ.)**

Solution. For water $\varepsilon/\varepsilon_0 = 80 \Rightarrow \varepsilon = 80\varepsilon_0$, $\sigma = 10^{-3}$ mho/m, $\varepsilon_0 = 9 \times 10^{-12}$ F/m.

$$\therefore \quad \frac{\sigma}{\varepsilon} = \frac{10^{-3}}{80 \times 9 \times 10^{-12}} = 1.39 \times 10^6$$

(i) Frequency, $\nu = 200$ Hz $\Rightarrow \omega = 2\pi \times 200$. Therefore

$$\frac{\sigma}{\omega\varepsilon} = \frac{1.39 \times 10^6}{2\pi \times 200} = 1106$$

(ii) Frequency, $\nu = 100$ MHz $\Rightarrow \omega = 2\pi \times 100 \times 10^6 = 2\pi \times 10^8$.

$$\therefore \quad \frac{\sigma}{\omega\varepsilon} = \frac{1.39 \times 10^6}{2\pi \times 10^8} = 0.0022$$

Since $\sigma/\omega\varepsilon > 100$ *for* 200 Hz, water behaves as **a conductor**. But, *for* frequency 100 MHz, $\sigma/\omega\varepsilon < 100$. So, water behaves **as a dielectric.**

▶ **Example 7.** *A plane e.m. travels in vacuum with the field $\vec{B}$ amplitude 3×10^{-6} T. On being incident normally on a surface, it gets totally reflected. What is the radiation pressure exerted on the surface?*

Solution. If $|\langle\vec{S}\rangle|$ be the intensity of the incident e.m. wave, then the radiation pressure

$$P = \frac{2|\langle\vec{S}\rangle|}{c}, \quad c = \text{speed of e.m. wave.}$$

$$\text{But,} \quad |\langle\vec{S}\rangle| = \sqrt{\frac{\varepsilon_0}{\mu_0}} \cdot \frac{E_m^2}{2} = \frac{1}{2}\sqrt{\frac{\mu_0}{\varepsilon_0}} H_m^2 = \frac{c}{2\mu_0} B_m^2$$

$$\therefore \quad P = \frac{(3 \times 10^6)^2}{4\pi \times 10^{-7}} \quad (\because \mu_0 = 4\pi \times 10^{-7})$$

$$= \mathbf{0.72 \times 10^{-5}\ Pa.}$$

Note. Usually $P = |\langle\vec{S}\rangle|/c$. But here the surface is totally reflecting and the velocity of incident wave changes from $+c$ to $-c$ making the time rate of change of momentum of photons double for an absorbing surface. Hence the factor 2.

▶ **Example 8.** *Write appropriate solutions of Maxwell's equations if $\vec{J}(\vec{r}, t) = 0$ and the only charge is a point charge fixed at the origin.* **(Calcutta Hons.)**

Solution. If q be the point charge *fixed* at the origin, the electric field at a distance $\vec{r}$ is given by

$$\vec{E}(\vec{r}) = \frac{q}{4\pi\varepsilon_0} \frac{\vec{r}}{r^3} \tag{8.19.1}$$

$$\therefore \quad \vec{\nabla} \times \vec{E} = 0.$$

$$\text{But,} \quad \vec{\nabla} \times \vec{E} = -\partial\vec{B}/\partial t. \quad \therefore \partial\vec{B}/\partial t = 0$$

which implies $\vec{H}$ is independent of time.

Again $\vec{J} = 0 \Rightarrow \vec{\nabla} \times \vec{H} = \partial \vec{D}/\partial t = 0$, since $\vec{E}$ given by (8.19.1) is independent of t. Also, since $\vec{\nabla} \cdot \vec{B} = 0$, $\vec{\nabla} \cdot \vec{H} = 0$. Since $\vec{\nabla} \times \vec{H} = 0 = \vec{\nabla} \cdot \vec{H}$,

$$\vec{H} = 0 \tag{8.19.2}$$

there being no current source for the magnetic field.

Eq. (8.19.1) and (8.19.2) constitute the solutions of Maxwell's equations.

Review Questions

1. Write down Maxwell's equations in electromagnetic theory. Explain the physical significance of each of them. **(Himachal Univ.)**
2. State Maxwell's equations in electromagnetic theory and derive them from the basic laws of physics they represent. **(Guru Nanak Univ.)**
3. Using Maxwell's relations, derive the equation of electromagnetic waves in a medium having permeability μ, permittivity ε but conductivity $\sigma = 0$ (non-conducting medium). **(Kurukshetra Univ.)**
4. Derive Maxwell's equations for electromagnetic waves in a conducting medium, writing down the Maxwell's relations of electromagnetic theory. Which is the dissipative term in the wave equation? **(Punjab Univ.)**
5. State Maxwell's equations in electromagnetic theory. Show by mathematical treatment that electromagnetic waves are transerse in nature. **(Calcutta Univ.)**
6. Using Maxwell's equations in electromagnetic theory, derive the wave equation for a plane polarized electromagnetic wave, having permeability μ and permittivity ε for (a) a charge-free non-conducting medium : $\rho = 0$, $\sigma = 0$, and (b) a charge-free conducting medium : $\rho = 0$, $\sigma \neq 0$. **(Guru Nanak and Himachal Univ.)**
7. Write the wave equation for plane polarized electromagnetic waves in a dielectric medium having finite values of μ and ε but $\sigma = 0$, and write the solutions to these equations.
   What is the phase difference between $\vec{E}$ and $\vec{H}$? Explain. **(Punjab Univ.)**
8. Define Poynting vector. What does it represent? Show that Poynting vector measures the flow of energy per unit area per second in an electromagnetic wave. **(Kurukshetra Univ.)**
9. State and derive Poynting's theorem or derive expressions for the flow of electromagnetic energy in a medium. **(Himachal Univ.)**

10. (a) Derive the equation for a plane polarized electromagnetic wave in a conducting medium having finite values of μ, ε and σ.
   (b) Show that in a conducting medium, the displacement current leads the conduction current by 90°. **(Guru Nanak Univ.)**
   (c) The amplitude of the wave gets attenuated with distance inside the conducting medium. **(Himachal Univ.)**

11. Define skin depth of a conductor. Prove that as the frequency increases, skin depth decreases. Explain skin effect and the electric shielding properties of a conductor. **(Punjabi and Punjab Univ.)**

12. Define characteristic impedance of a dielectric to electromagnetic waves and calculate the characteristic (intrinsic) impedance of free space. Show that electric and magnetic fields are in phase. **(Guru Nanak Univ.)**

13. (a) Calculate the impedance of a conducting medium to electromagnetic waves. Show that when the impedance is small the conductor behaves as a short circuit to electric field.
   (b) Prove that for a conductor, the magnetic field energy increases as the electric field energy decreases. **(Guru Nanak Univ.)**

14. Explain the reflection and transmission of an electromagnetic wave incident normally on a plane between media of impedance Z_1 and Z_2. Find the expressions for the reflection and transmission coefficients. **(Punjabi Univ.)**

15. Write short explanatory notes on :
   (a) Contribution of Maxwell to electromagnetic wave theory
   (b) Displacement current
   (c) Poynting vector and Poynting theorem
   (d) Velocity of electromagnetic wave
   (e) Skin depth and skin effect
   (f) Characteristic impedance of dielectric.

Short Answer and Objective Questions

1. Write down Maxwell's equations in electromagnetic theory, indicating the meaning of the symbols.

2. What is meant by displacement current? Show that it has the dimensions of current. How was Ampere's law modified by the introduction of displacement current?

3. Distinguish between the conduction and the displacement currents. **(Himachal Univ.)**

4. What is the chief contribution of Maxwell to electromagnetic wave theory?

5. A changing electric field induces magnetic field. Write the mathematical relation to this effect. **(Himachal Univ.)**

6. Prove that the velocity of an electromagnetic wave in a medium of permeability μ, permittivity ε and conductivity $\sigma = 0$ is given by $c = \sqrt{1/\mu\varepsilon}$. **(Guru Nanak Univ.)**

7. Write down the characteristics of electromagnetic waves in free space. **(Himachal Univ.)**

8. According to electromagnetic theory, what vectors vary at right angles to the direction of propagation of the light wave?

9. A light wave is characterized by its electric vector $\vec{E}$, although it has a magnetic vector $\vec{B}$. Explain why? **(Punjab Univ.)**

10. Why does the electric field $\vec{E}$ of electromagnetic wave determine the polarisation rather than the magnetic field? **(Himachal Univ.)**

11. Define Pynting vector. What does it represent? State Poynting's theorem. **(Kurukshetra Univ.)**

12. The electric and magnetic fields are varying sinusoidally in time as

$$\vec{E} = \vec{E}_0 \sin \omega t; \quad \vec{H} = \vec{H}_0 \sin \omega t.$$

Show that the average value of Poynting vector is given by

$$\langle S \rangle = \frac{\vec{E}_0}{\sqrt{2}} \times \frac{\vec{H}_0}{\sqrt{2}} = \vec{E}_{\text{rms}} \times \vec{H}_{rms} = \frac{1}{2}(\vec{E}_0 \times \vec{H}_0)$$ **(Punjabi Univ.)**

13. Show that for a plane electromagnetic wave, the average value of Poynting vector is given by

$$S_{av} = \tfrac{1}{2} c\varepsilon E_0^2 = \tfrac{1}{2} c\mu H_0^2.$$ **(Himachal Univ.)**

14. Show that for a plane electromagnetic wave, the flow of energy in a homogeneous isotropic medium is in the direction of wave propagation. **(Punjab Univ.)**

15. Show that the energy flux for a travelling wave in free space is simply the energy density times the velocity of the wave.

16. Show that in a conductor, the average value of Poynting vector is given by

$$S_{av} = \tfrac{1}{2} E_0 H_0 \cos 45° = \tfrac{1}{2} H_0^2 \text{Re}\,(Z_c)$$ **(Punjabi Univ.)**

17. When is a medium a conductor or a dielectric? Show that in a 'good' conductor the magnetic field lags the electric field by 45°.

18. (a) Show that high frequency electromagnetic waves propagate only a small distance inside a conductor.
(b) Why a two-way conversation with a submerged submarine is impossible?

19. Show that in an electromagnetic wave, the electrostatic energy per unit volume is equal to magnetic energy per unit volume and the total energy in the dielectric $= \varepsilon_r \times$ total energy density in free space.

20. Show that : (a) refractive index of a dielectric, $n = \sqrt{\varepsilon_r}$. Hence find the refractive index of one medium with respect to another. (b) refractive index of a conducting medium, $n = (1 - j)\sqrt{\sigma/2\omega\varepsilon_0}$.

21. Prove that a perfect conductor is a perfect reflector of electromagnetic waves. **(Guru Nanak Univ.)**

22. Explain the phenomenon of dispersion of electromagnetic waves in a conductor. **(Guru Nanak Univ.)**

23. Maxwell's equation involving $\partial\vec{B}/\partial t$ is obtained from
(a) Ampere's law (b) Biot-Savart law
(c) Gauss's law (d) Faraday's law. **(Himachal Univ.)**

24. The velocity of plane electromagnetic wave in vacuum is given by
(a) $c = \dfrac{1}{\sqrt{\mu_0\varepsilon_0}}$ (b) $c = \sqrt{\mu_0\varepsilon_0}$ (c) $c = \sqrt{\mu_0/\varepsilon_0}$ (d) $c = \sqrt{\varepsilon_0/\mu_0}$.

25. The SI unit of Poynting vector is
(a) Wm^2 (b) Wm (c) Wm^{-1} (d) $\text{Js}^{-1}\text{m}^{-2}$. **(Himachal Univ.)**

26. Skin depth for a conductor in reference to electromagnetic waves varies
(a) inversely as square root of frequency
(b) inversely as the frequency
(c) directly as the frequency
(d) directly as the square of frequency.

27. During the propagation of electromagnetic waves in a medium
(a) Electric energy density is double the magnetic energy density.
(b) Electric energy density is half the magnetic energy density.
(c) Both electric and magnetic energy densities are equal.
(d) None of the above statements is true.

28. The characteristic impedance for a perfect conductor to electromagnetic waves is
(a) zero (b) 1 (c) ∞ (d) none of these.
(Guru Nanak Univ.)

29. The phase difference between electric and magnetic field vector in a conductor is
(a) π rad (b) $\pi/2$ rad (c) $\pi/4$ rad (d) zero.
(Himachal Univ.)

30. Which one of the following has Coulomb as the unit?
(a) $\oint \vec{H}\cdot d\vec{l}$ (b) $\oint \vec{E}\cdot d\vec{l}$ (c) $\iint \vec{D}\cdot d\vec{S}$ (d) $\iint \vec{E}\cdot d\vec{S}$.
(Himachal Univ.)

31. The magnitude of impedance in a conductor to electromagnetic waves is
(a) $Z_c = \sqrt{\sigma/\omega\mu}$ (b) $Z_c = \sqrt{\mu\sigma/\omega}$ (c) $Z_c = \sqrt{\omega\sigma/\mu}$ (d) $Z_c = \sqrt{\mu\omega/\sigma}$.
(Himachal Univ.)

32. The dimensions of $\partial\vec{D}/\partial t$ is
(a) Coulomb/metre^2 (b) Coulomb
(c) Ampere/metre^2 (d) None of these.

33. For a plane electromagnetic wave, the energy-flow in a homogeneous isotropic medium is
(a) in the direction of wave propagation
(b) opposite to the direction of wave propagation
(c) at right angles to the direction of propagation
(d) none of the above.

Problems

1. Find the velocity in vacuo of electromagnetic waves. Given, for vacuum, $\mu_0 = 4\pi \times 10^{-7}$ kg m/C^2, $\varepsilon_0 = 1/(36\pi \times 10^9)$ C^{2}s^2/kg.m^3. [3×10^8 ms^{-1}]

2. A radio station radiates power uniformly, over a hemisphere concentric with the station, of 10^5 Watt. Find (i) the magnitude of Poynting vector, (ii) the amplitudes of electric and magnetic fields at a point 10 km from the station ($\varepsilon_0 = 9 \times 10^{-12}$ Fm^{-1}, $\mu_0 = 4\pi \times 10^{-7}$H/m).
[(i) 1.59×10^{-4} W/m^2; (ii) 0.343 V/m, 9.1×10^{-4} Amp.turn/m]

3. The power radiated by the sun is 3.8×10^{26} W. If the average distance between the sun and the earth is 1.5×10^{11} m, find the solar constant (i.e., average solar energy received per sec per sq. m of earth's surface). The radius of the sun $= 7 \times 10^8$ m. [1.34×10^3 W/m^{2}s]

4. The energy received by earth's surface from the sun is 1.33 kW/m^2. Find the electric field associated with sunlight as found on earth's surface. Assume $\varepsilon = 9 \times 10^{-12}$ SI unit, and sunlight to be essentially monochromatic of $\lambda = 6000$ A. [992.6 V/m]

5. Calculate the skin depth for a frequency of 10^6. Given, $\mu = 4\pi \times 10^{-7}$ Am^{-2} and $\sigma = 5.8 \times 10^7$ mho/m. **(Himachal Univ.)** [6.61×10^{-5} m]

6. Discuss the behaviour of copper to electromagnetic waves of frequency 0.5×10^{16} Hz and 7.0×10^{20} Hz. Given the conductivity of copper, $\sigma = 5.8 \times 10^7$ mho/m and the permittivity $\varepsilon = 9 \times 10^{-12}$ C^2/Nm^2. **(Guru Nanak Univ.)**

   [conductor for lower freq; dielectric for higher freq.]

7. Show that radar waves of frequency 1 MHz get reflected from the metallic surface of the aeroplane for which $\mu = 4\pi \times 10^{-7}$ H/m and $\sigma = 5.8 \times 10^7$ mno/m. **(Punjab Univ.)**

8. In a medium of dielectric constant 5, the maximum displacement current is equal to the maximum conduction current at a frequency of 1 MHz. What is the conductivity of the medium? **(Calcutta Univ.)** [2.78×10^{-4} mho/m]

9. A laser beam has a diemeter of 2 mm. Calculate the amplitude of the electric field in the beam in vacuum. Given: the power of the laser is 1.5 mW. [600 V/m]

10. An electric field in free space is given by

$$E_x = E_y = 0$$
$$E_z = E_0 \cos kx . \cos \omega t.$$

   Find the components of the magnetic field vector $\vec{H}$. **(Lucknow Univ.)**

   [$H_x = H_z = 0;\ H_y = -\dfrac{kE_0}{\mu_0 \omega} \sin kx \cdot \sin \omega t$]

11. Find from Poynting flow the r.m.s. value of magnetic intensity in air at a distance of 1 m from a radiating source of power 10 kW. **(Calcutta Univ.)** [1.44 A/m]

Chapter 9

Fourier Analysis

9.1 Introduction

Fourier analysis is a very powerful mathematical technique with an enormous application not only in diverse fields of physics like acoustics, optics, electricity and quantum mechanics, but also in other branches of physical science such as electronics, engineering, communication sciences etc. In fact, Fourier analysis is an imperative need for the development of any branch of science where the sine and cosine oscillations play a prominent role. It deals with the addition of a number of sinusoidal oscillations to get a resultant and conversely, starting from a given resultant, it determines the sine and cosine oscillations that were compounded to obtain the given resultant. While the former is known as **Fourier synthesis**, the latter is called **Fourier analysis**.

9.1.1 Some terminologies

We shall first deal with some of the terminologies that are often used in Fourier analysis.

A quantity y is said to oscillate *sinusoidally* or *cosinusoidally* with time t, according as

$$y = a \sin 2\pi\nu t \tag{9.1.1}$$

$$\text{or,} \quad y = a \cos 2\pi\nu t \tag{9.1.2}$$

The constant $a = (y)_{\max}$ is called the **amplitude** and the constant ν is called the **frequency** of oscillation.

The time T for one complete oscillation is called the **time period** or simply the **period** of oscillation.

$$T = 1/\nu \tag{9.1.3}$$

Sometimes, instead of ν, a new parameter ω, defined as

$$\omega = 2\pi\nu \tag{9.1.4}$$

is introduced, and this parameter ω is called the **angular frequency**.

• The sinusoidal oscillation (9.1.1) lags behind the cosinusoidal oscillation (9.1.2) by a quarter-period, i.e., $T/4$. But, if represented graphically, it would show however, the same shape.

9.2 Harmonic oscillations

Consider the following two oscilations :

$$y = a\sin(2\pi\nu t + \theta) \tag{9.2.1}$$

$$\text{or,}\quad y = a\cos(2\pi\nu t + \phi) \tag{9.2.2}$$

Compared to the pure sine and cosine oscillations of (9.1.1) and (9.1.2), these oscillations are shifted in phase by θ and ϕ respectively. They are ahead of the pure oscillations by $\theta/2\pi\nu$ and $\phi/2\pi\nu$ respectively, since

$$2\pi\nu t + \theta = 0 \quad \text{at} \quad t = -\theta/2\pi\nu$$

$$\text{and} \quad 2\pi\nu t + \phi = 0 \quad \text{at} \quad t = -\phi/2\pi\nu.$$

• Equations (9.2.1) and (9.2.2) represent the same oscillation if $\theta = \pi/2 + \phi$.

We shall now represent (9.2.1) [or (9.2.2)] in another way as shown below. From (9.2.1), we get

$$\begin{aligned} y &= a\sin(2\pi\nu t + \theta) \\ &= a\sin\theta\cos 2\pi\nu t + a\cos\theta\sin 2\pi\nu t \\ &= A\cos 2\pi\nu t + B\sin 2\pi\nu t \end{aligned} \tag{9.2.3}$$

where $A = a\sin\theta$ and $B = a\cos\theta$.

$$\therefore \quad a = \sqrt{A^2 + B^2}, \tag{9.2.4}$$

Similarly, we get

$$\begin{aligned} y &= a\cos(2\pi\nu t + \phi) \\ &= a\cos\phi\cos 2\pi\nu t + a\sin\phi\sin 2\pi\nu t \\ &= A\cos 2\pi\nu t + B\sin 2\pi\nu t \end{aligned} \tag{9.2.5}$$

where $A = a\cos\phi$ and $B = a\sin\phi$.

$$\therefore \quad a = \sqrt{A^2 + B^2} \tag{9.2.6}$$

Equations (9.2.1) to (9.2.3) and (9.2.5), all describe the same dependence of y on t. Such an oscillation is called a **harmonic oscillation**. A is called the **cosine amplitude** and B the sine amplitude and a, given by (9.2.4) or (9.2.6), is known as the **amplitude** of the harmonic oscillation.

9.2.1 Harmonic synthesis and analysis

If two or more harmonic oscillations of the *same frequency* are added, then their resultant is another harmonic oscillation of the same frequency.

Let us add the oscillations

$$y_1 = a_1 \cos 2\pi\nu t + b_1 \sin 2\pi\nu t$$

$$\text{and} \quad y_2 = a_2 \cos 2\pi\nu t + b_2 \sin 2\pi\nu t$$

$$\therefore \quad y_1 + y_2 = (a_1 + a_2) \cos 2\pi\nu t + (b_1 + b_2) \sin 2\pi\nu t \tag{9.2.7}$$

So, $y_1 + y_2$ is also *harmonic* with $(a_1 + a_2)$ as cosine amplitude and $(b_1 + b_2)$ as the sine amplitude.

However, if two or more harmonic oscillations of *different frequencies* are added, the resultant will *not* be harmonic. This can be easily verified either analytically or graphically.

A *notable* feature that emerges is that if the oscillations that are being added have different frequencies but all multiplies of some highest common factor, say ν_1, then the resultant would be *periodic in time*, with a period $T = 1/\nu_1$, that is

$$y(t) = y(t + T) \quad \text{for all} \quad t;$$

ν_1 is known as the *fundamental frequency.*

The process of adding a number of harmonic oscillations to form a resultant is called the **harmonic synthesis**. For N different frequencies, we get

$$y = \sum_{n=1}^{N} [a_n \cos(2\pi\nu_n t) + b_n \sin(2\pi\nu_n t)] \tag{9.2.8}$$

The quantities a_n's and b_n's are the cosine and sine amplitudes at frequencies ν_n's. When a fundamental frequency ν_1 exists, then

$$\nu_n = n\nu_1$$

and ν_n is called the *nth harmonic frequency.*

The reverse process is called **harmonic analysis** where y is given as a function of t, either by a graph or by providing a tabulated set of values at equi-spaced short time-intervals, and from the given data, the values of the frequencies and the cosine and sine amplitudes, as given in (9.2.8), are to be deduced. Fourier gave us, through various formulae, the method of carrying out the analysis systematically. The harmonic analysis, executed by Fourier's method, is called **Fourier analysis**.

9.2.2 Fourier's contribution

Fourier's contribution lies in showing how to deal with an infinite number of frequencies rather than a finite one. He did this by two different approaches. We discuss them in turn in what follows.

First approach — Consider the following equation.

$$y = \sum_{n=1}^{\infty} [a_n \cos(2\pi n\nu_1 t) + b_n \sin(2\pi n\nu_1 t)] \tag{9.2.9}$$

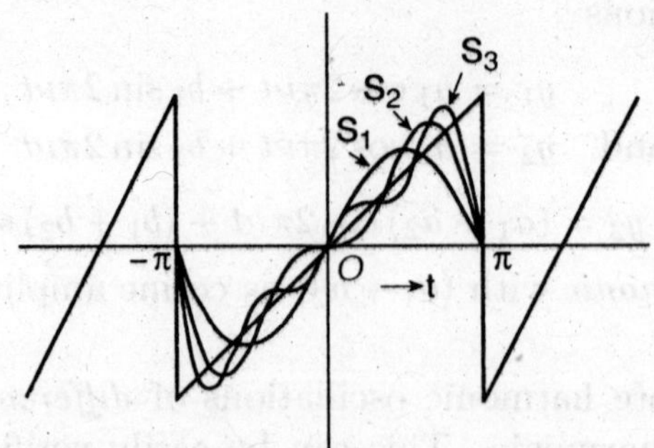

Fig 9.1

The above implies an *infinite set* of cosine and sine amplitudes, unlike (9.2.8) where N is *finite*. The value of y at a given t is obtained by first finding the partial sum S_N of N-number of harmonics, as in (9.2.8), and subsequently defining y as the limiting value of S_N as $N \to \infty$.

As an illustration, let $a_n = 0$ and $b_n = 1/n$. So

$$a_1 = a_2 = a_3 = \cdots = a_n = 0$$

$$\text{and} \quad b_1 = 1, \;\; b_2 = \tfrac{1}{2}, \;\; b_3 = \tfrac{1}{3}, \cdots \text{ and so on.}$$

In the graph (Fig. 9.1), the partial sum S_N for $N = 1, 2, 3$ has been represented. Note that as N increases, the resultant approaches more and more the *sawtooth-oscillation.*

Second approach — Let us now consider the expression

$$y = \int_0^\infty [a(\nu)\cos 2\pi\nu t + b(\nu)\sin 2\pi\nu t]\, d\nu \tag{9.2.10}$$

This implies that now the harmonic oscillations at all frequencies have been added, unlike the previous *infinite set* of discrete frequencies ν_1, $2\nu_1$, $3\nu_1, \cdots$, we have an *infinite continuum* of frequency from $\nu = 0$ to $\nu = \infty$.

Note. The harmonic amplitude a_n's and b_n's of (9.2.9) are now replaced by functions $a(\nu)$ and $b(\nu)$ whose values are defined for every ν. Let $d\nu$ be a small frequency-interval defined as

$$d\nu = \left(\nu + \tfrac{1}{2}\delta\nu\right) - \left(\nu - \tfrac{1}{2}\delta\nu\right)$$

and, in this interval, we now have an infinite number of contributing frequencies; it is assumed that they combine to approximate to a harmonic oscillation of frequency ν and having $a(\nu)\delta\nu$ and $b(\nu)\delta\nu$ as the cosine and sine amplitudes respectively. So, in this interval $\delta\nu$, the resultant is approximately given as a function of time, by

$$dy = [a(\nu)\cos(2\pi\nu t) + b(\nu)\sin(2\pi\nu t)]\,\delta\nu,$$

the time however must not be too large.

The integral of (9.2.10) is obtained by adding over all the frequency-intervals and taking the limit $\delta\nu \to 0$. The resultant y has a *continuous amplitude spectrum*; $a(\nu)$ is the *cosine amplitude spectrum* and $b(\nu)$ the *sine amplitude spectrum* of y.

9.3 Fourier series

We now begin the discussion of Fourier series with a statement of *Fourier's theorem* as under.

Fourier's theorem — Any single-valued periodic function $f(x)$ defined in the closed interval $[-\pi, \pi]$ may be represented over the interval by the following trigonometric series.

$$f(x) = \tfrac{1}{2}a_0 + \sum_{n=1}^{\infty} (a_n \cos nx + b_n \sin nx), \tag{9.3.1}$$

provided the *expansion coefficients*, a_n and b_n, are determined by **Euler's formulae** :

$$a_n = \frac{1}{\pi}\int_{-\pi}^{\pi} f(x)\cos nx\, dx, \quad (n = 0, 1, 2, \cdots) \tag{9.3.2}$$

$$\text{and} \quad b_n = \frac{1}{\pi}\int_{-\pi}^{\pi} f(x)\sin nx\, dx, \quad (n = 1, 2, 3, \cdots) \tag{9.3.3}$$

The expression on the right hand side of (9.3.1) is called the **Fourier series** of $f(x)$ and the a_n, b_n the Fourier series *cosine coefficient* and *sine coefficient* respectively. Equations (9.3.2) and (9.3.3), called the **Euler formulae**, determine a_n and b_n. The integrals of (9.3.2) and (9.3.3) are known as **Fourier integrals**.

Fourier discussed a number of special cases of the above theorem but could *not* develop any logical proof of it. In fact, the theorem as stated earlier is *not valid*. The first step towards a logical proof of the theorem was provided by Dirichlet who proved the following theorem called *Dirichlet's theorem*.

Dirichlet's theorem — If for the interval $[-\pi, \pi]$, the function $f(x)$ is *bounded*, *single-valued*, having at most a *finite number of extremums* and only a *finite number of finite discontinuities* (i.e., piecewise continuous), in the interval and *if*

$$f(x + 2\pi) = f(x) \quad \text{for values of } x \text{ outside } [-\pi, \pi],$$

$$\text{then,} \quad S_p(x) = \frac{1}{2}a_0 + \sum_{n=1}^{p} [a_n \cos nx + b_n \sin nx] \tag{9.3.4}$$

converges to $f(x)$ as $p \to \infty$ at values of x for which $f(x)$ is continuous and to $\frac{1}{2}[f(x+0) + f(x-0)]$ at points of discontinuity. The $f(x+0)$ and $f(x-0)$ correspond to the right hand and the left hand respectively. It is assumed that a_n's and b_n's are given by Euler's formulae.

The above *conditions* imposed on the function $f(x)$ by Dirichlet's theorem are known as **Dirichlet's conditions**. We note down the **conditions** again hereunder for an wider understanding.

1. *The function* $f(x)$ *should be bounded*, i.e., it should satisfy the condition $|f(x)| \le c$ for all values of x in the interval, c being some constant.

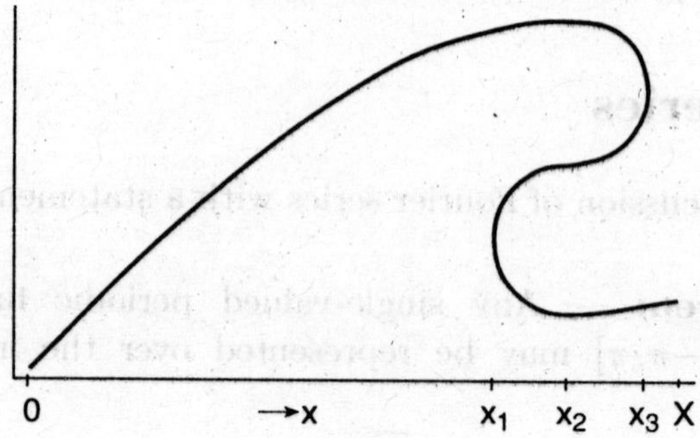

Fig 9.2

2. *The function* $f(x)$ *should be single valued.* Fig. 9.2 shows a multiple valued function where at $x = x_2$, the function has three possible values and Fourier expansion is *not possible* in the interval $x_1 < x < x_3$. But expansion is *possible* in the interval $0 < x < x_1$ and also in $x_3 < x < X$.

3. *The function can have only a finite number of extrema (maxima and minima).* Fig. 9.3a shows a plot of $\cos(1/x)$ against x and in the interval $0 < x < x_1$ (x_1 is any positive number), the function $\cos(1/x)$ has infinite number of maxima and minima. So, Fourier expansion is *not possible* in that interval. However, expansion is *possible* in the interval $x_1 < x < x_2$ where x_1, x_2 are any two positive numbers.

4. *The function must be piecewise continuous, i.e., it can only have a finite number of finite discontinuities.* Fig. 9.3b shows a piecewise continuous function. But functions like $f(x) = \tan x$ has, in the interval $0 < x < \pi$, an infinite discontinuity at $x = \pi/2$. Fourier expansion is therefore *not* possible.

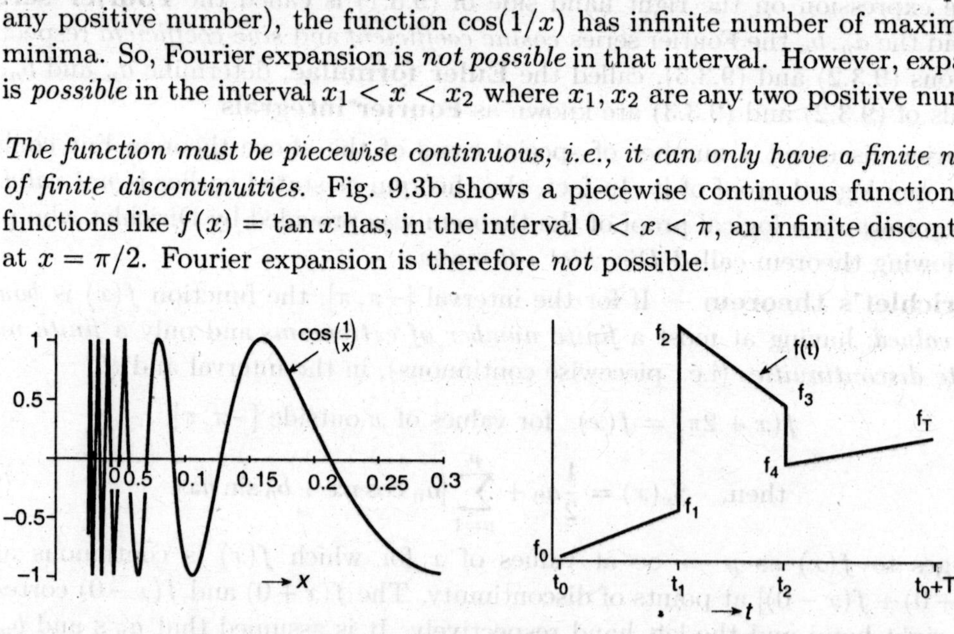

Fig. 9.3

The Dirichlet's conditions are sufficient but not necessary. Most of the functions representing solutions of physical problems satisfy Dirichlet's conditions.

The term $\frac{1}{2}a_0$ in Fourier series has been included to allow the possibility that $f(x)$ oscillates about some mean value other than zero, and it is written $\frac{1}{2}a_0$, rather than a_0, simply because a_0 would then be given by the relation (9.3.2) along with the cosine coefficient, since cosine of zero is 1.

$$\therefore \quad a_0 = \frac{1}{\pi}\int_{-\pi}^{\pi} f(x)dx \tag{9.3.5}$$

$$\therefore \quad \frac{1}{2}a_0 = \frac{1}{2\pi}\int_{-\pi}^{\pi} f(x)dx = \text{mean value of } f(x) = \overline{f(x)}.$$

• Fourier's theorem *does not* apply at the discontinuities of $f(x)$. At a discontinuity of $f(x)$, the value of the Fourier series is the *mean* of the values of $f(x)$ on both sides of the discontinuity.

• Equation (9.3.1) may also be written as shown under.

$$f(x) = \tfrac{1}{2}A_0 + \sum_{n=1}^{\infty} A_n \cos(nx - \alpha_n),$$

where, we have, $\quad A_n = \sqrt{a_n^2 + b_n^2}; \;\; \tan\alpha_n = b_n/a_n;$ and $A_0 = a_0$.

9.3.1 Fourier coefficients

We shall, in this section, justify the forms of a_0, a_n and b_n as given by (9.3.5), (9.3.2) and (9.3.3). We first state the following results that may be easily verified by simple integration.

1. $$\int_{-\pi}^{\pi} \cos nx \cos mx \, dx = \begin{cases} 0 & \text{for } n \neq m \\ 2\pi & \text{for } n = m = 0 \\ \pi & \text{for } n = m > 0. \end{cases}$$

2. $$\int_{-\pi}^{\pi} \sin nx \sin mx \, dx = \begin{cases} 0 & \text{for } n \neq m \\ 0 & \text{for } n = m = 0 \\ \pi & \text{for } n = m > 0. \end{cases}$$

3. $$\int_{-\pi}^{\pi} \sin nx \cos mx \, dx = 0 \quad \text{for all } n \text{ and } m.$$

4. $$\int_{-\pi}^{\pi} \sin nx \, dx = 0 \quad \text{for all } n.$$

5. $$\int_{-\pi}^{\pi} \cos nx \, dx = \begin{cases} 0 & \text{for } n > 0 \\ 2\pi & \text{for } n = 0 \end{cases}$$

Now, multiplying $f(x) = \frac{1}{2}a_0 + \sum\limits_{n=1}^{\infty}(a_n \cos nx + b_n \sin nx)$ by $\cos mx \, dx$ and integrating from $x = -\pi$ to $x = \pi$, we obtain

$$\int_{-\pi}^{\pi} f(x)\cos mx\,dx = \frac{1}{2}a_0\int_{-\pi}^{\pi}\cos mx\,dx + \int_{-\pi}^{\pi}\cos mx\left\{\sum_{n=1}^{\infty} a_n\cos nx\right\}dx$$
$$+\int_{-\pi}^{\pi}\cos mx\left\{\sum_{n=1}^{\infty} b_n\sin nx\right\}dx$$
$$= \frac{1}{2}a_0\int_{-\pi}^{\pi}\cos mx\,dx + \sum_{n=1}^{\infty}\left\{a_n\int_{-\pi}^{\pi}\cos mx\cos nx\,dx\right\}$$
$$+\sum_{n=1}^{\infty}\left\{b_n\int_{-\pi}^{\pi}\cos mx\sin nx\,dx\right\} \quad (9.3.6)$$

[We have assumed here that we may interchange the order of the integral and the summation notations, and then integrate the series term-by-term].

I. When $m = 0$, we have from (9.3.6), using the result of integrations **1–5** as stated at the very beginning of the section,

$$\int_{-\pi}^{\pi} f(x)dx = \tfrac{1}{2}a_0 \times 2\pi$$
$$\therefore \quad a_0 = \frac{1}{\pi}\int_{-\pi}^{\pi} f(x)dx$$

II. When $m = n > 0$, we have from (9.3.6), using the result of integration **1**,

$$\int_{-\pi}^{\pi} f(x)\cos nx\,dx = a_n\int_{-\pi}^{\pi}\cos nx\cos nx\,dx = \pi a_n$$
$$\therefore \quad a_n = \frac{1}{\pi}\int_{-\pi}^{\pi} f(x)\cos nx\,dx$$

To obtain the Fourier coefficients b_n, we similarly multiply

$$f(x) = \frac{1}{2}a_0 + \sum_{n=1}^{\infty}(a_n\cos nx + b_n\sin nx),$$

by $\sin mx\,dx$ and integrte, proceeding exactly as before, from $x = -\pi$ to $x = \pi$ to get finally

$$\int_{-\pi}^{\pi} f(x)\sin nx\,dx = b_n\int_{-\pi}^{\pi}\sin nx\sin nx\,dx = \pi b_n \quad (m = n > 0)$$
$$\therefore \quad b_n = \frac{1}{\pi}\int_{-\pi}^{\pi} f(x)\sin nx\,dx$$

• We have assumed here implicitly in the above derivation that $f(x)$ is *continuous* in $[-\pi, \pi]$. If however $f(x)$ is *discontinuous* at $x = x_0$, then the Fourier coefficients must be evaluated by summing the integrals over the continuous parts. For instance,

$$a_n = \frac{1}{\pi}\int_{-\pi}^{x_0} f(x)\cos nx\,dx + \frac{1}{\pi}\int_{x_0}^{\pi} f(x)\cos nx\,dx$$

and similarly for the coefficient b_n and a_0.

9.3.2 Fourier cosine and sine series : Symmetry

Even-function symmetry — The function $f(x)$ is an *even function* if $f(x) = f(-x)$, in the range $-\pi < x \leq \pi$. Then the Fourier coefficients, by virtue of the properties of definite integrals, are as follows.

$$a_0 = \frac{1}{\pi}\int_{-\pi}^{\pi} f(x)dx = \frac{2}{\pi}\int_0^{\pi} f(x)dx$$

$$a_n = \frac{1}{\pi}\int_{-\pi}^{\pi} f(x)\cos nx\, dx = \frac{2}{\pi}\int_0^{\pi} f(x)\cos nx\, dx$$

$$b_n = \frac{1}{\pi}\int_{-\pi}^{\pi} f(x)\sin nx\, dx = 0, \quad \text{for all } n\ (n = 1, 2, \cdots)$$

So, if $f(x)$ is an **even function**, its Fourier series reduces to a **cosine series**, along with the constant $\frac{1}{2}a_0$, i.e.,

$$\boxed{f(x) = \frac{1}{2}a_0 + \sum_{n=1}^{\infty} a_n \cos nx}$$

all the sine terms are totally absent.

Odd-function symmetry — Similarly, $f(x)$ is an *odd function*, if $f(x) = -f(-x)$, in the same range, and

$$a_0 = \frac{1}{\pi}\int_{-\pi}^{\pi} f(x)dx = 0$$

$$a_n = \frac{1}{\pi}\int_{-\pi}^{\pi} f(x)\cos nx\, dx = 0, \quad \text{for all } n.$$

$$b_n = \frac{1}{\pi}\int_{-\pi}^{\pi} f(x)\sin nx\, dx$$

$$= \frac{2}{\pi}\int_0^{\pi} f(x)\sin nx\, dx \quad (n = 1, 2, 3, \cdots)$$

Thus, if $f(x)$ is an **odd function**, its Fourier series reduces to a **sine series**, i.e.,

$$\boxed{f(x) = \sum_{n=1}^{\infty} b_n \sin nx}$$

all the cosine terms and the constant are absent.

• For example, x^2, $\cos x$, $\sec x$ are all even functions. Graphically, an even function is symmetrical about the y-axis, i.e., $f(x)$ axis. Here y-axis is the mirror for the reflection of the curve.

Similarly, x^3, $\sin x$, $\tan x$ are all odd functions. And graphically, an odd function is symmetrical about the origin.

9.3.3 Complex form of Fourier series

The complex form of Fourier series is obtained by expressing '$\cos nx$' and '$\sin nx$' in exponential form, that is,

$$f(x) = \tfrac{1}{2}a_0 + \sum_{n=1}^{\infty}(a_n \cos nx + b_n \sin nx)$$

$$= \tfrac{1}{2}a_0 + \sum_{n=1}^{\infty} a_n \left[\frac{e^{jnx} + e^{-inx}}{2}\right] + \sum_{n=1}^{\infty} b_n \left[\frac{e^{jnx} - e^{-inx}}{2j}\right]$$

$$= \sum_{-\infty}^{\infty} C_n e^{jnx}, \text{ where} \tag{9.3.7}$$

$$C_0 = \tfrac{1}{2}a_0$$
$$C_n = \tfrac{1}{2}(a_n - jb_n)$$
$$C_{-n} = \tfrac{1}{2}(a_n + jb_n)$$

Eq. (9.3.7) is the *complex form* of the Fourier series.

Multiplying both sides of (9.3.7) by e^{-jmx} and integrating with respect to x,

$$\int_{-\pi}^{\pi} f(x)e^{-jmx}dx = \sum_{-\infty}^{\infty} C_n \int_{-\pi}^{\pi} e^{j(n-m)x}dx$$

$$= \sum_{-\infty}^{\infty} C_n \times 2\pi\delta_{mn}$$

$$= 2\pi C_n$$

$$\therefore \quad C_n = \frac{1}{2\pi}\int_{-\pi}^{\pi} f(x)e^{-jnx}dx$$

$$C_{-n} = \frac{1}{2\pi}\int_{-\pi}^{\pi} f(x)e^{jnx}dx$$

$$C_0 = \frac{1}{2\pi}\int_{-\pi}^{\pi} f(x)dx = \tfrac{1}{2}a_0$$

• In some problems, it becomes more convenient to use the above complex form, rather than the trigonometric one, of the Fourier series.

9.3.4 Change in interval of expansion

So far, the expansion has been restricted to the interval $[-\pi, \pi]$. But it becomes necessary in many physical problems to develop Fourier series valid over a wider range, say $[-l, l]$. How then to obtain such an expansion?

Let we take $$f(x) = \tfrac{1}{2}a_0 + \sum_{n=1}^{\infty}(a_n \cos\phi x + b_n \sin\phi x)$$

To determine ϕ such that $f(x) = f(x + 2l)$. This is possible if, plainly, $\phi = n\pi/l$. And then, we obtain

$$f(x) = \tfrac{1}{2}a_0 + \sum_{n=1}^{\infty}\left[a_n \cos\frac{n\pi x}{l} + b_n \sin\frac{n\pi x}{l}\right]$$

for $-l \leq x \leq l$, if $f(x)$ satisfies the Dirichlet's condition in the given interval.

The corresponding Euler coefficients are

$$a_n = \frac{1}{l}\int_{-l}^{l} f(x)\cos\frac{n\pi x}{l}\,dx, \quad n = 0, 1, 2, \cdots$$

$$b_n = \frac{1}{l}\int_{-l}^{l} f(x)\sin\frac{n\pi x}{l}\,dx, \quad n = 1, 2, 3, \cdots$$

$$a_0 = \frac{1}{l}\int_{-l}^{l} f(x)dx$$

Half-range series — Occasionally, a function $f(x)$ is required to be expanded in a Fourier series in the range 0 to l, which is half the range of $2l$. Since the nature of the function outside the range $0 < x < l$ is immaterial, the function is extended in the range $-l < x < 0$ so that the function is even or odd. Then, $f(x)$ will contain either the cosine or sine series.

If $f(x)$ is to be expanded in a *half-range cosine series* in the interval $0 < x < l$, we take $f(x) = f(-x)$ in the range $-l < x < 0$. The function is then **even** over the entire period $(-l, l)$.

$$\text{Then,} \quad f(x) = \tfrac{1}{2}a_0 + \sum_{n=1}^{\infty} a_n \cos\frac{n\pi x}{l}, \quad \text{where}$$

$$a_0 = \frac{2}{l}\int_0^l f(x)dx, \quad \text{and}$$

$$a_n = \frac{2}{l}\int_0^l f(x)\cos\frac{n\pi x}{l}dx$$

If however $f(x)$ is to be expanded in a *half-range sine series* in the internal $0 < x < l$, we choose $f(x) = -f(x)$ in the range $-l < x < 0$ making it **odd** over the period $(-l, l)$. Then, we have

$$f(x) = \sum_{n=1}^{\infty} b_n \sin\frac{n\pi x}{l}, \quad \text{where } b_n = \frac{2}{l}\int_0^l f(x)\sin\frac{n\pi x}{l}dx$$

9.4 Applications of Fourier series

We shall now consider some examples in order that a working knowledge of the Fourier series expansion can be obtained. It is assumed here that the Fourier expansions developed converge to the original functions, and as more and more number of terms are included, the accuracy of the expansion steadily increases.

9.4.1 Sawtooth wave : Expansion of $f(x) = x, -\pi \le x \le \pi$

The graphical representation of the function $f(x)$ in the interval $[-\pi, \pi]$ and its periodic extension outside $[-\pi, \pi]$ are shown in Fig. 9.4

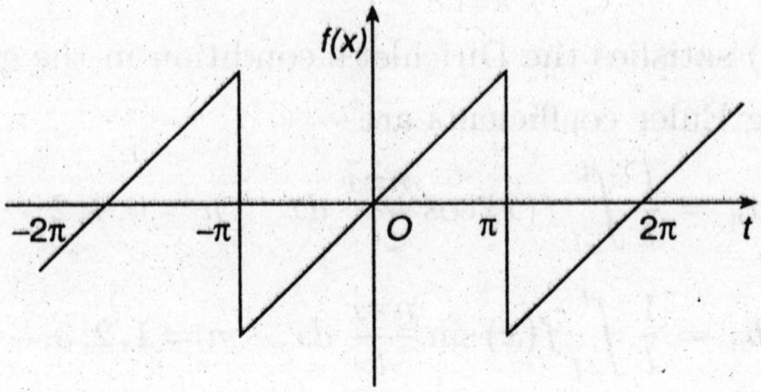

Fig 9.4

$$\text{Now,} \quad a_0 = \frac{1}{\pi}\int_{-\pi}^{\pi} f(x)dx = \frac{1}{\pi}\int_{-\pi}^{\pi} x\,dx = 0$$

$$\text{Also,} \quad a_{n>0} = \frac{1}{n}\int_{-\pi}^{\pi} f(x)\cos nx\,dx$$

$$= \frac{1}{\pi}\int_{-\pi}^{\pi} x\cos nx\,dx = 0$$

$$\text{and,} \quad b_{n>0} = \frac{1}{\pi}\int_{-\pi}^{\pi} f(x)\sin nx\,dx$$

$$= \frac{1}{\pi}\int_{-\pi}^{\pi} x\sin nx\,dx$$

$$= -\frac{2}{n}\cos n\pi = -\frac{2}{n}(-1)^{n+1}$$

The Fourier expansion is thus given by

$$f(x) = x = \frac{1}{2}a_0 + \sum_{n=1}^{\infty}[a_n\cos nx + b_n\sin nx]$$

$$= 2\sum_{n=1}^{\infty}\frac{(-1)^{n+1}\sin nx}{n}$$

The above function corresponds to what is usually called the **sawtooth wave.**

9.4.2 Square wave : Expansion of $f(x) = \begin{cases} 0 & -\pi \le x < 0 \\ h & 0 \le x < \pi \end{cases}$

The graphical representation of the function $f(x)$ in $[-\pi, \pi]$ and its periodic extension outside $[-\pi, \pi]$ are shown in the figure (Fig. 9.5).

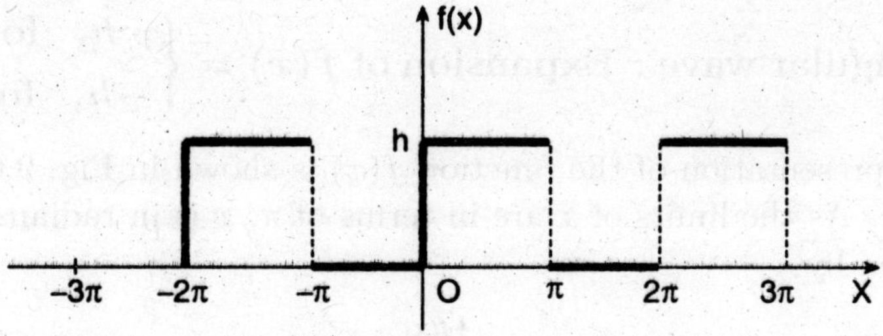

Fig 9.5

$$\text{Now,} \quad a_0 = \frac{1}{\pi}\int_{-\pi}^{0} f(x)dx + \frac{1}{\pi}\int_{0}^{\pi} f(x)dx$$

$$= \frac{1}{\pi}\int_{0}^{\pi} h\,dx = h$$

$$\text{Also,} \quad a_{n>1} = \frac{1}{\pi}\int_{-\pi}^{0} f(x)\cos nx\,dx + \frac{1}{\pi}\int_{0}^{\pi} f(x)\cos nx\,dx$$

$$= \frac{1}{\pi}\int_{0}^{\pi} h\cos nx\,dx$$

$$= \frac{h}{\pi}\int_{0}^{\pi} \cos nx\,dx = 0$$

$$\text{and} \quad b_{n>1} = \frac{1}{\pi}\int_{-\pi}^{0} f(x)\sin nx\,dx + \frac{1}{\pi}\int_{0}^{\pi} f(x)\sin nx\,dx$$

$$= \frac{h}{\pi}\int_{0}^{\pi} \sin nx\,dx$$

$$= \frac{h}{\pi}\left[-\frac{\cos nx}{n}\right]_0^{\pi}$$

$$= \begin{cases} 0, & \text{for } n \text{ even} \\ \frac{2h}{n\pi}, & \text{for } n \text{ odd} \end{cases}$$

∴ The Fourier expansion is therefore given by

$$f(x) = \frac{1}{2}a_0 + \sum_{n=1}^{\infty}(a_n\cos nx + b_n\sin nx)$$

$$= \frac{h}{2} + \frac{2h}{\pi}\sum_{n=1}^{\infty}\frac{\sin nx}{n} \quad (n \to \text{odd})$$

This function corresponds to what is called the **square wave**.

9.4.3 Rectangular wave : Expansion of $f(x) = \begin{cases} h, & \text{for } 0 \leq x \leq \pi \\ -h, & \text{for } \pi \leq x \leq 2\pi \end{cases}$

The graphical representation of the function $f(x)$ is shown in Fig. 9.6. It represents a rectangular wave. As the limits of x are in terms of π, x is in radians. So, the Fourier expansion is given by

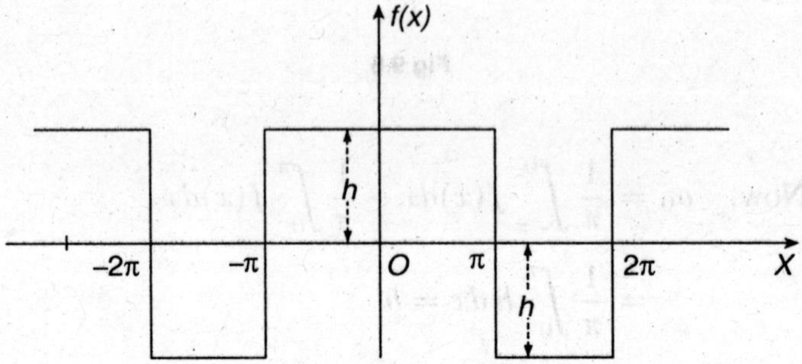

Fig 9.6

$$f(x) = \frac{a_0}{2} + \sum_{n=1}^{\infty}(a_n \cos nx + b_n \sin nx), \quad \text{where}$$

$$a_0 = \frac{1}{\pi}\int_0^{2\pi} f(x)dx = \frac{1}{\pi}\int_0^{\pi} h\,dx + \frac{1}{\pi}\int_{\pi}^{2\pi} -h\,dx = \frac{h}{\pi}[\pi - \pi] = 0$$

$$a_{n>0} = \frac{1}{\pi}\int_0^{2\pi} f(x)\cos nx\,dx = \frac{1}{\pi}\left[\int_0^{\pi} h\cos nx\,dx - \int_{\pi}^{2\pi} h\cos nx\,dx\right]$$

$$= \frac{h}{\pi}\left[\int_0^{\pi} \cos nx\,dx - \int_0^{2\pi} \cos nx\,dx\right] = 0$$

$$b_{n>0} = \frac{1}{\pi}\int_0^{2\pi} f(x)\sin nx\,dx = \frac{h}{\pi}\left[\int_0^{\pi} \sin nx\,dx - \int_{\pi}^{2\pi} \sin nx\,dx\right]$$

$$= \frac{h}{n\pi}\left\{[-\cos nx]_0^{\pi} + [\cos nx]_{\pi}^{2\pi}\right\} = \frac{2h}{n\pi}(1 - \cos n\pi)$$

$$\therefore \quad b_n = \begin{vmatrix} 0, & \text{for } n \text{ even} \\ \dfrac{4h}{n\pi}, & \text{for } n \text{ odd} \end{vmatrix}$$

So, the Fourier expansion of $f(x)$ is given by

$$f(x) = \frac{4h}{\pi}\sum_{n=1}^{\infty}\frac{\sin nx}{n} \quad (n \to \text{odd})$$

$$= \frac{4h}{\pi}\left[\sin x + \frac{\sin 3x}{3} + \frac{\sin 5x}{5} + \cdots\right]$$

This shows that (i) the frequency of the component vibrations are *odd multiples of the fundamental* freuqency, (ii) all components have the *same phase* and (iii) as frequencies increase in the ratio 1 : 2 : 5 : ⋯, the *amplitude decreases* in the ratio $1 : 1/3 : 1/5 : \cdots$.

9.4.4 Half-wave rectifier : Expansion of $f(x) = \sin x$, when $0 \leq x < \pi$, and $f(x) = 0$, when $\pi \leq x \leq 2\pi$

The graphical representation of the function $f(x)$ is shown in the figure (Fig. 9.7).

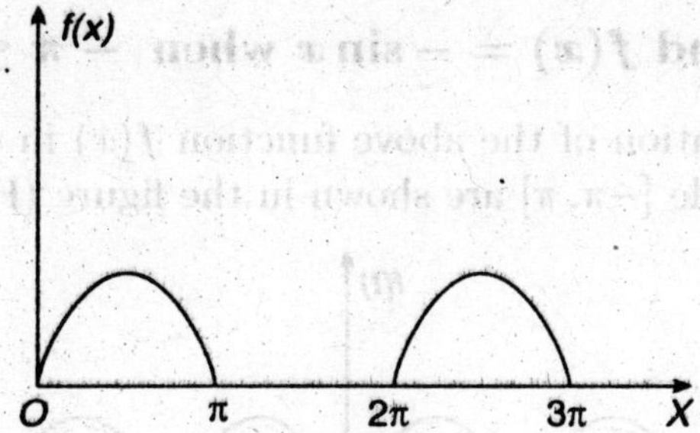

Fig 9.7

$$\text{Now,} \quad a_0 = \frac{1}{\pi}\int_0^{\pi} f(x)dx + \frac{1}{\pi}\int_{\pi}^{2\pi} f(x)dx$$

$$= \frac{1}{\pi}\int_0^{\pi} \sin x\, dx$$

$$= \frac{2}{\pi}$$

$$\text{Also,} \quad a_{n>1} = \frac{1}{\pi}\int_0^{\pi} f(x)\cos nx\, dx + \frac{1}{\pi}\int_0^{2\pi} f(x)\cos nx\, dx$$

$$= \frac{1}{\pi}\int_0^{\pi} \sin x \cos nx\, dx$$

$$= \begin{cases} 0, & \text{for } n\text{-odd} \\ -\frac{2}{\pi}\cdot\frac{1}{n^2-1}, & \text{for } n\text{-even} \end{cases}$$

$$\text{and} \quad b_{n>1} = \frac{2}{\pi}\int_0^{\pi} \sin x \sin nx\, dx$$

$$= \begin{cases} 0, \text{ for all } n, \text{ except } 1 \\ \frac{1}{2}, \text{ for } n = 1 \end{cases}$$

The Fourier expansion of the function $f(x)$ is thus given by

$$f(x) = \frac{1}{2}a_0 + \sum_{n=1}^{\infty}(a_n \cos nx + b_n \sin nx)$$

$$= \frac{1}{\pi} + \frac{1}{2}\sin x - \frac{2}{\pi}\sum_{n=\text{even}} \frac{\cos nx}{n^2 - 1}$$

This function corresponds to what is called the **half-wave rectifier**.

9.4.5 Full-wave rectifier : Expansion of $f(x) = \sin x$, when $0 \le x \le \pi$, and $f(x) = -\sin x$ when $-\pi \le x \le 0$

The graphical representation of the above function $f(x)$ in the inerval $[-\pi, \pi]$ and its periodic extension outside $[-\pi, \pi]$ are shown in the figure (Fig. 9.8).

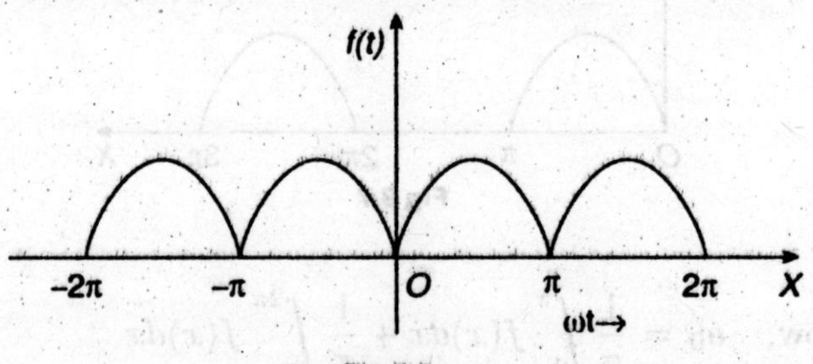

Fig 9.8

$$\text{Now,} \quad a_0 = \frac{1}{\pi}\int_{-\pi}^{0} f(x)dx + \frac{1}{\pi}\int_{0}^{\pi} f(x)dx$$

$$= -\frac{1}{\pi}\int_{-\pi}^{0} \sin x\, dx + \frac{1}{\pi}\int_{0}^{\pi} \sin x\, dx$$

$$= \frac{2}{\pi}\int_{0}^{\pi} \sin x\, dx = \frac{4}{\pi}$$

$$\text{Also,} \quad a_{n>0} = \frac{1}{\pi}\int_{-\pi}^{0} -\sin x \cos nx\, dx + \frac{1}{\pi}\int_{0}^{\pi} \sin x \cos nx\, dx$$

$$= \frac{2}{\pi}\int_{0}^{\pi} \sin x \cos nx\, dx$$

$$= \begin{cases} 0, & \text{for } n\text{-odd} \\ -\frac{4}{\pi}\cdot\frac{1}{n^2-1}, & \text{for } n\text{-even} \end{cases}$$

$$\text{and,} \quad b_{n>0} = \frac{2}{\pi}\int_{0}^{\pi} \sin x \sin nx\, dx = 0$$

∴ The Fourier expansion of the function $f(x)$ is thus given by

$$f(x) = \frac{1}{2}a_0 + \sum_{n=1}^{\infty}(a_n \cos nx + b_n \sin nx)$$

$$= \frac{2}{\pi} - \frac{4}{\pi}\sum_{n=0}^{\infty}\frac{\cos nx}{n^2-1} \quad (n \text{ even})$$

This function corresponds to what is called the **full wave rectifier.**

9.4.6 Riemann zeta function : Summation of series

The Riemann Zeta function is defined, in general, by

$$\xi(k) = \sum_{n=1}^{\infty}\frac{1}{n^k} = \frac{1}{1^k} + \frac{1}{2^k} + \frac{1}{3^k} + \cdots \text{ to } \infty \tag{9.4.1}$$

$$\therefore \quad \xi(2) = \sum_{n=1}^{\infty}\frac{1}{n^2} = \frac{1}{1^2} + \frac{1}{2^2} + \frac{1}{3^2} + \cdots \text{ to } \infty \tag{9.4.2}$$

$$\text{and,} \quad \xi(4) = \sum_{n=1}^{\infty}\frac{1}{n^4} = \frac{1}{1^4} + \frac{1}{2^4} + \frac{1}{3^4} + \cdots \text{ to } \infty \tag{9.4.3}$$

We shall evaluate the series (9.4.2) and (9.4.3) in turn by the application of Fourier expansion.

Let $f(x) = x^2$, $-\pi \le x \le \pi$. The graphical representation of the function $f(x)$ in the interval $[-\pi, \pi]$ and also its periodic extension outside $[-\pi, \pi]$ are shown in Fig. 9.9.

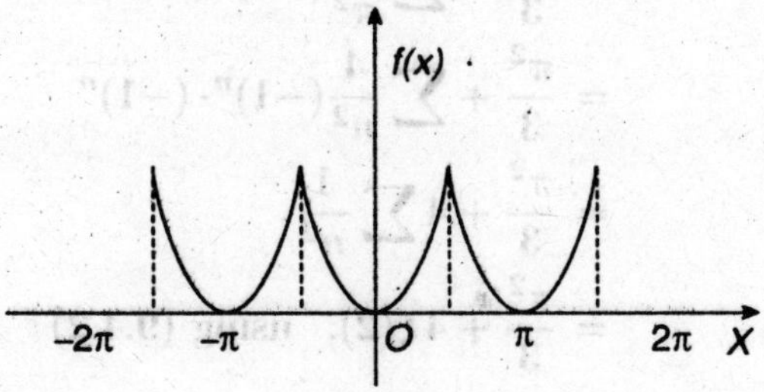

Fig 9.9

Since $f(-x) = (-x)^2 = x^2 = f(x)$, it is an *even* function and so $b_n = 0$.

$$\therefore \quad x^2 = \tfrac{1}{2}a_0 + \sum a_n \cos nx$$

$$\text{where,} \quad a_n = \frac{1}{\pi}\int_{-\pi}^{\pi} f(x)\cos nx\,dx = \frac{1}{\pi}\int_{-\pi}^{\pi} x^2\cos nx\,dx$$

$$= \frac{2}{\pi}\int_0^{\pi} x^2 \cos nx\,dx$$

$$= \frac{2}{\pi}\left[\frac{x^2}{n}\sin nx\right]_0^{\pi} - \frac{2}{\pi}\left[2\int_0^{\pi}\frac{x}{n}\sin nx\,dx\right]$$

$$= -\frac{4}{\pi}\int_0^{\pi} x\sin nx\,dx$$

$$= -\frac{4}{n\pi}\left[-\frac{x}{n}\cos nx\right]_0^{\pi} - \frac{4}{n\pi}\int_0^{\pi}\frac{1}{n}\cos nx\,dx$$

$$= \frac{4}{n^2}\cos n\pi - \frac{4}{n\pi}\left[\frac{1}{n}\cdot\frac{1}{n}\sin nx\right]_0^{\pi}$$

$$= \frac{4}{n^2}\cos n\pi = \frac{4}{n^2}(-1)^n$$

Also, $a_0 = \frac{1}{\pi}\int_{-\pi}^{\pi} f(x)dx$

$$= \frac{1}{\pi}\cdot 2\int_0^{\pi} f(x)dx \quad [\because f(x) = x^2 \text{ (even)}, f(-x) = f(x)]$$

$$= \frac{1}{\pi}2\int_0^{\pi} x^2dx = \frac{2}{\pi}\cdot\left[\frac{x^3}{3}\right]_0^{\pi} = \frac{2}{3}\pi^2$$

$$\therefore \quad f(x) = x^2 = \tfrac{1}{2}a_0 + \sum a_n\cos nx$$

$$= \frac{\pi^2}{3} + \sum_{n=1}^{\infty}\frac{4}{n^2}(-1)^n\cos nx \qquad (9.4.4)$$

Let us now put $x = \pi$. Therefore, we have, from (9.4.4)

$$\pi^2 = \frac{\pi^2}{3} + \sum\frac{4}{\pi^2}(-1)^n\cos n\pi$$

$$= \frac{\pi^2}{3} + \sum\frac{4}{n^2}(-1)^n\cdot(-1)^n$$

$$= \frac{\pi^2}{3} + 4\sum\frac{1}{n^2}$$

$$= \frac{\pi^2}{3} + 4\xi(2), \text{ using (9.4.2)}$$

$$\therefore \quad \xi(2) = \frac{1}{4}\left(\pi^2 - \frac{\pi^2}{3}\right) = \frac{\pi^2}{6}$$

$$\therefore \quad \boxed{\xi(2) = \sum\frac{1}{n^2} = \frac{\pi^2}{6}} \qquad (9.4.5)$$

To evaluate $\xi(4)$: Let us put $f(x) = x^4$

$$a_0 = \frac{1}{\pi}\int_{-\pi}^{\pi} x^4dx = \frac{2}{\pi}\int_0^{\pi} x^4dx = 2\cdot\frac{\pi^4}{5}$$

Also, $b_n = 0$, since $f(x)$ is an *even* function.

$$\therefore \quad f(x) = x^4 = \tfrac{1}{2}a_0 + \sum a_n \cos nx$$

$$= \frac{\pi^4}{5} + \sum a_n \cos nx$$

$$\text{But} \quad a_n = \frac{1}{\pi}\int_{-\pi}^{\pi} f(x)\cos nx\, dx$$

$$= \frac{2}{\pi}\int_0^{\pi} x^4 \cos nx\, dx$$

$$= \frac{2}{\pi}(-1)^n\left[\frac{4\pi^3}{n^2} - \frac{24\pi}{n^4}\right]$$

$$= (-1)^n\left[\frac{8\pi^2}{n^2} - \frac{48}{n^4}\right]$$

$$\therefore \quad x^4 = \frac{\pi^4}{5} + \sum(-1)^n\left(\frac{8\pi^2}{n^2} - \frac{48}{n^4}\right)\cos nx \tag{9.4.6}$$

Let us now put $x = \pi$. We have, therefore, from (9.4.6),

$$\pi^4 = \frac{\pi^4}{5} + \sum(-1)^n\left(\frac{8\pi^2}{n^2} - \frac{48}{n^4}\right)\cos n\pi$$

$$= \frac{\pi^4}{5} + \sum(-1)^n\left(\frac{8\pi^2}{n^2} - \frac{48}{n^4}\right)\cdot(-1)^n$$

$$= \frac{\pi^4}{5} + \sum\left(\frac{8\pi^2}{n^2} - \frac{48}{n^4}\right)$$

$$\therefore \quad \frac{4}{5}\pi^4 = \sum\left(\frac{8\pi^2}{n^2} - \frac{48}{n^4}\right)$$

$$= 8\pi^2\sum\frac{1}{n^2} - 48\sum\frac{1}{n^4}$$

$$= 8\pi^2\xi(2) - 48\xi(4), \quad \text{using (9.4.2) and (9.4.3).}$$

$$\therefore \quad 48\xi(4) = 8\pi^2\xi(2) - \frac{4}{5}\pi^4$$

$$= \frac{8}{6}\pi^4 - 4\cdot\frac{\pi^4}{5} = \frac{8}{15}\pi^4, \quad \text{using (9.4.5)}$$

$$\therefore \quad \boxed{\xi(4) = \frac{\pi^4}{90}} \tag{9.4.7}$$

9.5 Illustrated Examples

By way of applications of Fourier series, we have already given a number of important illustrated examples in Art 9.4.1 to Art 9.4.6. In this section we provide some more illustrated examples.

▶ **Example 1.** *Obtain the Fourier series of the function*

$$f(x) = \begin{cases} \cos x & \text{when } 0 < x \leq \pi \\ -\cos x & \text{when } -\pi \leq x < 0 \end{cases}$$

Solution. The function defined above is a periodic function, the period being $-\pi$ to π. It is shown graphically in Fig. 9.10. It is an *odd* function of x in $-\pi \leq x < \pi$ and in this interval has a finite discontinuity.

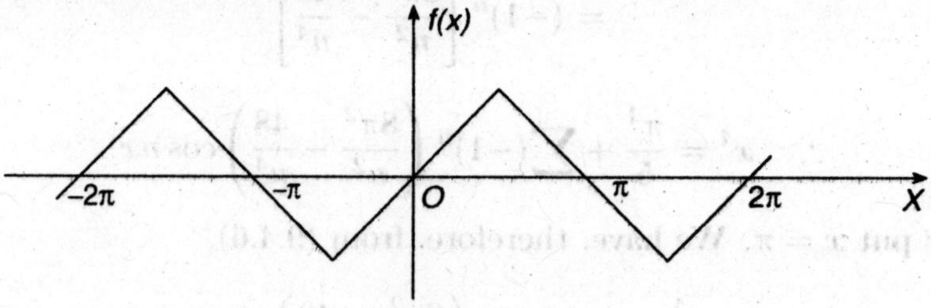

Fig 9.10

Hence, we have

$$f(x) = \frac{1}{2}a_0 + \sum_{n=1}^{\infty}(a_n \cos nx + b_n \sin nx)$$

where $a_n = 0$, the function being an *odd* one.

$$\text{Now,} \quad a_0 = \frac{1}{\pi}\int_{-\pi}^{0} -\cos x\, dx + \frac{1}{\pi}\int_{0}^{\pi} \cos x\, dx = 0$$

$$\text{and} \quad b_n = \frac{1}{\pi}\int_{-\pi}^{0} -\cos x \sin nx\, dx + \frac{1}{\pi}\int_{0}^{\pi} \cos x \sin nx\, dx$$

$$= \frac{1}{2\pi}\left[\frac{\cos(n+1)x}{n+1} + \frac{\cos(n-1)x}{n-1}\right]_{-\pi}^{0}$$

$$-\frac{1}{2\pi}\left[\frac{\cos(n+1)x}{n+1} + \frac{\cos(n-1)x}{n-1}\right]_{0}^{\pi}$$

$$= \frac{1}{\pi}\left[\frac{1-(1)^{n+1}}{n+1} + \frac{1-(-1)^{n-1}}{n-1}\right]$$

$$= \begin{cases} 0, & \text{for } n \text{ odd} \\ \dfrac{4n}{\pi(n^2-1)} & \text{for } n \text{ even.} \end{cases}$$

The required Fourier series is thus given by

$$f(x) = \frac{8}{\pi}\left(\frac{1}{1.3}\sin 2x + \frac{2}{3.5}\sin 4x + \frac{3}{5.7}\sin 6x + \cdots\right)$$
$$= \frac{4}{\pi}\sum_{n}^{\infty}\frac{n}{n^2-1}\sin nx \quad (n \text{ even})$$

This series represents $f(x)$ everywhere in $-\pi \le x \le \pi$ where $f(x)$ is continuous. At $x = 0$, the series has a sum equal to zero.

▶ **Example 2.** *Expand in Fourier series the function* $f(x)=x$ *in interval* $-1<x<1$.

Solution. Here a change in interval is involved.

$$\therefore \quad f(x) = x = \frac{1}{2}a_0 + \sum_{n=1}^{\infty}(a_n\cos n\pi x + b_n \sin n\pi x)$$

$$\text{Now,} \quad a_0 = \int_{-1}^{1} f(x)dx = \int_{-1}^{1} x\,dx = \left.\frac{x^2}{2}\right|_{-1}^{1} = 0$$

$$\text{And,} \quad a_n = \int_{-1}^{1} f(x)\cos n\pi x\,dx$$
$$= \int_{-1}^{1} x\cos n\pi x\,dx = 0, \quad \text{for all } n.$$

$$\text{Also,} \quad b_n = \int_{-1}^{1} f(x)\sin n\pi x\,dx$$
$$= \int_{-1}^{1} x\sin n\pi x\,dx$$
$$= -\frac{2}{\pi n}(-1)^n$$

$$\therefore \quad f(x) = \frac{2}{\pi}\left(\sin \pi x - \tfrac{1}{2}\sin 2\pi x + \tfrac{1}{3}\sin 3\pi x - \cdots\right)$$
$$= -\frac{2}{\pi}\sum_{n=1}^{\infty}(-1)^n\frac{\sin n\pi x}{n}$$

▶ **Example 3.** *Find the Fourier series for the following function*

$$f(x) = \begin{cases} -k & \textit{when} \quad -\pi < x < 0 \\ k & \textit{when} \quad 0 < x < \pi \end{cases}$$

Hence show that

$$\frac{\pi}{4} = 1 - \frac{1}{3} + \frac{1}{5} - \frac{1}{7} + \frac{1}{9} - \cdots \qquad \textbf{(Calcutta Hons.)}$$

Solution. We have : $f(x) = \frac{1}{2}a_0 + \sum(a_n \cos nx + b_n \sin nx)$

$$\text{Here,}\quad a_0 = \frac{1}{\pi}\int_{-\pi}^{\pi} f(x)dx$$

$$= \frac{1}{\pi}\int_{-\pi}^{0} f(x)dx + \frac{1}{\pi}\int_{0}^{\pi} f(x)dx$$

$$= \frac{1}{\pi}\int_{-\pi}^{0} -k\,dx + \frac{1}{\pi}\int_{0}^{\pi} k\,dx = 0$$

$$\text{And,}\quad a_n = \frac{1}{\pi}\int_{-\pi}^{\pi} f(x)\cos nx\,dx$$

$$= \frac{1}{\pi}\int_{-\pi}^{0} -k\cos nx\,dx + \frac{1}{\pi}\int_{0}^{\pi} k\cos nx\,dx = 0$$

$$\text{Also,}\quad b_n = \frac{1}{\pi}\int_{-\pi}^{\pi} f(x)\sin nx\,dx$$

$$= \frac{1}{\pi}\int_{-\pi}^{0} -k\sin nx\,dx + \frac{1}{\pi}\int_{0}^{\pi} k\sin nx\,dx$$

$$= \frac{2k}{n\pi}(1-\cos nx)$$

$$= \begin{cases} 0, & \text{for } n \text{ even} \\ \frac{4k}{n\pi}, & \text{for } n \text{ odd} \end{cases}$$

$$\therefore\quad f(x) = \sum_{n\,(\text{odd})}^{\infty} \frac{4k}{n\pi}\sin nx$$

$$= \frac{4k}{\pi}\left[\sin x + \frac{1}{3}\sin 3x + \frac{1}{5}\sin 5x + \cdots\right]$$

Let us now put $x = \pi/2$ in the above relation. We thus get

$$f\left(\frac{\pi}{2}\right) = k = \frac{4k}{\pi}\left[1 - \frac{1}{3} + \frac{1}{5} - \frac{1}{7} + \cdots\right]$$

$$\therefore\quad \frac{\pi}{4} = 1 - \frac{1}{3} + \frac{1}{5} - \frac{1}{7} + \cdots$$

► **Example 4.** *Give the Fourier series expansion for the function $f(x)$ given by*

$$f(x) = \begin{cases} -\frac{2}{\pi}(x+\pi), & \textit{when } -\pi \le x \le -\frac{1}{2}\pi \\ \frac{2}{\pi}x, & \textit{when } -\frac{1}{2}\pi \le x \le \frac{1}{2}\pi \\ -\frac{2}{\pi}(x-\pi), & \textit{when } \frac{1}{2}\pi \le x \le \pi \end{cases}$$

Also, $f(x+2\pi) = f(x)$, *for* $-\infty < x < \infty$.

Solution. The function $f(x)$ is a continuous periodic function having a period $-\pi$ to π.

$$\therefore \quad f(x) = \frac{1}{2}a_0 + \sum_{n=1}^{\infty}(a_n \cos nx + b_n \sin nx)$$

Here,
$$a_0 = \frac{1}{\pi}\int_{-\pi}^{\pi} f(x)dx$$
$$= \frac{1}{\pi}\int_{-\pi}^{-\frac{1}{2}\pi} -\frac{2}{\pi}(x+\pi)dx + \frac{1}{\pi}\int_{-\frac{1}{2}\pi}^{\frac{1}{2}\pi} \frac{2}{\pi}x\,dx + \frac{1}{\pi}\int_{\frac{1}{2}\pi}^{\pi} -\frac{2}{\pi}(x-\pi)dx$$
$$= \frac{2}{\pi^2}\left[\int_{-\pi}^{-\frac{1}{2}\pi} -(x+\pi)dx + \int_{-\frac{1}{2}\pi}^{\frac{1}{2}\pi} x\,dx + \int_{\frac{1}{2}\pi}^{\pi} -(x-\pi)dx\right]$$
$$= \frac{2}{\pi^2}\left[-\frac{(x+\pi)^2}{2}\right]_{-\pi}^{-\frac{1}{2}\pi} + \frac{2}{\pi^2}\left[\frac{x^2}{2}\right]_{-\frac{1}{2}\pi}^{\frac{1}{2}\pi} + \frac{2}{\pi^2}\left[-\frac{(x-\pi)^2}{2}\right]_{\frac{1}{2}\pi}^{\pi}$$
$$= \frac{1}{\pi^2}\left[-\frac{\pi^2}{4} + \frac{\pi^2}{4} - \frac{\pi^2}{4} + \frac{\pi^2}{4}\right] = 0$$

The axis of the curve thus corresponds to the x-axis.

Now,
$$a_n = \frac{1}{\pi}\int_{-\pi}^{\pi} f(x)\cos nx\,dx$$
$$= \frac{1}{\pi}\int_{-\pi}^{-\frac{1}{2}\pi} -\frac{2}{\pi}(x+\pi)\cos nx\,dx + \frac{1}{\pi}\int_{-\frac{1}{2}\pi}^{\frac{1}{2}\pi} \frac{2}{\pi}x\cos nx\,dx$$
$$+\frac{1}{\pi}\int_{+\frac{1}{2}\pi}^{\pi} -\frac{2}{\pi}(x-\pi)\cos nx\,dx$$
$= 0$, on evaluation of the integrals.

And,
$$b_n = \frac{1}{\pi}\int_{-\pi}^{\pi} f(x)\sin nx\,dx$$
$$= \frac{1}{\pi}\int_{-\pi}^{-\frac{1}{2}\pi} -\frac{2}{\pi}(x+\pi)\sin nx\,dx + \frac{1}{\pi}\int_{-\frac{1}{2}\pi}^{\frac{1}{2}\pi} \frac{2}{\pi}x\sin nx\,dx$$
$$+\frac{1}{\pi}\int_{+\frac{1}{2}\pi}^{\pi} -\frac{2}{\pi}(x-\pi)\sin nx\,dx$$
$$= \begin{cases} 0, & \text{for } n \text{ even} \\ \dfrac{8}{n^2\pi^2}\sin\dfrac{n\pi}{2}, & \text{for } n \text{ odd.} \end{cases}$$

$$\therefore \quad f(x) = \frac{8}{\pi^2}\left[\sin\pi - \frac{1}{3^2}\sin 3\pi + \frac{1}{5^2}\sin 5\pi - \cdots\right]$$

For the curve representing $f(x)$, see Fig. 9.6.

► **Example 5.** *Express $f(x)=x$ as a half-range sine series in the internal $0<x<2$.*

(Burdwan Hons.)

Solution. Here, the function $f(x) = x = \sum\limits_{n=1}^{\infty} b_n \sin \dfrac{n\pi x}{2}$

$$\text{where}\quad b_n = \frac{2}{2}\int_0^2 f(x)\sin\frac{n\pi x}{2}dx$$

$$= \int_0^2 x\sin\frac{n\pi x}{2}dx$$

$$= \left[x\cdot\frac{-\cos(n\pi x/2)}{n\pi/2} - \frac{-\sin(n\pi x/2)}{(n\pi/2)^2}\right]_0^2$$

integrating by parts,

$$= -\frac{4}{n\pi}\cos n\pi$$

$$= \frac{4}{n\pi},\quad \text{when } n = \text{odd}$$

$$= -\frac{4}{n\pi},\quad \text{when } n = \text{even}$$

$$\therefore\quad f(x) = x = -\frac{4}{\pi}\sum_{n=1}^{\infty}\frac{(-1)^n}{n}\sin\frac{n\pi x}{2}$$

$$= \frac{4}{\pi}\left[\sin\frac{\pi x}{2} - \frac{1}{2}\sin\pi x + \frac{1}{3}\sin\frac{3}{2}\pi x - \cdots\right]$$

► **Example 6.** *Express in a Fourier series the periodic function $f(x)$ with period $2l$, where*

$$f(x) = |x| \ \textit{for}\ -l < x < l$$

and hence show that

$$\frac{1}{1^2} + \frac{1}{3^2} + \frac{1}{5^2} + \cdots = \frac{\pi^3}{8}$$

(Calcutta and Burdwan Hons.)

Solution. We have : $f(x) = \frac{1}{2}a_0 + \sum\limits_{n=1}^{\infty}\left(a_n\cos\dfrac{n\pi x}{l} + b_n\sin\dfrac{n\pi x}{l}\right)$.

Since $f(x) = |x|$ is an **even** function, $b_n = 0$.

$$\therefore\quad f(x) = \frac{1}{2}a_0 + \sum_{n=1}^{\infty} a_n\cos\frac{n\pi x}{l}$$

$$\text{Now,}\quad a_0 = \frac{1}{l}\int_{-l}^{l} f(x)dx = \frac{2}{l}\int_0^l x\,dx = \frac{2}{l}\cdot\frac{l^2}{2} = l$$

Also, $a_n = \frac{1}{l}\int_{-l}^{l} f(x)\cos\frac{n\pi x}{l}dx = \frac{2}{l}\int_0^l x\cos\frac{n\pi x}{l}dx$

$$= \frac{2}{l}\left[x\frac{\sin(n\pi x/l)}{n\pi/l} - \frac{-\cos(n\pi x/l)}{(n\pi/l)^2}\right]_0^l, \text{ integrating by parts.}$$

$$= \frac{2l}{n^2\pi^2}(\cos n\pi - 1)$$

$= 0$, when n is even

$= -\frac{4l}{n^2\pi^2}$ when n is odd.

$$\therefore \quad f(x) = |x| = \frac{l}{2} - \frac{4l}{\pi^2}\sum_{n=\text{odd}}^{\infty}\frac{1}{n^2}\cos(n\pi x/l)$$

$$= \frac{l}{2} - \frac{4l}{\pi^2}\left[\cos\frac{\pi x}{l} + \frac{1}{3^2}\cos\frac{3\pi x}{l} + \frac{1}{5^2}\cos\frac{5\pi x}{l} + \cdots\right]$$

Putting in the above relation $x = 0$, we get

$$0 = \frac{l}{2} - \frac{4l}{\pi^2}\left[\frac{1}{1^2} + \frac{1}{3^2} + \frac{1}{5^2} + \cdots\right]$$

$$\Rightarrow \quad \frac{1}{1^2} + \frac{1}{3^2} + \frac{1}{5^2} + \cdots = \frac{l}{2} \times \frac{\pi^2}{4l} = \boldsymbol{\frac{\pi^2}{8}}$$

▶ **Example 7.** *Show that the function*

$$f(x) = \begin{cases} 0 & \text{for } -\pi \le x < 0 \\ x & \text{for } \quad 0 \le x < \pi \end{cases}$$

can be expanded in Fourier series as

$$f(x) = \frac{\pi}{4} - \frac{2}{\pi}\sum_{n=1}^{\infty}\frac{\cos(2n-1)x}{(2n-1)^2} - \sum_{n=1}^{\infty}(-1)^n\frac{\sin nx}{n}$$

From the expansion show that

$$1 + \frac{1}{3^2} + \frac{1}{5^2} + \cdots = \frac{\pi^2}{8}$$ **(Calcutta Hons.)**

Solution. We have $f(x) = \frac{1}{2}a_0 + \sum_{n=1}^{\infty}(a_n\cos nx + b_n\sin nx)$

Now, $a_0 = \frac{1}{\pi}\int_{-\pi}^{\pi} f(x)dx = \frac{1}{\pi}\left[\int_{-\pi}^{0} 0\cdot dx + \int_0^{\pi} x\,dx\right] = \frac{\pi}{2}$

Also, $a_n = \frac{1}{\pi}\int_{-\pi}^{\pi} f(x)\cos nx\,dx = \frac{1}{\pi}\int_0^{\pi} x\cos nx\,dx$

$$= \frac{1}{\pi}\left[x\frac{\sin nx}{n} - \frac{-\cos nx}{n^2}\right]_0^{\pi}, \text{ integrating by parts.}$$

$$= \frac{1}{\pi n^2}(\cos n\pi - 1)$$

And, $$b_n = \frac{1}{\pi}\int_{-\pi}^{\pi} f(x)\sin nx\,dx$$

$$= \frac{1}{\pi}\int_{0}^{\pi} x\sin nx\,dx$$

$$= \frac{1}{\pi}\left[x\frac{-\cos nx}{n} - \frac{-\sin nx}{n^2}\right]_0^{\pi}$$

$$= -\frac{1}{n}\cos n\pi$$

$$= \left.\begin{array}{ll}\frac{1}{n}, & \text{when } n = \text{odd}\\ -\frac{1}{n}, & \text{when } n = \text{even}\end{array}\right\} \text{ i.e., } = (-1)^n(1/n)$$

$$\therefore \quad f(x) = \frac{\pi}{4} + \frac{1}{\pi}\sum_{n=1}^{\infty}\frac{\cos n\pi - 1}{n^2}\cos nx - \sum_{n=1}^{\infty}\frac{(-1)^n}{n}\sin nx$$

$$= \frac{\pi}{4} + \frac{1}{\pi}\left[-\frac{2\cos x}{1^2} - \frac{2\cos 3x}{3^2} - \frac{2\cos 5x}{5^2} - \cdots\right] - \sum_{n=1}^{\infty}\frac{(-1)^n}{n}\sin nx$$

$$= \frac{\pi}{4} - \frac{2}{\pi}\sum_{n=1}^{\infty}\frac{\cos(2n-1)x}{(2n-1)^2} - \sum_{n=1}^{\infty}\frac{(-1)^n}{n}\sin nx$$

Substituting $x = 0$, in the above relation, we obtain

$$f(0) = 0 = \frac{\pi}{4} - \frac{2}{\pi}\left(\frac{1}{1^2} + \frac{1}{3^2} + \frac{1}{5^2} + \cdots\right)$$

$$\text{or,} \quad 1 + \frac{1}{3^2} + \frac{1}{5^2} + \cdots = \frac{\pi}{4}\cdot\frac{\pi}{2} = \frac{\pi^2}{8}$$

► **Example 8.** *Assuming that the Fourier expansion of the function $f(x)$ is uniformly convergent, show that*

$$\frac{1}{\pi}\int_{-\pi}^{\pi}[f(x)]^2 dx = \frac{1}{2}a_0^2 + \sum_{n=1}^{\infty}(a_n^2 + b_n^2)$$

Solution. We have

$$\frac{1}{\pi}\int_{-\pi}^{\pi}[f(x)]^2 dx = \frac{1}{\pi}\int_{-\pi}^{\pi}\left[\frac{1}{2}a_0 + \sum_{n=1}^{\infty}a_n\cos nx + \sum_{n=1}^{\infty}b_n\sin nx\right]^2 dx$$

$$= \frac{1}{\pi}\left[\frac{1}{4}a_0^2\int dx + \sum_{n=1}^{\infty}a_n^2\int\cos^2 nx\,dx + \sum_{n=1}^{\infty}b_n^2\int\sin^2 nx\,dx\right.$$

$$+ \sum_{n=1}^{\infty}2\frac{1}{2}a_0a_n\int\cos nx\,dx + \sum_{n=1}^{\infty}2\frac{1}{2}a_0b_n\int\sin nx\,dx$$

$$\left. + \sum_{r=1}^{\infty}\sum_{s=1}^{\infty}-2a_ra_s\int\cos rx\sin sx\,dx\right]$$

$$= \frac{1}{\pi}\left[\frac{1}{4}a_0^2 \cdot 2\pi + \sum_{n=1}^{\infty} \pi a_n^2 + \sum_{n=1}^{\infty} \pi b_n^2 + 0 + 0 + 0\right]$$

$$= \frac{1}{2}a_0^2 + \sum_{n=1}^{\infty}(a_n^2 + b_n^2).$$

The above relation is called **Parseval relation.**

▶ **Example 9.** *Find a Fourier series to represent $x - x^2$ from $x = -\pi$ to $x = \pi$.*

(Andhra and Gauhati Univ.)

Solution. Let $f(x) = x - x^2 = \frac{1}{2}a_0 + \sum_{n=1}^{\infty} a_n \cos nx + \sum_{n=1}^{\infty} b_n \sin nx.$

Now, $$a_0 = \frac{1}{\pi}\int_{-\pi}^{\pi}(x - x^2)dx$$

$$= \frac{1}{\pi}\left|\frac{x^2}{2} - \frac{x^3}{3}\right|_{-\pi}^{\pi} = -\frac{2\pi^2}{3}$$

$$a_n = \frac{1}{\pi}\int_{-\pi}^{\pi}(x - x^2)\cos nx\,dx$$

$$= \frac{1}{\pi}\left[(x - x^2)\frac{\sin nx}{n} - (1 - 2x)\left(-\frac{\cos nx}{n^2}\right) + (-2)\left(-\frac{\sin nx}{n^3}\right)\right]_{-\pi}^{\pi}$$

integrating by parts.

$$= -\frac{4(-1)^n}{n^2} \qquad [\because \cos n\pi = (-1)^n]$$

$$\therefore \quad a_1 = 4/1^2, \; a_2 = -4/2^2, \; a_3 = 4/3^2, \; a_4 = -4/4^2, \cdots.$$

$$b_n = \frac{1}{\pi}\int_{-\pi}^{\pi}(x - x^2)\sin nx\,dx$$

$$= \frac{1}{\pi}\left[(x - x^2)\left(-\frac{\cos nx}{n}\right) - (1 - 2x)\left(-\frac{\sin nx}{n^2}\right) + (-2)\left(\frac{\cos nx}{n^3}\right)\right]_{-\pi}^{\pi}$$

$$= \frac{-2(-1)^n}{n} \qquad [\because \sin n\pi = 0]$$

$$\therefore \quad b_1 = 2/1, \; b_2 = -2/2, \; b_3 = 2/3, \; b_4 = -2/4, \cdots.$$

We have, therefore

$$f(x) = x - x^2$$

$$= -\frac{\pi^2}{3} + 4\left[\frac{\cos x}{1^2} - \frac{\cos 2x}{2^2} + \frac{\cos 3x}{3^2} - \cdots\right] + 2\left[\frac{\sin x}{1} - \frac{\sin 2x}{2} + \frac{\sin 3x}{3} - \cdots\right]$$

which is the required Fourier series.

• **Note** that if we put $x = 0$ in the above series, we obtain

$$0 = -\frac{\pi^2}{3} + 4\left(\frac{1}{1^2} - \frac{1}{2^2} + \frac{1}{3^2} - \frac{1}{4^2} + \cdots\right)$$

$$\therefore \quad \frac{1}{1^2} - \frac{1}{2^2} + \frac{1}{3^2} - \frac{1}{4^2} + \cdots = \frac{\pi^2}{12}$$

which is an interesting series.

▶ **Example 10.** *Show that, if a real function $f(x)$ be expanded into complex Fourier series,*

$$\sum_{-\infty}^{\infty} C_n e^{jnx}$$

then, C_{-n} is the complex conjugate of C_n.

Solution. We have

$$C_n = \frac{1}{2\pi}\int_{-\pi}^{\pi} f(x)e^{jnx}dx$$

$$\therefore \quad C_{-n} = \frac{1}{2\pi}\int_{-\pi}^{\pi} f(x)e^{-jnx}dx = C_n^*$$

where C_n^* denotes the complex conjugate of C_n.

Review Questions

1. What is a Fourier series? What is the advantage of Fourier expansion over that of Taylor?

2. State under what conditions, a function $f(x)$ can be Fourier expanded into a convergent trigonometric series. How can Fourier expansion be applied to periodic functions?

3. State and explain Fourier's theorem. Analyse, with its help, a square periodic wave given by the following form

$$y = \begin{cases} A\,(\text{const.}), & \text{for } 0 \le t \le \frac{T}{2} \\ 0, & \text{for } \frac{T}{2} \le t \le T \end{cases}$$

What harmonics are present? **(Calcutta Hons.)**

4. Expand the function $f(x) = x$ in Fourier series in the interval $-\pi \le x \le \pi$.

5. State Fourier's theorem. A periodic square wave has the displacement curve

$$x = \begin{cases} f(t) = \text{const.}, & 0 \le t \le \frac{1}{2}T \\ \quad\;\; = 0, & \frac{1}{2}T \le t \le T. \end{cases}$$

Find the Fourier series for x. **(Burdwan Hons.)**

6. Expand $f(x) = x^2$ in Fourier series and hence show that

$$\frac{\pi^2}{12} = \frac{1}{1^2} - \frac{1}{2^2} + \frac{1}{3^2} - \frac{1}{4^2} + \cdots \text{ to } \infty$$

7. Develop the Fourier expansion for

$$f(t) = \begin{cases} 0, & \text{for } -\pi \le \omega t \le 0 \\ \sin \omega t, & \text{for } \quad 0 \le \omega t \le \pi \end{cases}$$

8. Expand $f(x) = 1$ for values of x lying between 0 and π in a Fourier series. Hence show that

$$\frac{\pi}{4} = 1 - \frac{1}{3} + \frac{1}{5} - \frac{1}{7} + \cdots$$ **(Calcutta Hons.)**

9. A *triangular wave* is represented by

$$f(x) = \begin{cases} x, & \text{for } \quad 0 < x < \pi \\ -x, & \text{for } -\pi < x < 0 \end{cases}$$

Develop the Fourier expansion for the function $f(x)$.

10. Expand $f(x) = x^2$ in the interval $0 < x < 2\pi$ in a Fourier series when the period is 2π and show that

$$f(x) = \frac{4\pi^2}{3} + \sum_{n=1}^{\infty} \left(\frac{4}{n^2} \cos nx - \frac{4\pi}{n} \sin nx \right)$$ **(Calcutta Hons.)**

11. Find the Fourier series for the function

$$f(x) = \begin{cases} -x, & \text{for } -4 \le x \le 0 \\ x, & \text{for } \quad 0 \le x \le 4 \end{cases}$$ **(Burdwan Hons.)**

12. Using Fourier series, show that

$$\frac{\pi^2}{6} = \sum_{n=1}^{\infty} \frac{1}{n^2}; \quad \frac{\pi^2}{90} = \sum_{n=1}^{\infty} \frac{1}{n^4}; \quad \frac{\pi^2}{8} = \sum_{n=1}^{\infty} \frac{1}{(2n-1)^2}$$

13. Find the Fourier series expansion for the function $f(x)$, if

$$f(x) = \begin{cases} \pi, & \text{when } -\pi < x \le \frac{1}{2}\pi \\ 0, & \text{when } \frac{1}{2}\pi < x \le \pi \end{cases}$$

14. State Fourier's theorem. Give the conditions of its validity. What values of the function do we obtain at the discontinuities? Illustrate with a simple example.

15. Expand a *square wave*, a *sawtooth wave* and a *symmetrical triangular wave* into a Fourier series.

16. Expand the function $f(x) = |x|$ betwen $-\pi$ and π.

17. A rectified full wave is given by the relation

$$i = \begin{cases} i_0 \sin \omega t, & 0 \le \omega t \le \pi \\ -i_0 \sin \omega t, & \pi \le \omega t \le 2\pi \end{cases}$$

Show that the Fourier series in this case would consist only of *even harmonics*. Evaluate the constant term.

18. In a circuit with a half-wave rectifier, the current is

$$i = \begin{cases} i_0 \sin \omega t, & \text{from } \omega t = 0 \text{ to } \omega t = \pi \\ 0, & \text{from } \omega t = \pi \text{ to } \omega t = 2\pi \end{cases}$$

Give the first four terms of the equivalent Fourier series.

19. (a) If the Fourier series expansion of a function $f(x)$ contains only sine terms, what kind of function is $f(x)$?

(b) Let $F(x)$ have a Fourier series expansion

$$F(x) = \sum_{n=1}^{\infty} a_n \cos nx + \sum_{n=1}^{\infty} b_n \sin nx;$$

then, prove that $\langle F^2(x) \rangle = \dfrac{1}{2\pi} \displaystyle\int_{-\pi}^{\pi} F^2(x) dx = \sum_{n=1}^{\infty} \dfrac{a_n^2 + b_n^2}{2}$

(Calcutta Hons.)

20. By expanding in Fourier series show that

$$\frac{1}{1^2} - \frac{1}{2^2} + \frac{1}{3^2} - \frac{1}{4^2} + \cdots = \frac{\pi^2}{12}$$

(Mysore and Osmania Univ.)

21. Obtain Fourier series for the function $f(x)$ given by

$$f(x) = \begin{cases} 1 + \frac{2x}{\pi}, & -\pi \le x \le 0, \\ 1 - \frac{2x}{\pi}, & 0 \le x \le \pi \end{cases}$$ **(Madras Univ.)**

Deduce that $\dfrac{1}{1^2} + \dfrac{1}{3^2} + \dfrac{1}{5^2} + \dfrac{1}{7^2} + \cdots = \dfrac{\pi^2}{8}$ **(Mangalore Univ.)**

22. Enumerate some of the important features of a Fourier expansion.

23. Completely analyse the square wave periodic function. **(Ranchi Univ.)**

24. Apply Fourier's theorem to analyse a rectangular and a sawtooth wave. **(Kurukshetra Univ.)**

25. Can you expand $f(x) = \tan x$ in a Fourier series? Explain. **(Calcutta Hons.)**

Chapter 10

Ultrasonics and Building Acoustics

10.1 Introduction

The limits of audibility of human ear lie approximately (slight variation is noted among individuals) between 20 and 20,000 cycles per second—the so-called **sonics.** *Sound waves having frequencies above the upper limit of human audibility,* i.e., *greater than* 20 kHz *and up to* 1 GHz *are known as* **ultrasonic waves,** while those below the lower audibility limit (i.e., < 20 Hz) are called **infrasonics**. Sound waves with frequencies above 1 GHz are called **hypersonics.** Formerly, ultrasonics were known as **supersonics**. But in modern acoustical technology, the term *supersonics* refers to speeds higher than that of sound.

The ultrasonic waves travel with the velocity of sound so that their wavelength $\lambda = 330\,\text{m}/20 \times 10^3 = 1.65 \times 10^{-2}\,\text{m} = 1.65\,\text{cm}$ or less. The lowest wavelength that has been generated by the highest acoustic frequency is of the length of visible light.

The extremely small wavelength of ultrasound has been exploited in many branches of science and industry, such as non-destructive industrial testing, destruction of bacteria, underwater signalling etc. Apart from these applications, the study of ultrasonics is of much academic interest in as much as newer phenomena, undetected in the audible range, are observed in the ultrasonic range. Ultrasonics therefore form an important chapter in modern acoustics.

10.2 Production of ultrasonic waves

There are a number of methods for the production of ultrasonic waves. Of them, the most common and important ones are : (i) the use of *magnetostriction* and (ii) the *piezoelectric phenomenon.*

Magnetostriction transducers—Magnetic field causes changes in the dimensions of ferromagnetic substances. This effect is known as **magnetostriction**. Compared to other ferromagnetics, the variation in length is most in nickel. Thus, if a high frequency alternating field is applied parallel to the length of a nickel rod placed in a solenoid, its length varies each time the field grows. The change in length is independent of the sign of the field but depends only on its magnitude and the nature of the material. This results in the setting up of vibrations in the rod that acts as a transducer, converting magnetic energy into sound energy. The vibrations have a frequency double that of the applied field. Ordinarily, the amplitude of vibrations of the rod is small. But if the natural frequency of the rod be in unison with the applied frequency of the field, resonance occurs when the amplitude of vibration is considerably increased. If the applied frequency is of the order of ultrasonic frequency, the nickel rod would excite high frequency longitudinal waves or ultrasound in the medium in which it is placed.

Oscillator circuit—A simple magnetostriction oscillator circuit is illustrated in Fig. 10.1. R is a nickel rod clamped at its centre and carries at the ends two magnetising coils L_1 and L_2, insulated from R. The coil L_1 and the tuning capacitor C form a tank circuit. Its resonant frequency is equal to the fundamental frequency of longitudinal vibration of R. The tank circuit is connected to the collector of the transistor T. L_2 is connected to the base of the transistor via the coupling capacitor C_1. The collector voltage V_{CC} produces a current in L_1. The resulting magnetic field changes the length of the rod R and consequently the flux through L_2 inducing an e.m.f. in it. The induced e.m.f. enhances the forward bias of base-emitter junction. So the collector current increases and provides a positive feedback to sustain the oscillation. Resistances R_1, R_2, R_E along with V_{CC} determine the *dc* operating point of the transistor. C_E is the *ac* by-pass capacitor. Since the nickel rod is free at its both ends, its natural frequency n is given by

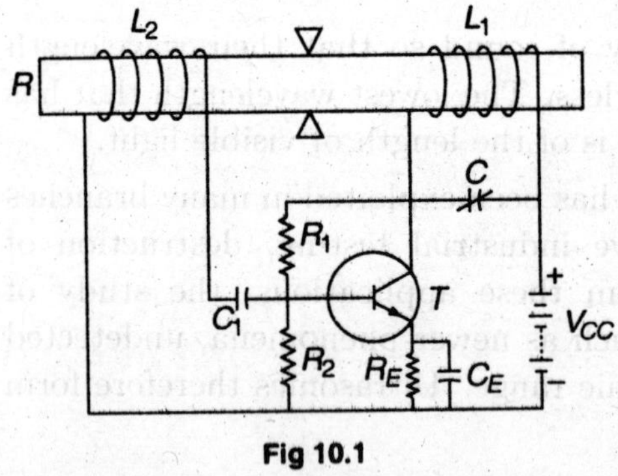

Fig 10.1

$$n = \frac{1}{2l}\sqrt{\frac{Y}{\rho}}$$

where l is the length of the rod, Y the Young's modulus and ρ the density of nickel. When the natural frequency of the rod, serving as transducer, coincides with that of the circuit, the rod is thrown into resonant vibration.

The magnetostriction device for the production of ultrasonic waves is suitable in the range 15–100 kHz. For higher frequencies, piezoelectric oscillator is used.

Piezoelectric transducer—Certain crystals, notably quartz (Fig. 10.2), exhibit **piezoelectric effect**, i.e., develop when stretched or compressed along certain axis (called *mechanical axis*) an electric potential difference along a perpendicular axis (called *electrical axis*). The converse effect is also true and is called **electrostriction**, i.e., when an alternating p.d. is applied along the electric axis, the crystal is set into elastic vibrations along the mechanical axis. If there is unison between the natural frequency of the crystal and that of electrical oscillations, the crystal vibrates with large amplitude, and this is utilised in producing ultrasonic waves. The alternating potential difference is obtained by means of a valve oscillator.

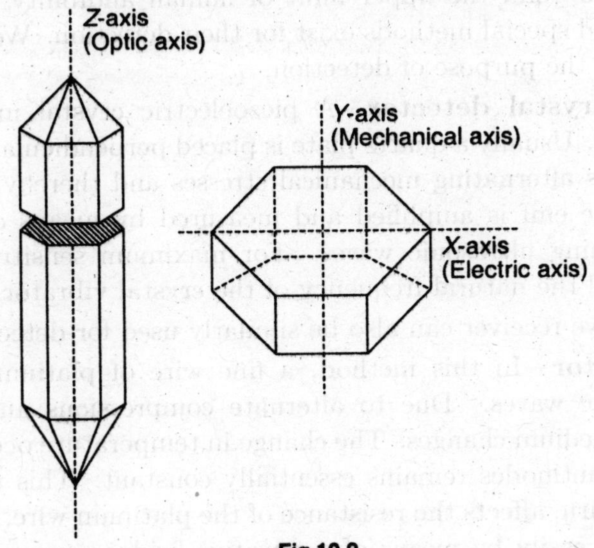

Fig 10.2

Oscillator circuit—The simple oscillator circuit of a piezoelectric generator is the modified *Hartley circuit,* as shown (Fig. 10.3), consisting of a tuned oscillator circuit (tank circuit) with an inductance L (L_1, L_2) and a variable capacitor C in parallel. The electrical oscillations maintain the mechanical vibration of the quartz crystal Q connected across the variable capacitor C.

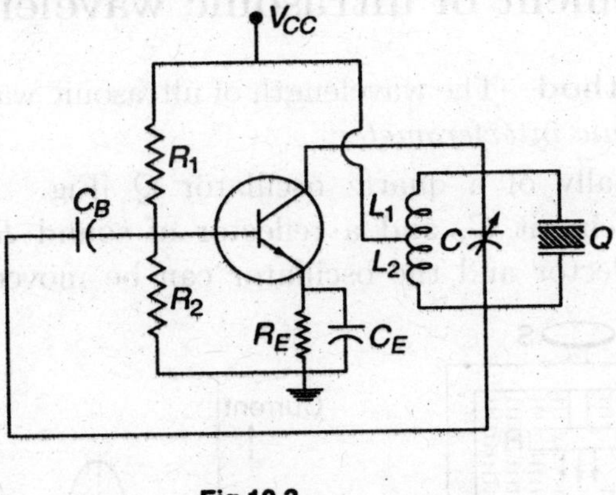

Fig 10.3

C is adjusted to tune the frequency of the resonant circuit to the parallel resonance frequency of the quartz crystal to set it into mechanical vibrations to produce ultrasonics. Feedback voltage for maintaining oscillations is obtained by connecting the positive terminal of collector supply V_{CC} to a tap in the inductor of tank circuit.

This device for the production of ultrasonic waves is suitable in the frequency range of 100–10 MHz.

10.3 Detection of ultrasonics

Since ultrasonics is beyond the upper limit of human audibility, we cannot directly detect ultrasonics and special methods exist for their detection. We shall mention here only *two* of them for the purpose of detection.

Piezoelectric crystal detector—A piezoelectric crystal may also serve as a detector of US-waves. Usually a quartz plate is placed perpendicularly to the ultrasonic beam which develops alternating mechanical stresses and thereby an alternating emf across its faces. The emf is amplified and measured by means of an output meter to detect the oncoming ultrasonic waves. For maximum sensitivity, the ultrasonic frequency must equal the natural frequency of the crystal vibrator.

A magnetostrictive receiver can also be similarly used for detecting the US-waves.

Thermal detector—In this method, a fine wire of platinum is moved in the medium of ultrasonic waves. Due to alternate compressions and rarefactions, the temperature of the medium changes. The change in temperature occurs at nodes, while the temperature at antinodes remains essentially constant. This temporal change in temperature, in its turn, affects the resistance of the platinum wire, placed at the node, and can be detected easily by means of a sensitive bridge arrangement (Wheatstone bridge). Hence the existence of ultrasonic waves are also detected.

Among the other methods of ultrasonic detection, mention may made of (i) **Kundt's tube method** and (ii) **sensitive flame method** in the discussin of which we shall not enter here.

10.4 Measurement of ultrasonic wavelength

Interferometric method—The wavelength of ultrasonic waves can be measured with the help of an *ultrasonic interferometer.*

It consists essentially of a quartz oscillator Q [Fig. 10.4(a)], along with the associated electronic circuit G, and a reflector of sound R placed parallel to each other. Both the reflector and the oscillator can be moved up and down with the

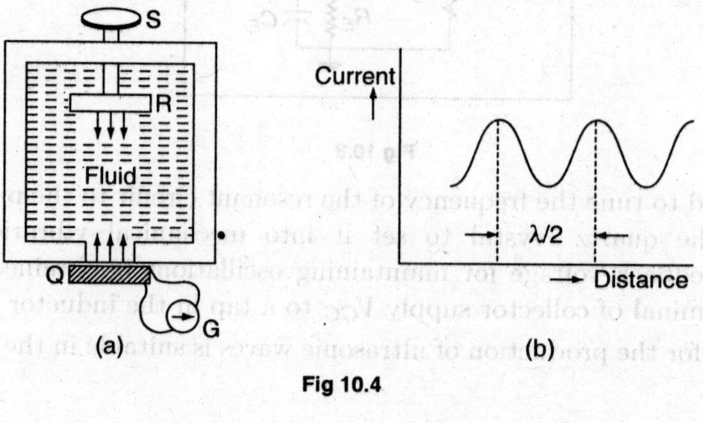

Fig 10.4

help of micrometer screw S. A standing wave pattern would be generated between them by the superposition of the incident and the reflected ultrasonic beams whenever the separation between them is an integral multiple of $\lambda/2$ where λ is the wavelength. The fluid medium vibrates as a *resonance column*. The distance between Q and R is increased by means of the screws and the corresponding plate current for various increments are noted and a *curve* is obtained [Fig. 10.4(b)] by plotting the current against the distance. The wavelength λ is thus known from the difference of the two adjacent positions of the reflector for maximum plate current.

- The method may be used to determine the ultrasonic velocity c in liquids or gases. For $c = n\lambda$, where n is the frequency of vibration of the crystal.

Diffraction grating method—The phenomenon of diffraction of light by ultrasonic waves passing through a liquid was first demonstrated in 1932 by P. Debye and his co-workers.

When ultrasonic waves pass through a medium, a periodic variation of pressure (compressions and rarefactions) occurs and consequently the *density of the medium varies from layer to layer*. The effect is very *prominent* in liquid. Thus ultrasonic standing waves could be formed in the intermediate regions of high and low density caused in the nodal and antinodal planes. As a result, to a beam of monochromatic light travelling in it at right angles to the sound, these act like parallel rulings of a diffraction grating (**acoustical grating**). The wavelength λ_s of sound wave corresponds to the *grating element d* of the usual transmission grating.

If θ be the angle of diffraction for the nth order maxima,

$$\boxed{\lambda_s \sin\theta = n\lambda}$$

where λ is the wavelength of the incident light. The value of λ_s can thus be obtained from the above expression.

Experimental set-up—The experimental set-up is illustrated diagrammatically in Fig. 10.5. G is a *liquid cell* of glass containing a *quartz oscillator* Q at one end and a *reflector* P at the other. The position of P is so adjusted that stationary waves are formed between Q and P. The cell G is illuminated by a parallel beam

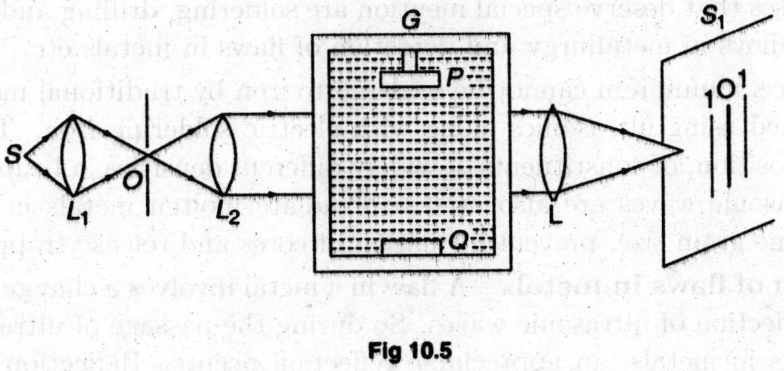

Fig 10.5

of monochromatic light in a direction perpendicular to that of the sound waves. S_1 is the screen on which the emerging light from the liquid cell is focussed by a condensing lens L. Diffraction pattern is seen on the screen, symmetrical about the central image O of the slit. The diffraction angle θ is measured by means of a spectrometer.

10.5 Effects and applications of ultrasonics

The applications of ultrasonics are so many and so varied that it is difficult to provide an exhaustive list. Ultrasonics are highly energetic waves and *mechanical, chemical, heating* and *biological effects* are produced in many substances when they are exposed to a strong beam of ultrasonics. Because of extremely small wavelength ($\sim$ cm or less), they can be transmitted over long distances without appreciable loss of energy. When they propagate in a liquid bath, stationary wave pattern is formed due to reflection of waves from the other end. It leads to the variation in density of liquid from layer to layer along the direction of propagation. This behaves as a plane diffraction grating that can diffract light.

The applications of ultrasonics in various fields are based on their above characteristic features. We describe hereunder some of the *important* applications.

1. Particles of colloidal materials in liquids are shattered forming extremely fine emulsions. Polymerised molecules disintegrate, the chain of starch molecules splits into fragments. The particles of suspended liquid coagulate. Smoke coagulates to such large sized particles that they cannot remain suspended in air. Smokes and dust particles are thus removed from air and other gases (*antipollutant*). Many microorganisms, even tadpoles, frogs and fishes, can be killed by ultrasound. They can also be used for sterilisation of equipments and freeing milk from bacteria, for cleaning utensils, washing clothes, removing dust and soot from chimney.

2. In the *medical field* as well, ultrasonics are finding ever-increasing applications. The body parts affected by neuralgic or rheumatic pains on being exposed to ultrasonics get sufficient relief from pains. Abnormal growth in brain and other internal body organs, certain tumours that are difficult to detect by X-rays can be detected by ultrasonics. **Ultrasonography** is now routinely used in medical diagnosis.

3. In *industry*, the importance of ultrasonic waves can hardly be overestimated. Some of the uses that deserve special mention are soldering, drilling and metal cutting, formation of alloys in metallurgy and detection of flaws in metals etc.

For instance, aluminium cannot be soldered to iron by traditional methods but can be accomplished using ultrasonics along with electric soldering iron. To get alloys of uniform composition, of constituents of widely different densities, a beam of ultrasonics is used. Ultrasonic waves are also used to irradiate molten metals in the process of cooling to refine grain size, prevent formation of cores and release trapped gases.

Detection of flaws in metals—A flaw in a metal involves a change in the medium and causes reflection of ultrasonic waves. So during the passage of ultrasonics through holes or cracks in metals, an appreciable reflection occurs. Reflection also occurs at the back surface of the metal. The reflected pulses are picked up by receivers, suitably amplified and applied to one set of plates of CRO. The transmitted signal, reflected signal from the flaw and that from the back surface of metal each produces a peak. The position of the second peak on the time-base of CRO gives the distance of flaw. The experimental set-up (Fig. 10.6) is illustrated and is self-explantory.

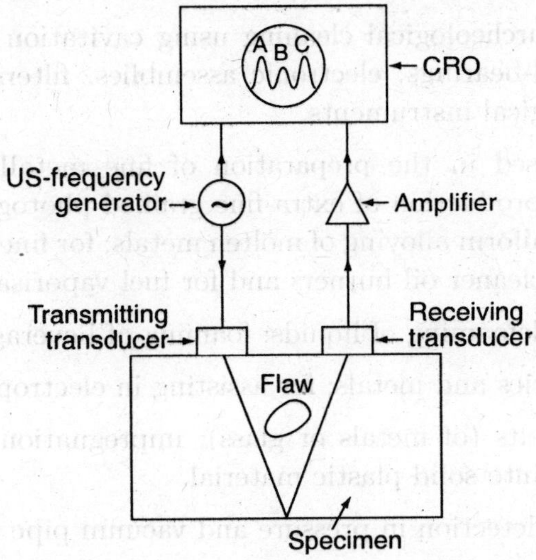

Fig 10.6

4. Among the other uses of ultrasonics, **sonar** for the detection of submarines, icebergs and other objects in the ocean, finding the depth of the sea, direction signalling etc. are of importance.

Sonar—The term 'Sonar' is an acronym of '*sound navigation and ranging*' and is employed for underwater detection, ranging and depth measurement. The process is similar to that used in radar. A sharp ultrasonic beam is directed in various directions into the sea and are picked up on return after reflection. The reflection of waves from any direction ensures the presence of some underwater reflecting body; the time interval gives the distance of it and the change in frequency of the echo signal due to Doppler effect can give the velocity of the body and its direction.

Depth of sea—Ultrasonics are highly energetic and show little diffraction. They can therefore be used for finding the depth of the sea. The interval between the transmission of ultrasonic wave and the reception of the reflected wave back from the sea bed is recorded. Let this interval be t and d be the depth of the sea there. If v be the known velocity of US-waves, then

$$d = \text{depth of sea} = v \times \frac{t}{2}$$

Direction signalling—The frequencies of US-waves are very high and this implies that their wavelengths are very small. The US-waves can therefore be concentrated into a sharp beam and direction signalling can be readily made.

List of Applications of Ultrasonics

1. Non-destructive testing, i.e., detection of defects in metals; corrosion inspection and thickness measurements.
2. Industrial and archeological cleaning using cavitation : antiques, minute parts of watches, ball-bearings, electronic assemblies, filters, gauges, jet engine fuel nozzles and surgical instruments.
3. Cavitation is used in the preparation of fine metallic dispersion of colloidal sulphur; in the production of extra-fine grained photographic emulsions; for fine structure and uniform alloying of molten metals; for finer fuel (and hence cheaper) in burners and cleaner oil burners and for fuel vaporisation in carburettors.
4. Degassing and defoaming of liquids; foaming of beverages.
5. Welding of plastics and metals; for assisting in electroplating.
6. Degassing of melts (of metals or glass); impregnation of porous materials and metal insertion into solid plastic material.
7. Air or gas leak detection in pressure and vacuum pipe systems.
8. Dyeing and manufacturing of paints.
9. Detection of bearing wear, hydraulic or pneumatic turbulence; friction or corona discharge.
10. Matching of fragile and hard materials including glass and ceramics; soldering and brazing; reducing friction during wire drawing or metal rolling and in determination of hardness.
11. Separation of mixtures into their components; separation of paper fibres and for molecular dispersion; accelerating chemical reactions for aging of liquors, tanning of hides and extractions etc.
12. Burglar alarms; in non-contact counting and sensing devices, measuring fluid flow and particle sizes; location of submarines, schools of fish and oyester beds in seas, and to measure the depth of the sea.
13. Level sensing devices for the control of propellants in rockets and for detecting water in nuclear submarines.
14. Viscometer for measuring viscosity of liquids.
15. Study of optical diffraction; ultrasonic light modulation to reproduce, record and play back TV-pictures and phototube amplifying systems.
16. Measurement of moduli of elasticity of solids and plastics at high temperatures; determining the degree of roasting of porcelain articles; study of concrete solidification process, and vulcanisation of rubbers.

17. Atomisation of medicine for easier inhalation, location of stones in the kidney and for destroying mould and bacteria.
18. Diagnosis of retinal detachment, lens dislocation, intraocular tumours and measurement of axial length of the eye (done to an accuracy of 0.2 mm).
19. Ultrasonically aided extractor to remove foreign bodies from the eye.
20. Studying brain lesions, tumours and foetus in the womb.
21. Prediction of the possibility of thrombosis.

10.6 Building acoustics : Introduction

Compared to other branches of sound, the *building acoustics* is of rather recent origin and concerns primarily with the constructions of an enclosure of high acoustical properties, i.e., *to provide the best audible sound of whether lecture or musical concert and the like such as drama, recitation etc. to the audience.* Till comparatively recently, it was enveloped in great mystery and uncertainty. The Fogg Art Museum hall of Harvard University was acoustically so defective that the speakers' words were *not intelligible* to the audience. It was the pioneering investigations (1903–11) of Professor Wallce Sabine of the Department of Physics, Harvard University, that gave the study of building acoustics a truly scientific turn. It was later followed by Watson, Knudsen, Fletcher, Steinberg, Richardson and others so much so that it is now recognised as an important course of study in *Architectural Engineering.*

Acoustical requisites for a good auditorium—Sabine prescribed the following essential requisites for an auditorium with good acoustical properties.

1. The sound should be **sufficiently loud** or intense in every part of the enclosure.

2. The *quality of sound* of speech and music *must remain unaltered*, i.e., relative intensities of several components of a complex sound be maintained everywhere in the auditorium.

3. There should be *no echoes* inside more than what is necessary to upkeep the continuity of the sound.

4. The successive sounds of speech or music etc. must remain *distinctly intelligible* and must be *free from* one another and from the *extraneous noises* to avoid confusion. The successive syllables articulated must be clear and distinct without any overlapping of syllables.

5. There should be *no undesirable concentration of sound* in any part of the auditorium and depletion at any other.

6. There should be *no unpleasant reinforcement of any articulation* by the object present inside the auditorium and *no resonance* within the building and no *echelon effect.*

7. The reverberation should be proper, *neither too large nor too small.* The *reverbration time* should be 1–2 seconds for music and 0.5 – 1 sec for speech.

The above conditions are necessarily dependent on the specific purpose for which the auditorium is meant, e.g., lecture theatre, motion picture theatre or an auditorium for orchestral or vocal music etc. But, undoubtedly, some of the above requisities are essential for the good acoustics of any hall.

10.7 Growth and decay of sound

When a pulse of sound is emitted inside an enclosure, a wave of compression spreads out in all directions and, in the absence of reflecting bodies, would be heard as a sharp single impression. As a matter of fact, however, the pulses are reflected a large number of times by the walls, ceilings of the enclosure and other reflecting surfaces, and a series of waves of diminishing amplitude passes the listner's ear until quieted by dissipation in air, ceiling, walls and floor. The result of the process of this *multiple reflections* is that after the first impulse of the sound is heard, the sound rolls over and is heard for a short interval of time until the intensity falls below the *threshold of audibility.*

Let now a *continuous source*, instead of pulses, be started in an enclosure. The waves would spread out and the listener becomes conscious of the start of sound when the first of these reaches him and also of an addition of energy corresponding to each reflected wave. This energy is subsequently added to the energy of the incident wave still being emitted from the source. Had there been no dissipation of energy, the resulting intensity would increase indefinitely. The loss of energy by absorption during every reflection makes the intensity reach a *steady* average value when the rate of energy-absorption equals the rate of emission. If now the source is *cut-off*, this intensity in listener's ear does *not* instantaneously fall to zero but gradually diminishes as the reflected waves reach the ear even sometime after the direct waves. The rise and fall of intensity is shown in Fig. 10.7(a). If many impulses are sent *in succession*, the effect of withdrawing them would be in accordance with Fig. 10.7(b).

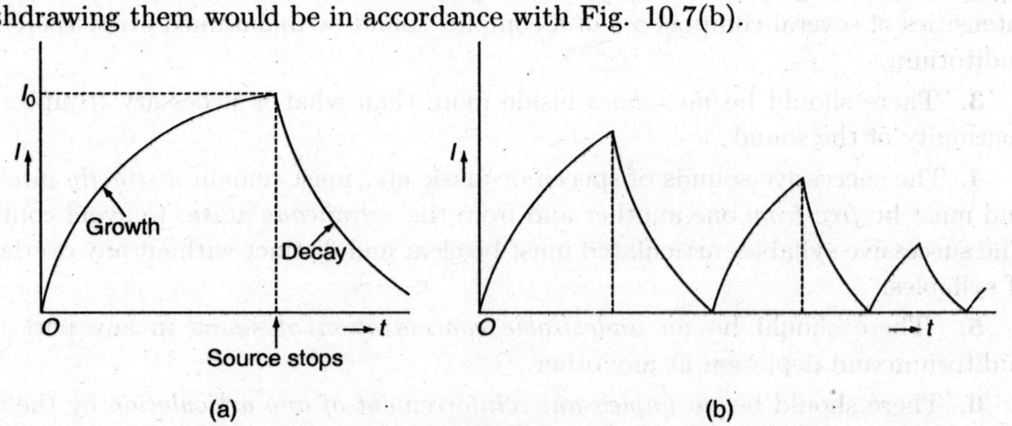

Fig 10.7

Reverberation and time of reverberation—This phenomenon of the prolongation or persistence of audible sound in an enclosure by repeated reflections for some interval of time after the source is stopped is known as **reverberation**.

The time taken for the rollings of the sound to die away, i.e., the duration for which the sound persists, reckoned from the time the original pulse was generated or, in case of a continuous note, the time when the source is cut-off, is known as the **time of reverberation** which is a very important quantity in determining the good acoustic quality of an auditorium. It may as well be defined as the time taken by the sound to fall below the threshold of audibility measured from the instant when the source stopped emitting.

10.8 Optimum reverberation time : Reverberation formula

It is the *reverberation* time that forms the *central problem* in constructing an auditorium of good acoustic properties.

The reverberation time in any hall should neither be too large nor too small. If it is too long, the reflected sound will not be reduced sufficiently in intensity and will therefore interfere with the succeeding syllables coming directly from the source and thus cause confusion. If however the time is too short, absorption of the room is too great and the weak sources cannot be supported. The feeling of continuity is also thereby affected. The time of reverberation should therefore have an optimum value, called the **optimum reverberation time**. The room for which the reverberation time is large is called a **live room** and that for which it is small is a **dead room**.

Reverberation formula : An empirical formula for the time of optimum reverberation T was suggested by Sabine and is given by

$$\boxed{T = \frac{0.05V}{a}}$$

where V is the volume of the enclosure in cubic feet and a is the absorption power of the hall.

Sabine's formula was deduced theoretically, by Jager (1911) and others, on the basis of the following *two main* **assumptions.**

1. That an average acoustic energy density, i.e., energy per unit volume p could be conceived, i.e., sound energy is uniformly distributed in the enclosure.

2. That the dissipative influences, other than absorption, are negligible; that is, energy is lost by way of absorption only and the loss can be treated as continuous.

The *other assumptions* include (i) that the source emits sound uniformly in all directions at a constant rate independent of the energy density in the enclosure, (ii) that there is no interference of sound waves, and (iii) that the absorption coefficients are independent of incident sound intensity.

Following Jager, we shall first calculate on the basis of the above assumptions, the energy falling on a given area of the walls per second.

Let V be the volume of the enclosure and ρ be the average acoustical energy density at any instant t. Consider a volume element dV (Fig. 10.8) in the medium, distant r from an element dS of the wall-surface and making an angle θ with the normal to dS. The acoustical energy in dV is thus ρdV. The area of a sphere of radius r surrounding dV is $4\pi r^2$. So the amount of energy radiated by dV and received normally by the surface area dS would thus be

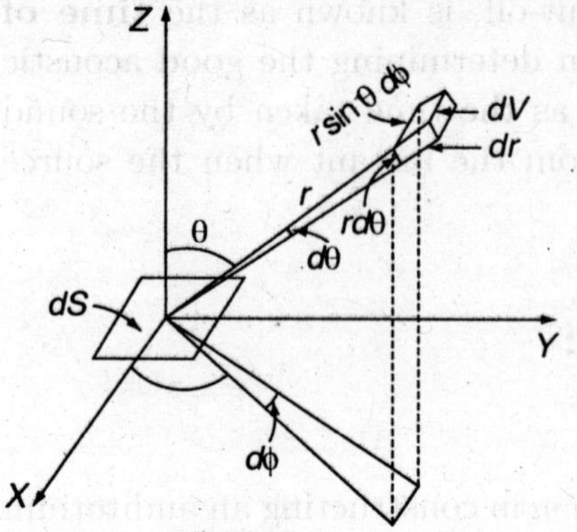

Fig 10.8

$$\rho dV \frac{dS\cos\theta}{4\pi r^2}$$

Now, in spherical polar co-ordinates (r, θ, ϕ), the volume element dV is given by,

$$dV = r\sin\theta d\phi \cdot r\, d\theta \cdot dr$$

Since the angle θ may vary from 0 to $\pi/2$ and ϕ from 0 to 2π, the total energy received by the surface dS is

$$\Delta E = \int_0^{\pi/2} \rho \frac{r\sin\theta \cdot r d\theta \cdot dr}{4\pi r^2} dS\cos\theta \int_0^{2\pi} d\phi$$

$$= \frac{\rho dS}{4\pi} dr \int_0^{\pi/2} \sin\theta\cos\theta\, d\theta \int_0^{2\pi} d\phi = \frac{\rho dS}{4} dr \qquad (10.8.1)$$

This energy falls on dS in the time $dt = dr/c$, where c is the velocity of sound in the medium.

$\therefore$ Total energy falling on dS *per second* is

$$\boxed{\frac{\Delta E}{dr/c} = \frac{\rho c dS}{4}} \qquad (10.8.2)$$

If $\alpha_1, \alpha_2 \cdots, \alpha_n$ be the *absorption coefficients*, that is, *the fraction of incident energy that is not reflected*, from the areas $dS_1, dS_2, \cdots, dS_n$ forming the interior walls of the hall, the rate of absorption of energy by the walls is

$$\frac{1}{4}\rho c \sum_{n=1}^{n} \alpha_n dS_n = \frac{1}{4}\rho c a$$

$$\text{where} \quad a = \sum \alpha_n dS_n, \qquad (10.8.3)$$

the total absorption or the **absorption power** of the hall.

But the rate of increase of acoustic energy inside the hall is

$$\frac{d}{dt}(\rho V) = V\frac{d\rho}{dt} \qquad (10.8.4)$$

So, if E_0 be the energy produced per second by the source, then we have

$$E_0 - \frac{1}{4}\rho c a = V\frac{d\rho}{dt} \qquad (10.8.5)$$

Solving the differential equation (10.8.5), we obtain the average acoustic density

$$\rho = \frac{4E_0}{ca}\left(1 - e^{-cat/4V}\right) \qquad (10.8.6)$$

The equation (10.8.6) is known as the *equation for the* **growth of sound** and is analogous to that of growth of current in an inductive a.c. circuit. It shows that when a source of sound is operated in an enclosure, the energy density gradually increases and finally reaches exponentially a steady maximum value $\rho_0 = 4E/ca$.

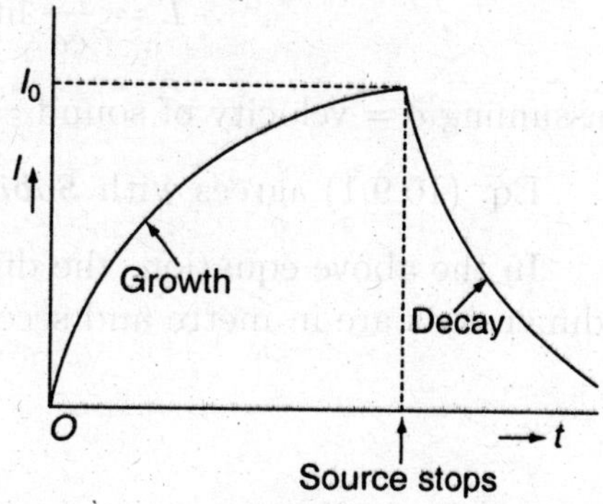

Fig 10.9

The *time constant* of the growth of the intensity, $\tau = 4V/ca$. If the absorption a is small τ is large and conversely, so that a fairly long time is required for the intensity to grow to a certain percentage of its maximum value.

If however the source is suddenly cut-off when the sound intensity has attained its maximum, the energy decays and the differential **decay equation** is, from (10.8.5), given by

$$\frac{1}{4}\rho ca + V\frac{d\rho}{dt} = 0$$

$$\text{Solving,} \quad \rho = \rho_0 e^{-cat/4V} = \rho_0 e^{-t/\tau} \qquad (10.8.7)$$

where ρ_0 is the *steady-state energy density* and $\tau = 4V/ca$, the *time constant* for the decay. The equation clearly demonstrates that the energy density does *not* immediately fall to zero when the source is stopped. And this indeed is what we call **reverberation**. The growth and decay of energy density with time is shown in Fig. 10.9.

10.9 Calculation of optimum reverberation time (Live room)

At the very outset, we shall *define* the term *optimum reverberation time*. According to Sabine, *the standard* or **optimum reverberation time is defined as the time required for the sound to be reduced from its initial intensity to one millionth of it**, or in other words, *it is the time during which the steady acoustic energy density reduces by* 60 *decibel* (*dB*).

From the decay equation (10.8.7), therefore,

$$\frac{\rho}{\rho_0} = 10^{-6} = e^{-caT/4V}$$

where T is the *standard* or the *optimum reverberation time.*

$$\therefore \quad T = \frac{4V}{ca} \ln (10^6) = \frac{4V \times 2.303 \times 6}{1120a} = \frac{0.05V}{a} \qquad (10.9.1)$$

assuming $c =$ velocity of sound $= 1120$ ft/s.

Eq. (10.9.1) agrees with ***Sabine's empirical formula*** for reverberation time.

In the above equation, the dimensions are in feet and second. When, however, the dimensions are in metre and second, $c = 330$ m/s and

$$\boxed{T = 0.165\frac{V}{a}} \qquad (10.9.2)$$

Sabine's formula is in good agreement with experiments for small absorption coefficients < 0.4 and hence for large T. A room where T is large is called a **live room.** This indicates that the assumptions made are essentially correct.

In acoustic engineering, V is usually expressed in cubic foot (cft) and a in *Sabin.* A **Sabin** is the unit of absorption coefficient. *The absorption coefficient of a surface is said to be* **1 Sabin** *if it absorbs sound at the same rate as does* 1 sq. ft. *of a perfect absorber such as an open window* (O.W.). Hence *Sabin* is also known as *open window unit* (O.W.U).

As T depends on the absorption coefficients of the materials in the room, it is possible by a correct choice of the absorbing surfaces, to arrange for any desired reverberation time. Different types of sound like speech, music etc. require different optimum reverberation times.

10.10 Absorption coefficient : Its measurement

The absorption coefficient of any material is defined as *the ratio of acoustic energy absorbed by its surface to that of the incident total sound energy on the surface.*

$$\therefore \quad \text{Absorption coefficient,} \quad a = \frac{\text{Sound energy absorbed}}{\text{Total sound energy incident}}.$$

An open window behaves as a *perfect absorber* since all the sound waves incident on it, passes through it. Hence, *the absorption coefficient of a material is defined as the rate of sound energy absorbed by a certain area of the surface to that absorbed by an open window of equal area.* Thus, the absorption coefficient of carpet is 0.1 means that 1 m^2 of the open window absorbs the same amount of energy as that absorbed by 10 m^2 of carpet $\Rightarrow a_{\text{carpet}} = 1/10 = 0.1$. Its unit is open window unit (written O.W.U.) or Sabins.

The absorption coeffcients of a number of common materials are provided in Table 10.1 in the next page.

Table 10.1 Absorption coefficient of common materials

Material	Absorption coefficient
Brick wall (30 cm thick)	0.03
Brick wall painted	0.016
Wooden floor	0.06
Glass	0.02
Carpets	0.15 − 0.30
Ordinary chair	0.17
Human body	4.3 − 4.7
Hair felt (2.5 cm thick)	0.58
Open window	1.00
Plaster	0.03
Felt	0.71

Measurement of absorption coefficient—Of the usual methods for determining the absorption coefficient of a material, the *two* that deserve special mention are :

1. Based on determining T, the *standard reverberation time.*
2. Based on the *decay equation* of sound.

We shall describe here the *first method* based on standard reverberation time only.

Let α_1 be the absorption coefficient of a material of area S_1. If T_1 and T_2 be the standard reverberation times, with and without the *large sample* of the material under study then, from Sabine's formula,

$$\frac{1}{T_2} = \frac{\sum \alpha_i S_i}{0.16V}; \quad \frac{1}{T_1} = \frac{\sum \alpha_i S_i + \alpha_1 S_1}{0.16V}$$
$$= \frac{\sum \alpha_i S_i}{0.16V} + \frac{\alpha_1 S_1}{0.16V}$$

where V represents the volume of the enclosure.

$$\therefore \frac{1}{T_1} - \frac{1}{T_2} = \frac{\alpha_1 S_1}{0.16V}$$

$$\Rightarrow \alpha_1 = \frac{0.16V}{S_1}\left(\frac{1}{T_1} - \frac{1}{T_2}\right) \tag{10.10.1}$$

Therefore, α_1 can be readily evaluated from a knowledge of the terms on the right hand side of (10.10.1).

10.11 Criticism of Sabine's formula : Eyring's formula

An examination of the expression $T = 0.05V/a$ or $T = 0.165V/a$ for the optimum reverberation time shows that it *cannot* be applied for a ***perfectly dead room*** as $a = 1$ for such a room and T becomes $0.05V$ or $0.165V$ whereas it should actually be *zero* for there is *no* reflection within a perfectly dead room. Hence Sabine's formula is *valid only for live rooms*, that is, with small values of a. The error in Sabine's formula increases as a increases beyond 0.2.

Sabine assumed a *continuous* absorption and an existence of a *uniform* energy density in the room, both of which are open to objections. Erying argued that absorption of sound is *discontinuous* and due to the standing waves in the room, the energy density *cannot* be *uniform*. On the basis of these arguments Eyring deduced the following relation for the optimum reverberation time.

$$T = \frac{0.05V}{-S\ln(1-\alpha)} \quad \text{(in dimensions of ft and sec)}$$

$$= \frac{0.16V}{-S\ln(1-\alpha)} \quad \text{(in dimensions of m and sec)}$$

where $S = \sum_n dS_n$ and $\alpha = a/S$, the **average** *absorption coefficient.*

10.12 Acoustic design of auditoriums

We have already indicated the requisites an accoustically good auditorium should have. While special consideration has to be made about reverberation times, the fulfilment of other conditions, namely

(a) *adequate* **loudness**,

(b) **uniform distribution** *of sound*,

(c) **optimum reverberation**,

(e) *exclusion of* **extraneous sound**, and

(f) *elimination of* **echelon effect**

are also to be ensured by proper design of the enclosure.

We shall now discuss the above factors one by one in what follows.

Adequate loudness—Adequate loudness in every portion of the hall is an important factor for satisfactory hearing. Loudness may be increased by using large *sounding boards* behind the speaker and directed to the audience. Uniform loudness may be obtained by large polished wooden parabolic reflectors with the source at the focus. Low ceilings are also of great help in this context in reflecting the sound energy towards the audience. In case loudspeaker is used for providing additional sound energy, it should be arranged a little above the speaker's head.

Uniform distribution of sound—As mentioned above, *parabolic reflectors* serve this purpose. Concave, cylindrical or spherical reflectors should be *discouraged* as they generate focussing effect. The ceiling and floor must also be *flat*.

Optimum reverberation—If the reverberation is high there is overlapping and confusion of successive sounds. If however it is low, the chances of confusion may be minimised, but the hall appears dead. Hence halls must be so designed with absorbers like asbestos, cork etc. that the reverberation is *optimum*. A few open windows also increase the absorption. Heavy curtains put up in the hall also absorb sound. A good audience is equivalent to 5 sq. ft. of a perfect absorber and may be utilized for the purpose.

Freedom from resonance—Sometimes window-panes and walls lacking in rigidity may be thrown into vibrations and create other sounds resulting in resonance for some note of audio frequency. This changes the intensity of note and may produce undesirable effect. Also, due to interference between the original sound and the created sound, the original sound may get distorted. Enclosed air in the hall may also cause resonance. Such resonant vibrations should be properly damped. To avoid resonance, no two dimensions of the hall should be the same. A value of 2 : 1 and 1.2 : 1 for length-width ratio has been found suitable.

Exclusion of extraneous sound (noise)—In an acoustially good hall, no noise i.e., unwanted sound should reach from outside. Sound from outside may enter the hall through doors, windows or metallic structures or produced by machineries. The extraneous sound may thus be classed as :

(a) **air-borne** which enters through doors, windows and ventilators, (b) **solid-borne** or **structure-borne** which are carried by solid substances like metallic structures and machineries. Using absorbent-pads in the machinery and placing curtains near them, their noise can be reduced. Sound entering through conducting structures can be avoided by interposing layers of felt or cork or some other insulators of sound. The air-borne sound can be avoided only by affecting the ventilation by using double or triple doors or windows and using felt baffles in ventilators.

Elimination of Echelon effect—If there is a large number of any regular spacing of reflecting surfaces of a structure such as a set of railings and the stairs, a musical note due to the regular succession of echoes, called **echelon effect**, may be produced to make the original sound confused or unintelligible. This can be eliminated by breaking the regularity of equal spacing at intervals.

Apart from the factors discussed so far, the optimum volume per seat in the enclosure (which is a variable with the type of enclosure) should also be ensured. As regards the seat arrangement, proper care must be taken so that they are close to the source and properly distributed as the audience greatly absorbs sound energy. To ensure free flow of sound from the source to the audience, the seats should be arranged in a plane *inclined by about* 8° (degrees).

For noise reduction and sound proofing, a knowledge of the loudness level of sound is essential. Here in Table 10.2 we give a list of some common sounds and their loudness level in decibel (dB) unit.

Table 10.2 Some common sounds and their loudness

No.	Sound	Loudness	No.	Sound	Loudness
1.	Threshold of audibility	0 dB	4.	Ordinary conversation	60 dB
2.	Normal breathing	10 dB	5.	Heavy traffic	80 dB
3.	Whispering	20 dB	6.	Thunder	110 dB

10.13 Transmission loss of sound

When sound is produced in an auditorium, it moves outward in spherical waves to strike the boundaries of the hall. In doing so, the waves are reflected, absorbed and transmitted in varying degrees, depending on the characteristics of the boundaries and the frequency of the sound.

Transmission loss—The loss of accoustic energy across a wall or a barrier is what is called *transmission loss* and is measured as *the loss in intensity of sound expressed in dB.*

$$\therefore \quad \text{Transmission loss } (T.L.) = 10 \ln \ (1/T_1).$$

$$\begin{aligned} \text{where} \quad T_1 &= \text{transmission coefficient of wall} \\ &= \text{transmitted intensity/incident intensity} \\ &= I_t/I_i \end{aligned}$$

$$\therefore \quad \boxed{\boldsymbol{T.L. = 10 \ln \left(\frac{I_t}{I_i}\right)}}$$

• Transmission loss depends on a number of factors. For instance, it varies with (i) the *frequency* of sound, (ii) the *materials used* in construction and also (iii) the *methods used* in construction. Larger the transmission loss, greater is the sound insulation. If the sound levels of either side of the barrier (or wall) be respectively X (dB) and Y (dB), then the transmission loss of the barrier is simply their difference, i.e., $(X - Y)$ in dB.

10.14 Sound absorbers

Sound absorbers or sound absorbing materials *are* those characteristic *materials* that are *used to increase the absorption and reduce the reflection* of sound waves in a room or hall. These materials should be (i) *cheap*, (ii) easily *available.* (iii) *light weight* but structurally sufficiently *strong*, (iv) *water proof*, (v) *economic in maintenance*, (vi) adequately *fire-proof* and (vii) with *high sound absorbing capacity* and the like.

These materials are broadly classified as :

1. Porous absorbers, e.g., fibre boards, soft plasters, wood wool, glass silk, asbestos fibre spray etc.

2. Cavity resonators i.e., chambers with a small opening.

3. Panel absorbers, where the absorbing material is fixed on a timber framing with an air space between the framing and the wall e.g., hard board panels, rigid plastic boards, gypsum boards, etc.

4. Composite absorbers, when the absorbers of Type 1, 2 and 3 are all combined in a single unit.

10.15 Illustrated Examples

▶ **Example 1.** *Calculate the reverberation time in an auditorium of volume measuring* 60 × 90 × 20 ft, *the absorbing materials consisting of* 9500 sq.ft. *of plaster* ($\alpha_1 = 0.03$), 7000 sq.ft. *of wood* ($\alpha_2 = 0.06$), 500 sq.ft. *of glass* ($\alpha_3 = 0.03$), 600 *seats* ($\alpha_4 = 0.3$) *and an audience of* 400 ($\alpha_5 = 4$).

Solution. The total absorption in Sabin is

Plaster	···	9500×0.03	=	285
Wood	···	7000×0.06	=	420
Glass	···	500×0.03	=	15
Seats	···	600×0.3	=	180
Audience	···	400×(4 − 0.3)	=	1480
		Total	=	2380

∴ From Sabine's formula, the reverberation time

$$T = 0.05 \times 60 \times 90 \times 20/2380$$

$$\simeq \mathbf{2.5\ s} \text{ (rather high)}$$

▶ **Example 2.** *The volume of a room is* $1200\,\text{m}^3$. *The surface area of the wall is* $220\,\text{m}^2$, *the floor area is* $120\,\text{m}^2$. *The average absorption coefficients are* 0.03 *for walls,* 0.80 *for ceilings and* 0.06 *for floor. Find the average absorption coefficient and reverberation time.*

Solution. The average absorption coefficient $\bar{\alpha}$ is given by

$$\bar{\alpha} = \frac{\alpha_1 S_1 + \alpha_2 S_2 + \alpha_3 S_3}{S_1 + S_2 + S_3} = \frac{0.03 \times 220 + 0.80 \times 120 + 0.06 \times 120}{220 + 120 + 120}$$

$$= \mathbf{0.24}$$

$\therefore$ Total absorption of the room $= \bar{\alpha}S$

$$= 0.24(220 + 120 + 120)$$

$$= 0.24 \times 460 \text{ metric Sabin}$$

$\therefore$ Reverberation time, T is given by

$$T = \frac{0.165V}{\bar{\alpha}S} = \frac{0.165 \times 1200}{0.24 \times 460} = \mathbf{1.80\ s}$$

► **Example 3.** *A window having an area* 1.4 m^2 *opens on a street. The intensity level of the sound at the window is* 60 dB. *How much acoustic power enters the window via the sound waves? Standard intensity level* $= 10^{-12}$ W/m^2.

Solution. Intensity level in $dB = 10\log_{10}(I/I_0)$ where I_0 = standard intensity level. Therefore, by the problem,

$$60 = 10\log_{10}(I/10^{-12})$$

$$\Rightarrow \quad I = 10^{-6} \text{ W/m}^2$$

$$\therefore \quad \text{Acoustic power} = \text{Intensity} \times \text{area}$$

$$= 10^{-6} \times 1.4$$

$$= \mathbf{1.4 \times 10^{-6}\ Watt}$$

► **Example 4.** *A rectangular enclosure with dimensions* 30 × 15 × 10 ft *has inner walls a mean absorption coefficient* 0.1. *Calculate the reverberation time and also the maximum intensity in the enclosure if there is a sound source of output* 15 μW *inside the enclosure.*

Solution. Volume of the enclosure, $V = 30 \times 15 \times 10 = 4500$ cft. The area of the inner walls is given by

$$S = 2(30 \times 15 + 15 \times 10 + 30 \times 10) = 1800 \text{ ft}^2$$

Total wall absorption, $A = 1800 \times 0.1 = 180$ sabin.

$$\therefore \quad \text{Reverberation time } T = \frac{0.05 \times V}{A} = \frac{0.05 \times 4500}{180} = \mathbf{1.23\ s}$$

Maximum intensity for a source of power P (= 15μW) is

$$I_m = \frac{P}{A} = \frac{15\ \mu\text{W}}{180} = \mathbf{0.083\ \mu W/m^2}$$

Review Questions

1. What is meant by ultrasonics? Is it the same as supersonics? State the different methods for the production of ultrasonics.

2. Describe, in details, the method of producing ultrasonic waves utilising the phenomenon of (i) magnetostriction and (ii) piezoelectricity.
3. Discuss how would you detect and measure the wavelength of ultrasonic waves.
4. Enumerate some of the effects of ultrasonic waves and apply an appropriate one for detecting flaws and inhomogeneities in solids materials.
5. What is a sonar? Write a short explanatory note on it.
6. What is reverberation? Explain reberberation time. What are the conditions to be satisfied for an acoustically acceptable auditorium?
7. What is absorption coefficient of a material? What is the unit in which it is measured?
8. Derive and discuss Sabine's formula relating the reverberation time in a room with its volume, the velocity of sound and the total absorption of the room.
9. Write a short note an Eyring's formula for reverberation time and compare it with that of Sabine. What is the role of interference in the acoustics of buildings?
10. Give a detailed account of the various applications of ultrasonics in the fields of (i) industry and (ii) medicine.

Short Answer and Objective Questions

1. What are magnetostrictive and piezoelectric effects?
2. Define ultrasonics and explain the term 'Sonar'. How are ultrasonics used for Sonar exploration?
3. Describe some of the applications of ultrasonic waves.
4. Define reverberation time. What is its unit?
5. Explain what do you mean by 'absorption coefficient' in sound.
6. What factors determine the reverberation time of a hall ? Should it be too high or too low? Explain.
7. Ultrasonic waves

   (a) are sound waves of very long wavelength,

   (b) move faster than the sound waves,

   (c) are sound waves of very high frequencies,

   (d) move with the same velocity as sound waves.

8. The walls of a hall meant for music concerts should

   (a) reflect sound (b) absorb sound (c) transmit sound (d) amplify sound.

9. If 50 dB and 10 dB are the sound intensity levels measured on either side of a wall, then the transmission loss of the wall is

   (a) 40 dB (b) 60 db (c) 50 dB (d) none.

10. The reverberation time is given by

    (a) Sabine's formula (b) Eyring's formula

    (c) Both (a) and (b) (d) None of (a) and (b).

Problems

1. A quartz crystal of thickness 0.001 m is vibrating at resonance. Find the fundamental frequency, given $Y = 7.9 \times 10^{10}\,\text{Nm}^{-2}$ and $\rho = 2650\,\text{kgm}^{-3}$ for quartz. [2730 kHz]

2. A cinema hall has a volume of 7500 m^3. It is required to have a vibration time of 1.5 sec. Calculate what should be the total absorption in the hall. [825 O.W.U.]

3. A lecture hall has a volume 12×10^4 m^3. It has a reverberation time of 1.5s. What is the average absorbing power of the surface? The total absorbing surface area is 25,000 sq. m. [0.51 W.]

4. Find the reverberation time of an auditorium of volume 1500 m^3 having the capacity of an audience of 120 when (i) the auditorium is empty, (ii) the auditorium is full and (iii) the audience occupying only cushioned seat, from the following data.

Surface	Area	Coefficient of absorption
Plastered wall	112 m^2	0.03
Wooden floor	130 m^2	0.06
Plastered ceiling	170 m^2	0.04
Wooden door	20 m^2	0.06
Cushioned chair	100 (No.)	1.0
Audience	120 (persons)	4.7

[2.08s; 0.362s; 0.420s]

5. An auditorium of volume 7×10^4 cft has a reverberation time 0.98s when empty. What will be the reverberation time when 250 people form the audience inside it? Assume the absorption coefficient per person = 0.4 and average absorbing area per person = 6 sq. ft. [0.78 s]

6. An audiorium of volume 2,500 m^3 has a total absorption equivalent to 300 m^2 open windows. If there are 10 persons, each of 4 sabin absorption, find (i) the reverberation time and (ii) the maximum intensity, when two persons on the average talk inside, each with 5 μW output. [1.32 s; 0.033 $\mu W/m^2$]

7. The interior walls of an enclosure of dimensions $250 \times 60 \times 30$ ft have a mean absorption coefficient 0.2. The floor and the ceiling have absorption coefficients 0.3 and 0.4 respectively. Find the reverberation time and the power output of a source to produce an intensity level of 70 dB. [1.55 s; 13.2 mW]

Part II
Optics (Geometrical and Physical)

Chapter 1

Refraction at Spherical Surfaces

1.1 Introduction

Refraction of light at spherical surfaces is of fundamental importance in the studies of geometrical optics. Various relations involving refraction in single spherical surface and lenses can be obtained. Its importance therefore can hardly be overestimated. In deriving the relations, however, the *sign convention* is to be settled and a number of *assumptions* are to be made which are laid down as under.

Sign convention—In all the figures (exception is rare), the incident light would proceed from *left to right.* While in case of a lens, *distances* are measured *from* the *optical centre*, it is measured from the *pole* for a refracting (or reflecting) surface. *Distances in the direction of travel of light are considered as* **positive**, *those measured against are counted as* **negative**. In case of transverse distances, they are taken as *positive* when measured *upward*, *negative* when *downward* with respect to the *principal axis.* The convention of signs is thus in accordance with the convention of coordinate geometry.

Assumptions—In deriving the various relations involving refraction at spherical surfaces, a number of assumptions is made. These are :

1. The *aperture* of the surface is *small* and the *object* is a *point object on the principal axis.*

2. The *incident ray* from the object strikes the surface at a point very *close to the pole* so that the angles of incidence and refraction are both very small.

3. The incident ray, the refracted ray and the normal at the point of incidence make *small angles* (i.e., oblique incidence and oblique refraction) with the principal axis (paraxial rays).

1.2 Fermat's principle

Fermat's *principle of least time* states that *when a ray of light passes from one point to another through a number of media, it always takes up the one for which the time taken is a minimum.*

The above statement of Fermat's principle however is *not* complete in itself. Although, it is true in some cases, there are situations in which it is found that the path followed by a ray of light is *maximum*, while in many others it is *neither a maximum, nor a minimum* but is *stationary.* For instance, in case of reflection and refraction in plane surfaces, the time taken is indeed *minimum.* But in the formation of images by a lens where various rays from an object point travel over different paths to meet at the image point, the time taken by each of them is *stationary* or *constant.* The same is true for refraction through prisms.

Analytically, the law is therefore stated as follows : *For a non-homogeneous anisotropic medium where the optical density varies from point to point, the optical path l of a ray AB, given by*

$$l = \int_A^B \mu \, dl$$

is an **extremum**, i.e., either *minimum, maximum* or *stationary.*

1.3 Fermat's principle and the laws of reflection

Let the plane mirror M_1M_2 and the plane $PQRS$ be normal to each other (Fig. 1.1). A ray of light AO incident at the point O on the mirror gets reflected along OB. So AO is the *incident* ray, OB the *reflected ray*, O being the *point of incidence.* Draw ON *normal* at O to the plane M_1M_2. Thus AO, OB and ON all lie in the plane $PQRS$.

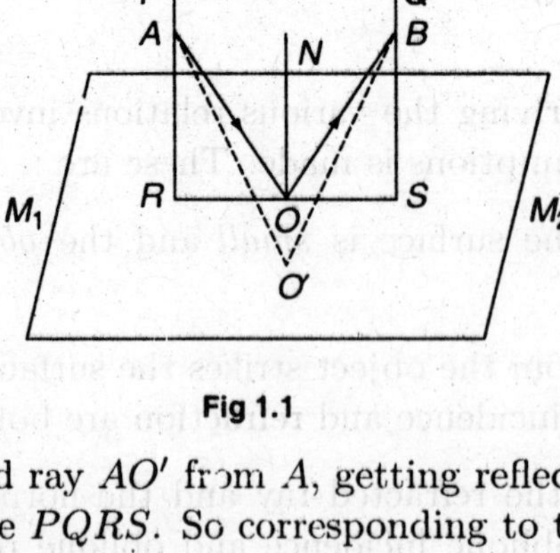

Fig 1.1

Let us next have a second ray AO' from A, getting reflected from the mirror at O' where OO' is $\perp^r$ to the plane $PQRS$. So corresponding to the *incident ray* AO', $O'B$ is the *reflected ray.*

Now, since $AO' > AO$ and $O'B > OB$, the path $(AO' + O'B) > (AO + OB)$, which violates Fermat's principle. So O' must coincide with O for the path to be *minimum.*

Thus, *the incident ray AO, the reflected ray OB and the normal ON at the point of incidence, all lie in the same plane*, which is the **first law of reflection**.

To derive the **second law of reflection**, let the point of incidence O on the mirror be such that the time t taken by light to go from $A \to B$ *via* O is *minimum*.

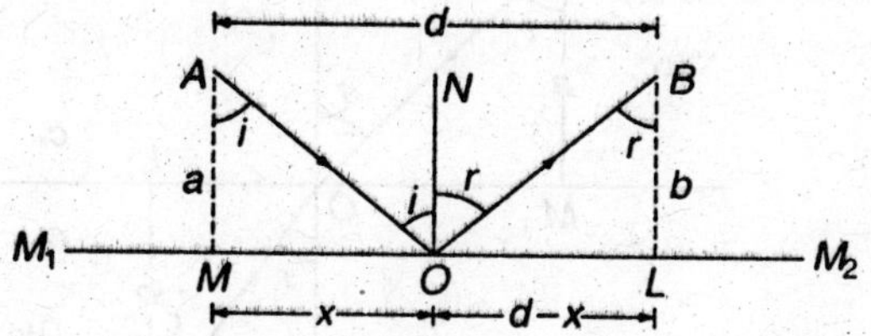

Fig 1.2

$$\text{Now,}\quad t = \frac{AO+OB}{c},\quad c \text{ being the velocity of light in vacuo (or air).}$$

$$= \frac{\sqrt{a^2+x^2}+\sqrt{b^2+(d-x)^2}}{c} \quad \text{(Fig. 1.2)} \qquad (1.3.1)$$

where d is the separation between the feet M, L of the perpendiculars drawn from A and B respectively onto the mirror M_1M_2; x, a, b are as shown in Fig. 1.2.

If O be shifted slightly along M_1M_2, x would change by dx, say and a different path and a different t-value would be involved for light to go to B. But A, B are two *fixed points* implying t would be the same for the new path. But, by Fermat's principle, the time of traverse should be **extremum**, $\Rightarrow$ $dt/dx = 0$. Differentiating (1.3.1)

$$\therefore \quad \frac{dt}{dx} = \frac{1}{c}\frac{d}{dx}\left[\sqrt{a^2+x^2}+\sqrt{b^2+(d-x)^2}\right] = 0$$

$$\Rightarrow \quad \frac{1}{c}\left[\frac{x}{\sqrt{a^2+x^2}} - \frac{d-x}{\sqrt{b^2-(d-x)^2}}\right] = 0$$

$$\Rightarrow \quad \frac{x}{\sqrt{a^2+x^2}} = \frac{d-x}{\sqrt{b^2-(d-x)^2}}$$

$$\text{or,} \quad \sin i = \sin r \Rightarrow \boxed{i = r}$$

where i is the angle of incidence and r the angle of refraction.

So the path of light from A to B *via* the plane mirror is traversed in least time provided $i = r$, i.e., the *angle of incidence* = the *angle of reflection*.

• The second derivative of LHS of (1.3.1) is a positive quantity for $i = r$ so that the condition $i = r$ makes AOB the path of least time.

1.4 Fermat's principle and the laws of refraction

Let AOB be the path of a ray (Fig. 1.3) from a medium of index μ_1 into a medium of index μ_2 ($\mu_2 > \mu_1$) *via* the point O situated on the interface of the two media. So AO

is the *incident ray* at an angle of incidence i and OB the *refracted ray* at an angle of refraction r, O the *point of incidence* and ON the *normal* to the interface at O.

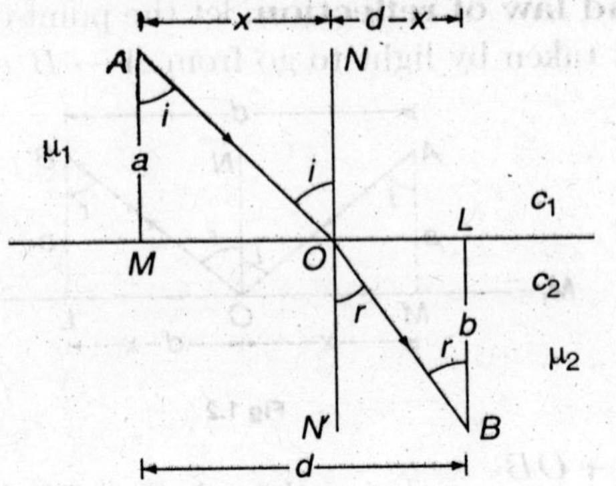

Fig 1.3

If c_1, c_2 be the velocities of light in the media of index μ_1, μ_2 respectively, the time light takes to move from $A \to B$ is

$$t = \frac{AO}{c_1} + \frac{OB}{c_2} = \frac{\sqrt{a^2+x^2}}{c_1} + \frac{\sqrt{b^2+(d-x)^2}}{c_2}. \tag{1.4.1}$$

where a, x, d are as shown in Fig. 1.3.

By Fermat's principle, t is to be minimum, maximum or stationary i.e., $dt/dx = 0$.

Now, from (1.4.1),

$$\frac{dt}{dx} = \frac{1}{c_1}\frac{x}{\sqrt{a^2+x^2}} - \frac{1}{c_2}\frac{(d-x)}{\sqrt{b^2+(d-x)^2}} = 0$$

$$\text{or,} \quad \frac{\sin i}{c_1} = \frac{\sin r}{c_2} \tag{1.4.2}$$

$$\text{But,} \quad \mu_1 = \frac{c}{c_1} \Rightarrow c_1 = \frac{c}{\mu_1}, \quad \mu_2 = \frac{c}{c_2} \Rightarrow c_2 = \frac{c}{\mu_2}$$

where c is the velocity of light in vacuo.

$\therefore$ From (1.4.2),

$$\boxed{\mu_1 \sin i = \mu_2 \sin r} \tag{1.4.3}$$

which is the well-known **Snell's law**. It determines the actual path of the ray.

- It is easily seen that the second derivative of (1.4.1) is *positive* so that the actual path of the ray is *minimum*.
- Similar to reflection, here also it can be shown that the incident ray, the reflected ray and the normal at the point of incidence lie in the same plane.

1.5 Refraction at a convex spherical surface

Case 1. Real image formation

Let R be the radius of curvature of a convex spherical surface separating two media of indices μ_1 and μ_2 ($\mu_2 > \mu_1$). C is the centre of curvature (Fig. 1.4), and P the pole of the spherical surface.

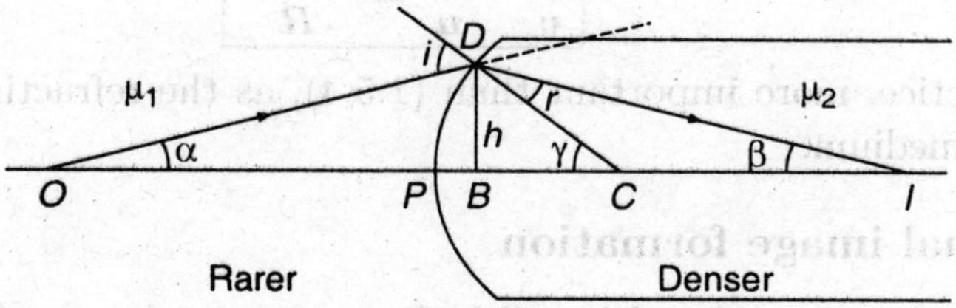

Fig 1.4

Let O be a point object on the principal axis in the rarer medium of index μ_1, distant u from the pole P in front of the convex surface ($u > R$). The incident ray OD, after refraction at D, bends towards the normal CD to the surface at D into the medium of index μ_2 to produce a real image at I, distant v from the pole.

If i and r be respectively the angles of incidence and refraction and α, β, γ the angles made with the principal axis by the incident ray, refracted ray and the normal to the spherical surface at D in order, then

$$\mu_1 \sin i = \mu_2 \sin r$$

$$\Rightarrow \quad \mu_1 i = \mu_2 r \tag{1.5.1}$$

as the aperture of the surface is very small and so also are i and r, the rays being paraxial.

Also, by geometry, we have

$$\left.\begin{aligned} i &= \alpha + \gamma \\ \text{and } \gamma &= r + \beta \\ \Rightarrow r &= \gamma - \beta \end{aligned}\right\} \tag{1.5.2}$$

$\therefore$ From (1.5.1), using (1.5.2),

$$\mu_1(\alpha + \gamma) = \mu_2(\gamma - \beta)$$

$$\Rightarrow \quad \mu_2\beta + \mu_1\alpha = (\mu_2 - \mu_1)\gamma. \tag{1.5.3}$$

But since the angles are small, we may write $\alpha = h/u$; $\beta = h/v$; $\gamma = h/R$, where $h = DB$, the normal to the axis from D.

$\therefore$ From (1.5.3), imposing the sign convention,

$$\mu_2\frac{h}{v} - \mu_1\frac{h}{u} = \frac{(\mu_2 - \mu_1)h}{R}$$

$$\Rightarrow \quad \boxed{\frac{\mu_2}{v} - \frac{\mu_1}{u} = \frac{\mu_2 - \mu_1}{R}} \tag{1.5.4}$$

which is the ***Gaussian formula*** for a single refracting surface.

In case the rarer medium is air, $\mu_1 = 1$, when let $\mu_2 = \mu$. So, from (1.5.4),

$$\boxed{\frac{\mu}{v} - \frac{1}{u} = \frac{\mu - 1}{R}} \tag{1.5.5}$$

which is, in practice, more important than (1.5.4), as the refracting surface is placed generally in air medium.

Case 2. Virtual image formation

Let, corresponding to the point object O in front of the convex surface on the principal axis in the rarer medium of index μ_1, I be the virtual image (Fig. 1.5) formed in the same medium of index μ_1, also on the axis. As before, let u be the distance of O from the pole P ($u < R$) and v the distance I, also from the pole.

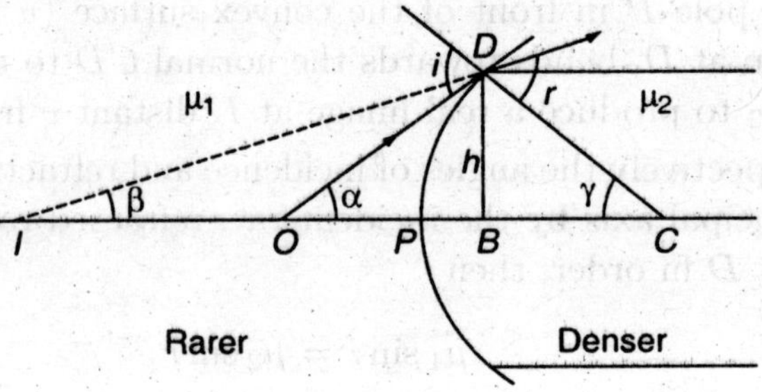

Fig 1.5

From geometry, again, the angle of incidence i and the angle of refraction r are given by

$$i = \alpha + \gamma; \quad r = \beta + \gamma.$$

where α, β and γ are the angles as shown in Fig. 1.5.

So from (1.5.1), we obtain

$$\mu_1(\alpha + \gamma) = \mu_2(\beta + \gamma) \quad \Rightarrow \quad -\mu_2\beta + \mu_1\alpha = (\mu_2 - \mu_1)\gamma$$

$$\Rightarrow \quad -\mu_2\frac{h}{v} + \mu_1\frac{h}{u} = \frac{(\mu_2 - \mu_1)h}{R} \quad (\because \ \alpha = h/u \text{ etc.})$$

where h is the perpendicular dropped from the point of incidence D onto the axis.

Imposing the sign convention,

$$-\mu_2\frac{h}{-v} + \mu_1\frac{h}{-u} = \frac{(\mu_2 - \mu_1)h}{R}$$

$$\therefore \quad \boxed{\frac{\mu_2}{v} - \frac{\mu_1}{u} = \frac{\mu_2 - \mu_1}{R}} \tag{1.5.6}$$

which is the required ***Gaussian formula***, being the same as the relation (1.5.4).

1.6 Refraction at a concave spherical surface

Let OD be the incident ray from the axial point object O. On refraction at D, it gets bent towards the normal CD (C is the centre of curvature of spherical surface) to produce the axial point image I, which is virtual (Fig. 1.6).

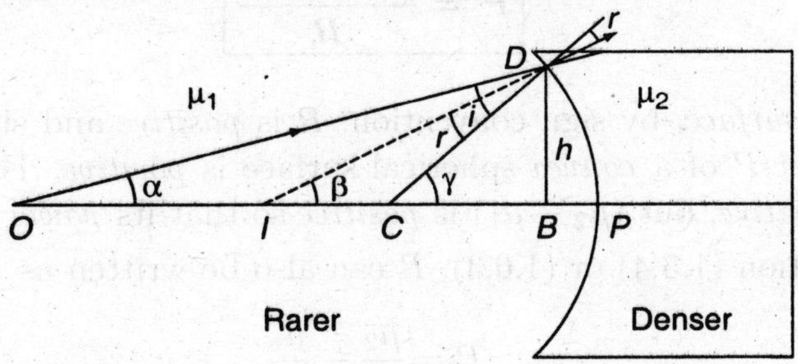

Fig 1.6

From geometry, we have

$$\alpha + i = \gamma \quad \Rightarrow \quad i = \gamma - \alpha;$$
$$\text{and} \quad \beta + r = \gamma \quad \Rightarrow \quad r = \gamma - \beta.$$

But, from the Snell's law,

$$\mu_1 \sin i = \mu_2 \sin r \quad \Rightarrow \quad \mu_1 i = \mu_2 r \tag{1.6.1}$$

since i and r are very small angles.

From (1.6.1), putting the values of i and r, we obtain,

$$\mu_1(\gamma - \alpha) = \mu_2(\gamma - \beta)$$

$$\Rightarrow \quad \mu_2\beta - \mu_1\alpha = (\mu_2 - \mu_1)\gamma. \tag{1.6.2}$$

But $\alpha = h/u$, $\beta = h/v$ and $\gamma = h/R$ where h is the length of perpendicular from D on the axis and R the radius of curvature. Imposing the sign convention, therefore, we get from (1.6.2), since u, v, R are all *negative*,

$$\mu_2\frac{h}{v} - \mu_1\frac{h}{u} = \frac{(\mu_2 - \mu_1)h}{R}$$

$$\therefore \quad \boxed{\frac{\mu_2}{v} - \frac{\mu_1}{u} = \frac{\mu_2 - \mu_1}{R}} \tag{1.6.3}$$

which is the *same* relation as obtained in refraction at a convex spherical surface.

1.7 Power, principal foci and principal planes

Some important terms related to a spherical surface are defined and discussed here before proceeding further.

Power of a spherical surface—Let a spherical surface of radius of curvature R be placed in a medium of index μ_1 and let μ_2 be the index of its material ($\mu_2 > \mu_1$). Then, the **power P** of the refracting spherical surface is defined as

$$\boxed{P = \frac{(\mu_2 - \mu_1)}{R}}$$

For a *convex surface*, by sign convention, R is *positive* and since $(\mu_2 - \mu_1)$ is also *positive*, the *power* P of a *convex* spherical surface is *positive*. For a *concave surface*, however, R is *negative*, but $(\mu_2 - \mu_1)$ is *positive* so that its *power* P is *negative*.

From the relation (1.5.4) or (1.6.3), P can also be written as

$$P = \frac{\mu_2}{v} - \frac{\mu_1}{u}$$

where u and v are respectively the object distance and image distance from the pole of the spherical surface.

If the first medium is air, $\mu_1 = 1$ and $\mu_2 = \mu$, say so that we have

$$P = \frac{\mu}{v} - \frac{1}{u}$$

If R, u and v are expressed in *metres*, the *power* is said to be in **diopter** (D).

Principal foci—The Gaussian relation

$$\frac{\mu_2}{v} - \frac{\mu_1}{u} = \frac{\mu_2 - \mu_1}{R}$$

defines *two* very important *points* on the *principal axis*, known respectively as the **first principal focus** F_1 and the **second principal focus** F_2. These two points are called **principal foci.**

First principal focus—*The first* **first principal focus** F_1 *is a point on the principal axis such that rays moving towards the point or diverging from it* (Fig. 1.7) *become parallel to the principal axis, after refraction.*

The distance of F_1 from the pole O of the refracting surface is called the **first focal length** f_1 of the surface. If an object be at F_1, then the image will be formed at *infinity*, i.e., if $u = f_1$, $v = \infty$.

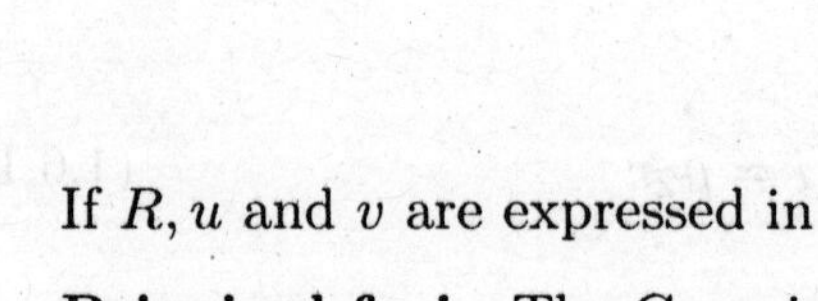

Fig. 1.7

$$\therefore \quad \frac{\mu_2}{\infty} - \frac{\mu_1}{f_1} = \frac{\mu_2 - \mu_1}{R} \Rightarrow f_1 = -\frac{\mu_1 R}{\mu_2 - \mu_1} \tag{1.7.1}$$

Second principal focus—*The* **second principal focus** F_2 *is a point on the principal axis such that the rays incident parallel to the principal axis will converge to or appear to diverge from* (Fig. 1.8) *this point, after refraction.*

The distance of F_2 from the pole O of the refracting surface is called the **second focal length** f_2 of the surface. If the object is at *infinity* i.e., $u = \infty$, then the image is at F_2, i.e., $v = f_2$.

$$\therefore \quad \frac{\mu_2}{f_2} - \frac{\mu_1}{\infty} = \frac{\mu_2 - \mu_1}{R} \quad \Rightarrow \quad f_2 = \frac{\mu_2 R}{\mu_2 - \mu_1} \qquad (1.7.2)$$

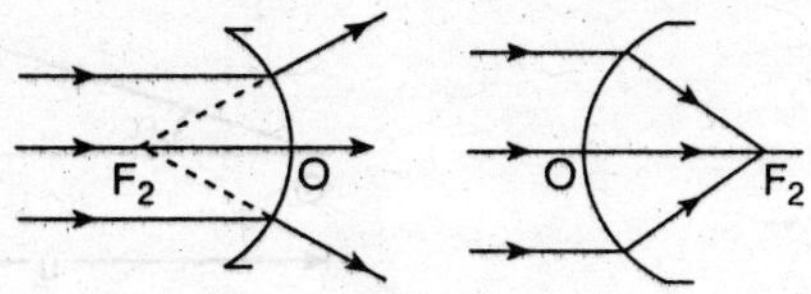

Fig. 1.8

If the first medium is air, $\mu = 1$ and $\mu_2 = \mu$, say and we have, from (1.7.1) and (1.7.2) respectively

$$\boxed{f_1 = -\frac{R}{\mu - 1}} \quad \text{and} \quad \boxed{f_2 = \frac{\mu R}{\mu - 1}} \qquad (1.7.3)$$

From (1.7.1) and (1.7.2), we get

$$\frac{f_2}{f_1} = -\frac{\mu_2}{\mu_1} \qquad (1.7.4)$$

which shows that the *two principal focal lengths are always of opposite sign.*

Principal planes—The two planes perpendicular to the principal axis, one passing through the first principal focus F_1 and the other through the second principal focus F_2, are called the **principal planes**. The one through F_1 is the **first principal plane** and the other through F_2, the **second principal plane** of the refracting surface.

• From (1.7.4),

$$\boxed{\mu_1 f_2 + \mu_2 f_1 = 0}$$

and if the first medium be air, $\mu = 1$ and $\mu_2 = \mu$, say, then

$$f_2 + \mu f_1 = 0 \qquad (1.7.5)$$

which *distinguishes* the two focal lengths.

1.8 Illustrated Examples

▶ **Example 1.** *An object is situated at the optical axis of a convex spherical surface. A ray of light starting from the object subtends an angle of* 5° *with the optical axis and intersects the spherical surface at a height of* 2.5 cm. *If the radius of curvature of the surface is* +5 cm *and the index of the material* 1.5, *find the angle which the refracted ray makes with the optical axis.* **(Himachal Univ.)**

Solution. Let u = object distance, v = image distance, R = the radius of curvature of the surface and μ the index of the material.

Here $\mu = 1.5$, $R = 5.0$ cm, $\alpha = 5°$, $AB = 2.5$ cm (Fig. 1.9), $\beta = ?$, $BO = PO = -u$ and $BI = PI = v$, O being the axial object, I the axial image.

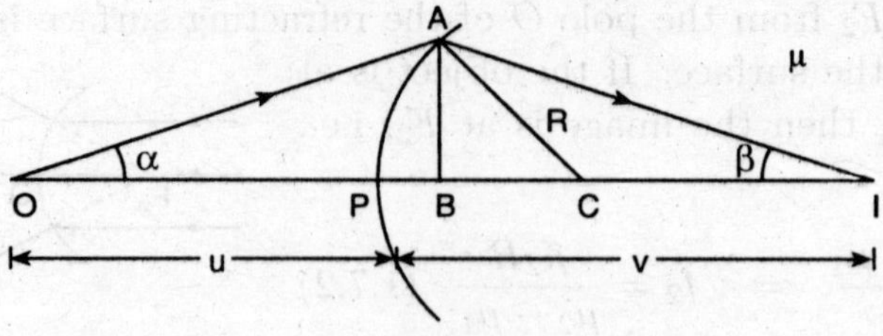

Fig. 1.9

$$\therefore \quad \tan\alpha = \frac{2.5}{-u} \Rightarrow u = -\frac{2.5}{\tan 5°} = -\frac{2.5}{0.0875} = -28.57 \text{ cm}$$

Since refraction occurs from a rarer medium (air) to a denser medium

$$\frac{\mu}{v} - \frac{1}{u} = \frac{\mu - 1}{R} \quad \Rightarrow \quad \frac{\mu}{v} = \frac{1.5 - 1}{5.0} - \frac{1}{28.57} = 0.1 - 0.035$$

$$\therefore \quad v = \frac{\mu}{0.1 - 0.035} = \frac{1.5}{0.065} = 23.08 \text{ cm}$$

$$\therefore \quad \tan\beta = \frac{2.5}{23.08} = 0.1083 \quad \Rightarrow \quad \beta = \mathbf{6.2°}$$

► **Example 2.** *Show that for refraction at a concave spherical surface (separating glass-air medium), the object distance should be greater than three times the radius of curvature of the refracting surface for the image to be real.*

Solution. Here $\mu_1 = 1.5$, $\mu_2 = 1$, $R = -$ve, $u = -$ve. So, from the relation

$$\frac{\mu_2}{v} - \frac{\mu_1}{u} = \frac{\mu_2 - \mu_1}{R}$$

$$\frac{1}{v} - \frac{1.5}{-u} = \frac{1 - 1.5}{-R}$$

$$\Rightarrow \quad \frac{1}{v} = \frac{1}{2R} - \frac{3}{2u}$$

For the image to be real, v is to be positive and, from above, the condition is

$$\frac{1}{2R} > \frac{3}{2u} \quad \Rightarrow \quad \boldsymbol{u > 3R}$$

► **Example 3.** *A small air bubble in a sphere of glass,* 3 mm *in dia, appears to be* 1 cm *from the surface when looked along a diameter. Find the true position of the air bubble.*

Solution. For refraction at a curved surface, we have

$$\frac{\mu_2}{v} - \frac{\mu_1}{u} = \frac{\mu_2 - \mu_1}{R}$$

Here refraction is at a concave surface. So $v = -1$ cm, $R = -1.5$ cm, $\mu_2 = 1$ (air), $\mu_1 = 1.5$

$$\therefore \quad \frac{1}{-1} - \frac{1.5}{u} = \frac{1-1.5}{-1.5}$$

$$\therefore \quad \mathbf{u = -1.12\ cm}$$

▶ **Example 4.** *When seen normally through a flat surface, the greatest thickness of a hemispherical glass block appears to be* 4.8 cm, *but when seen normally from the curved surface, it appears to be* 6.0 cm *thick. The actual thickness is* 7.2 cm. *Find* (i) *the refractive index of glass and* (ii) *the radius of curvature of the surface.*

(Calcutta Hons.)

Solution. (i) When *viewed normally through the flat surface*, the object O (Fig. 1.10) appears to be at I (its image) as refraction occurs at flat surface from a denser medium of index μ_1 to a rarer medium of index μ_2.

Now, $\mu_2 = 1$, $R = \infty$, $u = OP = 7.2$ cm and $v = IP = 4.8$ cm

$\therefore$ From the relation,

$$\frac{\mu_2}{v} - \frac{\mu_1}{u} = \frac{\mu_2 - \mu_1}{R}$$

$$\frac{1}{4.8} - \frac{\mu_1}{7.2} = 0, \quad \text{since } R = \infty$$

$$\therefore \quad \mu_1 = \frac{7.2}{4.8} = \frac{3}{2} = \mathbf{1.5}$$

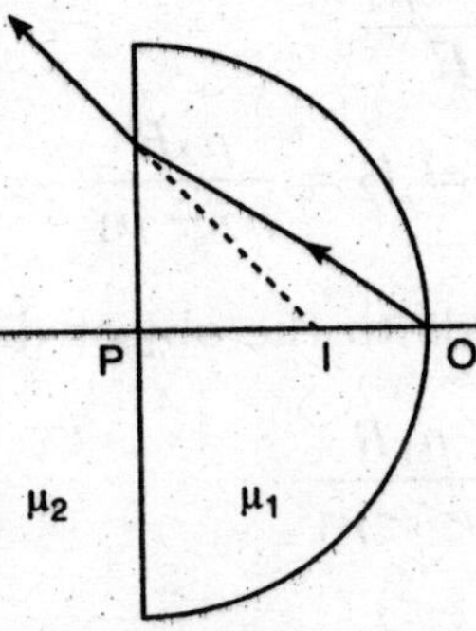

Fig. 1.10

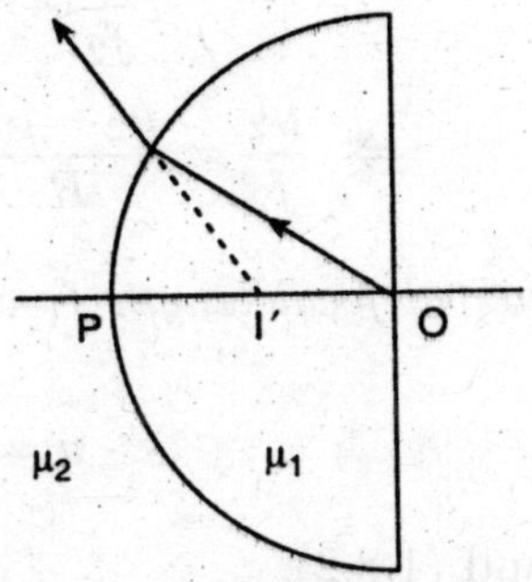

Fig. 1.11

(ii) In this case (Fig. 1.11), the object is *seen normally through the curved surface* and appears to be at I' (its image).

Here, therefore, $u = OP = 7.2$ cm, $v = I'P = 6.0$ cm, and $R = ?$

$$\therefore \quad \frac{\mu_2}{v} - \frac{\mu_1}{u} = \frac{\mu_2 - \mu_1}{R} \text{ gives : } \frac{1}{6} - \frac{1.5}{7.2} = \frac{1-1.5}{R}$$

$$\Rightarrow \quad -\frac{1}{24} = -\frac{0.5}{R}$$

$$\therefore \quad R = 24 \times 0.5 = \mathbf{12\ cm}$$

► **Example 5.** *The eye can be regarded as a single spherical refracting surface of radius of curvature of cornea* 7.8 mm, *separating two media of indices* 1.00 *and* 1.34. *Calculate the distance from the refracting surface at which a parallel beam of light will come to focus.*

(Delhi Hons.)

Solution. Here, $\mu_1 = 1.00$, $\mu_2 = 1.34$, $R = 0.78$ cm, $u = -\infty$. So, from the relation,

$$\frac{\mu_2}{v} - \frac{\mu_1}{u} = \frac{\mu_2 - \mu_1}{R},$$

$$\frac{1.34}{v} - \frac{1.00}{-\infty} = \frac{1.34 - 1.00}{0.78}$$

$$\therefore\ v = \frac{1.34 \times 0.78}{0.34} = \mathbf{3.07\ cm}$$

► **Example 6.** *Show that* $(f_1/u) + (f_2/v) = 1$ *where* f_1, f_2 *are the principal focal lengths of a spherical refracting surface, u and v having their usual significance.*

(Visweswariah Univ.)

Solution. The *second principal focal length* f_2 corresponds to the image distance v for object distance $u = \infty$. So from Gauss' relation

$$\frac{\mu_2}{f_2} - \frac{\mu_1}{\infty} = \frac{\mu_2 - \mu_1}{R}$$

$$\Rightarrow\ \frac{\mu_2}{f_2} = \frac{\mu_2 - \mu_1}{R} \quad \Rightarrow\ v = f_2 = \frac{\mu_2 R}{\mu_2 - \mu_1} \tag{1.8.1}$$

The *first principal focal length* f_1 corresponds to the object distance u for which $v = \infty$.

$$\therefore\ u = f_1 = -\frac{\mu_1 R}{\mu_2 - \mu_1} \tag{1.8.2}$$

$\therefore$ From (1.8.1) and (1.8.2),

$$\frac{\boldsymbol{f_2}}{\boldsymbol{v}} + \frac{\boldsymbol{f_1}}{\boldsymbol{u}} = \frac{\mu_2 R}{v(\mu_2 - \mu_1)} - \frac{\mu_1 R}{u(\mu_2 - \mu_1)} = \frac{R}{\mu_2 - \mu_1}\left[\frac{\mu_2}{v} - \frac{\mu_1}{u}\right]$$

$$= \frac{R}{\mu_2 - \mu_1} \cdot \frac{\mu_2 - \mu_1}{R}, \text{ using Gauss' relation,}$$

$$= \mathbf{1}$$

► **Example 7.** *A glass dumbell of length* 0.5 m *and index* 1.5 *has spherical ends of* 0.05 m *radius of curvature. If a point object is situated in air at a distance of* 0.2 m *from the end of the dumbell along the axis, calculate the position of the image formed by refraction at one end only.*

Solution. Here $u = -0.2$ m, $R = 0.05$ m, $\mu_2 = 1.5$, $\mu_1 = 1$, $v = ?$ (Fig. 1.12)

$$\text{Since,} \quad \frac{\mu_2}{v} - \frac{\mu_1}{u} = \frac{\mu_2 - \mu_1}{R}$$

$$\therefore \quad \frac{1.5}{v} + \frac{1}{0.2} = \frac{1.5 - 1}{0.05} \quad \Rightarrow \quad v = 0.3 \text{ m}$$

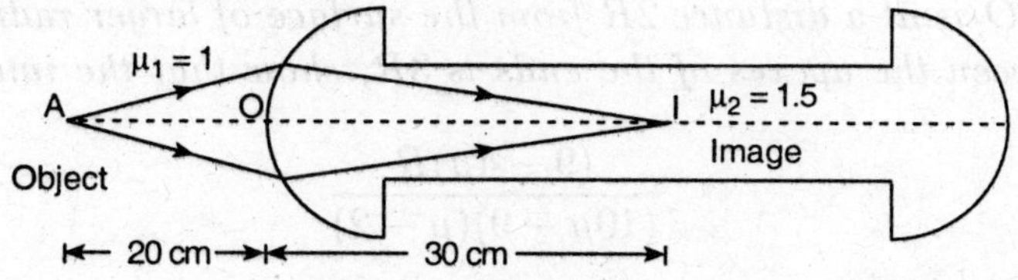

Fig. 1.12

As v is positive, the image is formed **0.3 m** to the **right** of the vertex O of the dumbell, as shown in Fig. 1.12.

▶ **Example 8.** *A transparent sphere is of radius* 15 cm *and refractive index* $\mu = 1.5$. *Find the position of its paraxial focal point.*

Solution. For refraction at the first surface, we get

$$\frac{\mu_1}{v_1} - \frac{1}{u_1} = \frac{\mu_1 - 1}{R_1}$$

with usual meanings of the symbols used.

Here $u_1 = -\infty$, $R_1 = +15$ cm, $\mu = 1.5$. So, we have

$$\text{image distance,} \quad v_1 = O_1D = 3R = 45 \text{ cm}$$

For refraction at the second surface, D is the virtual object. So, the object distance

$$u = O_2D = O_1D - O_1O_2 = 45 - 30 = 15 \text{ cm}$$

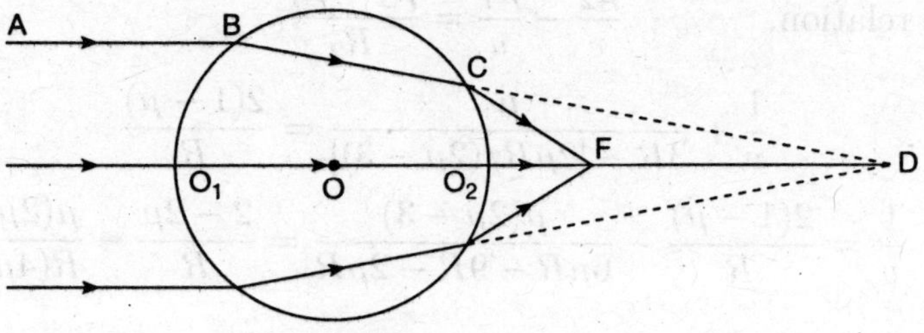

Fig. 1.13

Radius of curvature of the second surface, $R_2 = -15$ cm.

$$\therefore \text{ From the relation :} \quad \frac{\mu_2}{v} - \frac{\mu_1}{u} = \frac{\mu_2 - \mu_1}{R}$$

and remembering $\mu_2 = 1$, $\mu_1 = 1.5$, $u = 15$ cm and $R = R_2 = -15$ cm here,

$$\frac{1}{v} - \frac{1.5}{15} = \frac{1-1.5}{-15} \Rightarrow v = +7.5 \text{ cm}$$

$\therefore$ Distance of paraxial focal point F (Fig. 1.13) from O_2 is $O_2F =$ **+ 7.5 cm.**

► **Example 9.** *A glass rod has ends as shown in* Fig. 1.14. *The index of glass is* μ. *The object point O is at a distance* $2R$ *from the surface of larger radius of curvature. If the distance between the apexes of the ends is* $3R$, *show that the image of O is formed at a distance*

$$\frac{(9-4\mu)R}{(10\mu-9)(\mu-2)}$$

from the right-hand vertex.

Solution. For refraction at the first surface, we have

$$\frac{\mu_2}{v_1} - \frac{\mu_1}{u} = \frac{\mu_2 - \mu_1}{R_1}$$

Here $\mu_2 = \mu$, $\mu_1 = 1$, $u = -2R$, $R_1 = +R$. So, from above, we obtain

$$\frac{\mu}{v_1} + \frac{1}{2R} = \frac{\mu - 1}{R} \Rightarrow v_1 = \frac{2\mu R}{2\mu - 3}$$

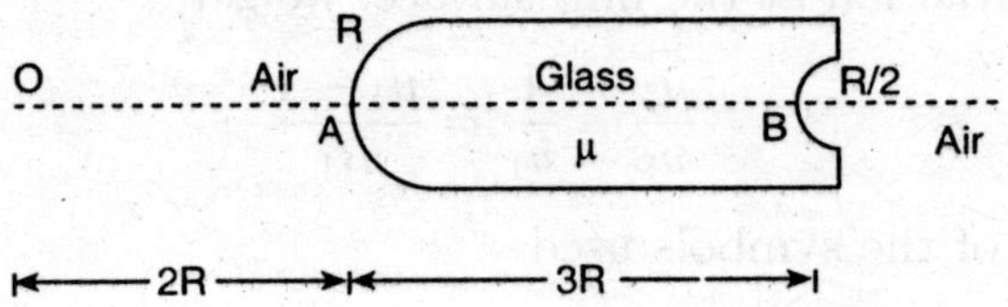

Fig. 1.14

For refraction at the second surface,

$$\mu_2 = 1, \quad \mu_1 = \mu \text{ and } R_2 = +R/2 \text{ and } u = -(3R - v_1) = -\left[3R - \frac{2\mu R}{2\mu - 3}\right]$$

$\therefore$ From the relation, $$\frac{\mu_2}{v} - \frac{\mu_1}{u} = \frac{\mu_2 - \mu_1}{R_2}$$

$$\frac{1}{v} + \frac{\mu}{3R - [2\mu R/(2\mu - 3)]} = \frac{2(1-\mu)}{R}$$

$$\therefore \quad \frac{1}{v} = \frac{2(1-\mu)}{R} - \frac{\mu(2\mu-3)}{6\mu R - 9R - 2\mu R} = \frac{2 - 2\mu}{R} - \frac{\mu(2\mu - 3)}{R(4\mu - 9)}$$

$$= \frac{-10\mu^2 + 29\mu - 18}{R(4\mu - 9)} = \frac{-(10\mu - 9)(\mu - 2)}{R(4\mu - 9)}$$

$$\therefore \quad v = \frac{-R(4\mu - 9)}{(10\mu - 9)(\mu - 2)} = \boldsymbol{\frac{(9 - 4\mu)R}{(10\mu - 9)(\mu - 2)}}$$

Review Questions

1. Explain Fermat's principle and using this principle, establish the laws of reflection of light. **(Burdwan Univ)**

2. Using Fermat's principle establish (a) the laws of reflection on a plane surface and (b) the condition of total internal reflection. **(Calcutta and Gorakhpur Univ.)**

3. Derive the following relation for refraction at a single convex spherical surface:
$$\frac{\mu}{v} - \frac{1}{u} = \frac{\mu - 1}{R}$$
(Calcutta Univ.)

4. Derive a relation connecting u, v, R and μ when refraction occurs at a concave spherical surface.

5. Establish the relation, for refraction at a single spherical surface,
$$\frac{\mu_2}{v} - \frac{\mu_1}{u} = \frac{\mu_2 - \mu_1}{R},$$
where the symbols have their usual significance. **(Delhi Univ.)**

6. Explain the following quantities related to a spherical surface : (a) power, (b) principal foci and (c) the principal planes.

7. Explain the terms : first principal focus and second principal focus and show that
$$f_1 v + f_2 u = uv.$$
with usual meanings of the symbols.

Short Answer and Objective Questions

1. State the assumptions made in deriving the various relations for refraction at a spherical surface.

2. State and explain Fermat's principle. **(Visweswariah Univ.)**

3. Derive Snell's law using Fermat's principle.

4. Write down the Gaussian formula for refraction at a single spherical surface.

5. Explain the term 'power' of a spherical surface. What is its unit?

6. What do you mean by principal foci and principal planes of a spherical surface?

7. Correct expression for power P of a spherical surface is given by

(a) $\frac{\mu_2 - \mu_1}{R}$ (b) $\frac{1}{v} - \frac{\mu}{u}$ (c) $\frac{\mu_2}{v} - \frac{\mu_1}{u}$ (d) none

8. For refraction at a concave spherical surface (separating glass-air medium), the object distance should be greater then n times the radius of curvature of the refracting surface, for real image. The value of n is

(a) 1 (b) 2 (c) 3 (d) 4

9. A double convex air-bubble in water will behave like a

(a) concave lens (b) convex lens (c) plane glass (d) none.

Problems

1. A concave surface of radius of curvature 30 cm separates two media of indices 4/3 and 1.5. An object is kept at a distance of 20 cm from the surface in the first medium. Calculate the position of the image. [0.2455 m]

2. A mark placed on the surface of a glass sphere is viewed through glass from a position directly opposite. If the diameter of the sphere is 0.1 m and the index of glass is 1.5, find the position of the image. [−0.20 m]

3. The incident face of a glass block ($\mu = 1$) is bounded by a concave surface of radius 0.03 m. A small object 2×10^{-2} m high is situated in air at a distance 0.1 m from the vertex. Find (a) the first and second focal lengths, (b) the image distance, (c) the lateral magnification and (d) the power of the surface.

[0.06 m, −0.09 m; −0.0562 m; 0.375; 16.67 D]

4. When viewed normally through the flat surface, the greatest thickness of a plano-convex surface appears to be 2.0 cm but when viewed normally through the curved surface, it appears to be 2.5 cm. The actual thickness is 3.0 cm. Find (a) the index of the material, (b) the radius of curvature of the curved surface.

[1.5; 5 cm]

5. A convex surface of radius of curvature 0.2 m separates two media of indices 1.33 and 1.50. An object is kept in the first medium at a distance of 2.4 m from the refracting surface. Find the position and nature of image.

[+5.07 m from refracting surface; real]

6. If in Example 6 (worked out), the object is 5 cm from the dumbell, what is the position of the image? [−15 cm]

7. A small filament is at the centre of a hollow glass sphere of inner and outer radii 8 cm and 9 cm respectively. The index of glass is 1.5. Find the position of the image of the filament when viewed from outside the sphere.

   [at the centre of sphere]

8. A glass dumbell 200 cm long and of index 1.5 has ends of radius of curvature 10 cm. Find the position of the image due to refraction at one end only, when the object is at a distance 40 cm from one end. [60 cm]

9. Obtain the condition for the formation of a real image when paraxial rays are refracted at a concave spherical water-air interface. Radius of curvature of the surface is R. [$\mu > 4R$]

10. Two concave refracting surfaces of equal radii of curvature and index 1.5 face each other in air. A point object O is placed midway between the centre and one of the vertices. Find the separation between the images of O formed by each refracting surface. [$0.114R$]

11. A parallel beam of light travelling in water ($\mu = 4/3$) is refracted by a sphercial air bubble of radius 2 mm, situated in water. Find the position of the image due to refraction at the first surface and position of the final image.

    Also, draw a ray diagram showing the positions of both images.

    [6 mm to left of first surface, 5 mm to left of second surface]

12. A small object is located 0.1 m in front of a convex surface of radius 0.03 m made on a glass block of $\mu = 1.66$. Find the position of the image. The object and the glass are in air. **(Calcutta Hons.)** [0.138 m inside glass from convex surface]

Chapter 2

Refraction Through Lenses

2.1 Convex lens equation

We shall deduce here the convex lens equation connecting the focal length, the radii of curvature of its surfaces, and the indices of the material of the lens and the surrounding medium.

Let a thin double convex lens of index μ_2 separate it from a medium of index μ_1 on either side. Let A, B be its curved surfaces having R_1 and R_2 as the radius of curvature respectively (Fig. 2.1). O is an axial point object in front of the face A and distant u from it. From O, let OA be a ray passing through the optical centre of the lens. It would thus suffer no deviation. OD is an *oblique ray* from O and is incident at D on the face A of the lens. It would have met the axis at I_1 if the lens material were continuous. So I_1 is the image of O, produced by refraction at face A only, and is distant v_1 from it.

$$\therefore \quad \frac{\mu_2}{v_1} - \frac{\mu_1}{u} = \frac{\mu_2 - \mu_1}{R_1} \tag{2.1.1}$$

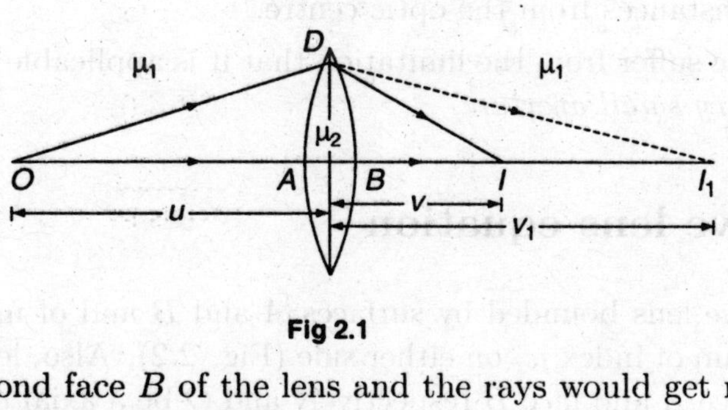

Fig 2.1

But there is a second face B of the lens and the rays would get refracted by it. I_1 would act as the object for this surface and the final image would be at the axial point I, distant v and in the medium of index μ_1.

$$\therefore \quad \frac{\mu_1}{v} - \frac{\mu_2}{v_1} = \frac{\mu_1 - \mu_2}{R_2} = \frac{\mu_2 - \mu_1}{-R_2} \tag{2.1.2}$$

Adding (2.1.1) and (2.1.2).

$$\frac{\mu_1}{v} - \frac{\mu_1}{u} = (\mu_2 - \mu_1)\left(\frac{1}{R_1} - \frac{1}{R_2}\right)$$

$$\Rightarrow \quad \frac{1}{v} - \frac{1}{u} = \left(\frac{\mu_2 - \mu_1}{\mu_1}\right)\left(\frac{1}{R_1} - \frac{1}{R_2}\right) \tag{2.1.3}$$

If the object be at infinity i.e., $u = \infty$, then $v = f$.

$$\therefore \quad \frac{1}{f} = \left(\frac{\mu_2 - \mu_1}{\mu_1}\right)\left(\frac{1}{R_1} - \frac{1}{R_2}\right). \tag{2.1.4}$$

If the lens is surrounded by air $\mu_1 = 1$ and $\mu_2 = \mu$, say, so that

$$\boxed{\therefore \quad \frac{1}{v} - \frac{1}{u} = \frac{1}{f} = (\mu - 1)\left(\frac{1}{R_1} - \frac{1}{R_2}\right)}$$

using (2.1.3) and (2.1.4).

The above is the *convex lens equation.*

The more familiar form, however is given by

$$\boxed{\frac{1}{v} - \frac{1}{u} = \frac{1}{f}}$$

• In deriving the equation, the following *approximations* have been made : (i) the *aperture* of the lens is so *small* that the *angles* made by the oblique rays with optical axis are also *small,* (ii) the *lens* is *very thin* so that distances measured from the surface can be taken as distances from the optic centre.

• The formulae suffer from the limitation that it is applicable to a lens which is *very thin* and has a *very small aperture.*

2.2 Concave lens equation

Let a thin concave lens bounded by surfaces A and B and of material of index μ_2 be placed in a medium of index μ_1 on either side (Fig. 2.2). Also, let R_1, R_2 be the radius of curvature of face A and face B respectively and O be a axial point object in front of A, distant u from C, the optical centre. The oblique ray OD from O suffers refraction at the concave surface and diverges along E, appearing to come from I_1 which is the virtual image of O formed by refraction at the surface A and is distant v_1 from the optical centre.

$$\therefore \quad \frac{\mu_2}{v_1} - \frac{\mu_1}{u} = \frac{\mu_2 - \mu_1}{R_1} \tag{2.2.1}$$

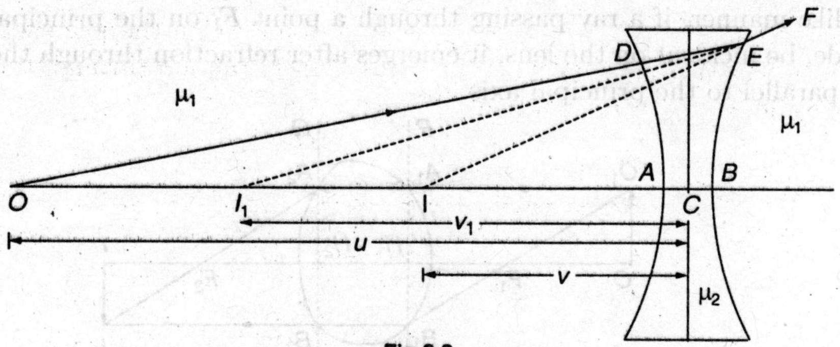

Fig 2.2

The rays would now strike the second face B of the lens to be deviated along EF to form at a distance v, the final axial point image I for which I_1 would act as the object. I lies in the medium of index μ_1.

$$\therefore \quad \frac{\mu_1}{v} - \frac{\mu_2}{v_1} = \frac{\mu_1 - \mu_2}{R_2} = \frac{\mu_2 - \mu_1}{-R_2} \tag{2.2.2}$$

Adding (2.2.1) and (2.2.2), we obtain

$$\frac{\mu_1}{v} - \frac{\mu_1}{u} = (\mu_2 - \mu_1)\left(\frac{1}{R_1} - \frac{1}{R_2}\right)$$

$$\Rightarrow \frac{1}{v} - \frac{1}{u} = \left(\frac{\mu_2 - \mu_1}{\mu_1}\right)\left(\frac{1}{R_1} - \frac{1}{R_2}\right)$$

If the lens be in air, $\mu_1 = 1$ and $\mu_2 = \mu$, say so that

$$\boxed{\frac{1}{v} - \frac{1}{u} = \frac{1}{f} = (\mu - 1)\left(\frac{1}{R_1} - \frac{1}{R_2}\right)}$$

which is very similar to that for a thin convex lens.

- This equation suffers from and is limited by the same factors as discussed in Art 2.1.

2.3 Cardinal points of a coaxial optical system

There are **six** *cardinal points* of a *coaxial optical system* or a lens system and these are : (i) **two** *focal points*, (ii) **two** *principal* or *unit points* and (iii) **two** *nodal points.* They are of cardinal importance in that they possess some unique properties, a knowledge of which is essential in understaning the optical system in its totality.

(i) Two focal points—A ray of light, incident on the lens system in a direction parallel to the principal axis, on its object side, would be refracted by the lens system such that on emergence it passes through a point F_2, on the principal axis, on the image side.

In a like manner, if a ray passing through a point F_1 on the principal axis, on the object side, be incident on the lens, it emerges after refraction through the lens system, as a ray parallel to the principal axis.

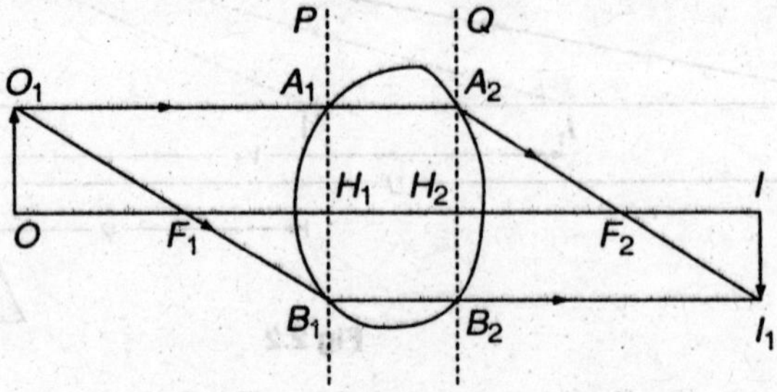

Fig 2.3

The points F_1 and F_2 (Fig. 2.3) are known respectively as the *first* and the *second focal points.*

Definitions : *The* **first focal point** (F_1) *is a point on the principal axis of a lens such that any ray of light passing through it is rendered parallel to the principal axis, after refraction by the lens system.*

The **second focal point** (F_2) *is a point on the principal axis of a lens system such that any ray of light parallel to the principal axis, after refraction by the lens system, passes through it.*

• The *planes* passing through F_1 and F_2, and *perpendicular* to the *principal axis* are termed as the **first** and the **second focal planes** respectively.

(ii) Two principal points—These are *two conjugate points situated on the principal axis for which the lens system has positive unit linear magnification.* This means that *if an object is placed at one principal point, an erect image of the same size will be produced at the other principal point.* In Fig. 2.3. H_1 and H_2 are respectively the **first** and the **second principal points**. The principal points are also known as **unit points.**

The distance of the first focal point from the first principal point, i.e., F_1H_1 is called the **first focal length, f_1** of the lens, and the distance of the second focal point from the second principal point, i.e, F_2H_2 is known as the **second focal length, f_2**.

If the *same* medium exists on either side of the lens, then $F_1H_1 = F_2H_2$, i.e., $f_1 = f_2$, the *first focal length equals the second focal length,* but otherwise not.

• The planes P and Q that pass through the points H_1 and H_2 respectively and are normal to the principal axis are known as **first** and **second principal planes** or **unit planes.**

(iii) Two nodal points—*The nodal points are two conjugate points on the principal axis having unit positive angular magnification.* This means that *a ray of light directed to one of the pair of points would, after refraction from the lens system, emerge out from the other in a direction parallel to the incident ray.*

Let P, Q be the *first* and the *second principal plane* of a lens system, and AF_1 and BF_2 the *first* and *second focal plane* respectively (Fig. 2.4). So a ray of light AA_1 from A on the first focal plane, parallel to the axis, will proceed to F_2, the second focal point, from the point B_1 on the second principal plane Q.

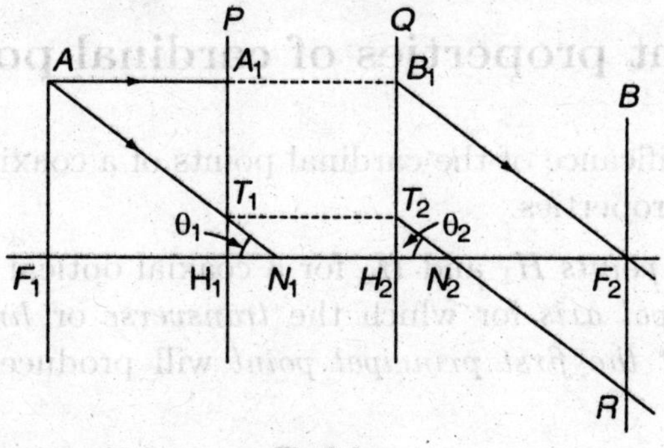

Fig 2.4

AT_1 is another ray from A, parallel to B_1F_2—the emergent ray. It strikes the first principal plane at T_1 and emerges from the point T_2 on the second principal plane such that $T_1H_1 = T_2H_2$ and $T_2R \parallel B_1F_2$. The points N_1 and N_2, the points of intersection of the incident ray AT_1 and the conjugate ray T_2R with the axis respectively, are the positions of the two **nodal points**.

Properties of nodal points—We have

$$\tan\theta_1 = \tan\theta_2 \quad \Rightarrow \quad \theta_1 = \theta_2$$

Since $T_1H_1 = T_2H_2$ and $\theta_1 = \theta_2$, $\Delta T_1H_1N_1$ and $\Delta T_2H_2N_2$ are congruent $\Rightarrow H_1N_1 = H_2N_2$.

$$\therefore \quad H_1N_1 + N_1H_2 = H_2N_2 + N_1H_2 \quad \Rightarrow \quad \boxed{H_1H_2 = N_1N_2.}$$

This gives the important information that *the distance between the two nodal points is equal to the distance between the two principal points.*

Again, $AF_1 = B_1H_2$ and $\angle AN_1F_1 = \angle B_1F_2H_2$. So the right-angled ΔAF_1N_1 and $\Delta B_1H_2F_2$ are congruent $\Rightarrow F_1N_1 = H_2F_2$.

$$\therefore \quad F_1H_1 + H_1N_1 = H_2F_2 \quad \Rightarrow \quad H_1N_1 = H_2F_2 - F_1H_1$$

$$\text{But,} \quad F_1H_1 = -f_1 \quad \text{and} \quad H_2F_2 = +f_2$$

where f_1 and f_2 are respectively the first and the second focal length of the lens system.

$$\therefore \quad H_1N_1 = H_2N_2 = f_2 + f_1$$

If the system is surrounded by the same medium, $f_2 = -f_1$.

$$\therefore \quad \boxed{H_1N_1 = H_2N_2 = 0.}$$

The principal (or unit) *points coincide with the nodal points if the medium on both sides of the system is the same.*

• The planes passing through the nodal points and normal to the principal axis are known as the **nodal planes**.

2.4 Significant properties of cardinal points summed up

To highlight the significance of the cardinal points of a coaxial optical system, we sum up their important properties.

1. The *principal points* H_1 and H_2 for a coaxial optical system are *two conjugate points* on the *principal axis* for which the *transverse* or *linear magnification is* $+1$. Hence, the *object at the first principal point* will produce an *image at the second principal point.*

2. The *first focal point* F_1 is a *point* on the *principal axis* of a lens such that a *ray* passing *through it* is rendered *parallel* to the *principal axis,* after refraction.

3. The *second focal point* F_2 is a *point* on the *principal axis* such that a *ray parallel* to the *principal axis,* after refraction, *passes through it.*

4. The *nodal points* N_1, N_2 are a *pair of conjugate points* on the *principal axis* for which the *angular magnification is* $+1$. In other words, *a ray directed to any one of the pair emerges out, after refraction through the optical system, from the second point in a direction parallel to the incident ray.*

5. The distance N_1N_2 between the *two nodal points* is equal to the *distance* H_1H_2 between the *two principal points* i.e., $N_1N_2 = H_1H_2$.

6. The *principal points* will *coincide* with the *nodal points* if the medium on both sides of the system is the *same* i.e., $H_1N_1 = H_2N_2 = 0$.

7. The distance of the *first focal point* F_1 from the *first principal point* H_1 is called the *first focal length* f_1 of the lens; and the distance of the *second focal point* F_2 from the *second principal point* H_2 is called the *second focal length* f_2, i.e., $F_1H_1 = f_1$ and $F_2H_2 = f_2$. If the medium on the two side of the lens is the same, then $f_1 = f_2$ (numerically).

2.5 Equivalent lens : Principal plane of a co-axial system by deviation method

Deviation by a thin lens—The function of a lens is to *deviate* the ray of light incident on it. We shall now find an expression for this deviation δ.

Let a monochromatic ray, parallel to the principal axis [Fig. 2.5(a)], be incident on a thin lens L of focal length f at a height h above the axis. As the ray is parallel to the principal axis, it will pass after refraction through the second focus F_2.

$\therefore$ Deviation suffered by the ray, $\delta = \tan\delta = h/f$ ($\because$ ray is paraxial, δ is small).

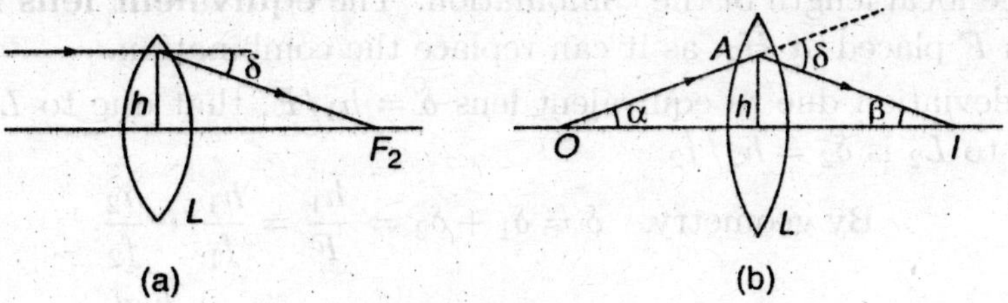

Fig 2.5

In Fig. 2.5(b), O is an axial point object, its corresponding image is I (also axial). OA is incident ray, AI the refracted ray.

$\therefore$ Deviation suffered by OA is given by

$$\delta = \alpha + \beta = \frac{h}{-u} + \frac{h}{v} = h\left[\frac{1}{v} - \frac{1}{u}\right] = \frac{h}{f}$$

Equivalent lens—If a combination of *two* thin lenses, with a *common* axis and separated by some *finite* distance, can be replaced by a *single* thin lens which, when placed at an appropriate fixed point, will produce an *image of the same size but* **not**, *in general, in the same position*, then the *single lens* is said to be an **equivalent lens** of the combination and its focal length the **equivalent focal length**.

• The *equivalence* of the equivalent lens is in a restricted sense in that it produces the same linear or transverse magnification as does the combination in respect of small objects placed normal to the axis, but generally the position of the image is different.

Equivalent focal length—We shall now derive an expression for the equivalent focal length F in terms of the focal lengths f_1 and f_2 of the two co-axial lenses and their distance of separation d.

Let L_1, L_2 be the lenses and PA a ray parallel to the axis incident at A on L_1 at a height $h_1 = C_1A$ from the axis (Fig. 2.6). Ray PA will suffer deviation by an angle δ_1 and proceed towards F_1, the principal focus. But in doing so, it meets the lens L_2 at B at a height $h_2 = C_2B$ from the axis, and on refraction by L_2, it gets further deviated by an angle δ_2 to meet the axis finally at F_2.

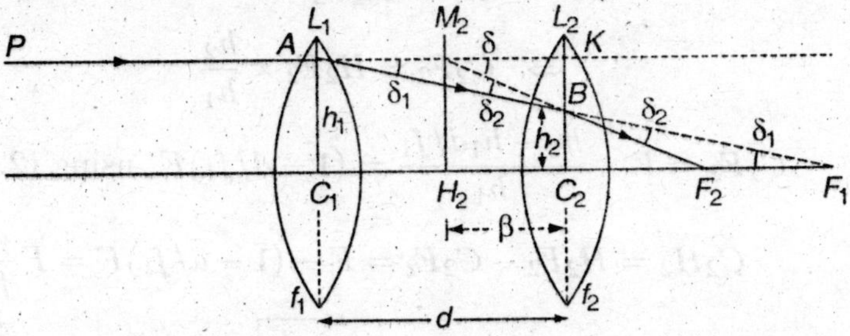

Fig 2.6

Since PA is parallel to the axis and meets, after refraction through both the lenses, the principal axis at F_2, F_2 must be the *second principal focus* of the combination. So

$H_2F_2 = F$, the focal length of the combination. The **equivalent lens** is thus a lens of **focal length** F placed at H_2, as it can replace the combination.

Now the deviation due to equivalent lens $\delta = h_1/F$, that due to L_1 is $\delta_1 = h_1/f_1$ and that due to L_2 is $\delta_2 = h_2/f_2$.

$$\text{By geometry,} \quad \delta = \delta_1 + \delta_2 \Rightarrow \frac{h_1}{F} = \frac{h_1}{f_1} + \frac{h_2}{f_2} \tag{2.5.1}$$

$$\text{But,} \quad h_2 = C_2B = C_2K - BK = h_1 - \frac{h_1 d}{f_1} \tag{2.5.2}$$

$$(\because BK/d = \tan\delta_1 = \delta_1)$$

$\therefore$ From (2.5.1), using (2.5.2), we obtain

$$\frac{h_1}{F} = \frac{h_1}{f_1} + \frac{h_1 - h_1 d/f_1}{f_2}$$

$$\Rightarrow \boxed{\frac{1}{F} = \frac{1}{f_1} + \frac{1}{f_2} - \frac{d}{f_1 f_2}} \tag{2.5.3}$$

which is an important relation giving the *equivalent focal length* F of the combination in terms of f_1, f_2 and d.

• From (2.5.3), $F = f_1f_2/(f_1 + f_2 - d) = -f_1f_2/\Delta$ where $\Delta = d - (f_1 + f_2)$ is called the **optical separation** or **optical interval** between the lenses.

• When the two thin lens are in contact $d = 0$. So (2.5.3) assumes the form

$$\boxed{\frac{1}{F} = \frac{1}{f_1} + \frac{1}{f_2}}$$

Position of equivalent lens—This can be clearly expressed as C_2H_2 or C_1H_2 and is symbolised by $\boldsymbol{\beta}$.

From similar $\Delta M_2H_2F_2$ and ΔBC_2F_2,

$$\frac{h_2}{h_1} = \frac{C_2F_2}{H_2F_2}$$

$$\Rightarrow C_2F_2 = H_2F_2 \times \frac{h_2}{h_1}$$

$$\therefore \quad C_2F_2 = F \times \frac{h_1 - h_1 d/f_1}{h_1} = (1 - d/f_1)F, \text{ using (2.5.2)}$$

$$\therefore \quad C_2H_2 = H_2F_2 - C_2F_2 = F - (1 - d/f_1)F = F\frac{d}{f_1}$$

$$\boxed{\therefore \beta = -F\frac{d}{f_1}}$$

since by sign convention β is $-ve$. So, the *equivalent lens must be placed at a distance* Fd/f_1 *in front* of the second lens L_2.

Location of principal planes—We can easily locate the **second principal plane** of the lens system. The plane M_2H_2, whence the deviation of the incident parallel ray seems to occur, is the *second principal plane*, H_2 being the **second principal point**. The second principal focus F_2 is at a distance F to its right.

Now to locate the **first principal plane**. Let a ray PA be incident, as before, parallel to the axis of the co-axial system, but from the right, on lens L_2 (Fig. 2.7).

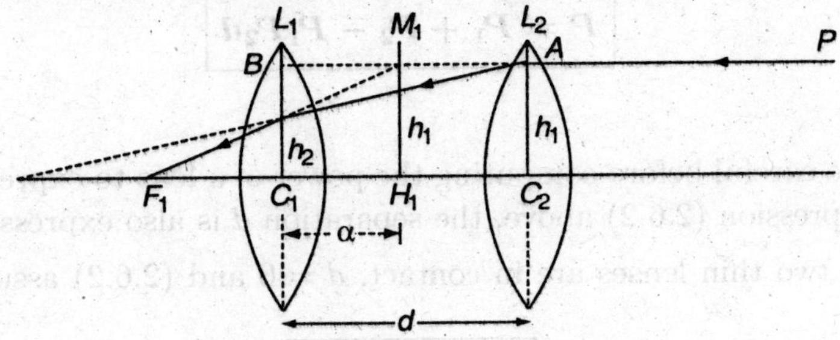

Fig 2.7

Similar procedure would now give us

$$F = +\frac{f_1 f_2}{f_1 + f_2 - d}; \quad C_1H_1 = F\frac{d}{f_2} = \frac{f_1 d}{f_1 + f_2 - d} = \alpha, \text{ the usual symbol.}$$

$$\therefore \quad \alpha = \frac{f_1 d}{f_1 + f_2 - d} = +\frac{Fd}{f_2}$$

$$\text{and} \quad C_2H_1 = C_2C_1 - C_1H_1 = d - \frac{Fd}{f_2} = d\left(1 - \frac{F}{f_2}\right)$$

The first principal focus F_1 is to the left of H_1 at a distance F.

• From above, it is seen that the *equivalent focal length is independent of the direction in which light enters the system.* The equivalent lens placed in a fixed position **cannot** reproduce the behaviour of the system in all respects. It should be placed at M_2H_2 when light enters from the left and at M_1H_1 when the entry is from the right.

• Note that $\boldsymbol{\alpha f_2 = -\beta f_1}$ which is a rather **useful** relation.

2.6 Power of equivalent lens of a coaxial system

The **power** of a lens is defined as the *reciprocal* of its *focal length* expressed in **metre**. Its unit is **Dioptre.**

$$\therefore \quad \text{Power} = \frac{1}{\text{focal length (in m)}}$$

Let F be the focal length of the equivalent lens of a combination of two lenses of focal lengths f_1 and f_2 respectively and separated by a distance d. Then,

$$\frac{1}{F} = \frac{1}{f_1} + \frac{1}{f_2} - \frac{d}{f_1 f_2}. \tag{2.6.1}$$

If P_1 be the power of lens 1 and P_2 that of the lens 2 and P the power of the combination thereof, then

$$P = 1/F, \quad P_1 = 1/f_1, \quad P_2 = 1/f_2.$$

$\therefore$ From (2.6.1), we obtain

$$\boxed{P = P_1 + P_2 - P_1 P_2 d} \qquad (2.6.2)$$

in dioptre (**D**).

One must be careful before computing the power of a lens to express its *focal length in metre.* In expression (2.6.2) above, the separation d is also expressed *in metre.*

• When the two thin lenses are in contact, $d = 0$ and (2.6.2) assumes the form as given below.

$$\boxed{P = P_1 + P_2}$$

2.7 Illustrated Examples

► **Example 1.** *Show that the focal length of a equiconvex lens made of crown glass of* $\mu = 1.5$ *is equal to the radius of curvature of either surface. If the radius is changed from* 0.05 m *to* 0.06 m, *show that the change in power is about* 3.3 dioptre.

(Punjab Univ.)

Solution. By the problem, $R_1 = -R_2 = R$ and $\mu = 1.5$ $\therefore$ From the relation,

$$\frac{1}{f} = (\mu - 1)\left(\frac{1}{R_1} - \frac{1}{R_2}\right)$$

$$\frac{1}{f} = (1.5 - 1)\frac{2}{R} = 0.5 \times 2/R = \frac{1}{R}$$

$$\Rightarrow \quad f = R$$

$\therefore$ When $R = 0.05$ m, $f_1 = 0.05$ m; when $R = 0.06$ m, $f_2 = 0.06$ m.

$\therefore$ Power in the first case, $P_1 = \dfrac{1}{f_1} = \dfrac{1}{0.05} = 20$ Dioptre

Power in the second case, $P_2 = \dfrac{1}{f_2} = \dfrac{1}{0.06} = 16.7$ Dioptre.

$\therefore$ Change in power, $\Delta P = P_1 - P_2 = 20 - 16.7 = \mathbf{3.3\ D}$

► **Example 2.** *A convex lens of focal length* 0.25 m *and refractive index* 1.5 *is immersed in water of index* 1.33. *Calculate the change in focal length of the lens due to immersion.*

Solution. Before immersion, $\mu_2 = 1.5$, $\mu_1 = 1$. So, the focal length f is given by

$$\frac{1}{f} = (\mu - 1)\left(\frac{1}{R_1} - \frac{1}{R_2}\right) = (\mu - 1)k,$$

$$= 0.5k \quad \text{where } k = (1/R_1 - 1/R_2) \tag{2.7.1}$$

If f_1 be the new focal length due to immersion in water,

$$\frac{1}{f_1} = \left(\frac{\mu_2 - \mu_1}{\mu_1}\right)k = \frac{1.50 - 1.33}{1.33}k = \frac{0.17}{1.33}k \tag{2.7.2}$$

$$\therefore \quad \frac{f_1}{f} = \frac{0.5 \times 1.33}{0.17} \Rightarrow f_1 = \frac{0.5 \times 1.33 \times 0.25}{0.17} = +0.978 \text{ m}$$

The positive sign shows that the lens is a *converging one* and the change in focal length = 0.978 − 0.250 = **0.728 m.**

▶ **Example 3.** *Two thin convex lenses A and B having focal lengths* 0.05 m *and* 0.02 m *are placed co-axially and separated by a distance of* 0.03 m. *Find the equivalent focal length and the positions of principal points.* **(Delhi and Guru Nanak Univ.)**

Solution. We have : $\dfrac{1}{F} = \dfrac{1}{f_1} + \dfrac{1}{f_2} - \dfrac{d}{f_1 f_2}$,

$$F = \frac{f_1 f_2}{f_1 + f_2 - d} = \frac{0.05 \times 0.02}{0.05 + 0.02 - 0.03} = \mathbf{0.025 \text{ m}}$$

which is the equivalent focal length.

Let α = distance of the first principal point H_1 from lens A. Therefore,

$$\alpha = +\frac{Fd}{f_2} = \frac{0.025 \times 0.03}{0.02} = \mathbf{0.0375 \text{ m}}$$

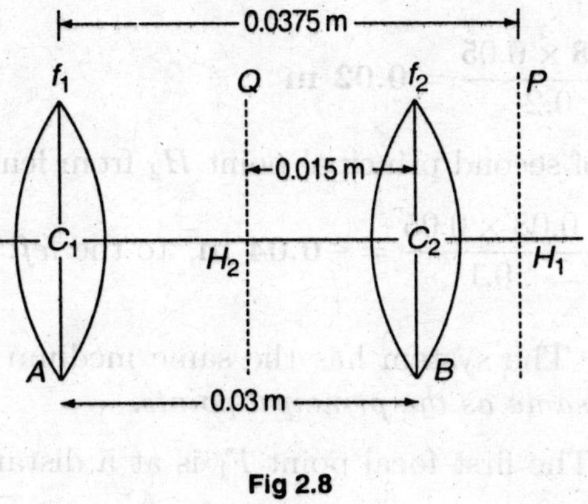

Fig 2.8

The distance of the second principal point H_2 from lens B is given by

$$\beta = -\frac{Fd}{f_1} = -\frac{0.025 \times 0.03}{0.05} = \mathbf{-0.015\ m}$$

The negative sign shows that H_2 is to the *left* of lens B.

► **Example 4.** *Two thin converging lenses of powers* 5D *and* 4D *are placed coaxially* 10 cm *apart. Find the focal length and the power of the combination.* **(Delhi Univ.)**

Solution. Here, P_1 = power of the first lens = +5 D; P_2 = power of the second lens = +4 D. Distance between the lenses = 10 cm = 0.1 m. So, the power P of the combination is

$$P = P_1 + P_2 - P_1P_2d$$

$$= 5 + 4 - 5 \times 4 \times 0.1$$

$$= 9 - 2 = \mathbf{+7D}$$

$\therefore$ Focal length of the combination, $F = \frac{1}{P} = \frac{1}{7} = \mathbf{0.1429\ m}$

► **Example 5.** *Two convex lenses A and B of focal lengths* 10 cm *and* 20 cm *are placed* 5 cm *apart in air. Find the cardinal points of the optical system.* **(Kurukshetra Univ.)**

Solution. Here the equivalent focal length

$$F = \frac{f_1 f_2}{f_1 + f_2 - d} = \frac{0.1 \times 0.2}{0.1 + 0.2 - 0.05} = +0.08\ \text{m}$$

Principal points : If α is the distance of the first principal point H_1 from lens A (Fig. 2.9) of focal length f_1, then

$$\alpha = \frac{Fd}{f_2} = \frac{0.08 \times 0.05}{0.2} = \mathbf{0.02\ m}$$

β = distance of second principal point H_2 from lens B, of focal length f_2,

$$= -\frac{Fd}{f_1} = -\frac{0.08 \times 0.05}{0.1} = \mathbf{-0.04\ m}, \text{ to the } \textit{left} \text{ of the lens } B.$$

Nodal points : The system has the same medium (air) on either side. So the *nodal points* are the *same as the principal points.*

Focal points : The first focal point F_1 is at a distance $(F_1H_1 - \alpha)$ to the *left* of lens A.

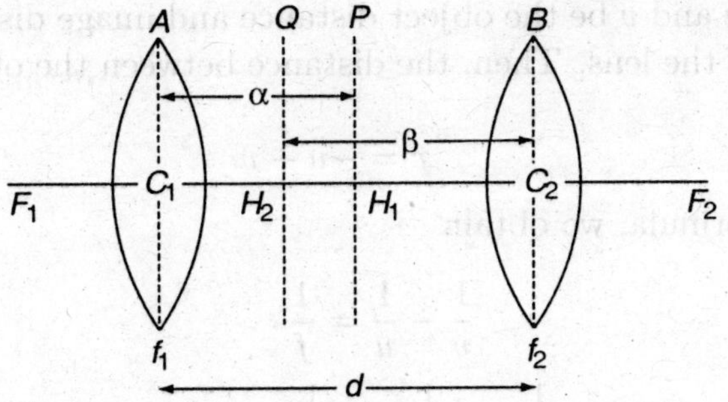

Fig 2.9

Now, we have

$$F_1H_1 - \alpha = 0.08 - 0.02 = \mathbf{0.06\ m}$$

The second focal point F_2 is to the *right* of lens B at a distance

$$F_2H_2 - \beta = 0.08 - 0.04 = \mathbf{0.04\ m}$$

▶ **Example 6.** *A convex lens of focal length f produces an image of an object. If the distance between the object and the screen be d and the magnification of the image be m, then prove that*

$$f = \frac{md}{(1+m)^2}$$ **(Calcutta Univ.)**

Solution. Multiplying the convex lens-equation $1/v - 1/u = 1/f$ by the image distance v,

$$1 - \frac{v}{u} = \frac{v}{f} \Rightarrow 1 - m = \frac{v}{f} \quad (\because m = -v/u) \tag{2.7.3}$$

Similarly, multiplying the same equation by u, the object distance

$$-\frac{1}{m} - 1 = \frac{u}{f} \tag{2.7.4}$$

Subtacting (2.7.4) from (2.7.3), we obtain

$$1 + m + \frac{1}{m} + 1 = \frac{v-u}{f} = \frac{d}{f} \quad (\because d = v - u)$$

$$\text{or,} \quad \frac{m^2 + 2m + 1}{m} = \frac{d}{f}$$

$$\Rightarrow \boldsymbol{f = \frac{md}{m^2 + 2m + 1} = \frac{md}{(1+m)^2}}$$

▶ **Example 7.** *Show that the minimum distance between an object and its real image in a convex lens is four times the focal length of the lens.* **(Burdwan Univ.)**

Solution. Let u and v be the object distance and image distance respectively from the optical centre of the lens. Then, the distance between the object and its real image is given by

$$x = -u + v \tag{2.7.5}$$

From the lens formula, we obtain

$$\frac{1}{v} - \frac{1}{u} = \frac{1}{f}$$

$$\therefore \quad \frac{1}{v} - \frac{1}{v-x} = \frac{1}{f}, \quad \text{using (2.7.5)}$$

$$\Rightarrow \quad v(v-x) = -xf$$

$$\text{or,} \quad v^2 - vx + xf = 0$$

$$\therefore \quad v = \frac{x \pm \sqrt{x^2 - 4xf}}{2}$$

For real values of v, the condition is thus $x^2 \geq 4xf \quad \Rightarrow \quad x \geq 4f$.

$\therefore$ **Minimum value** of $x = \mathbf{4f}$

Review Questions

1. Deduce the relation for a thin convex lens connecting the focal length, the radii of curvature of the surface, the indices of the material of the lens and the surrounding medium. **(Calcutta and Burdwan Univ)**

2. Obtain the relation between the distance of an object and its image formed by a thin concave lens in terms of the radii of curvature of the lens. **(Guru Nanak Univ.)**

3. What are the cardinal points of a coaxial optical system? What are their properties? **(Cochin and Kurukshetra Univ.)**

4. Show how the principal planes of a coaxial system can be located using the theory of the method of deviation. **(Bombay Univ.)**

5. Calculate the focal length of a combination of two thin lenses of focal lengths f_1 and f_2 placed at a distance d apart. Also derive the expressions giving the positions of the two principal points. **(Calcutta and Madras Univ.)**

6. From the expression of focal length of a co-axial system, obtain an expression for the power of an equivalent lens, corresponding to two thin lenses of known powers kept coaxially in air and separated by a distance.

Short Answer and Objective Questions

1. What approximations are used and what limitations do they impose on the usual formula for a thin convex lens? **(Delhi and Punjab Univ.)**

2. What are the cardinal points of a co-axial optical system? What are their properties? **(Himachal Univ.)**

3. What do you mean by an equivalent lens? Explain critically. **(Calcutta and Visweswariah Univ.)**

4. Explain the significance of the principal points and the principal foci.

5. If two convex lenses of power P_1 and P_2, separated by a distance d in air, form an optical system, what is its equivalent power?

6. What are nodal points? What is the property of the nodal points? **(Punjab Univ.)**

7. Power of a lens is measured in
   (a) metre (b) cm (c) km (d) dioptre

8. Two thin lenses are in contact and the combination has focal length 80 cm. If the focal length of one of the lenses is 20 cm, the power of the other lens is
   (a) 1.66 D (b) 4.00 D (c) -1.00 D (d) -3.75 D

9. A convex lens of focal length 40 cm is in contact with a convex lens of focal length 25 cm. The power of the combination is
   (a) -1.5 D (b) -6.5 D (c) $+6.5$ D (d) none

10. A lens of power +2D is in contact with a lens of power -1D. The combination behaves as a
    (a) converging lens of $f = 50$ cm (b) diverging lens of $f = 100$ cm
    (c) converging lens of $f = 100$ cm (d) converging lens of $f = 200$ cm.

11. The radius of the convex surface of a plano-convex lens is 10 cm and the focal length of the lens is 30 cm. The index of the material of the lens is
    (a) 1.5 (b) 1.66 (c) 1.33 (d) none

12. The distance between the two nodal points equals the distance between the
    (a) two conjugate points (b) two focal points
    (c) two principal points (d) none of these.

13. A convex lens of focal length f produces a real image, n times the size of an object. The distance of the object from the lens is

(a) $(n-1)f$ (b) $(n+1)f$ (c) $f\dfrac{(n+1)}{u}$ (d) $f\dfrac{(n-1)}{v}$

Problems

1. Find the focal length of a plano-convex lens, the radius of the curved surface being 30 cm and $\mu = 1.50$. [60 cm]

2. A convex lens of focal length 15 cm and refractive index 1.50 is placed (a) in water of $\mu = 1.33$ and (b) in carbon bisulphide, $\mu = 1.65$. Find the focal length and the behaviour of the lens in each case. [+0.587 m, convex; −0.825 m, concave]

3. An equi-convex lens of index 1.50 and focal length 0.1 m is held with its axis vertical and its lower surface immersed in water, the upper surface being in air. At what distance from the lens a vertical beam of parallel rays falling on the lens will be focussed? How will the position of focus be altered if the lens is wholly immersed in water of refractive index 1.33? [0.1985 m; 0.391 m]

4. A mark placed on the surface of a glass sphere is viewed through glass from a position directly opposite. If the diameter of the sphere is 0.1 m and the index of glass is 1.5, find the position of the image. [−0.20 m; 0.391 m]

5. Two thin converging lenses A, B of focal lengths 0.2 m and 0.3 m are placed co-axially 0.10 m apart in air. An object is located at 0.6 m in front of the lens A of smaller focal length. Find the positions of two principal planes and that of the image. [0.05 m to right of A, −0.075 m to left of B; 0.12 m from second p. plane]

6. The focal length of a convex lens for a certain monochromatic light is 0.3 m and refractive index of the material is 1.5. What will be the focal length for another monochromatic light whose refractive index is higher than the first by 2%. **(Calcutta Univ.)** [0.283 m]

7. Two similar thin convex lens, of focal length 10 cm each, are placed co-axially 5 cm apart. Find the equivalent focal length and the positions of principal points. Also find the position of the object for which the image is formed at infinity.

[$F = 6.67$ cm; $\alpha = +3.33$ cm, $\beta = -3.33$ cm; $u = -3.33$ cm]

8. Two thin converging lenses A and B of focal lengths 30 cm and 40 cm respectively, are placed co-axially 20 cm apart. An object is located at a distance of 48 cm

from the first lens. Find (i) the position of the image and (ii) the positions of the principal points. **(Delhi Univ.)**

[$\alpha = +12$ cm, $\beta = -16$ cm; final image is 24 cm to the right of B]

9. A convex lens of focal length 20 cm ($\mu = 1.50$) is totally immersed in water ($\mu = 1.33$). Find its focal length in water. **(Delhi Univ.)** [96 cm.]

10. A co-axial lens system placed in air has two lenses of focal lengths $3F$ and F separated by a distance $2F$. Find the positions of the cardinal points.

    [$\alpha = +3F$, $\beta = -F$; $F_1 = \frac{3}{2}F$ from the first lens, $F_2 = \frac{1}{2}F$ from the second lens; nodal points coincide with principal points P_1, P_2]

11. A thin converging lens and a thin diverging lens are placed co-axially in air at a distance of 5 cm. If the focal length of each is 10 cm, find for the combination (a) the focal length, (b) the power and (c) the position of principal points.

    [+ 20 cm; + 5D; $\alpha = \beta = -10$ cm]

12. An object placed 90 cm from a convex lens produces a real image on a screen placed 45 cm from the lens. On placing a concave lens in contact with the convex lens a real image is formed on the screen by moving the screen 75 cm further off. Find the focal length of the concave lens. **(Burdwan Univ.)** [-72 cm]

Chapter 3

Matrix Method in Geometrical Optics

Matrix method in geometrical optics is very useful in deducing various relations involving refractions and translations in optical systems, without taking recourse to tedious ray tracings. The object distance and image distance relation in lenses, the determination of cardinal points in thick lenses, the equivalent focal length of two thin lenses etc. all can be simply and readily obtained with relief by the matrix method. This interesting method of handling the ray-tracing equations was formulated in the beginning of the 1930s by T. Smith, but was not widely appreciated for almost 30 years. In particular, the analysis can now be done by feeding the matrix data directly to computers and a rebirth of interest in matrix-approach occurred. The basic requirements for applying the method are (i) to define what are called *optical coordinates of a ray* and (ii) an exposure to the method of matrix.

3.1 Coordinates of a ray : Effect of translation

Let z-axis be the axis of a cylindrically symmetric optical system. The paraxial rays passing through the axis are all confined to a plane and a ray such as PQ, inclined to the axis, may be specified by its distance x from the axis and the angle α it makes with the axis (Fig. 3.1). While x changes as the ray proceeds in straight line in a homogeneous medium, the angle α *does not.* For instance, at $P : x = x_1$ and $\alpha = \alpha_1$.

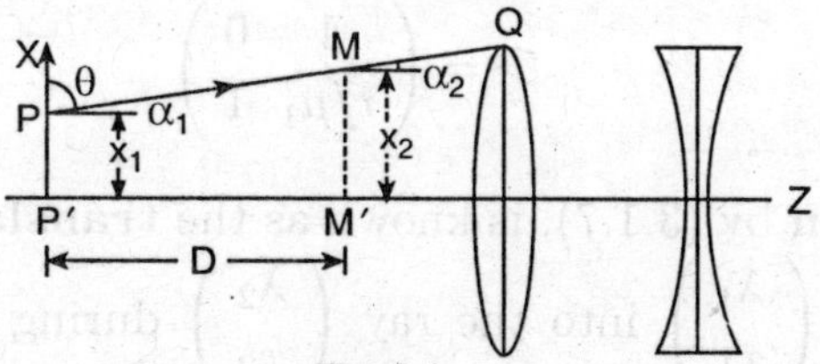

Fig. 3.1

At $M : x = x_2$, $\alpha = \alpha_2$, say, and $\alpha_1 = \alpha_2$. The quantities (x_1, α_1) represent the **optical coordinates of the ray** at P. Instead of α, we may also use the angle θ which the ray makes with x-axis. Then,

$$\lambda_1 = \mu_1 \cos\theta$$

where λ_1 is the *optical direction cosine* of the ray at P, μ_1 being the index of the medium. So, the optical coordinates of the ray at P may as well be expressed as (λ_1, x_1).

Effect of translation—Let a ray PQ move, in a homogeneous medium of index μ_1 with (x_1, α_1) and (x_2, α_2) as its coordinates at P and M respectively (Fig. 3.1). The medium being homogeneous, the ray moves in straight line and $\alpha_1 = \alpha_2$.

Now, $x_1 = PP'$ and $x_2 = MM'$ and let $P'M' = t$, the translation of the ray.

$$\therefore \quad x_2 = x_1 + t\tan\alpha_1 = x_1 + \alpha_1 t \tag{3.1.1}$$

since for a paraxial ray α_1 is very small.

In terms of λ, we may write

$$\lambda_1 = \mu_1\cos(90° - \alpha_1) = \mu_1 \sin\alpha_1 = \mu_1\alpha_1 \tag{3.1.2}$$

$$\lambda_2 = \mu_2\cos(90° - \alpha_2) = \mu_2 \sin\alpha_2 = \mu_2\alpha_2 \tag{3.1.3}$$

$$\therefore \quad \lambda_2 = \lambda_1 \quad (\because \alpha_1 = \alpha_2;\ \mu_2 = \mu_1) \tag{3.1.4}$$

$$\therefore \quad \boxed{x_2 = x_1 + \frac{t}{\mu_1}\lambda_1} \tag{3.1.5}$$

from (3.1.1), using (3.1.2).

The two equations (3.1.4) and (3.1.5) may be given the matrix form

$$\begin{pmatrix} \lambda_2 \\ x_2 \end{pmatrix} = \begin{pmatrix} 1 & 0 \\ t/\mu_1 & 1 \end{pmatrix} \begin{pmatrix} \lambda_1 \\ x_1 \end{pmatrix} \tag{3.1.6}$$

The column matrices $\begin{pmatrix} \lambda_1 \\ x_1 \end{pmatrix}$ and $\begin{pmatrix} \lambda_2 \\ x_2 \end{pmatrix}$ may be looked upon as the rays at points P and M, and the effect or process of translation over a distance t in a medium of index μ_1 is given by the 2×2 matrix

$$\mathcal{T} = \begin{pmatrix} 1 & 0 \\ t/\mu_1 & 1 \end{pmatrix} \tag{3.1.7}$$

The matrix $\mathcal{T}$, given by (3.1.7), is known as the **translation or transfer matrix**. It transforms the ray $\begin{pmatrix} \lambda_1 \\ x_1 \end{pmatrix}$ into the ray $\begin{pmatrix} \lambda_2 \\ x_2 \end{pmatrix}$ during translation through t, in a homogeneous medium of index μ_1.

• **Note** that the determinant $\mathcal{T}$, i.e.,

$$\det \mathcal{T} = \begin{vmatrix} 1 & 0 \\ t/\mu_1 & 1 \end{vmatrix} = 1$$

3.2 Refraction matrix

We shall now determine the matrix that will represent the *effect of refraction* through a spherical surface.

Let DP be a convex spherical refracting surface of radius of curvature R, separating two media of index μ_1 and μ_2 ($\mu_2 > \mu_1$), C being the centre of curvature (Fig. 3.2). AD is a ray incident at an angle i at D on the spherical surface to be refracted along DB, making an angle of refraction r. So, by Snell's law,

$$\mu_1 \sin i = \mu_2 \sin r \quad \Rightarrow \quad \mu_1 i = \mu_2 r \tag{3.2.1}$$

since the angles i and r are small.

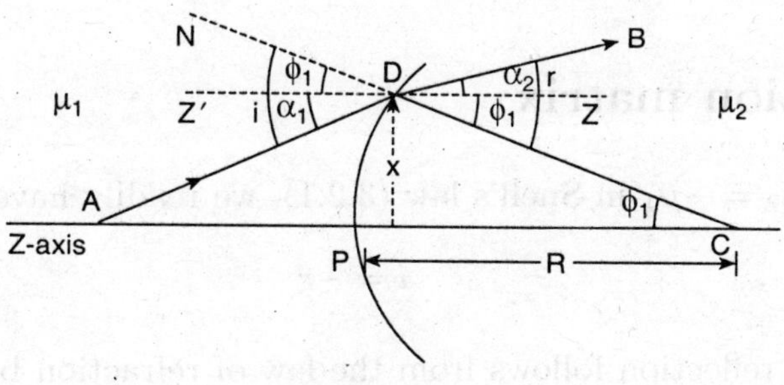

Fig. 3.2

If ϕ_1 be the inclination of the normal CDN with z-axis, and α_1, α_2 the angles made by the incident and refracted rays with z-axis, then

$$i = \phi_1 + \alpha_1; \quad r = \phi_1 + \alpha_2$$

Therefore, from equation (3.2.1)

$$\begin{aligned} \mu_1(\phi_1 + \alpha_1) &= \mu_2(\phi_1 + \alpha_2) \\ \Rightarrow \quad \mu_2\alpha_2 &= \mu_1\alpha_1 - (\mu_2 - \mu_1)\phi_1 \\ &= \mu_1\alpha_1 - \frac{\mu_2 - \mu_1}{R}x \quad (\because \text{ small, } \phi_1 = x/R) \\ \therefore \quad \lambda_2 &= \lambda_1 - \frac{\mu_2 - \mu_1}{R}x, \qquad \text{(using (3.1.2) and (3.1.3))} \\ &\Rightarrow \lambda_2 = \lambda_1 - Px \end{aligned} \tag{3.2.2}$$

where $P = (\mu_2 - \mu_1)/R$, the **power of the refracting surface**.

Also, $x_2 = x_1 = x$, since the height of the ray before and after refraction at D is the same.

So, the matrix for the refracted ray may be written as

$$\begin{pmatrix} \lambda_2 \\ x_2 \end{pmatrix} = \begin{pmatrix} 1 & -P \\ 0 & 1 \end{pmatrix} \begin{pmatrix} \lambda_1 \\ x_1 \end{pmatrix} \tag{3.2.3}$$

Thus, the *effect or process of refraction* is given by the 2×2 matrix, $\mathcal{R}$, given by

$$\mathcal{R} = \begin{pmatrix} 1 & -P \\ 0 & 1 \end{pmatrix} \tag{3.2.4}$$

and the matrix $\mathcal{R}$ is known as *refraction matrix*. $\mathcal{R}$ transforms the incident ray $\begin{pmatrix} \lambda_1 \\ x_1 \end{pmatrix}$ into the refracted ray $\begin{pmatrix} \lambda_2 \\ x_2 \end{pmatrix}$ during refraction at the surface.

As before, we have $\det \mathcal{R} = \begin{vmatrix} 1 & -P \\ 0 & 1 \end{vmatrix} = 1$

3.3 Reflection matrix

On substituting $\mu_2 = -\mu_1$ in Snell's law (3.2.1), we readily have

$$i = -r \tag{3.3.1}$$

So, the law of reflection follows from the law of refraction by replacing μ_2 by $-\mu_1$ and interpreting the angle of refraction as the angle of reflection. The negative sign in the law of reflection is consistent with the sign convention.

Thus, the **reflection matrix** for a curved surface may be obtained from the corresponding refraction matrix $\mathcal{R}$ simply by replacing μ_2 by $-\mu_1$. So, the reflection matrix is given by

$$\begin{pmatrix} 1 & -P \\ 0 & 1 \end{pmatrix} = \begin{pmatrix} 1 & -2\mu_1/R \\ 0 & 1 \end{pmatrix}$$

where μ_1 is the refractive index of the medium in front of the reflector and R, the radius of curvature of the surface.

3.4 System matrix

We have just seen how refraction through a spherical surface can be characterised by a (2×2) matrix. In general, an optical system consists of a number of lenses and hence can be characterised by refraction and translation matrices.

If $\begin{pmatrix}\lambda_1\\x_1\end{pmatrix}$ specifies a ray on entering an optical system and the same is specified by $\begin{pmatrix}\lambda_2\\x_2\end{pmatrix}$ when it leaves the system, then, in general,

$$\begin{pmatrix}\lambda_2\\x_2\end{pmatrix}=\begin{pmatrix}b & -a\\-d & c\end{pmatrix}\begin{pmatrix}\lambda_1\\x_1\end{pmatrix}$$

where the 2×2 matrix

$$\mathcal{A}=\begin{pmatrix}b & -a\\-d & c\end{pmatrix}$$

is known as the **system matrix**, being determined entirely by the optical system under reference. The optical system consists of refraction and translation and thus the *system matrix* $\mathcal{A}$ *is the product of refraction and translation matrices* $(\mathcal{R}\times\mathcal{T})$, and is sometimes called **translation-refraction matrix**.

• To some elements of system matrix, negative sign has been given due entirely for convenience and would be clear as we proceed.

• The product of matrices is the product of the determinant of the matrices. Thus

$$\det\mathcal{A}=1\ \Rightarrow\ bc-ad=1$$

3.5 System matrix in terms of magnification

We shall, in this section, express system matrix in terms of magnification.

Let O be an object on the axis, distant t_1 from the *first* refracting surface PP', and I the paraxial image, distant t_2 from the the *last* refracting surface, QQ' (Fig. 3.3).

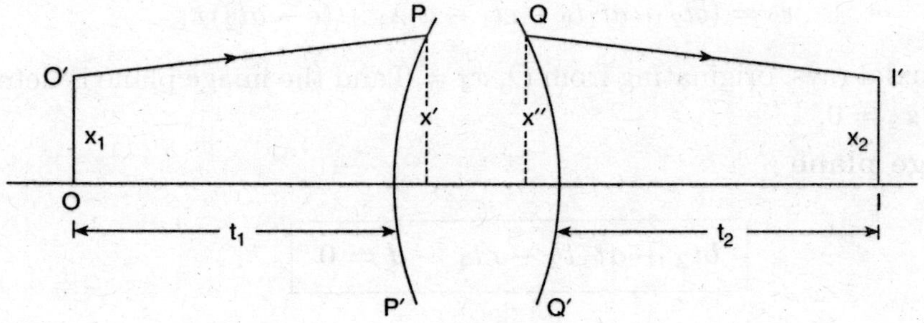

Fig. 3.3

The sign convention dictates that t_1 is *positive* and t_2 *negative*, implying the image is *real*. If however the image is formed to the left of the last refracting surface, t_2 is *negative* and the image is *virtual*.

While the **object plane** is the *plane containing the object and normal to the axis*, the **image plane** is the one *containing the image and normal to the axis*.

System matrix—Let a ray $O'P$ start from O', an object in the object plane and emerge as QI' from the last refracting surface forming thereby an image I' in the image plane. Also, let x_1, x_2 be the distances of O' and I' respectively from the axis. Let (λ_1, x_1), (λ', x'), (λ'', x'') and (λ_2, x_2) be the coordinates of the ray at O', P, Q and I' respectively.

Matrix corresponding to translation t_1 is

$$\begin{pmatrix} \lambda' \\ x' \end{pmatrix} = \begin{pmatrix} 1 & 0 \\ -t_1 & 1 \end{pmatrix} \begin{pmatrix} \lambda_1 \\ x_1 \end{pmatrix} \qquad (\because \ \mu_1 = 1 \text{ for air})$$

System matrix of the optical system is

$$\begin{pmatrix} \lambda'' \\ x'' \end{pmatrix} = \begin{pmatrix} b & -a \\ -d & c \end{pmatrix} \begin{pmatrix} \lambda' \\ x' \end{pmatrix}$$

Matrix corresponding to translation t_2 is

$$\begin{pmatrix} \lambda_2 \\ x_2 \end{pmatrix} = \begin{pmatrix} 1 & 0 \\ t_2 & 1 \end{pmatrix} \begin{pmatrix} \lambda'' \\ x'' \end{pmatrix} \qquad (\because \ \mu_2 = 1 \text{ for air})$$

$$\therefore \quad \begin{pmatrix} \lambda_2 \\ x_2 \end{pmatrix} = \begin{pmatrix} 1 & 0 \\ t_2 & 1 \end{pmatrix} \begin{pmatrix} b & -a \\ -d & c \end{pmatrix} \begin{pmatrix} 1 & 0 \\ -t_1 & 1 \end{pmatrix} \begin{pmatrix} \lambda_1 \\ x_1 \end{pmatrix}$$

$$= \begin{pmatrix} b & -a \\ bt_2 - d & -at_2 + c \end{pmatrix} \begin{pmatrix} 1 & 0 \\ -t_1 & 1 \end{pmatrix} \begin{pmatrix} \lambda_1 \\ x_1 \end{pmatrix}$$

$$= \begin{pmatrix} b + at_1 & -a \\ bt_2 - d + at_1 t_2 - ct_1 & -at_2 + c \end{pmatrix} \begin{pmatrix} \lambda_1 \\ x_1 \end{pmatrix} \qquad (3.5.1)$$

$$\therefore \quad x_2 = (bt_2 + at_1 t_2 - ct_1 - d)\lambda_1 + (c - at_2)x_1$$

For the paraxial rays, originating from O, $x_1 = 0$ and the image plane is determined by considering $x_2 = 0$.

$\therefore$ For **image plane** :

$$\boxed{\boldsymbol{bt_2 + at_1 t_2 - ct_1 - d = 0}} \qquad (3.5.2)$$

which is rather an *important relation.*

From (3.5.1), using (3.5.2), the system matrix is

$$\begin{pmatrix} \lambda_2 \\ x_2 \end{pmatrix} = \begin{pmatrix} b + at_1 & -a \\ 0 & c - at_2 \end{pmatrix} \begin{pmatrix} \lambda_1 \\ x_1 \end{pmatrix} \qquad (3.5.3)$$

For $x_1 \neq 0$, $x_2 = (c - at_2)x_1 \Rightarrow$ **magnification,** $m = x_2/x_1 = c - at_2$.

Since the determinant of system matrix is 1, we have from (3.5.3)

$$(b + at_1)(c - at_2) = 1$$

$$\Rightarrow \quad b + at_1 = \frac{1}{c - at_2} = \frac{1}{m}$$

∴ System matrix, in terms of magnification, is

$$\begin{pmatrix} \lambda_2 \\ x_2 \end{pmatrix} = \begin{pmatrix} 1/m & -a \\ 0 & m \end{pmatrix} \begin{pmatrix} \lambda_1 \\ x_1 \end{pmatrix}$$

3.6 System matrix for refraction at a convex spherical surface

Let a convex spherical surface of radius of curvature R separate two media of indices μ_1, μ_2 ($\mu_2 > \mu_1$), C being the centre of curvature of the surface (Fig. 3.4). A ray OA starts from O and is incident at A, with i as the angle of incidence. On refraction along AI, with r as the angle of refraction, it forms an image at I.

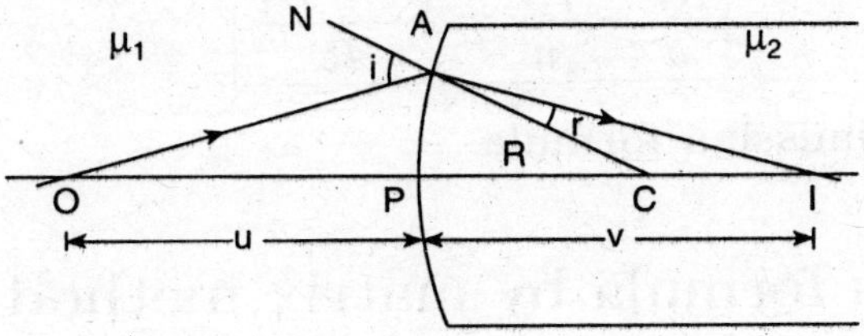

Fig. 3.4

Let P be the *power* of the refracting surface, so that

$$P = \frac{\mu_2 - \mu_1}{R}$$

Let u and v be the object distance and image distance respectively, measured from the pole P.

∴ **System matrix** of the system is

$$\mathcal{A} = \begin{pmatrix} 1 & 0 \\ v/\mu_2 & 1 \end{pmatrix} \begin{pmatrix} 1 & -P \\ 0 & 1 \end{pmatrix} \begin{pmatrix} 1 & 0 \\ u/\mu_1 & 1 \end{pmatrix}$$

$$= \begin{pmatrix} 1 & 0 \\ v/\mu_2 & 1 \end{pmatrix} \begin{pmatrix} 1 - (u/\mu_1)P & -P \\ u/\mu_1 & 1 \end{pmatrix}$$

$$= \begin{pmatrix} 1 - (u/\mu_1)P & -P \\ (1 - uP/\mu_1)v/\mu_2 + u/\mu_1 & -vP/\mu_2 + 1 \end{pmatrix} \quad (3.6.1)$$

Application—From the above system matrix, we may readily obtain the *Gaussion formula* for refraction at a single spherical (convex) surface.

The matrix equation, corresponding to the system matrix (3.6.1), gives

$$x_2 = \left\{\frac{v}{\mu_2}\left(1 - \frac{u}{\mu_1}P\right) + \frac{u}{\mu_1}\right\}\lambda_1 + \left(1 - \frac{v}{\mu_2}P\right)x_1$$

For paraxial rays originating from O, $x_1 = 0$ and the *image plane* is determined by $x_2 = 0$.

$$\therefore \quad \frac{v}{\mu_2}\left(1 - \frac{u}{\mu_1}P\right) + \frac{u}{\mu_1} = 0$$

$$\Rightarrow \quad \frac{v}{\mu_2} + \frac{u}{\mu_1} = \frac{uv}{\mu_1\mu_2}P$$

$$\text{or,} \quad \frac{\mu_1}{u} + \frac{\mu_2}{v} = \frac{\mu_2 - \mu_1}{R},$$

on dividing throughout by $uv/\mu_1\mu_2$ and substituting for P.

Now, by sign convention, v and R are *positive* and u *negative.*

$$\therefore \quad \boxed{\frac{\mu_2}{v} - \frac{\mu_1}{u} = \frac{\mu_2 - \mu_1}{R}}$$

which is the well-known **Gaussian formula.**

3.7 Lens-maker's formula by matrix method

Let a thin convex lens L of index μ be placed in air and R_1, R_2 the radii of curvature of the first and the second surface respectively (Fig. 3.5). Also, let O be a paraxial point, distant t_1 from the first surface. A ray of light OP, after refraction by the lens, travels along QI to form a paraxial image at I, distant t_2 from the second surface.

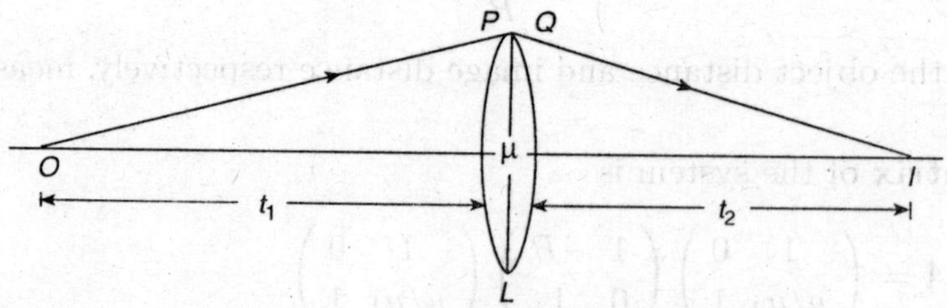

Fig. 3.5

By sign convention, R_1 is *positive,* R_2 *negative,* t_2 *positive* and t_1 *negative.*

$$\therefore \quad P_1 = \text{power of first surface} = (\mu - 1)/R_1$$

$$P_2 = \text{power of second surface} = -(\mu - 1)/R_2$$

∴ **System matrix** for the thin lens is

$$A = \begin{pmatrix} b & -a \\ -d & c \end{pmatrix} = \begin{pmatrix} 1 & -P_2 \\ 0 & 1 \end{pmatrix}\begin{pmatrix} 1 & -P_1 \\ 0 & 1 \end{pmatrix}$$

$$\Rightarrow \begin{pmatrix} b & -a \\ -d & c \end{pmatrix} = \begin{pmatrix} 1 & -P_1 - P_2 \\ 0 & 1 \end{pmatrix}$$

$$\therefore \; a = P_1 + P_2, \;\; b = 1, \;\; c = 1 \text{ and } d = 0.$$

From the equation of image plane for refracting surfaces,

$$bt_2 - at_1t_2 - ct_1 - d = 0$$

$$\text{or,} \;\; t_2 - (P_1 + P_2)t_1t_2 - t_1 = 0$$

Dividing throughout by t_1t_2 and substituting for P's

$$\frac{1}{t_1} - \frac{1}{t_2} = -(\mu - 1)\left(\frac{1}{R_1} - \frac{1}{R_2}\right)$$

$$\text{or,} \;\; \frac{1}{t_2} - \frac{1}{t_1} = (\mu - 1)\left(\frac{1}{R_1} - \frac{1}{R_2}\right) = \frac{1}{f} \qquad (3.7.1)$$

where t_1 and t_2 are the distances of the object and the image. Equation (3.7.1) is the well-known **Lens-maker's formula.**

• The lens is thin. So, $t_1 = u$, the object distance measured from the optical centre. Also, $t_2 = v$, for the same reason. So, (3.7.1) may also be written as

$$\boxed{\frac{1}{v} - \frac{1}{u} = \frac{1}{f}}$$

which is none else t! he conventional ***thin lens formula.***

3.8 System matrix for a thick lens

Let the thick lens L of thickness t be bounded by two spherical surfaces A_1B_1 and A_2B_2 of radii of curvature R_1 and R_2 respectively, and μ the index of the material of the lens (Fig. 3.6).

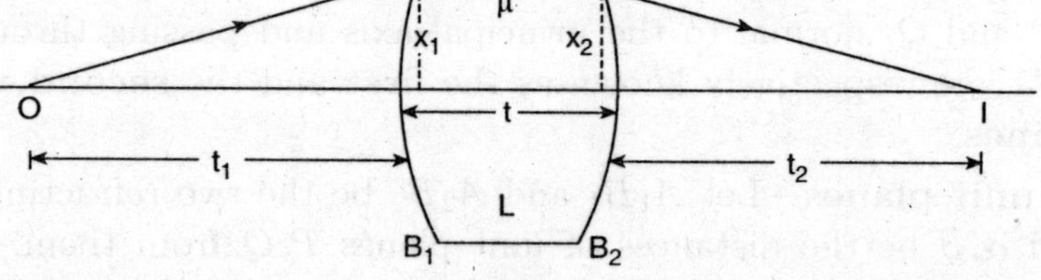

Fig. 3.6

A ray of light OP starts from the axial point O to strike the surface A_1B_1 at P and emerges, after refraction in the lens, at Q as QI to form an image on the axial point I. Let (λ_1, x_1) and (λ_2, x_2) be the coordinates of the rays at P and Q respectively. Apart from the two refractions at P and Q, the ray undergoes a translation through a distance t in the medium of index μ.

The powers P_1, P_2 of the surfaces A_1B_1 and A_2B_2 are :

$$P_1 = \frac{\mu - 1}{R_1}; \quad P_2 = \frac{1-\mu}{R_2} = -\frac{\mu - 1}{R_2}$$

$$\therefore \quad \begin{pmatrix} \lambda_2 \\ x_2 \end{pmatrix} = \begin{pmatrix} 1 & -P_2 \\ 0 & 1 \end{pmatrix} \begin{pmatrix} 1 & 0 \\ t/\mu & 1 \end{pmatrix} \begin{pmatrix} 1 & -P_1 \\ 0 & 1 \end{pmatrix} \begin{pmatrix} \lambda_1 \\ x_1 \end{pmatrix}$$

$\therefore$ **System matrix for *a* thick lens** is

$$\begin{aligned} \mathcal{A} = \begin{pmatrix} b & -a \\ -d & c \end{pmatrix} &= \begin{pmatrix} 1 & -P_2 \\ 0 & 1 \end{pmatrix} \begin{pmatrix} 1 & 0 \\ t/\mu & 1 \end{pmatrix} \begin{pmatrix} 1 & -P_1 \\ 0 & 1 \end{pmatrix} \\ &= \begin{pmatrix} 1 - P_2t/\mu & -P_2 \\ t/\mu & 1 \end{pmatrix} \begin{pmatrix} 1 & -P_1 \\ 0 & 1 \end{pmatrix} \\ &= \begin{pmatrix} 1 - P_2t/\mu & -P_1 - P_2(1 - tP_1/\mu) \\ t/\mu & 1 - tP_1/\mu \end{pmatrix} \end{aligned}$$

System matrix for a thin lens—If the lens is thin, $t \to 0$ and the system matrix $\mathcal{A}$ becomes

$$\mathcal{A} = \begin{pmatrix} b & -a \\ -d & c \end{pmatrix} = \begin{pmatrix} 1 & -(P_1 + P_2) \\ 0 & 1 \end{pmatrix}$$

which is the same as was obtained earlier.

3.8.1 Unit points and their distances

The **unit points** are the *two* points H_1 and H_2 on the principal axis, one in object space and the other in image space, for which the lens has *unit positive linear magnification.* They are also called **principal points** : H_1, the **first** and H_2 the **second** principal points (Fig. 3.7).

The planes P and Q, normal to the principal axis and passing through the unit points H_1 and H_2, are respectively known as the **first** and the **second unit planes** or **principal planes.**

Position of unit planes—Let A_1B_1 and A_2B_2 be the two refracting surfaces of a thick lens, and α, β be the distances of unit planes P, Q from them. The object plane and the image plane are similarly at distances t_1 and t_2 from A_1B_1 and A_2B_2 respectively.

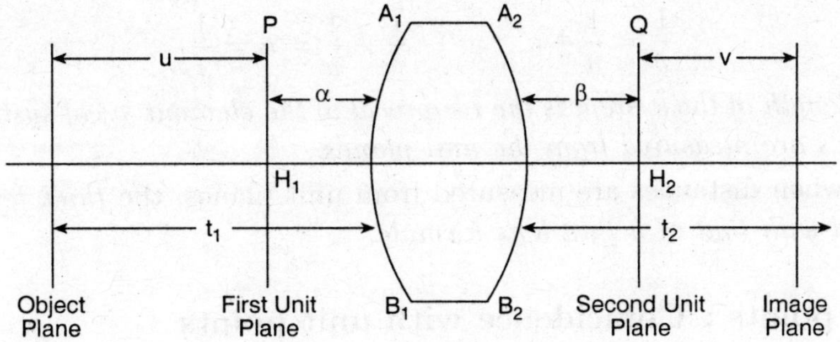

Fig. 3.7

By the system matrix for image plane

$$b + at_1 = \frac{1}{c - at_2} = \frac{1}{m} = 1 \tag{3.8.1}$$

If the distances are measured from *unit planes*,

$$b + a\alpha = 1 \quad \Rightarrow \quad \alpha = (1 - b)/a \tag{3.8.2}$$

$$\text{and} \quad c - a\beta = 1 \quad \Rightarrow \quad \beta = (c - 1)/a \tag{3.8.3}$$

By sign convention, here α is *negative* and β *positive.* The distances of unit points are thus known by the elements of the system matrix.

3.8.2 Focal length of an optical system

If the distances are measured from unit planes, then the focal length of a system is given by the reciprocal of the element 'a' of the system matrix.

Proof. Let u and v be the distances of the object plane from the first and the second unit plane (Fig. 3.7) respectively.

$$\therefore \quad t_1 = u + \alpha = u + (1 - b)/a \tag{3.8.4}$$

$$t_2 = v + \beta = v + (c - 1)/a \tag{3.8.5}$$

But, by the system matrix for the image plane,

$$t_2 = \frac{d + ct_1}{b + at_1} \tag{3.8.6}$$

Substituting the values of t_1, t_2 from (3.8.4) and (3.8.5) in (3.8.6), we obtain

$$v + \frac{c-1}{a} = \frac{d + c(u + (1-b)/a)}{b + a(u + (1-b)/a)} = \frac{ad + acu - c - bc}{a(1 + au)}$$

$$\therefore \quad v = \frac{ad + acu - c - bc}{a(1 + au)} - \frac{c-1}{a}$$

$$= \frac{au}{a(1 + au)}, \quad \text{simplifying and using } bc - ad = 1$$

$$\therefore \quad \frac{1}{v} = \frac{1}{u} + a \quad \Rightarrow \quad \frac{1}{v} - \frac{1}{u} = a = \frac{1}{f}$$

So, *the focal length of the system is the reciprocal of the element 'a' of system matrix, provided distances are measured from the unit planes.*

• Note that when distances are measured from unit planes, the *thick lens formula becomes identical with that of a thin lens formula.*

3.8.3 Nodal points : Coincidence with unit points

Apart from the pair of principal foci and unit points, there is another important pair of conjugate points on the axis of the system, called **nodal points**, which are the *points of positive* **unit angular magnification**. This implies that *a ray of light directed to the first nodal point* N_1 (Fig. 3.8), *after refraction by the optical system, will emerge so as to proceed or appear to proceed through or from the second nodal point* N_2, *in a direction parallel to that of the incident ray.* The planes passing through the nodal points and normal to the axis are called **nodal planes**.

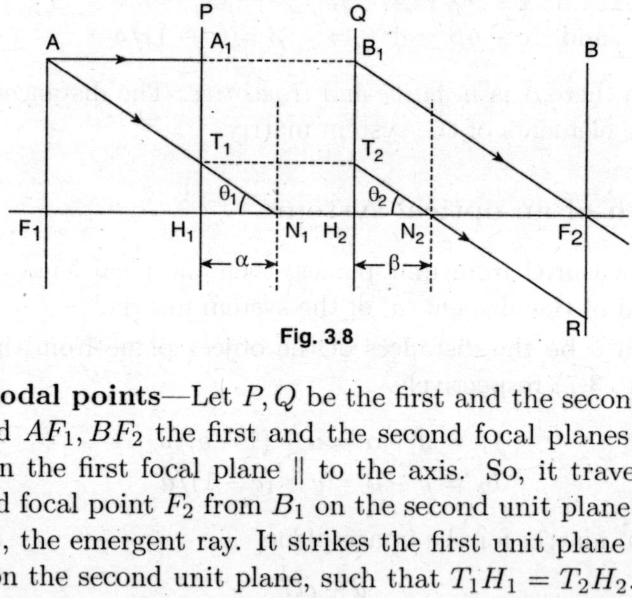

Fig. 3.8

Position of nodal points—Let P, Q be the first and the second unit planes of an optical system and AF_1, BF_2 the first and the second focal planes respectively. AA_1 is a ray from A on the first focal plane $\|$ to the axis. So, it travels, after refraction towards the second focal point F_2 from B_1 on the second unit plane Q. AT_1 is another ray from A $\|B_1F_2$, the emergent ray. It strikes the first unit plane at T_1 and emerges from T_2, a point on the second unit plane, such that $T_1H_1 = T_2H_2$. The ray T_2R will travel $\|B_1F_2$. The point of intersection of incident ray AT_1 and the axis, i.e., point N_1 is the *first nodal point.* Similarly, N_2, the point of intersection of the conjugate ray T_2R and the axis, is the *second nodal point.*

The definition of nodal points dictates $\theta_1 = \theta_2$. If the medium on both sides of the system be of index μ, $\lambda_1 = \lambda_2$ and for *axial* point $x = 0$.

Now, let α_1, β_1 be the distances of the nodal points N_1, N_2 respectively from unit planes.

$$\therefore \quad b + a\alpha_1 = 1 \quad \Rightarrow \quad \alpha_1 = (1-b)/a$$
$$c - a\beta_1 = 1 \quad \Rightarrow \quad \beta_1 = (c-1)/a$$

Comparing these with equations of unit planes

$$\alpha = \alpha_1, \quad \beta = \beta_1$$

This means that *the nodal planes coincide with the unit planes if the optical system has, on its either side, media of the same index.*

3.9 Focal length of a combination of two thin lenses

Let two thin lenses A, B of focal length f_1 and f_2 respectively be placed in air at a distance t apart. Also, let α, β be the distances of unit planes P, Q (Fig. 3.9) from the lens A and B respectively.

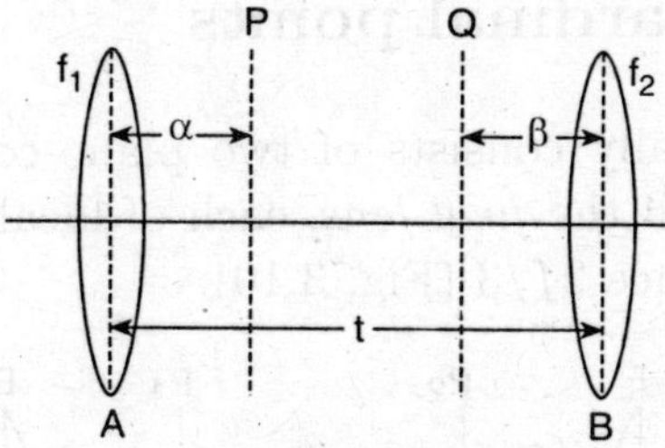

Fig. 3.9

The system matrix consists of the combination of two refraction matrices of the two lenses and the translation matrix of the ray through a distance t in air.

$$\therefore \quad A = \begin{pmatrix} b & -a \\ -d & c \end{pmatrix} = \begin{pmatrix} 1 & -1/f_2 \\ 0 & 1 \end{pmatrix} \begin{pmatrix} 1 & 0 \\ t & 1 \end{pmatrix} \begin{pmatrix} 1 & -1/f_1 \\ 0 & 1 \end{pmatrix}$$

$$= \begin{pmatrix} 1 - t/f_2 & -1/f_2 \\ t & 1 \end{pmatrix} \begin{pmatrix} 1 & -1/f_1 \\ 0 & 1 \end{pmatrix}$$

$$= \begin{pmatrix} 1 - t/f_2 & -1/f_1 + t/f_1 f_2 - 1/f_2 \\ t & -t/f_1 + 1 \end{pmatrix}$$

The matrix element a is the reciprocal of the focal length of the system. So, the **focal length of the combination**, F is given by

$$\frac{1}{F} = \frac{1}{f_1} + \frac{1}{f_2} - \frac{t}{f_1 f_2} \tag{3.9.1}$$

Distance of the first unit plane, P is

$$\alpha = \frac{1-b}{a} = \frac{1 - 1 + t/f_2}{1/F} = +\frac{tF}{f_2},$$

the positive sign implies that the plane is on the *right* of the lens A of focal length f_1.

Distance of the second unit plane, Q is

$$\beta = \frac{c-1}{a} = \frac{1 - t/f_1 - 1}{1/F} = -\frac{tF}{f_1},$$

the negative sign indicates that the unit plane Q is on the *left* of the lens B of focal length f_2.

- If the two thin lenses are *in contact,* $t = 0$ so that from (3.9.1),

$$\frac{1}{F} = \frac{1}{f_1} + \frac{1}{f_2}$$

3.10 Ramsden's eyepiece : Equivalent focal length and cardinal points

Ramsden's eyepiece essentially consists of two plano-convex lenses E and F, called respectively the *eye lens* and the *field lens,* each of focal length f and separated from each other in air by a distance $2f/3$ (Fig. 3.10).

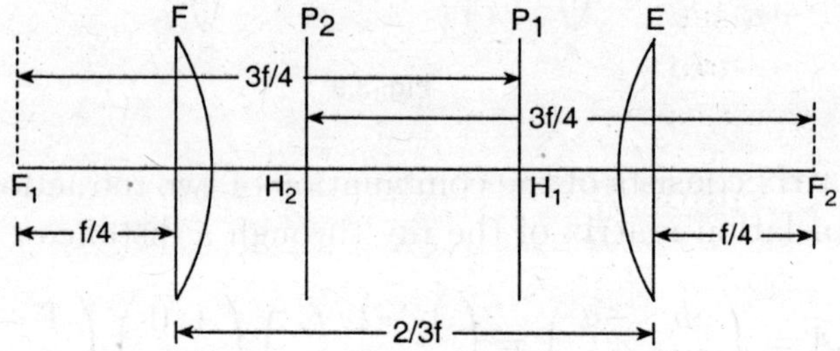

Fig. 3.10

We may therefore write the system matrix as

$$\mathcal{A} = \begin{pmatrix} b & -a \\ -d & c \end{pmatrix} = \begin{pmatrix} 1 & -1/f \\ 0 & 1 \end{pmatrix} \begin{pmatrix} 1 & 0 \\ 2f/3 & 1 \end{pmatrix} \begin{pmatrix} 1 & -1/f \\ 0 & 1 \end{pmatrix}$$

$$= \begin{pmatrix} 1 - 2/3 & -1/f \\ 2f/3 & 1 \end{pmatrix} \begin{pmatrix} 1 & -1/f \\ 0 & 1 \end{pmatrix}$$

$$= \begin{pmatrix} 1/3 & -1/3f - 1/f \\ 2f/3 & -2/3 + 1 \end{pmatrix}$$

$$\therefore \quad a = \frac{1}{3f} + \frac{1}{f} = \frac{4}{3f}; \quad b = \frac{1}{3}; \quad c = \frac{1}{3}; \quad d = -\frac{2f}{3}$$

$\therefore$ **Focal length of equivalent lens is**

$$F = \frac{1}{a} = \frac{3}{4}f$$

Cardinal points : A. Unit planes—Let α be the distance of the first unit plane P_1 from the lens E, and β be the same for the second unit plane P_2.

$$\therefore \quad \alpha = \frac{1-b}{a} = \frac{1-1/3}{4/3f} = \frac{2}{3} \times \frac{3f}{4} = \frac{f}{2}; \quad \beta = \frac{c-1}{a} = \frac{-2/3}{4/3f} = -\frac{f}{2}$$

So, while *the first unit plane* P_1 *is behind the field lens at a distance* $f/2$, *the second unit plane* P_2 *is behind the eye lens at a distance* $f/2$.

B. Nodal planes—As the system is in air, i.e., the same medium on either side of the optical system, the *nodal planes or points* coincide with the *unit planes or points.*

C. Focal planes—The *first focal plane* F_1 is at a distance $\frac{3}{4}f$ from *the first unit plane* i.e., *at* $(3f/4 - f/2) = \frac{1}{4}f$ to the *left of the field lens.*

The *second focal plane* F_2 is at a distance $\frac{1}{4}f$ *behind the eye lens away from the field lens.*

All the cardinal points are thus known.

3.11 Huygens' eyepiece : Equivalent focal length and cardinal points

Huygens' eyepiece consists of two phano-convex lenses—the *field lens F* and the *eye lens E*—of focal lengths $3f$ and f respectively and separated from each other in air by a distance $2f$. They are arranged (Fig. 3.11) with their convex faces towards the incident rays.

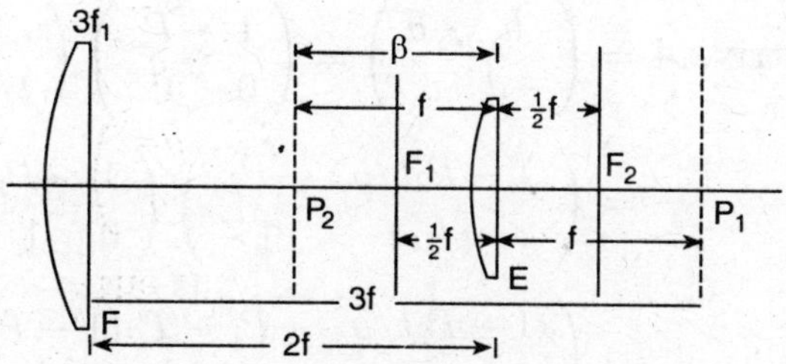

Fig. 3.11

We may thus write the system matrix

$$\mathcal{A} = \begin{pmatrix} b & -a \\ -d & c \end{pmatrix} = \begin{pmatrix} 1 & -1/f \\ 0 & 1 \end{pmatrix} \begin{pmatrix} 1 & 0 \\ 2f & 1 \end{pmatrix} \begin{pmatrix} 1 & -1/3f \\ 0 & 1 \end{pmatrix}$$

$$= \begin{pmatrix} 1 & -1/f \\ 0 & 1 \end{pmatrix} \begin{pmatrix} 1 & -1/3f \\ 2f & -\frac{2}{3}+1 \end{pmatrix}$$

$$= \begin{pmatrix} 1-2 & -1/3f - 1/3f \\ 2f & 1/3 \end{pmatrix} = \begin{pmatrix} -1 & -2/3f \\ 2f & 1/3 \end{pmatrix}$$

$$\therefore \quad a = \frac{2}{3f}; \quad b = -1; \quad c = \frac{1}{3}; \quad d = -2f$$

$\therefore$ **Focal length of equivalent lens is :** $\quad F = 1/a = \frac{3}{2}f.$

Let α be the distance of the **first unit (or principal) plane** P_1 from lens E and β be the same for the **second unit (or principal) plane** P_2.

$$\therefore \quad \alpha = \frac{1-b}{a} = \frac{(1+1)3f}{2} = 3f; \quad \beta = \frac{c-1}{a} = \frac{(1/3-1)3f}{2} = -f$$

Since the same air medium is on both side of the optical system, the **nodal planes** or **points** coincide with *unit planes* or *points.* The **focal planes** may also be found readily.

3.12 Illustrated Examples

► **Example 1.** *A convex lens of a material of index* 1.5 *has radii of curvature* 2 cm *and* −2 cm. *If the thickness of the lens is also* 2 cm, *find the system matrix, positions of principal points and focal points.* **(Punjab Univ.)**

Solution. Given : $R_1 = 2$ cm, $R_2 = -2$ cm, $t = 2$ cm and $n = 1.5$.

P_1 = power of first surface = $(\mu - 1)/R_1 = (1.5-1)/2 = 0.25\ \text{cm}^{-1}$

P_2 = power of second surface = $(1-\mu)/R_2 = (1 - 1.5/(-2) = 0.25\ \text{cm}^{-1}$.

$$\therefore \quad \text{System matrix, } \mathcal{A} = \begin{pmatrix} b & -a \\ -d & c \end{pmatrix} = \begin{pmatrix} 1 & -P_2 \\ 0 & 1 \end{pmatrix} \begin{pmatrix} 1 & 0 \\ t/\mu & 1 \end{pmatrix} \begin{pmatrix} 1 & -P_1 \\ 0 & 1 \end{pmatrix}$$

$$= \begin{pmatrix} 1 - P_2 t/\mu & -P_2 \\ t/\mu & 1 \end{pmatrix} \begin{pmatrix} 1 & -P_1 \\ 0 & 1 \end{pmatrix}$$

$$= \begin{pmatrix} 1 - P_2 t/\mu & -P_1 - P_2(1 - P_1 t/\mu) \\ t/\mu & 1 - P_1 t/\mu \end{pmatrix}$$

$$\therefore \quad a = P_1 + P_2(1 - P_1 t/\mu) = 0.25 + 0.25\left(1 - \frac{2 \times 0.25}{1.5}\right) = 0.417\ \text{cm}^{-1}$$

$$\therefore \quad \text{Focal length. } f = \frac{1}{a} = \frac{1}{0.417} = \mathbf{2.4\ cm}$$

$$b = 1 - \frac{P_2 t}{\mu} = 1 - \frac{0.25 \times 2}{1.5} = 0.667; \quad c = 1 - \frac{P_1 t}{\mu} = 1 - \frac{0.25 \times 2}{1.5} = 0.667$$

Distance of first principal point H_1 (Fig. 3.12) from first surface A_1B_1 is

$$\alpha = \frac{1-b}{a} = \frac{1-0.667}{0.417} = \mathbf{0.8\ cm}$$

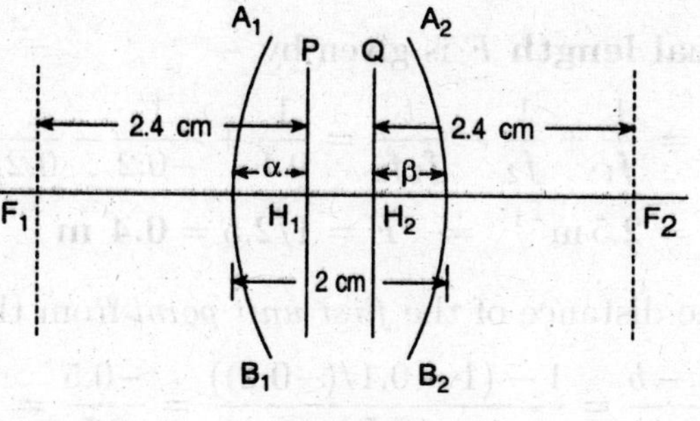

Fig. 3.12

Distance of second principal point H_2 from second surface A_2B_2 is

$$\beta = \frac{c-1}{a} = \frac{0.667-1}{0.417} = \mathbf{-0.8\ cm}$$

∴ First focal point F_1 is at a distance **2.4 cm** from H_1 to the *left of the first surface* and the second focal point F_2 is at a distance **2.4 cm** to the *right of second surface.*

▶ **Example 2.** *A convex lens of focal length* 20 cm *is separated from a concave lens of same focal length by a distance of* 10 cm. *Find the equivalent focal length and the positions of cardinal points.* **(Himachal Univ.)**

Solution. The system matrix of the combination of two lenses A and B (Fig. 3.13) of focal lengths f_1 and f_2 respectively and separated by t is

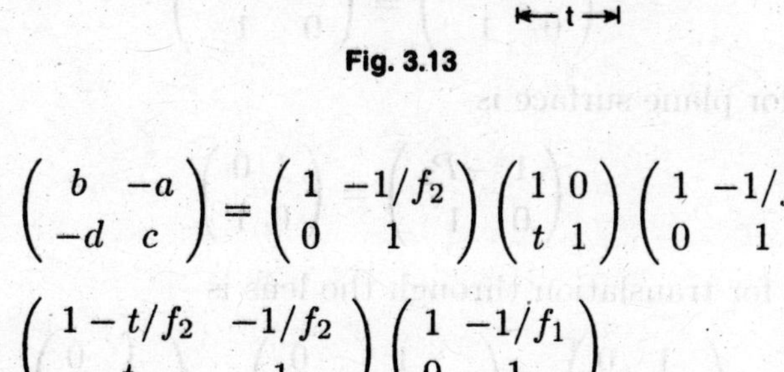

Fig. 3.13

$$\mathcal{A} = \begin{pmatrix} b & -a \\ -d & c \end{pmatrix} = \begin{pmatrix} 1 & -1/f_2 \\ 0 & 1 \end{pmatrix} \begin{pmatrix} 1 & 0 \\ t & 1 \end{pmatrix} \begin{pmatrix} 1 & -1/f_1 \\ 0 & 1 \end{pmatrix}$$

$$= \begin{pmatrix} 1-t/f_2 & -1/f_2 \\ t & 1 \end{pmatrix} \begin{pmatrix} 1 & -1/f_1 \\ 0 & 1 \end{pmatrix}$$

$$= \begin{pmatrix} 1-t/f_2 & -1/f_1 + t/f_1 f_2 - 1/f_2 \\ t & -t/f_1 + 1 \end{pmatrix}$$

$\therefore$ **Equivalent focal length F is given by**

$$\frac{1}{F} = a = \frac{1}{f_1} + \frac{1}{f_2} - \frac{t}{f_1 f_2} = \frac{1}{0.1} + \frac{1}{-0.2} - \frac{0.1}{0.2 \times (-0.2)}$$
$$= 2.5\,\mathrm{m}^{-1} \quad \Rightarrow \quad F = 1/2.5 = \mathbf{0.4\ m}$$

Unit points : The distance of the *first unit point* from the lens A is

$$\alpha = \frac{1-b}{a} = \frac{1-(1-0.1/(-0.2))}{2.5} = \frac{-0.5}{2.5} = \mathbf{-0.2\ m}$$

So the first unit point H_1 is at a distance **0.2 m** to the *left* of the lens A.

The distance of second unit point H_2 from the lens B is given by

$$\beta = \frac{c-1}{a} = \frac{-0.1/0.2+1-1}{2.5} = -\frac{0.5}{2.5} = \mathbf{-0.2\ m}$$

So, the second unit point is at a distance **0.2 m** to the *left* of the lens B.

Nodal points : The *nodal points N_1, N_2 coincide with the unit points H_1, H_2*, as the system has the same medium (air) on its either side.

Focal points : The *first focal point* F_1 is at a distance 0.4 m from H_1. So, $C_1F_1 = 0.4 + 0.2 = \mathbf{0.6\ m}$ from the lens A to its *left.*

The *second focal point* F_2 is at a distance 0.4 m from H_2. So, $C_2F_2 = 0.4 - 0.2 = \mathbf{0.2\ m}$ from the lens B to its *right.*

▶ **Example 3.** *Consider a plano-convex lens of a material of index* 1.5. *The convex surface has a radius of curvature of* 2.5 cm *and is facing the incident light. The centre thickness of the lens is* 0.6 cm. *Construct the system matrix.* **(Calcutta Hons.)**

Solution. P_1 = Power of front face $= (\mu - 1)/R_1 = (1.5-1)/2.5 = 0.2\ \mathrm{cm}^{-1}$.

P_2 = Power of back plane surface $= (1-\mu)/R_2 = (1-1.5)/\infty = 0$.

$\therefore$ Refraction matrix for front surface is

$$\begin{pmatrix} 1 & -P_1 \\ 0 & 1 \end{pmatrix} = \begin{pmatrix} 1 & -0.2 \\ 0 & 1 \end{pmatrix}$$

Refraction matrix for plane surface is

$$\begin{pmatrix} 1 & -P_2 \\ 0 & 1 \end{pmatrix} = \begin{pmatrix} 1 & 0 \\ 0 & 1 \end{pmatrix}$$

Translation matrix for translation through the lens is

$$\begin{pmatrix} 1 & 0 \\ t/\mu & 1 \end{pmatrix} = \begin{pmatrix} 1 & 0 \\ 0.6/1.5 & 1 \end{pmatrix} = \begin{pmatrix} 1 & 0 \\ 0.4 & 1 \end{pmatrix}$$

$$\therefore \textbf{ System matrix, } A = \begin{pmatrix} 1 & 0 \\ 0 & 1 \end{pmatrix} \begin{pmatrix} 1 & 0 \\ 0.4 & 1 \end{pmatrix} \begin{pmatrix} 1 & -0.2 \\ 0 & 1 \end{pmatrix}$$

▶ **Example 4.** *A sphere of glass ($\mu = 1.5$) is of 20 cm diameter. A parallel beam of light enters from one side. Apply matrix method to find where it will get focussed on the other side?* **(Punjab Univ.)**

Solution. Let AB and CF form the *parallel* beam (Fig. 3.14) incident on one side of the sphere $BDEF$ of dia 0.2 m (= 20 cm). The ray AB suffers refraction at B on the first surface BPF of power P_1 and emerges out at D of the surface DQE of power P_2 to meet the axis at I such that $QI = v$. To find the value of v.

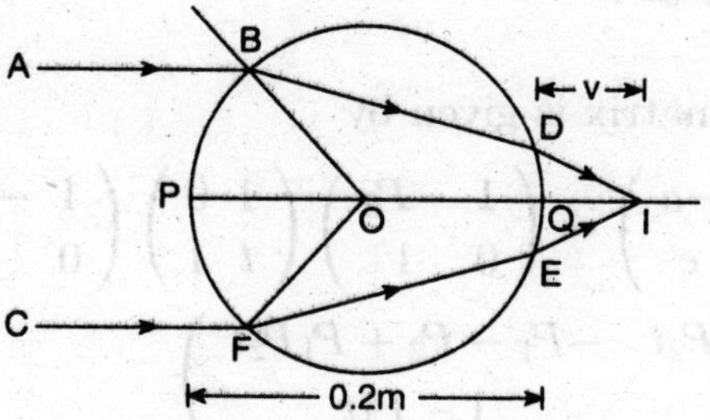

Fig. 3.14

From the given data, $\mu = 1.5$, $PQ = 0.2\,\text{m} \Rightarrow R = 0.1\,\text{m}$

$$\therefore \quad P_1 = \frac{\mu - 1}{R} = \frac{1.5 - 1}{0.1} = +5\,\text{m}^{-1}; \quad P_2 = \frac{1-\mu}{-R} = \frac{1-1.5}{-0.1} = +5\,\text{m}^{-1}.$$

$$\therefore \quad \text{System matrix, } \mathcal{A} = \begin{pmatrix} 1 & 0 \\ v & 1 \end{pmatrix}\begin{pmatrix} 1 & -P_2 \\ 0 & 1 \end{pmatrix}\begin{pmatrix} 1 & 0 \\ t/\mu & 1 \end{pmatrix}\begin{pmatrix} 1 & -P_1 \\ 0 & 1 \end{pmatrix}$$

$$= \begin{pmatrix} 1 & 0 \\ v & 1 \end{pmatrix}\begin{pmatrix} 1 & -5 \\ 0 & 1 \end{pmatrix}\begin{pmatrix} 1 & 0 \\ 2/15 & 1 \end{pmatrix}\begin{pmatrix} 1 & -5 \\ 0 & 1 \end{pmatrix}$$

$$= \begin{pmatrix} 1 & 0 \\ v & 1 \end{pmatrix}\begin{pmatrix} 1 & -5 \\ 0 & 1 \end{pmatrix}\begin{pmatrix} 1 & -5 \\ 2/15 & -2/3+1 \end{pmatrix}$$

$$= \begin{pmatrix} 1 & 0 \\ v & 1 \end{pmatrix}\begin{pmatrix} 1-2/3 & -5+10/3-5 \\ 2/15 & 1/3 \end{pmatrix}$$

$$= \begin{pmatrix} 1/3 & -20/3 \\ v/3+2/15 & -20v/3+1/3 \end{pmatrix}$$

$\therefore$ Matrix equation for the *image plane* is given by

$$\begin{pmatrix} \lambda_2 \\ x_2 \end{pmatrix} = \begin{pmatrix} 1/3 & -20/3 \\ v/3+2/15 & -20v/3+1/3 \end{pmatrix}\begin{pmatrix} \lambda_1 \\ x_1 \end{pmatrix}$$

$$\Rightarrow \quad x_2 = \left(\frac{1}{3}v + \frac{2}{15}\right)\lambda_1 + \left(\frac{1}{3} - \frac{20}{3}v\right)x_1.$$

As the *incident ray is parallel to the axis PQ*, $\lambda_1 = 0$ so that

$$x_2 = \left(\frac{1}{3} - \frac{20}{3}v\right) x_1$$

At the *focal plane*, $x_2 = 0 \Rightarrow \frac{1}{3} - \frac{20}{3}v = 0 \Rightarrow v = \mathbf{0.05\ m.}$

► **Example 5.** *Find the positions of two principal focal points and two nodal points for a combination of two convex lenses of focal lengths* 20 cm *and* 10 cm, *situated at a distance of* 10 cm *apart in air.* **(Calcutta Hons.)**

Solution. The system matrix is given by

$$\mathcal{A} = \begin{pmatrix} b & -a \\ -d & c \end{pmatrix} = \begin{pmatrix} 1 & -P_2 \\ 0 & 1 \end{pmatrix} \begin{pmatrix} 1 & 0 \\ t & 1 \end{pmatrix} \begin{pmatrix} 1 & -P_1 \\ 0 & 1 \end{pmatrix}, \quad \mu = 1$$

$$= \begin{pmatrix} 1 - P_2 t & -P_1 - P_2 + P_1 P_2 t \\ t & 1 - P_1 t \end{pmatrix}$$

$$\therefore \quad a = P_1 + P_2 - P_1 P_2 t = \frac{1}{20} + \frac{1}{10} - \frac{1}{20} \cdot \frac{1}{10} \cdot 10 = \frac{1}{10}\ \text{cm}^{-1}$$

$$b = 1 - P_2 t = 1 - \frac{1}{10} \cdot 10 = 0; \quad c = 1 - P_1 t = 1 - \frac{1}{20} \cdot 10 = 0.5$$

$$d = -t = -10\ \text{cm}.$$

$$\therefore \ \textbf{Equivalent focal length, } \boldsymbol{F = \frac{1}{a} = 10}\ \textbf{cm}$$

$\therefore$ Distance of *first principal point* from the *first lens* is

$$\alpha = \frac{1 - b}{a} = \frac{1 - 0}{1/10} = \mathbf{10\ cm}$$

Distance of *second principal point from the second lens is*

$$\beta = \frac{c - 1}{a} = \frac{0.5 - 1}{1/10} = \mathbf{-5\ cm}$$

The *nodal points coincide with the principal points* as the surrounding medium is the same, namely air.

Distance of *the first focal point from the first lens* is 10 cm − 10 cm = **0.**

Distance of *second focal point from the second lens* is 10 cm − 5 cm = **5 cm.**

► **Example 6.** *A ray of light, emerging from an axial point object, subtends an angle* $\alpha_1 = 6°$ *with the optic axis. At a height* $h = 2$ cm, *it intersects a convex spherical surface of radius* 5 cm, *the index* $\mu = 1.62$. *Find the angle the ray subtends with the optic axis after refraction.*

Solution. Here $\lambda_1 = \mu_1 \alpha_1 = \alpha_1 \ (\because \ \mu = 1) = +6° = +0.1047$ rad.

$$\lambda_2 = \mu_2 \alpha_1 = 1.62 \alpha_2; \ x_1 = 2\ \text{cm}.$$

$$\therefore \begin{pmatrix} \lambda_2 \\ x_2 \end{pmatrix} = \begin{pmatrix} 1 & -P \\ 0 & 1 \end{pmatrix} \begin{pmatrix} \lambda_1 \\ x_1 \end{pmatrix} \Rightarrow \begin{pmatrix} 1.62\alpha_2 \\ x_2 \end{pmatrix} = \begin{pmatrix} 1 & -(1.62-1)/5 \\ 0 & 1 \end{pmatrix} \begin{pmatrix} \alpha_1 \\ x_1 \end{pmatrix}$$

$$\text{or, } \begin{pmatrix} 1.62\alpha_2 \\ x_2 \end{pmatrix} = \begin{pmatrix} 1 & -0.124 \\ 0 & 1 \end{pmatrix} \begin{pmatrix} \alpha_1 \\ 2 \end{pmatrix}$$

$$\therefore \quad x_2 = x_1 = 2 \text{ cm}$$

$$1.62\alpha_2 = \alpha_1 - 2 \times 0.124 = 0.1047 - 0.248$$

$$\therefore \quad \alpha_2 = \frac{0.1047 - 0.248}{1.62} = -0.88 \text{ rad} = \mathbf{5^\circ \ (approx)}.$$

▶ **Example 7.** *A concave lens is placed at a distance of* 25 cm *in front of a concave mirror of focal length* 20 cm. *It is found that a pin placed at a distance of* 45 cm *in front of the lens coincides with its own inverted image formed by the combination. Using refraction matrix, find the focal length of the lens.* **(Calcutta Hons.)**

Solution. By the problem, the refracted rays from the lens fall on the mirror normally. So, the rays emerging from the lens appear to diverge from the centre of curvature C of the mirror (Fig. 3.15).

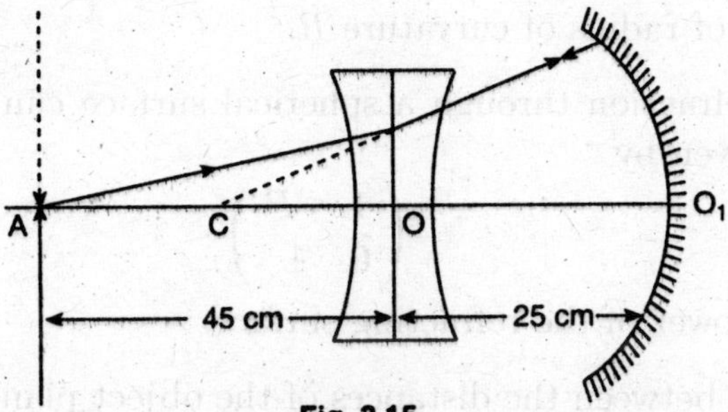

Fig. 3.15

$\therefore$ Image position of lens $\equiv$ Centre of curvature of mirror
But, object distance, $u = -OA = -45$ cm
and image distance, $v = -CO = -[CO_1 - OO_1] = -[2 \times 20 - 25] = -15$ cm

Let (λ_1, x_1) and (λ_2, x_2) be the optical co-ordinates of an object point and the corresponding image point for the lens.

$$\therefore \begin{pmatrix} \lambda_2 \\ x_2 \end{pmatrix} = \begin{pmatrix} 1 & 0 \\ v & 1 \end{pmatrix} \begin{pmatrix} 1 & -P \\ 0 & 1 \end{pmatrix} \begin{pmatrix} 1 & 0 \\ -u & 1 \end{pmatrix} \begin{pmatrix} \lambda_1 \\ x_1 \end{pmatrix}$$

$$= \begin{pmatrix} 1 & 0 \\ -15 & 1 \end{pmatrix} \begin{pmatrix} 1 & -P \\ 0 & 1 \end{pmatrix} \begin{pmatrix} 1 & 0 \\ 45 & 1 \end{pmatrix} \begin{pmatrix} \lambda_1 \\ x_1 \end{pmatrix}$$

$$= \begin{pmatrix} 1 & 0 \\ -15 & 1 \end{pmatrix} \begin{pmatrix} 1-45P & -P \\ 45 & 1 \end{pmatrix} \begin{pmatrix} \lambda_1 \\ x_1 \end{pmatrix}$$

$$= \begin{pmatrix} 1-45P & -P \\ -15(1-45P)+45 & 1+15P \end{pmatrix} \begin{pmatrix} \lambda_1 \\ x_1 \end{pmatrix}$$

$$\therefore \quad x_2 = [-15(1-45P)+45]\,\lambda_1 + (1+15P)x_1$$

For an *axial* object point, $x_1 = 0$ and as the image point also is *on the axis*, $x_2 = 0$.

$$\therefore \quad -15(1-45P)+45 = 0$$

$$\therefore \quad P = 2/45 \text{ cm}^{-1}, \text{ on simplification.}$$

$$\therefore \quad \textbf{Focal length,} \; F = \frac{1}{P} = \frac{45}{2} = \textbf{22.5 cm}$$

Review Questions

1. Determine the matrix that will represent the effect of translation in a medium of index μ.

2. Determine the matrix which will represent the effect of refraction through a spherical surface of radius of curvature R. **(Guru Nanak Univ.)**

3. Show that the refraction through a spherical surface can be characterised by a (2×2) matrix given by

$$\begin{pmatrix} 1 & -P \\ 0 & 1 \end{pmatrix}$$

where P is the power of the refracting surface. **(Punjab Univ.)**

4. Find the relation between the distances of the object plane and the image plane. Hence find the system matrix in terms of magnification.

5. Explain the terms : Translation matrix, refraction matrix and sytem matrix of an optical system for paraxial rays. Obtain the fundamental translation and refraction matrices for such a system.

6. Derive the following relation

$$\frac{\mu_2}{v} - \frac{\mu_1}{u} = \frac{\mu_2 - \mu_1}{R}$$

for refraction at a convex spherical surface by the matrix method. **(Calcutta Hons.)**

7. Derive the expression for focal length of a thin convex lens by the method of system matrix for it, under the condition of paraxial approximation. **(Kurukshetra Univ.)**

8. Obtain the system matrix for a thick lens and hence obtain the system matrix for a thin lens. **(Himachal Univ.)**

9. Obtain the system matrix for a thin lens and hence derive a relation for the focal length of a combination of two thin lenses separated by a distance. **(Kakatiya Univ.)**

10. Construct the system matrix for refraction of paraxial rays through a thick lens. Hence find the same for a thin lens and show that

$$\frac{1}{f} = (\mu - 1)\left(\frac{1}{R_1} - \frac{1}{R_2}\right),$$

the symbols having their usual meanings.

11. Define unit or principal points and determine their distances from the respective surfaces in terms of elements of system matrix.

12. Obtain the relation $\frac{1}{v} - \frac{1}{u} = \frac{1}{f}$ for a thin convex lens using the matrix method. **(Punjab Univ.)**

13. Show that if the distances are measured from unit planes, the focal length of the system is the reciprocal of the element 'a' of the system matrix.

14. Define nodal points and show that if the medium on either side of the optical system be of the same index, the nodal planes coincide with unit planes. Use the matrix method.

15. Calculate the focal length of the combination of two thin lenses of focal lengths f_1 and f_2 placed t distance apart in air. Also, find the positions of two principal (or unit) points. **(Himachal Univ.)**

16. Using matrix method, obtain the expression for equivalent focal length and positions of cardinal points of (a) a Ramsden's eyepiece and (b) a Huygens' eyepiece. **(Himachal Univ.)**

17. Show that the system matrix of an optical system may be written as

$$\begin{pmatrix} 1/m & -a \\ 0 & m \end{pmatrix}$$

where m is the linear magnification of the system.

18. Formulate the equations for the image plane and magnification of an optical system using the matrix method. Explain the terms : unit planes and nodal planes briefly. **(Kakatiya Univ.)**

19. Obtain Snell's law for paraxial rays and hence derive a matrix for the combination of refraction and translation for light rays passing through a thick lens. **(Andhra Univ.)**

Short Answer and Objective Questions

1. Explain, with a neat diagram, what you mean by the optical coordinates of a ray.
2. Write down the matrix for (a) translation and (b) refraction through a spherical surface.
3. What is meant by translation-refraction matrix? Explain. **(Punjabi Univ.)**
4. What is the power P of a refracting surface that separates two media with indices μ_1, μ_2 and having a radius of curvature R?
5. Write down the relation connecting the distance of the object plane and the image plane.
6. What is a system matrix? Write down the system matrix of an optical system in terms of magnification m.
7. What are unit points and nodal points? Under what condition the two coincide?
8. What is the determinant of the translation matrix? **(Himachal Univ.)**
9. Which one of the following elements of the system matrix is directly related to the focal length of the optical system?

   (a) a, (b) b, (c) c, (d) d. **(Himachal Univ.)**
10. The determinant of the translation matrix is

    (a) -1, (b) 0, (c) $+1$, (d) none of these.

Problems

1. Two thin lenses, one convex of focal length 0.2 m and the other concave of focal length 0.1 m are placed on a common axis 0.08 m apart. An object 0.01 m long is mounted at a distance of 0.4 m from the convex lens. Find the position and size of the image. [$v = -0.1455$ m; 0.0045 m]
2. A hemisphere of glass ($\mu = 3/2$) of 3 cm radius acts as an optical system. Using matrix method, find the position of the image when an object is placed 4 cm in front of the curved surface. [14 cm from plane surface towards object side]
3. Using matrix multiplication, find the new coordinates (x', y') in terms of the initial coordinates (x_0, y_0) :

$$\begin{pmatrix} 3 & 1 \\ 2 & 0 \end{pmatrix}\begin{pmatrix} 2 & 0 \\ -1 & 3 \end{pmatrix}\begin{pmatrix} x_0 \\ y_0 \end{pmatrix} = \begin{pmatrix} x' \\ y' \end{pmatrix}$$

[$x' = x_0 + 3y_0;\ y' = 4x_0$]

4. Two thin converging lenses of focal length 30 cm and 40 cm are placed co-axially 20 cm apart. An object is located at a distance of 48 cm from the first lens. Find (a) positions of principal points, (b) positions of focal points and (c) position of the image. [(a) $\alpha = +12$ cm, $\beta = -16$ cm; (b) $F_1 = 24$ cm left of first lens; $F_2 = 24$ cm to right of second lens; (c) 24 cm]

5. A small object is situated 0.1 m in front of a convex surface of radius 0.03 m and index 1.66, ground on a glass block. Find the image position if (a) the object and image are in air, (b) both are immersed in water ($\mu = 4/3$).

[−13.8 cm towards right of convex surface; 67.9 cm towards image side]

6. Using matrix method, find the position of cardinal points of a coaxial lens system, consisting of two lenses of focal length 15 cm and 5 cm in air, separated by 10 cm.

[$\alpha = +15$ cm, $\beta = -5$ cm; nodal points coincide with unit points, $F_1 = 7.5$ cm from first lens, $F_2 = 2.5$ cm from second lens]

7. Two thin convex lenses of focal length 0.12 m and 0.06 m are placed coaxially 0.08 m apart. Find the system matrix and positions of cardinal points of the system. [$a = 1/0.072$ m, $b = -0.333$ m, $c = 0.333$ m, $d = -0.08$ m; $\alpha = 0.096$ m, $\beta = -0.048$ m. Focal points : 0.024 m to right of lens 1 and 0.024 m to right of lens 2; Nodal points coincide with unit points]

8. A double convex lens is of material of index 1.5 and has a thickness 1 cm. The radii of curvature of the first and the second surfaces are 10 cm and 20 cm respectively. Find the system matrix and the positions of principal planes.

[$a = -91/1200$, $b = 61/60$, $c = 31/60$, $d = -2/3$; $\alpha = 20/91$ cm, $\beta = -40/91$ cm]

9. An object of height 1.5 cm is placed at a distance 40 cm in front of a convex lens of focal length +20 cm. Behind the convex lens, a concave lens of focal length −10 cm is placed at a distance of 15 cm from the convex lens. Using matrix method, find the size, position and nature of the image.

[1 cm, −16.66 cm (to left of 2nd lens), virtual and errrect]

10. A concave lens of focal length 30 cm is placed 10 cm in front of a convex lens of focal length 20 cm, construct the system matrix and find the positions of principal planes. [$a = 1/30\,\text{cm}^{-1}$, $b = 1/2$, $c = 4/3$, $d = -10$ cm; $\alpha = +15$ cm, $\beta = +10$ cm]

11. Obtain the system matrix for a lens placed in air and made of glass of $\mu = 1.5$ and having radii of curvature 50 cm each. **(Osmania Univ.)** $\left[\begin{pmatrix} 1 & -0.002 \\ 0 & 1 \end{pmatrix}\right]$

12. Calculate the position of the paraxial focal point of a lens in the form of a sphere of radius 20 cm and having a refractive index 1.6. [0.67 cm]

13. A thick convex lens has radii of corvature +6 cm and −6 cm and thickness 2 cm. The index of its material is 1.6 and is placed in air. Obtain the system matrix and determine its focal length and the positions of unit points.

$$\left[\begin{pmatrix} 0.875 & -0.1875 \\ 1.25 & 0.875 \end{pmatrix};\ f = 5.3\text{ cm};\ 0.67\text{ cm},\ -0.67\text{ cm}\right]$$

Chapter 4

Dispersion

4.1 Dispersion and dispersive power

The refractive index μ of a material is *not* a constant but varies with wavelength, i.e., it is different for different wavelengths of light. This aspect is contained in the very statement of Snell's law. The index is greater for violet rays than that for red rays. As a result, when a ray of composite (i.e., polychromatic) light such as white light is passed through a prism P the violet rays are deviated most and the red rays the least. This causes a ray of white light after its passage through a prism to break up (Fig. 4.1) into its constituent colours. This phenomenon is called **dispersion** and the pattern obtained thereof is called a **spectrum.**

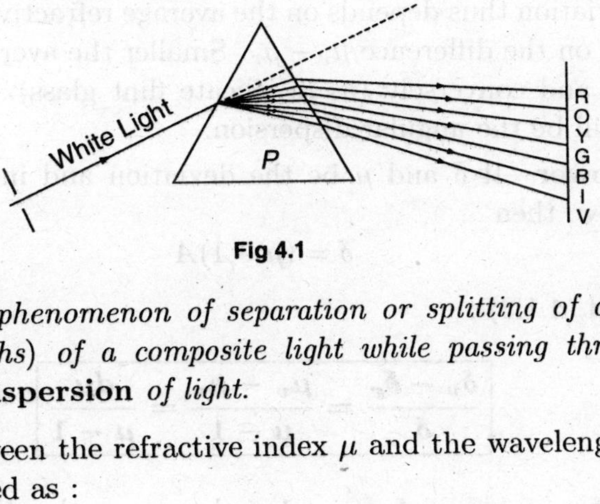

Fig 4.1

Dispersion—*The phenomenon of separation or splitting of different constituent colours (or, wavelengths) of a composite light while passing through a transparent medium is known as* **dispersion** *of light.*

• The relation between the refractive index μ and the wavelengt λ of light may be approximately described as :

$$\mu = A + B/\lambda^2$$

where A, B are constants known as the Cauchy's constants.

Angular dispersion—When a beam of white light passes through a prism of transparent material, light of different wavelengths suffer, as already indicated,

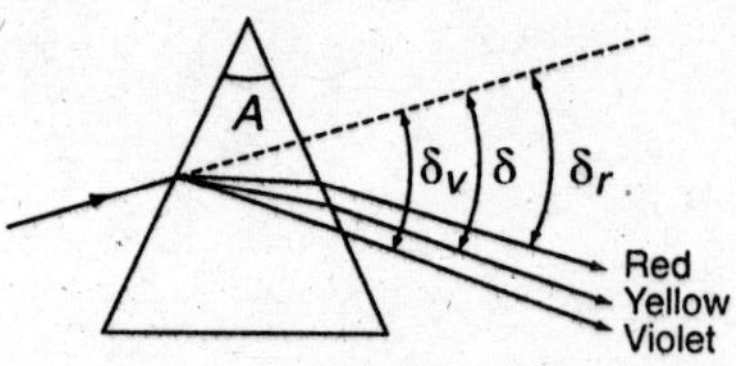

Fig 4.2

deviations by different amounts. The overall deviation of the beam is measured by the deviation of yellow light, since its deviation is roughly the average of all deviations (Fig. 4.2). Clearly, if δ_r and δ_v be the deviations for red and violet components, the angular divergence of the transmitted beam is given by $\delta_v - \delta_r$. This divergence is known as **angular dispersion.**

Definition—*The difference in deviations for the extreme violet and extreme red radiations, that is, $\delta_v - \delta_r$, is defined as the* **angular dispersion.**

If A, the angle of the prism, be very small and the prism be at minimum deviation

$$\mu = \frac{A+\delta}{A} = 1 + \frac{\delta}{A}$$

$$\Rightarrow \quad \boxed{\delta = (\mu - 1)A} \tag{4.1.1}$$

where δ is the angle of minimum deviation of the mean ray.

$$\therefore \quad \delta_v - \delta_r = (\mu_v - 1)A - (\mu_r - 1)A, \quad \text{using (4.1.1)}$$

$$\boxed{\delta_v - \delta_r = (\mu_v - \mu_r)A} \tag{4.1.2}$$

which gives the relation of angular dispersion in terms of refractive indices.

• The mean deviation thus depends on the average refractive index, and the angular dispersion depends on the difference $\mu_v - \mu_r$. Smaller the average value of μ (fluorite), smaller is $\mu_v - \mu_r$ and conversely (as in silicate flint glass). Thus, larger the mean deviation, larger will be the angular dispersion.

Dispersive power—If δ and μ be the deviation and index respectively for the mean ray, say yellow, then

$$\delta = (\mu - 1)A \tag{4.1.3}$$

$\therefore$ From (4.1.2) and (4.1.3) :

$$\boxed{\frac{\delta_v - \delta_r}{\delta} = \frac{\mu_v - \mu_r}{\mu - 1} = \frac{d\mu}{\mu - 1}} \tag{4.1.4}$$

The quantity $(\delta_v - \delta_r)/\delta$ or $d\mu/(\mu - 1)$ is known as the **dispersive power** of the material of the prism and is symbolised by ω. Note that it is *independent of the angle of prism.* Eq. (4.1.4) may be taken as the definition of *dispersive power.*

Definition—Dispersive power ω *may be defined as the ratio of the angular dispersion $\delta_v - \delta_r$ and the mean deviation δ when light gets dispersed at minimum deviation by a prism of small angle.*

It is given by the mathematical relation (4.1.4) above.

• Since our eye is sensitive to the spectral region between C and F lines (Fraunhofer) that lie in the red and blue region respectively of hydrogen spectrum, and since their mean refractive index is essentially the same as that of the D-lines of sodium in the yellow region, the *internationally accepted* expression for the dispersive power is

$$\omega = \frac{\mu_F - \mu_C}{\mu_D - 1}$$

4.2 Achromatic prisms : Deviation without dispersion

The dispersive powers of prisms are different for different materials. For instance, ω of flint glass is greater than that of crown glass. It is thus possible *to combine a crown glass prism with that of a flint glass with their bases on the opposite sides* (Fig. 4.3) *such that a white light ray, in passing through the combination, may suffer deviation without dispersion. Such a prismatic combination is known as an* **achromatic combination** *of prisms.*

The constituent colours of the composite white light are parallel on the emergence side and recombine to form white light again.

• Strictly speaking, the prismatic combination is *achromatic* only in respect of the extreme colours of C and F lines. In so far as the intermediate colours are concerned, it is only approximately true.

Deviation without dispersion : Condition—Let ABC and ACD be two prisms with small angles of refraction A and A' respectively. While ABC is of crown glass, ACD is of flint glass (Fig. 4.3).

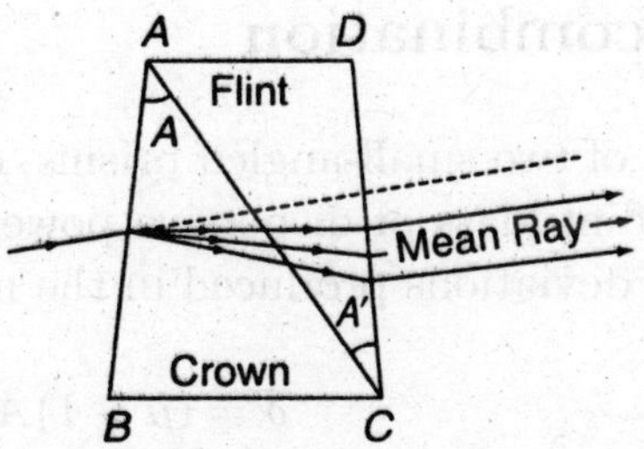

Fig 4.3

Also let δ_v, δ_r be the deviations and μ_v, μ_r the indices for the violet and red rays respectively for the prism ABC. The corresponding quantities for the prism ACD are simply primed. So, we may write, using (4.1.2),

$$\delta_v - \delta_r = (\mu_v - \mu_r)A$$

$$\delta'_v - \delta'_r = (\mu'_v - \mu'_r)A'$$

For *achromatism*, i.e., *zero angular dispersion*, therefore,

$$\delta_v - \delta_r = \delta'_v - \delta'_r$$

$$\Rightarrow \quad (\mu_v - \mu_r)A = (\mu'_v - \mu'_r)A'$$

$$\Rightarrow \frac{\mu_v - \mu_r}{\mu - 1}(\mu - 1)A = \frac{\mu'_v - \mu'_r}{\mu' - 1}(\mu' - 1)A'$$

$$\Rightarrow \quad \omega\delta = \omega'\delta'$$

$$\therefore \quad \boxed{\frac{\omega}{\omega'} = \frac{\delta'}{\delta}}$$

where ω and ω' are the dispersive powers of the materials of crown and flint glass prisms respectively.

So, the **condition** for a prismatic combination for achromatisms i.e., **no dispersion** is that *the dispersive powers should be inversely proportional to the deviations produced by them.*

But what is the net **deviation** produced by the combination?

For the prism ABC, the deviation $\delta = (\mu - 1)A$.

For the prism ACD, the deviation $\delta' = (\mu' - 1)A'$.

$\therefore$ Net deviation by the combination is :

$$\delta - \delta' = (\mu - 1)A - (\mu' - 1)A'$$

• The above treatment, it is to be noted, is valid only for a prism of small angle and for near-normal incidence.

4.3 Dispersion without deviation in prismatic combination

Let A and A' be the angles of two small-angled prisms, one of crown glass of dispersive power ω and the other of flint glass of dispersive power ω', be combined as shown in Fig. 4.4. If δ and δ' be the deviations produced in the mean ray of indices μ and μ' by the prisms, then

$$\delta = (\mu - 1)A; \quad \delta' = (\mu' - 1)A'$$

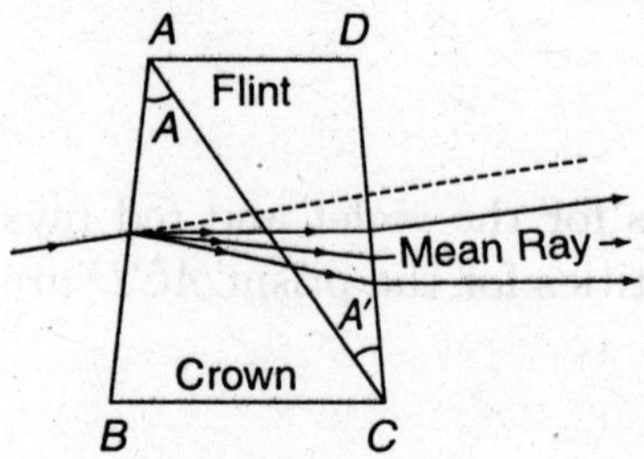

Fig 4.4

For the combination to show **no deviation**, we should have

$$\delta = \delta' \Rightarrow (\mu - 1)A = (\mu' - 1)A'$$

$$\therefore \quad \boxed{\frac{A}{A'} = \frac{\mu' - 1}{\mu - 1}}$$

which is the required **condition** *for no deviation.* Note that the angle A' of flint glass prism will be smaller than that of the crown glass.

Although there is no deivation, there is *some dispersion* which we shall now calculate.

Dispersion—Let δ_v and δ_r be the deviations for the violet and red rays for the crown glass prism and δ'_v and δ'_r the corresponding values for the flint glass prism. Also let μ_v and μ_r be the indices for the violet and red rays for crown glass and μ'_v and μ'_r the corresponding quantities for flint glass.

$$\therefore \quad \delta_v - \delta_r = (\mu_v - \mu_r)A = \text{angular dispersion of crown prism}$$

$$\delta'_v - \delta'_r = (\mu'_v - \mu'_r)A' = \text{angular dispersion of flint prism}$$

$$\begin{aligned}\therefore \quad \text{Resultant dispersion} &= (\delta_v - \delta_r) - (\delta'_v - \delta'_r) \\ &= (\mu_v - \mu_r)A - (\mu'_v - \mu'_r)A' \\ &= \frac{\mu_v - \mu_r}{\mu - 1}(\mu - 1)A - \frac{\mu'_v - \mu'_r}{\mu' - 1}(\mu' - 1)A' \\ &= \omega\delta - \omega'\delta' \\ &= (\omega - \omega')\delta \\ & (\because \delta = \delta', \textit{ no deviation})\end{aligned}$$

Thus although there is no deviation, there is a **net dispersion**, since $\omega \neq \omega'$.

But $\omega' > \omega \Rightarrow \omega - \omega'$ is *negative*, which means that the order of colours in the resultant dispersion by the combination is opposite to that in the dispersion due to crown glass prism.

4.3.1 Direct vision spectroscope

The *principle* on which this optical device is based is that an appropriate combination of prisms may produce *dispersion without deviation*. For a combination of prisms—one of crown glass, the other of flint glass—the condition is

$$(\mu - 1)A = (\mu' - 1)A'$$

as has already been proved in Art 4.3.

If m number of *crown* glass prisms and n number of *flint* glass prisms are used, then the condition for **no deviation** becomes

$$m(\mu - 1)A = n(\mu' - 1)A'$$

Construction—A direct vision spectroscope essentially consists of *alternate crown* and *flint* glass prisms cemented together with *canada balsam*, a transparent substance. The prisms are so arranged that the *alternate prisms have their refracting angles in opposite directions.* This ensures that the deviation produced by the flint glass prism is equal and opposite to that produced by crown glass prism.

Fig. 4.5 illustrates a simple *direct vision spectroscope* where 3 crown glass prisms (C's) and 2 flint glass prisms (F's) of appropriate angles are mounted alternately. These are properly and securedly held in position within a metallic tube and arranged with their refracting angles oppositely placed so that the middle part of the spectrum suffers no deviation. S is a narrow adjustable slit $\parallel$ to the refracting edge of the prisms. A pencil of white light is allowed to pass through S and is rendered parallel by means of an achromatic lens L_1 situated between S and the crown glass prism. The light in passing through the prism-system gets dispersed, the red and violet rays (R, V) being indicated in the figure. On emergence, the rays are brought to a focus in the focal plane of the converging lens L_2 and examined by the eyepiece E— a simple magnifier.

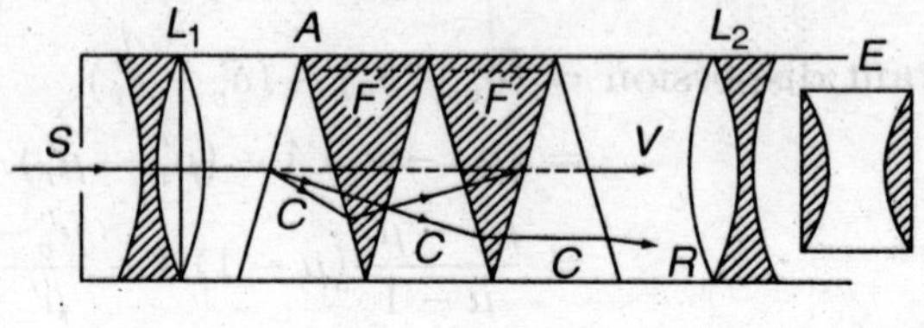

Fig 4.5

Qualitative studies of astral bodies can be made with a direct vision spectroscope. But its resolving power is too small for separating two very close lines.

4.4 Chromatic aberration of a lens

The index μ of every refracting medium is a function of the wavelength or the colour of light. In the visible region, the index is least for red and maximum for violet. The focal length of a lens depends on the refractive index μ and would thus be different for different colours or wavelengths. Since $\mu_v > \mu_r$, $f_v < f_r$. Thus a lens, for a given point object will *not produce a point image*, but a *series of coloured images* at different distances from the lens, corresponding to each constituent colour in the incident light beam.

This variation of the image distance from the lens, for a given object distance, with refractive index is a measure of the phenomenon called **longitudinal** *or* **axial aberration**.

There is another aspect of the variation in focal length with colour or wavelength of light *via* the refractive index. The magnification v/u of the image depends on the focal length. Since for a given u, v may have different values for different colours or wavelengths, *there will be a variation in the size of the image also and this variation measures what is called* **lateral chromatic aberration.**

We illustrate both the longitudinal and lateral chromatic aberration in Fig. 4.6 using an object having an extension. The object AB gives the image A_1B_1 corresponding to violet light and A_2B_2 corresponding to red light. The violet is thus closer to the lens than the red image. The distance $B_1B_2 = x$ on the *axis* measures the *longitudinal or axial chromatic aberration.* The distance $y = A_2B_2 - A_1B_1$ is the *measure of the lateral chromatic aberration.* The images of intermediate colours between violet and red lie between A_1B_1 and A_2B_2 and their size increases from violet to red.

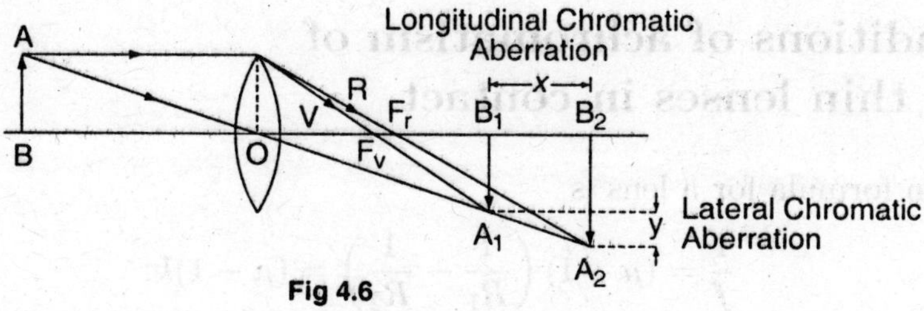

Fig 4.6

• There is *no* position where the images are in *sharp* focus. The effect due to axial and lateral aberrations is together known as **chromatic aberration**. In discussing the chromatic aberration we have deliberately avoided the *monochromatic aberrations*, which occur even if the light is of single colour or wavelength.

• If light is incident on the lens from left to right, the violet image is to the left of the red in a *convex lens*, as in Fig. 4.6 and the *chromatic aberration* is called *positive*. In a *concave lens*, however, in the same situation violet image will be to the right of the red image and the *aberration* is called *negative*. This hints at the fact that a proper combination of a convex and a concave lens may result in **no chromatic aberration.** Such a combination is called an **achromatic combination** or an **achromat** for the pair of wavelengths.

Calculation of chromatic aberration—Let us calculate the *longitudinal chromatic aberration* of a lens for an object at ∞.

Let f_r, μ_r and f_v, μ_v be the focal length and index for the red and violet rays respectively. Also, let f and μ be the corresponding quantities for the mean ray. So for red rays,

$$\frac{1}{f_r} = (\mu_r - 1)\left(\frac{1}{R_1} - \frac{1}{R_2}\right)$$
$$= \frac{\mu_r - 1}{\mu - 1}(\mu - 1)k \quad \text{where } k = (1/R_1 - 1/R_2)$$
$$= \frac{\mu_r - 1}{\mu - 1} \cdot \frac{1}{f}$$

Similarly, for the violet rays, we get

$$\frac{1}{f_v} = \frac{\mu_v - 1}{\mu - 1} \cdot \frac{1}{f}$$

$$\therefore \quad \frac{1}{f_v} - \frac{1}{f_r} = \frac{\mu_v - \mu_r}{\mu - 1} \cdot \frac{1}{f} = \frac{\omega}{f}$$

$$\Rightarrow \quad f_r - f_v = \frac{\omega}{f} f_v f_r = \frac{\omega}{f} f^2 = \omega f \quad \left(\because \; f_r f_v = f^2\right)$$

$$\therefore \quad \boxed{f_r - f_v = \omega f}$$

$\therefore$ **Longitudinal chromatic aberration** $(f_r - f_v)$ *is the product of the dispersive power* ω *and the focal length* f *for the mean ray.*

4.5 Conditions of achromatism of two thin lenses in contact

The refraction formula for a lens is

$$\frac{1}{f} = (\mu - 1)\left(\frac{1}{R_1} - \frac{1}{R_2}\right) = (\mu - 1)k, \tag{4.5.1}$$

$$\text{where } k = \left(\frac{1}{R_1} - \frac{1}{R_2}\right) = \frac{1}{\mu - 1}\frac{1}{f} \tag{4.5.2}$$

Differentiating (4.5.1) and putting the value of k from (4.5.2), we get

$$-\frac{1}{f^2}df = k \cdot d\mu$$

$$= \frac{d\mu}{\mu - 1} \cdot \frac{1}{f} = \frac{\omega}{f} \tag{4.5.3}$$

where $\quad d\mu = \mu_v - \mu_r$

and $\quad \dfrac{d\mu}{(\mu - 1)} = \omega,$

the dispersive power.

When two thin lenses of focal lengths f_1 and f_2 are in contact, the focal length F of the combination is given by

$$\frac{1}{F} = \frac{1}{f_1} + \frac{1}{f_2}$$

On differentiation, $\quad -\dfrac{1}{F^2}dF = -\dfrac{1}{f_1^2}df_1 - \dfrac{1}{f_2^2}df_2 \qquad (4.5.4)$

Now, for the extreme rays to be brought to one focus, i.e., to form an *achromatic combination*, the change in focal length dF of the combination must be *zero*. So, we have from (4.5.4),

$$\frac{df_1}{f_1^2} + \frac{df_2}{f_2^2} = 0 \tag{4.5.5}$$

$$\Rightarrow \boxed{\frac{\omega_1}{f_1} + \frac{\omega_2}{f_2} = 0}, \quad \text{using (4.5.3)} \tag{4.5.6}$$

where ω_1 and ω_2 are the dispersive powers of the materials of lens 1 and 2 respectively and are *positive*.

From (4.5.6), therefore, it follows that f_1 and f_2 should be of *opposite* sign. So for achromatism, the **conditions** are: (a) *the combination should consist of* **one convex** *and the* **other concave** *lens,* (b) *they must be of* **different materials** ($\because \omega_1 \neq \omega_2$) and (c) *satisfy the relation* (4.5.6).

4.6 Achromatism of two thin lenses separated by a distance

When two thin lenses of focal lengths f_1 and f_2 are placed t apart, the focal length F of the combination is given by

$$\frac{1}{F} = \frac{1}{f_1} + \frac{1}{f_2} - \frac{t}{f_1 f_2}$$

Differentiating the above expression we get,

$$-\frac{1}{F^2}dF = -\frac{1}{f_1^2}df_1 - \frac{1}{f_2^2}df_2 + t\left[\frac{df_1}{f_1^2 f_2} + \frac{df_2}{f_1 f_2^2}\right] \tag{4.6.1}$$

Now, for a single lens, we have

$$\frac{1}{f} = (\mu - 1)k,$$

$$\text{where } k = \left(\frac{1}{R_1} - \frac{1}{R_2}\right) = \frac{1}{\mu - 1}\cdot\frac{1}{f}$$

$$\therefore \quad -\frac{1}{f^2}df = d\mu \cdot k = \frac{d\mu}{\mu - 1}\cdot\frac{1}{f} = \frac{\omega}{f} = \frac{\text{dispersive power}}{\text{focal length}}$$

If ω_1, ω_2 be the dispersive powers for the lenses 1 and 2 respectively and ω that of the combination, then we get from (4.6.1),

$$\frac{\omega}{F} = \frac{\omega_1}{f_1} + \frac{\omega_2}{f_2} - t\left[\frac{\omega_1}{f_1 f_2} + \frac{\omega_2}{f_1 f_2}\right]$$

For *achromatism*, ω must be zero. This implies

$$\boxed{\frac{\omega_1}{f_1} + \frac{\omega_2}{f_2} = \frac{(\omega_1 + \omega_2)}{f_1 f_2}t} \tag{4.6.2}$$

which is the **required condition** for achromatism.

If the lenses are **in contact,** $t = 0$ and we get from (4.6.2),

$$\boxed{\frac{\omega_1}{f_1} + \frac{\omega_2}{f_2} = 0,}$$

the same relation as (4.5.6) obtained before.

If the two lenses are of the **same material** and placed t apart, then from (4.6.2), imposing $\omega_1 = \omega_2 = \omega$ say, we get

$$\frac{\omega}{f_1} + \frac{\omega}{f_2} = 2t\frac{\omega}{f_1 f_2}$$

$$\Rightarrow \quad f_1 + f_2 = 2t$$

$$\Rightarrow \quad \boxed{t = \tfrac{1}{2}(f_1 + f_2)} \tag{4.6.3}$$

which is rather an **important** relation that shows that *it is possible to form an achromatic combination of two lenses made of the same material when separated by a distance equal to half the sum of their focal lengths.*

There is also an added *advantage* in this case. The combination is ***truly achromatic***, for the expression (4.6.3) does *not* contain the dispersive power of the material of the lenses so that the aberration is eliminated for all colours. Also, since t is to be positive, either *both* the *lenses* should be *convex* or the *one with higher focal length* should be *convex*.

4.7 Rainbows : Primary and secondary

A *rainbow* is an interesting natural phenomenon consisting of curved patterns of spectral colour hanging in the sky. When sunlight falls on small droplets of water suspended in the atmosphere during, before, or after a rain, it suffers refraction, total internal reflection and dispersion. If the observer stands with his back towards the sun and the water droplets are in front, he or she may see a rainbow, sometimes even *two*. The *inner bow* is called the **primary rainbow** and the *outer one* the **secondary rainbow**.

While the *primary rainbow* originates due to *one* internal reflection of the rays before emerging out of the drop, the **secondary rainbow** is formed by rays undergoing *two* internal reflections before emerging out of the drop. The primary rainbow is brighter than the secondary one and its radius subtends a mean angle of about 41° at the observer's eye. The primary bow shows the solar spectrum pattern with *red on its outer edge* and *violet on the inner*. The secondary rainbow, on the other hand, is much less bright and its radius subtends a mean angle of about 52° at the observer's eye and the order of colours is reversed, i.e., *red in the inner circle, violet in the outer.*

4.7.1 Theory of formation of primary rainbow

To develop the theory, we shall first calculate the deviation suffered by a ray undergoing *one* internal reflection.

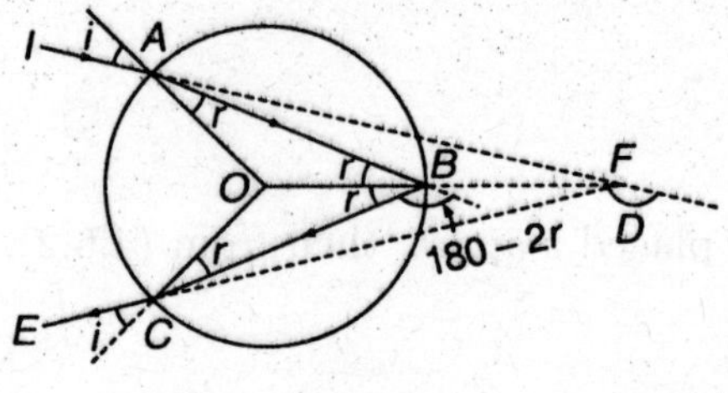

Fig 4.7

Let IA be the ray incident at an angle i on the surface of a drop (assumed spherical) of water of index μ in a plane passing through the centre O (Fig. 4.7). It suffers refraction at A at an angle r and an internal reflection at B to emerge finally at C by refraction along CE. Let the incident and emergent ray meet, on being extended, at F so that the ray IA suffers, due to *one* internal reflection, a deviation through an angle D.

Now, the deviation of the ray IA at A is $(i - r)$ and that at B is $(180° - 2r)$, and finally at C it is $(i - r)$. So, the total deviation

$$D = (180° - 2r) + 2(i - r) = 180° + 2i - 4r \tag{4.7.1}$$

Clearly,

$$D_{\max} = 180°, \text{ corresponding to } i = r = 0.$$

The minimum value of D is obtained by equating $dD/dr = 0$. Using (4.7.1),

$$\frac{dD}{dr} = 2\frac{di}{dr} - 4 = 0. \text{ So the condition for } D_{\min} \text{ is } \frac{di}{dr} = 2$$

Now, from Snell's law, $\sin i = \mu \sin r$ and we get on differentiation,

$$\cos i \frac{di}{dr} = \mu \cos r$$

$$\Rightarrow 2\cos i = \mu \cos r, \quad \text{for } D_{\min}$$

$$\Rightarrow 4\cos^2 i = \mu^2 \cos^2 r = \mu^2(1 - \sin^2 r)$$

$$\Rightarrow 4(1 - \sin^2 i) = \mu^2(1 - \sin^2 i/\mu^2)$$

$$\Rightarrow 3\sin^2 i = 4 - \mu^2$$

$$\therefore \quad \boxed{\sin i = \sqrt{(4 - \mu^2)/3}} \tag{4.7.2}$$

This i-value corresponds to the rays suffering *minimum deviation,* and it turns out that *at minimum deviation, the intensity of light is maximum.*

Now, $\mu_r = 1.329$ and $\mu_v = 1.343$ for water. From (4.7.2), using these values of μ's, the angle of incidence is 59.65° for red rays and 58.8° for violet. The corresponding angles of refraction are 40.5° and 39.6° respectively.

$$\therefore \quad D_r = 180° + 2 \times 59.65° - 4 \times 40.5° = 137.3°$$

$$D_v = 180° + 2 \times 58.8° - 4 \times 39.6° = 139.2°$$

So, it follows that the emergent rays that suffer *one* internal reflection will all be contained in a right circular cone of semi-vertical angle = (180° − 137.3°) = 42.7° for red rays and (180° − 139.2°) = 40.8° for violet rays and for other intermediate colours they will be sandwiched between them (Fig. 4.8). Note that the rays are packed near the surface of the cone than elsewhere within it. The reason is that the angle of deviation varies extremely slowly with the change in angle of incidence near the minimum deviation position.

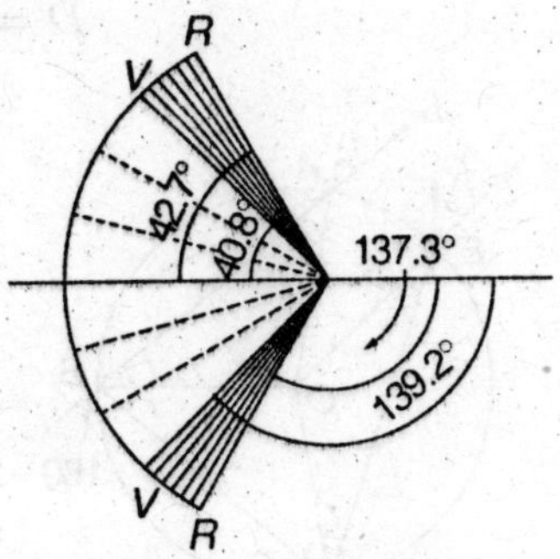

Fig 4.8

Consider the cone with PC (Fig. 4.9) as its axis and semi-vertical angle 42.7°, PC being parallel to the direction of sun rays. All the droplets like O_2 on the surface of this cone will return the light to P at 42.7° and they will be predominantly *red.* So, to an observer at P, the red rays will appear to come from a circle with O_2C as

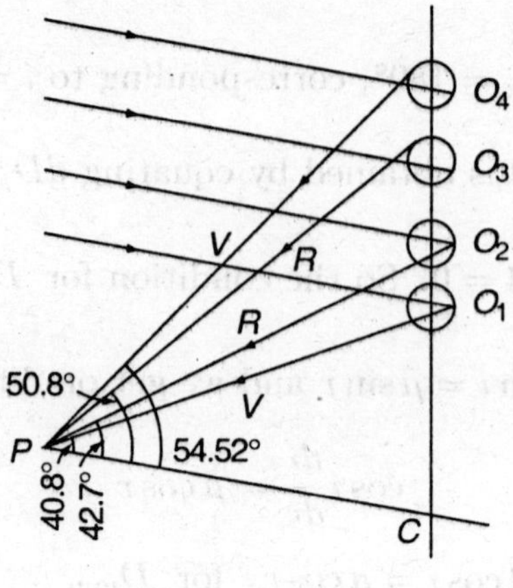

Fig 4.9

radius that subtends an angle of 42.7°. Similarly, the violet rays will appear to come to the observer from a circle with O_1C as radius that subtends an angle of 40.8°. From the ground level, the observer will usually see *only* (since the point C lies below the horizon) the different coloured *arcs* of the rainbow—*the red arc on the outer side of the cone* and the *violet one on the inner.* This is how a **primary rainbow** originates.

• A complete circular rainbow however can be made visible from an elevated position, e.g., from an aeroplane.

Formation of a secondary rainbow—As already indicated, the **secondary rainbow** originates if light suffers two internal reflections before coming out of the rain drop.

Fig. 4.10 illustrates the case of a ray undergoing *two* internal reflections. In this case, the total deviation of the incident ray IA is, by a similar method of computation,

$$D = 2(i - r) + 2(180° - 2r) = 360° + 2i - 6r$$

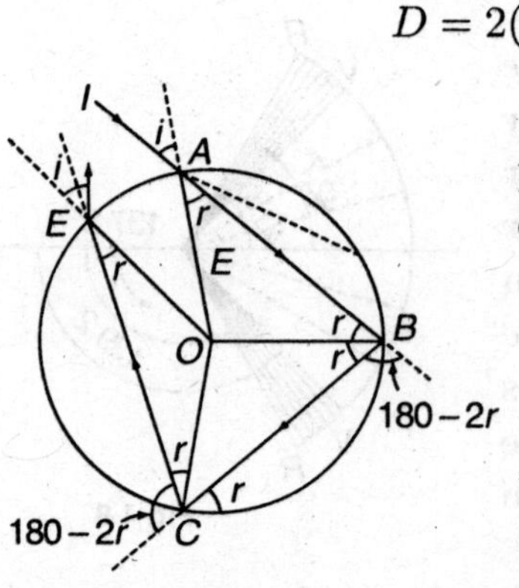

Fig 4.10

An analysis, similar to the one made in the case of primary rainbow, will show that the *minimum deviatioon* corresponds to an angle of incidence i, given by

$$\boxed{\sin i = \sqrt{9 - \mu^2/8}} \qquad (4.7.3)$$

On substitution of the values of μ_r and μ_v for water in the above relation and using Snell's law, one readily obtains, as before, i_r, i_v, r_r and r_v. Then, the computed minimum deviations are

$$D_r = 180° + 50.8°; \quad D_v = 180° + 54.52°$$

The rays would thus be packed in a cone whose external semi-vertical angle is 54.52° for violet rays and 50.8° for red rays (Fig. 4.11). The order of the colour would thus be *reversed—red on the inner side* and *violet on the outer.* The rays are densely packed in the peripheral part of the cone.

All droplets like O_4 that lie on a circle, with centre at C and the radius subtending an angle of 54.5° at the observer's eye P, will reflect *violet light* after *two* internal reflections. Similarly, all droplets like O_3 lying on a circle making an angle of 50.8° will reflect *red light* to the observer P. Thus, the *secondary rainbow* (Fig. 4.9) will be formed and would be *situated above the primary* bow. For a clearer understanding Fig. 4.12 is illustrative.

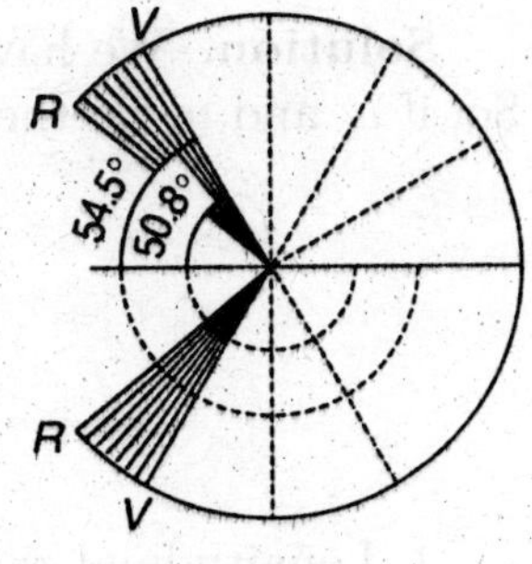

Fig 4.11

Note 1. It is interesting to see that the space between the primary and the secondary bow appears *darker* relative to the rest of the sky, since light reflected (either single or double internal reflection) from droplets of this space cannot reach the observer as the angle of deviation would have to be smaller than the minimum deviation.

Note 2. Two observers *do not* see the same rainbow! For a primary rainbow to be visible to the observer, the emergent rays must subtend a mean angle of 41° (for secondary, it is 52.5°) at the observer's eye. The position of drops that *can do* this depends on the position of the observer. Hence.

Note 3. The common centre C (Fig. 4.12) of the primary and the secondary rainbows is situated on the extension of the line from the sun to the observer's eye at P. We also have the fact that the angular radius of the outer circle (red) of the primary is about 42°. So as the *elevation* of the sun will *increase* above the horizon, the point C will get *shifted down* causing *smaller and smaller part of the bow to be visible.* An observer on the ground level cannot see the primary bow if the elevation of the sun exceeds 42°; for the secondary to be visible, the elevation of the sun must not exceed 54°. This is what is called the **effect of elevation of the sun** on the rainbows.

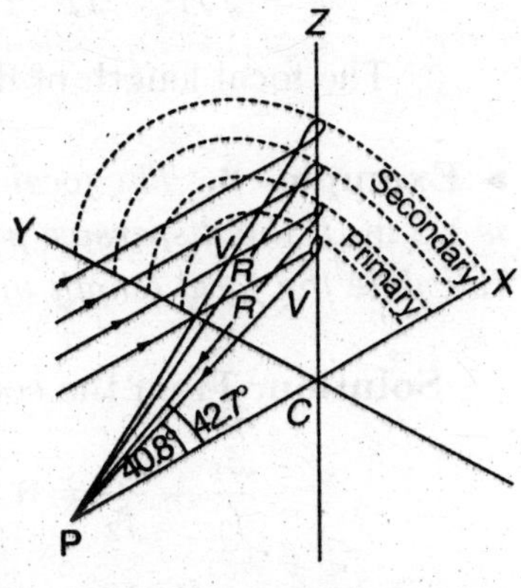

Fig 4.12

Note 4. If the observer of the rainbow moves in the upward direction, the line PC also would move *upward.* As a consequence, the common centre C of the rainbows will also *move up* and *more and more part* of the rainbow will *become visible.* This is what is called the **effect of elevation of the observer**. As already indicated, *one can see a completely circular rainbow* from an aircraft with the sun overhead.

4.8 Illustrated Examples

► **Example 1.** *A convex lens has a focal length* 15.50×10^{-2} m *for red colour. Its focal length for violet colour is* 14.45×10^{-2} m. *If an object is kept at a distance of* 0.4 m *from the lens, calculate the longitudinal chromatic aberration produced by the lens.* **(Meerut Univ.)**

Solution. We have $f_r = 15.50 \times 10^{-2}$ m, $f_v = 14.45 \times 10^{-2}$ m and $u = -0.4$ m. So, if v_r and v_v be the image distances for the red and violet light from the lens,

$$v_r = \frac{uf_r}{u+f_r} = \frac{-0.4 \times 15.50 \times 10^{-2}}{-0.4 + 15.50 \times 10^{-2}} = 0.2530 \text{ m}$$

$$\text{and} \quad v_v = \frac{-0.4 \times 14.45 \times 10^{-2}}{-0.4 + 14.45 \times 10^{-2}} = 0.2262 \text{ m}$$

∴ Longitudinal chromatic aberration = 0.2530 m − 0.2262 m = **0.0268 m**

► **Example 2.** *Calculate the focal length of a lens of flint glass of dispersive power* 0.4 *which will make an achromatic doublet with a converging lens of crown glass of focal length* 0.4 m *and dispersive power* 0.2. **(Kanpur Univ.)**

Solution. Here $f_1 = 0.4$ m, $\omega_1 = 0.2$, $\omega_2 = 0.4$, $f_2 = ?$
The condition to be satisfied for achromatism is

$$\frac{\omega_1}{f_1} + \frac{\omega_2}{f_2} = 0 \quad \Rightarrow \quad f_2 = -\frac{\omega_2}{\omega_1} f_1 = -\frac{0.4}{0.2} \times 0.4 = -0.8 \text{ m}$$

∴ The focal length of flint glass lens is **0.8 m** and it is a **concave lens.**

► **Example 3.** *The focal length of an achromatic combination of two lenses in contact is* 1.5 m. *If the dispersive powers of the materials of the two lenses are* 0.018 *and* 0.027, *calculate the focal length of the lenses.* **(Delhi and Calcutta Univ.)**

Solution. From the condition of achromatism, we have

$$\frac{\omega_1}{f_1} + \frac{\omega_2}{f_2} = 0 \quad \Rightarrow \quad \frac{1}{f_1} = -\frac{\omega_2}{\omega_1} \cdot \frac{1}{f_2} = -\frac{0.027}{0.018} \cdot \frac{1}{f_2} = -\frac{3}{2} \cdot \frac{1}{f_2} \qquad (4.8.1)$$

since here, $\omega_1 = 0.018$ and $\omega_2 = 0.027$.

$$\text{Also,} \quad \frac{1}{f_1} + \frac{1}{f_2} = \frac{1}{1.5} \Rightarrow -\frac{3}{2} \frac{1}{f_2} + \frac{1}{f_2} = \frac{1}{1.5}, \quad \text{using (4.8.1)}$$

$$\Rightarrow \frac{1}{f_2}\left(1 - \frac{3}{2}\right) = \frac{1}{1.5}$$

$$\Rightarrow \boldsymbol{f_2} = 1.5 \times (1 - 3/2) = \mathbf{-0.75 \text{ m}}$$

∴ From the relation (4.8.1), we get

$$\frac{1}{f_1} = -\frac{3}{2} \times \left(\frac{1}{-0.75}\right) = 2 \Rightarrow \boldsymbol{f_1} = \mathbf{0.5 \text{ m}}$$

The achromatic combination thus consists of a **convex lens** of focal length **0.5 m** in contact with a **concave lens** of focal length **0.75 m.**

▶ **Example 4.** *The mean indices of two specimens of glass are* 1.52 *and* 1.66 *respectively. The difference in indices for the same two lines of the spectrum are* 0.013 *for the first lens and* 0.022 *for the second. Find the focal length of lens of the second glass which when combined with a convex lens of* 0.5 m *focal length will make an object achoromatic for these two lines.* **(Calcutta Univ.)**

Solution. Assuming ω_1 and ω_2 to be the dispersive powers of the glass specimen 1 and 2, and taking the two lines as violet and red,

$$\omega_1 = \frac{\mu_v - \mu_r}{\mu - 1}; \quad \omega_2 = \frac{\mu'_v - \mu'_r}{\mu' - 1}$$

By the problem, $\mu_v - \mu_r = 0.013$ and $\mu = 1.52$, for the first lens and $\mu'_v - \mu'_r = 0.022$ and $\mu' = 1.66$, for the second lens.

$$\therefore \ \omega_1 = \frac{0.013}{1.52 - 1} = \frac{0.013}{0.52} = \frac{1}{40};$$

$$\omega_2 = \frac{0.022}{1.66 - 1} = \frac{0.022}{0.66} = \frac{1}{30}$$

Imposing the condition of achromatism,

$$\frac{\omega_1}{f_1} + \frac{\omega_2}{f_2} = 0, \quad \text{we get:} \ \frac{1}{40} \times \frac{1}{0.5} = -\frac{1}{30} \times \frac{1}{f_2}$$

$$\therefore \quad f_2 = -\frac{40 \times 0.5}{30} = -\frac{2}{3} = \mathbf{-0.667\ m}$$

The negative sign shows that the lens should be **concave.**

▶ **Example 5.** *The refractive indices of a flint glass for red and violet light are* 1.613 *and* 1.632 *respectively. What will be the angular dispersion produced by a thin prism of refracting angle* 5° *of flint glass?*

Solution. The deviation of the red light is given by

$$\delta_r = (\mu_r - 1)A$$

The deviation of the violet light is similarly given by

$$\delta_v = (\mu_v - 1)A$$

$$\therefore \quad \text{Angular dispersion} = \delta_v - \delta_r = (\mu_v - \mu_r)A$$

$$= (1.632 - 1.613) \times 5^\circ$$

$$= \mathbf{0.095^\circ}$$

▶ **Example 6.** *The dispersive power of crown glass is* 0.03 *and that of flint glass is* 0.05. *The indices for yellow light (mean) for these glasses are* 1.517 *and* 1.621 *respectively. An achromatic combination of prisms of crown and flint glasses produces a deviation of* 1° *for the yellow ray. Find the angles of the prisms.*

Solution. Let A and A' be the angles of the crown prism and the flint prism respectively.

For achromatic combination of the prisms, we have

$$\frac{\omega}{\omega'} = \frac{\delta'}{\delta} = \frac{(\mu'-1)A'}{(\mu-1)A}$$

$$\Rightarrow \frac{A'}{A} = \frac{(\mu-1)\omega}{(\mu'-1)\omega'}$$

where ω and ω' are the dispersive powers, δ an δ' the deviations of mean ray, and μ and μ' the indices of crown and flint prism respectively.

$$\therefore \quad \frac{A'}{A} = \frac{(1.517-1)\times 0.03}{(1.621-1)\times 0.05} = \frac{0.517\times 0.03}{0.621\times 0.05} = 0.50 \qquad (4.8.2)$$

Also, $\delta = (\mu-1)A$; $\delta' = (\mu'-1)A'$; and $\delta - \delta' = 1°$, by the problem.

$$\therefore \quad 1° = 0.517A - 0.621A' \qquad (4.8.3)$$

Solving (4.8.2) and (4.8.3), we obtain $\boldsymbol{A = 4.8°}$ (crown); $\boldsymbol{A' = 2.4°}$ (flint).

► **Example 7.** *Two lenses of focal lengths* 18 cm *and* 14 cm *are placed at a certain distance apart. Calculate the separation between them if they form a combination free from chromatic aberration.*

Solution. We have, $f_1 = 18$ cm and $f_2 = 14$ cm. If t be the required separation between the lenses for achromatism,

$$t = \frac{f_1+f_2}{2} = \frac{18+14}{2} = \mathbf{16\ cm}$$

► **Example 8.** *It is desired to make a converging achromatic lens of mean focal length* 30 cm *by using two lenses A and B of different materials. If the dispersive powers of A and B are in the ratio* 1 : 2, *find the focal length of each lens.* **(Mysore Univ.)**

Solution. By the problem, $\omega_1/\omega_2 = 1/2$. And from the condition of achromatism,

$$\frac{\omega_1}{f_1} + \frac{\omega_2}{f_2} = 0 \quad \Rightarrow \quad \frac{1}{f_2} = -\frac{\omega_1}{\omega_2}\cdot\frac{1}{f_1} = -\frac{1}{2}\cdot\frac{1}{f_1} \qquad (4.8.4)$$

We also have, $\dfrac{1}{f_1} + \dfrac{1}{f_2} = \dfrac{1}{F} = \dfrac{1}{30}$

$$\Rightarrow \quad \frac{1}{f_1} - \frac{1}{2f_1} = \frac{1}{30} \quad \Rightarrow \quad \frac{1}{2f_1} = \frac{1}{30} \quad \Rightarrow \quad f_1 = \mathbf{15\ cm}$$

$\therefore$ From (4.8.4), we obtain

$$\frac{1}{f_2} = -\frac{1}{2}\cdot\frac{1}{15} = -\frac{1}{30} \quad \Rightarrow \quad f_2 = \mathbf{-30\ cm}$$

Review Questions

1. What is meant by dispersion? Define angular dispersion and dispersive power. **(Calcutta and Punjab Univ.)**

2. What is an achromatic prism? Explain how two narrow-angled prisms of different dispersive powers be combined to produce deviation without dispersion.

3. Deduce the condition for a prism-combination to give dispersion without deviation. Calculate the resultant dispersion. **(Calcutta Univ.)**

4. Describe the theory, construction and the use of a direct vision spectroscope.

5. What is meant by chromatic aberration of a lens? Show that the longitudinal chromatic aberration = dispersive power × mean focal length for yellow light. **(Cochin and Burdwan Univ.)**

6. Find the condition of achromatism for two thin lenses in contact. **(Punjabi and Cochin Univ.)**

7. (a) Obtain the condition that must be satisfied in order that a combination of two lenses of focal lengths f_1 and f_2 at a distance d apart be achromatic. **(Poona and Kurukshetra Univ.)**

   (b) What will be the condition when the two lenses are of the same material? Is such a combination of lenses truly achromatic? **(Viswesaraiah Univ.)**

8. Explain the formation of a primary rainbow. In what order are the colours arranged? **(Guru Nanak Univ.)**

9. What is a secondary rainbow? Explain the formation of a secondary rainbow.

10. Write short explanatory notes on the following :

    (a) Dispersion and dispersive power

    (b) Direct vision spectroscope

    (c) Primary and secondary rainbows.

Short Answer and Objective Questions

1. What is dispersion? Define dispersive power.

2. What is meant by an achromatic combination of prisms?

3. Prove that $\omega = d\mu/(\mu - 1)$. To which colour does μ in the denominator of the expression correspond?

4. Is it possible to have (a) deviation without dispersion and (b) dispersion without deviation in a combination of prisms? Explain.

5. Write down the conditions, if any, in case (a) and (b) as cited in Q. 4 above, are at all possible. Name an instrument where we have the situation (b) above.

6. Explain with a diagram the longitudinal and lateral chromatic aberration in a lens.

7. What is achromatism? Write down (do not deduce) the condition of achromatism for two thin lenses (a) in contact and (b) separated by a distance t.

8. In case of (b) above, what will be the condition when the two lenses are of the same material? Is such a combination of lenses truly achromatic?

9. What are primary and secondary rainbows? In what order are the colours arranged? Is the order the same in both the above types of rainbows?

10. What is the effect of (a) elevation of the sun and (b) elevation of the observer, on the visibility of rainbows?

11. Why is it that the space between the primary and the secondary rainbows appear darker than the rest of the sky? **(Aligarh Univ)**

12. Will two observers see the same rainbow? Explain the matter. **(Punjab Univ.)**

13. Soap bubbles appear to be coloured due to the phenomenon of
    (a) polarisation (b) interference
    (c) dispersion (d) diffraction. **(Himachal Univ.)**

14. Dispersion is maximum for light of colour
    (a) violet (b) red (c) yellow (d) blue.

Problems

1. A compound achromatic lens of focal length 40 cm is to be made from two different glasses. Find the focal lengths of the component lenses if the dispersive power of crown glass is 0.21 and that of flint glass is 0.46. [+21.71 (cr); −47.62 (fl)]

2. The dispersive powers of crown and flint glasses are 0.03 and 0.05 respectively. If the difference in indices of blue and red colours is 0.014 for crown glass and 0.023 for flint glass, calculate the angles of the two prisms for a deviation (without disperson) of 10°. [53.6°; 32.6°]

3. For a crown and flint glass for C and F lines $\mu_C = 1.515$, $\mu_F = 1.523$ and $\mu'_C = 1.644$, $\mu'_F = 1.664$ respectively. Calculate the angle of flint glass prism which may be combined with crown glass prism having refracting angle 20° so that the combination is achromatic for C and F rays. [8.0°]

4. A telescope objective is made by cementing together two lenses, the cemented surfaces having the same radii of curvature. The combination is plano-convex and is of focal length 1.0 m. Find the radii of curvature of the component lenses. Given : refractive index and dispersive power of crown glass are 1.50 and 0.02 respectively and those of flint glass are 1.60 and 0.03. **(Punjab Univ.)**

5. An achromatic objective of focal length 30 cm is to comprise of two thin lenses cemented together with canada balsam and are made from hard crown and dense flint glasses respectively. The free surface of the divergent lens is plane. Calculate the radii of curvature of the surfaces of both the lenses. Given : $\mu_v = 1.5231$, $\mu_r = 1.5171$ for crown glass and $\mu'_v = 1.6347$, $\mu'_r = 1.6224$ for flint glass.
[$R_1 = -13.28$ cm, $R_2 = \infty$, (flint : concave);
$R_1 = +13.28$ cm, $R_2 = -12.65$ cm (crown : convex)]

6. A crown glass prism of angle 5° is to be combined with a flint glass prism such that the mean ray passes undeviated. Find (a) the angle of flint prism needed and (b) the angular dispersion produced by the combination when white light passes through it. The indices for red, yellow and violet light are 1.514, 1.517 and 1.523 respectively for crown glass, and the corresponding values for flint glass are 1.613, 1.620 and 1.632. [4.2°; 0.0348° (magnitude)]

7. A certain material has μ-values 1.56, 1.60 and 1.68 for red, yellow and violet light respectively. Calculate (a) the dispersive power, and (b) the angular dispersion produced by a thin prism of angle 6° of this material. [0.2; 0.72°]

8. A flint glass prism and a crown glass prism are combined so as to make the deviation of the mean ray zero. Given, μ (flint) = 1.620, μ (crown) = 1.518 for the mean ray. The angle of flint prism is 6°. What is the angle of crown prism? [7.2°]

9. The object glass of a telescope is an achromat of focal length 90 cm. If the magnitude of the dispersive powers of the two lenses are 0.024 and 0.036, calculate their focal lengths. **(I.A.S.)** [30 cm, −45 cm]

10. Two glasses have dispersive powers in the ratio 2 : 3. These glasses are to be used in the manufacture of an achromatic objective of focal length 20 cm. What are the focal lengths of the lenses? **(Osmania Univ.)** [6.67 cm, −10 cm]

11. Calculate the focal lengths of the crown and flint glass lenses which form an achromatic doublet of focal length 50 cm when placed in contact. The refractive indices for the C and F lines are given as under.

Crown : $\mu_C = 1.515$, $\mu_F = 1.523$; Flint : $\mu_C = 1.644$, $\mu_F = 1.664$.

[24.79 cm (crown convex), −44.19 cm (flint concave)]

12. A prism of crown glass of angle 5° and mean index 1.51 is combined with a flint glass prism of index 1.65 to produce no deviation. Find the angle of the flint glass prism and the net dispersion. [3.92°; −0.0284]

Chapter 5

Lens Aberrations and Eyepieces

5.1 Introduction

The simple formulae that we derived earlier for lenses were based on the assumptions that (i) the *aperture is small* and (ii) *only paraxial rays* (i.e., rays close to the axis) are used. For this reason, we approximated $\sin\theta = \theta$ and obtained a point image of a point object close to the axis. In practical optical systems, however, the aperatures are usually large enough to admit more light from consideration of brightness. So, non-paraxial rays also participate in the image formation and $\sin\theta = \theta$ is no longer true. The rays from a given axial point object do not converge to or appear to dverge from a point image. This is an *aberration*, i.e., *a departure* of the actual image from the prediction of *first order* ($\sin\theta = \theta$) *theory*.

For non-paraxial rays, however, a first estimate of error may be obtained by taking $\sin\theta = \theta - \theta^3/3!$. The theory developed by Seidal on this basis is called *third order theory* and he found *five* distinct *types* of departure or aberrations of image from the one predicted by the first order theory. These defects are called **Seidal aberrations**.

The five aberrations are : (1) **spherical aberration**, (2) **coma**, (3) **astigmatism**, (4) **curvaure** and (5) **distortion**. These aberrations are present even with strictly monochromatic light and are collectively called **monochromatic aberrations**.

The other kind of important aberration in lenses is called **chromatic aberration** in which images formed by refraction with white (or composite) light become coloured. It arises out of dispersion of white light leading to variation of focal length with wavelength. This also therefore occurs in lenses with small aperture and using paraxial rays. Chromatic aberration has already been discussed in Chapter : *Dispersion*.

The correction or minimization of these aberrations is rather complicated and details are beyond the scope of the book. We shall rest content by making a rapid and short survey of the phenomenon.

- Monochromatic aberrations are found in both lenses and curved mirrors.
- Chromatic aberration is absent from images formed by a mirror.

5.2 Spherical aberration in lens

Let a point object O (Fig. 5.1) be placed on the axis of a **converging lens** of wide aperture and rays be incident at different zones of the lens. The *paraxial rays* from the *central portion* of the lens are brought to a focus at I_c and those *away from the axis*—the *marginal rays*—are brought to a focus at I_m, I_m being *nearer* to the lens. Rays falling at intermediate zones of the lens are focussed between I_c and I_m.

This failure of the lens to form a point-image of an axial point-object due to variation in the focal length of a lens when one passes from its central region to the rim is called **spherical aberration.** I_cI_m is a measure of it.

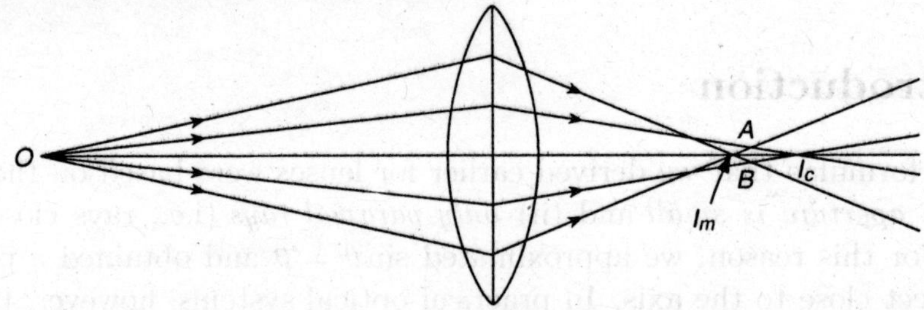

Fig. 5.1

Clearly, the image formed is *not sharp at any plane* on the axis. If a screen be placed normal to the axis at I_m, the *outer portions* of the object *are in focus* and we get a bright circular edge and a faint centre; at I_c the *central portions are only in focus* and we get a bright centre and a faint edge. As the screen is moved along the axis parallel to itself, it is seen that the *section* is *everywhere circular* and is *smallest* at AB where the paraxial and the marginal rays cross. *The smallest cross-section of almost uniform illumination, is called the* **circle of least confusion** and is the nearest approach to a point image. It thus corresponds to the *best image.*

The distance I_cI_m measures what is called the **axial** or **longitudinal spherical aberration**, and the *radius of the circle of least confusion gives the measure of* **lateral spherical aberration when the object is at infinity** i.e., parallel light is incident on the lens.

The paraxial focus I_c lies on the positive side of I_m, the marginal focus, and so the *longitudinal aberration* is called *positive*, and a *convex lens* is said to be *under-corrected.*

• In the above definition the object is taken at infinity, because the distance I_mI_c, varies with the position of object point O. So for comparison, it is always computed for the special case of parallel incident light.

Concave lens — The spherical aberration **for a concave lens** is illustrated in Fig. 5.2 When a parallel beam of light is incident on different zones of the lens, the paraxial rays from the central zone appear, after refraction, to diverge from the point F_c which is away from the lens than the point F_m where the marginal rays are brought to a focus. Thus F_c lies on the *negative* side of F_m and the spherical aberration is called *negative* and the concave lens is said to be *over-corrected.*

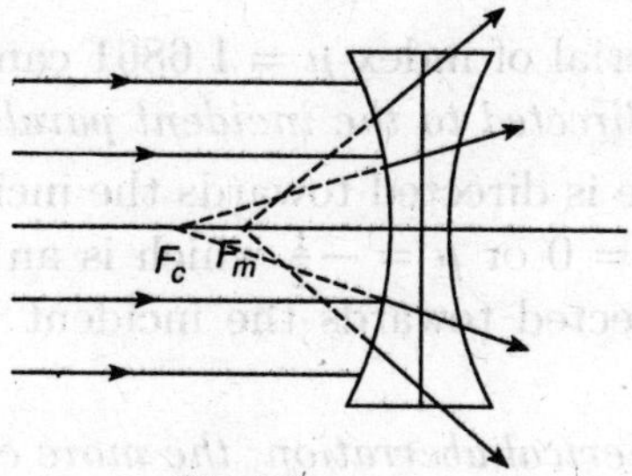

Fig. 5.2

Note. To have an idea of the magnitude, consider a equiconvex lens of $f = 10$ cm. The longitudinal spherical aberration for the region 1 cm from axis is 1.5 mm, i.e. 1.5% of f, the focal length.

5.2.1 Minimization of spherical aberration

Spherical aberration for a lens *cannot* altogether *be removed.* It can however be *minimized* by a number of ways.

1. By using stops — If the marginal or paraxial rays are cut-off, the rest come practically to a point focus to give a well-defined image. In a camera lens, the aperture is reduced to a small region around the centre of the lens, while in a telescope objective having large aperture, the central portion of the the lens is covered *using a stop* in front of the lens at the cost of reduced light intensity (Fig. 5.3).

Rayleigh showed that stopping of the central portion of a lens is superior to stopping of the marginal portion for, in this case, the resolving power of the lens is not reduced.

2. By a proper choice of radii of curvature of lens surfaces — It can be shown analytically that for minimum spherical aberration.

$$\frac{R_1}{R_2} = \frac{2\mu^2 - \mu - 4}{\mu(1+2\mu)} \quad \Rightarrow \quad \frac{R_1}{R_2} = -\frac{1}{6} \quad \text{for } \mu = 1.5.$$

where R's are the radii of curvature and μ the index of the lens, the more curved surface facing the more parallel beam. These lenses, designed to give minimum spherical aberration, are called **crossed lenses.** The negative sign indicates that the lens must either be double convex or double concave.

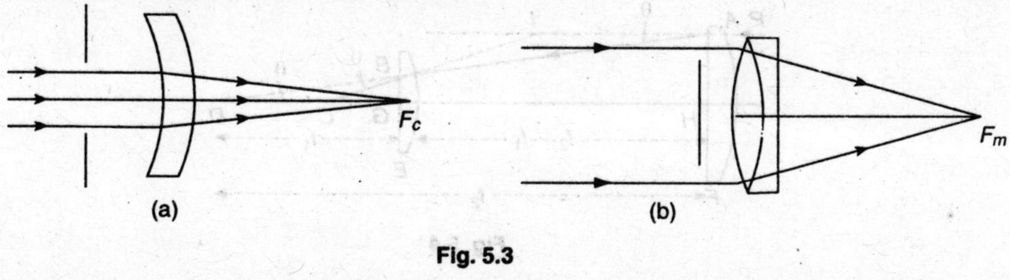

Fig. 5.3

• For instance, with a plano-convex lens with the curved surface directed towards the incident parallel rays, $R_2 = \infty \Rightarrow 2\mu^2 - \mu - 4 = 0$. Solving, $\mu = 1.6861$. Thus, a plano-convex lens of a material of index $\mu = 1.6861$ can *minimise spherical aberration when its curved surface is directed to the incident parallel rays.*

If however the plane face is directed towards the incident parallel rays, $R_1 = \infty \Rightarrow \mu(2\mu + 1) = 0 \Rightarrow$ either $\mu = 0$ or $\mu = -\frac{1}{2}$ which is an absurdity. So, a plano-convex lens with its plane face directed towards the incident parallel rays, cannot minimise spherical aberration.

Thus *for minimising spherical aberration, the more curved surface of the lens should be directed either* (i) *towards the incident or* (ii) *towards the emergent rays whichever is more parallel to the axis.*

3. By using a combination of lenses — The spherical aberration is greater when the angle of incidence is greater. Also, when the angle of incidence is greater, the deviation is greater. If therefore the total deviation is distributed amongst a number of refracting surfaces such that the deviation is small at each surface, the spherical aberration can be reduced. This can be achieved *by using a number of lenses and making equal deviation in each lens.*

(a) If two convergent lenses, separated by a distance d, are used and there is equal deviation due to refraction at all the surface, then the *spherical aberration is minimum* under the condition

$$f_2 \sim f_1 = d$$

where f_2, f_1 are the focal lengths of the lenses. This principle is used in **Huygens' eyepiece.**

(b) If a *convex lens* of *positive* spherical aberration be properly combined with a *concave lens of negative* aberration then the two abberations may cancel out. This device is however seldom used as it works only for a particular pair of object and image for which it has been devised.

4. By using aplanatic surfaces — This has been subsequently discussed in greater details.

• **Deduction of the condition for equal deviation** — The ray PA (Fig. 5.4) parallel to the axis is deviated through θ by refraction at A on the lens F. Another deviation through θ occurs in it by refraction at B on lens E before its final emergence along BC. Without the lens E, the emergence would have been along AD.

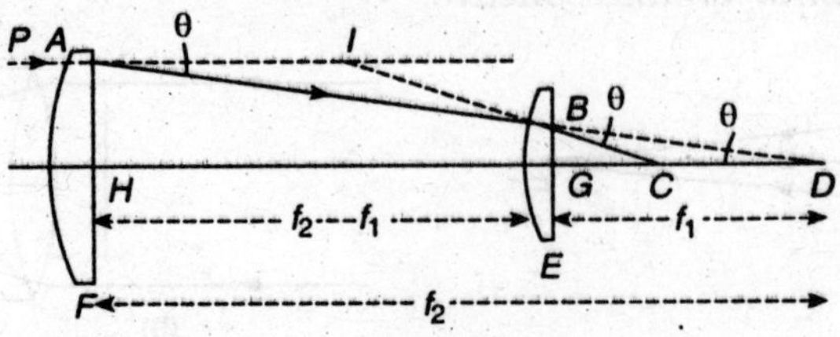

Fig. 5.4

From geometry, $\angle IAB = \angle BDC = \angle DBC = \theta \Rightarrow BC = CD$.

If B is *very near* G, then $BC = CG - CD = DG/2$.

But D acts as a virtual object for lens E of focal lenth f_1 and its image is formed at the point C.

$$\therefore \quad DG = +u, \quad CG = v = u/2$$

$$\therefore \quad \frac{1}{f_1} = \frac{1}{v} - \frac{1}{u} = \frac{2}{u} - \frac{1}{u} = \frac{1}{u} \Rightarrow f_1 = u = DG$$

Since PA is parallel to the axis, it would have converged to the point D, if there were no eye lens.

$\therefore HD = f_2 \Rightarrow HG = HD - GD = f_2 - f_1 = d$, separation between the lenses.

So, *for equal deviation, at the surface of the lens F and at that of the lens E, the separation between the lenses must equal the difference between their focal lengths.*

$$\therefore \quad \boxed{f_2 - f_1 = d}$$

5.3 Coma or Comatic error

In discussing *coma,* we assume that there is *no other* aberration present.

Comatic error deals with image formation of object points *off the axis.* Fig. 5.5 illustrates the *effect of coma* which arises because different zones of a lens produce different lateral magnifications.

Let PP' be the principal axis of the lens and Q an object point *off the axis.* Of the pencil of rays from Q, consider the principal ray that passes through the centre of the lens to be focussed at Q' called the *paraxial image of* Q. The hollow cone of rays 11 through the inner cone of the lens are focussed at Q'_1, slightly to the *left* of the image plane $P'Q'$, for the lens has a shorter focal length for non-paraxial rays. This cone of rays on emerging from Q'_1 will produce a *small bright circle* on the screen if placed at $P'Q'$, the dia of which depends on the distance between Q'_1 and $P'Q'$ and increases with the radius of the zone.

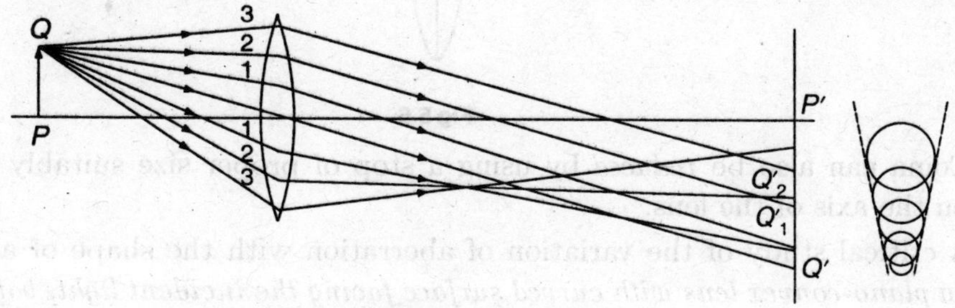

Fig 5.5

Similarly, the rays 22 passing through a zone of larger radius will be imaged at Q'_2, giving a *larger bright circle* on the screen at $P'Q'$. Rays 33 through the outer zone will give a still larger bright circle. This will lead to unequal magnifications for the different zones and the rays passing through various annular zones of the lens will give rise to *an image consisting of a series of overlapping circles* with the smallest point circle starting at Q' and the largest one touching P'. The *total effect, as shown, is to give the image a* **comet-like** *appearance and hence named* **coma**. The image looks brightest at Q', the intensity falling off upward in the direction of P'.

• Coma endows the image an asymmetrical shape and is highly objectionable in optical instruments using large apertures e.g., wide-angle photographic lenses, microscope objectives, etc.

• A lens corrected for spherical aberration produces a point-image of a point object situated on its principal axis. It still fails to give a point-image of a point object situated *off the axis.*

5.3.1 Reduction or removal of coma

1. Coma can be *completely eliminated* by designing a lens such that the lateral magnification is constant for all the zones. This is ensured by what is called **Abbe's sine condition** :

$$\boldsymbol{\mu y \sin\theta = \mu' y' \sin\theta'}$$

where y and y' are the lengths of the object and the image for the zone, θ and θ' the angles which the conjugate rays make with the axis and μ, μ' are the indices of the object and image space (Fig 5.6). It thus follows that the optical system is free from coma, if for all the zones.

$$\boldsymbol{\frac{\sin\theta}{\sin\theta'} = \frac{\mu' y'}{\mu y} = \text{const.}}$$

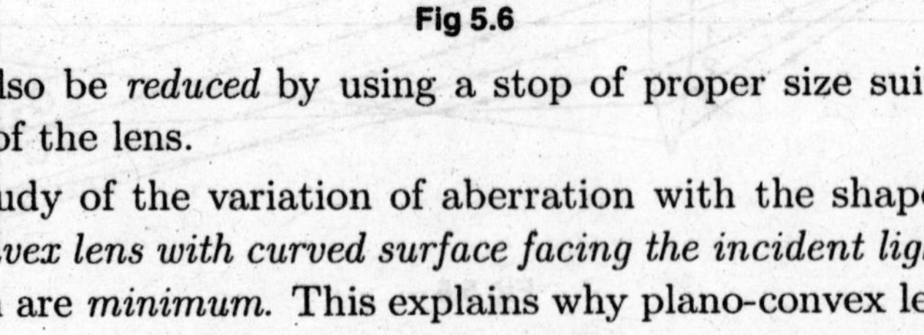

Fig 5.6

2. Coma can also be *reduced* by using a stop of proper size suitably located co-axially on the axis of the lens.

3. A critical study of the variation of aberration with the shape of a lens shows that *for a plano-convex lens with curved surface facing the incident light, both coma* and spherical aberration are *minimum.* This explains why plano-convex lenses are common in eyepieces.

4. The so-called **aplanatic lenses** are *free from coma* and spherical aberration *for* a particular pair of object and image points, called *aplanatic points.* But it is *not possible* to correct simultaneously both spherical aberration and coma for all the positions of the object.

5.4 Astigmatism

If spherical aberration and coma are totally eliminated, all points on or very close to the axis of the lens will form point images. However, if the object point be *well off the axis,* the rays from it on refraction through the lens *do not,* in fact, give a point image but a pair of mutually perpendicular lines separated by some distance. This failure leads to an aberration termed **astigmatism**, meaning '*not a point*'.

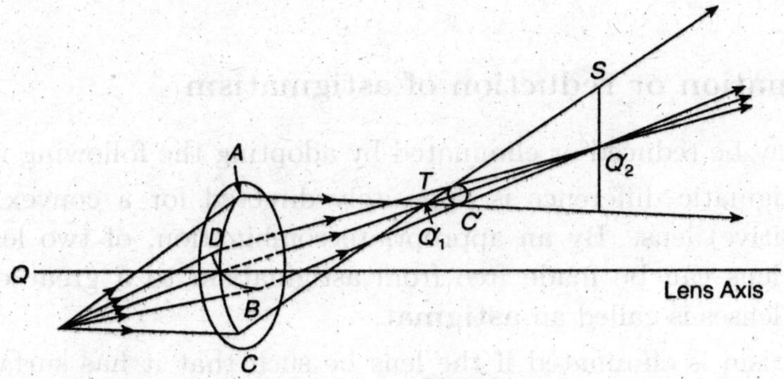

Fig. 5.7

Let the rays from a point object Q, considerably off the axis, fall on the lens $ABCD$ (Fig. 5.7). If there is no coma, the rays contained in the *vertical plane* (also called *meridional plane* QAC will give a point image Q_1' on the chief ray and those in the *horizontal plane* also called *sagittal plane* QBD give a point image at Q_2', also on the chief ray. As the latter rays making for Q_2' are still *converging* passing Q_1', the image at Q_1' called **primary image,** is a line T in the *horizontal plane* and normal to the chief ray. In a like manner, the rays in the vertical plane after intersecting at Q_1' *diverge* forming a line at S, called the **secondary image** that lies in the *vertical plane* and normal to both the chief ray and the primary image. Rays in oblique planes would also pass through the two line images T and S, called the *first* and the *second focal lines* respectively.

Nowhere, therefore, *sharp image* of Q is obtained. This type of aberration is called **astigmatism** that arises due to the fact that for an *off-axis* object point the focal length of the circular zone is *not* the same at every point of it — being a *minimum* in meridional section and *maximum* in the sagittal section. The section of the beam coming out of the lens is *elliptical* that degenerates into *straight lines* at the *primary* and *secondary images* and to a *circle* C' at a point more or less midway between them— the **circle of least confusion**–where, under the situation, the best possible image is obtained. The distance between T and S *along the chief ray* is called the **astigmatic difference** that increases with the obliquity of the rays.

• In the above discussion, we considered the image of one point only. If however the images of all the points of a plane object perpendicular to the axis are taken into consideration, each point will form its corresponding *primary* and *secondary image,* and also the *circle of least confusion.* The locus of primary images of these points is a *paraboloid surface* known as **primary image surface**, that of secondary images the **secondary image surface**, and the **surface of best focus** corresponds to *circles of least confunsion.*

• Astigmatism is called *positive* when the *primary* surface is formed *to the left of the secondary* surface, for a *convex lens.* For a *concave lens*, the situation is reversed— the *secondary surface* lies *to the left of the primary* surface and astigmatism is called *negative.*

5.4.1 Elimination or reduction of astigmatism

Astigmatism may be reduced or eliminated by adopting the following methods.

1. The astigmatic difference is oppositely directed for a convex (positive) and a concave (negative) lens. By an appropriate combination, of two lenses, therefore, the compound lens can be made free from astigmatism to a great extent. Such a combination of lenses is called an **astigmat**.

2. Astigmatism is eliminated if the lens be such that it has surfaces of different horizontal and vertical curvatures. Such a lens is known as a **toric lens**.

3. It is also possible to eliminate astigmatism by proper choice of elements of an optical system satisfying the following condition known as **Petzval condition :**

$$\boxed{\mu_1 f_1 + \mu_2 f_2 = 0.}$$

where μ_1, f_1 are the index and focal length of one lens and μ_2, f_2 are those of the other. The lenses may either be in contact or separated. To have the combination convergent, $f_1 > f_2$ where f_1 refers to *positive element* and the condition requires $\mu_1 > \mu_2$.

5.5 Curvature and distortion

Curvature—If an *extended plane object* is seen through a converging lens, the image formed by refraction through the lens is *not* flat. While the *central part* of the image is *in sharp focus* and *flat*, the *edges* are not. The edges are rather *quite blurred.* The reason is that the focal length of the lens for a paraxial beam $f_c > f_p$, the focal length for the oblique pencil. For a *virtual image*, as shown in Fig. 5.8(a), the image points well off the axis is drawn farther from the lens compared to those situated in the central paraxial part. The result is that the observed image is *curved outward.*

For a *real image*, however, the *curvature* of the image is *in the opposite direction* as illustrated in Fig. 5.8(b), i.e., the observed image is *curved inward.*

This aberration of the lens is known as the **curvature of the field** and occurs even though the aperture is reduced by using an appropriate stop to reduce spherical aberration, coma and astigmatism.

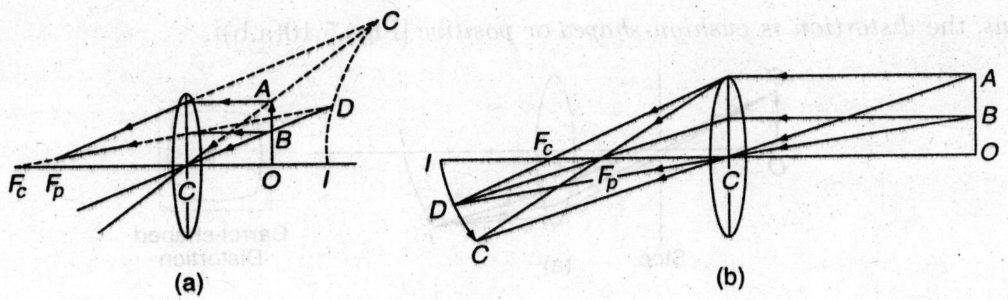

Fig 5.8

Curvature is closely associated with astigmatism. The elimination of astigmatism dictates that the primary and secondary image surfaces should be brought into coincidence. When this is done, the resulting astigmatic surface becomes usually *curved*. *To eliminate curvature* of the field, the primary and secondary image surfaces should have *equal and opposite curvature*. It is *not* possible to secure this with a *single lens* or a *contact doublet* of one positive and the other negative lens. It is only possible by a proper choice of elements of an optical system satisfying the **Petzral condition** and by using appropriate stop placed properly in front of the lens.

Distortion—For an extended object, the distances of the object points from the axis are different. Due to these differences, the lateral magnifications of the different parts are also different. The variations in this lateral magnification will render the image somewhat dissimilar from the object. This is what is called **distortion**.

In the case of virtual image in Fig. 5.8(a) the part DC of the image is more magnified than the part ID. But in Fig. 5.8(b) when real image is formed, ID is more magnified than the part DC. So, the image of a *rectangular object* say, a wire ganze, formed by a thick convex lens will appear **barrel-shaped** or *negative* when it is *real*, and **pin-cushion shaped** or *positive* when the image is *virtual*. Plainly, therefore, the *magnification of the central part* is much *more* than that of the *edges* in *negative distortion*. In *positive distortion*, however, the *magnification at the edges* is much *more* compared to that of the *central part* of the image (Fig. 5.9).

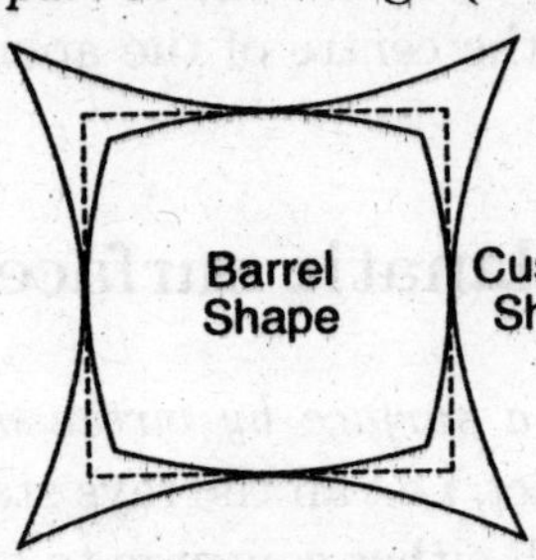

Fig 5.9

The type of distortion—positive or negative—even for a real image depends on the position of the stop. For instance, if it is placed *between the object and the lens*, the distortion is *barrel-shaped* or *negative*. If however the *stop* is *between* the *image and the lens*, the *distortion is cushion-shaped* or *positive* [Fig. 5.10(a,b)].

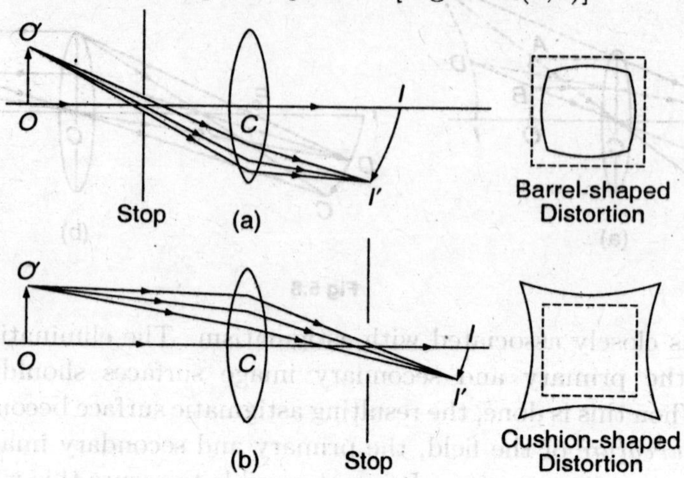

Fig 5.10

5.5.1 Elimination or reduction of distortion

Distortion can be eliminated by having a *compound lens* with *two elements and a stop* placed *appropriately* between them. In such a case, the *positive distortion* of the first lens is *nullified* by the *negative distortion* of the second lens.

By having an optical system with *two identical* lenses (Fig. 5.11) with a stop between them, the *positive pin-cushion distortion* of the first lens may be duly compensated by the *negative barrel shaped distortion* of the second lens, provided the *magnification* is −1. This method is employed in constructing a lens system in photographic cameras and projectors. Such a corrected lens-combination is called an **orthoscopic doublet** or *rapid rectilinear lens.*

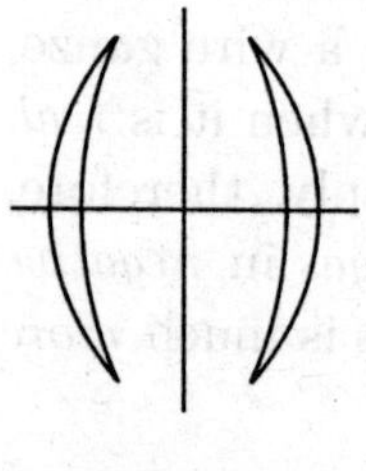

Fig 5.11

- A thin lens system shows no distortion if the optical centre is coincident with the centre of the aperture stop

5.6 Aplanatism : Aplanatic surface and points

Aplanatism *is the property of a surface by virtue of which it gives rise to an axial point image of an axial point object*, i.e., all the rays starting from an axial object point, after refraction at the surface, will either converge to or appear to diverge from another single axial point—the image point.

The *surface possessing this property of aplanatism is called an* **aplanatic surface**. The *two conjugate points–the object point and the image point*–are called the **aplanatic points**.

A *lens* possessing such a pair of conjugate points, free from both spherical aberration and coma, is called an **aplanatic lens**.

Aplanatism of a spherical refracting surface — We shall now show that a *spherical refracting surface is aplanatic with respect to a certain position of the object on the axis.*

Consider a spherical refracting surface with centre C, radius of curvature R (Fig. 5.12) and index μ. Let O be a point object on the axis, distant R/μ from C. The ray OA from O makes an angle θ_1 with the axis and after refraction at A bends away from the normal CN, along AB. If produced backward, BA meets the axis at I making an angle θ_2 with the axis. So I is the image of O.

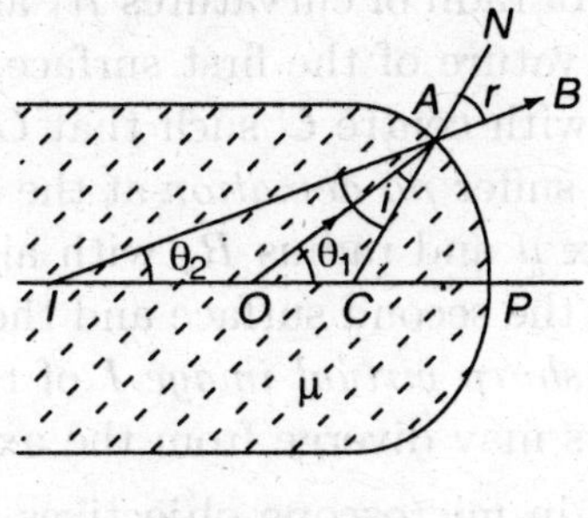

Fig 5.12

If i and r be the angles of incidence and refraction

$$\mu \sin i = \sin r \quad \Rightarrow \quad \sin i/\sin r = 1/\mu.$$

But from ΔAOC, $$OC/\sin i = AC/\sin\theta_1$$

$$\Rightarrow \quad \sin i/\sin\theta_1 = OC/AC = (R/\mu)/R = 1/\mu$$

$$\therefore \quad \frac{\sin i}{\sin r} = \frac{\sin i}{\sin\theta_1} \quad \Rightarrow \quad r = \theta_1$$

From ΔAOI, $$\theta_2 + r - i = \theta_1 \quad \Rightarrow \quad \theta_2 = i \quad (\because\ r = \theta_1)$$

From ΔACI, $$IC/\sin r = CA/\sin\theta_2$$

$$\Rightarrow \sin r/\sin\theta_2 = IC/CA = \sin r/\sin i = \mu$$

$$\therefore \quad IC = \mu CA = \mu R$$

Thus, *if the object is at a distance R/μ from the centre of curvature of the spherical refracting surface, its image is formed at a distance μR, irrespective of the slope angles θ_1 and θ_2.*

So, there shall be *no spherical aberration* and since the **sine condition** : $\sin\theta_1/\sin\theta_2 = \sin r/\sin\theta_2 = \mu =$ constant, is also *satisfied* there shall be *no comatic aberration* as well. So the refracting surface is aplanatic for O and I which are called its *aplanatic points.* Note that I, the image point, is *virtual.*

Aplanatic lens—The above property of spherical refracting surface may be used to construct an **aplanatic lens.**

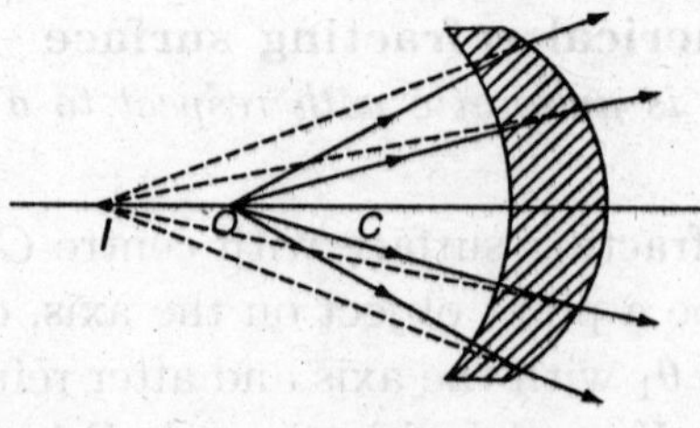

Fig 5.13

Consider a meniscus lens of radii of curvatures R_1 and R_2 (Fig. 5.13) and an object point at O, the centre of curvature of the first surface. Let O and I be the *aplanatic points* of the second surface with centre C such that $OC = R_2/\mu$ and $IC = R_2\mu$. So, all the rays diverging from O suffer *no deviation* at the first surface. O may be thought to be in the medium of index μ and radius R_2 with air as surrounding medium. The refraction now takes place at the second surface and the rays appear to diverge from I. The lens would thus form a *sharp virtual image* I of the object O, *irrespective of the slope angles* at which the rays may diverge from the axis.

Aplanatic lens is utilized in microscope objectives where a pencil of light as wide as possible is required from the small object to enter the objective to ensure that the image could be sufficiently bright.

5.6.1 High power oil immersion microscope objective

In high power microscopes, say with an overall magnification of 500, the objective alone gives a magnification $\sim$ 50 and hence must satisfy the following *conditions* :

1. It must be *free from chromatic* and *spherical aberrations.*
2. *Much light should enter* as the object to be viewed is too small.

For satisfying the above conditions, Abbe devised a rather complicated objective called **homogeneous oil immersion objective.**.

Oil immersion objective—It consists of a combination of a large number of lenses in order to be free from chromatic and spherical aberrations. The lowermost lens L_1 (Fig. 5.14) is a *plano-hemispherical convex* lens, O being the object (specimen). A drop of **cedar wood oil** having the same refractive index ($\mu = 1.51$) as that of the material of the lens is placed on the microscope slide between the plane surface of the lens and the object O which is thus virtually embedded optically within the lens L_1. The rays from O thus suffer no refraction till they reach the spherical surface with centre C and radius R of lens L_1.

The vertical focussing movement consists in bringing the specimen O, in contact with the oil, at a distance R/μ from C such that all rays diverging from O appear to come from its conjugate aplanatic point I_1 at a distance μR from C, independent of the angle of incidence. This arrangement is called **homogeneous immersion**.

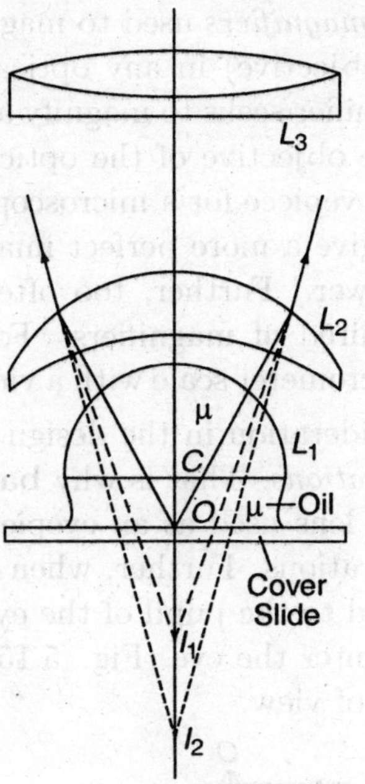

Fig 5.14

The second lens L_2 is an aplanatic meniscus so adjusted that I_1 lies at the centre of curvature of its lower surface and the final image, free from spherical aberration and coma, appears at I_2—the conjugate aplanatic point corresponding to I_1 for the second surface of lens L_2. The arrangement changes the wide angle beam of light from O into a narrow angle pencil, originating from I_2.

Introducing more aplanatic lenses in succession, the divergence of the rays from the object can be further reduced without any aberration. The rays are finally rendered *convergent* by using *two achromatic lenses* that also serve to *compensate the chromatic aberration* introduced by the first two lenses. Such an objective is called an **oil-immersion objective**.

- A really good microscope objective consists of 10 such aplanatic lenses and corrects the chromatic aberration for *three* colours and spherical aberration for *two*.

- There is a limit to introducing the number of more and more aplanatic lenses as the chromatic aberration becomes troublesome.

- The advantage of immersion objective over a dry objective is that it *collects a wider cone of light* from the specimen (object).

- Reflecting surfaces may as well act as aplanatic surfaces, e.g., a *paraboloidal mirror* is an *aplanatic surface* having its forcus and infinity as the two *aplanatic points. Ellipsoidal* mirror is also *aplanatic surface* with its two foci as *aplanatic* points.

5.7 Eyepieces (or ocular)

Eyepieces are fundamentally *magnifiers* used to magnify an image formed by a lens or a lens system preceding it (objective) in any optical instrument. The only *essential difference* is that while a magnifier seeks to magnify a real object, an eyepiece magnifies the real image formed by the objective of the optical instrument. Whereas a simple magnifier may be used as an eyepiece for a microscope or a telescope, special eyepieces are devised and designed to give a more perfect image than what can be obtained by a single lens of the same power. Further, too often eyepieces are to perform some *additional functions* not required of magnifiers. For instance, it may be desired to introduce a *cross-wire* or a micrometer scale with a view *to make measurements possible.*

The most important consideration in the design of eye-pieces is the *correction for chromatic and spherical aberrations.* That is why **basically an eyepiece consists of two lenses.** A single convex lens used as an eyepiece would have suffered from both chromatic and spherical aberrations. Further, when the eye is placed close to the lens and the dia is larger compared to the pupil of the eye, rays passing through the outer positions of the lens do not enter the eye (Fig. 5.15) resulting in a lower intensity of the image and a limited field of view.

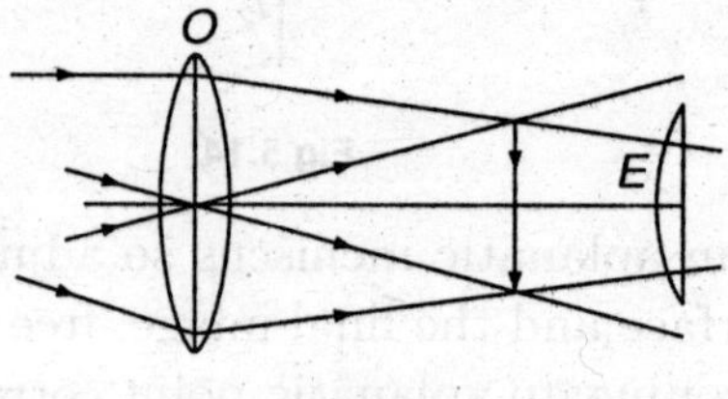

Fig 5.15

To overcome these difficulties, two lenses—one called **field lens** F and the other called **eye lens** E—are used to design an eyepiece. The field lens introduced between the objective and the eye lens is of a relatively *larger aperture* than the eye lens. The image formed by the objective is made to coincide with the principal plane of the field lens whose function is to turn the rays passing through its outer portions towards the principal axis. The rays are thus made to pass through the middle of the eye lens. As a result, all the rays from different parts of the wide area enter the observer's eye and *the field of view is increased.* Without it, the rays of outer part, shown dotted in Fig. 5.16, would have been missed by the eye.

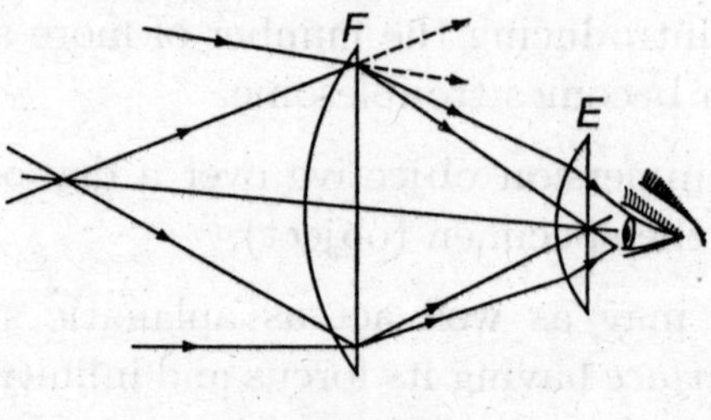

Fig 5.16

By using two lenses, the total deviation of the rays is spread over *four* surfaces to *reduce spherical aberration.* The *separation* between the lenses is so *adjusted* that the *chromatic aberration is also minimum.*

The two most widely used eyepieces are the **Ramsden's eyepiece** and the **Huygens' eyepiece**, each with its own merits and demerits. Each consists of two plano-convex lenses obviously to minimize spherical aberration, but the ratio of the focal lengths of the components and their separation are different.

We shall now describe each of them with greater depth separately and make finally a comparison of the two types of eyepieces, for the sake of completeness of our studies of the eyepieces.

5.8 Ramsden's eyepiece

Ramsden's eyepiece essentially consists of *two plano-convex* lenses, *each of focal length* f, placed *coaxially* so that their *curved surfaces face each other,* being *separated by* a distance equal to $\frac{2}{3}f$ (Fig. 5.17). Designed by Ramsden, it is commonly used in all optical *instruments* where *accurate measurements* are desired and is provided with cross-wires. The eyepiece is placed beyond the image formed by the objective.

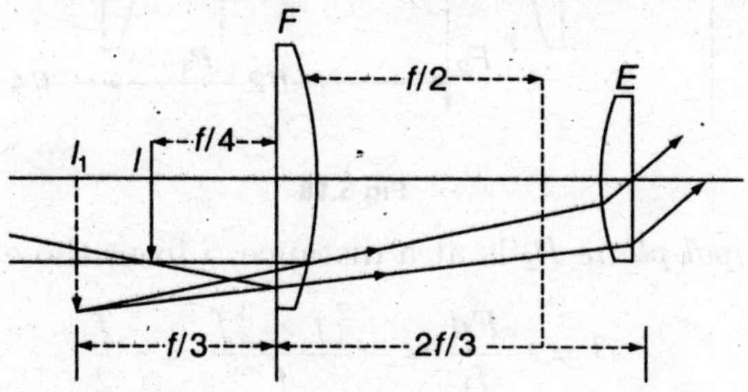

Fig 5.17

Focal length of eyepiece—If F be the focal length of the eyepiece, i.e., equivalent focal length of the lenses,

$$\frac{1}{F}=\frac{1}{f_1}+\frac{1}{f_2}-\frac{d}{f_1 f_2}=\frac{1}{f}+\frac{1}{f}-\frac{\frac{2}{3}f}{f\cdot f}=+\frac{4}{3f}\Rightarrow F=+\frac{3}{4}f$$

So, the *equivalent lens is of focal length* $\frac{3}{4}f$ *and must be placed behind the field lens at a distance*

$$\alpha=\frac{Fd}{f}=\frac{\frac{3}{4}f\times\frac{2}{3}f}{f}=\frac{f}{2}$$

⇒ *the equivalent lens is between the field lens and the eye lens.*

Since the eyepiece is of focal length $\frac{3}{4}f$, the image of the object due to objective must be at a distance $\left(\frac{3}{4}f-\frac{1}{2}f\right)$ or $f/4$ *in front* of the *field lens,* at I. This image will behave as an object for the eyepiece and the final image will be at *infinity.* So the *cross-wires* are to be placed at the image position I, i.e., *at* $f/4$ distance *in front of the field lens* F. This is a great *advantage* of Ramsden's eyepiece.

To minimize the chromatic aberration, both the lenses are made of a proper combination of a crown and a flint glass. The spherical aberration is also small here, since both the lenses are plano-convex with their convex surfaces facing each other.

Position of principal planes—The *first principal plane* of the eyepiece, i.e., the equivalent lens, is at P_1 *behind* the field lens at a distance α given by

$$\alpha = \frac{Fd}{f_2} = \frac{\frac{2}{3}f \times \frac{3}{4}f}{f} = +\frac{f}{2}$$

So, the equivalent lens is to be placed at P_1 *between* the field lens F and the eye lens E, distant $f/2$ behind the field lens, as illustrated in Fig. 5.18. The point H_1 where the plane intersects the principal axis is the **first principal point**.

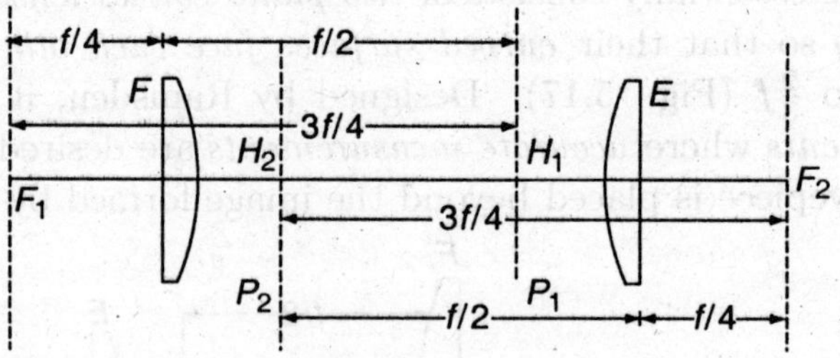

Fig 5.18

The *second principal plane* P_2 is at a distance β from the eye lens E given by

$$\beta = -\frac{Fd}{f_1} = -\frac{\frac{2}{3}f \times \frac{3}{4}f}{f} = -\frac{f}{2}$$

The second principal plane P_2 is thus *behind* the eye lens E at a distance $f/2$. The point H_2 where P_2 intersects the principal axis is the **second principal point.**

Nodal points—The two *nodal points* here *coincide* with the *principal points* for the system is in the same medium, namely air.

Position of focal points—The (equivalent) focal length of the eyepiece is $\frac{3}{4}f$. So, the *first focal point* F_1 is at a distance $\frac{3}{4}f$ from the *first principal point* H_1, i.e., at $\left(\frac{3}{4}f - \frac{1}{2}f\right) = f/4$ to the *left* of the *field lens* F.

The *second focal point* F_2 is at a distance $\frac{3}{4}f$ from the *second principal point* H_2, i.e., at $f/4$ *behind* the *eye lens* E away from field lens F.

• Ramsden's eyepiece is called a **positive eyepiece** since here the image acts as a real object for the eye lens and the final image is at infinity. It is an *advantage* of Ramsden's eyepiece (over Huygens') in that the cross-wires may be replaced by measuring scales and magnification as produced by eyepiece is the same for the cross-wires (or scales) and the image.

Merits and demerits—Ramsden's eyepiece has both *merits* and *demerits* as would be clear from the following points.

1. The *field lens* is *not* very *efficient*, but the *field of view* is *fairly wide.*

2. It is *not* entirely *achromatic.* However, it is minimized by using achromatic combinations for both the field lens and the eye lens.

3. *Spherical aberration* is *minimized* by the use of two lenses so that the deviations occurring at four different surfaces, are small. Use of two plano-convex lenses with convex surfaces facing each other also reduces it further. *Coma* and *distortion* are also *sufficiently small.*

4. It is usually *used* in optical instruments where *accurate quantitative measurements* are desired, for it is a *positive eyepiece* and *cross-wires* are *outside* being in front of the field lens.

5. It can be used as a *simple microscope* as the focal plane is to the left of the field lens and is thus real.

• **Gauss modified** this eyepiece (Fig. 5.19) to *illuminate* the *field of view* by introducing a glass plate G at 45° to its axis. Light from a source S enters through an opening in the side of the holder tube and gets reflected along the axis illuminating the cross-wires.

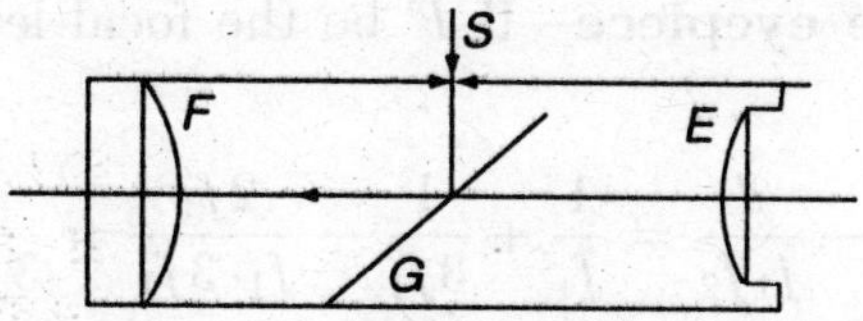

Fig 5.19

This eyepiece is used in telescopes fitted with spectrometer and the telescope is to be correctly adjusted for parallel rays and the correct adjustment is ensured when the cross-wires and their images are both in sharp focus. The illuminated field of view, by the modification introduced, is of great help in the matter.

• It may be argued that if the two lenses were separated by $\frac{2}{3}f$, instead of f, the eyepiece could have been *free* from *chromatic aberration.* However this is done to avoid *magnification of dust particles* on the surface of field lens that would make the *image blurred.*

5.9 Huygens' eyepiece

Huygens' eyepiece essentially consists of *two plano-convex lenses* of the *same kind* of glass, usually of crown, of focal lengths f_1 and f_2 $(= 3f_1)$ placed *co-axially* at a certain distance d $(= 2f_1)$ *apart,* the *plane surfaces* of both being *nearer the observer.* Devised to eliminate or minimize the spherical aberration, apart from chromatic, the lenses are so arranged that a ray, initially parallel and close to the axis, suffers *equal deviation at both the lens surface.*

As obtained earlier in Art. 5.2.1, *for equal deviation, at the surface of the field lens and at that of the eye lens, the separation between the lenses must equal the difference between their focal lengths.*

$$\therefore \quad f_2 - f_1 = d \tag{5.9.1}$$

For correcting the *chromatic* aberration, the required condition is

$$\frac{f_2 + f_1}{2} = d \tag{5.9.2}$$

For *simultaneous minimization* of both *spherical* and *chromatic aberrations*, we have from (5.9.1) and (5.9.2),

$$f_2 - f_1 = \tfrac{1}{2}(f_2 + f_1) \quad \Rightarrow \quad f_2 = 3f_1$$

$$\therefore \quad d = f_2 - f_1 = 3f_1 - f_1 = 2f_1$$

Construction—Thus the *focal length* of the *field lens* should be *three times the focal length* of the *eye lens* and the *separation* of the lenses should be *equal* to *double the focal length of the eye lens.* It was on this principle that Huygen constructed his eyepiece. The path of the rays through the eyepiece is shown in Fig. 5.20.

Focal length of the eyepiece—If F be the focal length of equivalent lens i.e., the eyepiece,

$$\frac{1}{F} = \frac{1}{f_1} + \frac{1}{f_2} - \frac{d}{f_1 f_2} = \frac{1}{f_1} + \frac{1}{3f_1} - \frac{2f_1}{f_1 \cdot 3f_1} = \frac{2}{3f_1} \quad \Rightarrow \quad F = \frac{3}{2}f_1$$

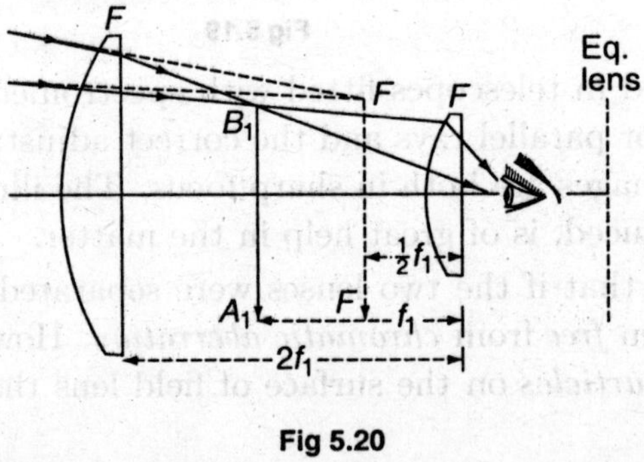

Fig 5.20

Position of principal points—The equivalent lens should be placed at a distance α from the field lens, given by

$$\alpha = \frac{Fd}{f_1} = \frac{\frac{3}{2}f_1 \cdot 2f_1}{f_1} = 3f_1$$

But the separation between the lenses is $2f_1$. So, the *position of equivalent lens* is $3f_1 - 2f_1 = +f_1$, i.e., it should be placed *away from eye lens* or, in other words, *behind it* by a distance $(= +f_1)$ equal to its *focal length,* as indicated by dotted line P_1 in Fig. 5.21.

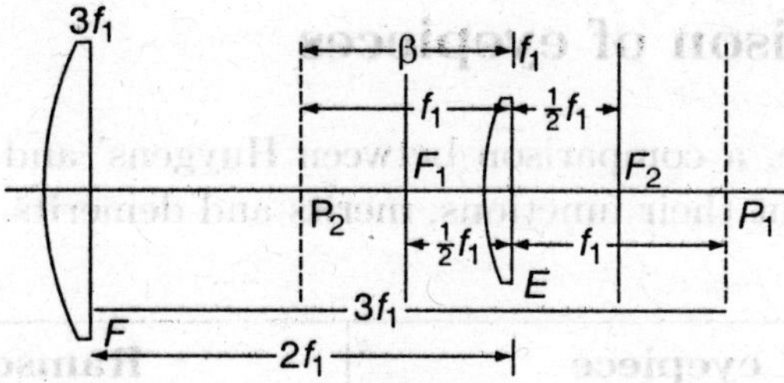

Fig 5.21

The *first principal point* P_1 is at a distance $\alpha = 3f_1$ from the field lens. The *second principal point* P_2 lies at a distance β from the eye lens towards the field lens.

$$\therefore \beta = -\frac{Fd}{f_2} = -\frac{\frac{3}{2}f_1 \cdot 2f_1}{3f_1} = -f_1$$

Position of focal points—The *first focal point F* is at a distance $\frac{3}{2}f_1$ from P_1, the first principal point or, at a distance $f_1/2\left(=\frac{3}{2}f_1 - f_1\right)$ from the eye lens on the side of the field lens.

The *second focal point* F_2 is at a distance $\frac{3}{2}f_1$ from the second principal point P_2, i.e., at a distance $\left(\frac{3}{2}f_1 - f_1\right)$ or $\frac{1}{2}f_1$ from the eye lens away from the field lens.

Position of cross-wires—The first principal focus F_1 of the eyepiece is at $f_1/2$ in front of the eye lens where the image AB due to the objective should be formed to get the final image at infinity. The rays from the objective are however intercepted by the field lens so that the image AB gets displaced to A_1B_1.

For the field lens of focal length $3f_1$, AB acts as the virtual object. So $u = \frac{3}{2}f_1$. If v be the distance of the image A_1B_1, then from lens formula,

$$\frac{1}{3f_1} = \frac{1}{v} - \frac{1}{\frac{3}{2}f_1} \Rightarrow \frac{1}{v} = \frac{1}{3f_1} + \frac{2}{3f_1} = +\frac{1}{f_1} \Rightarrow v = f_1.$$

$\therefore$ A_1B_1 is *midway* between the lenses. This fixes the position of the cross-wires. The image observed by the eye is distinct but slightly curved without distortion.

- As the image formed by the objective lies on the *negative side* (behind the field lens), Huygens' eyepiece is known as **negative eyepiece**.

- This eyepiece is called *theoretically perfect* since here both the chromatic and the spherical aberrations are simultaneously minimized with the focal length of the field lens three times that of the eye lens and separating them by a distance twice the focal length of the eye lens.

- *Quantitative measurements* with Huygens' eyepiece are *not* possible since the measurements are *not reliable.* For, the cross-wires are seen only through the eye lens as their position is between the field lens and the eye lens. The eye lens is a single lens that suffers from both spherical and chromatic aberrations. This results in an unequal magnification of the image and the cross-wires. This is why *Ramsden's eyepiece is preferred to Huygens'*. In Ramsden's eyepiece the cross-wires are in front of the eyepiece and hence the magnification of cross-wires and the image is the same.

5.10 Comparison of eyepieces

In the following table, a comparison between Huygens' and Ramsden's eyepieces has been made to highlight their functions, merits and demerits.

Huygens' eyepiece	Ramsden's eyepiece
1. $f_1 = 3f$; $f_2 = f$; $d = 2f$; $F = \frac{3}{2}f$	**1.** $f_1 = f$; $f_2 = f$; $d = \frac{2}{3}f$; $F = \frac{3}{4}f$.
2. Its first focal plane lies between the lenses.	**2.** First focal plane is in front of the two lenses.
3. Ordinarily, therefore, cross-wires cannot be used. To place cross-wires *between field lens and eye lens* involves mechanical difficulty, and they are *adjusted permanently for* persons of *normal vision* only.	**3.** Cross-wire or transparent scales can be used—a *great advantage.* No difficulty to place them as they are to be *placed outside, in front of field lens.* Can be *adjusted* for persons of *varying vision.*
4. Can magnify only virtual objects and hence called a *negative eyepiece.*	**4.** Can magnify real objects as a simple magnifier and hence called a *positive eyepeice.*
5. *Lateral chromatic aberration* is *eliminated* as $d = \frac{1}{2}(f_1 + f_2)$, but *some longitudinal colour* is present.	**5.** As $d \neq \frac{1}{2}(f_1 + f_2)$, it *shows lateral chromatic aberration,* but reduced with achromats. *Longitudinal colour* is half that of Huygens'.
6. There is some spherical aberration although the eyepiece *satisfies* the *condition of minimum spherical aberration.*	**6.** By *designing plano-convex* components and *positioning* them *properly, spherical aberration* is *reduced to* 1/5th that of Huygens', but separation *does not* conform to *minimum spherical aberration.*
7. Field of view fair up to 50°.	**7.** Field of view is of the same order.
8. Some *astigmatism* and pin cushion *distortion* present but entirely *free from coma.*	**8.** *Distortion* is *half* as great. But there is *no coma.*
9. *Eye-relief* is rather *too small* for any comfort.	**9.** *Eye-relief is* 50% *greater; another* distinct *advantage.*
10. Gives no *trustworthy quantitative measurement* due to unequal magnification of image and cross-wires.	**10.** Gives *accurate quantitative measurements* since the image and the cross-wires are equally magnified.

5.11 Illustrated Examples

▶ **Example 1.** *A convergent doublet of separated lenses, corrected for spherical aberration, has an equivalent focal length of* 10 cm. *The lenses are separated by* 2 cm. *What are the focal lengths of the component lenses?* **(Gorakhpur Univ.)**

Solution. If f_1, f_2 be the focal lengths of the component lenses, then, as spherical aberration is corrected,

$$d = f_1 - f_2 = 2 \text{ cm} \quad \Rightarrow \quad f_1 = (f_2 + 2) \text{ cm} \tag{5.11.1}$$

The equivalent focal length, $F = \dfrac{f_1 f_2}{f_1 + f_2 - d} = 10$ cm, by the problem.

$$\text{or,} \quad \frac{(f_2+2)f_2}{(f_2+2)+f_2-2} = 10, \quad \text{using (5.11.1).}$$

Solving we get : $f_2 = \mathbf{18\ cm}$ and $f_1 = \mathbf{20\ cm.}$

▶ **Example 2.** *Two thin lenses of focal lengths* 16 cm *and* 12 cm *form a combination which is corrected for spherical aberration. Find the distance between the principal points of the combination.*

Solution. If d be the separation between the lenses, then correction for spherical aberation dictates that

$$d = f_1 - f_2 = 16 - 12 = 4 \text{ cm}.$$

Also, the equivalent focal length,

$$F = \frac{f_1 f_2}{f_1 + f_2 - d} = \frac{16 \times 12}{16 + 12 - 4} = 8 \text{ cm}.$$

Distance of first principal point H_1 from the first lens L_1 (Fig. 5.22) is

$$L_1H_1 = +\frac{Fd}{f_2} = +\frac{8 \times 4}{12} = +\frac{8}{3} \text{ cm (on right of } L_1)$$

Distance of second principal point H_2 from the second lens L_2 is given by

$$L_2H_2 = -\frac{Fd}{f_1} = -\frac{8 \times 4}{16} = -2 \text{ cm (on left of } L_2)$$

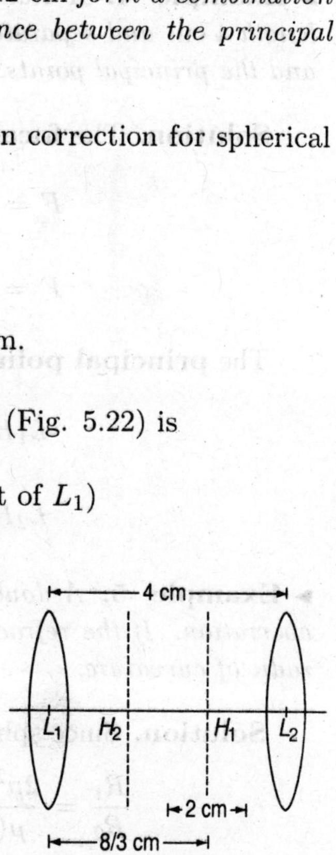

Fig 5.22

∴ Distance between the principal points is given by

$$\begin{aligned} H_2H_1 &= H_2L_2 - H_1L_2 \\ &= H_2L_2 - (L_1L_2 - L_1H_1) \\ &= 2 - (4 - 8/3) = \mathbf{2/3\ cm.} \end{aligned}$$

► **Example 3.** *The lenses in a Huygens' eyepiece employed in a microscope have focal lengths* 2 cm *and* 4 cm. *Find the separation between the lenses and locate the cardinal points.* **(Agra Univ.)**

Solution. Here the correction for chromatic aberration is most important. So the **separation** between the lenses is given by

$$d = \tfrac{1}{2}(f_1 + f_2) = \tfrac{1}{2}(4 + 2) = \mathbf{3\ cm.}$$

Equivalent focal length of eyepiece, $F = \dfrac{f_1 f_2}{f_1 + f_2 - d} = \dfrac{4 \times 2}{4 + 2 - 3} = \dfrac{8}{3}$ cm.

Positions of **principal planes** are :

$$L_1H_1 = +\frac{Fd}{f_2} = \frac{(8/3) \times 3}{2} = \mathbf{+4\ cm}$$

$$L_2H_2 = -\frac{Fd}{f_1} = -\frac{(8/3) \times 3}{4} = \mathbf{-2\ cm.}$$

The system is in air. So the **nodal points** *coincide with the principal planes.*

► **Example 4.** *A Ramsden's eyepiece consists of two positive lenses, each of focal length* 4 cm *and separated by a distance of* 3 cm. *Find the focal length of the eyepiece and the principal points.*

Solution. The **focal length** of the eyepiece F is given by the relation

$$F = \frac{f_1 f_2}{f_1 + f_2 - d}; \quad \text{Here } d = 3 \text{ cm}; \ f_1 = f_2 = 4 \text{ cm.}$$

$$\therefore \quad F = \frac{4 \times 4}{4 + 4 - 3} = \frac{16}{5} = \mathbf{3.2\ cm.}$$

The **principal points** are given by

$$L_1H_1 = \frac{Fd}{f_2} = \frac{(16/5) \times 3}{4} = \frac{4}{5} \times 3 = \mathbf{2.4\ cm}$$

$$L_2H_2 = -\frac{Fd}{f_1} = -\frac{(16/5) \times 3}{4} = \mathbf{-2.4\ cm.}$$

► **Example 5.** *A double convex lens of focal length* 20 cm *shows minimum spherical aberration. If the refractive index of the material of the lens is* $\mu = 1.5$, *calculate its radii of curvature.*

Solution. Since spherical aberration is minimum,

$$\frac{R_1}{R_2} = \frac{2\mu^2 - \mu - 4}{\mu(2\mu + 1)} = \frac{2 \times (1.5)^2 - 1.5 - 4}{1.5(2 \times 1.5 + 1)} = -\frac{1}{6} \quad \therefore \ R_2 = -6R_1$$

But, $\dfrac{1}{f} = (\mu - 1)\left(\dfrac{1}{R_1} - \dfrac{1}{R_2}\right) \Rightarrow \dfrac{1}{20} = 0.5\left(\dfrac{1}{R_1} + \dfrac{1}{6R_1}\right)$

$$\therefore \quad \frac{7}{6R_1} = \frac{1}{20 \times 0.5} \Rightarrow R_1 = \frac{7 \times 20 \times 0.5}{6} = \mathbf{11.67\ cm}$$

$$\therefore \quad R_2 = -6R_1 = -6 \times 11.67 = \mathbf{-70cm}$$

▶ **Example 6.** *The equivalent focal length of a convergent doublet of separated lenses in air is 30 cm. The doublet satisfies the conditions for no chromatic error and minimum spherical aberration. Calculate the focal lengths of the component lenses (of the same material) and the also their separation.*

Solution. Let f_1, f_2 be the required forcal lengths and d their separation in air.

For achromatism $d = \frac{1}{2}(f_1 + f_2)$; for minimum spherical aberration, $d = f_1 - f_2$. Solving these two equations,

$$\therefore \quad f_1 = \frac{3d}{2}; \quad f_2 = \frac{d}{2}$$

Again, since equivalent focal length is 30 cm,

$$\frac{1}{30} = \frac{1}{f_1} + \frac{1}{f_2} - \frac{d}{f_1 f_2} = \frac{2}{3d} + \frac{2}{d} - \frac{4}{3d} = \frac{4}{3d}$$

$$\therefore \quad d = \mathbf{40\ cm}$$

$$f_1 = \tfrac{3}{2} \times 40 = \mathbf{60\ cm}$$

$$f_2 = \tfrac{1}{2} \times 40 = \mathbf{20\ cm}$$

Review Questions

1. What is spherical aberration of a lens? Describe, without mathematical details, how this can be removed. **(Guru Nanak Univ.)**

2. Explain the defects : Coma and astigmatism. How are these minimized? **(Shivaji Univ.)**

3. Explain what do you understand by curvature of field and distortion. How are these minimized? **(Cochin Univ.)**

4. What is meant by an 'aplanatic surface'? Show that a spherical refracting surface is aplanatic with respect to a certain position of the object. Give practical applications of the same. **(Calcutta Hons.)**

5. Describe the high power microscope objective based upon the use of aplanatic points. **(Punjabi Univ.)**

6. What is an eyepiece? Explain why an eyepiece should consist of two lenses. **(Visweswaria Univ.)**

7. Give the construction and working of a Ramsden's eyepiece. **(Guru Nanak Univ.)**

8. Calculate and show in a diagram the positions of principal (unit) points, the focal points and the nodal points of a Ramsden's eyepiece. **(Himachal Univ.)**

9. Give the merits and demerits of a Ramsden's eyepiece. Why is it called a positive eyepiece? **(Guru Nanak Univ.)**

10. With the help of a neat diagram, give the construction and working of a Huygens' eyepiece. **(Kurukshetra Univ.)**

11. Calculate the focal length and the position of cross-wires in a Huygens' eyepiece and show by a diagram the positions of cardinal points. **(Punjab Univ.)**

12. Compare a Ramsden's eyepiece with a Huygens' eyepiece. **(Calcutta and Cochin Univ.)**

13. What are lateral and longitudinal spherical aberrations, and the circle of least confusion? **(Poona Univ.)**

14. What are Siedal aberrations? Briefly explain them and discuss the methods of minimising them.

Short Answer and Objective Questions

1. What is spherical aberration of a lens? Explain with diagram using a convex lens.

2. Why cannot spherical aberration be removed in a single lens? **(Meerut Univ.)**

3. Justify : A plano-convex lens ($\mu = 1.5$) is nearly as good as crossed lens in the construction of optical instruments.

4. What is coma? State Abbe's sine condition, with short explanation, for its removal.

5. What is meant by an aplanatic surface? What are aplanatic points? **(Calcutta Hons.)**

6. What is an eyepiece? Why should it consist of two lenses? **(Guru Nanak Univ.)**

7. Why is Ramsden's eyepiece called a positive eyepiece and Huygens' eyepiece a negative eyepiece?

8. What modification was introduced to Ramsden's eyepiece by Gauss and why? **(Himachal Univ.)**

9. Why can we not use Huygens' eyepiece for quantitative measurements?
(Guru Nanak Univ.)

10. Why does one call Huygens' eyepiece theoretically perfect and negative?
(Himachal Univ.)

11. What are the functions of field lens and eye lens in an eyepiece?

12. Why are the two lenses of equal focal lengths in Ramsden's eyepiece are separated by $\frac{2}{3}f$ and not f although the latter would make it free from chromatic aberration?

13. Why do we prefer a Ramsden's eyepiece to a Huygens' eyepiece?
(Himachal Univ.)

14. What are the aplanatic points of a paraboloidal mirror? Where and why is it used? What is ellipsoidal mirror? What is its importance?

15. Huygens' eyepiece consists of two
    (a) Plano-convex lenses of focal lenths $3f$ and f, separated by $2f/3$,
    (b) Plano-concave lenses of focal lengths $3f$ and f, separated by $2f$,
    (c) Plano-convex lenses of focal lengths $3f$ and f, separated by $2f$,
    (d) Plano-convex lenses of same focal lengths f, separated by $3f/2$.
(Himachal Univ.)

16. In Ramsden's eyepiece
    (a) there are no cross-wires,
    (b) the cross-wires are placed betwen the field and the eye lens,
    (c) the cross-wires are placed outside in front of field lens,
    (d) none of the above is true.

Problems

1. Two thin lenses of focal lengths 6 cm and 2 cm are placed co-axially at such a distance apart that the spherical aberration is minimum. Calculate the distance of separation between the lenses. [4 cm]

2. Two thin lenses having focal lengths 6 cm and 2 cm are placed co-axially to form an eyepiece. If the distance between the lenses be 4 cm, find the focal length of the eyepiece and the positions of principal planes and focal planes.
[3 cm; $\alpha = 6$ cm, $\beta = -2$ cm]

3. Two thin lenses of focal lenths f_1 and f_2 and separated by a distance d form a co-axial system of equivalent focal length $F = 50$ cm. The combination acts as an achromat and satisfies the condition of minimum spherical aberration. Find the values of f_1, f_2 and d. [100 cm; 33.3 cm; 66.67 cm]

4. For an infinitely distant object, a lens of $\mu = 1.5$ gives minimum spherical aberration when the ratio of the radii of curvature is given by

$$\frac{R_1}{R_2} = \frac{2\mu^2 - \mu - 4}{\mu(1 + 2\mu)}$$

Calculate the minimum spherical aberration (parallel incident light) for (i) an equiconvex lens and (ii) a plano-convex lens with convex surface facing object. [1.5 mm; 1.1 mm]

5. Light from the sun is falling upon the Ramsden's eyepiece. Locate the position of the image thus formed and also find the point from which the distance of the image is to be measured. The focal length of each lens of the eyepiece is 3.0 cm. **(Punjab Univ.)** [2.25 cm from second principal plane which is 1.5 cm behind eye lens]

6. From the condition for no chromatic aberration and minimum spherical aberration of a combination of two separate thin lenses, design a combination of equivalent focal length 5 cm. **(Gorakhpur Univ.)** [Two plano-convex lenses of $f = 10$ cm and 3.3 cm kept 6.7 cm apart]

7. Calculate the radii of aplanatic surfaces for a refracting sphere of $\mu = 1.60$ and $R = 3.2$ cm. Also calculate the magnification of the image in area. **(Meerut Univ.)** [2.0 cm, 5.12 cm; $\mu^6 (= 1.6^6)$]

Chapter 6

Speed of Light

6.1 Introduction

The determination of the speed of light c in air or vacuum is of great significance that can hardly be overestimated. The speed of electromagnetic radiation is the same as that of light so that the frequency n of radiowaves can be computed from its wavelength λ using the relation $c = n\lambda$ and conversely. The ratio of corresponding units on electromagnetic and electrostatic system being equal *numerically* to the speed light, it lends support to the electromagnetic nature of light. Further, the relation of equivalence between mass and energy is given by $E = mc^2 = \text{mass} \times (\text{speed of light})^2$ and the mass m of a body moving with a velocity v is related to its rest mass m_0 by $m = m_0/\sqrt{1 - v^2/c^2}$. Both involve the speed of light c. The latter relation provides the important information that the highest speed of a material particle is equal to that of light c. In addition, by the quantum theory, the energy associated with a wave of frequency ν is $h\nu$ and the momentum $h\nu/c$, where h is Planck's constant. Again, what is known as the *refractive index* of the medium is nothing but the ratio of the speed of light in vacuum to that in the medium concerned.

In the context of the above, the determination of the speed of light in vacuum, a *fundamental constant*, is of paramount **importance** in physics.

- The most striking fact about the speed of light is that if an object moves with this speed in one frame, it has the same speed in any other frame.

6.2 Michelson's rotating mirror method

This is one of the *accurate* and *modern* methods of measuring the velocity of light.

Set up—The set up of the apparatus of Michelson's method is illustrated in Fig. 6.1. S is a strong point source of light and A a regular *octagonal* (eight-sided) rotating mirror. Light from S, on passing through a fine slit, strikes the lower part 1 of the face of A and gets reflected to the *plane* mirror M_1 which in its turn reflects it to another *plane* mirror M_2. From M_2, after reflection, light falls on a large *concave* mirror C_1. It then travels about 35 km to strike the *concave* mirror C_2 having the same dimensions as those of C_1. Finally, it is brought to focus on another *concave* mirror C_3, whence it returns as a parallel beam back to C_1 and is reflected from the plane mirror M_3 that lies at right angles to and above M_2, and passes on to the mirror M_4. M_4, in its turn, reflects it on to the upper part 5 of the face of the rotating mirror A and the final image is formed after refraction through a right angled prism P at the focal plane of a telescope T.

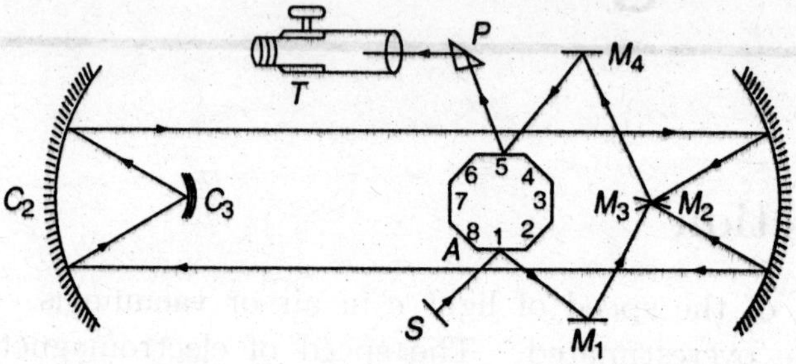

Fig 6.1

Action or Working—With the above set up of the optical system, the eyepiece of T is adjusted so that the image falls on the intersection of cross-wires. The rotating mirror A is then allowed to rotate in a regulated fashion till at last the image again falls on the intersection of the cross-wires. This *will occur only when one face of the rotating octahedral mirror A will exactly occupy the next face when light travels from A to the distant mirror and back.*

The speed of rotation of the rotating mirror is measured *stroboscopically* with an electrically maintained tuning fork of known frequency.

Calculations—Let the distance between the two stations C_1 and C_2 be d so that the total distance travelled by light during the time t taken by one face of the rotating mirror to be exactly succeeded by the next. But $t = 1/8n$ sec, where n is the number of rotations per sec of the rotating octahedal mirror.

$$\therefore \text{ Speed of light, } \quad c = \frac{2d}{1/8n} = 16nd$$

If however m be the number of faces of the rotating mirror, we get

$$c = \frac{2d}{1/mn} = 2mnd$$

The mean value of the speed of light obtained from a series of similar experiments by Michelson and co-workers was **2.99797 × 10⁸ m/s**.

Advantages and limitations—Michelson's method for the determination of speed of light has a number of *advantages.* It also suffers from a number of *limitations.*

The *advantages* are : (a) the distance between the two stations was sufficiently large; (b) it *does not* require the measurement (null method) of any displacement of the image as in Focault's method (not included); (c) the image remains at the point of intersection of the cross-wires only for a *certain* speed of rotation and appears or disappears abruptly; (d) the *brightness* of the image is enough to ensure accurate measurements.

The *limitations* of the method are : (a) it is difficult to keep the speed of rotation *constant* for a long time. In fact, it can be maintained for a *few seconds only* in which observation is to be completed; (b) at very high speeds of rotation, there is a *risk of breakage* of the sides of the rotating mirror and Michelson reduced the risk by increasing the number of sides of the mirror.

6.2.1 Modified Michelson's method of Pease and Pearson

Although Michelson's method is one of the most accurate methods for the determination of the speed of light, it *cannot* be used to find the speed directly in vacuum. Pease and Pearson therefore modified the method to measure the speed of light directly in vacuum.

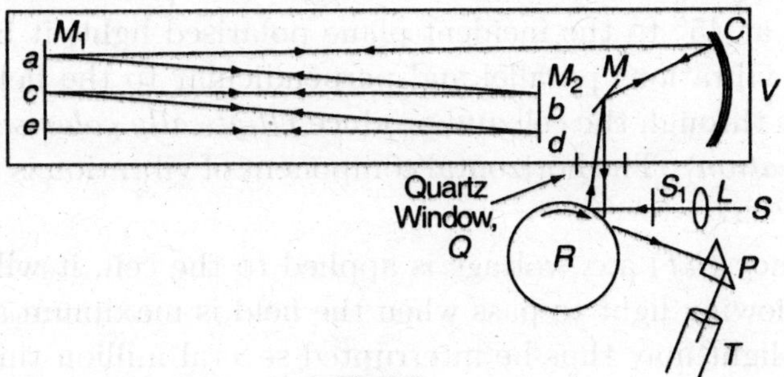

Fig 6.2

Set up—The experimental set up of Pease and Pearson's method is illustrated in Fig. 6.2. S is an *arc lamp* whence light is focussed on to the *slit* S_1 with the lens L. It then strikes the upper half of one of the faces of a multifaced *rotating cylinder* R. All the 32 faces of R are plane polished. On being reflected from R, light enters into a 1.6 km long *vacuum tube* V of 1 m dia *via* a *quartz window* Q, to fall on the plane mirror M. The light reflected by M next strikes the concave mirror C, of 1 m dia and focal length of about 15 m, and travels to the opposite end of the tube V to fall on the plane mirror M_1, 60 cm in dia. It suffers a reflection at a on M_1 to return to the slightly inclined similar mirror M_2 whence back to M_1 and again to M_2 by successive reflections. Finally the light strikes M_1 *normally* at e to retrace its path to go back to the concave mirror C. On reflection from M, it strikes the lower half of the *same face* of the rotating cylinder R, to be observed finally through the telescope T after being totally reflected from the prism P.

6.3 Kerr cell method of finding the speed of light

This method of finding the speed of light is essentially based on the use of a Kerr cell as an *electro-optic shutter*. But what is a *Kerr cell* ?

Kerr cell—A *Kerr cell* utilizes the property known as the *Kerr electro-optic effect*. A Kerr cell essentially consists of a small glass vessel with two parallel metallic electrodes sealed into it and immersed in an appropriate insulating liquid such as pure nitrobenze (Fig. 6.3). If a high voltage be placed across the electrodes, the cell acts as a *uniaxial crystal* with *optic axis parallel to the field* **(Kerr effect)**.

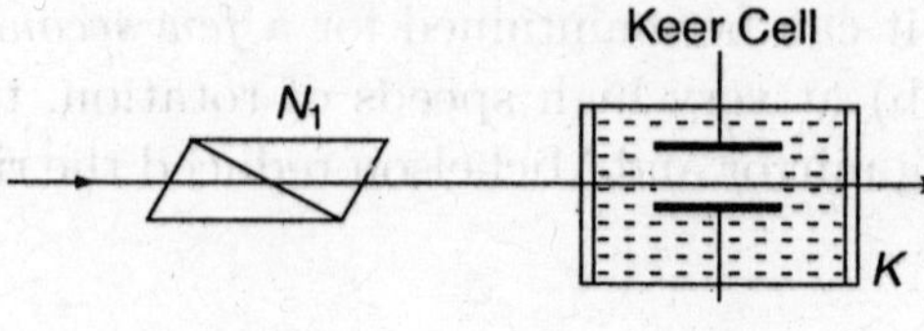

Fig 6.3

Let a Kerr cell K be placed between two crossed Nicols N_1 and N_2, N_1 acting as a *polariser* and N_2 as an *analyzer*. When the field is absent, no light will be transmitted (since crossed) by the analyser N_2. But when the field is 'on', the liquid inside the cell becomes doubly refracting so that some light now passes through N_2. If the Kerr cell K is oriented at 45° to the incident plane polarised light, it breaks into *two equal* components with vibrations parallel and perpendicular to the field. They travel with different velocities through the cell and produce *elliptically polarised* light on emergence (Chapter: *Polarisation*). The horizontal component of vibration is allowed transmission through the analyser.

If high frequency (hf) a.c. voltage is applied to the cell, it will act as an **electro-optic shutter** allowing light to pass when the field is maximum and stop it when the field is zero. The light may thus be interrupted several million times a second.

Speed of light by Kerr cell method—This was first used by Karolus and Mittelstaedt in 1925 to provide very accurate values for light speed. S is the light source (Fig. 6.4) whence light is focussed by means of a lens L_1 on to the Kerr cell K_1, sandwiched between the crossed Nicols N_1, N_2. Light, on emergence from N_2, is again rendered parallel by the lens L_2 to fall upon the plane mirror M.

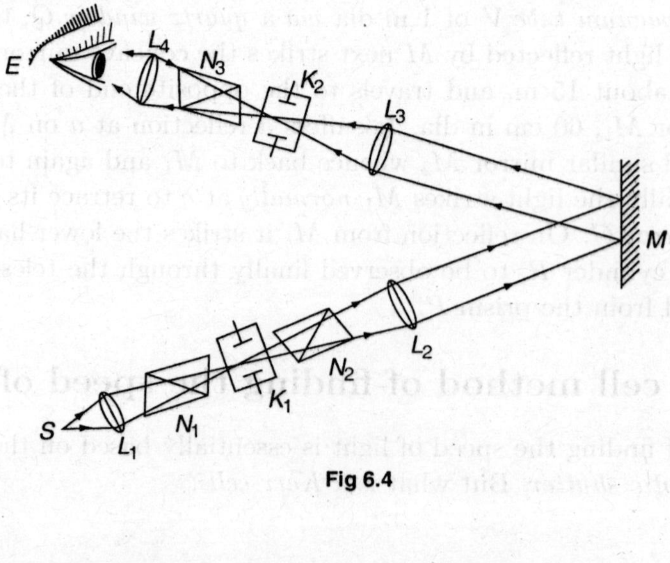

Fig 6.4

The reflected parallel beam from M is focussed on to an *identical* Kerr cell K_2, using the lens L_3, and on emergence falls on the Nicol N_3, being crossed with N_2. The light is finally received by the eye E through the lens L_4.

Action and calculation—As N_1 and N_2 are crossed, *no light* comes out of N_2 and so no light is visible to the eye. When an hf a.c. voltage is simultaneously applied to K_1 and K_2, the inside liquids become doubly refracting at the instant the voltage is maximum. Some light will now be transmitted by N_2. If during the time light takes to reach M and on reflection to reach K_2, the voltage passes through its *zero* value, *no light* will pass through N_2 as N_2 and N_3 are crossed. This implies that an *eclipse* will occur.

Let n be the frequency of the a.c. voltage used. Then it will fall from maximum to zero in time $1/4n$. If d be the effective distance the light travels between the cells K_1 and K_2, then

$$\text{Speed of light, } c = \frac{d}{1/4n} = 4nd$$

The *limitation* of the method however is that it is practically too difficult to exactly match the characteristics of the two identical Kerr cells.

Anderson in 1941, therefore, modified the above method to employ only one Kerr cell and also substituted a photocell in place of the eye, for a better judgement of light and its eclipse.

6.4 Anderson's method of finding the speed of light

The essential *set-up* of Anderson's method for the determination of velocity of light using Kerr cell as an *electro-optic shutter* is illustrated in Fig. 6.5.

S is a mercury lamp whence light, rendered parallel by means of the lens L_1, is allowed to pass through the Kerr cell K, sandwiched between two crossed Nicols N_1 and N_2. The light emerging from N_2 is focussed on the diaphragm D_1 by the lens L_2. D_1 is provided with cross-wires as shown. From D_1 light strikes a *semi-silvered* glass plate M_1 that divides the light into *two* parts : (i) a part is reflected to fall on the plane mirror M_2 after passing through the diaphragm D_2 and the lens L_3; (ii) the other part is transmitted and on reflection from the plane mirror M_3 falls *normally* on the *concave* mirror M_4. Both the beams retrace their paths and on passing through the semi-silvered glass plate M_1 are focussed on to the *photocell* P by means of the lens L_4. The position of M_2 is so adjusted that a phase diference π occurs between the two beams. The photocell would thus show a *minimum response* to the hf-variations in light intensity.

Next, the concave mirror M_4 is substituted by another concave mirror M_4' of focal length 5 m. M_4' will reflect light on to the plane mirror M_5 and thence to mirror M_6 where it falls *normally* and hence retraces its path. The position of M_2 is again adjusted for *minimum response* of photocell. The purpose of D_2 is to *match* the light intensities of the beams that travel different paths; closer the matching, more accurate would be the minimum intensity position.

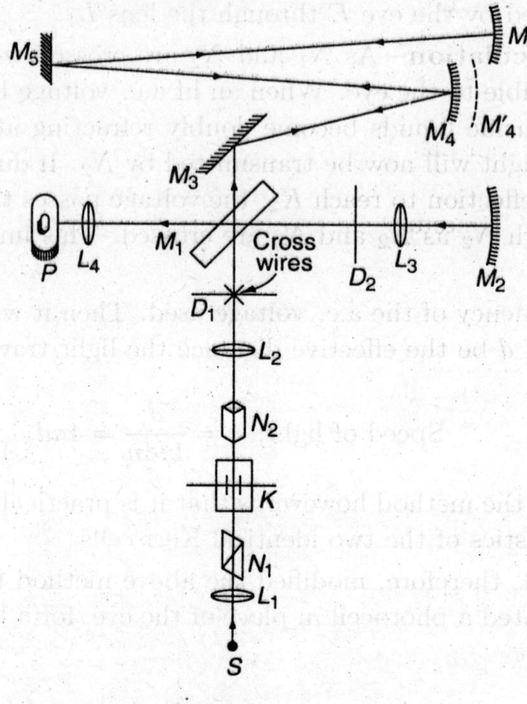

Fig 6.5

Calculations—Let x be the distance from M_1 to M_4 (also termed '*short path*') and y that from M_1 to M_2.

$$\therefore \quad 2x - 2y = m\lambda/2 \qquad (6.4.1)$$

where m is an integer and λ the wavelenth corresponding to hf used to modulate the light beam.

If s be the distance of M_4 from M_6 (also termed '*long path*'), δs the change in path when M_4 is replaced by M_4' and δy the shift of the plane mirror M_2, then

$$(2s + 2\delta s + 2x) - (2y - 2\delta y) = (2n + m)\frac{\lambda}{2}$$

$$\Rightarrow \quad 2(s + \delta s) + 2\delta y = n\lambda$$

In Anderson's experiments, $n = 11$; δs was measured by a micro-calliper. So λ can be computed. If N be the known frequency of the oscillator, then

$$\text{Speed of light } \; c = N\lambda$$

Anderson made nearly 3,000 observations and after introducing correction for vacuum obtained the value

$$\mathbf{c = 299776 \pm 6 \text{ km/s}}$$

Advantages—As already mentioned in Art. 6.4, the advantages of the method are (i) to avoid the use of two Kerr cells and (ii) the use of a photocell instead of the eye. However, the main source of error is the difficultly to arrange for both the beams in using the same position of the photocell.

- The focal length of the concave mirror M_4 is so chosen that the image of D_1 is formed in front of the photocell. This image is *inverted.* When M_4 is replaced by M_4' of larger focal length, the light is reflected on to M_6 and on return journey suffers a second reflection at M_4' and the image of D_1 formed in front of P is now *errect.* The cross-wire in D_1 enables the mirrors to be adjusted for the image to be formed in the same position (overlapping).
- The m-value as used in (6.4.1) is an integer and is made equal to unity.

6.5 Illustrated Examples

▶ **Example 1.** *In a Kerr-cell method for finding the velocity of light, the response of the photocell was minimum when the path difference was* 171.8 m *and frequency of modulation* 9.6 MHz. *If the value of integral was* 11, *calculate the speed of light.*

(Mysore Univ.)

Solution. Here the given path difference = 171.8 m; and $n = 11$.

$$\therefore \quad 2 \times 171.8 = 11 \times \lambda$$

$$\therefore \quad \lambda = \frac{2 \times 171.8}{11}$$

Also, the frequency of modulation $N = 9.6$ MHz $= 9.6 \times 10^6$ Hz. So, we have from the relation $c = N\lambda$, the required

$$\text{Speed of light, } c = 9.6 \times 10^6 \times 2 \times 171.8/11$$
$$= \mathbf{2.9986 \times 10^8 \ m/s}$$

▶ **Example 2.** *Calculate the time taken by a beam to travel through a glass pane of thickness* 1 mm *and refractive index* 1.5. *The speed of light through air is* 3×10^8 m/s.

Solution. From the definition of refractive index,

$$1.5 = \frac{\text{velocity of light in air, } c}{\text{velocity of light in the medium, } v} \quad \Rightarrow \quad v \text{ (glass)} = (3 \times 10^8/1.5) \text{ m/s}$$

Thickness of glass, $d = 1$ mm $= 10^{-3}$ m. So the time taken by the beam to travel through the medium is

$$t = \frac{d}{v} = \frac{10^{-3}}{3 \times 10^8/1.5} = \frac{1.5 \times 10^{-3}}{3 \times 10^8} = \mathbf{5 \times 10^{-12} \ s}$$

► **Example 3.** *In a Michelson's experiment, the octagonal mirror was making* 250 r.p.s. *and the distance between the two stations* C_1 *and* C_2 *was* 75 km. *Find the speed of light.*

Solution. Here $n = 250$ rev/s; $d = 75$ km. So, the velocity of light is given by

$$c = 2 \times 8nd = 16nd = 16 \times \times 250 \times 75 \text{ km/s}$$
$$= 3{,}00000 \text{ km/s}$$
$$= \mathbf{3 \times 10^8 \text{ m/s}}$$

Review Questions

1. Describe a modern method for measuring the speed of light in air and discuss its merits and demerits.
2. Describe and explain Michelson's rotating mirror method of finding the speed of light and compare its merits and demerits with those of other methods. **(Calcutta and Burdwan Univ)**
3. Describe Pease and Pearson's modified Michelson's method for an accurate measurement of the speed of light.
4. What is a Kerr cell? Describe the Kerr cell method for the determination of the speed of light in vacuum. **(Shivaji Univ.)**
5. Describe Anderson's method for determining the speed of light. What is the value obtained by him? And the present day accepted value? **(Delhi Univ.)**

Short Answer and Objective Questions

1. What is the present day accepted value for the velocity of light in vacuum?
2. Is the velocity of light in vacuum a fundamental constant? Why?
3. Why is this quantity (speed of light) of great significance in Physics? **(Punjab Univ.)**
4. What is Kerr effect? What is a Kerr cell? **(Shivaji Univ.)**
5. Why a Kerr cell is called an electro-optic shutter? Explain.

6. How is refractive index of a medium related to the velocity of light in vacuum?

7. Could the velocity of a material particle be increased indefinitely? Explain.

8. What is the advantage of using a polygonal mirror with longer number of faces in Michelson's method of measuring the speed of light?

9. Write down the advantages and limitations of Michelson's rotating mirror method for the determination of the velocity of light.

10. Is the speed of light in vacuum different for lights of different colours? Does the speed of light in some media depend on the direction of propagation? If so, name such a medium.

11. In vacuum, the speed of light depends on

    (a) wavelength (b) intensity (c) frequency (d) none of these

12. The colour of light that travels through glass with minimum speed is

    (a) red (b) green (c) yellow (d) violet

13. Light travels through a glass plate of thickness t and index μ. If c be speed of light in vacuum, the time taken by light to travel the plate is

    (a) $t/\mu c$ (b) $t\mu c$ (c) $\mu t/c$ (d) tc/μ

14. The speeds of red light and yellow light are exactly the same

    (a) in vacuum but not in air

    (b) in air but not in vacuum

    (c) in both vacuum and air

    (d) neither in air nor in vacuum

Problems

1. In Michelson's method, the octagonal mirror makes 625 revolutions per second. If the distance between the plane mirror and the rotating mirror is 30 km, calculate the velocity of light. [3×10^8 m/s]

2. Light travels to a target and back in 0.5 s in carbon disulphide. Calculate the distance of the target if the refractive index of carbon disulphide is 1.46.

   (Punjab Univ.) [51.37×10^6 m]

3. A certain monochromatic radiation has a wavelength of 5000 A in water. What is the wavelength in (i) vacuum and (ii) carbon disulphide? Given, μ for water $= 1.33$ and μ for $CS_2 = 1.628$. [6667 A; 4102A]

4. In the rotating mirror method used to measure the speed of light which travelled from Mt. Wilson to Mt. San Antonio, no shift of the observed source of light was obtained with an 8-face mirror rotated at a speed of 530 r.p.s. If the measured distance of the light travelled (in one direction) was 354.3×10^2 m, calculate the speed of light. [$\simeq 3 \times 10^8$ ms^{-1}]

5. In Michelson's experiment, the distance travelled by light between two reflections from the rotating mirror (having a shape of a regular octagon) is 4.8 km. What should be the minimum angular speed of the mirror (other than zero) so that the image is formed at the position where a non-rotating mirror forms it? [7.8×10^3 rev/s]

Chapter 7

Wave Theory

7.1 Introduction

The wave theory of light, in contradistinction to the corpuscular theory of Newton, was first proposed by the Dutch physicist Huygens. According to this theory, light is a form of **wave motion** generated by the mechanical vibrations of an all-pervading homogeneous medium known as *ether.* Ether however was a *hypothetical medium* which was endowed with properties of both *elasticity* and *inertia* in order that a wave motion can propagate through it. The mechanical disturbance travels from the point of its origin in all directions through the medium and the different colours of light were attributed to different wavelengths. Huygens assumed that light waves were longitudinal in character. But this supposition was later proved to be wrong. Light waves were explained by Fresnal and Young to be *transverse* in nature.

Now the wave velocity v, in an elastic medium of elasticity E and density ρ, is given by

$$v = \sqrt{E/\rho}$$

and ether appeared to be a problem child right from its birth. In the light of exceedingly high value of the velocity of light, it follows from the above relation that ether must possess a very high elasticity and extremely low density. These two properties are antagonistic to each other. So ether hypothesis was untenable. Happily in 1860, the English physicist Clark Maxwell put forward his celebrated *electromagnetic theory of light* according to which light waves are *transverse electromagnetic waves* composed of varying coupled electric and magnetic fields perpendicular to each other and also to the direction of propagation of the wave. The velocity of propagation of an electromagnetic wave is given by $\sqrt{1/\mu\varepsilon}$, where μ is the permeability and ε the permittivity of the medium. Since ε and μ have definite values for vacuum, it was unnecessary to assume the existence of the hypothetical ether medium.

The wave theory of light was highly successful in explaining most of the optical phenomena.

7.2 Huygens' principle of wave propagation

According to Huygens, a light source in a homogeneous hypothetical medium, ether, sends out light waves in all directions and it is these waves that carry energy with them to be transmitted in all directions.

Let S be a point source of light that sends out energy in all directions in the form of waves. After a short interval of time Δt, all the particles of the medium will be on the surface AB and be in the *same state of vibration*, i.e., the particle will be in the *same phase.* Plainly, AB is drawn with S as centre and radius $c\Delta t$ where c is the velocity of propagation of the waves. The surface AB is termed as the **primary wavefront**.

Wavefront—*A* **wavefront** *is defined as a surface on which the phase of disturbance is the same, i.e., the disturbance is in the same phase* at any given instant of time.

It is easy to see that in a *homogeneous medium,* for a *point source* of light, the *wavefront* is a **sphere** [Fig. 7.1(a)]. If the source is at a great distance, then a small portion of the spherical wavefront may be looked upon as a plane as illustrated in [Fig. 7.1(b)]. Thus rays of light diverging from or converging to a point give rise to a **spherical wavefront**, and a parallel beam of light gives rise to a **plane wavefront. The direction along which the disturbance propagates in a homogeneous medium is called the ray**. So, a *ray* is always *normal to* the *wavefront.*

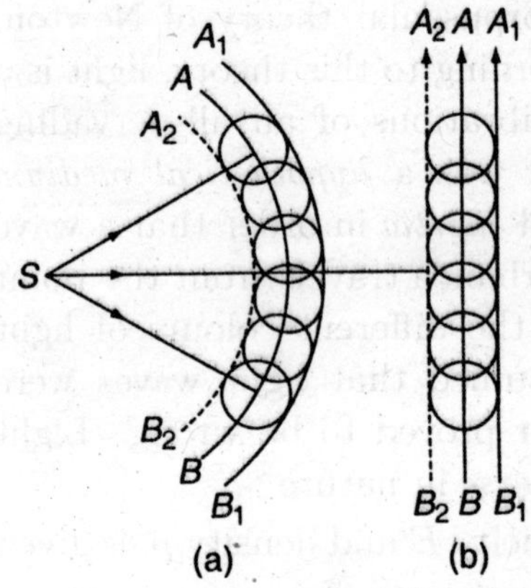

Fig 7.1

Now, an important question arises. Suppose, we know the wavefront AB at an instant t. How to construct the wavefront at a latter time $t + \delta t$? The answer to this question is the **Huygens' principle**. According to this principle, in an isotropic homogeneous medium *each point on the primary wavefront is to be looked upon as the centre of a new disturbance that sends out secondary wavelets. So, a series of secondary wavelets are sent out in all directions that propagate through space with the same velocity in an isotropic homogeneous medium.* After an interval of time, the envelope of the secondary wavelets gives rise to the *secondary wavefront.*

But how to locate the position of the new wavefront after an interval Δt sec? For this, a number of points are taken on AB and with each such point as centre and radius $c\Delta t$, where c is the velocity of light wave in the medium, spheres are drawn in turn. Then these spheres represent the secondary wavelets, starting from those points respectively. The surface A_1B_1 touching all such spheres in the **forward direction** is the *new wavefront.* This is how light wave propagates according to Huygens' principle.

The principle *assumes* that *wavelets are effective only at points where they are touched by their enveloping surface.*

• In case the medium is not homogeneous and isotropic, appropriate velocity is to be used in different points and directions to construct the new wavefront.

• Huygen's principle is open to *two objections* : (i) Why does *not* a wavefront propagate *backward*, giving rise to A_2B_2? (ii) What is the justification of *ignoring* arbitrarily the *large part* of the *secondary waves* not tangent to the enveloping surface?

Subsequently Fresnel, by taking into account the superposition and interference of the wavelets, showed that they are effective at the points of tangency only. Kirchhoff showed mathematically that the amplitude due to a wavelet in a direction making an angle θ with the ray is proportional to $\frac{1}{2}(1 + \cos\theta)$. For back-wave, $\theta = \pi$ which gives zero amplitude. Hence wavefront does not propagate backward. The quantity $(1 + \cos\theta)$ is often referred to as *obliquity factor*.

7.3 Reflection of light by wave theory

Let PQ be the reflecting surface and AB the plane wavefront incident on it (Fig. 7.2). When the wavefront strikes the reflector at A making an angle i with it (wavefront and reflector being perpendicular to the plane of the paper), the angle i is the *angle of incidence*, being equal to the angle between the normal and the incident ray (direction of arrows) at A. Now A becomes the source of secondary wavelets that tend to spread out in surrounding space. By the time the disturbance from B reaches the reflector at C, the secondary waves from A will have a radius BC. Draw a sphere with A as centre and a radius equal to BC. A plane through C, touching the sphere at D, will then give the reflected wavefront CD.

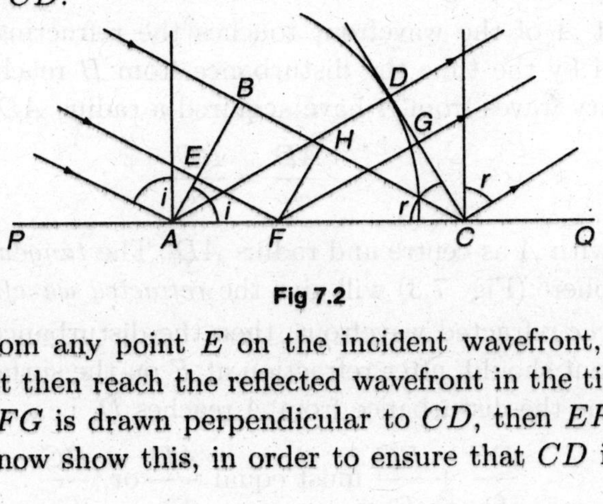

Fig 7.2

The disturbance from any point E on the incident wavefront, after reflection by the reflector at F, must then reach the reflected wavefront in the time the disturbance from A reaches D. If FG is drawn perpendicular to CD, then $EF + FG$ must equal AD or BC. We shall now show this, in order to ensure that CD is the *true reflected wavefront.*

Draw $FH \perp BC$ and take right-angled ΔABC and ΔADC. Since $AD = BC$ and AC is common,

$$\Delta ABC \equiv \Delta ADC \;\Rightarrow\; \angle BAC = \angle DCA = r \tag{7.3.1}$$

Again, $FH \parallel AB \;\Rightarrow\; \angle HFC = \angle BAC = \angle DCA = \angle GCF$. Now in right-angled ΔHFC and ΔFGC, $\angle GCF = \angle HFC$, FC is common.

$$\therefore \quad \Delta FHC \equiv \Delta FGC \;\Rightarrow\; FG = HC.$$

$\therefore\; EF + FG = BH + HC = BC$. $\therefore$ CD *is the true reflected wavefront.*

Now, from (7.3.1), $i = \angle BAC = \angle DCA = r \Rightarrow i = r$.

Angle of incidence = Angle of reflection.

Also, plainly the incident ray, the reflected ray and the normal at the point of incidence, all are in the same plane.

Hence the **laws of reflection** are *proved* from the wave theory.

7.4 Refraction by wave theory

Let a plane wavefront AB be incident on the plane refracting surface PQ that separates two media a and b (Fig. 7.3) in which the velocities of the waves are c_1 and c_2 ($c_1 > c_2$).

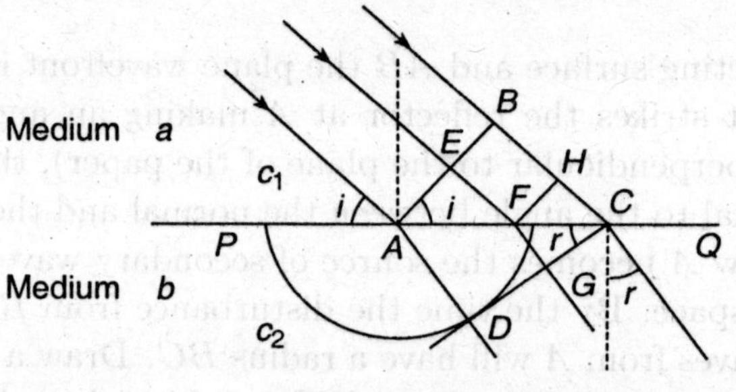

Fig 7.3

When the point A of the wavefront touches the refracting surface, the point B is away from it and by the time the disturbance from B reaches C at the surface of separation, secondary waves from A have acquired a radius AD in the medium b.

$$\therefore \quad \frac{AD}{c_2} = \frac{BC}{c_1} \tag{7.4.1}$$

Draw a sphere with A as centre and radius AD. The *tangent plane* CD through C and touching the sphere (Fig. 7.3) will give the *refracted wavefront.*

If CD be the *true* refracted wavefront, then the disturbance from any point E on the incident wavefront should, after refraction at F on the surface of separation, reach G on CD at the time the disturbance from A reaches D, i.e.,

$$\frac{EF}{c_1} + \frac{FG}{c_2} \text{ must equal } \frac{AD}{c_2} \text{ or, } \frac{BC}{c_1}$$

We shall now show this. Draw $FH \perp BC$. But, we have, $BC = BH + HC$

$$\Rightarrow \quad \frac{BC}{c_1} = \frac{BH}{c_1} + \frac{HC}{c_1} = \frac{EF}{c_1} + \frac{HC}{c_1} \quad (\because BH = EF).$$

Suffice, then if we can show: $HC/c_1 = FG/c_2$. Now, ΔABC and ΔFHC are similar. Also, ΔADC and ΔFGC are similar.

$$\therefore \quad \frac{FC}{AC} = \frac{HC}{BC} \quad \text{and} \quad \frac{FC}{AC} = \frac{FG}{AD} \tag{7.4.2}$$

$\therefore$ From (7.4.1) and (7.4.2), $HC/BC = FG/AD \Rightarrow AD/BC = FG/HC$.

$\therefore$ Using (7.4.1),

$$\frac{c_2}{c_1} = \frac{FG}{HC} \Rightarrow \frac{HC}{c_1} = \frac{FG}{c_2}$$

So, it is shown that CD *is the true refracted wavefront.*

Also, $\angle BAC = i =$ angle of incidence; $\angle ACD = r =$ angle of refraction.

$$\therefore \quad \frac{\sin i}{\sin r} = \frac{\sin \angle BAC}{\sin \angle ACD} = \frac{BC}{AC} \cdot \frac{AC}{AD} = \frac{BC}{AD} = \frac{c_1}{c_2} = \text{constant} \tag{7.4.3}$$

using the relation (7.4.1).

This is **Snell's law.** The constant is called the *refractive index*, ${}_a\mu_b$, of the medium b with respect to the medium a.

Plainly, from figure, the incident ray, the refracted ray and the normal at the point of incidence all lie in the same plane. Thus, the *laws of refraction* are established from the wave theory of light.

• From (7.4.3), we may write the expression for ${}_a\mu_b$ as

$${}_a\mu_b = \frac{\text{velocity of light in medium } a}{\text{velocity of light in medium } b}$$

If the medium a is vacuum ${}_a\mu_b$ is written as μ_b or simply μ, the *refractive index* of the medium b.

$$\therefore \quad \mu \text{ (of a medium)} = \frac{\text{velocity of light in vacuum (or air)}}{\text{velocity of light in the medium}}$$

The refractive index of a medium is the ratio of the velocity of light in vacuum to the velocity of light in the medium.

• The refractive index μ of a medium does *not* depend on the *frequency* of the wave for when light travels through a medium, it is the *wavelength* that *changes* but the *frequency* remains *unaltered.*

• The *geometrical path* of light should not be confused with the *optical path.* Corresponding to a **geometrical path** x in a medium of index μ, the **optical path** $= \mu x$.

7.5 Reflection of a spherical wave at a plane surface

Let LOS be a *spherical* wavefront that diverges from the point source P to be incident at O on a *plane reflector AOB* (Fig. 7.4). In the absence of the reflector, the wavefront after a time Δt would have been in the position AMB. So, $OM = v \cdot \Delta t$ where v is the velocity of light in the incident medium.

The presence of the reflector however causes the secondary wavelets generated at L, O and S on the wavefront LOS to be at A, N and B respectively after time Δt.

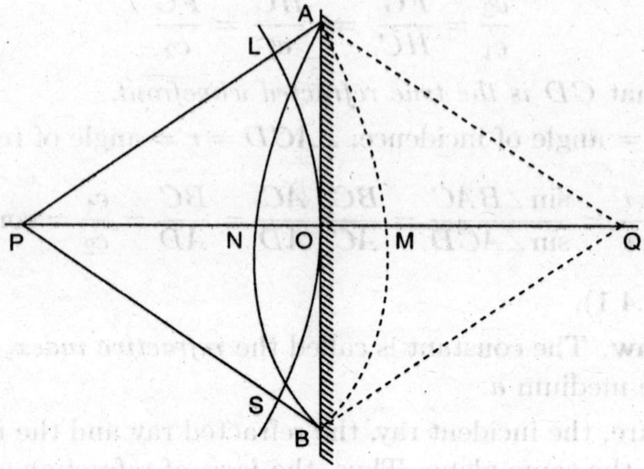

Fig. 7.4

Thus the *common envelope* of these secondary waves, i.e., ANB will be the *reflected wavefront* after time Δt. We thus have

$$OM = ON = AL = BS = v \cdot \Delta t$$

Since the medium is *homogeneous, the reflected wavefront* ANB will also be *spherical* and appear to diverge from Q. Q is thus the *image of* P.

From geometry of Fig. 7.4, we obtain,

$$AO^2 = ON(2QN - ON)$$

$$\simeq ON \cdot 2QN$$

$$\therefore \quad ON = AO^2/2QN$$

$$\text{and similarly} \quad OM = AO^2/2PM.$$

Since, $ON = OM$, we have, from above,

$$MP = QN$$

$$\Rightarrow \quad OM + OP = ON + OQ$$

$$\Rightarrow \quad OP = OQ \quad (\because OM = ON)$$

$\therefore$ **Object distance = Image distance**

for a plane mirror.

7.6 Reflection of a spherical wave at a spherical surface

Let a spherical wavefront LBS diverge from a point object P and be incident on a **concave** spherical reflector LAS. The aperture of the reflector is small; its pole is at A and the centre of curvature is at C (Fig. 7.5).

In a time-interval Δt, say, the secondary wave from B reaches A, when those from L and S arrive, after reflectioon from the reflector, at N and M respectively.

$$\therefore \quad AB = LN = SM = v \cdot \Delta t$$

where v is the velocity of light in the incident medium.

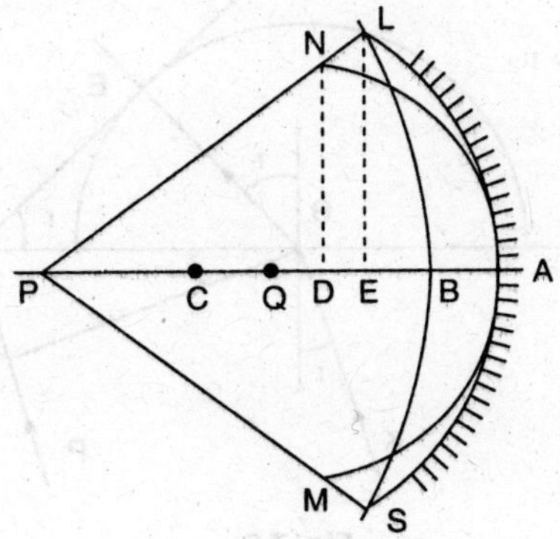

Fig. 7.5

NAM, the common envelope of the reflected secondary waves represents the *reflected wavefront* after time Δt. Like LBS, the reflected wavefront NAM will also be *spherical* since the object and image spaces are homogeneous, and converge at Q to form a *real image*. Draw $LE \perp AP$ and $ND \perp AP$. LE and ND are essentially equal.

Now, from geometry, we obtain

$$BE = LE^2/2BP = LE^2/(-2u) \tag{7.6.1}$$

$$(\because\ BP \simeq AP = -u)$$

$$AE = LE^2/2AC = LE^2/(-2r) \tag{7.6.2}$$

$$AD = DN^2/2AQ = DN^2/(-2v) \tag{7.6.3}$$

Now, from figure, we obtain,

$$AD = AE + ED = AE + AB = AE + (AE - BE)$$

$$\text{or,} \quad AD + BE = 2AE \quad (\because\ ED = LN = AB, \text{ practically})$$

$$\text{or,} \quad \frac{DN^2}{-2v} + \frac{LE^2}{-2u} = 2 \cdot \frac{LE^2}{-2r}$$

$$\text{or,} \quad \boxed{\frac{1}{v} + \frac{1}{u} = \frac{2}{r}}$$

which is the usual relation for reflection of light in a *concave* spherical reflector.

• The students are advised to work out the corresponding relation for reflection in a *convex* spherical surface.

7.7 Total internal reflection from wave theory

Let AB be a plane wavefront incident on the surface of separation BC of the two media of indices μ_1 and μ_2 ($\mu_1 > \mu_2$). KB and PA, both normal to the incident wavefront AB, are the incident rays on BC (Fig. 7.6).

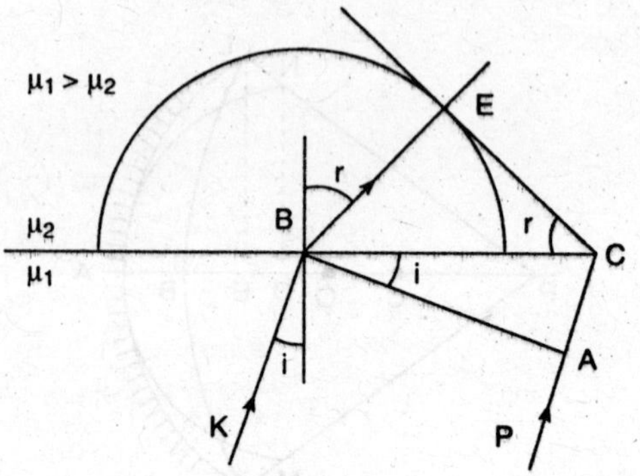

Fig. 7.6

The secondary wave generated at A at time $t = 0$ arrives at C during the time Δt, when that generated at B at $t = 0$ reaches E in the rarer medium of index μ_2. If v_1 and v_2 be the light velocities in medium of indices μ_1 and μ_2 respectively,

$$v_1 \Delta t = AC; \quad v_2 \Delta t = BE$$

$$\therefore \quad BE/AC = v_2/v_1 \Rightarrow BE = (v_2/v_1)AC = (v_2/v_1)BC \sin i \tag{7.7.1}$$

$$\text{Also,} \quad AC = (v_1/v_2)BE = (v_1/v_2)BC \sin r \tag{7.7.2}$$

Three *different* **cases** may now arise.

Case 1. Let the angle of incidence be such that

$$(v_2/v_1)\sin i < 1 \Rightarrow BE < BC \quad \text{(From (7.7.1))}$$

$$\text{But,} \quad BE = BC \sin r \Rightarrow \sin r < 1 \Rightarrow r < 90^\circ$$

So, real tangent can be drawn from C to the secondary wave generated at B in time Δt.

$$\text{Since } v_1 < v_2, \quad AC < BE \Rightarrow (AC/BC) < (BE/BC)$$

$$\Rightarrow \sin i < \sin r \Rightarrow r > i$$

and the refracted ray BE corresponds to the incident ray KB.

Case 2. Let now the angle of incidence $i = \theta$, such that $(v_2/v_1)\sin\theta = 1 \Rightarrow \sin\theta = (v_1/v_2) = (c/v_2)/(c/v_1) = \mu_2/\mu_1$, where c is the speed of light in vacuum.

$$\text{and } BE = BC, \quad \text{using (7.7.1).}$$

Substituting $BE = BC$ in (7.7.2), $BC \sin r = BC \Rightarrow \sin r = 1 \Rightarrow r = 90°$. So when the angle of incidence θ is such that $\sin\theta = \mu_2/\mu_1$, the corresponding angle of refraction $r = 90°$. The tangent plane CE to the secondary wavelet at B, generated at $t = \Delta t$ would be perpendicular to BC (Fg. 7.7). The angle of incidence θ, whose corresponding angle of refraction is 90° is called the *critical angle* of the two media.

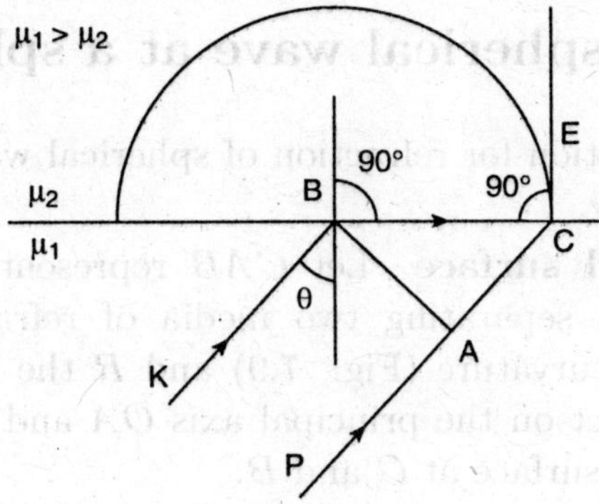

Fig. 7.7

Case 3. Let $i = i_1$ $(i_1 > \theta)$ so that $(v_2/v_1)\sin i_1 > 1$. ∴ From (7.7.1) :

$$BE > BC \Rightarrow BC \sin r > BC \Rightarrow \sin r > 1$$

which is an *absurdity.* So, in this case, *no refraction* can occur.

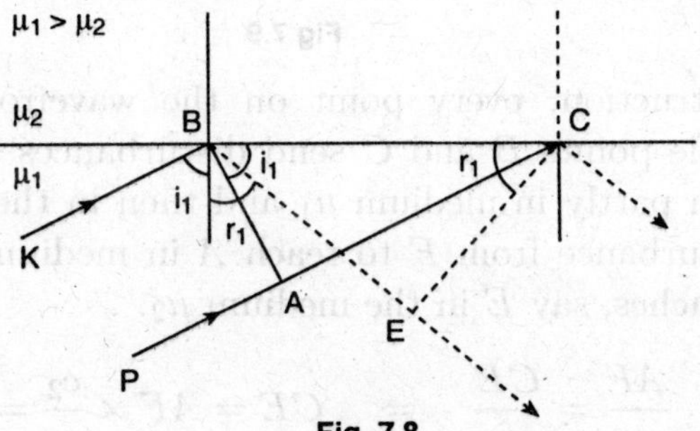

Fig. 7.8

The secondary wave from B reaches E in time Δt, when that from A reaches (Fig. 7.8) C. So $BE = AC$. The envelope CE then will be the position of the wavefront after time Δt, i.e., reflected wavefront. From Fig. 7.8, the right angled

$$\Delta ABC \equiv \Delta BCE \Rightarrow \angle ABC = \angle BCE$$

$$\therefore \quad \angle i_1 = \angle r_1$$

i.e., the angle i_1 made by the incident wavefront AB with the surface BC equals r_1, the angle made by the reflected wavefront CE with the surface BC, when the angle of incidence i_1 exceeds the critical angle θ. So the waves is **totally reflected internally** when the angle of incidence exceeds the critical angle of the two media.

• **Note** that the following *two conditions* must be satisfied for total reflection.

1. The *ray must travel from a denser to a rarer medium.*

2. The *angle of incidence* in the denser medium *must be greater than the critical angle* for the pair of media.

7.8 Refraction of spherical wave at a spherical surface

We shall now derive the relation for refraction of spherical wave at a spherical surface according to the wave theory.

1. Concave spherical surface—Let CAB represent a portion of a concave spherical refracting surface, separating two media of refracting indces μ_1 and μ_2 ($\mu_1 > \mu_2$), P its centre of curvature (Fig. 7.9) and R the radius of curvature. Let O be a luminous point object on the principal axis OA and BFC the portion of the spherical wave touching the surface at C and B.

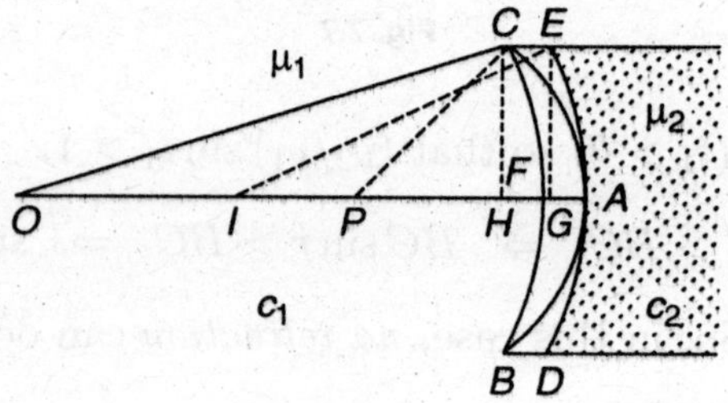

Fig 7.9

By Huygen's construction, every point on the wavefront BFC is a source of secondary waves. While points B and C send disturbances into the medium μ_2, the other points send them partly in medium μ_1 and then in the medium μ_2. If t be the time taken by the disturbance from F to reach A in medium μ_1, then in time t, the disturbance from C reaches, say E in the medium μ_2.

$$\therefore \quad \frac{AF}{c_1} = \frac{CE}{c_2} \quad \Rightarrow \quad CE = AF \times \frac{c_2}{c_1} = \frac{AF}{\mu} \tag{7.8.1}$$

where c_1 and c_2 are the wave velocities in medium μ_1 and medium μ_2 respectively so that $c_1/c_2 = \mu_2/\mu_1$.

In the same time t, the disturbance from B reaches D in the medium. Thus the disturbance from all points on the incident wavefront after time t will be on the wavefront DAE, the *refracted wavefront* originating from I. I is thus the *virtual image* of O.

Draw $CH \perp OA$ and $EG \perp OA$. Clearly, $CH = EG = y$, say.

Now, $AO = u$, $AI = v$ and $AP = R$. Also, from the geometry of a circle,

$$CH^2 \simeq AH \times 2R \quad \Rightarrow \quad AH = \frac{CH^2}{2R} = \frac{y^2}{2R} \tag{7.8.2}$$

$$\text{Similarly,} \quad FH = \frac{y^2}{2u}, \quad AG = \frac{y^2}{2v} \tag{7.8.3}$$

From figure (7.9), $CE = HG$ and from (7.8.1), $\mu_1 AF = \mu_2 CE = \mu_2 HG$

$$\therefore \quad \mu_1(AH - HF) = \mu_2(AH - AG)$$

$$\Rightarrow \quad \mu_1 \frac{y^2}{2R} - \mu_1 \frac{y^2}{2u} = \mu_2 \frac{y^2}{2R} - \mu_2 \frac{y^2}{2v}, \text{ using (7.8.2) and (7.8.3)}$$

$$\Rightarrow \quad \boxed{\frac{\mu_2}{v} - \frac{\mu_1}{u} = \frac{\mu_2 - \mu_1}{R}}$$

- If the first medium be air $\mu_1 = 1$ and taking $\mu_2 = \mu$, we get

$$\frac{\mu}{v} - \frac{1}{u} = \frac{\mu - 1}{R}.$$

Here, by sign convention, u, v, R are all negative. So, finally also, we have

$$\boxed{\frac{\mu}{v} - \frac{1}{u} = \frac{\mu - 1}{R}} \tag{7.8.4}$$

as the required relation.

2. Convex spherical surface—Let BAC be the portion of a convex refracting spherical surface, of index μ_2 separating it from a medium of index μ_1 ($\mu_2 > \mu_1$) and with centre of curvature P (Fig. 7.10). If c_1 and c_2 be the velocities of the wave in the medium of index μ_1 and μ_2 respectively, then $\mu_2/\mu_1 = c_1/c_2$ ($c_1 > c_2$). EAF is the trace of a wavefront diverging from the luminous point source O. By Huygens' construction, each point on the wavefront EAF is a source of secondary waves. While points E and F send disturbances in medium μ_1, other points send them partly in medium μ_1 and partly in the medium μ_2. Let t be the time taken by disturbance from F to reach C in medium μ_1, then, in the same time, the disturbance from A will reach, say D in the medium μ_2.

$$\therefore \quad \frac{CF}{c_1} = \frac{AD}{c_2} \quad \Rightarrow \quad CF = \frac{c_1}{c_2} AD = \frac{\mu_2}{\mu_1} AD \tag{7.8.5}$$

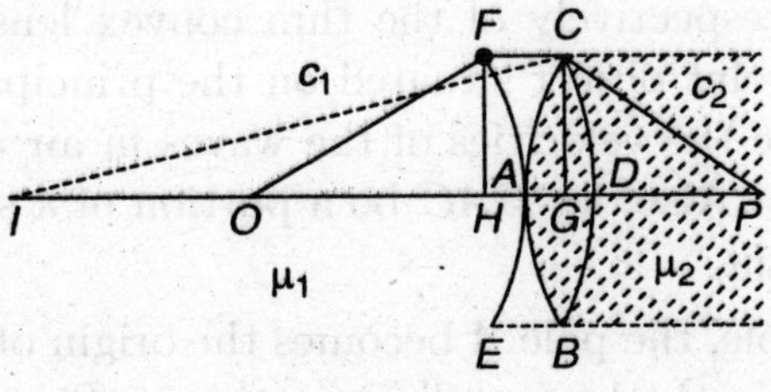

Fig 7.10

The disturbances from other points on incident wavefront EF will reach the wavefront CDB originating from I. I is thus the *virtual image* of O formed by refraction.

Draw $FH \perp OA$ and $CG \perp OA$. Plainly, $HG = CF = (\mu_2/\mu_1)AD$, using (7.5.6).

$$\Rightarrow \quad AH + AG = (\mu_2/\mu_1)(AG + GD)$$

$$\Rightarrow \quad (\mu_2 - \mu_1)AG = \mu_1 AH - \mu_2 GD \tag{7.8.6}$$

Now, $\quad CG = FH = y$, say, $AO = u$, $AI = v$ and $AP = R$.

From the geometry of a circle, we have

$$AH = \frac{y^2}{2u}, \quad GD = \frac{y^2}{2v} \quad \text{and} \quad AG = \frac{y^2}{2R}$$

Substituting these values in (7.8.6),

$$(\mu_2 - \mu_1)\frac{y^2}{2R} = \mu_1\frac{y^2}{2u} - \mu_2\frac{y^2}{2v}$$

$$\Rightarrow \quad \boxed{\frac{\mu_2 - \mu_1}{R} = \frac{\mu_1}{u} - \frac{\mu_2}{v}}$$

By sign convention, both u and v are negative, and R is positive. So, the above relation finally becomes

$$\boxed{\frac{\mu_2}{v} - \frac{\mu_1}{u} = \frac{\mu_2 - \mu_1}{R}}$$

When $\mu_1 = 1$ (i.e., air), and $\mu_2 = \mu$, say, the relation takes the usual form

$$\boxed{\frac{\mu - 1}{R} = \frac{\mu}{v} - \frac{1}{u}}$$

7.9 Lens-maker's formula from wave theory

We shall now derive the formula for thin lens, say, a convex lens, from the wave theory of light.

Convex lens—Let A and D be the poles of the two spherical surfaces of radii of curvature R_1 and R_2 respectively of the thin convex lens L of index μ (Fig. 7.11). Let O be a luminous point object situated on the principal axis, giving out spherical waves. Let c_1 and c_2 be the velocities of the waves in air and in the medium of index μ respectively ($c_1 > c_2$). Also, let BAC be a portion of a spherical wavefront touching the pole A at any instant.

By Huygens' principle, the pole A becomes the origin of secondary waves that move over the thickness AD in the lens medium in time t. So $t = AD/c_2$. The disturbance from points C and B travels in air and reaches say F and E respectively following paths CF and BE in time t. So $t = CF/c_1$ and FDE is the spherical wavefront *after*

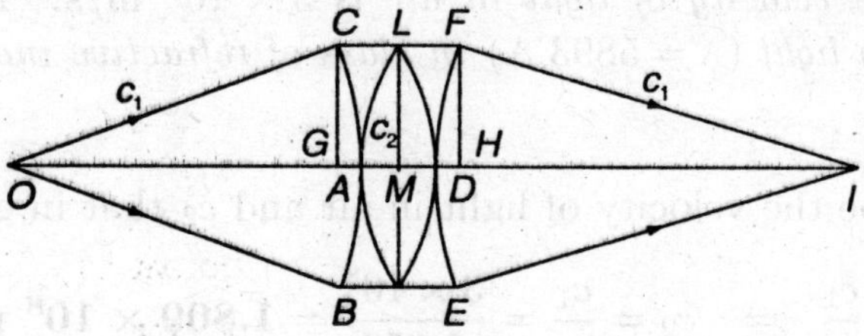

Fig 7.11

refraction through the lens, and is *convergent* with centre I, which is the *real image* of the object O.

$$\text{Now,} \quad \frac{AD}{c_2} = \frac{CF}{c_1}(= t) \quad \Rightarrow \quad CF = \frac{c_1}{c_2}AD = \mu AD \tag{7.9.1}$$

From C, L and F drop perpendiculars CG, LM and FH respectively on the axis.

$$\text{Plainly,} \quad GH = CF = \mu AD, \quad \text{using (7.9.1)}$$

$$\Rightarrow \quad GA + AD + DH = \mu AD$$

$$\Rightarrow \quad GA + DH = (\mu - 1)AD = (\mu - 1)(AM + MD) \tag{7.9.2}$$

Now, $AO = u$, $ID = v$ and $CG = LM = FH = y$, say. From geometry of circle,

$$GA = \frac{y^2}{2u}, \quad AM = \frac{y^2}{2R_1}, \quad MD = \frac{y^2}{2R_2} \quad \text{and} \quad DH = \frac{y^2}{2v}$$

Substituting these values in (7.9.2)

$$\frac{y^2}{2u} + \frac{y^2}{2v} = (\mu - 1)\left(\frac{y^2}{2R_1} + \frac{y^2}{2R_2}\right)$$

$$\Rightarrow \quad \frac{1}{u} + \frac{1}{v} = (\mu - 1)\left(\frac{1}{R_1} + \frac{1}{R_2}\right)$$

By sign convention, v and R_1 are positive, u and R_2 are negative.

$$\therefore \quad \frac{1}{v} - \frac{1}{u} = (\mu - 1)\left(\frac{1}{R_1} - \frac{1}{R_2}\right) \tag{7.9.3}$$

When $u = \infty$, (i.e., object at infinity), $v = f$ (i.e., image is at the focus). So from (7.9.3),

$$\boxed{\frac{1}{f} = (\mu - 1)\left(\frac{1}{R_1} - \frac{1}{R_2}\right)}$$

which is the required **lens-maker's formula.**

- By arguing in a similar line, the formula corresponding to thin concave lens may as well be derived without difficulty.

7.10 Illustrated Examples

► **Example 1.** *The velocity of light in air is* 3×10^8 m/s. *Find the velocity and wavelength of sodium light* ($\lambda = 5893$ A) *in glass of refractive index* 1.658.

(Burdwan Univ.)

Solution. If c_1 be the velocity of light in air and c_2 that in glass of index μ, then

$$\mu = \frac{c_1}{c_2} \quad \Rightarrow \quad c_2 = \frac{c_1}{\mu} = \frac{3 \times 10^8}{1.658} = \mathbf{1.809 \times 10^8\ m/s.}$$

If $\lambda_1 = 5893$ A for sodium light in air and λ_2 that in glass, then

$$c_1 = n\lambda_1, \quad c_2 = n\lambda_2, \quad (n = \text{frequency})$$

$$\Rightarrow \quad \frac{\lambda_1}{\lambda_2} = \frac{c_1}{c_2} = \mu$$

$\therefore$ Wavelength of sodium light in glass is

$$\lambda_2 = \frac{\lambda_1}{\mu} = \frac{5893}{1.658} = \mathbf{3554\ A}$$

► **Example 2.** *A light wave of frequency* 6×10^{10} Hz *is passed through a liquid. The wavelength measured in liquid is* 3×10^{-3} m. *Find* (i) *wavelength in vacuum,* (ii) *refractive index of the liquid and* (iii) *speed of light in liquid. Given : velocity in vacuum* $= 3 \times 10^8$ m/s. **(Punjabi Univ.)**

Solution. (i) If λ_1, λ_2 be the wavelengths in vacuum and liquid respectively and n the frequency of light then,

$$c_1 = n\lambda_1, \quad c_2 = n\lambda_2$$

where c_1 = velocity of light in vacuo and c_2 = velocity of light in liquid.

$$\therefore \quad c_2 = 6 \times 10^{10} \times 3 \times 10^{-3} = \mathbf{18 \times 10^7\ m}$$

$$\text{Also,} \quad \frac{\lambda_1}{\lambda_2} = \frac{c_1}{c_2} \quad \Rightarrow \quad \lambda_1 = \frac{c_1}{c_2}\lambda_2 = \frac{3 \times 10^8}{18 \times 10^7} \times 3 \times 10^{-3}$$

$$= \mathbf{5 \times 10^{-3}\ m}$$

$$\text{Refractive index,} \quad \mu = \frac{c_1}{c_2} = \frac{3 \times 10^8}{18 \times 10^7} = \mathbf{1.66}$$

► **Example 3.** *White light is a mixture of wavelenths between* 400 nm *and* 800 nm. *If such light goes through water having a refractive index* 1.33, *what would be the limits of wavelength in that medium?*

Solution. If light of wavelength λ_0 in vacuum goes into a medium of index μ, the wavelength in the medium becomes $\lambda = \lambda_0/\mu$.

$\therefore$ Corresponding to $\lambda_0 = 400$ nm, $\lambda = 400/1.33 = 300$ nm

Corresponding to $\lambda_0 = 800$ nm, $\lambda = 800/1.33 = 600$ nm.

Thus, the required limits of wavelength in water are **300 nm** and **600 nm.**

▶ **Example 4.** *It is found that the optical paths of a monochromatic light are identical when it goes through* 3.50 cm *in water or* 3.00 cm *of glass. Calculate the refractive index of glass when that of water is* 1.33.

Solution. If light goes through a distance d in a medium of index μ, its optical path $= \mu d$.

By the problem, here, we have

$$\mu_{\text{glass}} \times 3.00 \text{ cm} = \mu_{\text{water}} \times 3.50 \text{ cm} = 1.33 \times 3.50 \text{ cm}$$

$$\therefore \quad \mu_{\text{glass}} = \frac{1.33 \times 3.50}{3.00} = \mathbf{1.55}$$

▶ **Example 5.** *The index of fused quartz is* 1.472 *for light of* $\lambda = 400$ nm *and is* 1.452 *for light of* $\lambda = 760$ nm. *Find the speeds of light of these wavelengths in fused quartz.*

Solution. Let v_1 be the speed corresponding to $\lambda_1 = 400$ nm and v_2 that corresponding to $\lambda_2 = 760$ nm in fused quartz.

$\therefore$ From definition of refractive index, we get

$$1.472 = \frac{\text{velocity of light in vac.}}{\text{velocity of light in quartz}} = \frac{3 \times 10^8 \text{ m}}{v_1}$$

and $\quad 1.452 = \dfrac{3 \times 10^8 \text{ m}}{v_2}$. Solving for v_1 and v_2,

$$\mathbf{v_1 = 2.04 \times 10^8 \text{ m/s}; \quad v_2 = 2.07 \times 10^8 \text{ m/s}}$$

▶ **Example 6.** *Show that the function* $\psi(x, t) = f(x - vt)$ *is a solution of the differential wave equation.*

Solution. We have : $\quad \psi(x, t) = f(x - vt)$

$$\therefore \quad \frac{\partial \psi}{\partial x} = f'(x - vt); \quad \frac{\partial^2 \psi}{\partial x^2} = f''(x - vt).$$

$$\text{Also,} \quad \frac{\partial \psi}{\partial t} = -vf'(x - vt); \quad \frac{\partial^2 \psi}{\partial t^2} = v^2 f''(x - vt)$$

$$\therefore \quad \frac{\partial^2 \psi}{\partial t^2} = v^2 \frac{\partial^2 \psi}{\partial x^2}. \text{ Hence}$$

Review Questions

1. Briefly outline the wave theory of light. What is a wavefront?

2. Explain clearly Huygens' principle for the wave propagation of light. **(Guru Nanak and Meerut Univ.)**

3. Apply Huygens' principle to explain the laws of reflection and refraction of a plane wave at a plane surface.

4. Explain the laws of refraction of light on the basis of the wave theory. What is meant by refractive index and how is it related to the velocities of light in different media? **(Punjabi and Guru Nanak Univ.)**

5. Derive a relation for refraction of a spherical wave at a spherical surface according to the wave theory of light.

6. Derive the lens formula for a thin convex lens on the wave theory of light. **(Guru Nanak and Punjab Univ.)**

7. Account for the formation of images by refraction through lenses on wave theory and prove the relation
$$\frac{1}{v}-\frac{1}{u}=(\mu-1)\left(\frac{1}{R_1}-\frac{1}{R_2}\right)$$
where the symbols have their usual significance. **(Meerut Univ.)**

8. Derive the refraction formula for a thin concave lens on the basis of wave theory. **(Guru Nanak Univ.)**

9. On the basis of the wave theory, explain the phenomenon of total internal reflection and obtain the value of critical angle.

10. Prove the concave mimor equation
$$\frac{1}{v}+\frac{1}{u}=\frac{2}{r}$$
by applying the wave theory to study the reflection of spherical waves from a concave mirror.

Short Answer and Objective Questions

1. 'Light is dualistic in nature'. — Explain.

2. Explain what is meant by wavefront? What is a plane wave? What is the angle between the wavefront and the ray at a point?

3. State clearly the Huygens' principle of wave propagation.

4. What is meant by refractive index of an optical medium? How is it related to the velocity of light in the medium? What is the possible minimum value for it?

5. Distinguish between optical and geometrical paths. **(Himachal Univ.)**

6. Does refractive index of a medium depend on the frequency of the wave? Explain. **(Guru Nanak Univ.)**

7. Light is a
   (a) wave phenomenon
   (b) both particle and wave phenomenon
   (c) particle phenomenon
   (d) neither particle nor wave phenomenon.

8. Speed of light depends on
   (a) elasticity of medium only
   (b) inertia of medium only
   (c) elasticity as well as inertia
   (d) neither elasticity nor inertia.

9. When light is refracted, which of the following does not change?
   (a) Wavelength (b) Frequency (c) Velocity (d) Amplitude.

10. The wavefronts of light coming from a distant source of unknown shape are nearly
   (a) Plane (b) Cylindrical (c) Spherical (d) None of these.

Problems

1. The optical path of a monochromatic light is the same if it goes through 2 cm of glass or 2.25 cm of water. If the refractive index of glass is 1.50, what is the index of water? [1.33]

2. If the refractive index of glass is 1.50 and velocity of light in air is 3×10^8 m/s, find the velocity of light in glass. [2×10^8 m/s]

3. A beam of wavelength 0.16 m strikes the surface of a refracting medium of $\mu =$ 1.33. What will be the wavelength in the refracting medium? [0.12 m]

4. A ray of light strikes at an angle of incidence 55°. If the reflected and refracted rays are perpendicular to each other, find the refractive index of the medium. **(Punjabi Univ.)** [1.428]

5. Wavelength of sodium light in air is 589 nm. Find (a) the frequency in air, (b) its wavelength in water ($\mu = 1.33$), (c) its frequency in water and (d) its speed in water. [5.09×10^{14} Hz; 443 nm; 5.09×10^{14} Hz; 2.25×10^{8} m/s]

6. The speed of yellow light in a certain liquid is 2.4×10^{8} m/s. Find the refractive index of the liquid. [1.25]

7. Find the speed of light in vacuum when that in air is 3×10^{8} m/s. Given : the absolute refractive index of air = 1.00029. [3.00087×10^{8} m/s]

Chapter 8

Interference of Light

8.1 Introduction

Light is a form of wave motion and the waves are transverse in character. What happens when two wave trains of light are superposed on each other? If two or more disturbances arrive simultaneously at a space point, then the resultant disturbance at that point of space is given by the *principle of superposition.*

Principle of superposition—It states that *the resultant displacement of a particle in a medium, which is under the action of two or more waves simultaneously, is the vector sum of the displacements of the same particle due to the individual waves* (assumed small).

Let y_1 be the displacement of a particle at a point at time t in a given direction due to the action of a (single) wave train, and y_2 that due to another wave train in the absence of the first. Then, according to the principle of superposition, the instantaneous resultant displacement of the particle due to the two waves *acting simultaneously* is

$$y = y_1 \pm y_2$$

according as the two displacements are in the *same* or *opposite direction.*

8.2 Interference of light : Young's experiment

As a consequence of this principle, when two light beams intersect each other, the propagation of each beam remains unaffected by (i.e., independent of) the presence of the other. In the region of their intersection, however, both the beams act simultaneously and a change in intensity is likely to occur.

When monochromatic light waves, coming from two appropriate sources, proceed almost in the same direction and superpose at a point, the intensity of light at that point will be maximum or minimum according as the waves at the point are in the same or opposite phase. This phenomenon is called *interference of light* and is rather important, since the phenomenon requires, for its explanation, the *wave nature* of light.

Interference of light—When two light waves superimpose, then the resultant amplitude (or intensity, which is proportional to square of the amplitude) in the region of superposition is different from the amplitude (or intensity) of individual waves. At places where say, the *crest* due to one falls on the *crest* of the other or say, a *trough* of one falls on the *trough* of the other, the resultant *amplitude increases* and hence also the *intensity* of light. If, however, the *trough* of one falls on the *crest* of the other, the resultant amplitude or intensity decreases. Similar modifications in intensity also occur for intermediate points between a crest and a trough.

This modification in the distribution of light intensity in the region of superposition of two (or more) *waves of light is called* the **interference of light.**

When the resultant amplitude is the sum of the amplitudes of the separate waves, the interference is said to be **constructive,** and when the resultant amplitude is equal to the difference of the amplitudes of the interfering waves, the interference is said to be **destructive.**

Young's experiment—Young demonstrated the phenomenon of interference by an arrangement shown in Fig. 8.1. S is a pinhole through which sunlight was passed. The sunlight was then allowed to go through two other close pinholes S_1, S_2—much away but equidistant from S—to be received finally on a screen. The two sets of spherical waves that emerged from S_1 and S_2 interfered with each other to produce few coloured fringes of varying intensity on the screen.

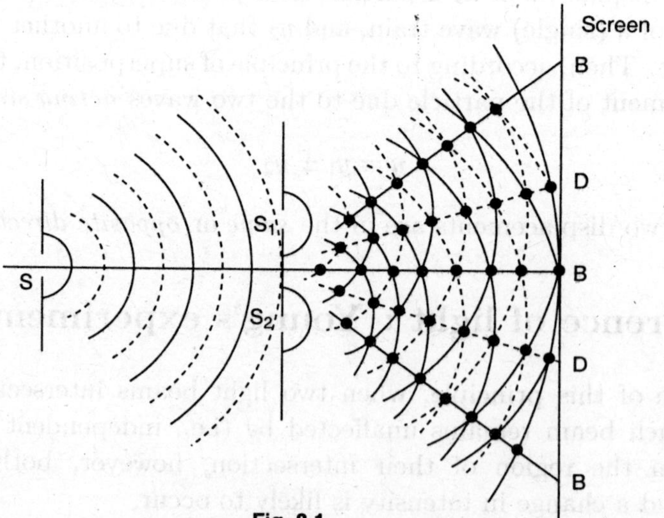

Fig. 8.1

• The improved version of the experiment consists in replacing the pinholes S_1, S_2 by narrow slits and the sunlight by monochromatic light. The intefering waves then are cylindrical and a number of alternate bright and dark fringes, parallel to the slit-length, are seen on the screen.

8.3 Conditions for interference of light

Two independent sources *cannot* produce interference. To obtain a *well-defined, observable* and *permanent* or *stationary* interference pattern, a number of conditions are to be satisfied.

1. Conditions for sustained interference—The required condictions are;

(a) *The two sources* i.e., *two interfering waves should be coherent. The phase difference between the waves must be zero or remain constant with time.*

Such sources are called **coherent** sources. If however the sources vibrate with varying phase difference, i.e., the phase changes rapidly, they are **incoherent.**

In a steady interference pattern, the intensity at any point which is given by

$$I = a_1^2 + a_2^2 + 2a_1a_2 \cos\delta$$

should be constant with time, *a*'s being the amplitudes of interfering waves. This implies that δ must be constant. But $\delta = \delta_1 + \delta_2$ where δ_1 is the initial phase difference and δ_2 is the phase difference at that point due to path difference between the waves; δ_2 is a constant as the path difference for a particular point remains constant in time. So for constant δ, the phase difference δ_1 between the waves must be constant.

(b) *The two interfering waves must have the same frequency* (or *wavelength*).

For light waves of different frequencies or wavelengths, the phase difference between them will continuously vary. Hence the resultant intensity at a point will change with time, being alternately maximum and minimum (like '*beats*' in sound) and will *not* be sustained.

(c) *If the interfering waves are polarised, they must be in the same state of polarisation.* This a necessary condition for interference of polarised light.

2. Conditions for good observation—For a good observation,

(a) *The separation between the two sources* (*d*) *should be small,* and

(b) *The distance between the screen and the two sources* (*D*) *should be large.*

For, we shall see shortly, the fringe-width $\beta = (D/d)\lambda$ and if d is small, the fringe-width is large and the fringes are separately visible. If however d is large, fringe-width will be small (since λ for light is extremely small) and the fringes may *not* be resolved as separate. For the same reason D also should be large.

(c) *The background should be dark.* Otherwise, the *visibility* will *not* be good and due to low intensity of fringes, they may *not* be visible at all.

3. Conditions for good contrast—For good contrast the conditions are:

(a) *The amplitudes of the interfering waves should be equal or very nearly equal.*

If a_1 and a_2 be the amplitudes of interfering beams, intensities of maxima and minima are proportional to $(a_1 + a_2)^2$ and $(a_1 \sim a_2)^2$ respectively. In case, $a_1 \gg a_2$, the intensities of maxima and minima will be practically the same and so the contrast will be *poor.* Hence for good contrast a_1 and a_2 must be nearly equal. Maximum contrast is obtained when $a_1 = a_2$, for then the dark fringes will have zero intensity.

(b) *The sources must be very narrow.* A broad source is equivalent to a large number of narrow sources juxtaposed. Each set of such two sources will give rise to its own pattern, leading to overlapping of a number of patterns. Contrast will thus be poor and the fringes may altogether disappear.

(c) *The two sources should be monochromatic* or *very nearly so.* Since the fringe-width $\beta\,(= D\lambda/d)$ depends on the wavelength λ, β for a given d and D will be constant and hence good intensity fringes will be seen. With composite light source, each will give rise to its own set of fringes with its own spacing. As a result, although the zero order maxima for all λ's will be in the same position, the other orders will overlap on each other. The net effect would be white light and uniform illumination. A few coloured fringes with poor contrast can only be observed near the centre.

• From above, it is apparent that **two real sources of light cannot act as coherent sources to produce any interference.** For (i) they will not emit light waves of the same frequency, nor have nearly the same amplitude; (ii) the phase of waves emitted by them will change independent of each other and hence the phase difference will not be constant in time.

• It is also clear from above that **we cannot obtain interference with wide slits.** A wide source is equivalent to a large number of narrow sources lying side by side, each pair producing its own fringe pattern which overlap leading to general illumination.

• It is interesting to see *what happens in double slit experiment if one slit is covered with green transparent paper and the other, say, with blue.* The two slits now give light waves of different wavelengths ($\lambda_b \neq \lambda_g$). The sources are no longer capable of producing interference pattern; they are *not* coherent.

• **Can we get interference pattern if slit-separation d be less than λ,** *the wavelength* ? Since $\beta = (D/d)\lambda$, if $d < \lambda$, then β is *extremely* large leading to *uniform illumination;* there will be *no* visible fringes.

8.4 Analytic treatment of interference (wave theory)

Let S be a monochromatic narrow light source (a slit) and S_1, S_2 two similar narrow parallel slits, very close to each other and equidistant from S (Fig. 8.2). The waves arriving at S_1 and S_2 from S will be in phase at all times. We shall find the resultant intensity of light at a point P on the screen XY placed parallel to S_1S_2 due to interference of light waves from S_1 and S_2.

Let a_1 and a_2 be the amplitudes at P due to waves from S_1 and S_2 respectively and δ be their phase difference on reaching P. If y_1 and y_2 be the displacements of the waves at P,

$$y_1 = a_1 \sin \omega t \tag{8.4.1}$$
$$y_2 = a_2 \sin(\omega t + \delta) \tag{8.4.2}$$

ω being their common frequency (angular).

By the principle of superposition, the resultant displacement at P is

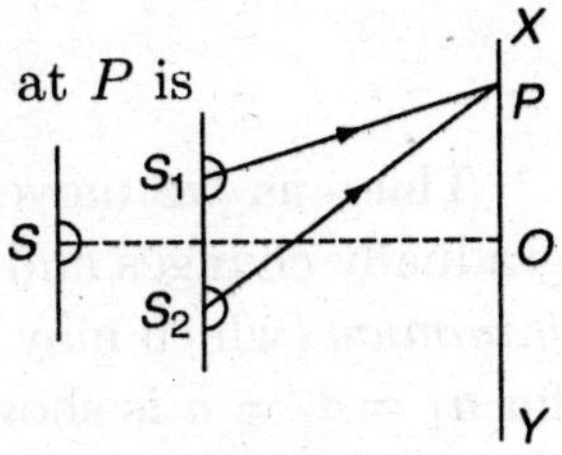

Fig 8.2

$$\begin{aligned} y = y_1 + y_2 &= a_1 \sin \omega t + a_2 \sin(\omega t + \delta) \\ &= a_1 \sin \omega t + a_2 \sin \omega t \cos \delta + a_2 \cos \omega t \sin \delta \\ &= \sin \omega t (a_1 + a_2 \cos \delta) + \cos \omega t (a_2 \sin \delta) \\ &= \sin \omega t \cdot R \cos \theta + \cos \omega t \cdot R \sin \theta \\ &= R \sin(\omega t + \theta) \end{aligned} \tag{8.4.3}$$

where $R \cos \theta = a_1 + a_2 \cos \delta$; $R \sin \theta = a_2 \sin \delta$, so that by squaring and adding

$$R^2 = a_1^2 + a_2^2 + 2a_1 a_2 \cos \delta \tag{8.4.4}$$

The resultant intensity I at P is proportional to R^2.

$$\therefore \ I = a_1^2 + a_2^2 + 2a_1 a_2 \cos \delta \tag{8.4.5}$$

taking, for the sake of simplicity, the constant of proportionality as 1.

The phase difference δ between the two waves at P is given by

$$\delta = \frac{2\pi}{\lambda} \times \text{path difference} = \frac{2\pi}{\lambda}(S_2P - S_1P)$$

The treatment shows that *the resultant intensity* (8.4.5) *at P is* **not** *just the sum of the intensities due to the separate waves, i.e., $I \neq a_1^2 + a_2^2$*; but it depends on δ. The two waves are said to *interfere*.

Maxima or minima—The intensity I is a **maximum** when $\cos \delta = +1$, that is,

$$\text{Phase difference, } \delta = 2n\pi, \qquad n = 0, 1, 2, \cdots$$
$$\text{or, Path difference, } \Delta = S_2P - S_1P = 2n\lambda/2, \ n = 0, 1, 2, \cdots$$

$\therefore$ *The intensity is* **maximum** *when the path difference of the point P from the two interfering waves is an* **even** *multiple of $\lambda/2$* and a bright band is obtained there. This is **constructive interference**.

The maximum intensity, $I_{\text{max}} = a_1^2 + a_2^2 + 2a_1a_2 = (a_1 + a_2)^2$, i.e., *the maximum intensity is greater than the sum of the two separate intensities* $(a_1^2 + a_2^2)$.

The intensity I is a **minimum** when $\cos \delta = -1$, that is,

$$\text{Phase difference, } \delta = (2n + 1)\pi, \qquad n = 0, 1, 2, \cdots$$
$$\text{or, Path difference, } \Delta = S_2P - S_1P = (2n + 1)\lambda/2, \ n = 0, 1, 2, \cdots$$

$\therefore$ *The intensity is* **minimum** *when the path difference of the point P from the two interfering waves is an* **odd** *multiple of $\lambda/2$* and a near dark band is obtained there. This is the **destructive interference**.

The minimum intensity, $I_{min} = a_1^2 + a_2^2 - 2a_1a_2 = (a_1 \sim a_2)^2$, i.e., *the minimum intensity is less than the sum of the two separate intensities* $(a_1^2 + a_2^2)$.

Special case—When the amplitudes of the two waves are equal, i.e., $a_1 = a_2 = a$ (say), then

$$I_{max} = a^2 + a^2 + 2a^2 = 4a^2$$
$$I_{min} = a^2 + a^2 - 2a^2 = 0.$$

Thus, as we move on the screen, the path difference Δ between the two waves gradually changes and a variation in intensity of light, being alternately *maximum* and *minimum* (which may be zero if $a_1 = a_2$), is observed. The intensity distribution curve for $a_1 = a_2 = a$ is shown in Fig. 8.3. This is known as the **interference pattern** and the alternate bright and dark regions the **interference fringes.**

• The central bright band at O corresponds to $\Delta = 0$, when $n = 0$. The next bright band corresponds to $n = 1$ and is known as the *first order* bright band, and so on.

• The above analysis is based on the fact that the two waves start from S_1 and S_2 in the *same phase.* Hence at P, they have a *constant phase difference* δ due to the difference in paths they describe. Such waves are called '*coherent*' and are in a position to *interfere.*

8.5 Interference and conservation of energy

At first sight, it may appear that the law of conservation of energy is violated in the phenomenon of interference, for the energy at the minima points is lost. But this is only *apparent.*

In the absence of interference, the intensity due to two waves should be $a^2 + a^2 = 2a^2$, but the intensity at the maxima points is $4a^2$ while that at minima points is *zero.* Fig. 8.3 shows the variation of intensity I as a function of phase difference δ.

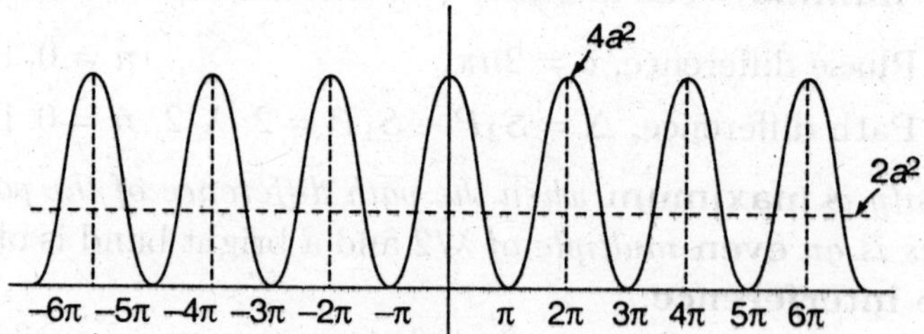

Fig 8.3

The average value of intensity in the range $\delta = 0$ to $\delta = 2\pi$ is

$$\bar{I} = \frac{\int_0^{2\pi} I\,d\delta}{\int_0^{2\pi} d\delta} = \frac{\int_0^{2\pi}(a_1^2 + a_2^2 + 2a_1a_2\cos\delta)d\delta}{2\pi} = \frac{\left[a_1^2\delta + a_2^2\delta + 2a_1a_2\sin\delta\right]_0^{2\pi}}{2\pi}$$

$$= a_1^2 + a_2^2 = 2a^2, \quad \text{when } a_1 = a_2 = a, \text{ say.}$$

Thus, *the average value of intensity equals the sum of the separate intensities.*

So the energy which apparently disappears at minima has actually been transferred to the maxima where the intensity is $4a^2 > 2a^2$, the sum of the intensity of the two beams acting separately. There is thus *no destruction of energy as a result of interference.* The **energy** is simply **redistributed**. *The formation of interference fringes is thus in accord with the law of conservation of energy.*

8.6 Fringe-width and shape of fringes in double slit

Let S be a narrow slit illuminated by monochromatic light of wavelength λ and S_1, S_2 two parallel narrow slits, very close to each other and equidistant from S whence waves arrive at S_1, S_2 in the same phase (Fig. 8.4). Beyond S_1, S_2 waves proceed to produce an interference pattern on the screen XY, placed parallel to S_1 and S_2.

Let MO be the perpendicular bisector of S_1S_2, which meets the screen at O. P is any point on the screen. Let S_1S_2, the slit separation, be d; MO, the slit-screen distance, be D and OP be x.

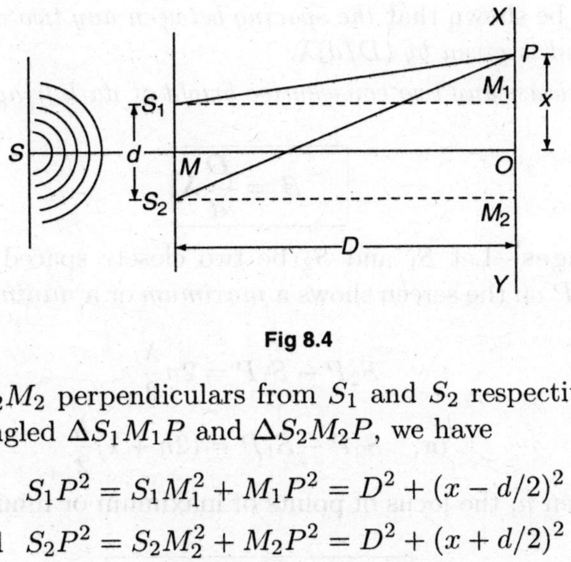

Fig 8.4

Draw S_1M_1 and S_2M_2 perpendiculars from S_1 and S_2 respectively on the screen. Now, from the right angled ΔS_1M_1P and ΔS_2M_2P, we have

$$S_1P^2 = S_1M_1^2 + M_1P^2 = D^2 + (x - d/2)^2 \tag{8.6.1}$$

$$\text{and} \quad S_2P^2 = S_2M_2^2 + M_2P^2 = D^2 + (x + d/2)^2 \tag{8.6.2}$$

$$\therefore \quad S_2P^2 - S_1P^2 = (x + d/2)^2 - (x - d/2)^2$$

$$\Rightarrow (S_2P - S_1P)(S_2P + S_1P) = 2xd$$

$$\text{or,} \quad (S_2P - S_1P)2D \simeq 2xd \quad (\because \ d \ll D)$$

$$\therefore \quad S_2P - S_1P = \text{path difference, } \Delta = x\frac{d}{D}$$

Spacing of fringes (Fringe-width)—For the nth **bright** fringe, we have

$$\Delta = 2n\frac{\lambda}{2} \quad (n = \text{integer, including } 0)$$

$$\Rightarrow \quad x\frac{d}{D} = 2n\frac{\lambda}{2} \quad \Rightarrow \quad x = 2n\frac{\lambda}{2}\frac{D}{d} \tag{8.6.3}$$

For the nth **dark** fringe, we have

$$\Delta = (2n+1)\frac{\lambda}{2} \qquad (n = \text{integer, including } 0)$$

$$\Rightarrow \quad x\frac{d}{D} = (2n+1)\frac{\lambda}{2} \quad \Rightarrow \quad x = (2n+1)\frac{\lambda}{2}\frac{D}{d} \tag{8.6.4}$$

Now, let x_n and x_{n+1} be the distances of nth and $(n+1)$th *bright fringe* from O.

$$\therefore \ x_n = 2n\frac{\lambda}{2}\frac{D}{d}; \qquad x_{n+1} = 2(n+1)\frac{\lambda}{2}\frac{D}{d}$$

$\therefore$ Spacing between nth and $(n+1)$th *bright fringes* is

$$\beta = x_{n+1} - x_n = \frac{D}{d}\left[2(n+1)\frac{\lambda}{2} - 2n\frac{\lambda}{2}\right] = \frac{D}{d}\lambda \tag{8.6.5}$$

which is *independent* of the order of the fringe n.

So, *the spacing between any two consecutive bright fringes is the same.*

It can similarly be shown that *the spacing between any two consecutive dark fringes is also the same and is given by* $(D/d)\lambda$.

The *spacing between any two consecutive bright* or *dark fringes is called the* **fringe-width, β**.

$$\boxed{\therefore \ \beta = \frac{D}{d}\lambda} \tag{8.6.6}$$

Shape of fringes—Let S_1 and S_2 be two closely spaced *coherent* sources [Fig. 8.5(a)]. The point P on the screen shows a *maximum* or a *minimum* intensity according as

$$S_2P - S_1P = 2n\frac{\lambda}{2}$$

$$\text{or,} \quad S_2P - S_1P = (2n+1)\frac{\lambda}{2}$$

Hence for a given n, the locus of points of maximum or minimum intensity is given by

$$\boxed{S_2P - S_1P = \text{constant}}$$

This represents a *hyperbola* with S_1 and S_2 as its foci. In space, the locus of points of maximum (or minimum) for a given n will thus be **confocal hyperboloids** and can be obtained by a revolution of the hyperbola about the line S_1S_2.

The fringes however are usually observed on a screen XY in a plane perpendicular to the plane of the figure and parallel to S_1S_2. So, the observed fringes are *sections of hyperboloids* by this plane, that is, they are *hyperbolae*. But the wavelength of light is small ($\sim 10^{-5}$ cm) and the value of $(S_2P - S_1P)$ is also of the same order. The hyperbolae therefore appear, in practice, as *straight lines* as shown in [Fig. 8.5(b)].

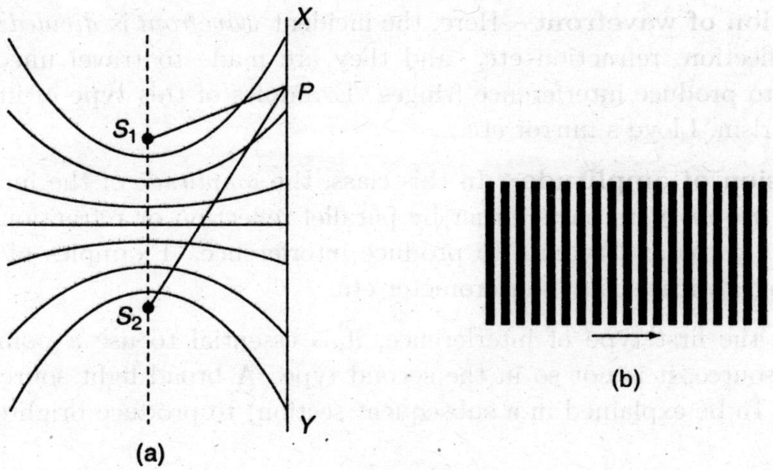

Fig 8.5

• Note that the fringe-width β is *directly proportional* to D and λ and *inversely proportional* to d. If D and d are constants, $\beta \propto \lambda$ so that *fringes produced by shorter wavelengths will be narrower* compared to those due to longer wavelengths.

8.7 White light fringes : Colour effect

The distance of nth bright fringe from the central one, in double-slit interference, is

$$x_n = n\frac{\lambda D}{d}$$

where d = inter-slit separation and D = source-screen distance.

Note that $x_n = f(\lambda)$, so that the fringes of different λ's (colours) will be in step only for the *central bright band* ($n = 0$). Now, for $n = 0$, $x_n = 0$ implying that the bright fringes of all wavelengths will coincide. The *central band will thus be white with white light.* For higher orders, however, $n > 0$ and x_n will be greater, greater the λ-value and conversely. As λ_r for red light is longer than λ_v for violet light, all bright bands, except the central one, will appear *coloured*; while red would be in the outermost position, violet will be in the innermost position.

When the path difference Δ between the interfering waves becomes large, the condition for constructive interference for one wavelength and that for destructive interference for another wavelength may be satisfied for the *same point.* In such a case, the resultant illumination is hardly distinguishable from white light. So, for white light fringes, the path difference should be kept very small and we shall then obtain few coloured fringes on either side of the central white fringe.

8.8 Types of interference

Interference may be divided into *two* main classes: (i) by the *division of wavefront* and (ii) by the *division of amplitude.*

1. Division of wavefront—Here, the incident *wavefront* is *divided* into *two* parts either by reflection, refraction etc. and they are made to travel unequal distances and reunite to produce interference fringes. Examples of this type of interference are Fresnel's biprism, Lloyd's mirror etc.

2. Division of amplitude—In this class, the *amplitude* of the incident beam is *divided* into *two or more* parts either by parallel reflection or refraction. They travel over different paths and reunite to produce interference. Examples of this class are Newton's rings, Michelson's interferometer etc.

While in the first type of interference, it is essential to use a point source or a narrow line source, it is not so in the second type. A broad light source is employed here (Why? To be explained in a subsequent section) to produce brighter fringes.

A. Interference by division of wavefront

8.9 Fresnel's Biprism

It is a device, made by Fresnel, to produce *two coherent sources* by the *division of wavefront* to obtain sustained interference.

It consists essentially of *two prisms, each of very small refracting angle,* placed base to base, forming a single obtuse angled prism. In practice, the prism is made from a single optically plane glass plate by proper grinding and polishing. The obtuse angle of the prism is about 179° and the other angles are about 30′ each.

Production of fringes : Set-up—The arrangement for obtaining inteference fringes is illustrated in Fig. 8.6.

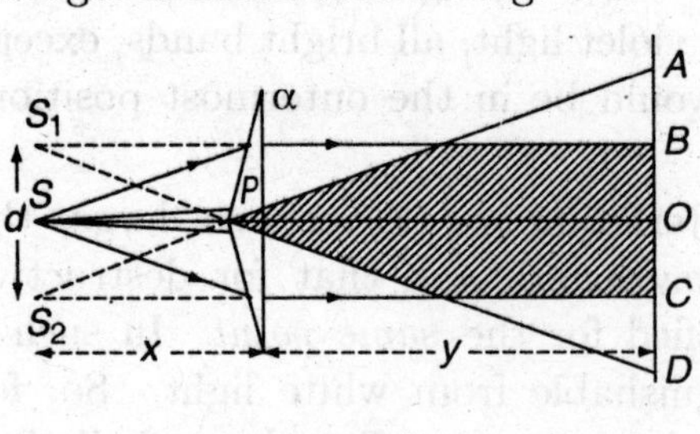

Fig 8.6

S is a *narrow vertical* slit and is illuminated by *monochromatic light.* The light emerging from S is allowed to fall *symmetrically* on the biprism P, placed at a small distance from the slit S and with its refracting edges parallel to S. The light beams emerging from the upper and lower halves of the biprism appear to diverge from two virtual images S_1 and S_2 respectively of the slit S. S_1 and S_2 thus act as *two coherent (virtual) sources.* The light cones BS_1D and AS_2C, diverging respectively from S_1 and S_2 are superposed, BC being the overlapping region where interference fringes are formed. They may be received on a screen AD placed perpendicular to the plane of paper or seen through an eyepiece. The fringes are *not localised* (non-localised fringes) in any particular plane and that is why the screen can be placed anywhere within a suitable distance.

8.9.1 Theory of biprism : Determination of unknown λ

The *theory* of biprism is essentially the same as that of the double slit, already discussed and would not be repeated.

The unknown wavelength λ of monochromatic light can be determined from biprism fringes, using the formula

$$\beta = \frac{D}{d}\lambda \quad \Rightarrow \quad \boxed{\lambda = \beta\frac{d}{D}} \tag{8.9.1}$$

where β is the fringe-width, d the distance between the virtual sources and D [$= (x + y)$] the separation between the slit S and the eyepiece at O, fitted with a micrometer.

The above formula has already been derived in Art. 8.6.

Procedure : ***Measurement of*** $\boldsymbol{\beta}$—The vertical crosswire is properly adjusted and made to coincide with any bright fringe. The corresponding reading of the micrometer is recorded. Next, by moving the micrometer, the crosswire is set on different bright fringes, one by one, and the micrometer readings recorded in each case. The mean distance between two successive fringes gives β.

Measurement of D—The slit-eyepiece separation can be read directly from the optical bench. On making the usual *bench error* correction to the above reading, D is obtained.

Measurement of d—For this, a suitable convex lens is mounted between the biprism and the eyepiece. The focal length of the lens should be less than one-fourth of the separation between the eyepiece and the biprism (Fig. 8.7). The lens is adjusted for two positions L_1 and L_2 for which sharp images of S_1 and S_2 are obtained in the field of view of the eyepiece. If c_1 and c_2 be the distances between the real images in the two cases, then from Fig. 8.7.

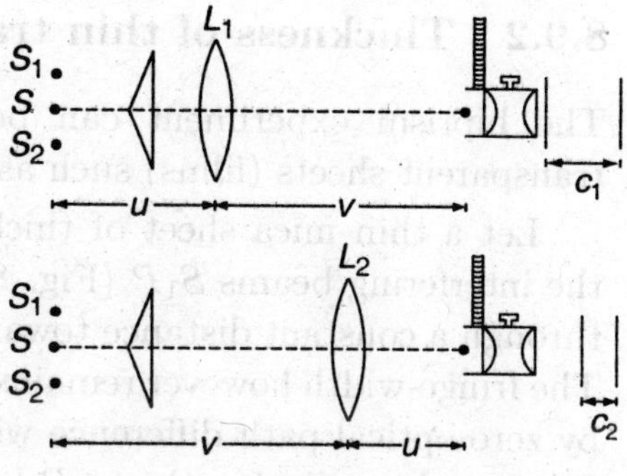

Fig 8.7

$$\frac{c_1}{d} = \frac{v}{u}; \quad \frac{c_2}{d} = \frac{u}{v} \quad \Rightarrow \quad \frac{c_1}{d} = \frac{d}{c_2}$$

$$\therefore \quad d^2 = c_1c_2 \quad \Rightarrow \quad d = \sqrt{c_1c_2}$$

So, by measuring c_1 and c_2, d can be readily evaluated.

By substituting the observed and computed values of β, d and D, the wavelength λ can be determined from (8.9.1).

• **How to locate the zero order** in the biprism fringe pattern when, with monochromatic light, all bright fringes look alike? With *white* light, however, only the *zero order* fringe will be *white*, all else coloured. It can then be readily located and the crosswire positioned there. Next, on replacing the white light by monochromatic one, the fringe appearing at the crosswire gives the *zero order* fringe.

• What happens in the fringe pattern if the **eyepiece is moved away from source**? Since the fringe-width $\beta = (D/d)\lambda$, β increases with D, the separation between the eyepiece and the source. The **fringes** thus become **wider.** The converse result follows when D decreases, i.e., the fringes become *narrower.*

• **How to measure the acute angle of a biprism?** The separation d between the virtual sources S_1, S_2 can be expressed, as follows, in terms of the refracting angle α of the prism, the index μ of its material and the slit-prism separation x. Since α is very small, the deviation of a ray produced by each of the prism is $(\mu - 1)\alpha$. If θ be the total deviation of the ray striking the centre of the biprism (Fig. 8.8),

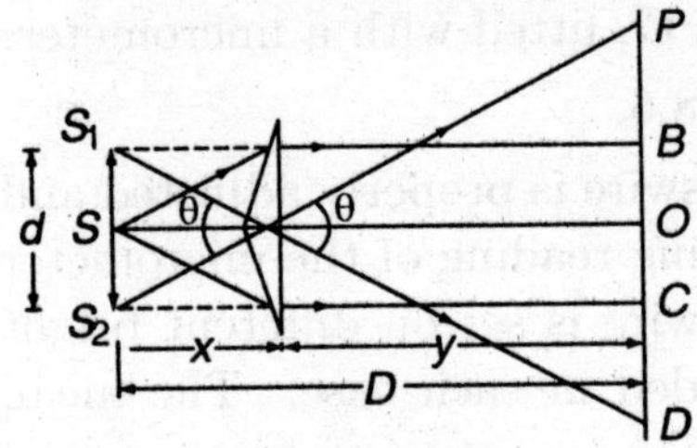

Fig 8.8

$$\tfrac{1}{2}\theta = (\mu - 1)\alpha; \text{ also } d = x\theta.$$

$$\therefore \quad \boxed{d = 2x(\mu - 1)\alpha}$$

Measuring d and x and knowing the index μ for the monochromatic light used, α can be calculated.

8.9.2 Thickness of thin transparent sheet (film)

The biprism experiment can be used to determine the thicknesses of *very thin* transparent sheets (films) such as of mica, glass etc.

Let a thin mica sheet of thickness t and index μ be placed in the path of one of the interfering beams S_1P (Fig. 8.9). The entire fringe sytem, as a result, gets shifted through a constant distance towards the path of the beam in which the sheet is placed. The fringe-width however remains unchanged. Let the central fringe at O, characterised by zero optical path difference with respect to the sources S_1 and S_2, be shifted to P. So, now the optical paths to P from S_1 and S_2 must be equal.

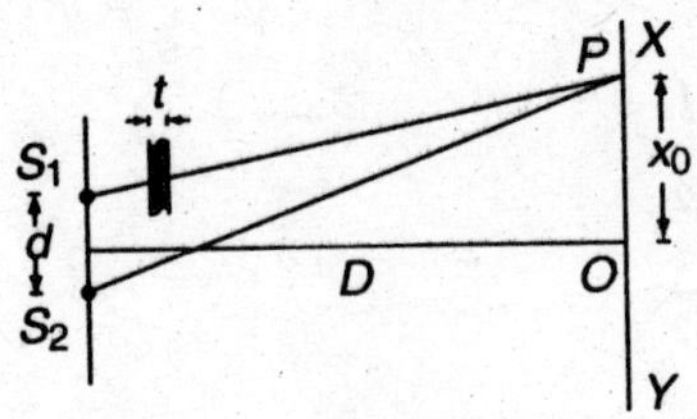

Fig 8.9

$$\therefore \quad (S_1P - t) + \mu t = S_2P$$

$$\Rightarrow S_2P - S_1P = (\mu - 1)t \qquad (8.9.2)$$

In the original fringe pattern (i.e., when the sheet was not interposed) P was the site of say nth bright fringe.

$$\therefore \quad S_2P - S_1P = n\lambda \qquad (8.9.3)$$

Thus, from equations (8.9.2) and (8.9.3), we obtain

$$(\mu - 1)t = n\lambda = n\beta\frac{d}{D}$$

$$(\because \beta = \text{fringe-width} = D/d\lambda \;\Rightarrow\; \lambda = \beta d/D)$$

But $n\beta$ = the amount of shift of the central bright fringe = x_0, say.

$$\therefore \quad (\mu - 1)t = x_0 \frac{d}{D}$$

$$\Rightarrow \quad \boxed{t = \frac{x_0 d}{(\mu - 1)D} = \frac{x_0}{\beta(\mu - 1)}\lambda} \quad (\because \beta = D\lambda/d) \qquad (8.9.4)$$

So, if μ and λ are known, the thickness t of the sheet can be readily evaluated by determining experimentally the fringe-width β and the shift x_0 of the central bright fringe.

With **monochromatic light,** all the bright fringes look alike and it is **not possible** to locate the central fringe or zero order to measure the shift x_0. This is usually done with white light when the central fringe is white, all else are coloured. Hence the position to which central fringe is shifted can be easily located. Finally, the shift is accurtely measured using monochromatic light.

- Relation (8.9.4) may also be exploited to find μ, provided λ and t are known or to find λ, if μ and t are known etc.
- Why should the *shift be towards the film*? From (8.9.4)

$$x_0 = \frac{D}{d}\left(\frac{c}{v} - 1\right)t$$

where c is the speed of light in vacuum. Light travels with smaller velocity v in denser medium (film), i.e., $c/v > 1 \Rightarrow x_0$ is positive. Hence the central fringe shifts towards the film.

8.10 Phase change on reflection : Stokes' treatment

If a light wave is reflected at the surface of an optically denser medium, it suffers a change in phase by π, i.e., an extra path difference of $\lambda/2$ occurs.

Stokes' treatment—Let a light wave AB of amplitude a be incident (Fig. 8.10) at B on the surface of separation PQ of two media 1 and 2, medium 2 being optically denser. The wave AB will be partly reflected along BC and partly transmitted along BD. If r and t be the fractions of amplitude reflected and transmitted respectively, the amplitude along BC is ar and that along BD is at.

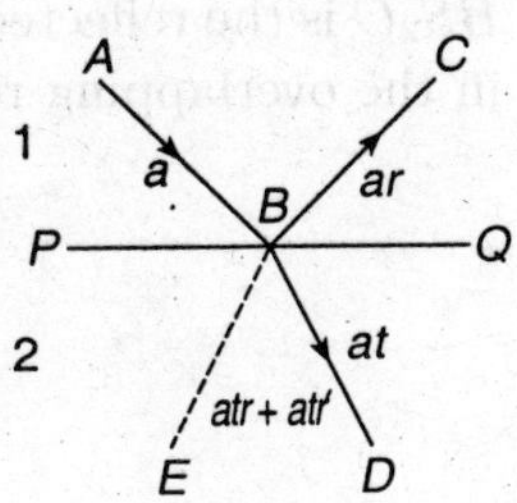

Fig 8.10

Suppose now that the direction of both BC and BD is reversed. On reversal, the wave BC gives a reflected wave of amplitude ar^2 along BA and a transmitted wave of amplitude art along BE. If r' and t' be the fractions of amplitude reflected and transmitted when the wave travels from a denser to a rarer medium, then the wave BD, on reversal, gives a transmitted wave of amplitude att' along BA and a reflected wave of amplitude atr' along BE.

Now, the *principle of reversibility* of light dictates that the reflected and transmitted waves BC and BD, on reversal, should give only the original ray BA of amplitude a.

$$\therefore \quad art + ar't = 0 \quad \Rightarrow \quad r' = -r \tag{8.10.1}$$

$$\text{and} \quad ar^2 + att' = a \quad \Rightarrow \quad tt' = 1 - r^2 \tag{8.10.2}$$

The negative sign in (8.10.1) indicates displacements in opposite directions, i.e., a phase of π at reflection either from rarer to denser medium or from denser to rarer medium.

This phase change in reflection is **demonstrated experimentally** in Lloyd's mirror experiment where interference occurs between waves direct from the source and those reflected from an optically denser medium. It is seen that the central fringe is *dark* instead of *bright*. Unless light suffers a phase change of π on reflection at the mirror, this is *not* possible. Thus **a phase change of π takes place on reflection at the surface of a denser medium.**

• It should be noted that *no such phase change* is found to be introduced if the *reflection takes place* at the surface of a *rarer medium.*

8.11 Lloyd's mirror

This is a simple device to obtain two *coherent* light sources to produce sustained interference.

It consists of just a plane mirror MN polished on the front surface and blackened at the back (to stop multiple reflections). The arrangement is illustrated in Fig. 8.11 where S_1 is a narrow slit, illuminated by monochromatic light, placed with its length parallel to the surface of the mirror. Light from S_1 falls on the mirror at nearly *grazing* incidence and the reflected beam appears to *diverge* from S_2, which is the virtual image of S_1. Thus S_1 and S_2 act as two coherent sources. BS_1E is the direct light cone and BS_2C is the reflected light cone. The two get superposed to produce interference fringes in the overlapping region BC on the screen XY.

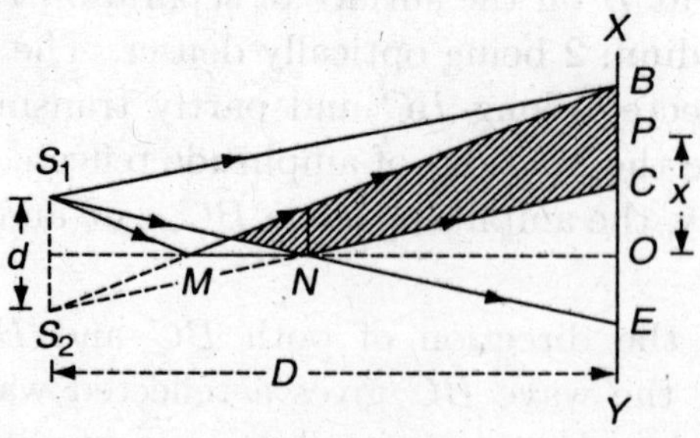

Fig 8.11

Central or zero order fringe—The central or zero order fringe is expected on the screen to be at O, on the the perpendicular bisector of S_1S_2. But it *is not,* as

O is outside the region of superposition. The zero order however can be made visible by (i) moving the screen towards MN, the mirror, such that O coincides with N, or (ii) by properly introducing a thin mica sheet in the path of light from so that the entire fringe pattern is *shifted downward.*

With white light, the central or zero order fringe is expected to be *bright*, but in fact it is *dark.* This is due to the phase change of π when light is reflected from the mirror (a denser medium), so that the path difference between the interfering waves at the position of zero order is *not* zero but $\lambda/2$. The **central fringe** is thus **dark.**

Determination of wavelength λ—As in double slit, here also the fringe-width β is given by

$$\beta = \frac{D}{d}\lambda \Rightarrow \lambda = \frac{\beta d}{D}$$

where d is the distance between the coherent sources and D that between the screen and source.

$\therefore$ λ can be evaluated by measuring the fringe-width β and knowing d and D.

The mirror is first mounted vertically on an optical bench and placed along the length of the bench. A narrow vertical slit is also properly mounted on the same bench and illuminated by a monochromatic source. The mirror is next rotated about an axis parallel to bench-length until distinct fringes appear in a micrometer eyepiece. This ensures that the plane of the mirror is exactly parallel to the length of the slit.

The fringe-width β is measured by the usual method using the eyepiece, d by displacement method using a convex lens, and D by a metre scale.

Achromatic fringes with white light—A system of bright and dark fringes, without any colour, obtained by white light is known as **achromatic fringes**.

With white light, we usually obtain a central bright fringe with a few coloured ones on either side, for the fringe-width $\beta\,(= D\lambda/d)$ is different for different λ's (colours). If however β could be made the same for all λ's, the maxima for each order for all λ's will coincide forming achromatic fringes. Necessarily, the condition for this is:

$$\frac{D}{d}\lambda = \text{constant} \;\Rightarrow\; \frac{\lambda}{d} = \text{constant}.$$

It is easily realised, in Lloyd's mirror, employing a normal spectrum produced by a plane diffraction grating as shown in Fig. 8.12.

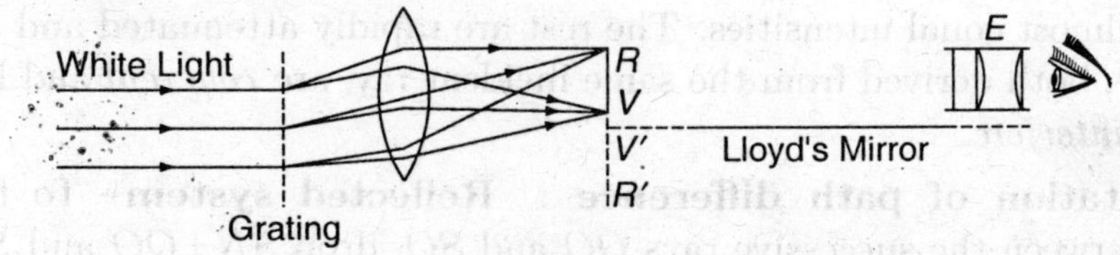

Fig 8.12

If a narrow spectrum RV is formed, say by a plane diffraction grating, the Lloyd's mirror will produce its reflected image $V'R'$. If we have,

$$\frac{\lambda_r}{RR'} = \frac{\lambda_v}{VV'}, \text{ then } \frac{\lambda}{d} = \text{constant, for all } \lambda\text{'s.}$$

Since in a grating spectrum, *the dispersion is accurately proportional to the wavelength, the condition,* λ/d = constant, *is simultaneously satisfied for all wavelengths.* If red light is used, red and black fringes of width $D\lambda/d$ are produced. If blue light is used blue and black fringes of the same fringe-width will be formed. So, if white light is used, black and white achromatic fringes will be produced.

Comparison of biprism and Lloyd mirror fringes—The main points of difference between the biprism and Lloyd's mirror fringes may be enumerated as under.

1. In biprism, the *complete pattern* of fringes is obtained, but in Lloyd's mirror *less than half of the complete pattern on one side of the central fringe* are visible; the *central fringe is itself invisible.*

2. While in biprism the *central fringe is bright,* it is *dark* in Lloyd's mirror.

3. Compared to the Lloyd's mirror, the *central fringe* in biprism is *less sharp.*

4. In biprism, fringes are produced by interference of *two coherent sources both* of which are *virtual;* in Lloyd's mirror, one coherent source is *real,* the other one is *virtual.*

B. Interference by division of amplitude

8.12 Interference in thin films (wedge-shaped)

Let a ray PQ of monochromatic light of wavelength λ, be incident at an angle i on a transparent wedge-shaped film, bounded by the surfaces AB and CD and enclosing an angle α, as shown in Fig. 9.14. At Q, the ray is partly reflected along QO and partly refracted along QR at an angle r. At R, it is again partly reflected along RS and partly refracted along RT_1. This process of *multiple reflections and refractions* are repeated at S, V etc. As a result, we have a set *reflected rays* QO, SO, $\cdots$ etc. and another set of *transmitted rays* $RT_1, VT_2, \cdots$ etc. At each point $Q, R, S, \cdots$ etc., only a fraction of light is reflected; the rest is refracted so that rays QO and SO, suffering one reflection each, have almost equal intensities. The rest are rapidly attenuated and are ignorable. QO and SO, both derived from the same incident ray, are *coherent* and hence are in a position *to interfere.*

Computation of path difference : Reflected system—To find the path difference between the successive rays QO and SO, drop $SN \perp QO$ and $SN_1 \perp QR$ and $SLM \perp CD$. Produce QR and SL to meet at M.

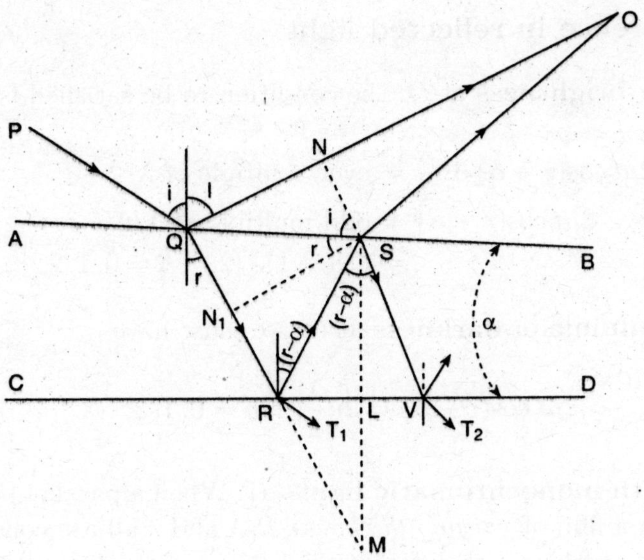

Fig. 8.13

From geometry, $RS = RM$ and $SL = LM = t$, the thickness of the film at S. So, the path difference between QO and SO is

$$\Delta = \text{Path } (QN_1 + N_1R + RS) \text{ in film } - \text{ path } QN \text{ in air}$$

$$= \mu(QN_1 + N_1R + RS) - QN \qquad (8.12.1)$$

μ being the index of the film.

But, $$\mu = \frac{\sin i}{\sin r} = \frac{QN/QS}{QN_1/QS} = \frac{QN}{QN_1}$$

$$\Rightarrow \quad QN = \mu \cdot QN_1$$

$\therefore$ From (8.12.1), $$\Delta = \mu(N_1R + RS) = \mu(N_1R + RM)$$

$$= \mu \cdot N_1M$$

$$= \mu \cdot SM\cos(r - \alpha)$$

$$\therefore \quad \boxed{\Delta = 2\mu t\cos(r - \alpha)} \qquad (8.12.2)$$

In computing the conditions for maxima and minima, one must remember that *a ray reflected at a surface backed by a denser medium suffers an abrupt phase change by* π, *equivalent to a path difference of* $\lambda/2$.

So, in **reflected system**, the net path difference between the above two rays (QO and SO) is

$$\boxed{\Delta = 2\mu t\cos(r - \alpha) \pm \frac{\lambda}{2}}$$

8.12.1 Interference in reflected light

$\therefore$ For **maxima** or **brightness** at O, the condition to be satisfied is

$$\begin{aligned} 2\mu t\cos(r-\alpha) \pm \frac{\lambda}{2} &= \text{even multiple of } \lambda/2 \\ \Rightarrow \quad 2\mu t\cos(r-\alpha) &= \text{odd multiple of } \lambda/2 \\ &= (2n+1)\lambda/2, \quad n = 0, 1, 2, \cdots, \end{aligned}$$

Similarly, for **minima** or **darkness** at O, we must have

$$2\mu t\cos(r-\alpha) = 2n\frac{\lambda}{2}, \quad n = 0, 1, 2, \cdots.$$

2. Fringes with monochromatic light—(i) When a parallel beam of monochromatic light falls on a film of *varying thickness*, μ, λ and r all are constants; so we shall get different orders of fringes for different thickness t of the film.

With increasing t, the order number of the fringe increases and a given order corresponds to the locus of points of equal t. The fringes are therefore called **fringes of equal thickness** or **FET** fringes.

At the thin edge of the film, t is essentially zero so that $\Delta = \lambda/2$. The thin edge would thus be dark, for all λ-values (lights of all colours).

(ii) For *extremely thin* film, $t \to 0$ for all parts of the film and the path difference between the reflected rays, $\Delta = \lambda/2$. The film surface then appears completely dark even with white light.

(iii) If the **film** surfaces are **plane-parallel** (i.e., $\alpha = 0$) and a parallel beam of monochromatic light is incident obliquely, μ, λ and t are all constants but r will vary with the inclination of the beam. The surface would thus be bright or dark according as r satisfies the conditions of brightness or darkness.

The fringe of a given order here corresponds to the locus of points of equal r. Such fringes are thus known as **fringes of equal inclination or FEI fringes.**

For *normally incident* parallel beam, $\cos r = 1$, so that the film surface will be uniformly bright or dark according as t satisfies the brightness or darkness condition.

2. Fringes with white light : Colour of thin films—When a parallel beam of white light falls on a thin wedge-shaped film, μ, λ, r will be different for light of different colours in the white light. As $\lambda_v < \lambda_r$, the first order bright fringe of violet light occurs at a smaller t of the film than the t-value for the corresponding bright fringe of red. We shall thus get *differently coloured fringes at different thickness.* This explains the *colour of thin films.* Such fringes are called **fringes of equal chromatic order** or **FECO** fringes.

At the *thin edge* of the film, t is essentially zero and the edge will be practically *dark* since $\Delta = \lambda/2$ for all λ's, as explained earlier.

It may so happen that we get another thickness where the brightness condition is *simultaneously* satisfied by two or more colours leading to a *coloured band* by overlapping. When the film thickness is sufficient, the overlapping is to such an extent that *uniform illumination results.*

8.12.2 Interference in transmitted light

A similar interference phenomenon occurs with rays transmitted through the film and emerging on the side CD of it.

Calculations, following a procedure similar to that adopted in reflected beam, yield for the path difference Δ between two successive rays and is given by

$$\boxed{\Delta = 2\mu t \cos(r - 2\alpha)} \tag{8.12.3}$$

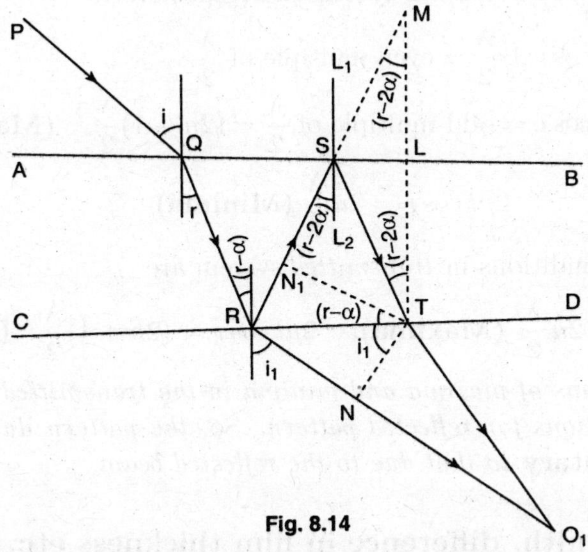

Fig. 8.14

Proof. In reference to Fig. 8.14, one part of the ray QR incident on the boundary CD emerges out along RO_1, the other part undergoes *two successive internal reflections* at R and S to emerge finally out of CD along TO_1.

Drop $TN_1 \perp RS$, $TN \perp RO_1$ and $TLM \perp AB$. RS and TL meet at M when produced. From geometry, $ST = SM$; $LT = LM = t =$ thickness of film at L. Also, since $NO_1 - TO_1$, the path difference of the two transmitted interfering rays is

$$\Delta = \mu(RN_1 + N_1S + ST) - RN \tag{8.12.4}$$

Now, $$\mu = \frac{\sin i_1}{\sin(r-\alpha)} = \frac{RN/RT}{RN_1/RT} = \frac{RN}{RN_1}$$

$$\Rightarrow \quad RN = \mu \cdot RN_1$$

$\therefore$ From (8.12.4) $\Delta = \mu(N_1S + SM) = \mu\cdot N_1M = \mu\cdot TM\cos(r - 2\alpha)$

$$= 2\mu t\cos(r - 2\alpha) \qquad (8.12.5)$$

Condition of maxima and minima—In the transmitted beam, the *reflection* at different points *occurs inside the film at the surface backed by rarer medium* (usually air) and so there is *no phase change by* π, unlike in the case of reflected system.

$\therefore$ For **maxima** or **brightness**, the condition is

$$2\mu t\cos(r - 2\alpha) = 2n\,\lambda/2 = n\lambda$$

And for **minima** or **darkness**,

$$2\mu t\cos(r - 2\alpha) = (2n + 1)\lambda/2$$

• If the wedge-angle of the film $\alpha = 0$ (i.e., a **parallel film**), then the conditions of *maxima* and *minima* in *reflected* system are respectively.

$$2\mu t\cos r \pm \frac{\lambda}{2} = \text{even multiple of } \frac{\lambda}{2}$$

$$\Rightarrow \quad 2\mu t\cos r = \text{odd multiple of } \frac{\lambda}{2} = (2n + 1)\frac{\lambda}{2} \quad \textbf{(Maxima)}$$

Similarly, $\quad 2\mu t\cos r = 2n\dfrac{\lambda}{2}$ **(Minima)**

Corresponding conditions in *transmitted* system are

$$2\mu t\cos r = 2n\frac{\lambda}{2} \ \textbf{(Maxima)}; \quad 2\mu t\cos r = (2n + 1)\frac{\lambda}{2} \ \textbf{(Minima)}$$

Thus the conditions of maxima and minima in the transmitted pattern are just the reverse of the conditions for reflected pattern. So, *the pattern due to the transmitted beam is* **complementary** *to that due to the reflected beam.*

8.12.3 Fringe width, difference in film thickness etc. in thin wedge-shaped film

Let BAC be a *thin* wedge-shaped film of a very small angle θ and index μ (Fig. 8.15). A parallel beam of light of wavelength λ falls on the upper surface and viewed by the reflected light. Then, as already discussed, alternate bright and dark bands will be observed.

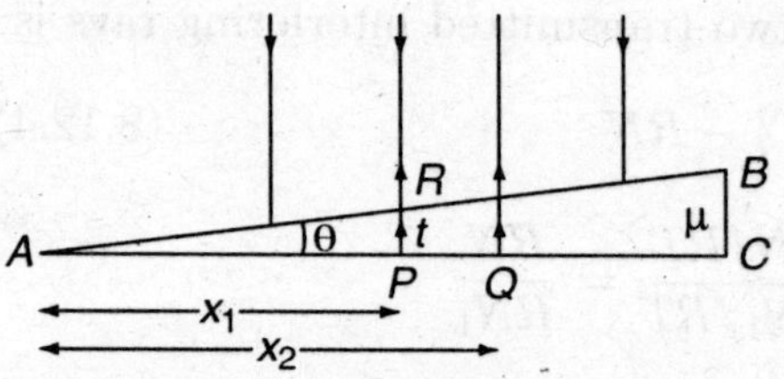

Fig 8.15

Fringe width — Let t be the thickness of any point P, distant x_1 from the end A of the wedge. For normal incidence, the net path difference between the light reflected from R on the upper surface and that reflected from P on the lower is $\Delta = 2\mu t - \lambda/2$, taking into account the extra path difference due to Stokes.

$\therefore$ P will be **dark** and a **dark band** will be obtained across the width, if

$$2\mu t - \lambda/2 = (2n-1)\lambda/2 \quad \Rightarrow \quad 2\mu t = n\lambda$$

But $t = x_1\theta$ ($\because$ θ very small). So, we obtain

$$2\mu x_1\theta = n\lambda. \tag{8.12.6}$$

If x_2 be the distance of $(n+m)$th dark fringe, then we have,

$$2\mu x_2\theta = (n+m)\lambda \tag{8.12.7}$$

$$\Rightarrow \quad 2\mu\theta(x_2 - x_1) = m\lambda.$$

$\therefore$ From (8.12.6) and (8.12.7), the fringe width,

$$\boxed{\boldsymbol{\beta = \frac{x_2 - x_1}{m} = \frac{\lambda}{2\mu\theta}}} \tag{8.12.8}$$

With **bright fringes** also, the condition for which is

$$2\mu t - \lambda/2 = n\lambda$$

$$\Rightarrow \quad 2\mu t = (2n+1)\lambda/2,$$

the same expression is obtained.

Difference of film thicknesses between two points—If t_1, t_2 be the thickness at P and Q respectively and m the number of fringes between these points, then $t_2 = x_2\theta$; $t_1 = x_1\theta$.

$$\therefore \quad \frac{t_2 - t_1}{\theta} = (x_2 - x_1)$$

$$\text{But} \quad (x_2 - x_1)/m = \lambda/2\mu\theta, \quad \text{from (8.12.8)}$$

$$\therefore \quad \frac{t_2 - t_1}{m\theta} = \frac{x_2 - x_1}{m} = \frac{\lambda}{2\mu\theta}$$

$$\therefore \quad \boxed{\boldsymbol{t_2 - t_1 = \lambda m/2\mu}}$$

Shape and location of fringes—We have seen that for a bright or a dark fringe, $t =$ constant. And in a wedge-shaped film t remains constant along lines parallel to the thin edge of the wedge. So the fringes, bright or dark, will also be straight lines $\parallel$ to the thin edge of the wedge, i.e., here fringes are of *equal thickness* (FET), each being the locus of points for which thickness $t =$ constant.

At the edege of the film, $t = 0 \Rightarrow$ the effective path difference $\Delta = \lambda/2$, a condition of minimum intensity. The *edge of the film is* thus **dark.**

And, as already observed, the interference fringes formed are *localised in the wedge itself* and can be easily seen by a low power microscope focussed on the wedge.

• With **white light,** each wavelength will produce its own interference fringes, the separation between two consecutive fringes being least for violet and greatest for red. At the edge, $t = 0$ and so $\Delta = \lambda/2$. Thus the intensity at the edge is minimum for all λ's. The edge will therefore be *dark.* As one proceeds away from the edge towards increasing thickness, few coloured bands of mixed colour is observed. With greater and greater thickness, the overlapping also increases to give eventually an uniform illumination.

8.13 Need of an extended source

An extended source is necessary to observe colours in thin films, for the narrow source limits the visibility of the film.

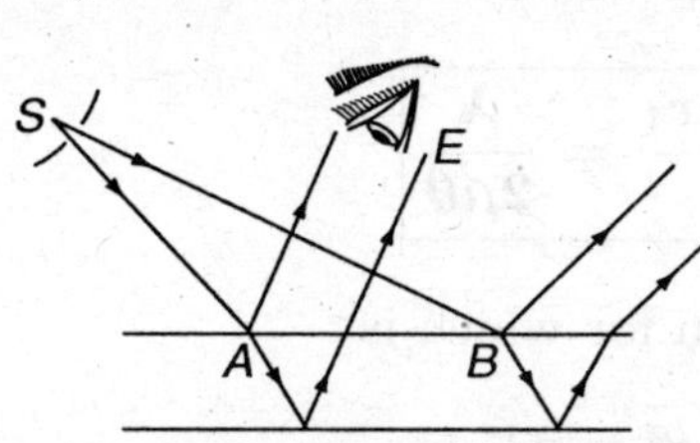

Fig 8.16

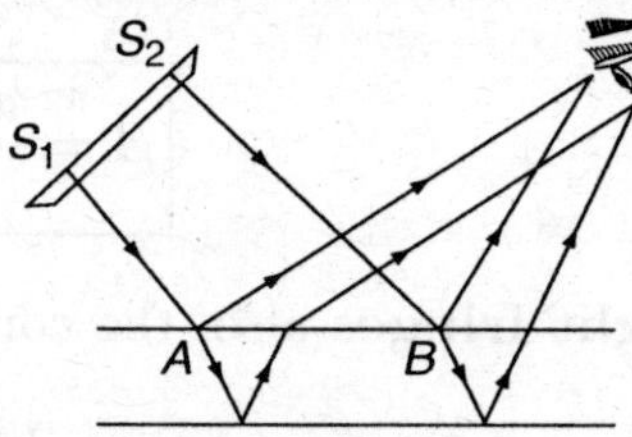

Fig 8.17

The situation is explained in reference to Fig. 8.16 and 8.17. Let SA be a ray that starts from a *narrow source* and be incident on the film. On reflection from the upper surface of the film and suffering one internal reflection, it enters the eye E. SB is another ray from S but is incident at a different angle. On reflection and internal reflection, it *cannot* reach the eye (Fig. 8.16). So, only a limited portion of the film becomes visible. To observe the whole of the film, the eye is to be continuously shifted from one position to another.

If however an *extended source* S_1S_2 is used (Fig. 8.17), the light reflected from a large part of the film reaches the eye E, placed in a suitable position. One can thus see the *entire film* simultaneously. This is why an extended source is used to view the film.

8.14 Film of thickness much less than wavelength

Consider an ulttra-thin *parallel film* of thickness $t \ll \lambda$. In this case, the geometrical path difference, $2\mu t \cos r$, between two successive reflected rays QR and Q_1R_1 (Fig. 8.18) becomes negligible. The only path difference is due to phase difference of π according to Stokes' law and is $\lambda/2$. Thus the rays QR and Q_1R_1 will produce destructive interference. But the destruction is *not* complete as they differ in amplitude. However, we show here that if the other rays due to multiple internal reflections are considered the destruction is complete.

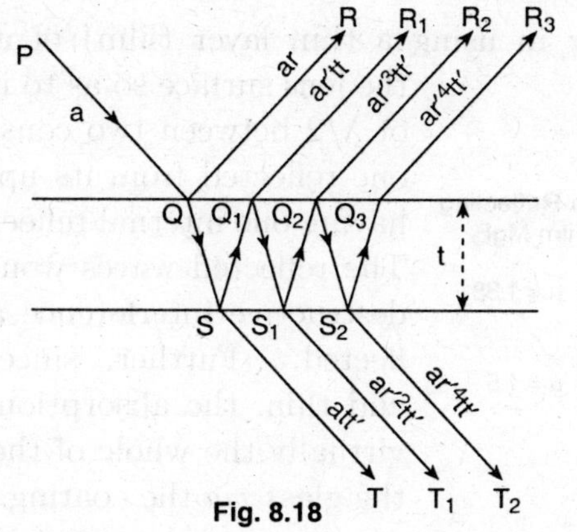

Fig. 8.18

The resultant amplitude of rays reflected along Q_1R_1, Q_2R_2, Q_3R_3, $\cdots$ etc. is

$$A = ar'tt'(1 + r'^2 + r'^4 + \cdots \text{ to } \infty)$$
$$= ar'tt'.\frac{1}{1 - r'^2}$$

where r and t are the reflection and transmission coefficients of the front surface, and r', t' the corresponding quantities of the lower surface.

But, according to Stokes' treatment $tt' = 1 - r^2$ and $r' = -r$. Therefore, $A = -ar$. So the net reflected amplitude including $QR = ar - ar = 0$. So, when the film is extremely thin it appears *perfectly dark* as it reflects no light and energy conservation principle dictates that *all light must be transmitted through the film.*

In so far as the transmitted rays are concerned, there is no phase change and the geometrical path difference between successive transmitted waves is negligible when $t \ll \lambda$. So the resultant transmitted amplitude is

$$A' = att' + att' \cdot r'^2 + att' \cdot r'^4 + \cdots$$
$$= att'(1 + r'^2 + r'^4 + \cdots)$$
$$= att'\frac{1}{1 - r'^2}$$
$$= a, \text{ the amplitude of incident wave} \quad (\because tt' = 1 - r'^2)$$

It is thus concluded that *an extremely thin transparent film transmits the whole of the incident light to the otherside.*

8.15 Non-reflecting film or anti-reflection coating

A part of light incident on lens-surfaces gets lost by reflection. While the loss may not be significant for a single lens surface, it is rather considerable in a multi-lens system. The elimination of such losses is thus important and is usually done by what is called *non-reflecting film* or *anti-reflection coating.*

It consists essentially in using a thin layer (film) of a transparent material on the lens surface so as to introduce a path-difference of $\lambda/2$ between two consecutive rays BC and EF, one reflected from its upper surface and the other having one internal reflection in the film (Fig. 8.19). The reflected waves would then undergo complete destructive interference and no light would be reflected. Further, since the film is transparent and thin, the absorption is also negligible so that virtually the whole of the incident light passes into the glass *via* the coating.

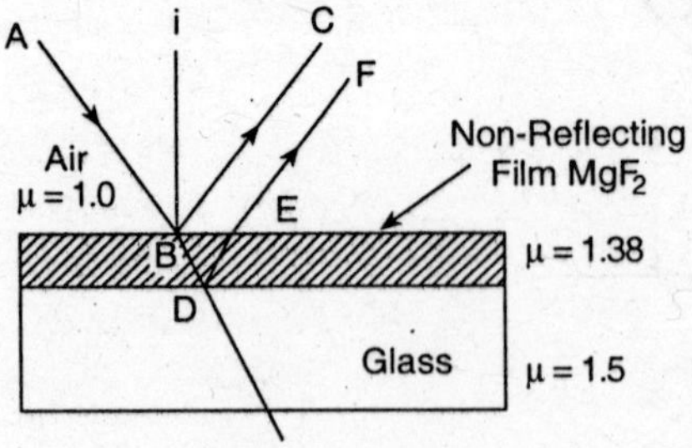

Fig. 8.19

The *choice of the material is important* in that its index should be intermediate between that of air and glass. The *thickness of the material is also critical* so as to ensure the destructive interference sought for.

Calculations—The reflectivity r at the boundary between two media of indices μ_1 and μ_2 for normal incidence is, by electromagnetic theory,

$$r = \left(\frac{\mu_2 - \mu_1}{\mu_2 + \mu_1}\right)^2 \tag{8.15.1}$$

So, for equality of amplitudes of waves along BC and EF

$$\left(\frac{\mu_c - \mu_a}{\mu_c + \mu_a}\right)^2 = \left(\frac{\mu_g - \mu_c}{\mu_g + \mu_c}\right)^2 \tag{8.15.2}$$

$$\text{or,} \quad \boldsymbol{\mu_c = \sqrt{\mu_a \, \mu_g}}, \quad \text{on simplification,} \tag{8.15.3}$$

where μ_a = index of air, μ_c = index of coating and μ_g = index of glass.

Rays BC and EF, both suffer sudden phase change of π. So, for desctructive interference, the geometric path difference must be $\lambda/2$.

$$\therefore \quad 2\mu_c t = \frac{\lambda}{2} \quad \Rightarrow \quad t = \frac{\lambda}{4\mu_c} \tag{8.15.4}$$

for normal incidence, t being the minimum thickness of coating.

Thus a coating of index given by (8.15.3) and of thickness given by (8.15.4) may act as an antireflection coating.

- Plainly, the coating material should be insoluble, scratch resistant and durable. MgF_2 and cryolite are often used.
- The technique of reducing the reflectivity of a surface is known as **blooming**.
- While a single layer coating is effective for a single wavelength or a narrow range of wavelengths, multilayer coatings prove efficient over a wide range of wavelengths.

8.16 Newton's rings

Newton's rings constitute a special instance of interference in a film of slowly varying thickness. It was Hooke (not Newton) which first observed it.

If a *plano-convex* lens of *large radius* of curvature be placed with the *convex surface* in contact with a *plane* glass plate, an air film of small thickness is formed between the two. The thickness of the film at the point of contact is zero and gradually increases outward. If monochromatic light be allowed to fall normally on the film and viewed normally with a low-power microscope in reflected light, **a system of alternate bright and dark concentric rings around the point of contact,** with their *centre dark*, is observed to be formed in the air film. These are called **Newton's rings.**

Experimental arrangement—The experimental arrangement for observing Newton's rings is illustrated in Fig. 8.20. L is the plano-convex lens, PC the glass plate on which L is placed with convex surface in contact, C being the point of contact. Light from an extended monochromatic source S falls on a glass plate G held at an angle of 45° with the vertical. The light reflected from G falls normally on the air-film enclosed between L and PC. Interference occurs between the rays reflected from the upper and the lower surfaces of the film giving rise to interference pattern in the form of circular rings *localised* in the film itself. The rings are seen by a low-power microscope M focussed on the film.

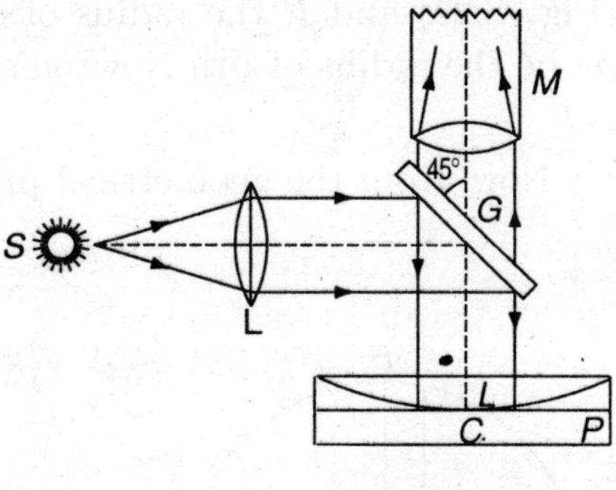

Fig 8.20

Formation of Newton's rings—Newton's rings are formed due to interference of light waves *reflected* from the upper and the lower surfaces of the air film. In Fig. 8.21, corresponding to the incident ray AB, the interfering rays are 1 and 2.

The rings being observed in **reflected light,** the effective path difference between the interfering rays is

$$\Delta = 2\mu t \cos r + \lambda/2$$

where μ = index of film (air), t the thickness of the film at B and r the inclination of the ray. The factor $\lambda/2$ is due to the phase change of π on reflection at lower surface of the film—the glass plate.

For *normal incidence,* $r = 0$ and *for air,* $\mu = 1$, so that

$$\Delta = 2t + \lambda/2$$

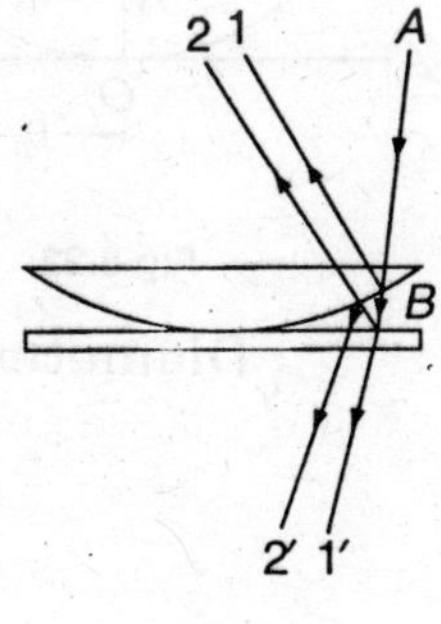

Fig 8.21

At the *point of contact,* $t = 0 \Rightarrow \Delta = \lambda/2$, which corresponds to the condition for **minimum** intensity. Hence the **central point is dark.**

The condition for nth order **maximum** intensity, i.e., **bright fringe** is

$$\Delta = n\lambda \quad \Rightarrow \quad 2t + \lambda/2 = n\lambda \quad \Rightarrow \quad 2t = (2n-1)\lambda/2.$$

The condition for nth order **minimum** intensity, i.e., **dark fringe** is

$$\Delta = (2n+1)\lambda/2 \quad \Rightarrow \quad 2t + \lambda/2 = (2n+1)\lambda/2 \quad \Rightarrow \quad 2t = 2n\lambda/2 = n\lambda$$

The above expressions show that, bright or dark, a fringe of any particular order n occurs for a constant t-value. In the air-film, t is constant along a *circle with centre at the point of contact.* So the *fringes* appear in the form of *concentric circles*—each circle corresponds to a definite t-value, and different circles to different t's. Since each fringe is the locus of constant thickness (t), Newton's rings are called **fringes of equal** (film) **thickness (FET)**.

Diameters of rings—We shall now find an expression for the diameters of the Newton's rings.

Let AB be the glass plate, LOL' the lens placed on it with O as the point of contact (Fig. 8.22) and R the radius of curvature of the curved surface of the lens. Also, let ρ_n be the radius of nth Newton's ring corresponding to a point P where the thickness is t.

Now, from the geometrical property of a circle, we obtain

$$PN^2 = ON \times NE$$

$$\Rightarrow \quad \rho_n^2 = t \times (2R - t) = 2Rt - t^2 \simeq 2Rt \quad (\because t \text{ is very small})$$

$$\therefore \quad 2t = \rho_n^2/R \tag{8.16.1}$$

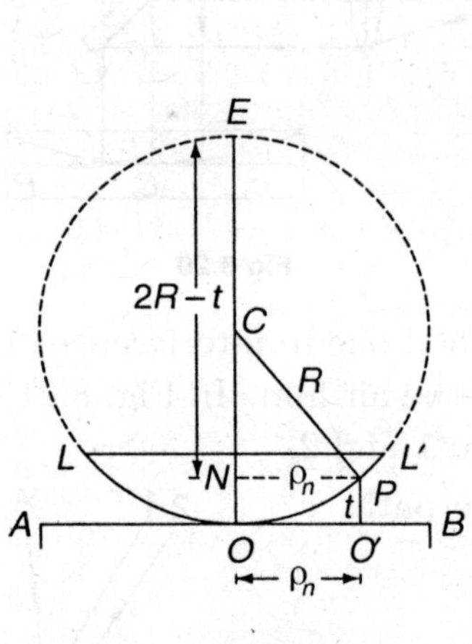

Fig 8.22

Since for a **bright ring,**

$$2\mu t = (2n-1)\lambda/2 \tag{8.16.2}$$

$$\rho_n^2/R = (2n-1)\lambda/2\mu, \quad \text{using (8.16.1) and (8.16.2)}$$

$$\Rightarrow \quad \boldsymbol{\rho_n^2 = (2n-1)\frac{\lambda R}{2\mu}}$$

$\therefore$ **Diameter of nth bright ring, D_n** is given by

$$D_n = 2\rho_n = 2\sqrt{2n-1}\sqrt{\lambda R/2\mu}$$

$$\propto \sqrt{2n-1} \tag{8.16.3}$$

Hence *the diameters of* **bright rings** *are proportional to the square roots of odd natural numbers,* i.e., diameters of the first few rings are in the ratio $1 : \sqrt{3} : \sqrt{5} : \sqrt{7} : \sqrt{9}$ etc., i.e., in the ratio 1 : 1.732 : 2.236 : 2.646 : 3 etc.

$\therefore$ Separation between successive rings are in the ratio: 0.732, 0.504, 0.410, 0.354 etc.

This shows that *as the order of the ring increases, the separation between them decreases* i.e., the rings gradually become narrower and more crowding of rings occur as one moves from the centre outward.

For a **dark ring,**

$$2\mu t = n\lambda \tag{8.16.4}$$

$$\text{or,} \quad \rho_n^2/R = n\lambda, \quad \text{using (8.16.1).}$$

∴ **Diameter of *n*th dark ring,** D_n is given by

$$D_n = 2\rho_n = 2\sqrt{nR\lambda} = \sqrt{4nR\lambda} = \sqrt{4R\lambda}\sqrt{n} \tag{8.16.5}$$

$$\Rightarrow \quad D_n \propto \sqrt{n}$$

So, the *diameters of* **dark rings** *are proportional to the square root of natural numbers.*

Fringe width — If D_n and D_{n+1} are the dia of two successive *bright* rings

$$D_{n+1}^2 - D_n^2 = \frac{4\lambda R}{\mu}$$

$$\Rightarrow \quad D_{n+1} - D_n = \frac{4\lambda R}{\mu(D_{n+1} + D_n)} \simeq \frac{4\lambda R}{\mu \cdot 2D_n}$$

$$\therefore \quad \textbf{Fringe width, } \boldsymbol{\beta} = \frac{D_{n+1} - D_n}{2} = \frac{\boldsymbol{\lambda R}}{\boldsymbol{\mu D_n}}$$

So, *the fringe width β decreases as the diameter D_n of the ring increases*

Newton's rings in intransmitted light—Newton's rings can be seen both in transmitted or refracted light. In Fig. 8.21, while 1 and 2 are reflected interfering rays, 1′ and 2′ are the transmitted interfering rays.

There are *two* essential differences in the transmitted system of rings relative to the reflected one. These are :

1. *The rings as seen in the transmitted light are exactly complementary to those observed in the reflected light.*

2. *The rings produced by transmitted light are much poorer in contrast relative to those formed by the reflected light.*

Let us explain the points. The ray 1 reflected at the upper surface of air-film suffers no phase change. But the ray reflected *internally* at the lower surface undergoes a phase change of π. So the two interfering reflected rays have a phase difference of π (over and above that due different optical paths covered by them). The ray 1′ gets transmitted directly through the air-film, but the ray 2′ undergoes two *internal* reflections, and hence a phase change of 2π, before its final emergence. Thus the condition for maxima in reflected light corresponds to minima in transmitted light and conversely. This leads to complementarity of transmitted and reflected ring systems.

Secondly, in transmitted light the ray 2′ differs from ray 1′ in amplitude ratio $\mathcal{R}^2$, where $\mathcal{R}$ is the reflection coefficient, because of its two reflections within the film. Thus, the interfering transmitted rays are of very unequal amplitudes. Therefore, the contrast of maxima and minima is poor, i.e., less distinct.

• **Newton's rings with white light** — With white incident light, the *central spot* will be *black* and few (8–10) *coloured rings* will surround it. The region beyond the coloured rings will be show a *general illumination* due to however overlapping of different coloured rings.

8.17 Applications of Newton's rings

We shall now deal with *two* of the most important applications of Newton's rings. These are :

1. Determination of wavelength—With the given extended source of monochromatic light, the apparatus is arranged, following the usual procedure, to produce Newton's rings. The eyepiece of the travelling microscope is first focussed on the cross-wire and then the microscope itself on the rings.

With the help of the microscope, the diameter D_n of the nth the dark ring is measured and is given by

$$D_n^2 = 4n\lambda R \quad \text{[Eq. (8.16.5)]}$$

Next, the diameter D_{n+m} of the $(n+m)$th dark ring is measured, and is given by the relation :

$$D_{n+m}^2 = 4(n+m)\lambda R$$

$$\therefore \quad D_{n+m}^2 - D_n^2 = 4m\lambda R \tag{8.17.1}$$

$$\Rightarrow \quad \boxed{\lambda = \frac{D_{n+m}^2 - D_n^2}{4mR}}$$

Hence λ can be readily calculated. While m is obtained by simple counting, the radius of curvature R of the lower surface of the lens is determined by using a spherometer or by Boys' optical method.

2. Determination of index of a liquid—The experiment is first performed with air film between the plano-convex lens and the optically plane glass plate kept in a container C (Fig. 8.23). As before, the diameter of the nth and $(n+m)$th dark rings are measured with the help of the travelling microscope.

$$\therefore \quad D_{n+m}^2 - D_n^2 = 4m\lambda R \tag{8.17.2}$$

Next, the experimental liquid is poured into the container, without disturbing the arrangement, such that the air film is replaced by a film of the liquid. The diameters of the nth dark ring and the $(n + m)$th dark ring are determined, as before, and let them be denoted by D'_n and D'_{n+m} respectively.

$$\therefore \quad D'^2_{n+m} - D'^2_n = 4m\lambda R/\mu$$

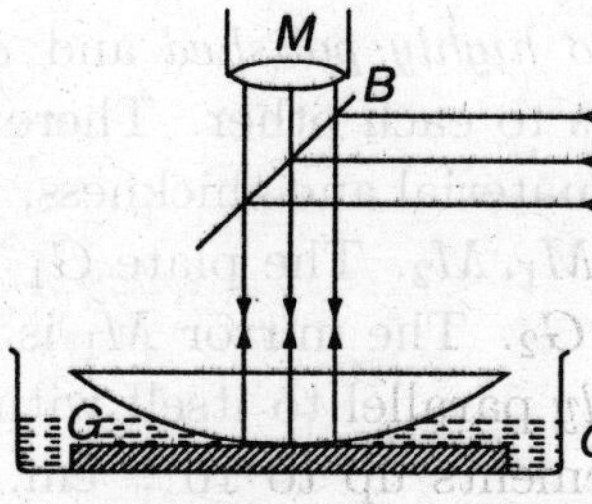

Fig 8.23

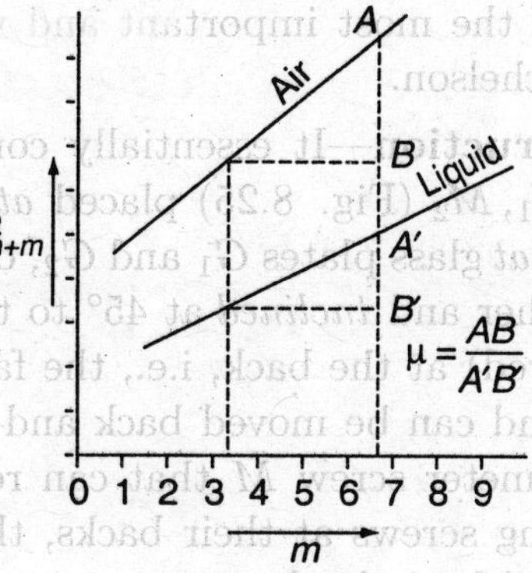

Fig 8.24

$$\therefore \quad \mu = \frac{4m\lambda R}{D'^2_{n+m} - D'^2_m} \tag{8.17.3}$$

So, μ can be calculated, if $m, \lambda, R, D'_n, D'_{n+m}$ are known.

• If, however, λ is not known, divide the relation (8.17.3) by (8.17.2) in order to get

$$\mu = \frac{D^2_{n+m} - D^2_n}{D'^2_{n+m} - D'^2_n}$$

Graphical method—The diameters of the dark rings are determined for various orders from say nth ring to $(n + m)$th ring, first with air and then with the liquid as the film.

A graph of D^2_{n+m} vs m is plotted (Fig. 8.24) where $m = 0, 1, 2, \ldots$ and we get two straight line curves—one for air and the other for the liquid. The ratio of the slopes of the two lines (air and liquid) gives the refractive index of the liquid.

$$\therefore \quad \mu = \frac{AB}{A'B'}$$

• **Newton's rings** may however be produced under many *different configurations* such as : (i) lens in contact with a **concave surface**, (ii) lens in contact with a **convex surface**, (iii) lens **separated from the plate** by some distance. For (i) and (ii) see 'Illustrated Examples' and for (iii) apply yourself.

8.18 Michelson's interferometer

Interferometers form a class of optical instruments based on the phenomenon of *interference*, and are widely used to determine wavelengths, refractive index etc. They have high resolving power.

One of the most important and widely known among the interferometers is that due to Michelson.

Construction—It essentially consists of *two highly polished* and *optically plane* mirrors M_1, M_2 (Fig. 8.25) placed *at right angles* to each other. There are *two* other *optically flat* glass plates G_1 and G_2, of the *same* material and thickness, placed *parallel* to each other and *inclined* at 45° to the mirrors M_1, M_2. The plate G_1 is *semisilvered* (half-silvered) at the back, i.e., the face towards G_2. The mirror M_1 is mounted on a carriage and can be moved back and forth *exactly* parallel to itself with the help of a fine micrometer screw M that can read displacements up to 10^{-5} cm. By means of the levelling screws at their backs, the mirrors M_1 and M_2 can be tilted both about vertical and horizontal axes so as to make them *exactly* at right angles to each other. The interference bands produced by the interferometer are observed in the field of view of the telescope T.

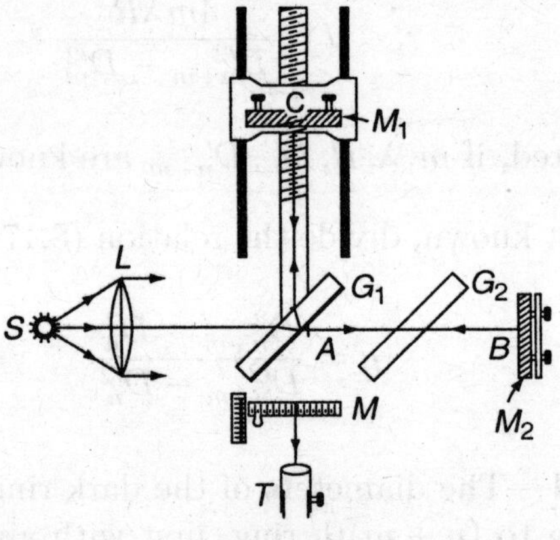

Fig 8.25

Working principle—Light from an *extended monochromatic* source S is rendered parallel by a collimating lens L and is allowed to fall on the plate G_1. It gets divided into two parts of equal intensities: one being *reflected* at the semisilvered back surface of G_1 to move along AC towards M_1, and the other being *transmitted* along AB towards M_2. Both the rays AC and AB fall *normally* on M_1, M_2 respectively and retrace their paths after reflection to enter finally into the telescope T along AT. Since the two rays are derived originally from the same single beam by division of amplitude, they are in a position to produce interference fringes in the field of view of the telescope T under suitable conditions.

What is the **role of the glass plate** G_2? For without it, the ray AC passes through G_1 *twice*, but the ray AB *none* at all. So the paths of the two rays AC and AB in glass are not equal. To *compensate* for this, an exactly similar glass plate G_2 and made of the same material is introduced in the path of AB. Hence G_2 is called a **compensator**. While G_2 is *not* essential in working with *monochromatic* light, it is a *must* for *white* light.

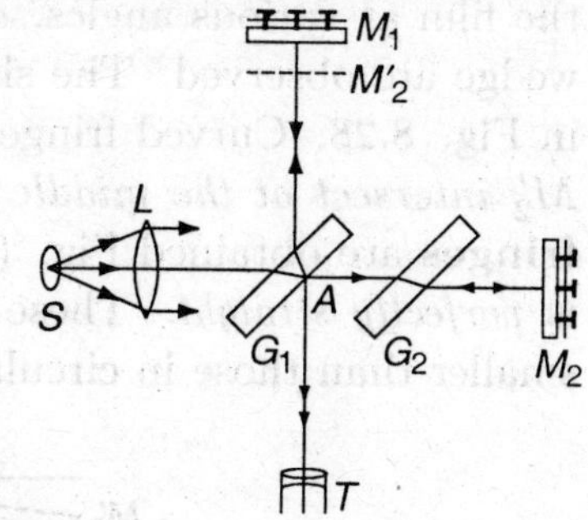

Fig 8.26

If one looks towards M_1 from T, one sees M_1 and also a virtual image M_2' of M_2 formed by G_1 (Fig. 8.24). The two interfering beams, received by telescope T, though come by way of reflections from M_1 and M_2 behave functionally as if reflected from M_1 and M_2'. So the interferometer is *optically equivalent* to an *air film* enclosed between M_1 and M_2'. Depending on the path difference and the angles between the mirrors M_1 and M_2' (virtual mirror), the fringes may be circular, straight etc.

Production of circular fringes—Circular fringes are produced in Michelson's interferometer with monochromatic light when M_1, M_2 are *exactly* perpendicular to each other. The production of such fringes is illustrated in and can be understood from Fig. 8.27.

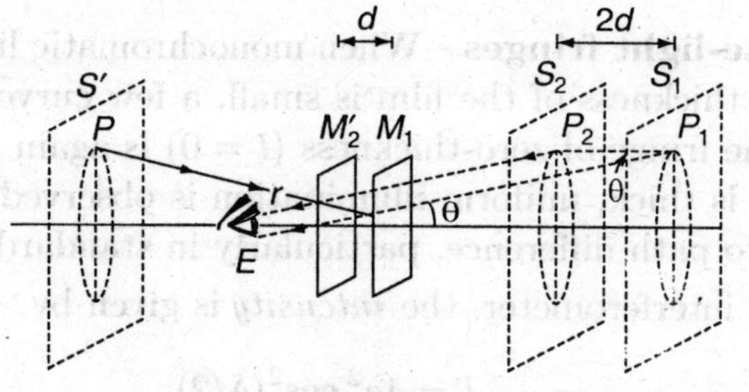

Fig 8.27

M_1 and M_2' (image of M_2) are parallel reflecting surfaces, the real extended source being replaced by its virtual image S', formed by reflection at G_1, behind the eye E. This virtual source S' again forms *two* virtual images S_1 and S_2 by M_1 and M_2' Light from a point P say, on the extended source appears to the eye E to come from the corresponding coherent points P_1 and P_2 on S_1 and S_2. If d be the separation between M_1 and M_2', then that between S_1 and S_2 will be $2d$. The path difference between the rays entering the eye is plainly $2d \cos\theta$ where θ is the angle the reflected beam makes with the normal.

If $2d\cos\theta = n\lambda$, $n = 0, 1, 2, \cdots$ etc., P appears **bright (maximum)**.

If $2d\cos\theta = (2n+1)\lambda/2$, $n = 0, 1, 2, \cdots$ etc. P appears **dark (minimum)**.

For a given value of d, n *and* λ, *the angle* θ *will be constant so that the locus of* P *is a circle with centre on the axis.* So, a series of alternate bright or dark **circular fringes** are observed. The fringe order, however, *decreases* as θ *increases*, i.e., as we move away from the centre. Since the interfering rays are *parallel*, these fringes are formed at *infinity* and hence observed through a telescope focussed for infinity.

Straight localised fringes—When M_2 is *not* exactly perpendicular to M_1, the enclosed air-film between M_1 and M_2' is *wedge-shaped.* Since the light is incident on the film at various angles, *curved fringes* with convexity towards the thin edge of the wedge are observed. The shape of the fringes for different path differences are shown in Fig. 8.28. Curved fringes are seen in Fig. (a) and (c). When the mirrors M_1 and M_2' *intersect at the middle* and the *film thickness is small*, practically **straight line fringes** are obtained Fig. (c). The fringe that corresponds to $t = 0$ is the one which is *perfectly straight.* These fringes are visible up to path differences relatively much smaller than those in circular fringes.

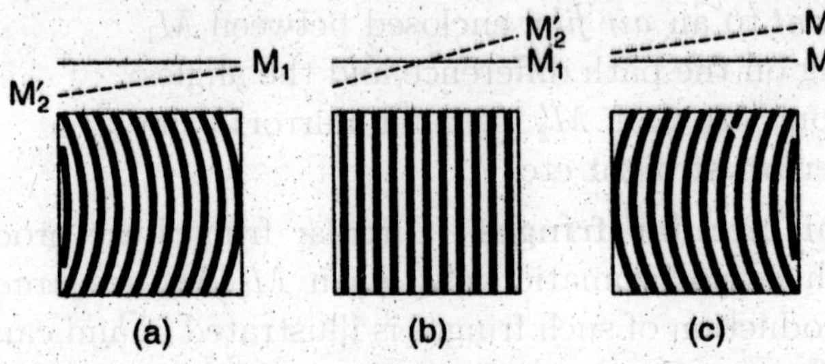

Fig 8.28

Localised white-light fringes—When monochromatic light source is replaced by white light and the thickness of the film is small, a few curved and coloured localised fringes are seen. The fringe of zero-thickness ($t = 0$) is again *perfectly achromatic* and *straight.* If the film is thick, uniform illumination is observed. White light fringes are useful in finding zero path difference, particularly in standardising a metre.

• In Michelson's interferometer, the *intensity* is given by

$$I = 4a^2 \cos^2(\delta/2)$$

$$\text{where} \quad \delta = \frac{2\pi}{\lambda}(2d\cos\theta)$$

d being the distance between M_1 and M_2'. I is *maximum* when $\delta = 2n\pi$, and *minimum* when

$$\delta = (2n+1)\pi, \quad n = 0, 1, 2, \ldots$$

8.19 Applications of Michelson's Interferometer

Michelson's interferometers are put to the following important uses.

1. Determination of *unknown wavelength* of light. **2.** Determination of the *wavelength-difference* of two very close spectral lines. **3.** Determination of *thickness* of thin transparent sheets. **4.** *Standardisation* of 'metre' in terms of standard wavelength.

We shall now discuss these applications in greater details one by one.

1. Determination of wavelength of monochromatic light—The interferometer is first adjusted for circular fringes in the field of view of the telescope. If mirrors

M_1, M_2 are equidistant from the glass plate G_1, the field of view will be perfectly dark. M_1 is next adjusted in position so as to obtain a particular bright fringe in the field of view of telescope with its centre coinciding with the crosswire. From this position when M_1 is moved away through a distance $\lambda/2$, the path-difference changes by λ and the position of a particular bright fringe is occupied by the next bright fringe. If n be the number of fringes that cross the centre of field when M_1 is moved from initial position x_1 to a final position x_2, then

$$x_2 - x_1 = n\frac{\lambda}{2} \quad \Rightarrow \lambda = \frac{2(x_2 - x_1)}{n}$$

The difference $(x_2 - x_1)$ can be measured correct up to 10^{-4} mm with the help of the micrometer screw and n is actually counted. λ is obtained by using the above formula.

2. Determination of difference of two very close wavelengths (resolution of spectral lines)—Let the source emit two very close wavelengths λ_1 and $\lambda_2 (\lambda_1 > \lambda_2)$, e.g., sodium D-lines, and the interferometer be adjusted to obtain circular fringes. There will be two sets of fringes—one corresponding to λ_1 and the other corresponding to λ_2. Since λ_1 is very close to λ_2, the two sets will practically coincide. As M_1 is moved slowly, the two sets will separate out slowly and when the dark fringe of λ_1 coincides with the bright fringe of λ_2, the result is maximum *indistinctness* (poor visibility). On further movement of M_1, the *next indistinct* position is reached after a distance x, say. In this position, if n fringes of λ_1 appear at the centre, then $(n + 1)$ fringes of λ_2 will appear at the centre of the field of view.

$$\therefore \quad x = n\frac{\lambda_1}{2} \quad \Rightarrow \quad n = \frac{2x}{\lambda_1};$$

$$\text{and} \quad x = (n+1)\frac{\lambda_2}{2} \quad \Rightarrow \quad n+1 = \frac{2x}{\lambda_2}$$

$$\therefore \quad (n+1) - n = 2x\left(\frac{1}{\lambda_2} - \frac{1}{\lambda_1}\right) = \frac{2x(\lambda_1 - \lambda_2)}{\lambda_1\lambda_2}$$

$$\therefore \quad \lambda_1 - \lambda_2 = \frac{\lambda_1\lambda_2}{2x} = \frac{\lambda_m^2}{2x}$$

where λ_m is the mean of the two wavelengths λ_1 and λ_2.

Thus measuring the distance x moved by mirror M_1 between two consecutive positions of maximum indistinctness, the difference of two close wavelengths can be determined.

3. Determination of thickness of a thin transparent sheet—The interferometer is set for localised white light fringes. And the cross-wires are made to coincide with the central dark fringe. The paths AB and AC are exactly equal.

The thin film is now introduced between the glass plate G_1 and the mirror M_2 such that light passes through it normally. The introduction of the sheet of thickness t and refractive index μ increases the optical path by $2(\mu-1)t$. The fringes will thus get shifted from their positions. The mirror M_1 is next moved till again a dark fringe of zero path coincides with the crosswire. The distance x through which M_1 is moved is noted.

The white light is next replaced by monochromatic light of wavelength λ and the mirror M_1 is moved slowly to count the number of fringes contained in x. Then,

$$2(\mu-1)t = n\lambda \quad \Rightarrow \quad t = \frac{n\lambda}{2(\mu-1)}$$

whence the thickness t can be evaluated from a knowledge μ, n and λ.

• It may appear that when the formula

$$2(\mu-1)t = n\lambda$$

(where n is the number of fringes by which the fringe system is displaced, λ the wavelength of light) is being used, localised white light fringes are unnecessary. In fact, it is not so, for the fringes with monochromatic light are exactly alike and it is difficult to count the number of fringes shifted, the shift being sudden on inserting the thin sheet.

• The same method and working formula may be used to *determine* the *refractive index* μ of the material of the sheet provided n, λ and t are known.

4. Standardisation of 'metre'—Michelson and Benoit *standardised* the *metre* in terms of wavelegth of *red* light of cadmium with the idea of reproducing the metre in case the original one gets lost.

The *principle* consists in counting the number of fringes crossing the field of view when the movable mirror is moved over the whole length of the metre. But there are *two problems* involved: (i) fringes *cannot* be obtained if path difference exceeds 24 cm or the mirror-displacement > 12 cm; (ii) a displacement of 10 cm involves transition of about 3×10^3 fringes, the accurate counting of which is *not possible* mechanically. The *difficulties,* however, were *removed* by using *nine* **etalons** which are intermediate standards, each twice as long as the previous one. The longest etalon was of 10 cm and the smallest one of 0.390625 cm.

Etalon—An etalon, as shown in Fig. 8.29, consists of two plane mirrors M_1, M_2 placed accurately parallel to each other and silvered at their front. The distance between them is the length l of the etalon. The mirrors can be made perfectly parallel using the screw T and spring S.

The *method* of standardisation consists of the following *three* consecutive steps.

1. Determination of the number of waves in the shortest etalon.

2. Comparison of the intermediate etalons, and

3. Comparison of the longest etalon with the standard metre.

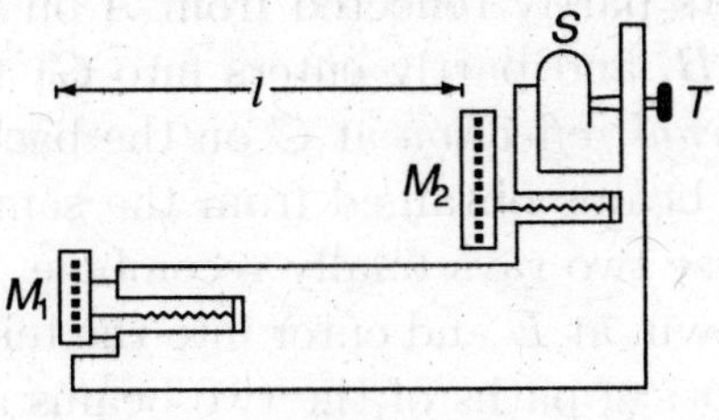

Fig 8.29

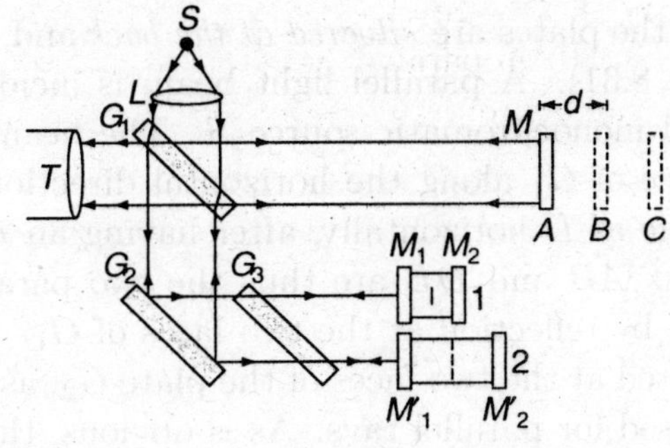

Fig 8.30

Step 1. For this purpose, Michelson's interferometer was modified to the form shown in Fig. 8.30. The two shortest etalons 1 and 2 were mounted with a field of view covering the four mirrors M_1, M_2 and M_1', M_2'. Using white light, the distances of M, M_1 and M_1' from the eye were made equal so that a dark fringe appeared in the field of view. M was next moved parallel to position B and when it is exactly coplanar with M_2, a *dark fringe* will appear again. AB was then exactly equal to the length l of the elaton M_1M_2. By replacing the white light with red light of Cd, M was slowly moved from A to B and the number of fringes crossing the reference line counted. The fraction of the fringe was also determined by rotating the compensating glass plate G_3, duly calibrated previously. The length of the first etalon was thus evaluated in terms of wavelength.

Step 2. Next, the smallest etalon was moved over its own length, without however counting the fringes till the white light dark fringe re-appeared in M_1. Finally, M was moved to the position C when again the reappearance of white light dark fringe occurred in M_2 and also in M_2'. Thus was obtained the number of waves contained in the length $M_1'M_2'$ of the second etalon. The second etalon was next compared with the third and in this way the number of waves in the longest etalon of 10 cm length was evaluated.

Step 3. The 10 cm long etalon is next taken as a substandard and is compared directly with the prototype standard metre.

The *metre* values obtained in terms of different λ's are as under.

1. **Red line :** 1 m = 1,553,163.5 λ_r where $\lambda_r = 6438.4722$ A
2. **Green line :** 1 m = 1,966,249.7 λ_g where $\lambda_g = 5985.8240$ A
3. **Blue line :** 1 m = 2,083,372.1 λ_b where $\lambda_b = 4799.9107$ A

8.20 Jamin's Interferometer

Jamin's interferometer is widely used in the determination of refractive indices of gases.

Construction—It consists essentially of two rectangular glass plates G_1, G_2 of *exactly the same thickness*, usually cut from the optically worked same piece of glass.

Both the plates are *silvered at the back* and are mounted *perfectly parallel* to each other (Fig. 8.31). A parallel light beam is incident on G_1 at A at an angle of 45° from a broad monochromatic source S. The beam gets partly reflected from A on the front surface of G_1 along the horizontal direction AB, and partly enters into G_1 to finally emerge at D horizontally, after having an *internal reflection* at C on the back surface of G_1. AB and DE are thus the two parallel beams obtained from the same parent beam by reflection at the two faces of G_1. These two rays finally recombine, on being reflected at the two faces of the plate G_2, as shown, at E and enter into the telescope T focussed for parallel rays. As is obvious, the optical paths of the two beams are equal and identical so as to produce *interference fringes* in the field of view of the telescope.

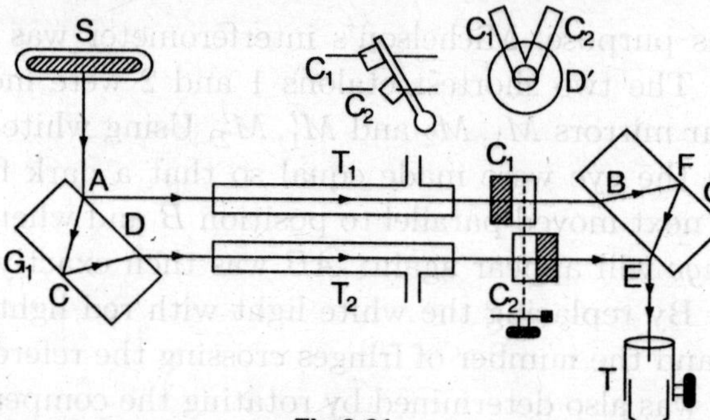

Fig 8.31

Application for finding the index of gases—In using the interferometer to determine the refractive index μ of a gas or air, at a given temperature and pressure, two glass tubes T_1 and T_2, exactly similar and closed with exactly similar glass plates, are introduced in the paths of the two beams. To start with, both T_1 and T_2 are exhausted and the eye is fixed on the fringes at the crosswire. Next, the experimental gas is slowly admitted into one of the tubes T_2, say. The presence of the gas in T_2 will introduce a path difference between the two beams and cause the fringes to move across the crosswire of the telescope.

Let n be the number of fringes that pass across the cross-wire at a desired temperature and pressure of the gas, t the tube-length and μ the index of the gas.

$$\therefore \text{ Optical path difference, } (\mu - 1)t = n\lambda.$$

If t, λ and n are known, μ of the gas at a known pressure and temperature can be determined.

Note. As the gas is admitted in one of the tubes, the optical path introduced is *not gradual* and so counting of n is difficult. To remove the difficulty Jamin used the so-called **Jamin's compensator**. The device consists of two thick glass plates C_1 and C_2, of the same thickness and cut from the same glass piece, having a small inclination with each other. They can be rotated together about a *horizontal axis* by a *single knob,* attached (Fig. 8.31) to a calibrated *circular disc* D', so as to place one in each path of the two beams. When C_1 and C_2 are *equally inclined* to the incident beam, the optical paths through them are equal. When rotated from this orientation, one path gets *shortened*, the other *lengthened*, causing a relative phase difference that compensates for the path difference in the two tubes due to admittance of gas in one. D' is calibrated by counting the fringes directly with monochromating light in terms of μ and λ. The compensator is most sensitive when the inclination between C_1 and C_2 is small.

C. Multiple beam interferometry

8.21 Multiple beam interferometers

8.21.1 Fabry-Perot interferometer

It is an interferometer of *high resolving power* in which fringes of constant inclination, called **Haidinger fringes**, are produced by transmitted light after multiple reflections in an air-film enclosed between two *parallel* and *highly reflecting* glass plates.

Construction—It consists essentially of two *optically plane* glass plates A, B (Fig. 8.32), placed *accurately parallel* to each other and separated by a distance d. The inner surfaces of A and B are *thinly silvered* so as to reflect 80-90% of the incident light. The *outer surfaces* of the plates are also *parallel but inclined to their respective inner surfaces*, (i.e., the plates are slightly prismatic) to avoid interference of the rays reflected at the outer unsilvered surface.

Monochromatic light from an extended source S_1 is made parallel by the collimating lens L_1. Each parallel ray suffers multiple reflections successively at the two silvered surfaces as shown. At each reflection, a small fraction of light is also transmitted so that each incident ray produces a group of coherent and parallel transmitted rays. There is a constant path difference between any two successive transmitted rays. A second converging lens L_2 brings these rays to focus at P in its focal plane where they interfere. Thus, the rays from all points of the source produce an interference pattern on a screen S_2 placed at the *focal plane* of L_2. This is known as **multiple beam interference**.

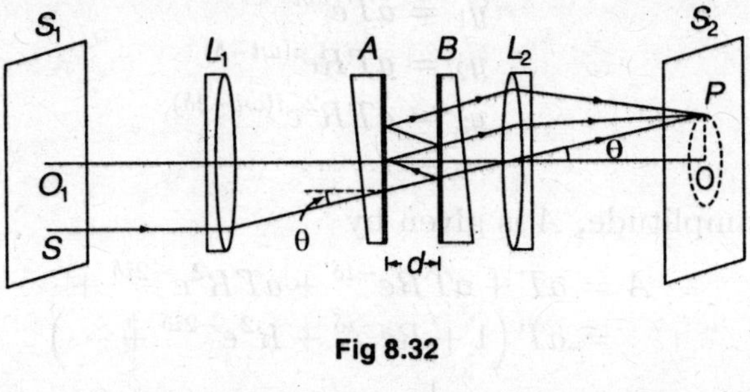

Fig 8.32

Formation of fringes—If θ be the inclination of a particular ray with the normal to the silvered surface of A and d the separation between the plates, then the path difference between two successive transmitted rays corresponding to the incident one is $2d\cos\theta$.

$\therefore$ For maxima, $2d\cos\theta = n\lambda$, $n = 0, 1, 2, \cdots$, etc. where n is the *order of the interference* and λ the wavelength of light.

The locus of points in the source giving rays of constant inclination θ is a *circle.* Thus with an extended source, *the interference pattern will be a series of bright concentric rings* (Fig. 8.33) *on a dark background.* Each of the rings will correspond to a particular θ.

Fig 8.33

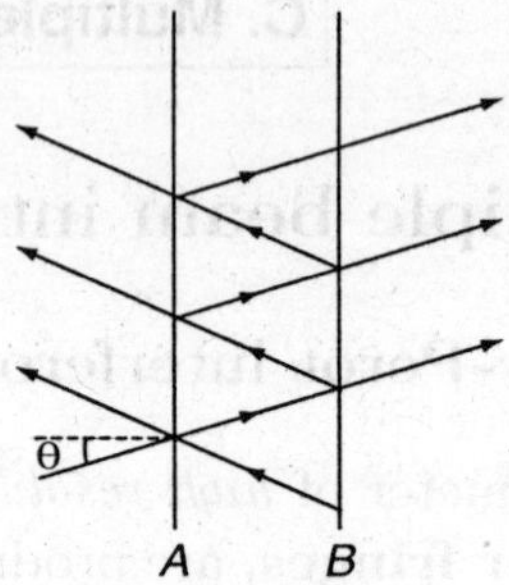

Fig 8.34

Intensity distribution of transmitted beam—It is clear from Fig. 8.34 that the intensity of successive parallel rays goes on decreasing, being dependent on reflection and transmission coefficients of silvered surface.

Let T and R be the fraction of the intensity of incident light which are respectively transmitted and reflected at each silvered surface. So $\sqrt{T}$ and $\sqrt{R}$ are the fractions of amplitudes transmitted and reflected respectively.

$\therefore$ Amplitudes of successive transmitted rays through the plates are :

$$aT,\ aTR,\ aTR^2, \cdots$$

The phase difference δ between any two rays reaching a point on the screen is

$$\delta = \frac{2\pi}{\lambda} \times \text{path diff.} = \frac{2\pi}{\lambda} 2d\cos\theta$$

Let $y = ae^{i\omega t}$ be the incident wave. Then the waves reaching a point on the screen are:

$$\begin{aligned} y_1 &= aTe^{i\omega t} \\ y_2 &= aTRe^{i(\omega t - \delta)} \\ y_3 &= aTR^2 e^{i(\omega t - 2\delta)} \\ \cdots &= \cdots\cdots \end{aligned}$$

$\therefore$ Resultant amplitude, A is given by

$$\begin{aligned} A &= aT + aTRe^{-i\delta} + aTR^2 e^{-2i\delta} + \cdots \\ &= aT\left(1 + Re^{-i\delta} + R^2 e^{-2i\delta} + \cdots\right) \\ &= aT\frac{1}{1 - Re^{-i\delta}} \end{aligned}$$

$$\therefore A^* = \text{conjugate of } A = aT\frac{1}{1 - Re^{i\delta}}$$

∴ Resultant intensity,

$$I = AA^* = \frac{a^2T^2}{(1-Re^{-i\delta})(1-Re^{i\delta})} = \frac{a^2T^2}{1+R^2-R(e^{-i\delta}+e^{i\delta})}$$

$$= \frac{a^2T^2}{1+R^2-2R\cos\delta} = \frac{a^2T^2}{(1-R)^2+2R(1-\cos\delta)}$$

$$= \frac{a^2T^2}{(1-R)^2+4R\sin^2\delta/2} = \frac{a^2T^2}{(1-R)^2}\left[\frac{1}{1+\frac{4R}{(1-R)^2}\sin^2\delta/2}\right] \quad (8.21.1)$$

The above expression (8.21.1) shows that the intensity is **maximum** when $\sin^2\delta/2 = 0 \Rightarrow \delta = 2n\pi$, where $n = 0, 1, 2, \cdots$

$$\therefore\ \boldsymbol{I_{\max} = \frac{a^2T^2}{(1-R)^2}} \quad (8.21.2)$$

And, the intensity will be **minimum** when

$$\sin^2\delta/2 = 1 \ \Rightarrow\ \delta = (2n+1)\pi, \quad n = 0, 1, 2, \cdots$$

$$\therefore \quad I_{\min} = \frac{a^2T^2}{(1-R)^2}\cdot\frac{1}{1+\frac{4R}{(1-R)^2}} \quad (8.21.3)$$

$$\therefore \quad \boldsymbol{I = I_{\max}\frac{1}{1+\frac{4R}{(1-R)^2}\sin^2\delta/2}} \quad (8.21.4)$$

the intensity expression for Fabry-Perot fringes.

Sharpness of fringes—A plot of I vs. δ for different values of R, the reflectivity, it is seen that the half-width is quite small (Fig. 8.35). The **half-width** is the width of the **peak** at $I = \frac{1}{2}I_{\max}$.

From (8.21.4), therefore, δ at half width is given by

$$1 + \frac{4R^2}{(1-R^2)^2}\sin^2(\delta/2) = 2$$

$$\Rightarrow \quad \sin^2(\delta/2) = \frac{(1-R^2)^2}{4R^2} \quad \Rightarrow \quad \sin(\delta/2) = (1-R^2)/2R$$

$$\therefore \quad \delta = 2\sin^{-1}\frac{1-R^2}{2R}$$

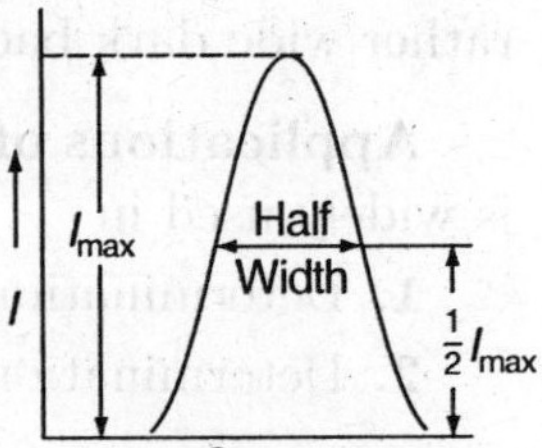

Fig 8.35

So, δ decreases with increasing R.

∴ When $R = 1$, $\delta = 0 \Rightarrow$ visibility is maximum. When $R = 0.9$, $\delta = 0.211$ radian.

In Michelson's interferometer, however, the intensity is given by

$$I = I_{max} \cos^2(\delta/2)$$

so that at half-width, i.e., $I = \frac{1}{2} I_{max}$,

$$\delta = 2\cos^{-1}(1/\sqrt{2}) = \pi/2 = 1.57 \text{ rad.}$$

This shows that *fringes obtained with a Fabry-Perot interferometer are comparatively much* **sharper** *than those obtained with a Michelson interferometer.*

A plot of I/I_{max} vs. δ for different R gives a set of curves as shown in Fig. 8.36. It is observed that higher the R, more rapid is the fall in intensity on both sides of a maximum. In other words, *more the reflectivity, sharper are the bright fringes.* Theoretically, sharpness is maximum for $R = 1$.

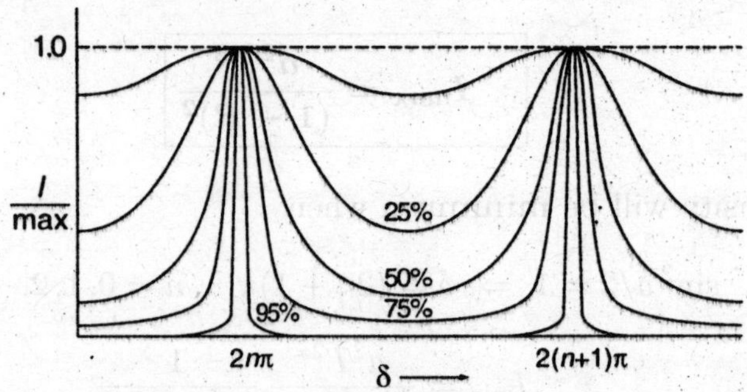

Fig 8.36

The **visibility** of the fringes is given by

$$V = \frac{I_{max} - I_{min}}{I_{max} + I_{min}} = \frac{2R}{1 + R^2} \qquad (8.21.5)$$

using (8.21.2) and (8.21.3) and simplifying.

Equation (8.21.5) shows that the visibility of fringes is a function of the reflection coefficient R only, and increases with increasing value of R.

For instance, when $R = 0.5$ (50%), $V = 0.8$ (80%). And when $R \to 100\%$, $V \to 1$, Hence, *more the reflectivity, more is the visibility* or *contrast.*

Thus, the Fabry-Perot interferometer gives a set of *concentric sharp rings* on a rather wide dark background.

Applications of Fabry-Perot interferometer—The Fabry-Perot interferometer is widely used in

1. Determination of **unknown wavelength** λ, and
2. Determination of **wavelength-difference** $\Delta\lambda$ of two close lines (*fine structure*)

Determination of λ—The condition for maxima in Fabry-Perot interferometer is

$$2d\cos\theta = n\lambda$$

n being the order of interference and λ the wavelength of light.

At the centre of the fringe pattern, $\theta = 0 \Rightarrow \cos\theta = 1$.

$$\therefore\ 2d = n_0\lambda$$

where n_0 is the order of interference at the centre.

If at the centre, we get a *bright* fringe of order n_1 for a separation d_1, and again a *bright* fringe of order n_2, for a separation d_2, then we have

$$2d_1 = n_1\lambda; \quad 2d_2 = n_2\lambda$$

$$\therefore \quad 2(d_2 - d_1) = (n_2 - n_1)\lambda = N\lambda, \text{ say}$$

where $N = n_2 - n_1$, i.e., the number of fringes crossing the centre of the fringe system, when the separation changes from d_1 to d_2.

$$\therefore \quad \boxed{\lambda = \frac{2(d_2 - d_1)}{N}}$$

Determination of $\Delta\lambda$ (fine structure)—Let the source emit two very close wavelengths λ_1 and λ_2. We shall now get two sets of fringes, one for each wavelength. Let mth order maximum for λ_1 coincide with nth order maximum for λ_2, with plate separation d_1, in a direction θ.

$$\therefore\ 2d_1\cos\theta = m\lambda_1 = n\lambda_2 \tag{8.21.6}$$

$$\Rightarrow \quad \frac{\lambda_1}{\lambda_2} = \frac{n}{m}$$

Now, if we move from the centre outward, the condition for the next order to coincide is given by

$$\frac{\lambda_1}{\lambda_2} = \frac{n-1}{m-1}$$

($\because$ order decreases outward)

Since m and n are rather *large*, $n/m \simeq (n-1)/(m-1)$ which implies that the maxima of λ_1 and λ_2 coincide over a *wide range of angle* covering the whole *field of view* appearing *distinct maximum*.

On increasing the plate separation to d_2, say, until the two sets of fringes again coincide showing *maximum distinct* field, then

$$2d_2\cos\theta = (m+p)\lambda_1 = (n+p+1)\lambda_2 \tag{8.21.7}$$

where p is the increase in order of λ_1 and $p+1$ that for λ_2.

$$\therefore\ 2(d_2 - d_1)\cos\theta = p\lambda_1 = (p+1)\lambda_2$$

using (8.21.6) and (8.21.7).

At the centre, $\cos\theta = 1$. Therefore, from above, we obtain

$$2(d_2 - d_1) = p\lambda_1 = (p+1)\lambda_2$$

$$\therefore \quad p(\lambda_1 - \lambda_2) = \lambda_2$$

$$\therefore \quad p = \frac{\lambda_2}{\lambda_1 - \lambda_2}$$

Substituting this p-value in (8.21.7), we obtain

$$2(d_2 - d_1) = \frac{\lambda_1\lambda_2}{\lambda_1 - \lambda_2} = \frac{\lambda_m^2}{\Delta\lambda}$$

$$\Rightarrow \quad \boxed{\Delta\lambda = \frac{\lambda_m^2}{2(d_2 - d_1)}}$$

where λ_m = mean of λ_1 and λ_2, and $(d_2 - d_1)$ = distance traversed by the movable plate for two consecutive positions of maximum distinctness of the field of view.

8.21.2 Lummer-Gehrcke plate

This is also a *high resolution* multiple beam interferometer and is widely used in studying *fine structure* of spectral lines. Unlike Fabry-Perot, here the coating of silver to increase reflectivity and, thereby to increase sharpness of fringes, has been dispensed with. For, the thick silvering will not only increase reflectivity but also absorption and hence reduce the intensity. Instead, therefore, the *phenomenon of total internal reflection* is utilised here.

Construction—It consists essentially of an *optically plane* and *accurately parallel* plate AB (Fig. 8.37) of glass or quartz, 10 – 15 cm long and 1-2 cm wide and a few mm thick. At one end of the plate, a small *totally reflecting prism* P is cemented.

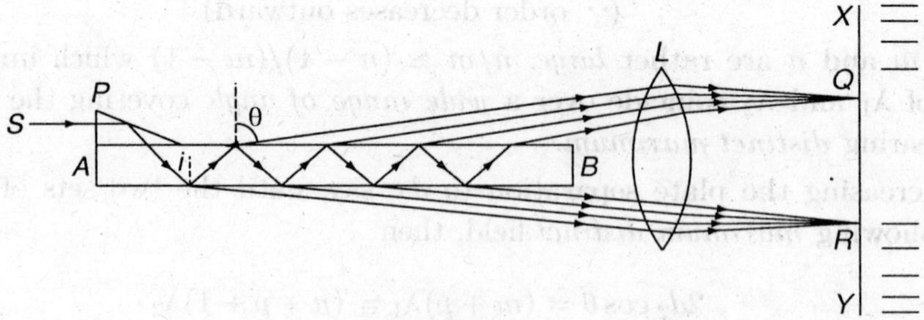

Fig 8.37

A beam of light from an *extended* source (not shown) enters into the plate through P, meeting the lower surface of the plate at an angle *slightly less* than the *critical angle* of the plate material. Only a small fraction of light thus emerges out, the rest being internally reflected many times successively at the two surfaces of the plate. The transmitted fractions at each reflection form a parallel bundle of rays on either side of the plate at grazing angles. Between any two consecutive rays in the bundle, there is a constant path difference. The parallel rays are brought to focus on a screen XY placed in the focal plane of a converging lens L. Two identical sets of fringes are obtained. The *fringes are bright, narrow maxima, separated by wide dark minima.* They are *approximately straight lines parallel to the plate surface* and are called **Haidinger fringes.**

When however the plate is made of *quartz,* it can be used even in *ultraviolet* which is usually absorbed by glass.

Theory in brief—If t be the thickness of the plate and μ the refractive index of its material for wavelength λ, then for maxima

$$2\mu t \cos i = n\lambda \tag{8.21.8}$$

where the integer n is the order of interference and i the angle of incidence.

But, $\quad \sin r = \mu \sin i \quad \Rightarrow \quad \cos i = \sqrt{1 - \sin^2 r/\mu^2}$

$\therefore$ From (8.21.8), we obtain

$$2t\sqrt{\mu^2 - \sin^2 r} = n\lambda$$

$$\Rightarrow \boxed{4t^2(\mu^2 - \sin^2 r) = n^2\lambda^2}$$

the *fundamental equation* for a Lummer-Gehrcke plate, especially used for observing the *fine structure* of spectral lines.

Unlike F-P interferometer, the number of reflections depends on the length of the plate. Also, due to large path difference, it is used for the study of *Zeeman splitting* (**Zeeman effect**) of spectral lines in a magnetic field.

8.22 Illustrated Examples

A. Interference by division of wavefront

► **Example 1.** *Calculate the fringe-width of interference pattern produced in Young's double slit experiment with two slits* 10^{-3} m *apart on a screen* 1 m *away. Wavelength of light is* 5893 A. **(Guru Nanak Univ.)**

Solution. Given: $D = 1$ m, $d = 10^{-3}$ m, $\lambda = 5893$ A $= 5893 \times 10^{-10}$ m.

$$\therefore \text{ Fringe width, } \beta = \frac{D}{d}\lambda = \frac{1}{10^{-3}} \times 5893 \times 10^{-10} \text{ m} = \mathbf{5893 \times 10^{-7} \text{ m.}}$$

▸ **Example 2.** *The inclined faces of a biprism* ($\mu = 1.5$) *make angles of* $1°$ *with the base of the prism. The slit is* 10 cm *from the biprism and is illuminated by light of* $\lambda =$ 5900 A. *Find the fringe-width observed at a distance of* 1 m *from the biprism.*
(Kurukshetra Univ.)

Solution. Given: $\mu = 1.5, \lambda = 5900\,\text{m} = 5900 \times 10^{-10}$ m; base angle, $\alpha = 1° = \pi/180$ radian; source-biprism distance, $x = 10\,\text{cm} = 0.1\,\text{m}$; screen-biprism distance, $y = 1.0\,\text{m}$. Fringe width, $\beta = ?$

From the relation:

$$\beta = \frac{\lambda(x+y)}{2(\mu-1)\alpha},$$

$$\beta = \frac{5900 \times 10^{-10}(0.1 + 1.0)}{2(1.5-1)\pi/180} = \frac{5900 \times 1.1 \times 10^{-10} \times 180}{2 \times 0.5 \times \pi} = \mathbf{0.000372\ m}$$

▸ **Example 3.** *Interference fringes are produced by biprism in the focal plane of a reading microscope which is* 1 m *away from the slit. A lens interposed between biprism and microscope gives two images of the slit in two positions:* 4.05 mm *in one position,* 2.90 mm *in the other. If the wavelength of light used is* $\lambda =$ 5893 A, *find the distance between consecutive bands.*
(Vidyasagar and Punjab Univ.)

Solution. Given, $\lambda = 5893$ A $= 5893 \times 10^{-10}$ m; $D = 1$ m; $c_1 = 4.05$ mm $= 4.05 \times 10^{-3}$ m; $c_2 = 2.9$ mm $= 2.9 \times 10^{-3}$ m.

$$\therefore\ d = \sqrt{c_1 c_2} = \sqrt{4.05 \times 2.9} \times 10^{-3} = 3.427 \times 10^{-3}\,\text{m}$$

$\therefore$ Distance between consecutive bands = fringe-width = β.

$$\therefore\ \beta = \frac{D}{d}\lambda = \frac{1.0 \times 5893 \times 10^{-10}}{3.427 \times 10^{-3}} = \mathbf{0.172 \times 10^{-3}\ m}$$

▸ **Example 4.** *When a thin sheet of transparent material of thickness* 7.2×10^{-6} m *is introduced in the path of one of the interfering beams, the central fringe shifts to a position occupied by the sixth bright fringe. If* $\lambda = 6 \times 10^{-7}$ m, *find the refractive index of the sheet material.*
(Guru Nanak Univ.)

Solution. Given: thickness $t = 7.2 \times 10^{-6}$ m; $\lambda = 6 \times 10^{-7}$ m; index $\mu = ?$

By the problem, shift in central fringe = 6β. From the relation, $x_0 = \beta(\mu - 1)t/\lambda$,

$$6\beta = \frac{\beta}{\lambda}(\mu-1)t = \frac{\beta}{6 \times 10^{-7}}(\mu - 1) \times 7.2 \times 10^{-6}$$

$$\therefore\ \mu - 1 = \frac{6 \times 6 \times 10^{-7}}{7.2 \times 10^{-6}} = 0.5 \quad \Rightarrow \quad \mu = \mathbf{1.5}$$

▶ **Example 5.** *In a biprism experiment, fringes were first observed with sodium light of wavelength* 5890 A *and fringe-width was measured to be* 0.347 mm. Na-*light was then replaced by white light and central fringe located. On introducing a thin glass sheet in half of the beam, the central fringe was shifted by* 2.143 mm. *Calculate the thickness of the sheet if* μ *of glass is* 1.542. **(Benaras Hindu Univ.)**

Solution. Given: $\lambda = 5890 \times 10^{-10}$ m; $\mu = 1.542$; $\beta = 0.342$ mm; $x_0 = 2.143$ mm. Let n be the number of fringes in x_0; then $n = 2.143/0.342$.

$$\text{But,}\quad (\mu - 1)t = n\lambda \quad \Rightarrow \quad t = \frac{n\lambda}{\mu - 1}$$

$$\therefore \text{ Thickness, } t = \frac{2.143 \times 5890 \times 10^{-10}}{0.342 \times 0.542} = \mathbf{6.71 \times 10^{-6}\ m.}$$

B. Interference by division of amplitude

▶ **Example 6.** *What should be the thickness of a non-reflecting layer to be deposited on glass surface corresponding to wavelength* 6000 A ? *Refractive index of the layer is* 1.35. **(Benaras Hindu Univ.)**

Solution. The condition for minimum thickness of a non-reflecting coating dictates:

$$t = \lambda/4\mu$$

where t = thickness, λ = wavelength of light and μ = refractive index of coating.

Given: $\mu = 1.35, \lambda = 6000\text{ A} = 6000 \times 10^{-10}\text{ m} = 6 \times 10^{-7}$ m and $t = ?$

$$\therefore\ t = \frac{6 \times 10^{-7}}{4 \times 1.35} = \mathbf{1.11 \times 10^{-7}\ m.}$$

▶ **Example 7.** *A soap film is illuminated by white light incident at an angle of* 30°. *The reflected light is examined by a spectroscope in which the dark band corresponding to* $\lambda = 6 \times 10^{-7}$ m *is found. Find the thickness of the film. Given* μ *of film* = 1.33. **(Kurukshetra Univ.)**

Solution. Given $\mu = 1.33$, $i = 30°, \lambda = 6 \times 10^{-7}$ m, t = ? If t corresponds to the minimum thickness of the film for dark band,

$$t = \frac{\lambda}{2\mu \cos r} = \frac{\lambda}{2\mu\sqrt{1 - \sin^2 r}}.$$

$$\text{But, } \mu = \sin i/\sin r \quad \Rightarrow \quad \sin r = \sin i/\mu = \sin 30°/1.33 = 0.5/1.33 = 0.3759$$

$$\therefore\ t = \frac{6 \times 10^{-7}}{2 \times 1.33 \times \sqrt{1 - (0.3759)^2}} = \mathbf{2.434 \times 10^{-7}\ m.}$$

► **Example 8.** *The fringes of equal thickness are formed when two glass plates are kept over each other with a small gap in between. If a parallel beam of light of wavelength* 6000 A *is used and fringe separation is* 3 mm, *then what is the angle between the plates in seconds?* **(Poona Univ.)**

Solution. Given, $\lambda = 6000\text{ A} = 6000 \times 10^{-10}$ m, β = fringe separation = 3 mm = 3×10^{-3} m, $\mu = \mu_{air} = 1$ and angle between the plates, $\theta = ?$

The relation,

$$\beta = \frac{\lambda}{2\mu\theta} \Rightarrow \theta = \frac{\lambda}{2\mu\beta}$$

$$\therefore\ \theta = \frac{6000 \times 10^{-10}}{2 \times 1 \times 3 \times 10^{-3}} = 10^{-4} \text{ radian} = \frac{10^{-4} \times 180}{\pi} \text{ degree}$$

$$= \frac{10^{-4} \times 180}{\pi} \times 3600 \text{ second} = \mathbf{20.62''}$$

► **Example 9.** *Show that the diameter D_n of the nth Newton's ring, when two surfaces of radii R_1 and R_2 are placed in contact, is given by*

$$\frac{1}{R_1} \pm \frac{1}{R_2} = \frac{4n\lambda}{D_n^2}.$$

(Kanpur Univ.)

Solution. Newton's rings result from interference between waves reflected from the upper and the lower surfaces of air-film of varying thickness enclosed between the two surfaces (Fig. 8.38).

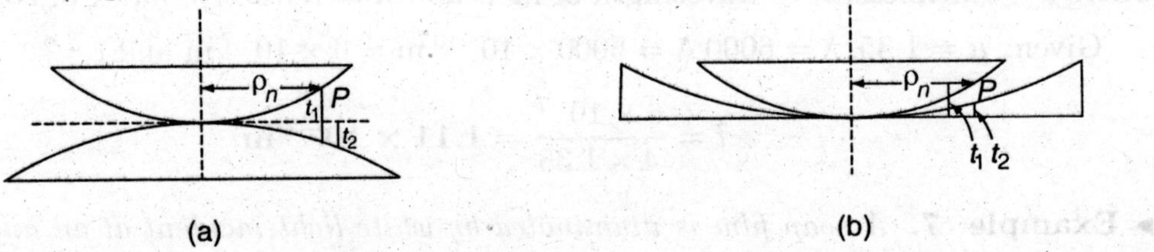

(a) (b)

Fig 8.38

If the radii of curvature of the surfaces be large and incidence of light is almost normal, then the *effective* path difference between interfering waves is

$$\Delta = 2t + \lambda/2$$

where t is the film-thickness at the point under reference.

$\therefore$ For a *dark* ring say,

$$2t + \frac{\lambda}{2} = (2n + 1)\frac{\lambda}{2} \quad \Rightarrow \quad 2t = n\lambda \qquad (8.22.1)$$

Let ρ_n be the radius of a ring corresponding to a point P, where thickness is t and R_1, R_2 be the (large) radii of curvature of the two surfaces.

$$\therefore\ t = t_1 \pm t_2 \quad \text{(Fig. (a) for + sign; Fig. (b) for − sign)}$$

$$= \frac{\rho^2}{2R_1} \pm \frac{\rho^2}{2R_2} \Rightarrow 2t = \rho_n^2\left(\frac{1}{R_1} \pm \frac{1}{R_2}\right).$$

Substituting this value of $2t$ in (8.22.1), we get

$$\rho_n^2\left(\frac{1}{R_1} \pm \frac{1}{R_2}\right) = n\lambda$$

$$\text{or,} \quad \frac{D_n^2}{4}\left(\frac{1}{R_1} \pm \frac{1}{R_2}\right) = n\lambda \quad (\because \text{ diameter } D_n = 2\rho_n)$$

$$\therefore \quad \frac{1}{R_1} \pm \frac{1}{R_2} = \frac{4n\lambda}{D_n^2}$$

▶ **Example 10.** *Newton's rings are observed in reflected light of* $\lambda = 5.9 \times 10^{-7}$ m. *The diameter of* 10*th dark ring is* 0.5 cm. *Find the radius of curvature of the lens and the thickness of the air film.* **(Calcutta, Delhi and Punjabi Univ.)**

Solution. Given, $\lambda = 5.9 \times 10^{-7}$ m, $D_{10} = 0.5$ cm $= 0.5 \times 10^{-2}$ m.

$$\text{But,} \quad D_n^2 = 4n\lambda R$$

$$\Rightarrow \quad R = \frac{D_n^2}{4n\lambda}$$

$$\therefore\ R = \frac{(0.5 \times 10^{-2})^2}{4 \times 10 \times 5.9 \times 10^{-7}} = \mathbf{1.059\,m}.$$

The relation,

$$2t = \frac{r^2}{R} \Rightarrow t = \frac{r^2}{2R} = \frac{(D/2)^2}{2R} = \frac{D^2}{8R}$$

$$\therefore\ t = \frac{D_n^2}{8R} = \frac{(0.5 \times 10^{-2})^2}{8 \times 1.059} = \mathbf{2.95 \times 10^{-6} m}$$

▶ **Example 11.** *In Newton's ring arrangement with a source emitting two wavelengths* $\lambda_1 = 6 \times 10^{-7}$ m, *and* $\lambda_2 = 5.9 \times 10^{-7}$ m, *it is found that nth dark ring due to one wavelength coincides with* $(n + 1)$*th dark ring due to the other. Find the diameter of nth dark ring if the radius of curvature of the lens is* 0.9 m. **(Punjab Univ.)**

Solution. By the problem, $(D_n^2)_{\lambda_1} = (D_{n+1})^2_{\lambda_2}$.

$$\therefore\ 4n\lambda_1 R = 4(n+1)\lambda_2 R \quad \Rightarrow \quad n\lambda_1 = (n+1)\lambda_2$$

$$\therefore \quad n \times 6 \times 10^{-7} = (n+1) \times 5.9 \times 10^{-7}. \quad \text{Solving, } n = 59$$

$$\therefore\ D_n = D_{59} = \sqrt{4 \times 59 \times 6 \times 10^{-7} \times 0.9} = \mathbf{1.1289 \times 10^{-2} m}$$

► **Example 12.** *Newton's rings are formed with reflected light of* $\lambda = 5.895 \times 10^{-7}$ m *with a liquid between the plane and the curved surface. The diameter of the 5th dark ring is* 0.3 cm *and the radius of curvature of the curved surface is* 1 m. *Calculate the refractive index of the liquid.* **(Punjabi Univ.)**

Solution. Given, $\lambda = 5.895 \times 10^{-7}$ m, $D_5 = 0.3\,\text{cm} = 0.3 \times 10^{-2}$ m and $R = 1$ m. Now, for the nth dark ring,

$$\frac{\mu D_n^2}{4R} = n\lambda \quad \Rightarrow \quad \mu = \frac{4Rn\lambda}{D_n^2}$$

$$\therefore \quad \mu = \frac{4 \times 1 \times 5 \times 5.895 \times 10^{-7}}{(0.3 \times 10^{-2})^2} = \mathbf{1.31}$$

► **Example 13.** *Newton's rings are formed with reflected light of wavelength* 5890×10^{-10} m *using a plano-convex lens and plane glass plate, with liquid between them. The diameter of the* 10*th ring is* 4.2 mm *and the radius of curvature of the curved surface of the lens is* 100 cm. *Find the refractive index of the liquid if the ring is (a) dark and (b) bright.* **(Guru Nanak Univ.)**

Solution. Given, $n = 10$, $R = 100$ cm $= 1$ m and $\lambda = 5890 \times 10^{-10}$ m, $D_{10} = 4.2$ mm $= 4.2 \times 10^{-3}$ m.

(a) If the ring is *dark,* we have

$$\mu = \frac{4Rn\lambda}{D_n^2} = \frac{4 \times 1 \times 10 \times 5890 \times 10^{-10}}{(4.2 \times 10^{-3})^2} = \mathbf{1.336}$$

(b) If the ring is *bright,* we have

$$\mu = \frac{4R(2n-1)\lambda}{2D_n^2} = \frac{4 \times 1 \times 19 \times 5890 \times 10^{-10}}{2 \times (4.2 \times 10^{-3})^2} = \mathbf{1.269}$$

C. Interferometry

► **Example 14.** *In Michelson's interferometer* 100 *fringes cross the field of view when the movable mirror is displaced through* 0.02948 mm. *Calculate the wavelength of monochromatic light used.* **(Delhi Univ.)**

Solution. By the problem, $n = 100, l = 0.02948$ mm $= 0.02948 \times 10^{-3}$ m and $\lambda = ?$
From the relation, $n\lambda/2 = l$, $\lambda = 2l/n$, we obtain

$$\lambda = \frac{2 \times 0.02948 \times 10^{-3}}{100} = 5896 \times 10^{-10}\,\text{m} = \mathbf{5896A}$$

► **Example 15.** *The movable mirror of Michelson's interferometer is moved through a distance of* 0.02603 mm. *If the wavelength used is* 5206 A, *find the number of fringes shifted across the cross-wire of the eyepiece of the telescope.*

Solution. Here $\lambda = 5206\text{ A} = 5206 \times 10^{-10}$ m; $l = 0.02603$ mm $= 0.02603 \times 10^{-3}$ m, and $n = ?$

$\therefore$ From the relation, $n\lambda/2 = l$, $n = 2l/\lambda$. We therefore get

$$n = \frac{2 \times 0.02603 \times 10^{-3}}{5206 \times 10^{-10}} = \mathbf{100}$$

▶ **Example 16.** *A Michelson interferometer is set for the white light straight fringes. When a mica sheet of thickness* 0.005 cm *is put in front of the fixed mirror, then to bring back the coloured fringes to their original position, the movable mirror is moved by* 0.0025 cm. *Calculate the refractive index of mica.* **(Guru Nanak Univ.)**

Solution. Let μ be the refractive index of mica sheet of thickness t, so that the increase in optical path is $(\mu - 1)t$. If l be the distance moved by the movable mirror to bring back the coloured fringes in position,

$$(\mu - 1)t = l \quad \Rightarrow \quad \mu - 1 = l/t$$

$$\therefore \quad \mu = 1 + \frac{l}{t} = 1 + \frac{0.0025}{0.005} = 1 + 0.5 = \mathbf{1.5}$$

▶ **Example 17.** *The wavelength of two components of D-lines of sodium are* 5890 A *and* 5896 A. *By how much distance one of the mirrors of Michelson's interferometer be moved so as to obtain consecutive positions of maximum distinctness.*

(Delhi and Himachal Univ.)

Solution. Here $\lambda_1 = 5890\text{ A} = 5890 \times 10^{-10}$ m; $\lambda_2 = 5896 \times 10^{-10}$ m. Let l be the distance over which one of the mirrors is to be moved in passing from the position of one maximum intensity to the next.

$$\therefore \quad \lambda_1 \sim \lambda_2 = \frac{\lambda_1 \lambda_2}{2l} \quad \Rightarrow \quad l = \frac{\lambda_1 \lambda_2}{2(\lambda_1 \sim \lambda_2)}$$

$$\therefore \quad l = \frac{5890 \times 10^{-10} \times 5896 \times 10^{-10}}{2(5896 - 5890) \times 10^{-10}} = \frac{5890 \times 5896 \times 10^{-10}}{2 \times 6} = \mathbf{0.2894 \times 10^{-3}\,m}$$

▶ **Example 18.** *In an experiment for determining the index of a gas by Michelson's interferometer, a shift of* 140 *fringes is observed when all the gas is removed from the tube. If the wavelength of light used is* 5460 A *and the length of the tube is* 20 cm, *calculate the refractive index of the gas.* **(Delhi Univ.)**

Solution. Here $\lambda = 5460\text{ A} = 5460 \times 10^{-10}$ m; $n = 140$, $l = 20$ cm $= 0.2$ m, $\mu = ?$

From the relation,

$$(\mu - 1)l = n\lambda/2$$

$$\mu - 1 = \frac{n\lambda}{2l} = \frac{140 \times 5460 \times 10^{-10}}{2 \times 0.2} = 0.000191$$

$$\therefore \quad \mu = 1 + 0.000191 = \mathbf{1.000191}$$

► **Example 19.** *White light is incident normally, on a Fabry-Perot interferometer with plate separation of* 4×10^{-6} m. *Calculate the wavelengths for which there are interference maxima in the transmitted beam in the range* 4000 A *to* 5000 A.

(Punjab Univ.)

Solution. Since light is incident normally, $\theta = 0 \Rightarrow \cos\theta = 1$. If d be the separation between the plates, then for maxima, the path difference $2d\cos\theta$ is given by

$$2d\cos\theta = n\lambda \quad \Rightarrow \quad 2d = n\lambda \quad (\because\ \theta = 0)$$

$$\therefore \quad \lambda = \frac{2d}{n} = \frac{2 \times 4 \times 10^{-6}}{n}\,\text{m}$$

For λ in the range 4000 A – 5000 A, i.e., 4×10^{-7} m to 5×10^{-7} m, the n-value for

$$\lambda = 4 \times 10^{-7}\,\text{m is}: \ n = \frac{2 \times 4 \times 10^{-6}}{4 \times 10^{-7}} = 20$$

$$\lambda = 5 \times 10^{-7}\,\text{m is}: \ n = \frac{2 \times 4 \times 10^{-6}}{5 \times 10^{-7}} = 16$$

$\therefore$ The relevant wavelengths correspond to $n = 16$ to $n = 20$.

$$\therefore \lambda_1 = \frac{2 \times 4 \times 10^{-6}}{16} = 5 \times 10^{-7}\ \text{m} = \mathbf{5000\ A}$$

$$\lambda_2 = \frac{2 \times 4 \times 10^{-6}}{17} = 4.706 \times 10^{-7}\ \text{m} = \mathbf{4706\ A}$$

$$\lambda_3 = \frac{2 \times 4 \times 10^{-6}}{18} = 4.444 \times 10^{-7}\ \text{m} = \mathbf{4444\ A}$$

$$\lambda_4 = \frac{2 \times 4 \times 10^{-6}}{19} = 4.211 \times 10^{-7}\ \text{m} = \mathbf{4211\ A}$$

$$\lambda_5 = \frac{2 \times 4 \times 10^{-6}}{20} = 4 \times 10^{-7}\,\text{m} = \mathbf{4000\ A}$$

► **Example 20.** *Two Fabry-Perot interferometers have equal plate separation but the coefficients of reflection are* 0.8 *and* 0.9. *Deduce the relative widths of the maxima in the two cases.*

Solution. In F-P interferometer, the intensity distribution is given by

$$I = \frac{I_{\max}}{1 + F\sin^2 \delta/2} \tag{8.22.2}$$

where $F = 4R/(1 - R^2)$, R being the coefficient of reflection.

Now (i) when $R = 0.8$, we obtain $F_1 = 4 \times 0.8/(1 - 0.8)^2 = 80$

Similarly, (ii) when $R = 0.9$, we have $F_2 = 4 \times 0.9/(1 - 0.9)^2 = 360$

If δ' be the fringe half-width, then from (8.22.1)

$$\frac{1}{2} = \frac{1}{1 + F \sin^2 \delta'/2} \quad (\because\ I = \tfrac{1}{2} I_{\max})$$

$$\Rightarrow\ 1 + F \sin^2 \delta'/2 = 2\ \Rightarrow \sin^2 \delta'/2 = 1/F$$

$$\therefore\quad \sin \delta'/2 = 1/\sqrt{F} \tag{8.22.3}$$

For large values of R, $1/\sqrt{F}$ is very small so that $\sin \delta'/2 = \delta/2$.

$$\therefore\ \delta' = 2/\sqrt{F},\ \text{using (8.22.2)} \tag{8.22.4}$$

If δ_1 and δ_2 be the relative widths of the maxima in the two cases, we have

$$\delta_1/\delta_2 = \sqrt{F_2/F_1} = \sqrt{360/80} = \sqrt{9/2} = \mathbf{3 : 1.41}$$

▶ **Example 21.** *In Jamin's interferometer, two evacuated tubes, each of length* 0.2 m *are placed in the two beams. A gas at a known temperature and pressure is admitted in one of the tubes and* 100 *fringes cross the centre of the field of view. Calculate the refractive index of the gas, if the wavelength of light used is* 5460 A.

Solution: Given: $n = 100$, $\lambda = 5460\text{A} = 5460 \times 10^{-10}\text{m}$ and $t = 0.2\text{m}$.

The path difference due to admittance of gas in one of the tubes $= (\mu - 1)t$.

$$\therefore\quad (\mu - 1)t = n\lambda$$

$$\Rightarrow\quad (\mu - 1) = \frac{n\lambda}{t} = \frac{100 \times 5460 \times 10^{10}}{0.2} = 0.000273.$$

$$\therefore\quad \text{Refractive index, } \mu = 1 + 0.000273 = \mathbf{1.000273}$$

Review Questions

Section A

1. Explain the phenomenon of interference of light. Describe and explain Young's experiment demonstrating interference of light.
(Burdwan and Vidyasagar Univ.)

2. What are coherent sources? How are they produced? Explain
(Cochin and Kanpur Univ.)

3. Explain analytically the interference of light on the basis of the wave theory. Explain briefly the phenomenon of interference in relation to the law of conservation of energy. **(Guru Nanak and Himachal Univ.)**

4. State and explain, in brief, the conditions for the interference of light for (i) observance, (ii) good contrast and (iii) stationary interference pattern. **(Benaras Hindu, Allahabad, Gauhati Univ.)**

5. Find an expression for fringe-width in case of Young's double slit expression. What is the shape of the fringes? **(Guru Nanak and Agra Univ.)**

6. Show that in a two slit interference pattern, the intensity at a point is given by
$$I = A + B\cos^2(kx/2)$$
where A, B and k are constants of the set up and x is the linear distance of this point from the central fringe.

7. Describe Fresnel's biprism. Explain the formation of interference fringes by it with a monochromatic source. Describe in detail how the wavelength of monochromatic light can be determined with its help. **(Burdwan and Vidyasagar Univ.)**

8. In a Fresnel's biprism arrangement, show that $d = 2x(\mu - 1)\alpha$ where x is the distance of the source from the biprism base and α the refracting angle of the prism.

9. Calculate the displacement of fringes when a thin transparent sheet is introduced in the path of one of the interfering beams in biprism. Show how this method is used for finding the thickness of a mica plate. **(Meerut and Indore Univ.)**

10. Explain the effect of introducing a thin plate of glass in the path of one of the interfering beams. Show how this method is used for finding the refractive index and the wavelength of light. **(Himachal and Guru Nanak Univ.)**

11. Give Stoke's treatment to explain the change of phase when reflection takes place at a denser medium. **(Himachal and Kanpur Univ.)**

12. Discuss the formation of fringes by Lloyd's single mirror and explain why the central fringe is black. Find an expression for the fringe-width. **(Guru Nanak Univ.)**

13. Give the theory and experimental arrangement of Lloyd's mirror to measure the wavelength of monochromatic light. **(Meerut and Rajasthan Univ.)**

14. Describe a method for obtaining achromatic fringes with Lloyd's mirror. In what respects do Lloyd's mirror fringes differ from those of biprism?

Section B

15. How does interference take place in thin films? Show that the reflected and transmitted interference patterns are complementary. **(Punjab and Himachal Univ.)**

16. What do you mean by non-reflecting or antireflection coating? Discuss its theory. **(Guru Nanak Univ.)**

17. Explain how interference fringes are formed by a thin wedge-shaped film, when examined by normally reflected light. How will you estimate the difference of film thickness between two points?

18. Describe the formation of Newton's rings by reflected light. Describe an arrangement to produce such rings. **(Calcutta and Vidyasagar Univ.)**

19. Derive the expression for the radius of nth dark as well as bright ring to show that while diameters of dark rings are proportional to the square root of simple numbers, those of bright rings are proportional to the square root of odd simple numbers. **(Calcutta and Kurukshetra Univ.)**

20. Derive an expression for the wavelength of monochromatic light source used in Newton's rings experiment in terms of diameters of rings and radius of curvature of the lens used. **(Vidyasagar and Burdwan Univ.)**

21. Explain the method of Newton's rings for determining the refractive index of rare liquids. Derive the necessary formula. **(Guru Nanak Univ.)**

22. Show that the diameter D_n of the nth Newton's ring, when two surfaces of radii R_1 and R_2 are placed in contact, is given by

$$\frac{1}{R_1} \pm \frac{1}{R_2} = \frac{4n\lambda}{D_n^2}$$

λ being the wavelength of light used. **(Kanpur Univ.)**

Section C

23. Describe the principle of construction and the working of a Michelson's interferometer. **(Himachal, Meerut and Guru Nanak Univ.)**

24. Explain how circular fringes are produced in Michelson's interferometer? Show that the radii of circular fringes are proportional to the square root of natural numbers. **(Punjabi Univ.)**

25. How do you obtain localised fringes and white light fringes in Michelson's interferometer?

26. Explain how Michelson's interferometer is used to determine (a) the wavelength of monochromatic light, (b) the wavelength difference of two very close wavelengths and (c) the refractive index and thickness of a thin transparent film. **(Meerut and Lucknow Univ.)**

27. Describe Michelson's interferometer. How will you use it to calibrate a metre in terms of a standard wavelength? **(Delhi Univ.)**

28. Describe Jamin's interferometer. How will you find the refractive index of a gas with its help?

29. Describe the principle, construction and production of interference fringes with Fabry-Perot interferometer. Discuss the intensity distribution of the fringes and find the ratio of I_{max} to I_{min}. **(Benaras Hindu Univ.)**

30. Explain the visibility and sharpness of the fringes obtained with Fabry-Perot interferometer.

31. Describe the construction and the working principle of a Lumer-Gehrcke plate. What is its special advantage? **(Kurukshetra Univ.)**

Short Answer and Objective Questions

Section A

1. What is meant by interference of light? What are coherent sources?

2. Enumerate the methods of production of coherent sources, with one example for each.

3. State in brief, the condition for observing a good and sustained interference pattern.

4. Is the conservation of energy principle violated in the phenomenon of interference? Explain.

5. Explain what happens in biprism experiment when (i) white light is used to illuminate the slit and (ii) the edge of the prism is not parallel to the slit. **(Guru Nanak Univ.)**

6. How will you locate the fringe of zero order in biprism experiment? Explain what happens when the eyepiece is moved away from the source. **(Poona Univ.)**

7. Explain why the central fringe in Lloyd's mirror is black. **(Kurukshetra Univ.)**

8. Compare the Lloyd's mirror fringes with those of the biprism.

9. Can two real source of light act as coherent sources? Explain. **(Guru Nanak and Punjab Univ.)**

10. Can we obtain interference with wide slits? Explain your answer.

11. In case of Young's double slit experiment, if one slit is covered with green transparent paper and the other with blue transparent one, what will be the effect on interference pattern? **(Guru Nanak Univ.)**

12. Can we obtain interference pattern if two coherent sources are separated by a distance less than a wavelength of light? Explain. **(Punjabi Univ.)**

13. Does the interference phenomenon give any information about the longitudinal or transverse nature of light? **(Guru Nanak Univ.)**

14. What becomes of energy of light waves whose destructive influence leads to dark lines in interference pattern?

15. What is the relation between the phase difference and the path difference?

16. Light waves from two coherent sources of same intensity I interfere. The intensity of light of maxima will be

    (a) I (b) $2I$ (c) I^2 (d) $4I^2$ **(Himachal Univ.)**

17. On reflection from a denser medium, the path difference introduced is

    (a) zero (b) $\lambda/2$ (c) λ (d) 2λ **(Himachal Univ.)**

18. In the phenomenon of interference, the law of conservation of energy is

    (a) never violated (b) always violated
    (c) sometime violated (d) violated depending on frequency

Section B

19. Why is a broad source of light necessary for observing colours in thin films, when a narrow source is required for biprism experiment? **(Kurukshetra Univ.)**

20. Show that the fringe-width decreases with the order in Newton's rings. **(Punjab Univ.)**

21. How does the interference pattern by reflection in thin films differ from that of refraction? Why? **(Himachal Univ.)**

22. Why do we see colours when white light falls on a thin film of transparent medium? **(Punjabi Univ.)**

23. What will happen in Newton's ring experiment if the glass plate is replaced by a plane mirror? **(Kurukshetra Univ.)**

24. How will you test the flatness of a surface by interference of light? **(Punjabi Univ.)**

25. Why do we prefer a convex lens of large radius of curvature in Newton's ring experiment?

26. Explain what will happen if white light is used instead of monochromatic light in Newton's experiment.

27. Explain what would happen in Newton's ring experiment when air in inter-space is replaced by a transparent liquid.

28. How from the study of colours of thin film, its thickness can be estimated?
(Guru Nanak and Punjabi Univ.)

29. Lenses coated with non-reflecting thin film appear purple in colour when seen in reflected light. Why? **(Himachal Univ.)**

Section C

30. Why the outer surfaces of the two plates of Fabry-Perot interferometer are not exactly parallel to the reflecting surfaces? **(Himachal Univ.)**

31. Give important applications of the Michelson's interferometer.
(Cochin and Himachal Univ.)

32. What are the shape of interference fringes obtained with Michelson's interferometer? **(Punjab Univ.)**

Problems

1. The inclined faces of a biprism of refractive index 1.50 make angle of 2° with the base. A slit illuminated by a monochromatic light is placed at a distance of 10 cm from the biprism. If the distance between the two dark fringes observed at a distance of 1 m from the biprism is 0.18 mm, find the wavelength of light used.
(Punjab Univ.) [5727 A]

2. In a typical biprism experiment, the fringe-width is 10^{-3} m for a wavelength of 5893 A. If $y/x = 20$, where y is the distance between the biprism and screen and x that between the slit and the biprism, calculate the refracting angle α of the biprism, if $\mu = 1.5$. [0.709°]

3. In a biprism experiment, the eyepiece is placed at a distance of 1.2 m from the source. The distance between the virtual sources were found to be 7.5×10^{-4} m. Find the wavelength of light if the eyepiece is to be moved transversely through a distance of 1.888×10^{-2} m for 20 fringes. **(Delhi Univ.)** [5900 A]

4. Calculate the separation between the coherent sources formed by a biprism whose inclined faces make angle of 2° with the base, the slit source being 10 cm away from the biprism and refractive index of the material of biprism 1.5.
(Kurukshetra Univ.) [0.00349 m]

5. Fringes are produced by a monochromatic light of $\lambda = 5.45 \times 10^{-7}$ m. A thin plate of glass of index 1.50 is then placed normally in the path of one of the interfering beams and the central bright band of the fringe system is found to move into the position previously occupied by the third bright band from the centre. Calculate the thickness of the plate. **(Punjab Univ.)** [3.27×10^{-6} m]

6. In a biprism experiment, fringes were first observed with sodium light of wavelength 5890 A and fringe-width was measured to be 0.347 mm. Sodium light was then replaced by white light and central fringe was located. On introducing a thin glass sheet in half of the beam, the central fringe was shifted by 2.143 mm. Calculate the thickness of the glass sheet, if the index of glass is 1.542.
(Benaras Hindu Univ.) [6.71×10^{-6} m]

7. When a thin piece of glass 3.4×10^{-4} cm thick is placed in the path of interfering beams in a biprism experiment, it is found that the central bright fringe shifts through a distance equal to the width of 4 fringes. Find the refractive index of the glass piece. **(Delhi Univ.)** [1.6424]

8. If a slit illuminated by sodium light (5893 A) is placed 0.15 cm from the plane of the Lloyd's mirror, what will be the distance between the consecutive bands formed 1 m away from the slit. **(Meerut Univ.)** [0.002 cm]

9. Two coherent sources whose intensity ratio is 81 : 1 produce interference fringes. Deduce the ratio of maximum to minimum intensity of the fringe system.
(Agra Univ.) [25 : 16]

10. In a double-slit arrangement, fringes are produced using light of $\lambda = 4800$ A. One slit is covered by thin glass plate of $\mu = 1.4$ and the other by another glass plate of the same thickness but of index 1.7. The result is a shift in the central bright fringe to the position of the original 5th bright fringe from the centre. Find the thickness of the plates. **(Gorakhpur Univ.)** [8.0×10^{-6} m]

11. Consider a non-reflecting film of refractive index 1.38. Assume that its thickness is 9×10^{-6} cm. Calculate the wavelength in the visible region for which the film will be non-reflecting. **(Kurukshetra Univ.)** [4968 A]

12. A wedge-shaped air film, having an angle of 40 seconds is illuminated by monochromatic light and fringes are observed vertically through a microscope. The distance measured between the consecutive bright fringes is 0.12×10^{-2} m. Calculate the wavelength of light used. **(Punjab Univ.)** [4656 A]

13. Using light of $\lambda = 5900$ A, it is found that in a thin film of air 7.4 fringes occur between two points. Deduce the difference of film thickness between them. [0.2183×10^{-5} m]

14. When a wedge-shaped air-film is viewed by a monochromatic source of light incident normally, the interference fringes 4 mm apart are observed. If the air space is filled with water of $\mu = 1.33$, how far apart will the fringes be observed? **(Kurukshetra Univ.)** [3×10^{-3} m]

15. In Newton's ring experiment, the diameter of 23rd ring was found to be 0.501 cm and that of 3rd ring was 0.181 cm. If the radius of curvature of the plano-convex lens is 0.5 m, calculate the wavelength of light used. **(Calcutta Univ.)** [5456 A]

16. In Newton's ring experiment, the diameter of the 4th and 12th rings are 0.4 cm and 0.7 cm respectively. Find the diameter of the 20th dark ring. [0.908×10^{-2} m]

17. A parallel beam of light of $\lambda = 5890$ A is incident on a thin glass plate of $\mu = 1.5$, such that the angle of refraction inside the plate is 60°. Calculate the *smallest* thickness of glass plate which will appear dark by reflection. **(Punjab Univ.)** [3.927×10^{-7} m]

18. In Newton's ring experiment, the radius of the 10th dark ring is found to be 4 mm when there is air between the convex lens and the glass plate. However, when air is replaced by a liquid, the radius of the ring shrinks from 4 mm to 3.3 mm. Find the refractive index of the liquid. [1.47]

19. A plano-convex lens of radius 3 m is placed on an optically flat glass plate and is illuminated by monochromatic light. The diameter of the 8th ring in the transmitted system is 0.72 cm. Calculate the wavelength of light used. **(Delhi Hons.)** [5760 A]

20. In a Michelson's interferometer, a thin plate is introduced in the path of one of the beams and it is found that 50 bands had crossed the line of observation. If the wavelength of light used is 5896 A and $\mu = 1.4$, determine the thickness of the plate. **(Punjab and Guru Nanak Univ.)** [4968 A]

21. In an experiment with Michelson's interferometer, the distance travelled by the mirror for two successive positions of maximum distinctness was 0.2945 mm. If the mean wavelength for the two components of sodium D-line is 5893 A, calculate the difference between the two wavelengths. **(Kanpur Univ.)** [5.896 A]

22. In an experiment with Jamin's interferometer for determining the refractive index of air, a shift of 150 fringes was observed when all the air was removed from the

tube placed in one of the optical paths. If the wavelength of light used is 4000 A in air and the length of the tube is 0.2 m, calculate the refractive index of air.

(Delhi Univ.) [1.0003]

23. Calculate the distance between two successive positions of the movable mirror of a Michelson's interferometer giving best fringes in the case of sodium light having lines of wavelengths 5890 A and 5896 A. **(Delhi Hons.)** [0.2894 mm]

24. In Jamin's interferometer, two evacuated tubes, each of length 0.2 m, are placed in the two beams. A gas at a known temperature and pressure is admitted in one of the tubes and 100 fringes cross the centre of the field of view. Calculate the refractive index of the gas, if the wavelength of light used is 5460 A. [1.000273]

25. Calculate the minimum plate separation in a Fabry-Perrot interferometer to obtain a free spectral range of 0.1 A in the wavelength region 5000 A. Also, calculate the smallest resolvable wavelength difference for the refractivity of 0.95.

[1.25 cm; 1.632×10^{-11} m]

Chapter 9

Diffraction of Light I

9.1 Introduction : Diffraction

A careful investigation reveals that light suffers some *deviation from* its *rectilinear path* whenever it passes close to the edge of opaque obstacles and narrow slits. As a result of this deviation some light *bends into the region of geometrical shadow.* The bending is however negligibly small if the dimensions of the obstacle or aperture are large compared to the wavelength of light. If however the size of the aperture or obstacle is comparable to the wavelength, the bending becomes somewhat more pronounced, although even in such cases only careful investigation can detect the phenomenon.

Diffraction. *The phenomenon of bending of light waves around obstacles or apertures of size comparable with the wavelength of light resulting in its spreading into the geometrical shadow of the object is known as* **diffraction**.

A form of wave motion, be it light or sound, exhibits diffraction. In fact, all wave motion shows diffraction under appropriate conditions. Diffraction of light gives rise to bright and dark fringes, called *diffraction bands,* outside the region of geometrical shadow. The correct interpretation of diffraction of light was provided by Fresnel who showed that *it is due to mutual interference of secondary wavelets originating from different parts of the wavefront* not blocked by the obstacle. Based on the principle of secondary wavelets and that of interference (superposition), he could calculate the position of fringes which agreed well with observation. It is to be noted that diffraction effects are observed only when a part of the wavefront is cut off by some obstacle.

All optical instruments use a limited portion of the incident wavefront and so some diffraction effects always accompany the image. For a proper understanding of the optical devices, therefore, the importance of diffraction can hardly be overestimated.

9.2 Distinction between interference and diffraction

Diffraction should be carefully distinguished from interference. The following are the *differences* between the two phenomena.

1. In interference, *interaction* occurs *between two separate wavefronts* originating from two *coherent* sources. In diffraction, however, *interaction* occurs *between the secondary wavelets* originating from different points *of the* exposed part of the *same wavefront.*

2. The *widths* of the fringes in interference *may* or *may not be equal.* But in diffraction, the widths of the fringes are *never equal.*

3. In interference, the regions of *minimum intensity* are *usually perfectly dark*, while in diffraction it is *not so.*

4. In interference, all the *maxima* are of *uniform intensity*, but in a diffraction pattern, the *bright bands* of maxima are of *varying intensities.*

9.3 Kinds of diffraction

There are two different kinds of diffraction. There are : (i) *Fresnel's diffraction* and (ii) *Fraunhofer's diffraction.*

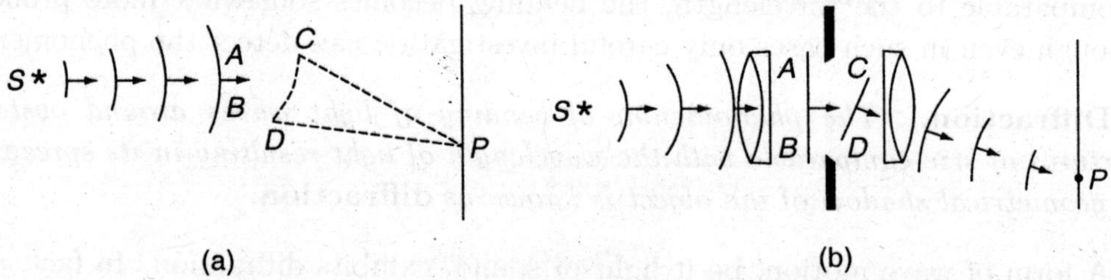

Fig 9.1

1. Fresnel's diffraction—In this type of diffraction, the *source* or the *screen* or *both* are placed *at finite* distances from the obstacle or aperture [Fig. 9.1(a)]. No lenses are used to render the rays parallel or convergent. The *incident wavefront AB* is *not plane,* but either *spherical* or *cylindrical.*

2. Fraunhofer's diffraction—In this type of diffraction, the *source* and the *screen* are *effectively at infinite distances* from the obstacle or aperture. This is made by using convex lenses—one for rendering the light from the source parallel before it falls on the aperture and the other to focus the light after diffraction on to the screen [Fig 9.1(b)]. The arrangement thus pushes both the source and the screen effectively to infinity. In this case, *the incident wavefront AB is plane.*

We shall study only the **Fresnel** type of **diffraction** in this chapter.

9.4 Fresnel's half-period zones

The whole of geometrical optics is based on the propagation of light in straight lines and the idea of light rays was thus developed. But it was indeed a difficult task to explain this on the basis of the wave theory. The solution was finally provided by Fresnel. He assumed that secondary wavelets originate from every point of the wavefront and interfere. The intensity at any point is the resultant of the intensities due to these secondary wavelets. With this idea, he not only could successfully explain the phenomenon but showed that *rectilinear propagation of light is only approximately true.*

We shall now discuss the Fresnel's method—the technique of **half-period zones**—for finding the effect of plane wavefront at some external point in front of it and use the technique to explain rectilinear propagation of light.

Fresnel's half-period zones—According to Huygens, each point on a wavefront sends out secondary wavelets which, Fresnel assumed, *interfere* among themselves to produce the resultant intensity at any point. To find this resultant intensity due to the entire wavefront, he divided the wavefront into a number of zones known as **Fresnel's half-period zones**. But how to construct the half-period zones?

Construction of half-period zones—Let a plane wavefront $ABCD$ of wavelength λ move from left to right and P be an external point at which the resultant (Fig. 9.2) intensity is to be determined. From P, draw a perpendicular. PO to the wavefront; the point O is known as the **pole** of the wavefront relative to P. Let $PO = p$ and with P as centre and radii $p + \lambda/2$, $p + 2\lambda/2$, $p + 3\lambda/2, \cdots$ etc. draw a series of *spheres* the sections of which by the wavefront are *concentric circles* $M_1, M_2, \cdots, M_n$, etc. with O as centre. The *area of* the first of the *innermost circle* is called the **first half-period zone**; the *annular area* between the *first and the second circles* is the **second half-period zone** and so on. The nth half-period zone is thus the annular area between the $(n-1)$th and the nth circles.

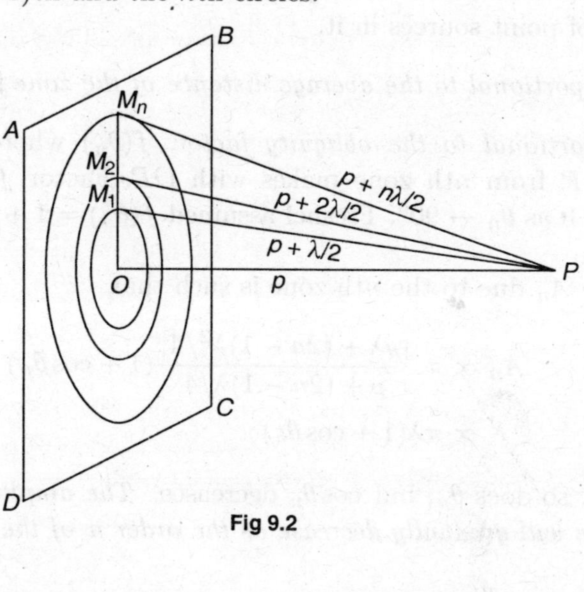

Fig 9.2

Area of a zone—The area of the nth half-period zone (Fig. 9.2) is

$$\begin{aligned}
S &= \pi(OM_n^2 - OM_{n-1}^2) \\
&= \pi\left[\left\{(p+n\lambda/2)^2 - p^2\right\} - \left\{(p+\overline{n-1}\,\lambda/2)^2 - p^2\right\}\right] \\
&= \pi\left[(pn\lambda + n^2\lambda^2/4) - \left\{p(n-1)\lambda + (n-1)^2\lambda^2/4\right\}\right] \\
&= \pi\left[p\lambda + \frac{\lambda^2}{4}\left\{n^2 - (n-1)^2\right\}\right] \\
&= \pi\left[p\lambda + \frac{\lambda^2}{4}(2n-1)\right] \\
&\simeq \pi p\lambda
\end{aligned}$$

since the term with λ^2 is ignorable compared to $p\lambda$, since $p \gg \lambda$.

Thus, *approximately, the area of the nth zone is independent of the order n of the zone, which implies that the area of each zone is nearly the same.* Strictly speaking, however, it increases slightly with n.

Average distance of nth zone from P—The average distance of the nth zone from P (Fig. 9.2) is

$$d_n = \tfrac{1}{2}[(p+n\lambda/2) + \{p+(n-1)\lambda/2\}] = p + (2n-1)\lambda/4$$

Amplitude, due to a zone, at P—The factors on which the amplitude of the disturbance at P due to a zone depends are (i) the *area of the zone,* (ii) the *average distance of the zone* and (iii) the *obliquity factor.* The **amplitude** is

(a) *directly proportional to the area of the zone* since more the area, more would be the number of point sources in it,

(b) *inversely proportional to the average distance of the zone* from P, and

(c) *directly proportional to the obliquity factor,* $f(\theta_n)$ where θ_n is the angle the direction of P from nth zone makes with OP. Factor $f(\theta_n) = 1$ for $\theta_n = 0$ and $f(\theta_n) \to 0$ as $\theta_n \to 90°$. Fresnel assumed $f(\theta_n) = 1 + \cos\theta_n$.

$\therefore$ Amplitude A_n due to the nth zone is such that

$$\begin{aligned}
A_n &\propto \pi\,\frac{[p\lambda + (2n-1)\lambda^2/4]}{p + (2n-1)\lambda/4}(1+\cos\theta_n) \\
&\propto \pi\lambda(1+\cos\theta_n)
\end{aligned}$$

As n increases, so does θ_n; but $\cos\theta_n$ decreases. *The amplitude of the disturbance at P due to a zone will gradually decrease as the order n of the zone increases.*

Resultant amplitude due to the entire wavefront—If $A_1, A_2, \cdots, A_n$ be the amplitudes at P due to the first, second, $\cdots$ etc. zone respectively, then they are in *descending order* of magnitude. Since, again, by very construction, the average distance of P from any two consecutive zones differs by $\lambda/2$ (half-period), the waves from two successice zones reach P in **opposite phase.**

$\therefore$ The resultant amplitude at P due to the whole wavefront $ABCD$ is

$$A = A_1 - A_2 + A_3 - A_4 + \cdots + (-1)^{n-1} A_n \tag{9.4.1}$$

Since A_2 is slightly *smaller* than A_1 but slightly *greater* than A_3, we have

$$A_2 = \tfrac{1}{2}(A_1 + A_3); \quad A_4 = \tfrac{1}{2}(A_3 + A_5); \cdots \text{ etc.} \tag{9.4.2}$$

to a close approximation.

$\therefore$ The equation (9.4.1) may be written as:

$$A = \frac{A_1}{2} + \left(\frac{A_1}{2} - A_2 + \frac{A_3}{2}\right) + \left(\frac{A_3}{2} - A_4 + \frac{A_5}{2}\right) + \cdots$$

the last term being $\frac{A_n}{2}$ if n is *odd* or $\frac{A_{n-1}}{2} - A_n$ if n is *even.*

Using the relations of (9.4.2), we obtain

$$A = \frac{A_1}{2} + \frac{A_n}{2} \cdots \text{ when } n \text{ is } odd$$

$$\text{and } A = \frac{A_1}{2} + \frac{A_{n-1}}{2} - A_n \cdots \text{ when } n \text{ is } even.$$

since the value of each bracketted term is essentially zero.

Now, n is extremely large and because of the obliquity factor $A_{n-1} = A_n = 0$.

$$\therefore \quad \boxed{A = \frac{1}{2} A_1}$$

So, *the amplitude due to the entire wavefront at a point in front of it is just half that due to the first half-period zone alone.* The contribution from other half-period zones are cancelled by *mutual interference.*

- As the intensity is proportional to the square of the amplitude, the intensity at P due to the entire wavefront is *one-fourth* of the intensity due to the *first half-period zone.*, i.e.,

$$I = A_1^2/4$$

- Since Fresnel assumed that the obliquity factor $f(\theta_n) = 1 + \cos\theta_n$, we have for waves travelling in backward direction, from first half-period zone, $\theta_n = \pi \Rightarrow f(\theta_n) = 0$. The resultant amplitude $A = \frac{1}{2}A_1$ at any point in the *backward direction* will be zero. This implies that *backward wave is absent.*

9.4.1 Rectilinear propagation of light

Fig. 9.3 shows the variation in the intensity of light at P with the number of exposed zones. The graph indicates that only a few central zones are effective and the intensity at P due to those zones is practically the same as due to the entire wavefront. So, if an obstacle is interposed between the wavefront and the point P is large enough to obstruct the effective central zones, the whole wavefront would become ineffective.

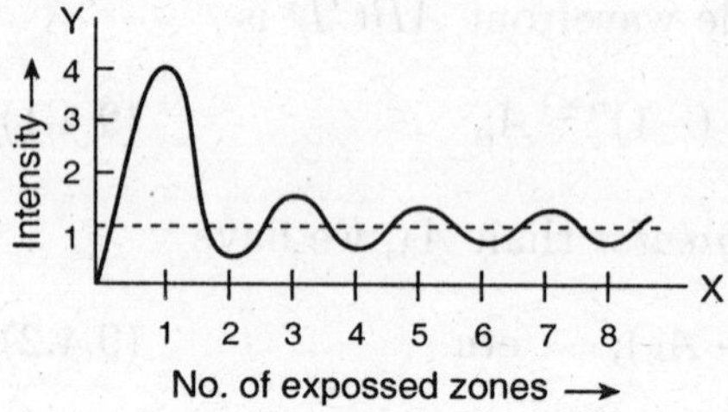

Fig. 9.3

Let a small *circular opaque* obstacle be placed at O (Fig. 9.2) so as to cover only the first half-period zone of the wavefront. Then, according to the above analysis (Art. 9.4), the resultant intensity at P due to the exposed wave front is proportional to $(A_2/2)^2$. Similarly, on gradual increase in size of the circular opaque obstacle, it would successively cover the first two, first three etc. half-period zones. This would result in the intensity at P to become proportional to $(A_3/2)^2$, $(A_4/2)^2$, etc. respectively. The illumination at P, therefore, would gradually diminish to become ultimately too small when the size of the obstacle is large enough to screen out an appreciable number of half-period zones. But the size of the half-period zones are small $(\pi p\lambda)$ so that a tiny obstacle is good enough to cover a large number of half-period zones, cutting off practically all light from the source. In other words, light propagates *approximately* in a straight line.

9.5 Zone plate : its theory

A *zone plate* is an optical device based on Fresnel's theory of half-period zeroes. It is a special diffracting screen designed to put a stop to the light from alternate half-period zones.

It is usually constructed by drawing a series of concentric circles on a white paper-sheet, the radii of the circles being proportional to the square-root of natural numbers, i.e., $\sqrt{1} : \sqrt{2} : \sqrt{3}$: etc. (Why? Wait till we develop the theory). The different zones thus formed are *alternately painted black*. A highly reduced photograph of this drawing is taken on a plate of glass. The negative thus obtained is the **zone plate.**

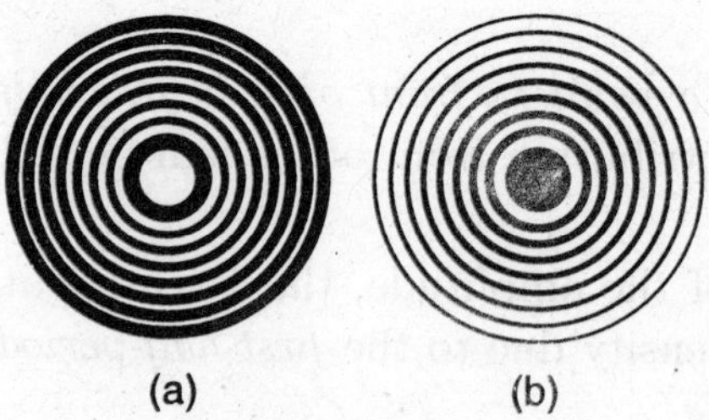

Fig 9.4

If *odd zones* are *transparent* and even zones are opaque, it is called a **positive zone plate.** If however *even zones* are *transparent* and odd zones are opaque, it is called a **negative zone** plate. The zone plate is illustrated in Fig. 9.4 where (a) shows positive plate and (b) shows negative plate.

A zone plate behaves as a **convergent lens** and produces image of a light-source on a screen placed at a suitable distance.

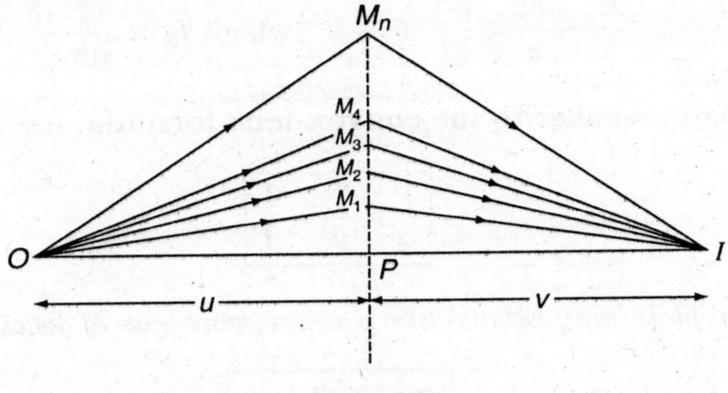

Fig 9.5

Theory—Let O be a point source of monochromatic light giving out spherical waves of wavelength λ whose effect at a point I on the screen is required (Fig. 9.5).

Consider an *imaginary plane* through P of a transparent medium, perpendicular to the plane of the paper and the line OI. The plane is divided into zones bounded by circles with centre at P and radii $PM_1 = r_1$, $PM_2 = r_2, \cdots, PM_n = r_n$ such that

$$OM_1 + IM_1 = OP + IP + \lambda/2 \tag{9.5.1}$$

$$OM_2 + IM_2 = OP + IP + 2(\lambda/2) \tag{9.5.2}$$

$$\cdots\cdots\cdots\cdots = \cdots\cdots\cdots\cdots\cdots\cdots$$

$$OM_n + IM_n = OP + IP + n(\lambda/2) \tag{9.5.3}$$

The annular rings thus formed are therefore *half period zones* for I, since the path-difference between corresponding points of two consecutive zones is $\lambda/2$.

Let $OP = u$ and $IP = v$, so that we have

$$OM_n = (u^2 + r_n^2)^{1/2} = u(1 + r_n^2/u^2)^{1/2} = u + r_n^2/2u, \quad (\because u \gg r_n.)$$

Similarly, we have $$IM_n = v + r_n^2/2v.$$

$\therefore$ From (9.5.3), we obtain

$$u + \frac{r_n^2}{2u} + v + \frac{r_n^2}{2v} = u + v + \frac{n\lambda}{2}$$

$$\Rightarrow \quad r_n^2\left(\frac{1}{u} + \frac{1}{v}\right) = n\lambda \tag{9.5.4}$$

($\Rightarrow r_n \propto \sqrt{n}$, since u, v, λ are constants and this justifies why the radii of concentric circles in preparing a zone plate are taken proportional to $\sqrt{1} : \sqrt{2} : \sqrt{3}$: etc.)

$$\therefore \quad \frac{1}{u} + \frac{1}{v} = \frac{n\lambda}{r_n^2} \tag{9.5.5}$$

Applying the sign convention, we have from (9.5.5),

$$\frac{1}{v}-\frac{1}{u}=\frac{n\lambda}{r_n^2}=\frac{1}{f_n},\quad \text{where } f_n=\frac{r_n^2}{n\lambda}. \tag{9.5.6}$$

Equation (9.5.6) is similar to the **convex-lens formula**, namely,

$$\boxed{\frac{1}{v}-\frac{1}{u}=\frac{1}{f}}$$

Thus the *zone plate may behave like a convergent lens of focal length*

$$\boxed{\boldsymbol{f_n=r_n^2/n\lambda}} \tag{9.5.7}$$

provided it has the *focussing action.*

• The expression (9.5.7) for the focal length f_n of a zone plate is λ-dependent, that is, a zone plate has different foci for different wavelengths.

Focussing action—In a zone plate, the alternate zones, say the *even* ones, are blocked. So, waves from the transparent zones (1st, 3rd, 5th etc.) differ in phase by λ and reach the point I in the same phase.

$\therefore$ Resultant amplitude A at I is

$$A = A_1 + A_3 + A_5 + ...$$

Thus, the resultant amplitude is now many times greater than $A_1/2$, the resultant amplitude due to all the zones. I is thus *extremely bright* and may be said to be the *image* of O. This explains the **focussing action** of zone plate.

9.5.1 Zone plate vs. Convex lens

A zone plate and a convex lens possess both similarity and dissimilarity. This aspect is discussed and elaborated below.

Similarity—A zone plate *behaves like a convex lens* and the focal length of both depends on the wavelength. Both of them, therefore, *suffer from chromatic aberration.* The *relation between the conjugate distances is also similar* in a zone plate and a convex lens.

Dissimilarity—However, there is no one-to-one correspondence between the two and, in fact, in many respects they differ from each other. For instance, (i) in contrast to one focus of a convex lens, the *zone plate,* for a given λ, *has multiple focal lengths* (n-dependence), thus forming for a point-source a series of point-images of diminishing intensity; (ii) also, while for a convex lens, all the rays reaching an image-point have the *same optical path,* it is not so for a zone plate where the *paths between the rays for two consecutive transparent zones differ by* λ; (iii) further, the *focal length in a zone plate* *decreases as* λ *increases; the focal length for red is less than that for violet* (i.e., $f_r < f_v$) for a zone plate, but for a convex lens $f_r > f_v$; the *order of colours* in chromatic effect of one is just the reverse of the other; (iv) the *thickness* of a convex lens *varies* from point to point, but that of a zone plate is *constant* everywhere; (v) also, in a convex lens, the rays are brought to focus by refraction, but the image is formed in a zone plate by diffraction and the image due to convex lens is more intense than that due to a zone plate.

9.5.2 Phase reversal zone plate

If instead of blocking the light from alternate zones, an additional optical path $\lambda/2$ is introduced between the light waves from consecutive zones, the amplitudes from each successive zone will be reinforced. Such a modified zone plate, first designed by R.W. Wood, is known as a **phase reversal zone plate**.

To prepare a *phase reversal zone plate*, the following procedures, as devised by Wood, are sequentially adopted.

1. A *glass plate* is *chemically cleaned* and *coated with* a thin layer of *gelatine* solution and carefully dried.

2. The coated plate is next *sensitized* by immersing it in a dilute solution of $K_2Cr_2O_7$ for a few seconds and *dried, but now in the dark.*

3. The plate is next placed in contact with an usual zone plate and *exposed to sunlight.* The light passes through the transparent zones to act upon the gelatine to make it insoluble in water. The gelatine in contact with opaque zones, however, remains water-soluble.

4. The glass plate thereafter is *immersed in water* to dissolve the gelatine of unexposed parts to such a depth that the optical paths from consecutive zones on the plate get an additional optical path of $\lambda/2$.

The phase reversal zone plate gives *four times more intense image* than an ordinary zone plate.

9.6 Diffraction at a straight edge

The set-up—Let S be a narrow slit (Fig. 9.6) placed perpendicular to the plane of the paper, A be the sharp straight edge of an opaque obstacle AB (e.g., the edge of a razor blade) and XY a screen, both placed parallel to the slit. Let the slit be illuminated with a monochromatic light of wavelength λ. Join SA and extend it to meet the screen at M.

Observation—With no diffraction at the edge A, uniform illumination would be obtained on the screen above M, and geometrical shadow (complete darkness) below it. But what, in fact, is observed are a few *bright and dark bands of unequal width, poor contrast and varying intensities that run parallel to the length of the slit in the region* *above and near M* (illuminated region). *However, the geometrical shadow,* (below M), the *intensity falls rapidly and becomes zero at a small but finite distance from* M. This is the **diffraction pattern** due to a straight edge.

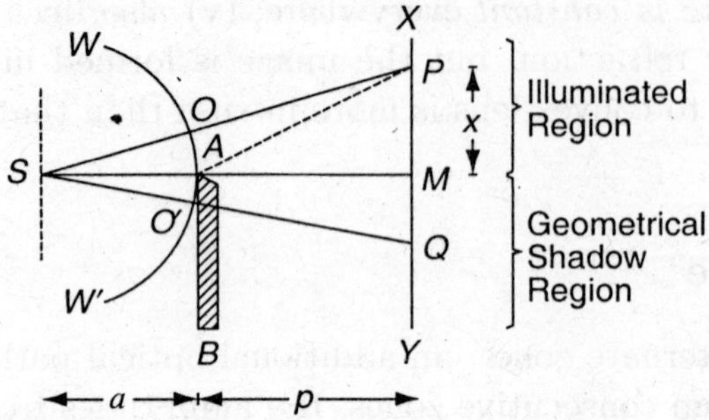

Fig 9.6

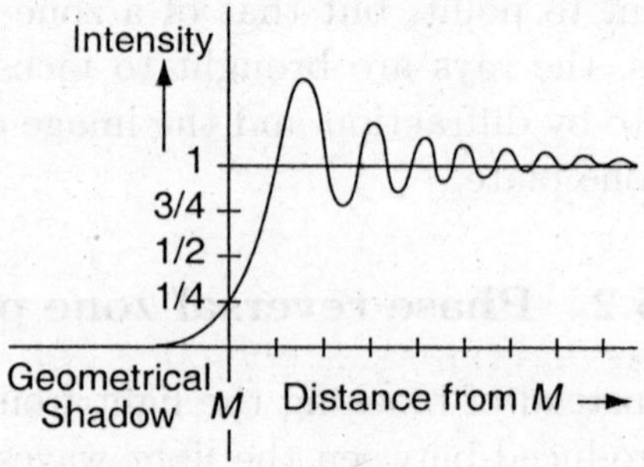

Fig 9.7

The *intensity distribution* of the pattern is illustrated graphically in Fig. 9.7 where it is seen that the intensity diminishes rapidly in the geometrical shadow, while a number of bright and dark bands are seen outside till uniform illumination sets in.

Explanation—Let WW' (Fig. 9.6) be the section of the cylindrical wavefront originating from S, and P be a point, distant x above M, on the screen. Join PS to intersect WW' at O, which therefore is the *pole* of the wavefront with respect to P. To find the intensity at P, the wavefront is to be divided, with O as pole, into *half-period elements.* The effect at P depends on the number of half-period elements contained in OA and the *upper-half* OW of the wavefront.

The effect at M, however, is due to the *upper-half* of wavefront *only,* so that the displacement at M is *half the displacement* that would have occurred if the entire wavefront were effective. Thus the intensity at M is *one-fourth* of that at a point far removed from M, where the whole wavefront is effective.

As one moves into the geometrical shadow, the pole O moves *via* A towards W' and half-period elements $1, 2, 3, \cdots$ are intercepted and intensity falls off gradually. On the contrary, if one moves away from M towards X, half-period elements $1, 2, 3, \cdots$ are exposed. The illumination at P is due to the *complete half of wave surface* OW plus the resultant of number of half-period elements contained in OA.

The amplitude at P will be maximum or minimum depending on if OA *contains an odd* or *even number of half-period elements.* The intensity at a maximum goes on decreasing, and at a minimum increasing, till finally at a large distance from M uniform illumination results. This is already illustrated in Fig. 9.7.

Mathematical theory—The number of half-period elements in AO (Fig. 9.6) depends on the path difference $AP - OP$. Putting $SA = a$, $AM = p$ and $MP = x$,

$$AP = \sqrt{AM^2 + MP^2} = \sqrt{p^2 + x^2} = p\left(1 + \frac{x^2}{p^2}\right)^{1/2} \simeq p\left(1 + \frac{x^2}{2p^2}\right) \qquad (9.6.1)$$

$$OP = SP - SO = SP - SA = (SM^2 + MP^2)^{1/2} - SA$$
$$= \left[(a+p)^2 + x^2\right]^{1/2} - a = (a+p)\left[1 + \frac{x^2}{(a+p)^2}\right]^{1/2} - a$$
$$\simeq (a+p) + \frac{x^2}{2(a+p)} - a = p + \frac{x^2}{2(a+p)}$$

$$\therefore\ AP - OP = \frac{x^2}{2p} - \frac{x^2}{2(a+p)} = \frac{ax^2}{2p(a+p)} \tag{9.6.2}$$

Now, P is a **maximum** if $AP - OP = (2n+1)\lambda/2$

$$\Rightarrow \quad \frac{ax^2}{2p(a+p)} = (2n+1)\frac{\lambda}{2}$$

$$\Rightarrow \quad x = \sqrt{\frac{p(a+p)(2n+1)\lambda}{a}} = k\sqrt{2n+1} \tag{9.6.3}$$

where k is a constant. The expression gives the position of the nth **maxima** in the diffraction pattern.

Similarly, P is a **minimum** if $AP - OP = n\lambda$. This implies that

$$\frac{ax^2}{2p(a+p)} = n\lambda, \text{ using (9.6.2)}$$

$$\Rightarrow \quad x = \sqrt{\frac{p(a+p)2n\lambda}{a}} = k\sqrt{2n} \tag{9.6.4}$$

which gives the position of nth **minima** in the pattern.

Distances of *successive maxima* from M are obtained by putting $n = 0, 1, 2$, etc. in (9.6.3).

$$\therefore\ x_1 = k;\ x_2 = k\sqrt{3};\ x_3 = k\sqrt{5};\ x_4 = k\sqrt{7}; \quad \text{etc.}$$

$\therefore$ Separation between successive maxima are

$$x_2 - x_1 = k(\sqrt{3} - 1) = 0.73k$$
$$x_3 - x_2 = k(\sqrt{5} - \sqrt{3}) = 0.50k$$
$$x_4 - x_3 = k(\sqrt{7} - \sqrt{5}) = 0.43k \text{ etc.}$$

which are in decreasing order. The bands are *not* therefore *equally spaced.*

Similarly, the separation between the successive minima also gradually decreases.

Determination of wavelength—A straight edge, e.g., a sharp razor blade is set with its edge parallel to the slit properly on the optical bench; the slit is illuminated with light of wavelength λ. Diffraction bands of unequal width and diminishing intensity are observed in the field of view of the micrometer eyepiece. The positions of the first maximum and the one most distant but clearly visible maximum, say 5th, are noted.

$$\therefore \quad x_1 = \sqrt{\frac{p(a+p)\lambda}{a}}; \quad x_5 = \sqrt{\frac{p(a+p) \times 11\lambda}{a}}$$

$$\therefore \quad x_5 - x_1 = \sqrt{\frac{p(a+p)\lambda}{a}}(\sqrt{11} - 1)$$

The distance p from the edge to the eyepiece and a that between the slit and the edge are measured. So from a knowledge of $(x_5 - x_1)$, we can find the unknown wavelength λ.

9.7 Fresnel's diffraction at a circular aperture

The set-up—Let AB be the narrow *circular* aperture placed in the path of light from a monochromatic point source S and the diffraction pattern be obtained on a screen XY held parallel to the slit. O is the centre of the slit and the line joining S and O meets, when extended, the screen at P normally (Fig. 9.8). Let P_1Q_1 be the section of a spherical wavefront that touches the aperture at any instant. The aperture will form the geometrical shadow above M and below N. The amplitude at P on the screen is the resultant of amplitudes of a large number of wavelets emanating from the wavefront AOB. With respect to P, the pole of the wavefront is at O. Let the wavefront be divided into half-period zones round O and the intensity at the axial point P be calculated.

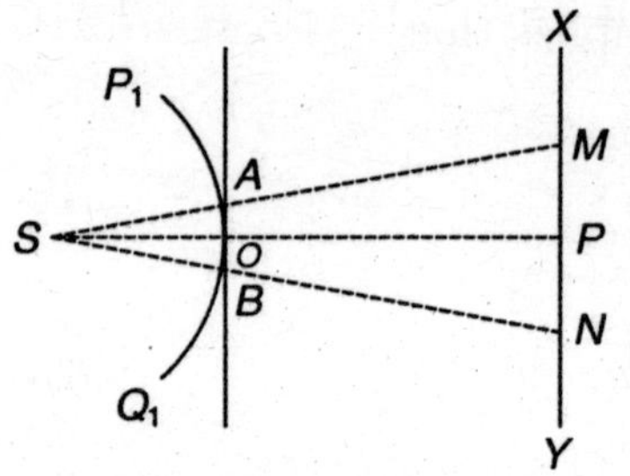

Fig 9.8

Intensity at axial point—If the size of the aperture be such that P exposes only the first zone, then the amplitude at P will be A_1, when the whole wavefront is exposed, the resultant amplitude at P is only $A_1/2$. The intensity at P is thus **four times** the intensity due to the entire wavefront. P is thus **extremely bright**—a position of *maximum* intensity.

If the screen is moved towards the aperture such that the first and the second zones are exposed then the amplitude at P is $A_1 - A_2$, i.e., almost *zero*. Thus the intensity at P is essentially *zero*—the position of *minimum*. If three zones are allowed, the resultant amplitude at P is $A_1 - A_2 + A_3$ which is again large making *intensity very high*. And so it goes on. So P gives a *maximum* or *minimum* according as the aperture allows an *odd* or *even* number of half-period zones. Hence the point P will be *alternately bright* or *dark*.

Distance of any point of maximum or minimum intensity—Let a be the distance of aperture AB from the source S and b the distance of screen XY (i.e., P)

from AB (Fig. 9.9). If r be the radius of the aperture, the path difference Δ for waves reaching P along SAP and SOP is

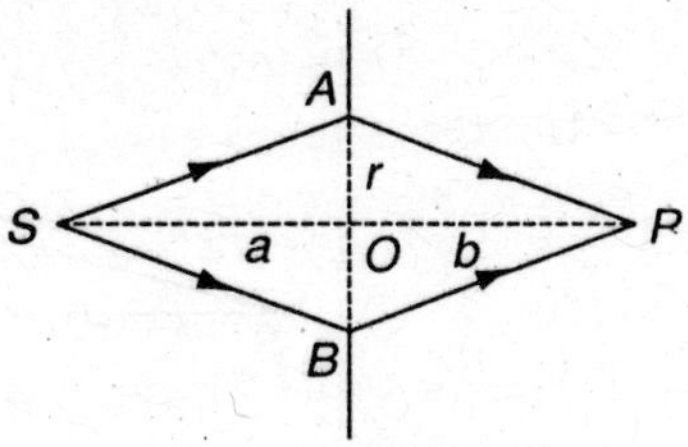

Fig 9.9

$$\begin{aligned}\Delta &= (SA + AP) - (SO + OP)\\ &= (a^2 + r^2)^{1/2} + (b^2 + r^2)^{1/2} - (a + b)\\ &= a(1 + r^2/a^2)^{1/2} + b(1 + r^2/b^2)^{1/2} - (a + b)\\ &= a(1 + r^2/2a^2) + b(1 + r^2/2b^2) - (a + b)\\ &= \frac{r^2}{2}\left(\frac{1}{a} + \frac{1}{b}\right)\end{aligned}$$

$$\therefore \quad \boxed{\frac{1}{a} + \frac{1}{b} = \frac{2\Delta}{r^2}} \tag{9.7.1}$$

If the position of the screen be such that n half-period zones are contained in the aperture, then the path difference $\Delta = n\lambda/2 \quad \Rightarrow 2\Delta = n\lambda$.

$\therefore$ From (9.7.1),

$$\frac{1}{a} + \frac{1}{b} = \frac{n\lambda}{r^2} \tag{9.7.2}$$

Now, P would be **maximum**, if n is **odd** and **minimum** if n is **even**.

In case the incident light is parallel, $a = \infty \quad \Rightarrow 1/a = 0$.

$$\therefore \text{ From (9.7.2)}, \quad \frac{1}{b} = \frac{n\lambda}{r^2} \quad \Rightarrow n = \frac{r^2}{b\lambda} = \frac{\pi r^2}{\pi b\lambda}$$

where πr^2 is the *area of aperture* and $\pi b\lambda$ *that of a half-period zone*, constructed for a point distant b from the wavefront.

Intensity at non-axial point—Let the screen be so placed that the intensity at P is *maximum* and the aperture contains five (*odd*) zones. If one moves away from the axis to the point P', then O' is the new pole where the line SP' meets the wavefront (Fig 9.10).

Let, in this position, the first four zones be fully exposed with about half of the 5th and 6th zones (Fig. 9.11). The resultant amplitude at P' is thus:

$$\begin{aligned}A_1 - A_2 + A_3 - A_4 + \tfrac{1}{2}A_5 - \tfrac{1}{2}A_6 &= \tfrac{1}{2}A_1 + \left[\tfrac{1}{2}(A_1 + A_3) - A_2\right]\\ &\quad + \left[\tfrac{1}{2}(A_3 + A_5) - A_4\right] - \tfrac{1}{2}A_6\\ &= \tfrac{1}{2}A_1 - \tfrac{1}{2}A_6\end{aligned}$$

which corresponds to a **minimum.**

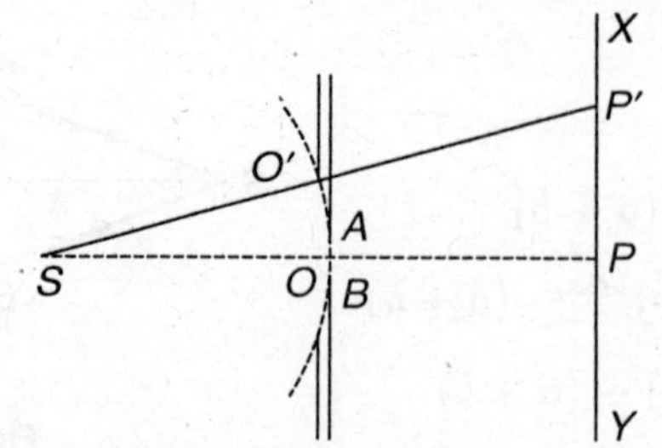

Fig 9.10

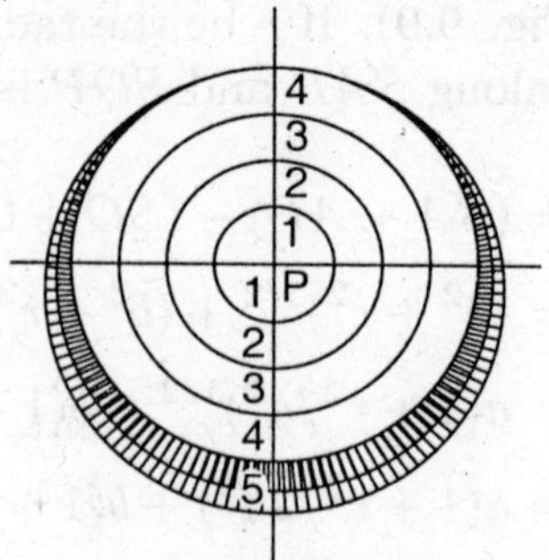

Fig 9.11

If one proceeds further to P'' (not shown) such that the first three zones are fully exposed with nearly half of the 4th, 5th, 6th and 7th zone, then similar computation for amplitude at P'' will show that this is a position of **maximum.**

Thus along PX, there is *a series of points with alternate maximum and minimum.* A rotation of the entire figure about PS as axis will make the points P', P'' etc. generate *bright and dark rings* of unequal width about P.

If the aperture be **large**, the rings will be visible only near the geometrical shadow with intensity rapidly falling off within the shadow. At a point well within, the intensity becomes zero. If, however, the aperture be **small** so as to contain only a fraction of the first half-period zone, with respect to P, the point P will be *bright.* And there will be *no ring* within the geometrical image of the aperture on the screen, and the bright point will be surrounded by alternate bright and dark rings, depending on whether the aperture contains an *even* or *odd* number of fractional zones.

9.8 Fresnel's diffraction at a rectangular aperture

Let a rectangular aperature of width AB be placed between a narrow slit S, illuminated with light of wavelength λ, and a screen XY, both being perpendicular to the plane of paper (Fig. 9.12). WW' is the section of cylindrical wavefront from S and incident on AB. The rectilinear propagation of light dictates that while the region $A'B'$ would be uniformly illuminated, those above A' and below B' should be completely dark. In fact, however, this does *not* occur; the illumination on the screen depends on the width AB and *three* situations may arise—*wide aperture, narrow aperture* and *very narrow aperture.*

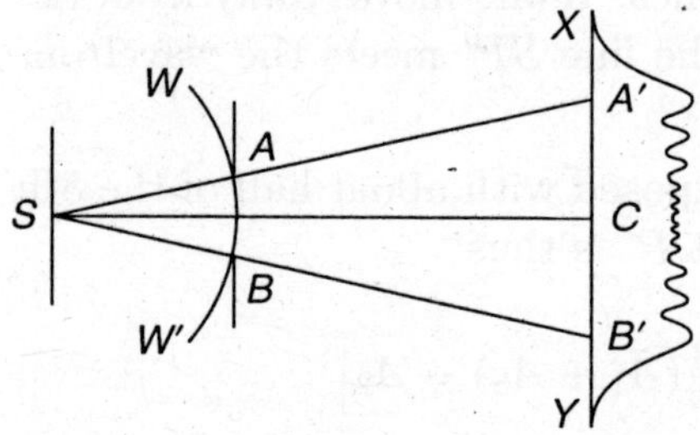

Fig 9.12

1. Wide aperture—In this case, there is uniform illumination at the centre C and some unequally spaced bands in the region $A'B'$. The contrast of the bands near A' and B' is poor. Above A' and below B' however the intensity rapidly falls to zero. But why?

Explantion—When the aperture AB is wide, it behaves essentially as two independent straight edges A and B, and hence we shall obtain, as in straight edge, unqually spaced bands with poor contrast near A' and B' within the region $A'B'$. These bands will merge into uniform illumination near the centre C. The intensity will show a rapid fall to zero in the region of geometrical shadow, i.e., above A' and below B'.

2. Narow aperture—In such a case, the centre C of the region $A'B'$ may appear *bright* or *dark*. Near A' and B' in the illuminated region we get unequally spaced bands with poor contrast. Above A' and below B', equally spaced bands are obtained.

Explanation—The point C is bright or dark according as AB contains an odd or even number of half-period strips of each half of the wavefront. For instance, if AB contains three half-period strips of each of the half wavefront, then the amplitude at C is

$$A = 2(A_1 - A_2 + A_3)$$

which indicates a *maximum* at the point C.

If, however, there are two half-period strips, then the amplitude at C would be

$$A = 2(A_1 - A_2),$$

which indicates that at C there is a *minimum.*

So, with the more and more opening up of the aperture, the centre of the illuminated region becomes *alternately bright and dark.*

Consider now the illumination near A' or B'. If AB contains three half-period strips, C will be, as seen already, bright. For a point above C, the pole of the wavefront is above O implying that the number of exposed half-period strips decreases in the upper half and increases in the lower half of the wavefront. Corresponding to a point P, where the number is 2 in upper half and 4 in the lower, say, then the resultant amplitude is

$$(A_1 - A_2) + (A_1 - A_2 + A_3 - A_4) \Rightarrow \text{minimum.}$$

At a still higher point, if the number be 1 and 5 say, then the amplitude is

$$A_1 + (A_1 - A_2 + A_3 - A_4 + A_5) \Rightarrow \text{maximum.}$$

So, in the region $A'B'$ we get bright and dark diffraction bands.

For a point such as Q (Fig 9.13), in the geometrical shadow, the pole is O' so that Q gets light only from few half-period strips contained in part AB of lower half of the wavefront. And

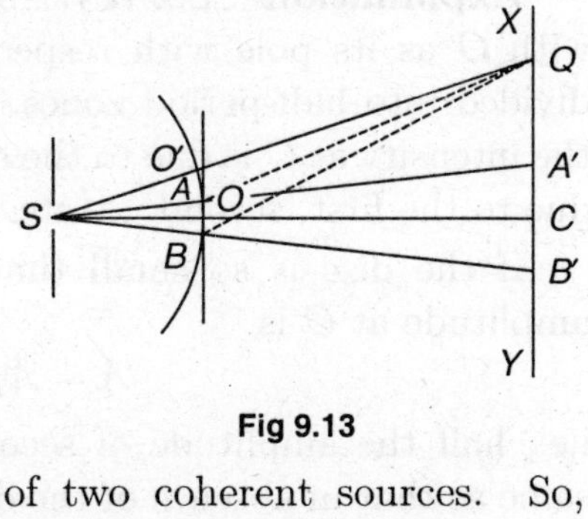

Fig 9.13

$$BQ - AQ = (2n+1)\lambda/2 \quad \Rightarrow \quad \text{maximum at } Q$$

$$BQ - AQ = 2n\lambda/2 \quad \Rightarrow \quad \text{minimum at } Q$$

The above conditions are similar to interference of two coherent sources. So, equispaced interference fringes are obtained above A' and (similarly) below B'.

3. Very narrow aperture—In this situation, the aperture contains a fraction of the first half-period strip in each half of the wavefront w.r.t. C. C is thus always bright. No bands are seen in region $A'B'$. In geometrical shadow, equi-spaced bands are observed.

Explanation—C is always bright since disturbances at C due to both halves of wavefront are in phase. As one goes towards A' or B' from C, the disturbance at any point is also due to a fraction of first half-period strip from each half-wavefront and are in phase. So in region $A'B'$, the intensity is always a maximum and no bands are observed.

Beyond A' and B', in the geometrical shadow, the number of half-period strips in AB may be more than one. So, the intensity at a point is *maximum* or *minimum* according as this number is *odd* or *even*. Thus we shall get *equispaced* dark and bright bands as in **Case 2.**

9.9 Diffraction at an opaque circular disc

Let AB be a small circular opaque disc placed in the path of light of a monochromatic point source S. The diffraction pattern is obtained on a screen XY perpendicular to the plane of the paper [Fig. 9.14(a)].

The pattern consists of a 'bright' spot at the centre C of the geometrical shadow $A'B'$, surrounded by a number of bright rings within the shadow. Outside $A'B'$, brighter and broader rings are obtained.

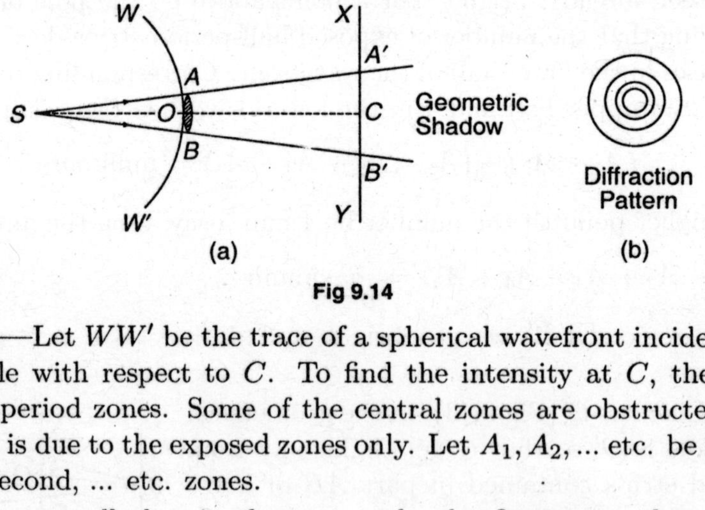

Fig 9.14

Explanation—Let WW' be the trace of a spherical wavefront incident on the disc with O as its pole with respect to C. To find the intensity at C, the wavefront is divided into half-period zones. Some of the central zones are obstructed by AB and the intensity at C is due to the exposed zones only. Let $A_1, A_2, ...$ etc. be the amplitude due to the first, second, ... etc. zones.

If the disc is so small that it obstructs only the first zone, then the resultant amplitude at C is

$$A = A_2 - A_3 + A_4 - A_5 + ... \quad \simeq \tfrac{1}{2}A_2$$

i.e., half the amplitude of second zone $\Rightarrow$ intensity $I \propto \frac{1}{4}A_2^2$. This is essentially the same as that in absence of the disc when $I \propto \frac{1}{4}A_1^2$. Thus, C is almost as bright as when the disc is absent.

If the disc-area increases so as to cover the first two zones (or equivalently, the screen moved forward to a position so that the disc covers the first two zones), the amplitude at C is

$$A = A_3 - A_4 + A_5 - A_6 + \dots \quad \simeq \tfrac{1}{2}A_3 \quad \Rightarrow \text{Intensity, } I \propto \tfrac{1}{4}A_3^2$$

So, at C, the intensity I is proportional to $A_2^2/4$, $A_3^2/4$, etc. as the disc covers 1, 2, etc. zones. Thus C is *always bright,* but the intensity gradually decreases as disc-size increases or screen moved towards the disc.

The concentric alternate bright and dark rings inside the geometrical shadow are due to interference between the exposed parts of the wavefront above A and below B. The rings outside the shadow are unequally spaced 'diffraction' rings formed as in a straight edge [Fig. 9.14(b)].

9.10 Illustrated Examples

▶ **Example 1.** *A screen is placed at a distance of* 100 cm *from a circular hole illuminated by a parallel beam of light of* $\lambda = 6400$ A. *Find the radius of the fourth half-period zone.* **(Jabalpur Univ.)**

Solution. The radius of the nth half-period zone is given by

$$r_n^2 = \left(p + n\frac{\lambda}{2}\right)^2 - p^2 = np\lambda + n^2\lambda^2/4 \simeq np\lambda$$

where p is the distance of a point from the pole of a plane wavefront.

Here $p = 100$ cm, $\lambda = 6400\text{A} = 6400 \times 10^{-8}$ cm and $n = 4$.

$$\therefore \; r_n = \sqrt{np\lambda} = \sqrt{4 \times 100 \times 6400 \times 10^{-8}} = \mathbf{0.16\ cm}$$

▶ **Example 2.** *The diameter of the first ring of a zone plate is* 1 mm. *If plane waves* ($\lambda = 5000$ A) *fall on the plate, where should the screen be placed so that light is focussed to the brightest spot?* **(Kanpur Univ.)**

Solution. Plane waves falling on the zone plate will be focussed to the brightest spot at its first focus, where the screen should be placed.

Now, the focal length f_n of the zone plate is given by

$$f_n = \frac{r_n^2}{n\lambda}$$

where r_n = radius of the nth zone.

If r_1 = radius of the first ($n = 1$) zone, then the first focal length $f_1 = r_1^2/\lambda$.

Here $r_1 = 0.50$ mm $= 0.05$ cm, $\lambda = 5000$ A $= 5000 \times 10^{-8}$ cm.

$$\therefore \; f_1 = \frac{0.05 \times 0.05}{5000 \times 10^{-8}} = \mathbf{50\ cm}$$

$\therefore$ The screen is to be placed at a distance of **50 cm** from the zone plate.

► **Example 3.** *An object is placed at* 20 cm *from a zone plate and the brightest image is situated at* 20 cm *from the zone plate with light of* $\lambda = 4000$ A. *Calculate the number of Fresnel's zones in a radius of* 1 cm *of the plate.* **(Allahabad Univ.)**

Solution. From the zone-plate formula, we have

$$\frac{1}{u} + \frac{1}{v} = \frac{n\lambda}{r_n^2} = \frac{1}{f}$$

Here $u = v = 20$ cm. So, $1/f = 1/20 + 1/20 \quad \Rightarrow f = 10$ cm.
Let the radius of nth zone be $r_n = 1$ cm. From the relation,

$$\frac{n\lambda}{r_n^2} = \frac{1}{f}, \qquad n = \frac{r_n^2}{f\lambda} = \frac{1^2}{10 \times 4000 \times 10^{-8}} = \mathbf{2500}$$

► **Example 4.** *The central circle of a zone plate has a radius of* 0.07 cm. *Light of wavelength* 5000 A *coming from* (i) *an object at infinity,* (ii) *an object* 147 cm *away from the plate falls on it. Find the position of the principal image in each case.* **(Agra Univ.)**

Solution. The focal length of the zone plate is

$$f = \frac{r_n^2}{n\lambda} = \frac{0.07^2 \text{ cm}^2}{1 \times 5000 \times 10^{-8} \text{ cm}} = 98 \text{ cm}$$

If u and v be the distances of object and principal image from the zone plate,

$$\frac{1}{u} + \frac{1}{v} = \frac{1}{f}$$

Case 1. Here $u = \infty$, $f = 98$ cm, $v = ?$ Therefore, from the above relation,

$$\frac{1}{\infty} + \frac{1}{v} = \frac{1}{98} \quad \Rightarrow \mathbf{v = 98 \text{ cm}}$$

Case 2. Here $u = 147$ cm, $f = 98$ cm, $v = ?$

$$\therefore \frac{1}{147} + \frac{1}{v} = \frac{1}{98} \quad \Rightarrow \mathbf{v = 294 \text{ cm}}$$

► **Example 5.** *A zone plate is made such that the radii of the circles defining the zones are the same as the radii of Newton's rings formed between a plane surface and a surface whose radius of curvature is* 150 cm. *Find the primary focal length of the zone plate.*

Solution. In an air film ($\mu = 1$), the radii r_n of Newton's (dark) rings are given by

$$r_n^2 = n\lambda R \quad (n = 1, 2, 3, \cdots)$$

where R = radius of curvature of the curved surface.

The primary focal length of the zone plate is given by

$$f = \frac{r_n^2}{n\lambda} \quad (n = 1, 2, \cdots) \Rightarrow r_n^2 = n\lambda f.$$

But, by the problem, we have $\quad n\lambda f = n\lambda R \quad \Rightarrow f = R = \mathbf{150\ cm}$

▶ **Example 6.** *Fringes are seen within the shadow of a thin wire held vertically in front of a narrow slit illuminated by monochromatic light. Find the diameter of the wire if the fringe-width at a distance of* 4 m *from the wire is* 1.8 mm *and the wavelength of light used is* 5400 A. **(Agra Univ.)**

Solution. The fringe width β in the diffraction pattern of fringes due to a narrow wire is

$$\beta = \frac{D}{d}\lambda \Rightarrow d = \frac{D}{\beta}\lambda$$

where D = screen to wire distance = 4 m = 400 cm here; d = dia of the wire. Here β = 1.8 mm = 0.18 cm, $\lambda = 5400 \times 10^{-8}$ cm.

$$\therefore d = \frac{400 \times 5400 \times 10^{-8}}{0.18} = \mathbf{0.12\ cm}$$

▶ **Example 7.** *When a circular aperture of diameter* 2 mm *is illuminated by a plane wave of light, the most intense point on the axis is at a distance of* 200 cm *from the aperture. Calculate the wavelength of light.* **(Gauhati Univ.)**

Solution. Let AB be the aperture (Fig. 9.15) and P the most intense axial point. For P to be most intense, only one half-period zone is to be allowed by AB.

For this, we have, $PA = p + \lambda/2$, where $p = PO$.

If r be the radius of the aperture,

$$r^2 = OA^2 = (p + \lambda/2)^2 - p^2 \simeq p\lambda$$

$$\therefore \lambda = \frac{r^2}{p} = \frac{0.1 \times 0.1}{200} = 5 \times 10^{-5} \text{ cm}$$

$$\doteq \mathbf{5000\ A}$$

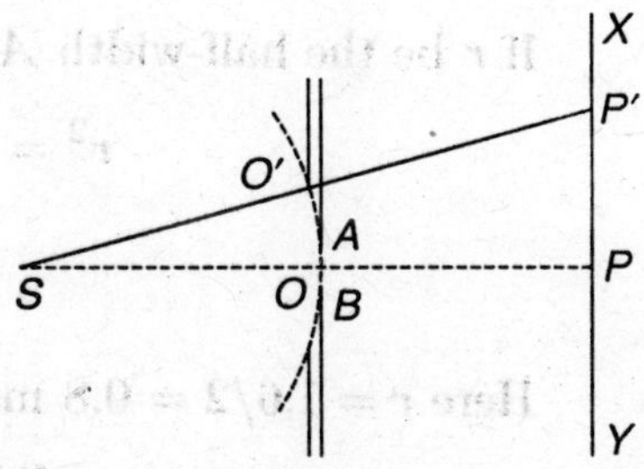

Fig 9.15

▶ **Example 8.** *Light of* $\lambda = 6000$ A *falls normally on a diaphragm with a round hole of* 6 mm *dia. A screen is placed behind the diaphragm* 3 m *away. How many Fresnel's zones can be accommodated in the diaphragm hole? Will the centre of the pattern on the screen be dark or bright? Explain.* **(Gorakhpur Univ.)**

Solution. Refer to the figure of the previous example. Let n be the required number of Fresnel's zones. Then

$$PA = p + n\frac{\lambda}{2}$$

If r be the radius of the hole, we have

$$r^2 = \left(p + n\frac{\lambda}{2}\right)^2 - p^2 \simeq pn\lambda,$$

ignoring the term $n^2\lambda^2/4$, which is indeed very small.

$$\therefore \quad n = \frac{r^2}{p\lambda}$$

Here $r = 3$ mm $= 0.3$ cm; $p = 3$ m $= 300$ cm and $\lambda = 6000 \times 10^{-8}$ cm.

$$\therefore n = \frac{0.3 \times 0.3}{300 \times 6000 \times 10^{-8}} = \mathbf{5}$$

Since the hole has *odd* number of Fresnel's zones, the centre of the pattern will be **bright.**

▶ **Example 9.** *A monochromatic beam of light on passing through a slit* 1.6 mm *wide falls on a screen held close to the slit. The screen is then gradually moved away and the middle of the patch of light on it becomes dark when the screen is* 50 cm *from the slit. Calculate the wavelength of light.* **(Kanpur Univ.)**

Solution. Refer to the figure of Ex. 7. The axial point P first becomes dark when the slit AB accommodates two half-period zones with respect to P.

$$\therefore PA = p + \lambda, \quad \text{where } p = OP.$$

If r be the half-width AO of the slit, then we have

$$r^2 = (p+\lambda)^2 - p^2 \simeq 2p\lambda, \quad (\text{ignoring } \lambda^2)$$

$$\therefore \quad \lambda = \frac{r^2}{2p}$$

Here $r = 1.6/2 = 0.8$ mm $= 0.08$ cm; $p = 50$ cm; $\lambda =$? Therefore,

$$\lambda = \frac{0.08 \times 0.08}{2 \times 50} = 6.4 \times 10^{-5} \text{ cm} = \mathbf{6400\ A}.$$

▶ **Example 10.** *Find the inner and outer radii of* 10*th half-period zone for a plane wavefront with respect to a point* 0.5 m *from it. The wavelength of light* $\lambda = 500$ nm.

Solution. The outer radius of nth half-period zone is

$$r_n \simeq \sqrt{np\lambda} \quad (\text{See Ex. 1})$$

The inner radius of nth half-period zone is

$$r_{n-1} \simeq \sqrt{(n-1)p\lambda}$$

By the problem, $n = 10$, $p = 0.5$ m, $\lambda = 500 \times 10^{-9}$ m.

$$\therefore \quad r_{10} = \sqrt{10 \times 0.5 \times 500 \times 10^{-9}} = \mathbf{1.58\ mm}$$

$$r_9 = \sqrt{9 \times 0.5 \times 500 \times 10^{-9}} = \mathbf{1.50\ mm}$$

▶ **Example 11.** *A parallel beam of monochromatic light is incident normally on a thin plate with a small circular hole of* 1 mm *dia. Find the wavelength of light used, if the screen has to be moved through* 10 cm *from its position where the centre is dark to the next similar position.*

Solution. For the first dark position (p_1), the hole must contain first two zones.

$$\therefore \; p_1^2 + r^2 = \left(p_1 + 2\frac{\lambda}{2}\right)^2 \simeq p_1^2 + 2p_1\lambda$$

$$\therefore \; p_1 = r^2/2\lambda \quad (r = \text{radius of the hole})$$

For the next dark position (p_2), the hole must contain first four zones.

$$\therefore \; p_2^2 + r^2 = \left(p_2 + 4\frac{\lambda}{2}\right)^2 \simeq p_2^2 + 4p_2\lambda$$

$$\therefore \; p_2 = r^2/4\lambda$$

$$\therefore \; p_1 - p_2 = r^2/2\lambda - r^2/4\lambda = r^2/4\lambda \;\Rightarrow\; \lambda = r^2/4(p_1 - p_2)$$

Here $r = 0.5$ mm, $p_1 - p_2 = 10$ cm. So, from the above relation, we get

$$\lambda = \frac{(0.5 \times 10^{-3})^2}{4 \times 10 \times 10^{-2}} \text{ m} = \mathbf{625\ nm}$$

▶ **Example 12.** *A plane wave of wavelength* 500 nm *is incident on a circular aperture of fixed radius* 0.50 mm*. Find the position of the brightest and darkest points on the axis.* **(Calcutta Hons.)**

Solution. If the distance p of the axial point be such that the aperture exposes only the first half-period zone, we shall get the **brightest** point.

$$\therefore \; p^2 + r^2 = \left(p + \frac{\lambda}{2}\right)^2$$

$$\text{or, } r^2 = p\lambda \;\Rightarrow\; p = \frac{r^2}{\lambda} = \frac{(0.5 \times 10^{-3})^2}{500 \times 10^{-9}} \text{ m} = \mathbf{0.5\ m}$$

For the *darkest* point on the axis, the aperture must expose two half-period zones.

$$\therefore \; p^2 + r^2 = \left(p + 2\frac{\lambda}{2}\right)^2 \simeq p^2 + 2p\lambda$$

$$\therefore \; p = \frac{r^2}{2\lambda} = \frac{(0.5 \times 10^{-3})^2}{2 \times 500 \times 10^{-9}} = \mathbf{0.25\ m}$$

Review Questions

1. What are Fresnel's half-period zones? Prove that the area of a half-period zone on a plane wavefront is essentially independent of the order of the zone. **(Lucknow Univ.)**

2. Show that the amplitude due to a large wavefront at a point in front of it is just half that due to the first half-period zone. Hence give Fresnel's explanation of the rectilinear propagation of monochromatic light. **(Allahabad Univ.)**

3. What is a zone plate? Give its theory; show that a zone plate has multiple foci. Compare a zone plate with a convex lens. What is meant by 'phase reversal zone plate'? **(Kanpur Univ.)**

4. Explain the rectiliner propagation of light on the basis of the wave theory. **(Calcutta Hons.)**

5. Explain the phenomenon of diffraction of light. What is the difference between Fresnel and Fraunhofer class of diffraction? Also, explain the difference between interference and diffraction of light. **(Saugar Univ.)**

6. Describe, with necessary theory, the Fresnel type of diffraction due to a straight edge. Show the intensity distribution in the diffraction pattern and comment on its features. How would you use it to determ the wavelength of monochromatic light? **(Meerut Univ.)**

7. A sharp razor blade is held vertically in a beam of monochromatic light coming from a narrow slit parallel to the edge of the blade. Discuss the positions and intensities of fringes observed on a screen behind the blade. **(Indore Univ.)**

8. Explain the method of dividing a cylindrical wavefront into half-period strips and find its effect at an external point. **(Agra Univ.)**

9. Describe and explain the diffraction pattern formed by a narrow wire illuminated by monochromatic light from a narrow slit parallel to the wire. How would you use the pattern to measure the thickness of the wire. What change do you expect if the wire is 'thick'? **(Kanpur Univ.)**

10. Describe the Fresnel's diffraction pattern when a monochromatic light is incident on a rectangular aperture. Discuss the case when the aperture is (i) wide, (ii) narrow and (iii) very narrow.

11. Discuss the Fresnel's diffraction at a small opaque disc illuminated by monochromatic light. Explain why the centre of the shadow of the disc is bright. What happens if the size of the disc is increased? **(Gauhati Univ.)**

12. Describe the diffraction effects due to a narrow circular aperture illuminated by monochromatic light from a point source. Show that if the screen on which diffraction pattern is obtained be moved towards the aperture, the intensity at the centre would be alternately maximum and minimum. **(Meerut Univ.)**

13. Write short explanatory notes on the following:

    (a) Fresnel's half-period zones

    (b) Rectilinear propagation of light by wave theory

    (c) Zone plate and its construction

    (d) Phase reversal zone plate

    (e) Fresnel and Fraunhofer class of diffraction.

Short Answer and Objective Questions

1. What do you mean by the term 'diffraction'? Can sound waves show diffraction phenomenon?
2. Why are the 'half-period zones' so called?
3. What is a zone plate? Write down the expression for the focal length of a zone plate? **(Cochin Univ.)**
4. How is a zone plate constructed? What is a phase reversal zone plate? **(Guru Nanak Univ.)**
5. In what essential respects the case of propagation of sound waves differ from those of light waves? **(Shivaji Univ.)**
6. What are the factors on which the amplitude of light waves from a half-period zone at an observation point depend? **(Himachal Univ.)**
7. Why do we get bright spot at the centre of the image of an opaque object when its size is extremely small? **(Punjab Univ.)**
8. Why is the diffraction effect in light difficult to observe?
9. What is the relation between the wavelength of the wave and size of the obstacle for diffraction to take place? **(Himachal Univ.)**
10. What is the cause of light streaks one sees while looking at a strong source of light with half-shut eyes? **(Himachal Univ.)**

11. The area of second half period zone A_2 and that of the first half period zone A_1 are related as

(a) $A_2 = 2A_1$ (b) $A_2 = \frac{1}{2}A_1$ (c) $A_2 = \sqrt{A_1}$ (d) $A_2 = A_1$

(Guru Nanak Univ.)

12. Which one of the following is essential for observing the diffraction?

(a) a screen (b) a narrow slit (c) two coherent sources (d) white light.

(Himachal Univ.)

Problems

1. Find the first three focal lengths of a zone plate for which the radius of the first zone is 0.3 mm, for light of $\lambda = 5000$ A. **(Allahabad Univ.)** [18, 6, 3.6 cm.]

2. What is the radius of the first zone in a zone plate if the focal length is 20 cm for light of $\lambda = 5000$ A? **(Rajasthan Univ.)** [0.0316 cm]

3. Find the radii of the first three transparent zones of a zone plate whose first focal length is 1 m for $\lambda = 5893$ A. [0.0767, 0.133, 0.172 cm]

4. A zone plate is illuminated with sodium light ($\lambda = 5896$ A) placed at a distance of 1.5 m. If the image of the point source is obained at a distance of 3 m on the other side, what will be the power of the equivalent lens which may replace the zone plate without disturbing the set up? Also calculate the radius of the first zone of the plate. **(Agra Univ.)** [+1D; 0.0768 cm]

5. A zone plate gives a series of images of a point source on its axis. If the strongest and second strongest images are at distances 30 cm and 6 cm respectively from the zone plate, both on the same side remote from the source, calculate the distance of the source from the zone plate, the principal focal length and the radius of the first zone ($\lambda = 5 \times 10^{-5}$ cm). [30 cm; 15 cm; 0.0274 cm]

6. In a diffraction experiment due to a straight edge, the distance between the slit-source ($\lambda = 6000$ A) and the straight edge is 6 m, and that between the straight edge and eyepiece is 4 m. Find the positions of the first three maxima and their separation. [0.200 cm, 0.346 cm, 0.447 cm; 0.146 cm, 0:101 cm]

7. A fine aperture is placed at a distance of 12 cm from a sharp razor blade edge held vertically at a distance of 25 cm from the screen. If the aperture be illuminated by light of $\lambda = 5890$ A, calculate the height of the 6th bright band above the line joining the edge to aperture. **(Kanpur Univ.)** [0.243 cm]

8. A very narrow vertical slit illuminated with light of $\lambda = 5.9 \times 10^{-5}$ cm casts a shadow of a vertical copper wire of 0.1 cm dia, and 20 cm away from the slit, on a screen 3 m away from the wire. Calculate the total number of bands inside the shadow. [9]

9. Light of $\lambda = 6000$ A passes through a narrow circular aperture having a radius of 0.09 cm. At what distance along the axis will the first maximum intensity be observed? [135 cm]

10. A circular aperture of 1.2 mm dia is illuminated with plane waves of monochromatic light. The diffracted light is received on a distant screen, which is gradually moved towards the aperture. The centre of the circular patch of light first becomes dark when the screen is 30 cm from the aperture. Find the wavelength of light used? **(Meerut Univ.)** [6000 A]

11. A parallel beam of $\lambda = 6000$ A is normally incident on a circular opening in an opaque screen. What should be the radius of the opening such that the intensity at a point 50 cm from the screen is 4 times the intensity on it when the screen is absent? Find the distance through which the screen should be moved towards the point to reduce the intensity at the point to zero. [0.055 cm; 25 cm]

12. A screen is placed at a distance of 60 cm from a circular aperture of 0.6 cm radius. Find the radius of the first half-period zone for light of $\lambda = 6000$ A. How many zones are contained in the circular aperture? **(Rajasthan Univ.)** [0.06 cm; 100]

13. A parallel beam of microwaves ($\lambda = 0.8$ cm) is incident on a circular aperture of adjustable radius. If a detector is placed on the axis of the aperture 80 cm behind it and the radius of the aperture is gradually increased, at what value of the radius would the response of the detector be maximum? **(Agra Univ.)** [8 cm]

14. Parallel monochromatic light of 6000 A falls normally on a circular aperture of diameter 0.12 cm and is viewed from the opposite side along a line through the centre of the hole and normal to its plane. Calculate the three largest distances from the hole at which the intensity is (i) maximum, (ii) zero.

[(i) 60, 20, 12 cm; (ii) 30, 15, 10 cm.]

15. Plane waves of light of $\lambda = 5000$ A fall normally, in one case on a small circular aperture, and in the other case on a small circular obstacle, each of radius 1.0 mm. In each case, light is observed at a point P distant 2 m from the aperture or obstacle. Show that the intensity of the point P in the case of aperture is approximately four times that in the case of the obstacle. **(Meerut Univ.)**

Chapter 10

Diffraction of Light II

We discussed the Fresnel type of diffraction phenomenon of light in the previous chapter. This chapter is devoted to a discussion of Fraunhofer type diffraction pattern in different cases such as single slit, double-slit, rectangular and circular slits, plane diffraction grating, concave grating etc.

10.1 Diffraction at a single slit

The case of diffraction by a single narrow slit is treated here as a *one-dimensional* (linear) problem.

Let a *parallel* beam of light of wavelength λ, i.e., a *planar* wavefront, be incident normally on a narrow slit AB (Fig. 10.1) of width a, placed perpendicular to the plane of paper. Each point of wavefront sends out secondary wavelets in all directions. The rays proceeding in the direction of incident rays are focussed at O' on the screen XY at the focal plane of the converging lens L. But those diffracted through an angle θ are focussed at P on the screen. We shall find the resultant intensity at P to get the *intensity distribution* in a single slit diffraction pattern.

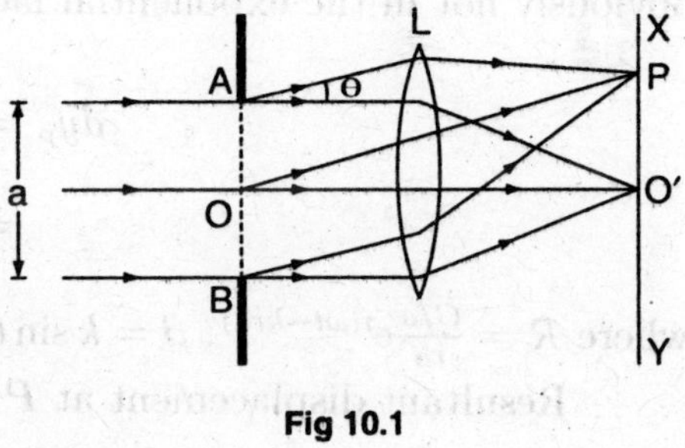

Fig 10.1

10.1.1 Intensity distribution function (IDF)

As the wavefront is planar, the energy passing per unit area per unit time is constant. So, the intensity C^2 associated with incident wavefront is a constant over the entire slit. Thus, the vibration of secondary sources at all points in the slit may be taken as :

$$y = (C/a)e^{i\omega t},$$

where C/a is the amplitude of vibration at the origin $O(0,0)$, the centre of the slit.

Consider (Fig. 10.2) a line element dz of the wavefront at $(0, z)$. If r be the distance of the point P from dz, the displacement at P due to dz at time t is

$$dy_P = \frac{C/a}{r} e^{i(\omega t - kr)} dz \tag{10.1.1}$$

where ω is the circular frequency and $k(= 2\pi/\lambda)$, the *propagation constant.*

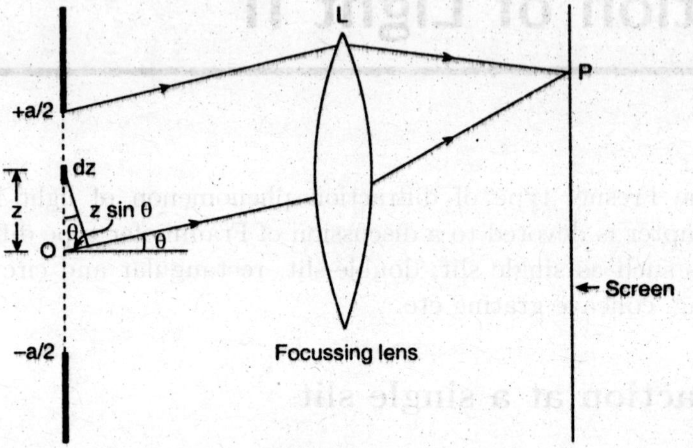

Fig. 10.2

We can express : r as $r = r_0 - z\sin\theta$, where r_0 is the value of r for $z = 0$. Since z and $\sin\theta$ are both small, r in the denominator of (10.1.1) may be replaced by r_0, but obviously not in the exponential factor.

$$\therefore \quad dy_P = \frac{C/a}{r_0} e^{i(\omega t - kr_0)} e^{ikz\sin\theta} dz$$

$$= \mathcal{R} e^{i\beta z} dz \tag{10.1.2}$$

where $\mathcal{R} = \frac{C/a}{r_0} e^{i(\omega t - kr_0)}$; $\beta = k\sin\theta = (2\pi/\lambda)\sin\theta$.

$\therefore$ Resultant displacement at P due to the whole slit is

$$y_P = \mathcal{R}\int_{-a/2}^{a/2} e^{i\beta z} dz = \frac{\mathcal{R}}{i\beta}\left(e^{i\beta a/2} - e^{-i\beta a/2}\right)$$

$$= \frac{\mathcal{R}}{i\beta} 2i\sin(\beta a/2) = a\mathcal{R}\frac{\sin(\beta a/2)}{(\beta a/2)}$$

$$= \frac{C}{r_0} e^{i(\omega t - kr_0)} \cdot \frac{\sin\alpha}{\alpha}$$

where we substitute $\alpha = \beta a/2 = \pi a\sin\theta/\lambda$.

Its complex conjugate is : $y_P^* = \frac{C}{r_0} e^{-i(\omega t - kr_0)} \cdot \frac{\sin\alpha}{\alpha}$

∴ Resultant intensity at P is given by

$$I_P = y_P y_P^* = \frac{C^2}{r_0^2} \cdot \frac{\sin^2\alpha}{\alpha^2}$$

When $\alpha \to 0$, $\sin\alpha/\alpha \to 1$ ∴ $\lim_{\alpha\to 0} I_P = I_0 = C^2/r_0^2$, the intensity at O' (Fig. 10.1).

$$\therefore \quad \boxed{I_P = I_0 \frac{\sin^2\alpha}{\alpha^2}} \qquad (10.1.3)$$

which is the **intensity distribution function** for a narrow single slit.

Maxima and minima—The directions of maxima or minima are given by

$$\frac{dI}{d\alpha} = 0$$

$$\Rightarrow \quad \frac{d}{d\alpha}\left[I_0\left(\frac{\sin\alpha}{\alpha}\right)^2 = 0,\right] \quad \text{using (10.1.3)}$$

$$\Rightarrow \quad I_0\left(\frac{2\sin\alpha}{\alpha}\right) \cdot \frac{\alpha\cos\alpha - \sin\alpha}{\alpha^2} = 0$$

$$\Rightarrow \quad \sin\alpha(\alpha\cos\alpha - \sin\alpha) = 0$$

Therefore, we have, either $\sin\alpha = 0$ or, $a = \tan\alpha$

But, it is found that $d^2I/d\alpha^2$ is *positive* for $\sin\alpha = 0$ and *negative* for $\alpha = \tan\alpha$.

∴ For **minima**, we have

$$\sin\alpha = 0$$

$$\Rightarrow \quad \boxed{\alpha = n\pi}$$

$$\text{or,} \quad \boxed{a\sin\theta = n\lambda} \qquad (10.1.4)$$

where $n = \pm 1, \pm 2, \pm 3, \cdots$ etc.

The equation (10.1.4) gives the direction of the first, second, $\cdots$ etc. minima corresponding to $n = \pm 1, \pm 2, \cdots$ etc. We have *excluded* $n = 0$, since when $\alpha \to 0$, $\mathrm{Lt}_{\alpha\to 0} \sin\alpha/\alpha = 1$, that gives a *maximum* ($I = I_0$).

For **maxima**, $\boxed{\alpha = \tan\alpha}$

which is a transcendental equation and can only be solved graphically by plotting the two curves :

$$y = \alpha \quad \text{and} \quad y = \tan\alpha \qquad (10.1.5)$$

and finding the point of intersection as shown in Fig. 10.3.

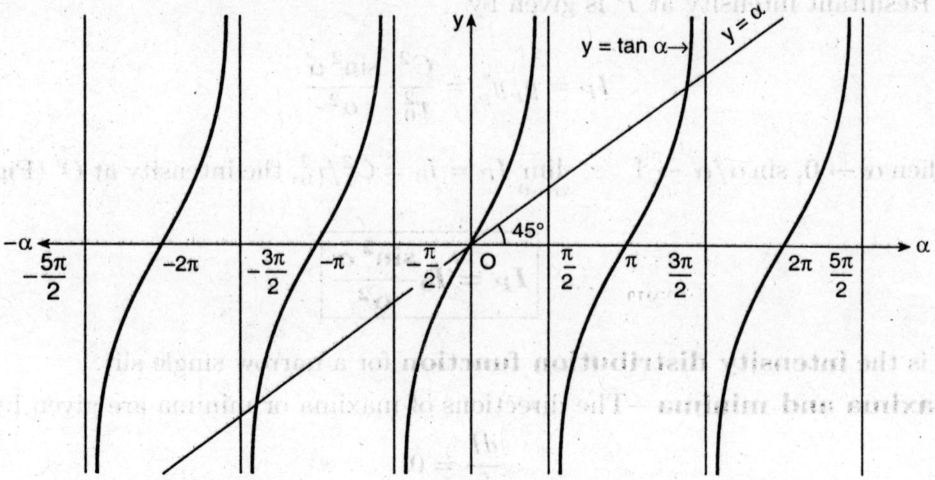

Fig. 10.3

The first of these is plainly a *straight line* through the origin inclined at 45° with the axes. The second is a *discontinuous curve* with a number of *branches*. The intersection of the two gives α-values that satisfy (10.1.5). The values are

$$\alpha = 0, \quad \pm 3\pi/2 \text{ (approx)}, \quad \pm 5\pi/2 \text{ (approx)}, \quad \cdots \text{ etc.}$$

or, **more exactly**, $\alpha = 0, \quad \pm 1.430\pi, \quad \pm 2.462\pi, \quad \cdots$ etc.

Central or principal maximum—This occurs when we have $\alpha = 0$. As $\alpha \to 0$, $\sin\alpha/\alpha = 1$, so that we obtain

$$I = I_0, \quad \text{the intensity of central maximum.}$$

Similarly, the intensities $I_1, I_2, I_3 \cdots$ etc. of the first, second, third $\cdots$ etc. *subsidiary* **(secondary)** *maxima* are

$$I_1 = I_0\left(\frac{\sin 3\pi/2}{3\pi/2}\right)^2 \simeq I_0\frac{4}{9\pi^2} = \frac{I_0}{22}$$

$$I_2 = I_0\left(\frac{\sin 5\pi/2}{5\pi/2}\right)^2 \simeq I_0\frac{4}{25\pi^2} \simeq \frac{I_0}{61}, \text{ etc.}$$

$$\therefore \quad I_0 : I_1 : I_2 : \cdots \text{ etc.} = 1 : 1/22 : 1/61 : \cdots \text{ etc.}$$

This shows that *most of the incident light is concentrated in the central maximum* that occurs in the direction given by $\alpha = 0 \Rightarrow \pi a \sin\theta/\lambda = 0 \Rightarrow \theta = 0$, i.e., *in the direction of incident light.* The secondary maxima are thus feeble relative to central maximum and their intensities diminish rapidly with increase in the order number.

Diffraction pattern—From the foregoing, it appears that the diffraction pattern consists of *a bright central maximum* in the direction of the incident light, along with *alternate minima* and *weak secondary maxima of rapidly diminishing intensity on either side of it.* The intensity distribution pattern is illustrated in Fig. 10.4.

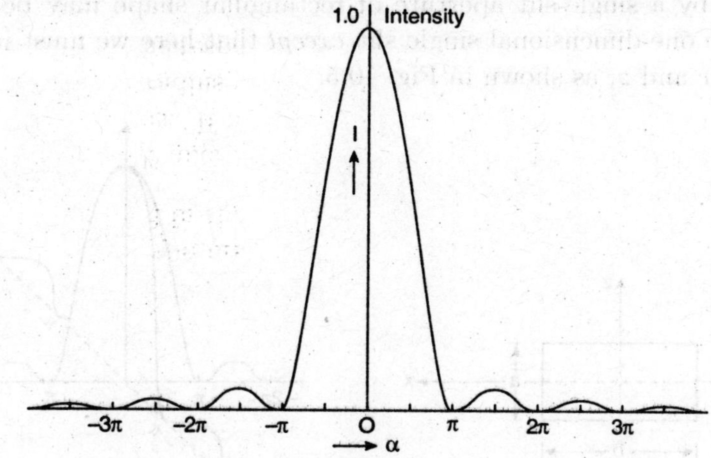

Fig. 10.4

The minima are at $\alpha = \pm\pi, \pm 2\pi, \pm 3\pi, \cdots$ etc. and the secondary maxima lie between two minima, but *do not fall exactly midway between them.* They are slightly displaced towards the centre of the pattern by an amount that decreases as the order increases.

Width of central maximum—We have the first minimum on either side of central maximum in a direction θ given by

$$a \sin\theta = \pm\lambda \;\Rightarrow\; \sin\theta = \pm\lambda/a$$
$$\text{or,} \quad \theta = \pm\lambda/a \quad (\because \theta \text{ is small})$$

which is plainly the **angular half-width** of central maximum.

$$\therefore \text{ Width of central maximum} = 2\lambda/a,$$

being *inversely proportional to the slit-width a.*

• **What happens when the slit is made narrower**—As the slit is narrowed, a gets decreased and consequently *the width of the central maximum* $(= 2\lambda/a)$ *increases.* In particular, when $a = \lambda$, we get, from the relation $a\sin\theta = \lambda$, $\theta = 90°$, i.e., the first minimum occurs at 90°, i.e., *the principal maximum fills the whole space.*

• **Difference between single-slit diffraction pattern and double-slit interference pattern**—While both the patterns consist of alternate maxima and minima, the two in fact differ widely. In the diffraction pattern, the *central maximum is brightest,* having λ/a as the angular half-width i.e., *narrower the slit, broader is the maximum. The subsidiary maxima* are on *either side* of it, situated *non-symmetrically* and are of *rapidly diminishing intensity.*

In the interference pattern, the *maxima and minima are equidistant* and *equally wide* and all the *maxima have* the *same intensity.* The *fringe width increases* with *decreasing separation* between the slits.

10.2 Rectangular aperture

The diffraction by a single-slit aperture of rectangular shape may be treated in the same way as the one-dimensional single slit *except* that here we must integrate in two dimensions say x and z, as shown in Fig. 10.5.

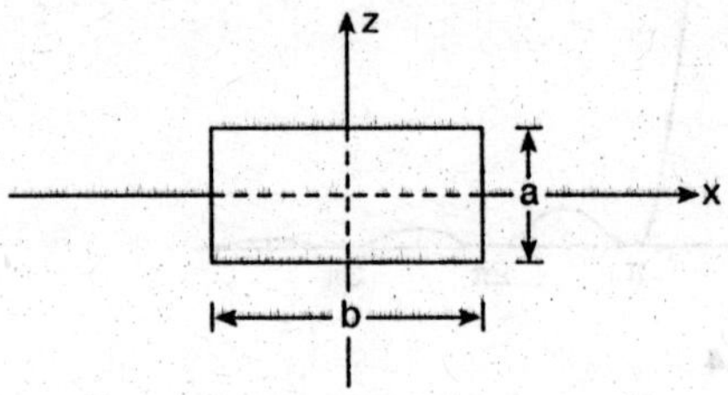

Fig. 10.5

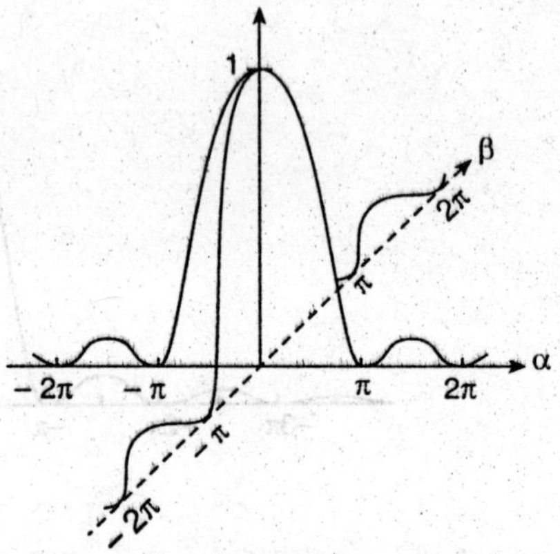

Fig. 10.6

The dimensions of the aperture are a and b and the angles θ and ϕ define the directions of the diffracted ray.

We leave this as an exercise for the reader to show that the **intensity distribution** in this case will be given by the product of two single-slit distribution functions :

$$I = I_0 \frac{\sin^2 \alpha}{\alpha^2} \cdot \frac{\sin^2 \beta}{\beta^2}$$

where $\alpha = \pi a \sin\theta/\lambda$ and $\beta = \pi b \sin\phi/\lambda$.

The resulting distribution pattern has lines of zero intensity defined by

$$\alpha = \pm\lambda, \pm 2\lambda, \pm 3\lambda, \cdots \text{ etc.}$$
$$\beta = \pm\lambda, \pm 2\lambda, \pm 3\lambda, \cdots \text{ etc.}$$

as is shown in Fig. 10.6.

As in the case of one-dimensional single slit, the scale of diffraction bears an inverse relationship to the scale of aperture.

10.3 Diffraction at a double-slit

Let a *parallel beam* of light, i.e., a *planar wavefront*, of wavelength λ be incident normally on two slits AB and CD (Fig. 10.7), each of width a and separated by an opaque space $BC = b$. The distance between any pair of *corresponding points* in the two slits is thus $(a + b)$ and is known as the *double-slit constant*, $d (= a + b)$.

We shall treat the case of diffraction by a narrow double slit also as a one-dimensional (linear) problem.

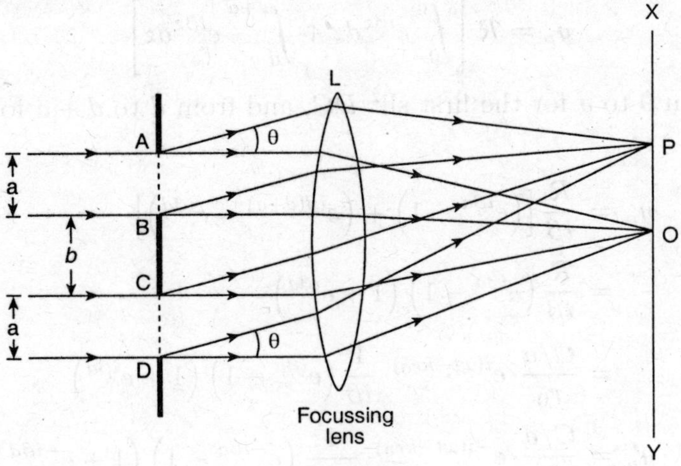

Fig. 10.7

Let the diffracted light be focussed by a convex lens L on a screen XY placed in the focal plane of the lens and parallel to the slits. The pattern obtained on the screen, as would be clear from the *intensity distribution function* (IDF), is the diffraction pattern due to a single slit on which is superposed a system of interference fringes. We shall therefore first find an expression for the intensity at P, on the screen, where the rays diffracted by an angle θ are focussed.

10.3.1 Intensity distribution function

Consider a line element dz (Fig. 10.8) of the planar wavefront, at the slit-plane, having coordinates $(0, z)$ with the *lower end* of the *lower slit* as origin $(0, 0)$. The energy C^2, associated with the wavefront, is uniformly distributed over both the slits and so the vibration of secondary sources in the slit-plane may be written as : $y = (C/a)e^{i\omega t}$, where C/a is the amplitude of vibration. If r be the distance of P from dz, the displacement at P due to element dz at time t is

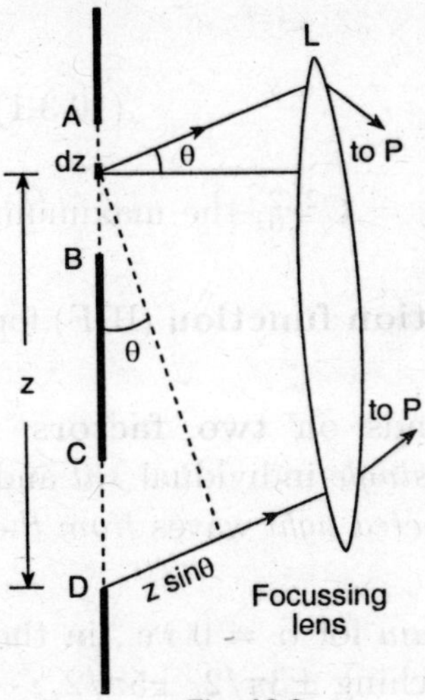

Fig. 10.8

$$dy_P = \frac{C/a}{r} e^{i(\omega t - kr)} dz$$

We can express r as : $r = r_0 - z \sin\theta$, where r_0 is the value of r for $z = 0$. Since z and $\sin\theta$ are both small, r in the denominator may be replaced by r_0 but obviously not in the exponential factor.

$$\therefore \quad dy_P = \frac{C/a}{r_0} e^{i(\omega t - kr_0)} e^{ikz \sin\theta} dz = \mathcal{R} e^{i\beta z} dz$$

where $\mathcal{R} = \frac{C/a}{r_0} e^{i(\omega t - kr_0)}$; $\beta = k \sin\theta = (2\pi/\lambda) \sin\theta$.

$\therefore$ The resultant displacement at P due to the whole wavefront incident on the two slits is

$$y_P = \mathcal{R}\left[\int_0^a e^{i\beta z}dz + \int_d^{d+a} e^{i\beta z}dz\right]$$

since z varies from 0 to a for the first slit DC, and from d to $d+a$ for the second slit BA.

$$\begin{aligned}
\therefore \quad y_P &= \frac{\mathcal{R}}{i\beta}\left[\left(e^{i\beta a}-1\right)+\left(e^{i\beta(d+a)}-e^{i\beta d}\right)\right] \\
&= \frac{\mathcal{R}}{i\beta}\left(e^{i\beta a}-1\right)\left(1+e^{i\beta d}\right) \\
&= \frac{C/a}{r_0}\cdot e^{i(\omega t - kr_0)}\cdot\frac{1}{i\beta}\left(e^{i\beta a}-1\right)\left(1+e^{i\beta d}\right) \\
\therefore \quad y_P^* &= \frac{C/a}{r_0}\cdot e^{-i(\omega t - kr_0)}\cdot\frac{1}{-i\beta}\left(e^{-i\beta a}-1\right)\left(1+e^{-i\beta d}\right)
\end{aligned}$$

$\therefore$ Resultant intensity at P is given by

$$\begin{aligned}
I_P = y_P y_P^* &= \frac{C^2}{r_0^2}\cdot\frac{1}{\beta^2 a^2}\left[2-\left(e^{i\beta a}+e^{-i\beta a}\right)\right]\left[2+e^{i\beta d}+e^{-i\beta d}\right] \\
&= \frac{C^2}{r_0^2}\cdot\frac{1}{\beta^2 a^2}\left[2-2\cos(\beta a)\right]\left[2+2\cos(\beta d)\right] \\
&= \frac{C^2}{r_0^2}\frac{4}{\beta^2 a^2}\left[1-\cos(\beta a)\right]\left[1+\cos(\beta d)\right] \\
&= \frac{C^2}{r_0^2}\frac{2\sin^2(\beta a/2)}{(\beta a/2)^2}\cdot 2\cos^2(\beta d/2)
\end{aligned}$$

$$\therefore \quad \boxed{I_p = 4I_0\frac{\sin^2\alpha}{\alpha^2}\cos^2\gamma} \qquad (10.3.1)$$

where $\alpha = \beta a/2 = \pi a\sin\theta/\lambda$, $\gamma = \beta d/2 = \pi d\sin\theta/\lambda$ and $I_o = C^2 r_0^2$, the maximum intensity of the single slit diffraction pattern.

Eq. (10.3.1) in the box is the required **intensity distribution function** (IDF) for a double slit diffraction pattern.

The intensity in the resultant pattern therefore depends on **two factors** : (i) $\sin^2\alpha/\alpha^2$ which gives the *diffraction pattern* due to *each single* individual *slit* and (ii) $\cos^2\gamma$ which gives the pattern due to *interference of diffracted light* waves *from the two slits.*

The **diffraction term $\sin^2\alpha/\alpha^2$** gives a *central maximum* for $\alpha = 0$ i.e., in the direction $\theta = 0$. The maxima also occur at values of α approaching $\pm 3\pi/2$, $\pm 5\pi/2$, $\cdots$ etc. These are *secondary diffraction maxima.*

The *minima* are obtained when $\sin^2\alpha/\alpha^2 = 0$ or $\sin\alpha = 0$, but $\alpha \neq 0$.

$$\therefore \quad \alpha = m\pi$$

$$\text{or,} \quad \frac{\pi a \sin\theta}{\lambda} = m\pi$$

$$\Rightarrow \quad a\sin\theta = m\lambda, \quad m = \pm 1, \pm 2, \cdots \ \textbf{except 0.}$$

These minima are known as *diffraction minima.*

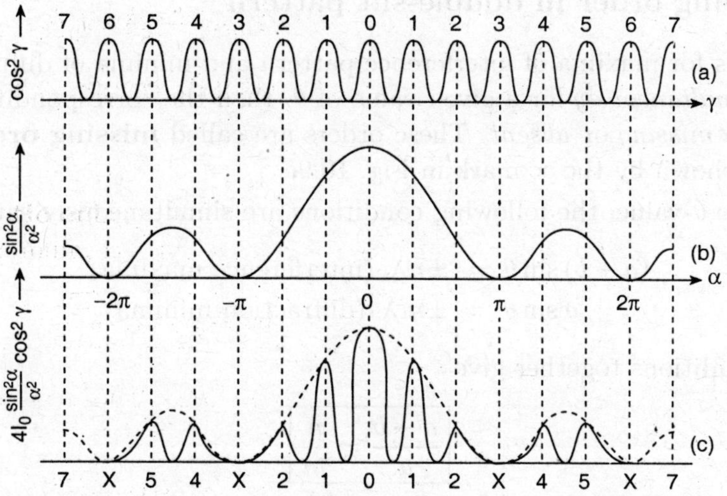

Fig. 10.9

So, based on *diffraction term*, the pattern consists of a central maximum in the direction $\theta = 0$ with alternate minima and secondary maxima of diminishing intensity on either side (Fig. 10.9(b)).

The **interference term $\cos^2\gamma$** gives a *set of equidistant* dark and bright fringes as in Young's double slit interference experiments (Fig. 10.9(a)). The **bright fringes** or **maxima** are in the direction given by

$$\cos^2\gamma = 1 \quad \Rightarrow \quad \gamma = n\pi$$

$$\text{or,} \quad \frac{\pi}{\lambda}(a+b)\sin\theta = n\pi$$

$$\Rightarrow \quad (a+b)\sin\theta = n\lambda, \quad n = 0, \pm 1, \pm 2, \cdots, \text{etc.}$$

corresponding to zero order, first order, second order, $\cdots$ etc. maxima. These are *interference maxima.*

The **positions of minima** due to $\cos^2\gamma$ are given by

$$\cos^2\gamma = 0 \quad \Rightarrow \quad \gamma = (2s+1)\frac{\pi}{2}$$

$$\text{or,} \quad \frac{\pi}{\lambda}(a+b)\sin\theta = (2s+1)\frac{\pi}{2}$$

$$\Rightarrow \quad (a+b)\sin\theta = (2s+1)\frac{\lambda}{2}$$

These are called the *interference minima.*

The *intensity distribution* in the resultant diffraction pattern is shown in Fig. 10.9(c). It is the plot of the product of the ordinates of the curve (b) with those of curve (a) and the constant $4I_0$. The whole pattern may be looked upon as consisting of interference fringes formed due to light from both the slits, and the intensities of these fringes are being controlled by diffraction at the individual slits.

10.3.2 Missing order in double-slit pattern

If the conditions for maxima of interference pattern and minima of diffraction pattern are satisfied *simultaneously* for a given value of θ, then the corresponding interference maxima will be *missing* or *absent.* These orders are called **missing orders** in double slit pattern as shown by the $\times$ mark in Fig. 10.9c.

Let for some θ-value, the following conditions are simultaneously satisfied.

$$\begin{aligned}(a+b)\sin\theta &= \pm n\lambda \text{ (interference maxima)}\\ a\sin\theta &= \pm m\lambda \text{ (diffraction minima)}\end{aligned}$$

The two conditions together give

$$\boxed{\frac{a+b}{a}=\frac{n}{m}}$$

If $b=a$, $n=2m=2,4,6,\cdots$, ($\because$ $m=1,2,\cdots$) which implies that 2nd, 4th, 6th, $\cdots$ etc. order of interference maxima will be missing, i.e., coincide with 1st, 2nd, 3rd, $\cdots$, etc. orders of diffraction minima.

If $b=2a$, $n=3m=3,6,0,\cdots\Rightarrow$ 3rd, 6th, 9th $\cdots$ order interference maxima will be missing, coinciding with the 1st, 2rd, 3rd, $\cdots$, etc. order of diffraction minima. The central diffraction maximum will then have *five* interference maxima. And so on.

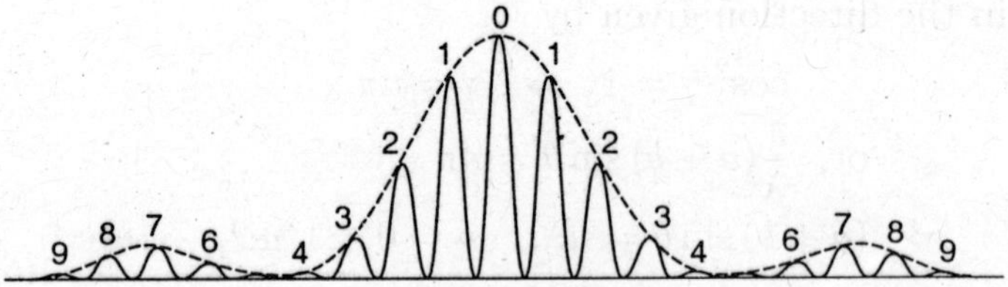

Fig. 10.10

Thus, as b increases to $a, 2a, 3a, \cdots$, the number of interference maxima within the central diffraction maximum increases due to missing orders.

For $b=4a$, the central diffraction maxima will contain *nine* interference maxima, as shown in Fig. 10.10.

• **What happens if $a+b=a$?** This means $b=0$ i.e., the two slits get joined and all the orders of interference maxima will be absent. The diffraction pattern observed will be that due to a *single slit* of width $2a$.

10.3.3 Effect of various factors on double-slit diffraction pattern

A number of factors such as slit-width, slit-separation, wavelength etc. may affect the double-slit diffraction pattern.

1. Effect of increasing slit-width—If the slit-width a is *increased*, the envelope of the fringe-pattern so changes that the *central peak* becomes *sharper*. The spacing between the fringes however does *not* change as it depends only on the slit-separation b. So, now within the central diffraction maximum, *less number of interference maxima* will fall. The converse occurs on *decreasing* the slit-width.

2. Effect of increasing slit-separation—If the slit-separation b is *increased*, keeping the slit-width a constant, the *fringes become closer together*, but the envelope of the pattern remains unaltered. So, *more number of interference maxima* will now fall within the central diffraction maximum.

3. Effect of increasing wavelength—If the wavelength λ of the incident light is *increased*, the envelope of the fringe-pattern becomes *broader*, and the fringes move *further apart*. The *converse effect* occurs on *decreasing* the wavelength.

10.3.4 Single slit vs. Double slit diffraction

A comparison of the diffraction patterns due to a single slit and a double slit is worth noting. The main points of difference between the two are :

1. The single slit diffraction pattern consists of a central brightest maximum along with minima and secondary (or subsidiary) maxima of rapidly diminishing intensities on either side.

The double-slit pattern, on the other hand, consists of equally spaced interference maxima and minima within the central maximum. The intensity of the central maximum in double-slit pattern is four times that of the central maximum ($4I_0$) due to a single slit diffraction. If however one of the slits is covered by an opaque screen, the pattern becomes similar to that due to a single slit.

2. The spacing of diffraction maxima and minima depends on the slit width a, but the spacing of the interference maxima and minima depends on both a and b, where b is the slit separation. The intensities of interference maxima are *not constant* but decreases to zero on either side of central maximum. The maxima reappear two or three times before the intensity becomes too low to be observed.

3. While in a double-slit diffraction pattern there arises, under certain condition, missing or absent orders—missing of interference maxima—there is *no* such phenomenon in the single-slit diffraction pattern.

10.4 Circular aperture

Let AB be a *narrow circular* aperture of diameter d and centred at C, placed perpendicular to the plane of the paper (Fig. 10.11) and also let a parallel beam of light of wavelength λ be incident *normally* on it.

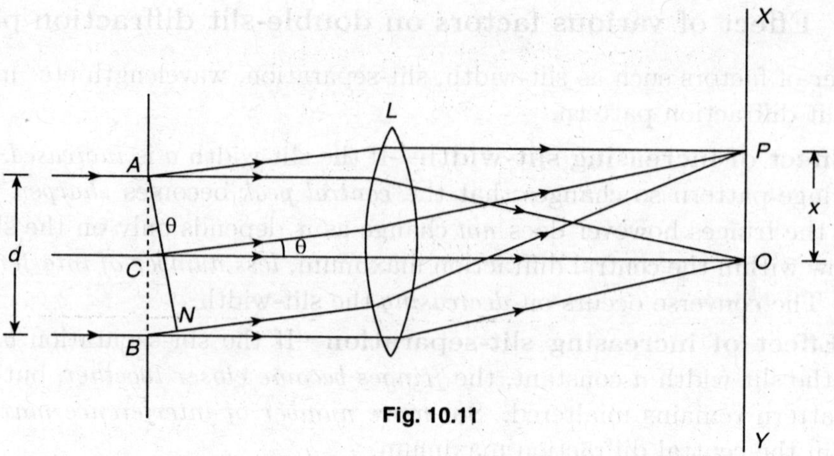

Fig. 10.11

The secondary waves travelling in the *direction* CO, the direction of the incident beam, are brought to a focus at O by the focussing lens L on the screeen XY, placed perpendicular to the paper in the focal plane of the lens. So, O corresponds to the *central maximum*, where all secondary waves, equidistant from C in the upper and lower half, travel the same distance and reinforce each other at O.

However, the rays undergoing diffraction at θ by the aperture will meet at a point P on the screen, distant x from O. The path-difference Δ between the secondary waves from the two extreme ends A and B of the aperture is

$$\Delta = BN = AB\sin\theta = d\sin\theta$$

If Δ is an even *mutiple* of $\lambda/2$, then with the same argument as in **single-slit pattern**, the point P will have *minimum* intensity.

$$\therefore \quad d\sin\theta = 2n\frac{\lambda}{2}, \quad n = 0, 1, 2, \cdots \text{ gives } \textbf{minimum}$$

If however Δ is an *odd multiple* of $\lambda/2$, then P will have maximum intensity.

$$\therefore \; d\sin\theta = (2n+1)\frac{\lambda}{2}, \quad n = 0, 1, 2, \cdots \text{ gives } \textbf{maximum}$$

The circular aperture is *symmetrical* about the axis and the resultant diffraction pattern may be obtained by rotating Fig. 10.11 about the same axis. P will then trace out a *circular ring* of uniform intensity, minimum or maximum, as dictated by the above conditions.

Thus, the diffraction pattern due to a circular aperture consists of *a central bright disc*, called **Airy's disc**, sourrounded by *alternate dark and bright concentric rings*, called the **Airy's rings** as shown in Fig. 10.12. While the intensity of the dark rings is zero, that of the bright ones decreases gradually outward from O.

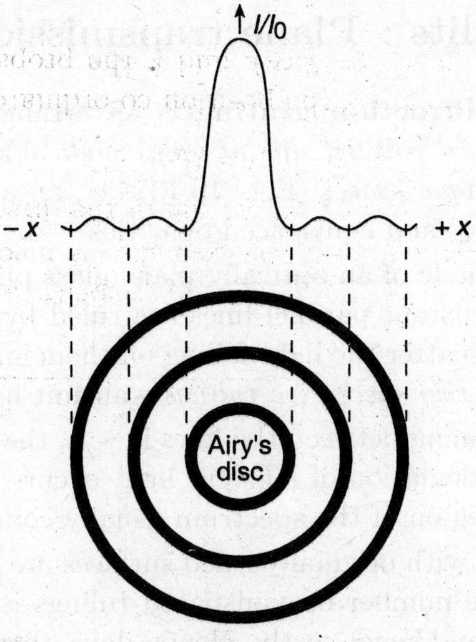

Fig. 10.12

Radius of Airy's disc—In case the focussing lens is very near the aperture or the screen is at a large distance from the lens,

$$\sin\theta = \theta = x/f \tag{10.4.1}$$

where f is the focal length of the lens.

But for the first secondary minimum

$$d\sin\theta = \lambda \;\Rightarrow\; \sin\theta = \theta = \lambda/d \tag{10.4.2}$$

From (10.4.1) and (10.4.2),

$$\therefore \quad \frac{x}{f} = \frac{\lambda}{d} \quad \Rightarrow \quad \boxed{x = \frac{\lambda}{d} f}$$

where x is the radius of the Airy's disc.

Actually, the radius of the first dark ring is slightly more than that given by the above boxed relation. Airy has shown that

$$\boxed{x = \frac{1.22\lambda}{d} f}$$

The intensity distribution of the bright and dark rings is similar to that given for a rectangular slit. Clearly, with increase in the diameter of the aperture, the radius of the central bright disc decreases.

- The diffraction of plane waves from a narrow circular aperture is of importance for its application to resolving power of telescopes and microscopes (*Chapter* 14), for a lens behaves like a circular aperture and gives a pattern of alternate, concentric dark and bright rings.

10.5 Multiple slits : Plane transmission grating

A *plane transmission* **diffraction grating** is an arrangement equivalent to a very *large number N* of *identical parallel slits of equal width* side by side and *separated* from one another *by equal opaque spaces* (Fig. 10.13) i.e., an array of alternate equispaced transparencies and equispaced opacities.

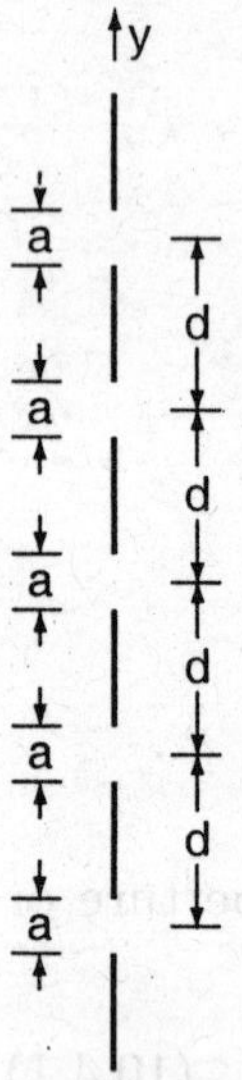

Fig 10.13

This is made of an optically-plane glass plate on which a large number of fine, equidistant parallel lines are ruled by means of a diamond point. The *rulings* scatter the light falling on them and *are opaque,* but the *spaces between any two successive rulings* transmit light and they *act as slits.*

If the spacing between the lines is $\sim \lambda$, the wavelength of light, then an appreciable deviation of incident light occurs. Gratings used for studying the visible region of the spectrum usually contain 10,000 lines per cm.

Gratings with originally ruled surfaces are only few since the process of making large number of equispaced rulings is difficult, tedious and time-consuming, and hence costly. Now-a-days, therefore, *replicas* of the original grating are prepared. For this, a thin layer of collodion solution is poured on the original grating surface and the solution is allowed to harden. The collodion film is then carefully removed from the grating surface and fixed between two glass plates. This serves as a *plane transmission grating.* A large number of replicas are commercially prepared in this way from a single original ruled surface.

Intensity distribution function—We shall evaluate the diffraction integral, in this case, in a manner similar to that used in the double-slit, using the same symbols and co-ordinate system. The total amplitude at any point P on the screen, therefore, is given by

$$y_p = \mathcal{R}\left[\int_0^a + \int_d^{d+a} + \int_{2d}^{2d+a} + \ldots + \int_{(N-1)d}^{(N-1)d+a} e^{i\beta z}\right] dz$$

$$= \frac{\mathcal{R}}{i\beta}\left(e^{i\beta a} - 1\right)\left[1 + e^{i\beta d} + e^{i\beta .2d} + \ldots + e^{i\beta(N-1)d}\right]$$

$$= \frac{\mathcal{R}}{i\beta}\left(i^{i\beta a} - 1\right)\frac{1 - e^{i\beta Nd}}{1 - e^{i\beta d}} = \frac{C}{r_0}\cdot\frac{1}{i\beta a}\left(e^{i\beta a} - 1\right)\frac{1 - e^{i\beta Nd}}{1 - e^{i\beta d}}\,.$$

$$\therefore \quad y_p^* = \frac{C}{r_0}\cdot\frac{1}{-i\beta a}\left(e^{-i\beta a} - 1\right)\frac{1 - e^{-i\beta Nd}}{1\ e^{-i\beta d}}$$

$\therefore$ Intensity at P is given by

$$I_p = y_p y_p^* = \frac{C^2}{r_0^2}\cdot\frac{1}{\beta^2 a^2}\frac{\left[2 - \left(e^{i\beta a} + e^{-i\beta a}\right)\right]\left[2 - \left(e^{i\beta Nd} + e^{-i\beta Nd}\right)\right]}{\left[2 - \left(e^{i\beta d} + e^{-i\beta d}\right)\right]}$$

$$= \frac{C^2}{r_0^2}\cdot\frac{1}{\beta^2 a^2}\frac{[2 - 2\cos(\beta a)][2 - 2\cos(N\beta d)]}{[2 - 2\cos(\beta d)]}$$

$$= I_0 \cdot \frac{4}{\beta^2 a^2} \frac{[1 - \cos(\beta a)][1 - \cos(N\beta d)]}{2[1 - \cos(\beta d)]}$$

$$= I_0 \frac{\sin^2(\beta a/2)}{(\beta a/2)^2} \cdot \frac{2\sin^2(N\beta d/2)}{2\sin^2(\beta d/2)}$$

$$\therefore \quad \boxed{I_p = I_0 \frac{\sin^2 \alpha}{\alpha^2} \cdot \frac{\sin^2(N\gamma)}{\sin^2 \gamma}} \tag{10.5.1}$$

where $\alpha = \beta a/2 = \pi a \sin\theta/\lambda$, $\gamma = \beta d/2 = \pi d \sin\theta/\lambda$ and $I_0 = C^2/r_0^2$, the maximum intensity of the single slit diffraction pattern.

Eq. (10.5.1) in the box is the required **intensity distribution function** (IDF) for diffraction grating.

• If in (10.5.1), we put $N = 2$, we get the intensity distribution function for a *double-slit*, for then

$$I_p = I_0 \frac{\sin^2 \alpha}{\alpha^2} \cdot \frac{\sin^2 2\gamma}{\sin^2 \gamma} = I_0 \frac{\sin^2 \alpha}{\alpha^2} \cdot \frac{2^2 \sin^2 \gamma \cos^2 \gamma}{\sin^2 \gamma}$$

$$= 4I_0 \frac{\sin^2 \alpha}{\alpha^2} \cdot \cos^2 \gamma$$

For diffraction grating, the resultant intensity I is therefore given by

$$\boxed{I = I_0 \frac{\sin^2 \alpha}{\alpha^2} \cdot \frac{\sin^2 N\gamma}{\sin^2 \gamma}} \tag{10.5.2}$$

The first factor $I_0 \sin^2 /\alpha^2$ is the **diffraction term** giving the diffraction pattern due to a single slit. The second factor $\sin^2 N\gamma/\sin^2 \gamma$ is the **interference term** giving the interference pattern due to N *identical* slits. We shall consider the intensity distribution due to the second factor first.

Principal maxima—When $\sin\gamma = 0$, i.e., $\gamma = \pm n\pi$, $n = 0, 1, 2, \cdots$ etc., we have,

$$\sin N\gamma = 0 \quad \Rightarrow \quad \sin N\gamma/\sin\gamma = 0/0 = \text{indeterminate}.$$

$$\text{But,} \quad \lim_{\gamma\to\pm n\pi} \frac{\sin N\gamma}{\sin\gamma} = \lim_{\gamma\to\pm n\pi} \frac{N\cos N\gamma}{\cos\gamma} = \frac{N\cos N(\pm n\pi)}{\cos(\pm n\pi)} = \pm N$$

$\therefore$ From (10.5.2),

$$I = I_0 \frac{\sin^2 \alpha}{\alpha^2} N^2$$

which is a *maximum*. These maxima are *most intense* and are known as **principal maxima.**

The *directions of principal maxima* are given by

$$\gamma = \pm n\pi$$

$$\Rightarrow \quad \frac{\pi(a+b)\sin\theta}{\lambda} = \pm n\pi$$

$$\Rightarrow \quad (a+b)\sin\theta = \pm n\lambda, \quad n = 0, 1, 2, ... \tag{10.5.3}$$

For $n = 0$, we get the **zero-order** spectrum. For $n = \pm 1, \pm 2, ...$ we get the *first, second, ...* order *principal maxima.* The significance of $\pm$ is that there are *two principal maxima for each order* situated on either side of the *zero-order* maximum.

• The positions of principal maxima, as given by (10.5.3), are the same as in double-slit diffraction pattern. Thus, whether $N = 2$ or $N > 2$, the positions of principal maxima *do not* change provided $(a+b)$, the *grating element* d, is the same.

Minima—When $\sin N\gamma = 0$, but $\sin\gamma \neq 0$, we have

$$\frac{\sin N\gamma}{\sin\gamma} = 0 \quad \Rightarrow \quad I = 0, \quad \text{from (10.5.2)}$$

and this corresponds to a *minimum.*

The *directions of the minima* are given by

$$\sin N\gamma = 0 \quad \Rightarrow \quad N\gamma = \pm m\pi$$

$$\text{or,} \quad N\frac{\pi}{\lambda}(a+b)\sin\theta = \pm m\pi$$

$$\Rightarrow \quad N(a+b)\sin\theta = \pm m\lambda \tag{10.5.4}$$

where m can take *all integral* values *excluding* $0, N, 2N, ...nN$, since these m-values make $\sin\gamma = 0$ and we obtain *principal maxima.*

$\therefore$ $m = 0$ gives a *principal maximum*
$m = 1, 2, 3, ..., (N-1)$ give *minima*
$m = N$ gives again a *principal maximum,* etc.

Secondary maxima—Since there are $(N-1)$ minima between two consecutive principal maxima, there *must be* $(N-2)$ other maxima between two principal maxima. These maxima are called **secondary maxima.**

The *positions of secondary maxima* are obtained by equating $dI/d\gamma$ to zero. From the intensity expression (10.5.2), therefore,

$$\frac{dI}{d\gamma} = I_0\frac{\sin^2\alpha}{\alpha^2}\cdot 2\left(\frac{\sin N\gamma}{\sin\gamma}\right)\cdot\frac{N\cos N\gamma\sin\gamma - \sin N\gamma\cos\gamma}{\sin^2\gamma} = 0$$

$$\Rightarrow \quad N\cos N\gamma\sin\gamma - \sin N\gamma\cos\gamma = 0$$

$$\Rightarrow \quad \tan N\gamma = N\tan\gamma \tag{10.5.5}$$

$$\therefore \quad \sin N\gamma = \frac{N\tan\gamma}{\sqrt{(1+N^2\tan^2\gamma)}}$$

$$\therefore \quad \frac{\sin^2 N\gamma}{\sin^2 \gamma} = \frac{N^2 \tan^2 \gamma}{(1 + N^2 \tan^2 \gamma) \sin^2 \gamma} = \frac{N^2}{\cos^2 \gamma + N^2 \sin^2 \gamma}$$

$$= \frac{N^2}{1 + (N^2 - 1) \sin^2 \gamma}$$

Thus, the **intensity** of secondary maxima is proportional to

$$\frac{N^2}{1 + (N^2 - 1) \sin^2 \gamma}$$

whereas that of principal maxima is proportional to N^2.

$$\therefore \quad \frac{\text{Intensity of secondary maxima}}{\text{Intensity of principal maxima}} = \frac{1}{1 + (N^2 - 1) \sin^2 \gamma}$$

This shows that *greater the N-value, weaker are the secondary maxima.* In practice, N is very large in a grating and consequently the secondary maxima are *not visible* in grating spectrum. The grating spectrum is illustrated graphically in Fig. 10.14. While (a) and (b) show the graphs of variation of intensity due to factors $\sin^2 \alpha/\alpha^2$ and $\sin^2 N\gamma/\sin^2 \gamma$ respectively, the resultant intensity curve is shown in (c).

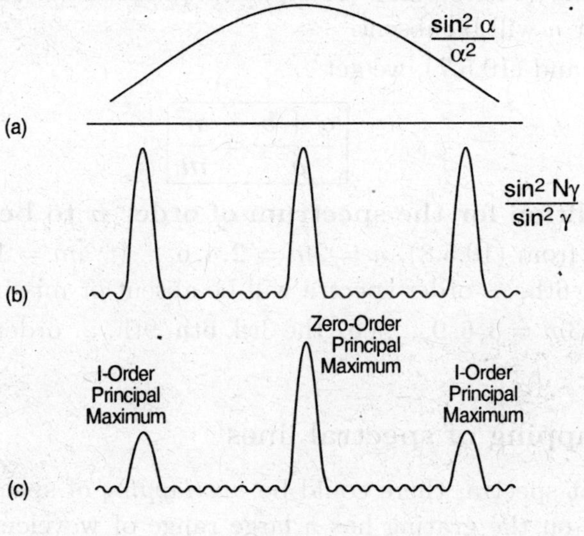

Fig 10.14

Multiple spectra : White light incidence—If light of wavelength λ be incident on a grating, the principal maxima are formed in a direction given by

$$(a + b) \sin \theta = \pm n\lambda; \quad n = 0, 1, 2, ...$$

This shows that for a given order n, the diffraction angle θ will vary with λ. *Longer the wavelength, greater will be the angle of diffraction.*

So, if the incident beam be white, then each order will contain principal maxima of *different wavelengths in different directions.* For example, principal maxima of all wavelengths corresponding to $n = 1$ will form the first-order spectra. The same would be true for $n = 2$ and so on.

However, corresponding to $n = 0$, the principal maxima *of all wavelengths* will be along the same direction given by $\theta = 0$. Hence the '***zero-order*** *maximum will be white,* with first-order spectra, second-order spectra etc. on either side of it.

10.5.1 Absent spectra in diffraction grating

We first deduce the condition for *absent spectra* or *missing spectra* in a plane transmission grating.

The principal maxima in the grating spectrum, with normal incidence, are obtained in the directions given by

$$(a + b) \sin \theta = n\lambda \tag{10.5.6}$$

where n is the order of the maximum.

The minima in a single-slit pattern are obtained in the directions given by

$$a \sin \theta = m\lambda; \quad m = 1, 2, 3, ... \tag{10.5.7}$$

If the equations (10.5.6) and (10.5.7) are satisfied *simultaneously,* a particular maximum of order n will be absent.

From (10.5.6) and (10.5.7), we get

$$\boxed{\frac{a + b}{a} = \frac{n}{m}} \tag{10.5.8}$$

which is the **condition for the spectrum of order n to be absent.**

If $b = a$, then from (10.5.8), $n = 2m = 2, 4, 6, ...$ ($\because \ m = 1, 2, 3, ...$) which implies that the 2nd, 4th, 6th, ... order spectra will be absent or missing.

If $b = 2a$, $n = 3m = 3, 6, 9, ...$, i.e., the 3rd, 6th, 9th, ... order spectra will be absent.

And so on.

10.5.2 Overlapping of spectral lines

Apart from absent spectra, there could be *overlapping* of spectral lines in grating. If the incident light on the grating has a large range of wavelengths, then the spectral lines of shorter wavelength and higher order may overlap on the spectral lines of longer wavelength and lower order.

If the angle of diffraction θ be the same for spectral line of λ_1 in first order, of λ_2 in the second order, of λ_3 in the third order etc., then

$$(a + b) \sin \theta = 1\lambda_1 = 2\lambda_2 = 3\lambda_3 = ...$$

leading to overlapping of spectral lines of wavelengths $\lambda_1, \lambda_2, \lambda_3, \cdots$, etc. in various appropriate orders.

So, overlapping of spectral lines occurs if : **$n\lambda$ = constant.**

For instance, red line 7000A in the 3rd order will overlap with green line 5250A in 4th order, violet line 4200A in 5th order etc.

There is however *no overlapping of spectral lines in the visible region* ranging from 4000A to 7200A since the diffracting angle of red end of spectrum in the first order is less than that for the violet end in the second order. To avoid unwelcome overlapping, suitable filters are utilised to absorb those wavelengths in the region under investigation.

• **Ghost lines**—In good gratings, the clear spaces are strictly periodic, but due to errors in the ruling process, there may occur irregularity in the periodicity of clear spaces. This results in the appearance of *spurious* or *ghost* lines other than the actual.

For instance, if d and d' be the grating constants in two different parts of the same grating, then the nth order principal maximum for wavelength λ is given by

$$d\sin\theta = n\lambda; \quad d'\sin\theta' = n\lambda$$

$$\Rightarrow \quad \sin\theta = (d'/d)\sin\theta'$$

10.5.3 Condition of minimum deviation of rays

For *oblique incidence of a parallel beam of light of wavelength* λ (Fig. 10.15), the grating equation is

$$(a+b)(\sin i + \sin\theta) = n\lambda \tag{10.5.9}$$

where $(a+b)$ is the grating element, i the angle of incidence, θ the angle of diffraction.

[**Proof :** Path difference Δ between the secondary waves through the corresponding points A and C is given by

$$\Delta = FC + CE = AC\sin i + AC\sin\theta = AC(\sin i + \sin\theta) = (a+b)(\sin i + \sin\theta)$$

For nth primary maximum, $\Delta = n\lambda \Rightarrow (a+b)(\sin i + \sin\theta) = n\lambda$]

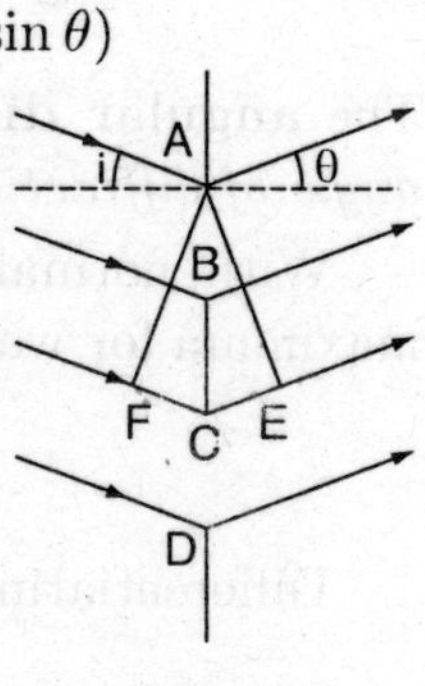

Fig 10.15

Eq. (10.5.9) may be given the form

$$2(a+b)\sin\frac{i+\theta}{2}\cos\frac{i-\theta}{2} = n\lambda$$

Now, the deviation suffered by the rays is $\delta = i + \theta$.

$$\therefore \quad 2(a+b)\sin(\delta/2)\cos\tfrac{1}{2}(i-\theta) = n\lambda$$

So, $\sin\delta/2$, and hence δ, will be minimum when

$$\cos\tfrac{1}{2}(i-\theta) \quad \text{is maximum}$$

$$\Rightarrow \quad i - \theta = 0 \quad \Rightarrow \quad i = \theta$$

which is the **required condition.**

Thus, *the rays forming the spectrum in a grating suffer minimum deviation when the angle of incidence equals the angle of diffraction.*

• Under the condition of minimum deviation, the grating equation becomes

$$2(a+b)\sin\delta_m/2 = n\lambda$$

where δ_m is the minimum deviation suffered by the rays.

10.5.4 Maximum number of orders available

With normal incidence, the principal maxima in the diffraction pattern of a grating are obtained in the direction given by

$$(a+b)\sin\theta = n\lambda$$

where n is the order of the principal maxima.

$$\therefore \quad n = \frac{(a+b)\sin\theta}{\lambda} \quad \Rightarrow \quad n_{\text{max}} = \frac{a+b}{\lambda}, \quad \text{when } \theta = \theta_{\text{max}} = 90^\circ$$

and this gives the maximum possible order in a grating.

If $(a+b) < 2\lambda$, where $\lambda =$ wavelength of incident light, we have

$$n_{\text{max}} < \frac{2\lambda}{\lambda} < 2$$

which implies that only the **first order spectrum** is obtained.

10.6 Angular dispersive power of a grating

The **angular dispersive power** *of a grating is defined as the rate of change of the angle of diffraction with the wavelength of light.* It is thus $d\theta/d\lambda$.

With normal incidence, the angle of diffraction θ for the nth order principal maximum for wavelength λ is given by the grating equation

$$(a+b)\sin\theta = n\lambda$$

Differentiating the above equation, with respect to λ, we get

$$(a+b)\cos\theta\frac{d\theta}{d\lambda} = n$$

$$\Rightarrow \quad \boxed{\frac{d\theta}{d\lambda} = \frac{n}{(a+b)\cos\theta} = \frac{nN}{\cos\theta}} \tag{10.6.1}$$

where $N = 1/(a+b) =$ number of lines per unit length of the grating.

Eq. (10.6.1) is the expression for the angular dispersive power of a grating.

The expression shows that the dispersive power is

1. *directly proportional* to the *order* n of the spectrum,

2. *directly proportional* to the number of *lines per unit length* N; in other words, *inversely proportional* to the *grating element* $(a+b)$, and

3. *inversely proportional* to $\cos\theta$, i.e., larger the θ-value, smaller is $\cos\theta$ and higher is the dispersive power.

• Since the angle of diffraction for red light is greater than that for violet, then by 3 above, the dispersion in the red region is greater than that in the violet.

• From the expression (10.6.1), it is clear that the angular spacing of any two spectral lines is double in the second order compared to the first.

• **Linear dispersion**—If the linear spacing of two spectral lines of wavelength λ and $\lambda + d\lambda$ be dx in the focal plane of telescope objective (or photographic plate), then

$$dx = f d\theta$$

where f is the focal length of the telescope objective.

$$\therefore \quad \frac{dx}{d\lambda} = f\frac{d\theta}{d\lambda} = f\frac{nN}{\cos\theta}$$

$$\Rightarrow \quad dx = f\frac{nN}{\cos\theta}d\lambda$$

The linear dispersion is important in the study of photographs of a spectrum.

• **Normal spectrum**—A spectrum is said to be *normal* if $d\theta \propto d\lambda$, i.e., $d\theta/d\lambda =$ constant. From (10.6.1), $d\theta/d\lambda = Nn/\cos\theta \simeq$ constant for small $\theta(< 6°)$. The *grating spectrum* is thus essentially *normal* as θ is usually small.

10.7 Concave reflection grating

So long we discussed plane transmission grating. In this section, we shall deal with what is known as *concave reflection grating.*

Rowland developed **concave reflection grating** *consisting of a polished spherical concave surface of metal like speculum on which are ruled fine parallel lines, equispaced along the chord of the arc joining the extreme rulings.* When incident light falls on such a grating, it gets diffracted and is automatically focussed without the use of lenses.

Theory—Let GG' be the trace of the concave grating (Fig. 10.16), C being its centre of curvature and A, B two corresponding points on the grating. So $AB = a+b$, the *grating element.* Let S be a narrow vertical slit illuminated by light of wavelength λ. SA and SB are two rays incident at angles i and $i + di$ respectively on the grating. The corresponding diffracted rays, diffracted at θ and $\theta + d\theta$, are AS' and BS'.

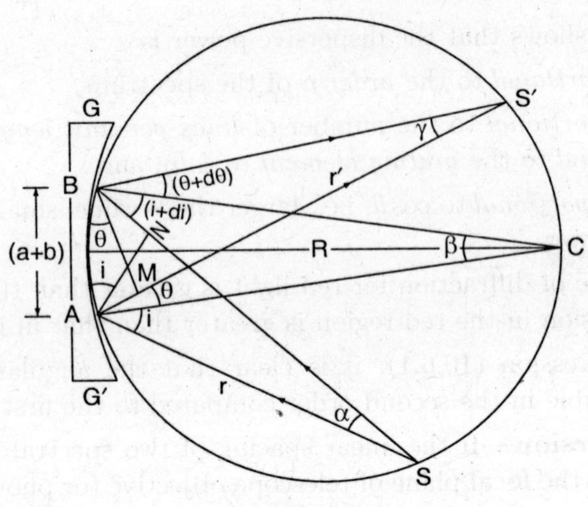

Fig 10.16

With S and S' as centres and SA and $S'B$ as radii draw arcs AN and BM which will be essentially straight. The path difference Δ between the rays SAS' and SBS' is given by

$$\begin{aligned}\Delta &= (SB + BS') - (SA + AS') = (SB - SA) - (AS' - BS') \\ &= BN - AM = AB\sin i - AB\sin\theta \\ &= AB(\sin i - \sin\theta) \\ &= (a+b)(\sin i - \sin\theta)\end{aligned}$$

The rays will *reinforce* each other to produce a *maximum* at S', if $\Delta = n\lambda$, i.e.,

$$(a+b)(\sin i - \sin\theta) = n\lambda \qquad (10.7.1)$$

All the diffracted rays of a given λ will be focussed at S', if the path difference for any such pair of corresponding rays be the same. So, the **focussing condition** is

$$(a+b)(\sin i - \sin\theta) = \text{const} \qquad (10.7.2)$$

$$\Rightarrow \quad \cos i\, di - \cos\theta\, d\theta = 0 \qquad (10.7.3)$$

If α, β and γ be the angles as shown in Fig. 10.16,

$$\alpha + i = \beta + i + di \quad \Rightarrow \quad di = \alpha - \beta$$

$$\beta + \theta = \gamma + \theta + d\theta \quad \Rightarrow \quad d\theta = \beta - \gamma$$

$\therefore$ From (10.7.3),

$$\cos i(\alpha - \beta) - \cos\theta(\beta - \gamma) = 0 \qquad (10.7.4)$$

Let $SA = r$, $AS' = r'$ and R be the radius of curvature of the concave grating.

$\therefore\ r\alpha = AN = (a+b)\cos i;\ R\beta = BA = a+b$ and $r'\gamma = BM = (a+b)\cos\theta$.

$\therefore\quad \alpha = (a+b)\cos i/r\,;\ \beta = (a+b)/R$ and $\gamma = (a+b)\cos\theta/r'$.

$\therefore$ From (10.7.4), using the above values for α, β and γ.

$$\cos i\left(\frac{\cos i}{r}-\frac{1}{R}\right)-\cos\theta\left(\frac{1}{R}-\frac{\cos\theta}{r'}\right)=0 \tag{10.7.5}$$

which is the **general equation** for the position of S'.

It follows from (10.7.5) that *if*

$$r = R\cos i, \text{ then } r' = R\cos\theta$$

$\therefore$ If S is situated on a circle of radius R, then S' also lies on the same circle.

So, *if the slit and the concave grating are placed on the circumference of a circle of diameter equal to the radius of curvature of the grating, then the spectra get focussed on the circumference of the same circle.* This circle is called the **Rowland circle.**

10.8 Mountings of concave grating

The concave grating can be used to obtain spectra in a number of different ways, called **mountings.** In all types of mountings, the grating is kept tangential to the Rowland circle, and the slit and eyepiece (or photographic plate) are placed on the same circle.

We shall describe here the following *three* principal mountings : **1. Rowland** *mounting,* **2. Paschen** *mounting,* and **3. Eagle** *mounting* and point out their relative merits and demerits.

1. Rowland mounting–The principle of this mounting is illustrated in Fig. 10.17. The mounting consists of two strong iron rails SX and SY, at right angles to each other in a horizontal plane. The grating G and the plate-holder P are kept on two carriages C_1 and C_2, capable of sliding along the rails and are joined by a rigid beam of length R, the radius of curvature of the grating. The slit S is placed at the junction of the two rails.

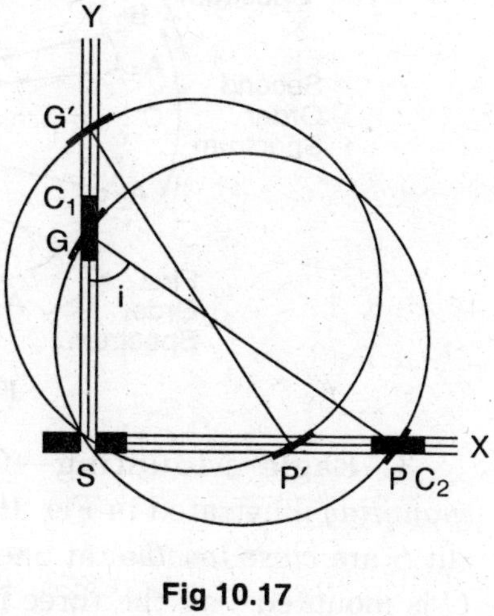

Fig 10.17

Now $\angle YSX = 90°$ and C_1C_2 = diameter of Rowland circle, G, S and P all lie on this circle. So with white light falling on the grating *via* the slit, it gets diffracted and spectra of different orders are focussed on the circle. A limited spectral portion however is used that falls on the photographic plate P which is *always normal* to G. As C_1C_2 is slided, G and P occupy new positions G' and P' respectively but lie always on the Rowland circle being diametrically opposite to each other. Spectra of various orders are automatically obtained.

As the angle of diffraction $\theta = 0$ along GP, the equation : $(a+b)(\sin i - \sin\theta) = n\lambda$, reduces to

$$(a+b)\sin i = n\lambda$$

But $\sin i = SP/PG = SP/R$, so that

$$(a+b)SP/R = n\lambda,$$

$\therefore\ SP \propto n$, for a given λ; also $SP \propto \lambda$ for a given n.

So, as P moves from S along SP, the first, second etc. order maxima occur at *equal intervals*. And the rail SP can be graduated to give the wavelengths directly.

Merits and demerits—The spectrum is *normal* and the rail SP can be calibrated to *read the wavelengths* of spectral lines *directly*. These are the merits of the mounting.

The demerits are : it requires *large space;* it is *expensive* involving considerable mechanism. It shows some *astigmatism* and the adjustment gets *disturbed* whenever spectral region is altered.

2. Paschen mounting—It is the common form of mounting and is illustrated in Fig. 10.18. The grating G is mounted at a *fixed* point on the Rowland circle on the circumference of which the slit S is also set and *fixed* so as to give the desired angle of incidence. The photographic plate is clamped to a circular steel frame along the Rowland circle. Spectra of different orders are *simultaneously* focussed on the photographic plate.

Merits and demerits—The mounting is very *rigid* and the *plate* can be *placed anywhere* along the frame to record *any portion of the spectrum* with *no change in adjustment.* With large film *all orders* can be *simultaneously* photographed.

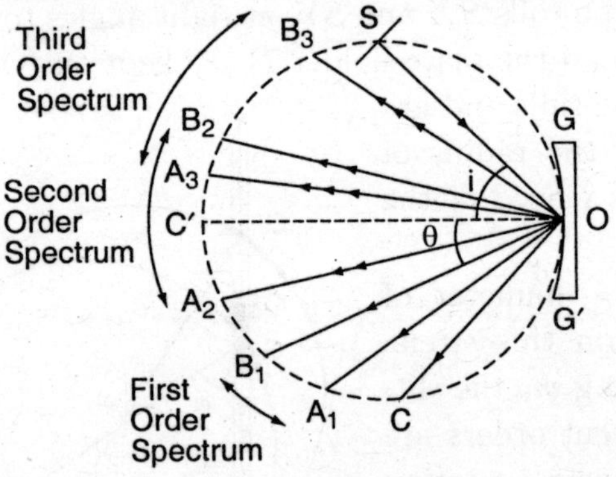

Fig 10.18

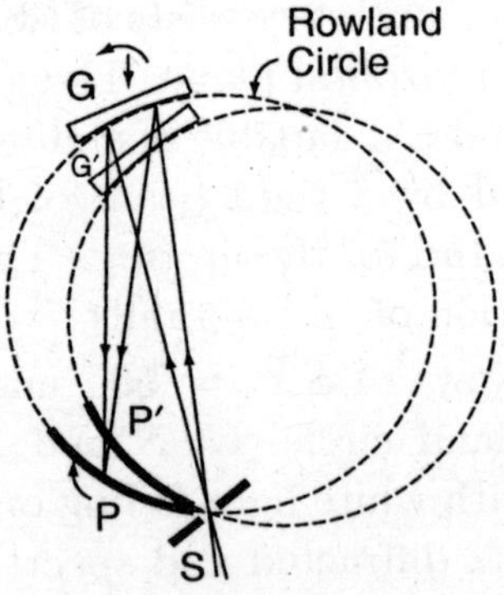

Fig 10.19

3. Eagle Mounting—One of the most widely used mountings is the *Eagle mounting* illustrated in Fig. 10.19. In this mounting, the photographic plate P and the slit S are *close together* at one end of a rigid bar; at the other end of the bar the grating G is mounted—all the three lying on the Rowland circle. Only a part of the spectrum which is diffracted back almost along the path of the incident beam is obtained on the plate. To get the different spectral region, G is moved along the bar and also rotated about a vertical axis to a new position G' such that the Rowland circle still passes through slit S. Plate P is also tilted to P' to have its surface along the new Rowland circle.

Merits and demerits—The mounting is *most compact* occuping much *less space.* It is thus widely used in vacuum spectrographs for studying the extreme UV-region. Further, *temperature control* is *easier* and *higher orders* can be obtained. *Astigmatism is* also comparatively *much less.*

The demerits are : spectrum is *not normal; cannot simultaneously focus* all spectral regions; it *involves* both *translation* and *rotation* of the grating along with the rotation of the plate.

• **Advantages of concave gratings over plane ones**—A concave grating requires no lenses for collimation or focussing and thus dispenses with chromatic aberration. It can be used in extreme UV-region where glass and even quartz becomes opaque.

Because of large radius of curvature, a concave grating has large dispersion enabling a far more accurate determination of wavelength. A concave grating, if used at grazing incidence, has highly intense spectral lines so that even weaker lines can be photographed.

10.9 Echelon grating of Michelson

Background—The resolving power of a grating is given by N where N is the total number of lines on the grating and n is the order of spectra. So, to increase the resolving power, one has to increase N and n. But it is difficult and tedious to rule a large number of lines on a grating surface and the diamond-point ruling fails as $N \to 10^5$. Again, with a ruled grating, one cannot get $m > 3$ as the intensity is very poor in higher orders. A resolving power $\sim$ 300,000 in ruled grating is thus a far cry.

In this background, Michelson in 1898 developed a special type of grating, called **Echelon grating** in which N is fairly small (~ 40) but n is made very high ($\sim 10^6$) with no loss in intensity. Such gratings can give a very high ($\sim$ 400,000) resolving power. Echelon grating may be of (i) *transmission type* and (ii) *reflection type.* The latter is more useful while the former is seldom used.

Construction—An echelon transmission grating consists of a number of *equally thick* and *parallel plates of optically worked* glass or quartz arranged *in the form of steps*, as shown in Fig. 10.20. In fact, all the plates are cut from an optically worked single glass or quartz plate. The step-formation is such that each plate projects the same distance (step-width) beyond the one below it. The step-width is $\sim$ 1 mm. Each *step* acts as *an aperture.*

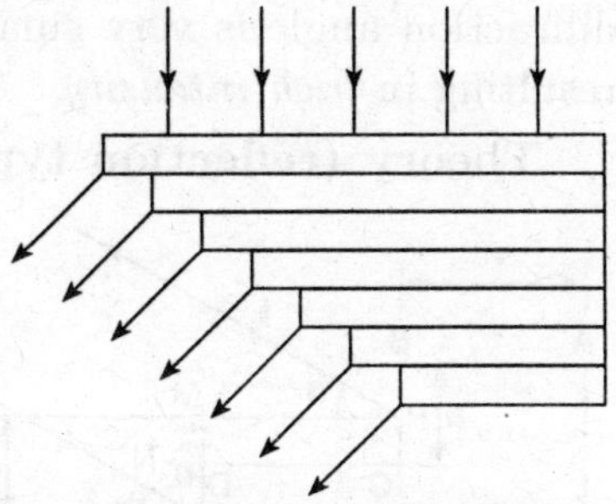

Fig. 10.20

When a parallel beam of light falls normally on the grating, each point on the step acts as a secondary source of disturbance to send out rays in all possible directions, to be collected by a telescope (not shown), as diffracted rays. As the width of each diffracting element (step) is *very large* compared to the wavelength λ of light, most of the incident light is diffracted *at a small angle.*

For the construction of reflection type echelon grating, the front face of the steps are aluminised for good reflection.

Theory (transmission type)—Let a be the width of each step and t the thickness of each plate of index μ (Fig. 10.21). A parallel beam of monochromatic light of wavelength λ is incident on the surface of the topmost plate. Consider two parallel rays diffracted at an angle θ from the *corresponding points* A and B.

The path difference Δ between these rays is given by

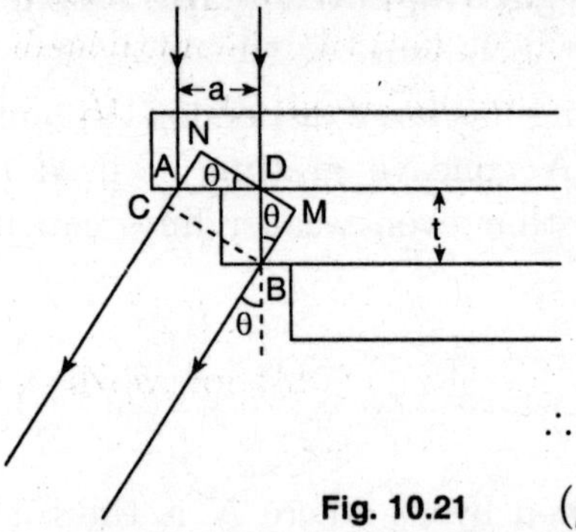

Fig. 10.21

$$\begin{aligned}\Delta &= \mu BD - AC = \mu BD - (CN - AN)\\ &= \mu BD - (BM - AN)\\ &= \mu BD - (BD\cos\theta - AD\sin\theta)\\ &= (\mu - 1)BD - AD\cdot\theta\\ &\qquad (\because\ \theta \text{ is small, } \cos\theta = 1,\ \sin\theta = \theta)\\ &= (\mu - 1)t + a\theta\end{aligned}$$

$\therefore$ **Grating equation**, giving the principal maximum, is

$$(\mu - 1)t + a\theta = n\lambda, \quad n = \text{order of spectrum.} \tag{10.9.1}$$

Resolving power : From (10.9.1), we have

$$n = \frac{(\mu - 1)t + a\theta}{\lambda} = \frac{(1.5 - 1)\times 1 \text{ cm}}{5\times 10^{-5}\text{ cm}} = \mathbf{10{,}000}$$

taking $\mu = 1.5$, $t = 1$ cm, $\theta \to 0$ (normal viewing) and $\lambda = 5\times 10^{-5}$ cm.

If the number of plates, $N = 40$, then the resolving power is given by

$$N \times n = 40 \times 10{,}000 = \mathbf{4{,}00{,}000}$$

Thus, a *very high resolving power* is obtained in echelon grating. Also, since diffraction angle is very small, almost all the light is focussed into one or two orders resulting in *high intensity.*

Theory (reflection type)—Let a be the width of each step and t the thickness of each plate. Consider two parallel rays QP and SR diffracted at an angle θ from the corresponding points P and R (Fig. 10.22).

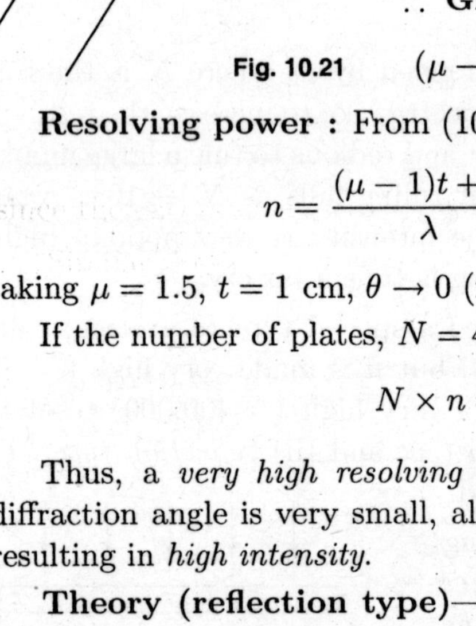

Fig. 10.22

Path difference between these rays is given by

$$\begin{aligned}\Delta &= MP + PL - RN\\ &= t + t\cos\theta - a\sin\theta\\ &= t + t - a\theta \quad (\because\ \theta \text{ small})\\ &= 2t - a\theta\end{aligned}$$

∴ **Grating equation**, giving the principal maximum, is

$$\boxed{2t - a\theta = n\lambda,} \tag{10.9.2}$$

where n = order of spectrum $= (2t - a\theta)/\lambda$.

When viewed along the normal, $\theta = 0$ and therefore, we have

$$n = 2t/\lambda$$

Resolving power—Taking $t = 1$ cm, $\lambda = 500$ nm, we obtain

$$n = \frac{2 \times 10^{-2}}{500 \times 10^{-9}} = \mathbf{40{,}000}$$

which is an extremely high value.

If the number of plates $N = 40$, then the resolving power is

$$N \times n = 40 \times 40{,}000 = \mathbf{16{,}00000}$$

Thus, an *extremely high resolving power* is obtained in echelon grating so that if the incident light is not truly monochromatic, two nearby spectral lines will be well resolved. It can thus be used to study the hyperfine structure of spectral lines (e.g., Zeeman effect). Reflection echelon grating is also used for an accurate determination of wsavelegth from far infrared to extreme ultraviolet region of spectrum.

- **Angular dispersive power**—For *reflection type*, we have

$$\theta = \frac{2t - n\lambda}{a}$$

$$\Rightarrow \quad \frac{d\theta}{d\lambda} = -\frac{n}{a}$$

which is the required expression for the angular dispersive power.

For *transmission type*,

$$\theta = \frac{n\lambda - (\mu - 1)t}{a} \quad \Rightarrow \quad \frac{d\theta}{d\lambda} = \frac{1}{a}\left[n - t\frac{d\mu}{d\lambda}\right]$$

which is the required expression, $d\mu/d\lambda$ being the rate of change of index of the material of the echelon with wavelength λ.

10.10 Illustrated Examples

▶ **Example 1.** *A convex lens of focal length* 20 cm *is placed after a slit of width* 0.6 mm. *If a plane wave of wavelength* 6000 A *falls on the slit normally, calculate the separation between the second minima on either side of the central maximum.*

(Punjab Univ.)

Solution. By the problem, $\lambda = 6000\,\text{A} = 6 \times 10^{-7}$ m; $f = 20$ cm $= 0.2$ m and slit-width, $a = 0.6$ mm $= 0.6 \times 10^{-3}$ m.

Let the second minima be in a direction making an angle θ_2 on either side of the central maximum.

$$\therefore \quad \sin\theta_2 = 2\frac{\lambda}{a} = \frac{2\times 6\times 10^{-7}}{0.6\times 10^{-3}} = 0.002$$

Let x be the distance of the second minimum from the central maximum. Then, since the diffraction pattern is on the screen at the focal plane of the lens placed close to the slit,

$$\sin\theta_2 = \tan\theta_2 = \frac{x}{f} \Rightarrow \frac{x}{f} = \frac{2\lambda}{a} \Rightarrow x = f\times 0.002 = 0.2\times 0.002 = 0.0004\,\text{m}$$

$\therefore$ Separation between the second minima on either side of central maximum $= 2\times 0.0004 =$ **0.0008 m.**

► **Example 2.** *A parallel beam of sodium light is allowed to be incident normally on a plane grating having* 4250 lines/cm *and a second order spectral line is observed to be deviated through* 30°. *Find the wavelength of the spectral line.* **(Visweswariah Univ.)**

Solution. By the problem, $n = 2$, $\theta_2 = 30°$, $N = 4250$ and $\lambda = ?$

Grating element, $a + b = 1/N = 1/4250$.

$\therefore$ From the relation, $(a+b)\sin\theta_2 = n\lambda$, we have

$$\lambda = \frac{(a+b)\sin\theta_2}{n} = \frac{0.5}{4250\times 2} = 5882\times 10^{-8}\text{ cm} = \mathbf{5882\ A.}$$

► **Example 3.** *In a plane transmission grating, the angle of diffraction for the second order maxima for wavelength* 5×10^{-5} m *is* 30°. *Calculate the number of* lines/cm *of the grating.* **(Kurukshetra Univ.)**

Solution. Here $n = 2$, $\theta = 30°$, $\lambda = 5\times 10^{-5}$ cm, $N =$ No. of lines/cm = ?

Grating element, $a + b = 1/N$. So, from the relation,

$$(a+b)\sin\theta = n\lambda, \quad (1/N)\sin 30° = 2\times 5\times 10^{-5}$$

$$\therefore \quad N = \frac{0.5}{2\times 5\times 10^{-5}} = \mathbf{5000\ cm}$$

► **Example 4.** *Show that in a diffraction grating with grating element* 1.5×10^{-6} m *and light of wavelength* 550 nm, *the third and higher order principal maxima are not visible.* **(Guru Nanak Univ.)**

Solution. Given, $a + b =$ grating element $= 1.5\times 10^{-6}$ m and $\lambda = 550$ nm $= 5500\times 10^{-10}$ m.

Let n be the maximum order of spectrum visible with the grating.

$$\therefore \quad (a+b)\sin\theta_{max} = n\lambda$$

where $\theta_{max} =$ maximum angle of diffraction $= 90°$, implying $\sin\theta_{max} = 1$.

$$\therefore \quad n = \frac{a+b}{\lambda} = \frac{1.5 \times 10^{-6} \text{ m}}{5500 \times 10^{-10} \text{ m}} = 2.72 < \mathbf{3}.$$

$\therefore$ **Third** and **higher orders** will **not** be **visible.**

► **Example 5.** *A diffraction grating used at normal incidence gives a line* (5400 A) *in certain order superposed on another line* (4050 A) *of the next higher order. If the angle of diffraction be* 30°, *how many* lines/cm, *are there on the grating?*

(Calcutta and Punjab Univ.)

Solution. Given $\lambda_1 = 5400$ A $= 5400 \times 10^{-10}$ m; $\lambda_2 = 4050$ A $= 4050 \times 10^{-10}$ m; $\theta_1 = \theta_2 = 30°$, $N = ?$ Also, by the problem, the two lines get superposed in nth and $(n+1)$th order.

$$\therefore \quad (a+b)\sin\theta_1 = n\lambda_1; \quad (a+b)\sin\theta_2 = (n+1)\lambda_2$$

$$\therefore \quad n\lambda_1 = (n+1)\lambda_2 \quad \Rightarrow \quad \frac{n+1}{n} = \frac{\lambda_1}{\lambda_2}$$

$$\therefore \quad 1 + \frac{1}{n} = \frac{5400}{4050} = 1 + \frac{1350}{4050} \quad \Rightarrow \quad n = \frac{4050}{1350} = 3$$

$$\therefore \quad (a+b)\sin 30° = 3 \times 5400 \times 10^{-10}$$

$$\therefore \quad N = \frac{1}{a+b} = \frac{0.5}{3 \times 5400 \times 10^{-10}} = \frac{10^8}{324} = 308642/\text{m} = \mathbf{3086/cm}$$

► **Example 6.** *A diffraction pattern is observed using a double-slit and light of wavelength* 5000 A. *If the slit-width is* 0.02 mm *and spacing between the slits is* 0.10 mm, *calculate* (i) *the distance between the central maximum and the first minimum of the fringe envelope and* (ii) *the distance between any two consecutive double-slit dark fringes. Given : the slit-screen separation* = 1 m. **(Indian Adm. Service)**

Solution. The given data are : $a = 0.02$ mm $= 2 \times 10^{-5}$ m; $b = 0.1$ mm $= 10^{-4}$ m; $\lambda = 5000\text{A} = 5 \times 10^{-7}$ m; $D = 1$ m. So $a + b = 1.2 \times 10^{-4}$ m; x_1 = distance between central maximum and first minimum of fringe envelope.

(i) Angular separation θ_1 between central maximum and first minimum is given by

$$\sin\theta_1 = \theta_1 = \lambda/2(a+b); \quad \text{also } \theta_1 = x_1/D$$

$$\therefore \quad \frac{x_1}{D} = \frac{\lambda}{2(a+b)} \Rightarrow x_1 = \frac{\lambda D}{2(a+b)} = \frac{5 \times 10^{-7} \times 1}{2 \times 1.2 \times 10^{-4}} = 2.08 \times 10^{-3} \text{ m} = \mathbf{2.08 \text{ mm}}$$

(ii) The angular separation between the two consecutive dark fringes is given by

$$\sin\theta_2 - \sin\theta_1 = \theta_2 - \theta_1 = \frac{3\lambda}{2(a+b)} - \frac{\lambda}{2(a+b)} = \frac{\lambda}{a+b}$$

If x_2 be the distance between two consecutive dark fringes,

$$\theta_2 - \theta_1 = \frac{x_2}{D} \Rightarrow x_2 = D(\theta_2 - \theta_1) = \frac{\lambda D}{a+b} = \frac{5 \times 10^{-7} \times 1}{1.2 \times 10^{-4}} = 4.16 \times 10^{-3} \text{ m} = \mathbf{4.16 \text{ mm.}}$$

► **Example 7.** *A Rowland grating (concave) has* 6000 *lines* per cm. *Find the angular separation between the two yellow lines of neon* (5882A *and* 5852A) *in the first order spectrum observed normally.* **(Kanpur Univ.)**

Solution. The equation for a concave grating is

$$(a+b)(\sin i - \sin\theta) = n\lambda \tag{10.10.1}$$

with usual meanings of the symbols.

Differentiating (10.10.1), with respect to λ,

$$-(a+b)\cos\theta\frac{d\theta}{d\lambda} = n$$

$$\Rightarrow \quad |d\theta| = \frac{n\,d\lambda}{(a+b)\cos\theta} = \frac{n\,d\lambda}{a+b} \quad (\because\ \theta = 0,\ \text{normal incidence})$$

Substituting the given values,

$$d\theta = \frac{1\times(5882-5852)\times 10^{-8}\ \text{cm}}{(1/6000)\ \text{cm}} = \mathbf{1.8\times 10^{-3}\ radian}$$

► **Example 8.** *In a Rowland (concave) grating, two lines of wavelengths* 4886.24A *and* 4887.37A *fall at linear positions* 12.372 mm *and* 14.268 mm. *What is the wavelength of the line that gives a reading* 11.638 mm? **(Agra Univ.)**

Solution. A Rowland (concave) grating gives a normal spectrum, i.e., $d\lambda/dx =$ constant.

For $d\lambda = 4887.37 - 4886.24 = 1.13$ A, $dx = 14.268 - 12.372 = 1.896$ mm.

$$\therefore \quad \frac{d\lambda}{dx} = \frac{1.13}{1.896} = 0.596\ \text{A/mm}$$

Let λ be the unknown wavelength. Then, we have

$$d\lambda = (4886.24 - \lambda)\ \text{A}$$

$$dx = 12.372 - 11.638 = 0.734\ \text{mm.}$$

$$\therefore \quad \frac{4886.24-\lambda}{0.734} = 0.596 \quad \Rightarrow \quad \lambda = \mathbf{4885.80\ A}$$

► **Example 9.** *Two spectral lines have wavelengths* λ *and* $\lambda + d\lambda$ $(d\lambda \ll \lambda)$ *respectively. Show that their angular separation in a grating spectrometer is given by*

$$d\theta = d\lambda/\sqrt{\{(a+b)/n^2\}^2 - \lambda^2}$$

where $(a+b)$ *is the grating element,* n *the order at which the lines are observed.*

(Meerut Univ.)

Solution. The grating equation for normal incidence is given by

$$(a+b)\sin\theta = n\lambda \tag{10.10.2}$$

Differentiating (10.10.2) with respect to λ

$$(a+b)\cos\theta\frac{d\theta}{d\lambda} = n$$

$$\therefore \quad \boldsymbol{d\theta} = \frac{nd\lambda}{(a+b)\cos\theta} = \frac{nd\lambda}{(a+b)\sqrt{1-\sin^2\theta}}$$

$$= \frac{nd\lambda}{(a+b)\sqrt{1-n^2\lambda^2/(a+b)^2}}, \quad \text{using (10.10.2)}$$

$$= \frac{nd\lambda}{\sqrt{(a+b)^2-n^2\lambda^2}} = \boldsymbol{\frac{d\lambda}{\sqrt{\{(a+b)/n\}^2-\lambda^2}}}.$$

► **Example 10.** *For a grating with grating element* $(a+b) = 18{,}000$ A, *obtain expression for angular dispersion in the first and third order spectra around* $\lambda = 5000$ A, *assuming normal incidence.* **(Gorokhpur Univ.)**

Solution. The angular dispersion produced by a grating is

$$\frac{d\theta}{d\lambda} = \frac{n}{(a+b)\cos\theta} = \frac{n}{(a+b)\sqrt{1-\sin^2\theta}} \tag{10.10.3}$$

But, from grating equation, in normal incidence

$$(a+b)\sin\theta = n\lambda \quad \Rightarrow \quad \sin\theta = n\lambda/(a+b)$$

$$\therefore \quad \frac{d\theta}{d\lambda} = \frac{n}{(a+b)\sqrt{1-n^2\lambda^2/(a+b)^2}} = \frac{n}{\sqrt{(a+b)^2-n^2\lambda^2}}$$

Here, $n = 1$, $\lambda = 5000$ A, $(a+b) = 18{,}000$ A. So

$$\frac{d\theta}{d\lambda} = \frac{1}{\sqrt{18000^2-5000^2}} = \mathbf{5.78 \times 10^{-5}\ radian/A}$$

Similarly, for $n = 3$, $d\theta/d\lambda =$ $\mathbf{3.0 \times 10^{-4}\ rad/A.}$

► **Example 11.** *A parallel beam of light of wavelength* 5200 A *is incident normally on a diffraction grating. The first order spectrum is formed in a direction making an angle* $\theta_1 = \sin^{-1} 0.25$ *with the normal. Find* (i) *the number of lines per metre on the grating,* (ii) *the angle* θ_2 *between the normal and the second order and* (iii) *the highest order.*

Solution. From the grating equation

$$(a+b)\sin\theta = n\lambda,$$

the number of lines per unit length, $N = \dfrac{1}{a+b}$

$$\therefore \quad \sin\theta = Nn\lambda$$

Here $\theta = \theta_1 = \sin^{-1} 0.25 \Rightarrow \sin\theta_1 = 0.25$; $n = 1$ and $\lambda = 5200\,\text{A} = 5.2 \times 10^{-7}$ m.

$$\therefore \quad \sin\theta_1 = N\lambda \Rightarrow N = \frac{0.25}{5.2 \times 10^{-7}} = \mathbf{5 \times 10^5/m}$$

$$\therefore \quad \sin\theta_2 = N \times 2 \times \lambda = 5 \times 10^5 \times 2 \times 5.2 \times 10^{-7} = 0.52 \quad \therefore \quad \theta_2 = \mathbf{31°}$$

Since $(\sin\theta)_{\max} = 1$, the highest possible order is given by

$$n = \frac{1}{N\lambda} = \frac{1}{5 \times 10^5 \times 5.2 \times 10^{-7}} = \mathbf{4}$$

▸ **Example 12.** *In Fraunhofer diffraction pattern due to a circular aperture, the screen is at a distance of* 1 m *from the aperture and the aperture is illuminated by light of* $\lambda = 5893$ A*. The diameter of the slit is* 0.1 mm*. Find the separation between the Airy's disc and the first secondary ring of minimum.*

Solution. Here $\lambda = 5893\,\text{A} = 5893 \times 10^{-8}$ cm; $f = 1\,\text{m} = 100$ cm, $d = 0.1\,\text{mm} = 0.01$ cm, and $x = ?$

For a circular aperture, we have

$$x = \frac{1.22\lambda}{d} f = \frac{1.22 \times 5893 \times 10^{-8} \times 100}{0.01} = \mathbf{0.7189\ cm.}$$

▸ **Example 13.** *How many orders would be visible if the wavelength of incident light is* 589 nm *and the number of lines in the grating is* 100/mm*?* **(Calcutta Hons.)**

Solution. Here $\lambda = 589$ nm $= 589 \times 10^{-9}$ m; $N = \dfrac{1}{a+b} = 100 \times 10^3$/m. $\theta = 90°$ (maximum) $\Rightarrow \sin\theta = 1$.

From $(a+b)\sin\theta = n\lambda$, $n = \dfrac{a+b}{\lambda}\sin\theta = \dfrac{1}{100 \times 10^{-3} \times 589 \times 10^{-9}} \times 1 = 16.32$

Thus **16** orders would be visible.

▸ **Example 14.** *For a grating with grating element* $(a+b) = 18000$ A*, obtain the expression for angular dispersion in the first and third order spectra, if* $\lambda = 5000$ A*.*

Solution. The angular dispersion produced by a grating is

$$\frac{d\theta}{d\lambda} = \frac{n}{(a+b)\cos\theta} = \frac{n}{(a+b)\sqrt{1-\sin^2\theta}}$$

From grating equation in normal incidence, we obtain

$$(a+b)\sin\theta = n\lambda \Rightarrow \sin\theta = n\lambda/(a+b)$$

$$\therefore \quad \frac{d\theta}{d\lambda} = \frac{n}{(a+b)\sqrt{1-n^2\lambda^2/(a+b)^2}} = \frac{n}{\sqrt{(a+b)^2-n^2\lambda^2}}$$

In first order, $n = 1$; here $\lambda = 5000$ A; $(a+b) = 18000$ A.

$$\therefore \quad \frac{d\theta}{d\lambda} = \frac{1}{\sqrt{18000^2 - 5000^2}} = \mathbf{5.78 \times 10^{-5}\ rad/A}$$

Similarly, in third order, $n = 3$ and we have

$$\frac{d\theta}{d\lambda} = \frac{3}{\sqrt{18000^2 - 5000^2 \times 3^2}} = \mathbf{3.0 \times 10^{-4}\ rad/A}$$

▶ **Example 15.** *Consider a grating of width* 5 cm *with slits of width* $a = 0.001$ mm *separated by a distance* 0.002 mm. *How many orders would be visible at* $\lambda = 550$ nm*? Find the width of the principal maximum; is there any missing order?*

Solution. Here $a = 0.001$ mm; $b = 0.002$ mm; $\lambda = 550 \times 10^{-9}$ m; $= 550 \times 10^{-6}$ mm. From the relation, $(a+b)\sin\theta = n\lambda$, the maximum number of orders visible is

$$n = \frac{(a+b)\sin\theta_{\max}}{\lambda} = \frac{(0.001 \times 0.002) \times 1}{550 \times 10^{-6}} = 5.45$$

$\therefore$ **Maximum** number of **order** visible = **5.**

Angular width of the principal maximum is

$$2d\theta = \frac{2\lambda}{N(a+b)\cos\theta}$$

Hee $N(a+b)$ = width of grating $= 5 \times 10^{-2}$ m and $\cos\theta = 1$ for principal maximum.

$$\therefore \quad 2d\theta = \frac{2 \times 550 \times 10^{-9}}{5 \times 10^{-2}} = \mathbf{2.2 \times 10^{-5}\ rad}$$

We have, again, $$\frac{a+b}{a} = \frac{0.001 + 0.002}{0.001} = \frac{3}{1}$$

$\therefore$ The **3rd, 6th, 9th** etc. order maxima will be missing.

▶ **Example 16.** *Fraunhofer double slit diffraction pattern is observed in the focal plane of lens of focal length* 0.5 m. *The wavelength of incident light is* 500 nm. *The distance between two maxima adjacent to zero order maximum is* 5 mm *and the fourth order maximum is missing. Find the width of each slit and the distance between their centres.* **(Calcutta Hons.)**

Solution. Since the 4th order is missing, we must have

$$\frac{a+b}{a} = 4 \;\Rightarrow\; b = 3a$$

Now for the first order maximum, $(a+b)\sin\theta_1 = \lambda$

$$\therefore \quad \theta_1 \simeq \frac{\lambda}{a+b} = \frac{\lambda}{4a}$$

$\therefore$ Distance between the two maxima adjacent to the zero order maximum is

$$x = f \times 2\theta_1 = f \times \frac{2\lambda}{4a} \quad (f = \text{focal length of the lens})$$

$$\therefore \quad a = f \times \frac{\lambda}{2x} = \frac{0.5 \times 500 \times 10^{-9}}{2 \times 5 \times 10^{-3}}\ \text{m} = \mathbf{0.025\ mm}$$

$$\therefore \quad b = 3a = 3 \times 0.025 = \mathbf{0.075\ mm}$$

Review Questions

1. Discuss Fraunhofer diffraction pattern due to a single slit. Find the expression for the width of the central maximum. **(Himachal Univ.)**
2. Discuss analytically the distribution of intensity in the diffraction pattern due to a single slit. **(Guru Nanak Univ.)**
3. Discuss the diffraction pattern produced by a circular aperture. What is its importance? **(Kurukshetra Univ.)**
4. Discuss Fraunhofer diffraction pattern due to a double-slit. Find the positions of maxima and minima. What is missing order in a double slit pattern. **(Kurukshetra Univ.)**
5. Using Frauhofer's theory of single slit diffraction, explain the diffraction by multiple (N) slits. **(Shivaji Univ.)**
6. Give a complete account of the phenomenon and the relevant theory of diffraction at N parallel slits. **(Punjab Univ.)**
7. What is a plane diffraction grating? Discuss its theory and derive the condition for the secondary maxima and minima. **(Calcutta and Visweswariah Univ.)**
8. Describe in full detail how you would use the plane transmission grating to determine the wavelength of light. **(Cochin Univ.)**
9. Given a spectrometer and monochromatic source of light, how will you find the grating constant? **(Mysore Univ.)**
10. Describe Rowland's concave reflection grating, and give its complete theory. **(Benares Hindu Univ.)**
11. Describe the three important mountings of a concave grating. **(Shivaji Univ.)**

12. Discuss Fraunhofer diffraction due to a single slit. Extend the theory to the case of a plane transmission grating. Explain what is meant by diffraction spectra of different orders and state the condition under which the grating spectra of even order are absent. **(Lucknow and Osmania Univ.)**

13. Discuss Fraunhofer diffraction at a narrow slit. Draw a curve indicating the distribution of light in the diffraction pattern. Is there any fundamental difference between interference and diffraction? Give reasons. **(Delhi Univ.)**

14. Discuss the phenomenon of Fraunhofer diffraction at a single slit and show that the intensities of successive maxima are nearly in the ratio

$$1 : \frac{4}{9\pi^2} : \frac{4}{25\pi^2} : \frac{4}{49\pi^2}.$$

(Kanpur Univ.)

15. Derive an expression for the intensity at a point in the Fraunhofer type of diffraction produced by two nearby parallel narrow slits illuminated by monochromatic light. Draw a diagram to indicate the distribution of intensity. **(Delhi Hons.)**

16. Give the theory of a concave reflection grating. Deduce the condition to obtain focussed spectra. Explain why a concave grating is preferred to a plane grating. What is a normal spectrum? **(Delhi Univ.)**

17. What is an echelon grating? Develop its theory. Discuss, in what respect, an echelon grating is superior to a plane diffraction grating. Indicate also its limitations.

Short Answer and Objective Questions

1. Derive an expression for the angular dispersion of a plane grating.

2. Distinguish between prism spectra and grating spectra.

3. Explain why a grating is designed to give only two orders. Why is a grating spectrum called a normal spectrum?

4. What happens in the diffraction pattern due to a single slit if the slit-width is gradually increased? **(Shivaji Univ.)**

5. Distinguish between a single slit and a double-slit diffraction pattern. **(Guru Nanak Univ.)**

6. How does the diffraction pattern due to a double-slit change if the distance between the slit centres is varied, keeping slit-width constant? **(Kurukshetra Univ.)**

7. Find the missing order in a double-slit diffraction pattern. **(Guru Nanak Univ.)**

8. What is a plane diffraction grating? What is grating element? What do you mean by opacities and transparencies? What is the usual order of the number of lines/cm of a grating? Why gratings with larger number of lines are preferred?

9. Derive the condition for missing order spectra for a diffraction grating. What particular order would be absent if the width of transparencies and opacities of a grating are equal? **(Punjab Univ.)**

10. What will be the effect on the diffraction pattern of a grating if the grating element is of the order of λ or less? **(Himachal Univ.)**

11. Draw a diagram indicating the intensity distribution in Fraunhofer diffraction at a single slit, with central maximum and one secondary maximum on each side of the central one. On the same diagram, show how the pattern changes if one more identical slit is placed parallel to the first a distance 4 times the slit width.

Problems

1. In an arrangement for Fraunhofer diffraction, we use a slit of width 0.2 mm and the first minimum is at 5 mm on either side of the central maximum. If the distance between the lens and the screen is 2 m, calculate the wavelength of light. **(Himachal Univ.)** [5×10^{-7} m]

2. A parallel beam of light of $\lambda = 5 \times 10^{-7}$ m is incident normally on a narrow slit of width 0.2 mm. The diffraction pattern is observed on a screen at the focal plane of a convex lens of $f = 50$ cm. Find the distance between the first two minima and the first two maxima on the screen. [0.125×10^{-3} m; 0.1288×10^{-2} m]

3. Find the missing orders in the Fraunhofer diffraction pattern of double-slit if the slit-width of each is 0.16 m and they are 0.8 mm apart. [6, 12, 18, etc.]

4. What is the highest order spectrum which may be observed with monochromatic light of wavelength 5000 A by means of a grating with 5000 lines/cm? **(Benares Hindu Univ.)** [4]

5. A diffraction grating has 6000 lines/cm. Show that, with white light as source, the second and the third order spectra overlap. Use the wavelength components 4000 A and 7000 A for studying the white light beam.

6. A wire grating is made of 200 wires/cm placed at equal distance apart. The diameter of each wire is 0.025 mm. Calculate the angle of diffraction for the third order spectra and also find the absent spectra, if any. Wavelength of light used = 6000 A. **(Punjab Univ.)** [$2°4'$; 2]

7. In a Fraunhofer diffraction pattern due to a narrow slit, a screen is placed 2 m away from the lens to obtain the pattern at its focal plane. Find the slit-width if the first minimum lies 5 mm on either side of central maximum when plane waves of 5000 A are incident on the slit. [0.2 mm]

8. A Rowland grating has 6000 lines to a centimeter. Calculate the angular separation of two mercury lines of wavelengths 5770 A and 5461 A in the first order spectrum. **(Delhi Univ.)** [$1°6'$]

9. A diffraction grating which has 4000 lines to a cm is used at normal incidence. Calculate the dispersive power of the grating in the third order spectrum in the wavelength region 5000 A. **(Delhi Univ.)** [15000]

10. A diffraction grating used at normal incidence gives a green line, $\lambda = 5400$ A in a certain order superimposed on the violet line $\lambda = 4500$ A of the next higher order if the angle of diffraction is 10°. How many lines are there per cm in the grating? **(Delhi Hons.)** [3986/cm]

11. A single slit of width 0.14 mm is illuminated normally by monochromatic light and diffraction bands are observed on a screen 2 m away. If the centre of the second dark band is 1.6 cm away from the middle of the central bright band, deduce the wavelength of light used. **(IAS)**

12. In double-slit Fraunhofer diffraction, calculate the fringe spacing on a screen 0.5 m away from the slits, if they are illuminated with blue light ($\lambda = 4800$ A). Slit separation $b = 0.1$ mm and slit width $a = 0.02$ mm. **(IAS)**

13. A convex lens of focal length 40 cm is employed to focus the diffraction pattern of a single slit of width 0.3 mm. Find the linear separation between the first order dark and the central band. Take $\lambda = 589$ nm. [0.7852 mm]

14. A parallel beam of light of $\lambda = 500$ nm is incident normally on a slit of width 0.2 mm. The diffrction is observed on a screen which is placed at the focal plane of a convex lens (close to the slit) of focal length 20 cm. Calculate the width of the central maximum and also the distance between the first two minima. **(Calcutta Hons.)** [1 mm; 0.5 mm]

Chapter 11

Resolving Power of Optical Instruments

11.1 Introduction

If two objects are very close to each other, or if they are at a very large distance from the eye, they may not be distinguishable as two separate objects. In other words, we say that the eye cannot *resolve* them. To resolve them, that is, to see them as *distinguishably separate,* optical instruments such as telescope, microscope etc. are employed. Similarly, to distinguish two very close sepctral lines (e.g., D-lines of sodium : $D_1 = 5890\,\text{A}$, $D_2 = 5896\,\text{A}$), optical instruments like prism, grating etc. are used.

While the theory of optical instruments is based on the laws of geometrical optics and rectilinear propagation of light, these laws work truly only approximately. As a matter of fact, when a beam of light from a point object or a line passes through the objective of a telescope, the lens serves as a circular aperture and gives rise to a diffraction pattern, instead of a point or a line image. The pattern consists of a bright central maximum and other secondary maxima, with minima in between, of rapidly decreasing intensity. So for an optical instrument, it would be possible to resolve two point objects provided the corresponding diffraction patterns are distinguishable from each other. *Qualitatively, the resolving power of an optical instrument is its ability to produce distinguishably separate patterns of two point objects.*

11.2 Rayleigh's criterion of resolution

To avoid subjectivity in distinguishing two patterns as separate, Lord Rayleigh gave a criterion of resolution.

According to Rayleigh, *two point objects* (or *sources*) *lying close to each other are said to be* **just** *resolved by an optical instrument if the central maxinum in the diffraction pattern of one falls on the first minimum in the pattern of the other.* This is known as **Rayleigh's criterion of resolution.** The criterion is equally applicable to *spectral lines of two very close wavelengths*, So, *two close lines in the spectrum will be resolved when the central maximum due to one wavelength falls on the first minimum due to the other.*

• The criterion, as stated above, in case of spectral lines amounts to this : *the least angular separation between the principal maxima of the diffraction patterns of the two spectral lines in a given order should be equal to half the angular width of either principal maximum.*

Illustration of Rayleigh's criterion—We shall now illustrate the criterion by considering the resolution of two wavelengths λ_1 and λ_2 by a grating.

The intensity curves of the patterns of two wavelengths is shown in Fig. 11.1(a) where the difference in wavelengths is rather large and their principal maxima A, B are separately visible. One can easily locate the existence of a distinct point of zero intensity lying between the two principal maxima. So the wavelengths λ_1 and λ_2 are *well-resolved.*

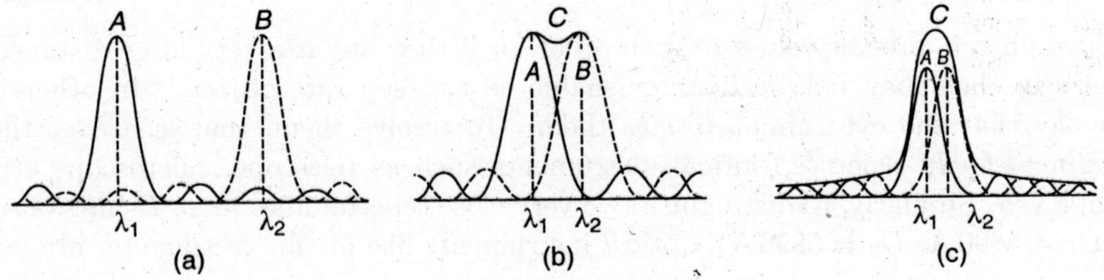

Fig 11.1

When the two wavelengths are comparatively much closer, the wavelength difference is such that the central maximum of each of the wavelengths coincides with the first minimum of the other as shown in Fig. 11.1(b). In this case, the resultant intensity curve C due to the two wavelengths shows a distinct dip in the middle of the two central maxima. This means that there is a *recognizable* diminution in intensity between the two central maxima giving the impression of the presence of *two different* wavelengths and, according to Rayleigh, the two wavelengths can be *just* distinguished and they are said to be *just resolved.*

Finally, consider the case when the difference in wavelengths is still smaller so that the central maxima corresponding to the two wavelengths λ_1, λ_2 also come closer still as shown in Fig. 11.1(c). The resultant intensity curve C now, unlike the previous case, is smooth without any dip. It thus gives an impression of the presence of a single wavelength, that is, the two wavelengths are *not resolved.*

11.3 Resolving power of a telescope

The purpose of a telescope is to see distant objects distinctly and the details which this optical device can provide depends on the *angle* subtended at the objective by two point objects and *not* on their *linear separation.*

The **resolving power** *of a telescope is defined as the reciprocal of the smallest angle subtended to its objective by two distinct object points that can be just recognised as separate ones by the telescope.*

Expression for resolving power—Let AB be the objective (Fig. 11.2) of the telescope of diameter D and O_1, O_2 (not shown) be two very close point-objects (say, two stars) from which parallel rays show an angular separation $d\theta$ at the objective. The objective lens (with the supporting ring) acts as a circular aperture and produces Fraunhofer diffration pattern as images of the point-objects O_1, O_2. Let P_1, P_2 be the positions of the central maxima of the two images. The patterns may be too close to each other with heavy overlapping to distinguish them as separate.

According to Rayleigh's criterion, however, the patterns will be *just resolved* if the central maximum of one falls on the first minimum of the other.

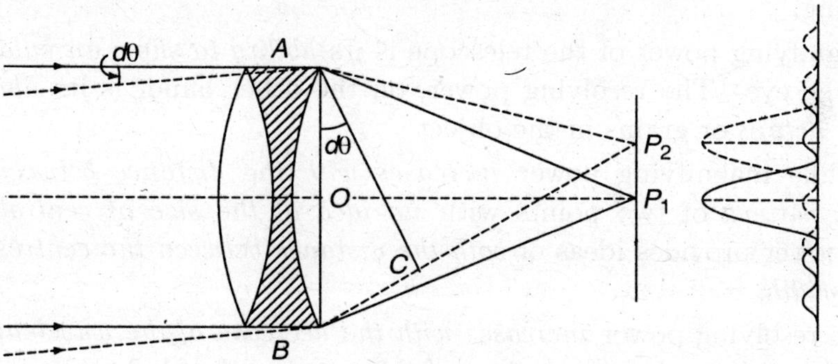

Fig 11.2

Now, the secondary wavelets travelling along AP_2 and BP_2 meet at P_2 and have a path difference

$$\Delta = (BP_2 - AP_2) = BC = AB \cdot \sin d\theta$$

$$= AB \cdot d\theta = D \cdot d\theta \quad (\because \text{ angle } d\theta \text{ is small})$$

If this $\Delta = \lambda$, the position of P_2 corresponds to the first minimum of the first image. So, for *just resolution,* the required condition is

$$D \cdot d\theta = \lambda \quad \Rightarrow \quad d\theta = \lambda / D \tag{11.3.1}$$

This condition holds for a rectangular aperture. For a circular aperture, however, Airy has shown by a rigourous treatment that the condition is

$$d\theta = \frac{1.22\lambda}{D} \tag{11.3.2}$$

where $d\theta$ is minimum resolvable angle between the two distant point objects. So, this $d\theta$ gives the **limit of resolution**. The *reciprocal of* $d\theta$ is the *resolving power* of the telescope.

$$\therefore \quad \textbf{Resolving power}, \ \frac{1}{d\theta} = \frac{D}{1.22\lambda} \tag{11.3.3}$$

The above expression for the resolving power of a telescope shows :

1. A telescope with a *larger diameter of objective* has a *higher resolving power and conversely.*

2. The *smaller the wavelength of light, higher* will be *the resolving power.*

• That P_2 corresponds to the first minimum may also be understood physically : Divide the whole wavefront AB into two halves AO and OB. Then the path-difference between secondary waves from corresponding points in the above two halves is $\lambda/2$. Thus, all the secondary waves from the two halves will interfere *destructively* with one another. P_2, therefore, will correspond to the *first minimum* of the *first image*

Magnifying power vs resolving power—The *magnifying power* of a telescope should be carefully distinguished from its *resolving power.* The following features are worth-noting.

1. Magnifying power of the telescope is its *ability to show an enlarged view of the object* to the eye. The resolving power, on the other hand, is its *ability to show the amount of details* or grains in the object.

2. While magnifying power *increases with the distance between the centres of diffraction pattern* of two points with *no idea of the size of central maximum,* the resolving power provides ideas of *both the distance between the centres of two maxima and their width.*

3. The resolving power *increases with the decrease of the wavelength* of light. But the magnifying power is *independent of the wavelength* of light.

4. The magnifying power of a telescope is given by F/f, i.e., the ratio of the focal length of the objective to that of the eyepiece. *Two telescopes* of the *same dia for the objective* will give *different magnifying powers* (as the focal length may be different), but the *same resolving power for a given wavelength* as the resolving power is given by the ratio D/λ.

• In general, **the magnifying power** of an optical instrument is defined as *the ratio of the angle subtended by the image at the eye, as seen through the instrument, to the angle subtended by the object at the centre of the unaided eye, when both the object and the image lie at the same distance from the eye.* For a microscope, this distance is the *least distance of distinct vision* and for a telescope, it is *infinity.*

The **resolving power** of an optical instrument, however, is defined as *the reciprocal of the smallest angle subtended at the objective by the two point objects that can* **just** *be distinguished as separate.*

11.4 Resolving power of a microscope

The *minimum distance by which two points* in the object *are separated so that their images formed by a microscope are just seen as separate is known as the* **limit of resolution** *of the microscope.*

The *reciprocal of the limit of resolution is called the* **resolving power**. The *smaller the limit of resolution, higher is the resolving power* and conversely.

Expression for resolving power—Let AB be the aperture of the microscope objective [Fig. 11.3(a)], O_1 and O_2 two self-luminous point objects separated by a distance x and very close to the objective and perpendicular to the axis. The periphery

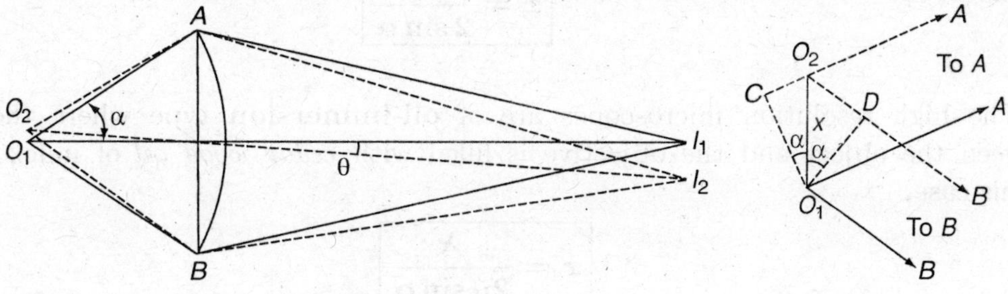

Fig 11.3

of the objective acts as a circular aperture so that the images of O_1 and O_2 are Fraunhofer diffraction patterns, each consisting of a central bright disc surrounded by a number of alternate dark and bright rings. I_1 corresponds to the central maximum of O_1 and I_2 to that of O_2.

By Rayleigh's criterion, in this case, the two objects O_1 and O_2 will be *just resolved* in the image plane if the first dark diffraction ring of the pattern at I_2 passes through I_1, the centre of the principal bright disc of O_1. This occurs when the path difference between the extreme rays O_2BI_1 and O_2AI_1 equals 1.22λ. We shall now compute for this path-difference.

With respect to Fig. 11.3(b) (a magnified view of O_1O_2 and the rays thereform), the

$$\textbf{Path difference } \Delta = (O_2B + BI_1) - (O_2A + AI_1)$$
$$= O_2B - O_2A \quad (\because \ BI_1 = AI_1)$$
$$= (O_2D + DB) - (CA - CO_2)$$
$$= O_2D + CO_2 \quad (\because \ DB = O_1B = O_1A = CA)$$

where $O_1C \perp CA$, $O_1D \perp O_2B$.

$$\therefore \ \Delta = O_2D + CO_2 = x \sin\alpha + x \sin\alpha = 2x \sin\alpha$$

where α is the semi-vertical angle of the cone of rays received by the objective AB.

$\therefore$ The two images will appear *just resolved* if

$$2x \sin\alpha = 1.22\lambda \ \Rightarrow \ x = \frac{1.22\lambda}{2 \sin\alpha}$$

which is the required *limit of resolution*.

We assumed in our derivation that the point objects are *self-luminous*, emitting light of wavelength λ. In case they are *not*, and are illuminated from an external source of light of wavelength λ, Abbe showed that the factor 1.22 may be dropped.

$$\therefore \quad \boxed{x = \frac{\lambda}{2 \sin \alpha}}$$

The high resolution microscopes are of **oil-immersion** type where the space between the object and the objective is filled with *cedar wood oil* of index μ, say. In this case,

$$\boxed{x = \frac{\lambda}{2\mu \sin \alpha}}$$

since in computing the path difference, multiplication by μ is now essential.

The quantity $\mu \sin \alpha$ is known as the **numerical aperature** (N.A.) of the microscope and is usually marked on the objective. It is a quantity of *great importance* in controlling the limit of resolution of the microscope. The *greater the N.A., smaller is x* and **higher is the resolution**, i.e., *finer is the detail that one can observe.*

Plainly, to decrease x between the points to be resolved, i.e., to increase the resolving power of the microscope:

1. The *wavelength* of the light should be *decreased.*

2. The *numerical aperture* should be *increased.*

$$\therefore \quad \boxed{\textbf{Limit of resolution, } x = \frac{\lambda}{2N.A.}}$$

11.5 Resolving power of a plane transmission grating

The resolving power of a grating measures the ability to distinguish two very close spectral lines λ and $\lambda + d\lambda$ and is defined by $\lambda/d\lambda$, $d\lambda$ being the smallest wavelength difference for which the spectral lines can be *just* resolved at wavelength λ.

Expression for resolving power—The nth order principal maximum of a spectral line of wavelength λ of a grating for normal incidence is given by

$$(a + b) \sin \theta = n\lambda \qquad (11.5.1)$$

where $(a + b)$ is the grating element and θ the diffraction angle corresponding to nth order spectrum.

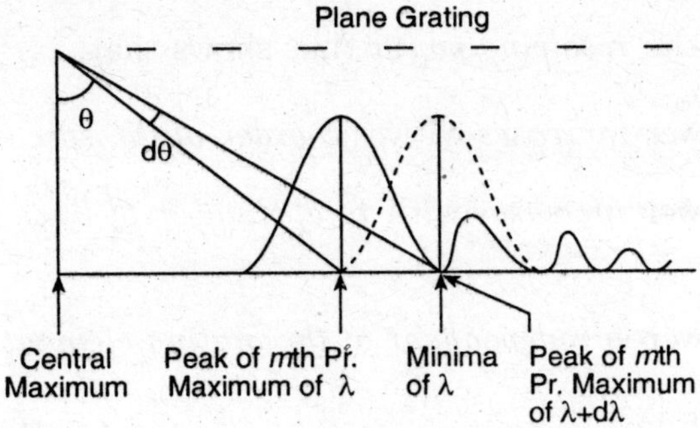

Fig. 11.4

The same for wavelength $\lambda + d\lambda$ will obviously be

$$(a+b)\sin(\theta + d\theta) = n(\lambda + d\lambda) \tag{11.5.2}$$

the corresponding diffraction angle (Fig. 11.4) being $(\theta + d\theta)$.

By Rayleigh's criterion, the two wavelengths will be *just resolved* if the position of nth order *principal maximum* for $\lambda + d\lambda$ corresponds to the *first minimum* adjacent to nth order principal maximum for λ. The condition of resolution thus dictates that the principal maximum of $\lambda + d\lambda$ in nth order should agree in direction with the first minimum for λ.

Now, the condition for *minimum* in a grating is given by

$$N(a+b)\sin\theta = m\lambda \tag{11.5.3}$$

where N is the total number of slits and m can have all *integral* values *except* $0, N, 2N, \cdots, nN$ since these m values satisfy the condition for *different maxima.*

So, the first minimum adjacent to nth principal maximum in the direction $(\theta + d\theta)$ is obtained by substituting $m = nN + 1$. So, the first minimum in the direction $(\theta + d\theta)$ satisfies the relation

$$N(a+b)\sin(\theta + d\theta) = (nN + 1)\lambda \tag{11.5.4}$$

Also, using (11.5.2)

$$N(a+b)\sin(\theta + d\theta) = nN(\lambda + d\lambda). \tag{11.5.5}$$

$\therefore$ For **just resolution,**

$$(nN+1)\lambda = nN(\lambda + d\lambda)$$

$$\Rightarrow \quad \lambda = nNd\lambda$$

$$\boxed{\therefore \quad \frac{\boldsymbol{\lambda}}{\boldsymbol{d\lambda}} = \boldsymbol{nN}} \tag{11.5.6}$$

which is the required expression for the resolving power (r.p.) of a grating.

The expression for the *resolving power* thus shows that

1. The resolving power *increases* with the *order of the spectrum* and conversely.

2. The resolving power *increases* with *total number of lines* in the effective part of grating.

3. The resolving power is *independent of the grating element.*

From equation (11.5.1)

$$n = \frac{(a+b)\sin\theta}{\lambda}$$

$$\therefore \quad \boxed{\frac{\boldsymbol{\lambda}}{\boldsymbol{d\lambda}} = \boldsymbol{nN} = \frac{\boldsymbol{N(a+b)\sin\theta}}{\boldsymbol{\lambda}}} \tag{11.5.7}$$

$$= W\frac{\sin\theta}{\lambda}$$

where $W = N(a+b)$, the total width of the ruled surface of the grating.

Relation (11.5.7) is often useful in working out numerical problems.

Dispersive power vs. resolving power—The (angular) *dispersive power* of a grating should be carefully distinguished from its *resolving power.*

The **angular dispersive power** *is defined as the change in the angle of diffraction corresponding to unit change in wavelength*, i.e., $d\theta/d\lambda$.

The expression for nth order principal maximum of a grating of element $(a+b)$ is, for normal incidence,

$$(a+b)\sin\theta = n\lambda \tag{11.5.8}$$

where λ is the wavelength of the incident light.

Differentiating (11.5.8),

$$(a+b)\cos\theta\, d\theta = n\, d\lambda.$$

$\therefore$ Angular dispersive power,

$$\boxed{\frac{\boldsymbol{d\theta}}{\boldsymbol{d\lambda}} = \frac{\boldsymbol{n}}{\boldsymbol{(a+b)\cos\theta}}} \tag{11.5.9}$$

The **angular dispersive** power of a *grating* thus

1. *increases* with the *increase in the order* of the spectrum.

2. *increases* with the *decrease* in the *grating element* $(a+b)$, i.e., with the *increase* in the *number of lines per unit length.*

3. *is independent of* N, the *total number of lines in the grating.*

Comparatively speaking, therefore,

1. While the dispersive power gives an idea of *angular separation* $d\theta$ between two spectral lines produced by the grating, the resolving power gives the *limit of just resolution* of two close wavelenths (or objects).

2. The dispersion is given by $d\theta/d\lambda = n/(a+b)\cos\theta$; the resolving power is measured by $\lambda/d\lambda = nN$.

3. When N (= total no. of lines) *increases, dispersive* power remains *unchanged*, but the *resolving power increases.*

4. If the *grating element* $(a+b)$ is *decreased*, the *dispersive power increases*, but the *resolving power* remains *unchanged.* So *a grating with a higher dispersive power does not necessarily have a higher resolving power.*

• If $\theta = 90°$, then from (11.5.7), we get

$$\left(\frac{\lambda}{d\lambda}\right)_{\max} = \frac{W}{\lambda} = \frac{N(a+b)}{\lambda}$$

the maximum value of r.p. for normal incidence.

• For oblique incidence, it can be shown that

$$\left(\frac{\lambda}{d\lambda}\right)_{\max} = \frac{2W}{\lambda} = \frac{2N(a+b)}{\lambda}.$$

11.6 Resolving power of a prism (spectrometer)

The **resolving power** *of a prism is its capacity to form separate spectral lines of two very close wavelengths* λ *and* $\lambda + d\lambda$ *and is measured by the ratio* of the mean *wavelength* λ to the *smallest difference in wavelength* $d\lambda$ that can be *just* distinguished as separate, i.e., by $\lambda/d\lambda$.

Expression for resolving power—Let ABC be the section of a prism (Fig. 11.5), BP a plane wavefront of wavelength λ incident on the prism placed in its position of minimum deviation. On refraction through the prism, the refracted wavefront CQ is produced which is focussed by the telescope objective L at its focal plane at I. I thus corresponds to the principal maximum for wavelength λ. It is assumed that light passes through the whole of the prism.

$$\therefore \quad PA + AQ = \mu \cdot BC \tag{11.6.1}$$

where μ is the refractive index of the prismatic material.

Equation (11.6.1) is obtained from Fermat's principle that for any wavelength, all actual optical paths between incident and emergent wavefronts are equal.

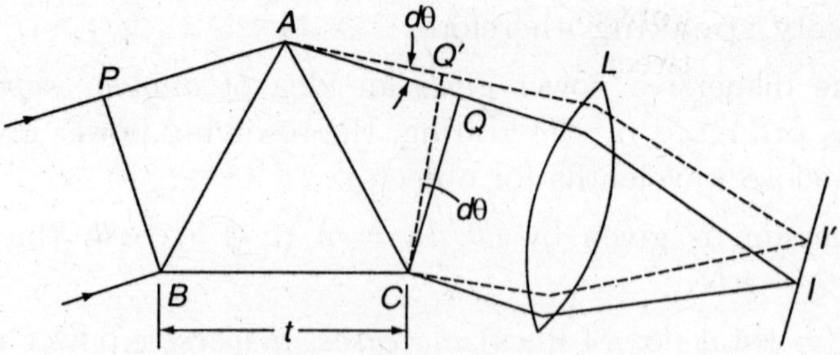

Fig 11.5

Let now another plane wavefront of wavelength $\lambda + d\lambda$ be incident on the prism along the *same path* (this is possible as $d\lambda \to 0$) to produce the refracted wavefront CQ'. Let $d\theta$ be the angle between the wavefronts CQ and CQ'. As before, CQ' is focussed by the objective of the telescope at its focal plane at the point I'. I' thus corresponds to the principal maximum for wavelength $\lambda + d\lambda$.

$$\therefore \quad PA + AQ' = (\mu - d\mu)BC \tag{11.6.2}$$

since μ decreases as λ increases.

Subracting (11.6.2) from (11.6.1), we obtain

$$AQ - AQ' = d\mu \cdot BC \quad \Rightarrow \quad QQ' = t \cdot d\mu \tag{11.6.3}$$

where $t = BC$, the width of the base of the prism.

If the refracted wavefront and also the diameter of the telescope objective be d, then

$$QQ' \simeq d \cdot d\theta \;\Rightarrow\; d \cdot d\theta = t d\mu \;\Rightarrow\; d\theta = \frac{t}{d} d\mu \tag{11.6.4}$$

If $d\theta$ be the limit of resolution of the telescope, we have

$$d\theta = \lambda/d \tag{11.6.5}$$

From (11.6.4) and (11.6.5), we get

$$\frac{\lambda}{d} = \frac{t}{d} d\mu \;\Rightarrow\; \boxed{\frac{\boldsymbol{\lambda}}{\boldsymbol{d\lambda}} = \boldsymbol{t}\frac{\boldsymbol{d\mu}}{\boldsymbol{d\lambda}}} \tag{11.6.6}$$

which is the required expression for the *resolving power* of a prism.

The expression shows that the *resolving power* of a prism *varies directly* as

1. t, the *width of the base* of the prism, and

2. $d\mu/d\lambda$, the *rate of change of refractive index with wavelength.*

• The difficulty in constructing a homogeneous prism of large base puts a limit to the maximum r.p. that can be obtained.

11.7 Grating spectra vs. prism spectra

It is interesting to make a comparative study of the grating spectrum and the prism spectrum. The important points of difference between the grating and the prism spectra are enumerated below.

1. While there is *only one spectrum* in the case of prism, we get a *number of spectra* situated on either side of the central maximum in grating.

2. In prisms, the *deviation* produced *depends* on (i) the *angle of the prism*, (ii) the *index of the* prismatic *material*, and (iii) the *wavelength of light*. Violet is deviated more than the red. In gratings however, the *deviation is proportional* (i) *directly to the wavelength* of light and (ii) *inversely to* the *grating element*. It is *independent* of the *material* of the grating.

3. The *dispersive power* of a prism is *higher*, *shorter* the *wavelength* and conversely. Violet end thus spreads out more so that violet part is less intense compared to red. The *dispersive power* of a grating is given by $n/(a+b)\cos\theta$ and since θ is usually small, the angular separation of two lines for a given spectrum is practically the same all along in a given order. But the dispersive power is *directly proportional* to n so that the angular separation in the second order is double that of the first, and also *increases* with *decreasing grating element*.

4. The *resolving power* of a prism is *small* and equals $t(d\mu/d\lambda)$, i.e., *directly propostional* to the *length of* the *prism-base*. It also *depends on the material* of the prism and the *wavelength* of light. The *resolving power* of a grating is *very large*, being given by the product of the total number of lines and the order of the spectrum (nN), but is *independent* of the *wavelength* and the *grating constant*.

While all the above points are in favour of grating spectrum, it suffers from a *serious disadvantage* that its *intensity is low* as most of the incident light is transmitted undeviated to produce the central band, and also because the part of the diffracted light gets distributed over various orders. In prism, there is one spectrum and the *intensity* of the spectrum is *excellent*.

11.8 Illustrated Examples

▶ **Example 1.** *Calculate the resolving power of a telescope having a diameter of* 200 inches *and* $\lambda = 6000$ A. **(Guru Nanak Univ.)**

Solution: Smallest angular separation of two objects that can be resolved is

$$d\theta = \frac{1.22\lambda}{D} = \frac{1.22 \times 6000 \times 10^{-10}}{200 \times 2.54/100} = \frac{1.22 \times 6 \times 10^{-5}}{2 \times 254} \text{ radian}$$

since here $\lambda = 6000\text{ A} = 6000 \times 10^{-10}$ m, $D = 200$ inches $= 200 \times 2.54/100$ m.

$$\therefore \quad \text{Resolving power} = \frac{1}{d\theta} = \frac{2 \times 254}{1.22 \times 6 \times 10^{-5}} = \mathbf{69.4 \times 10^5}$$

► **Example 2.** *The Mount Wilson Observatory telescope has an objective diameter* 2.54 m. *Assuming the mean wavelength of light to be* 5.5×10^{-5} cm, *estimate the smallest angular separation of two stars which can be resolved by it.*

(Punjab and Guru Nanak Univ.)

Solution: Here $\lambda = 5.5 \times 10^{-5}$ cm $= 5.5 \times 10^{-7}$ m; diameter of objective, $D = 2.54$ m.

$\therefore$ Smallest angular separation for resolution is

$$d\theta = \frac{1.22\lambda}{D} = \frac{1.22 \times 5.5 \times 10^{-7}}{2.54} = 2.64 \times 10^{-7} \text{ radian} = \mathbf{0.054 \text{ sec}}$$

► **Example 3.** *In a second order spectrum of plane diffraction grating, a certain spectral line appears at an angle of* $10°$, *while another line of wavelength* 0.5×10^{-10} m *greater appears at an angle* $3''$ *greater. Find the wavelength of the lines and the minimum grating width required to resolve them. Given :* $\sin 10° = 0.1736$ *and* $\cos 10° = 0.9848$.

(Delhi Hons.)

Solution: Here $n = 2$, $\lambda = ?$, $d\lambda = 0.5 \times 10^{-10}$ m and

$$d\theta = 3'' = \frac{3 \times \pi}{180 \times 60 \times 60} = 0.1455 \times 10^{-4} \text{ radian.}$$

$$\therefore \quad \frac{d\theta}{d\lambda} = \frac{0.1455 \times 10^{-4}}{0.5 \times 10^{-10}} = 2.91 \times 10^{5}$$

$$\text{But} \quad \frac{d\theta}{d\lambda} = \frac{n}{(a+b)\cos\theta}$$

$$\therefore \quad a + b = \frac{2}{2.91 \times 10^{5} \times \cos 10°} = \frac{2}{2.91 \times 10^{5} \times 0.9848} = 0.6979 \times 10^{-5} \text{ m}$$

If λ be the wavelength of first line and n be the order, we have

$$(a+b)\sin\theta = n\lambda \quad \Rightarrow \quad \lambda = \frac{(a+b)\sin\theta}{n} = \frac{0.6979 \times 10^{-5} \times \sin 10°}{2}$$

$$= \frac{0.6979 \times 10^{-5} \times 0.1736}{2} = \mathbf{6057.77 \text{ Å}}$$

$\therefore$ Wavelength of the second line $= 6057.77$ Å $+ 0.5$ A $=$ **6058.27 A**

If $N =$ total number of lines in the grating for *just* resolution of the lines,

$$\frac{\lambda}{d\lambda} = Nn \quad \Rightarrow \quad N = \frac{1}{n} \cdot \frac{\lambda}{d\lambda} = \frac{1}{2} \cdot \frac{6057.77}{0.5} = 6057.77$$

$$\therefore \quad \text{Grating width} = N(a+b) = 6057.77 \times 0.6979 \times 10^{-5}$$

$$= 0.04228 \text{ m} = \mathbf{4.228 \text{ cm}}$$

▶ **Example 4.** *Calculate angular dispersion in degrees per angstrom for a diffraction grating having* 14438 *lines per inch when used in the third order at* 4200 A.

(Punjab Univ.)

Solution: Here $n = 3$, $\lambda = 4200\text{ A} = 4200 \times 10^{-10}$ m, no. of lines per inch = 14438.

$$\therefore \quad a + b = \text{grating element} = \frac{2.54 \times 10^{-2}}{14438} = 0.1759 \times 10^{-5}\text{ m}$$

$$\therefore \quad \sin\theta = \frac{n\lambda}{a+b} = \frac{3 \times 4200 \times 10^{-10}}{0.1759 \times 10^{-5}} = 0.7163$$

$$\therefore \quad \cos\theta = \sqrt{1 - \sin^2\theta} = \sqrt{1 - (0.7163)^2} = 0.6978$$

$\therefore$ Angular dispersion is given by

$$\begin{aligned} d\theta &= \frac{n\,d\lambda}{(a+b)\cos\theta} \\ &= \frac{3 \times 1 \times 10^{-10}}{0.1759 \times 10^{-5} \times 0.6978} \\ &= 0.2444 \times 10^{-3}\text{ rad} \\ &= \mathbf{0.014^\circ} \end{aligned}$$

▶ **Example 5.** *Find the number of lines a grating should have to resolve the first order doublet having a wavelength difference of* 0.6 nm *at* 589.3 nm. **(Himachal Univ.)**

Solution: Here $\lambda = 589.3$ nm; $d\lambda = 0.6$ nm and $n = 1$.

Let N be the number of lines on the grating to resolve the doublet.

$$\frac{\lambda}{d\lambda} = Nn \quad \Rightarrow \quad N = \frac{1}{n} \cdot \frac{\lambda}{d\lambda} = \frac{\lambda}{d\lambda} = \frac{589.3}{0.6} = \mathbf{982}$$

▶ **Example 6.** *A diffraction grating just resolves two lines of* $\lambda = 5140$ A *and* 5140.5 A *in the first order. Will it be able to resolve the lines* 8037 A *and* 8037.5 A*?*

Solution: Here $d\lambda = 5140.5 - 5140 = 0.5$ A, $n = 1$, and

$$\lambda = \tfrac{1}{2}(\lambda_1 + \lambda_2) = \tfrac{1}{2}(5140 + 5140.5) = 5140.25\text{ A}$$

$\therefore$ Total number of lines in the grating

$$N = \frac{1}{n}\left(\frac{\lambda}{d\lambda}\right) = \frac{1}{1} \times \frac{5140.25}{0.5} = 10280.5$$

$\therefore$ Resolving power in the first order $= 1 \times 10280.5 = 10280.5$

Resolving power in second order = $2 \times 10280.5 = \mathbf{20561}$

For the second pair of lines $\lambda = \frac{1}{2}(8037+8037.5) = 8037.25$ A; $d\lambda = 8037.5-8037 = 0.5$ A

$$\therefore \text{ Resolving power} = \frac{\lambda}{d\lambda} = \frac{8037.25}{0.5} = \mathbf{16074.5}$$

As the required resolving power, 16074.5, is less than the available resolving power in the second order, the second pair of lines **will be resolved** in the second order.

► **Example 7.** *The thickness of the base of a flint glass prism of a spectrometer is* 1 cm *and the dispersion* $d\mu/d\lambda$ *for flint glass is* 1000 cm^{-1} *in the range* $\lambda = 6000$ A. *Find the smallest wavelength separation in this range which the spectrometer will be able to resolve.*

Solution: The resolving power of a prism is

$$\frac{\lambda}{d\lambda} = t\frac{d\mu}{d\lambda} \quad \Rightarrow \quad d\lambda = \frac{\lambda}{t(d\mu/d\lambda)}$$

$\therefore$ Minimum wavelength separation in the given range is given by

$$d\lambda = \frac{6000 \times 10^{-8} \text{ cm}}{1 \text{ cm} \times 1000 \text{ cm}^{-1}} = 6 \times 10^{-8} \text{ cm} = \mathbf{6A}$$

► **Example 8.** *The smallest object details than can be resolved by a microscope is* 3.8×10^{-5} cm *with sodium light of* $\lambda = 5893$ A. *Calculate* (a) *the numerical aperture of objective when used dry and* (b) *the same when an immersion oil of* $\mu = 1.5$ *is used.*

Solution: The limit of resolution of a microscope (when dry) is

$$d = \frac{\lambda}{2\sin\alpha} = \frac{\lambda}{2(N.A)}$$

where N.A. = numerical aperture.

(a) Numerical Aperture $= \lambda/2d = \dfrac{5893 \times 10^{-8} \text{ cm}}{2 \times 3.8 \times 10^{-5} \text{ cm}} = \mathbf{0.775}$

(b) Oil immersion $N.A = \mu \times$ dry $N.A. = 1.5 \times 0.775 = \mathbf{1.162}$

► **Example 9.** *The refractive index of glass for wavelengths* 656.3 nm *and* 527.0 nm *are* 1.6545 *and* 1.6635 *respectively. Find the length of the base of a* 60° *prism of this glass that can just resolve sodium D-lines of wavelengths* 589.0 nm *and* 589.6 nm.

Solution. Here $\lambda = \frac{1}{2}(589.6 + 589.0)$ nm = 589 nm; $d\lambda = 0.6$ nm

$$\text{and} \quad \frac{d\mu}{d\lambda} = \frac{1.6635 - 1.6545}{(656.3 - 527.0) \times 10^{-9}} = 69605 \text{ m}^{-1}$$

$$\text{Resolving power, } \frac{\lambda}{d\lambda} = t\frac{d\mu}{d\lambda} \Rightarrow t = \frac{\lambda}{d\lambda}\bigg/\frac{d\mu}{d\lambda}$$

$$\therefore\ t = \frac{589.3}{0.6} \times \frac{1}{69605} \text{ m} \simeq 0.014 \text{ m} = \mathbf{1.4\ cm}$$

▶ **Example 10.** *Find the angular limit of resolution of a telescope whose aperature is* 2 m *in diameter. If the focal length of the eyepiece be* 25 cm *and the aperture of human eye be* 2 mm, *find the minimum focal length of objective for the full utilisation of the resolving power of the telescope. The mean wavelength of light used is* 5000 A

Solution. The angular limit of resolution of a telescope is

$$\theta = \frac{1.22}{D}\lambda = \frac{1.22 \times 5000 \times 10^{-10} \text{ m}}{2\text{m}} = \mathbf{3.05 \times 10^{-7}\ rod}$$

For full utilization of resolving power,

r.p. of telescope × its normal magnifying power M = r.p. of the eye

$$\therefore\ M = \frac{\text{r.p. of eye}}{\text{r.p. of telescope}} = \frac{1.22\lambda/de}{1.22\lambda/D} = \frac{D}{d_e} = \frac{2}{2 \times 10^{-3}} = 1000$$

Also,
$$M = \frac{\text{focal lengh of objective}}{\text{focal length of eyepiece}} = \frac{f_O}{f_e}$$

$$\Rightarrow\ f_O = Mf_e = 1000 \times 2.5 \text{ cm} = 25 \text{ m}$$

$$\therefore\ \text{Required focal length} = \mathbf{25\ m}$$

Review Questions

1. State and explain Rayleigh's criterion of resolution and discuss its significance in studying the spectral lines. **(Cochin Univ.)**
2. Define resolving power and deduce an expression for the resolving power of a telescope. **(Kurukshetra Univ.)**
3. What is meant by the resolving power of a microscope? How is it related to the numerical aperture of its objective? **(Cochin Univ.)**
4. Derive an expression for the resolving power of a plane transmission grating. **(Calcutta and Benares Univ.)**
5. Distinguish between the angular dispersive pcwer and the resolving power of a grating. Deduce an equation for the dispersive power of a plane transmission grating. On what factors does it depend? **(Guru Nanak and Himachal Univ.)**

6. Explain what is meant by resolving power of a diffraction grating. Show that the resolving power is given by the product of the total number of rulings and the order of the spectrum.

7. Derive an expression for the resolving power of a prism and a telescope. **(Shivaji Univ.)**

8. State the difference between the grating spectra and the prism spectra. **(Calcutta and Vidyasagar Univ.)**

Short Answer and Objective Questions

1. Distinguish between magnifying power and resolving power. **(Punjab Univ.)**

2. State the Rayleigh's criterion of resolution.

3. Show that to obtain full resolution, the magnifying power of a telescope must not be less than the ratio of the diameter of the object-glass to the diameter of the pupil of the eye.

4. Explain the limit of resolution of a microscope. How is the resolving power related to the limit of resolution?

5. Distinguish between angular dispersive power and the resolving power of a grating.

6. Discuss the statement "A grating, having a higher dispersive power than another, does not necessarily has a higher resolving power". **(Guru Nanak Univ.)**

7. How will you increase the resolving power of a diffraction grating?

8. A grating with 1000 lines just resolves the D-lines of sodium (5896 Å and 5890 Å) in the first order. Explain what happens if half the portion of the grating is covered so that no light passes through this half?

9. A spectrometre, having a very large circular scale reading up to one second, is fitted with a telescope of 2.5 cm objective. Criticise its usefulness and suggest modification required. **(Punjab Univ.)**

10. When the number of lines per cm is increased in a plane diffraction grating
    (a) linear or angular dispersive power increases, resolving power decreases
    (b) linear or angular dispersive power decreases, resolving power increases
    (c) dispersive and resolving power both increase
    (d) dispersive and resolving power both decrease. **(Himachal Univ.)**

11. The resolving power of a microscope is limited by
   (a) the diameter of the objective (b) the wavelength of light
   (c) the kind of glass used (d) none of these is true.
12. The resolving power in the nth order of a grating having N number of total rulings is
   (a) nN (b) n/N (c) $(n+N)$ (d) none of these.
13. A telescope having larger diameter will have
   (a) a higher resolving power
   (b) a lower resolving power
   (c) any resolving power, independent of diameter
   (d) none of these is correct.

Problems

1. Calculate the useful magnifying power of a telescope of 1 inch objective, assuming the resolving power of the eye to be equal to 2 minutes and wavelength of light 6×10^{-5} cm. **(Punjab Univ.)** [20.2]
2. Calculate the magnifying power and the limit of resolution of a microscope from the following data : Numerical aperture = 0.12, wavelength of light $\lambda = 6 \times 10^{-7}$ m, distance of cross-wire from objective = 0.15 m and focal lengths of eye-piece and objective are equal and of value 0.025 m. [55, 0.25×10^{-5} m]
3. A diffraction grating having 4000 lines per cm is used at normal incidence. Calculate the dispersive power of the grating in the third order spectrum in the wavelength region 5000 A. [15,000]
4. Light is incident normally on a grating of width 0.5 cm and having 2500 lines. Find the angular separation for the principal maxima of two sodium lines in the first order. Are the two lines ($\lambda = 5890$ Å, $\lambda_2 = 5896$ Å) resolved? **(Punjab Univ.)** [3.139×10^{-4} rad; Yes]
5. A grating has grating constant 5×10^{-5} m. What is its resolving power in the second order, if the width of the grating is 0.02 m? [800]
6. The separation of sodium lines (mean wavelength $\lambda = 5893$ A) in the second order spectrum of a plane transmission grating having 5000 lines per cm is 2.5′ of an arc for normal incidence. Find the difference in wavelength between the two yellow lines. [5.876 A]

7. The refractive index for C and F lines for a certain glass is $\mu_C = 1.6545$ and $\mu_F = 1.6635$. The wavelengths of C and F lines in the solar spectrum are 6563 A and 5270 Å respectively. Find the base of 60° prism of this glass which will just resolve the two sodium lines of wave-lengths 5896 A and 5890 A. [1.41 cm]

8. A grating has 15 cm of the surface ruled with 5000 lines. Calculate its resolving power in the second order. [150×10^3]

9. Calculate the minimum number of lines per cm in a $\frac{1}{2}$ inch width grating in order that the D_1 and D_2 lines of sodium is resolved in the first order. [775]

10. A grating having a width 2″ is ruled with 15,000 lines per inch. Find the smallest wavelength separation that can be resolved in the second order at a mean wavelength 6000 Å. [0.10 A]

11. A plane transmission grating has 40,000 lines in all with grating element 12.5×10^{-8} cm. Find the maximum resolving power for which it can be used in the range $\lambda = 5000$ A. [80,000]

12. What must be the thickness of the base if a flint glass prism which would resolve, the D-lines of sodium? Given : $d\mu/d\lambda = 952\ \text{cm}^{-1}$. [1.03 cm]

13. The objective of the telescope at St. Palomer is 5.08 m. Calculate the least distance between two objects on the moon's surface that can be just resolved by it. Given : mean wavelength of moonlight = 5890 A; earth-moon distance $= 3.84 \times 10^5$ km. [54.3 m]

14. A prism spectrometer uses a prism of base 5 cm and is of material having $d\mu/d\lambda = 200$ in the range $\lambda = 5000$ A. Calculate the smallest difference of wavelength in the range resolvable by the spectrometer. [5 A]

15. Consider two gratings, one of width 2 cm has 2000 lines and the other of width 1 cm has 1000 lines. Compare their angular dispersion and resolving powers in the second order. [Dispersive power same; $(\text{r.p})_1 = 2(\text{r.p})_2$]

16. If sodium light (589.0 nm and 589.6 nm) is incident normally on a grating having 100 lines/mm with width of ruling 2 cm, calculate in the first order (a) chromatic dispersion and (b) resolvng power. Are the D-lines resolved in the first order?

(Calcutta Hons.) [0.2′, 200, Yes]

Chapter 12

Polarisation of Light

12.1 Introduction

The phenomena of *interference* and *diffraction* of light proved beyond doubt that light is a form of *wave motion*, but unfortunately could not decide if the wave is longitudinal or transverse in nature. This was established only after the discovery of the phenomenon of **polarisation** and here lies its importance. Polarisation required for its explanation that *light must be a transverse wave.* Electromagnetic theory of light (Chapter 15) considers *light* as a *transverse electromagnetic wave* consisting of vibrating electric and magnetic vectors at right angles to each other and also to the direction of propagation.

The *electric vector* is identified as the **light vector**. In ordinary light beam with thousands of waves, the light vectors of component waves will be in all possible directions on a plane, say $ABCD$ at right angles to that of propagation say, XY (Fig. 12.1). This is due to the random orientations of excited atoms or molecules in the source. *Such a light beam with electric vectors arranged symmetrically about the direction of propagation is known as* **natural** or **unpolarised light**.

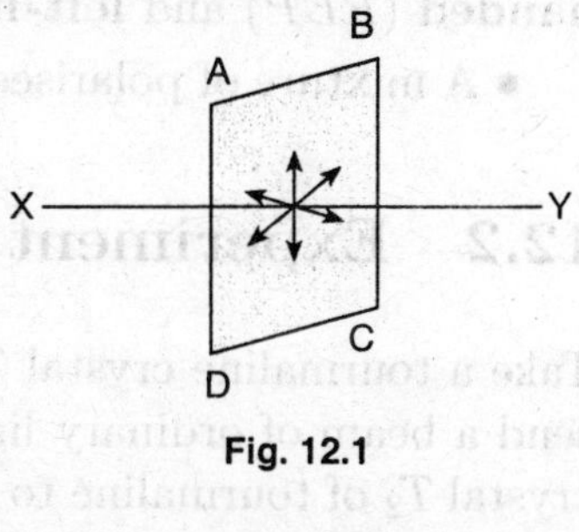

Fig. 12.1

Now, all these vibrations may be resolved into two rectangular components of *equal amplitude* but with rapidly and randomly *varying relative phase difference.* If by some means, one of these rectangular vibrations are cut off, we shall have vibrations of all the component waves confined in one definite direction. Such a light beam is said to be **plane polarised** or **linearly polarised**. The representations of unpolarised and plane polarised light are shown in Fig. 12.2. Unpolarised light consisting of vibrations in all directions with equal probability, in a plane normal to the direction of propagation, is usually represented by a *star of dots* and *arrows*, as shown in Fig. 12.2a.

In polarised light, when the *vibrations* are *in the plane of* (i.e., parallel to) *the paper*, it is represented by a straight *arrow* as shown in Fig. 12.2b. However, when the *vibrations* lie *in a direction perpendicular to the plane of the paper*, it is represented by a *dot* as shown in Fig. 12.2c.

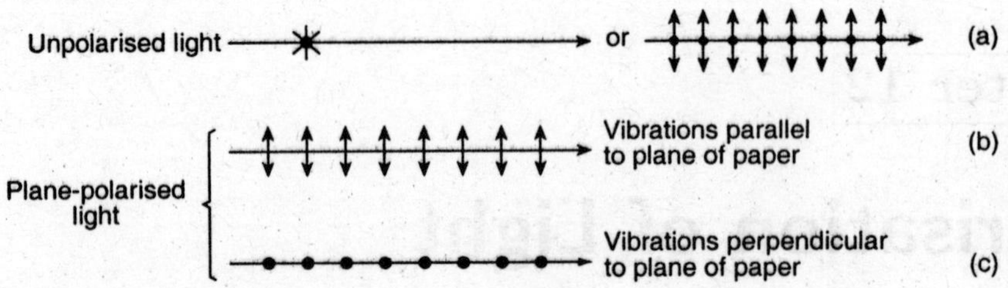

Fig. 12.2

• By superposition of two plane polarised waves, the resultant light vector, under appropriate conditions, may *rotate* in a plane perpendicular to the direction of propagation. If the magnitude of the resultant light vector remains constant, its tip appears to trace *a circle* at a fixed space. Such a light is called **circularly polarised** (*CP*). If the rotation, when viewed *against* the incoming light, is *clockwise* then the light is called **right circularly polarised** (*RCP*); if however the rotation is *anti-clockwise* it is said to be **left circularly polarised** (*LCP*).

In case the magnitude of the resultant light vector varies periodically between a maximum and a minimum, as it rotates, its tip appears to trace *an elliptic path.* Such a light is called **elliptically polarised** (*EP*) and like *CP*, this can also be **right-handed** (*REP*) and **left-handed** (*LEP*) under the same proviso.

• A mixture of polarised and unpolarised light is said to be *partially polarised light.*

12.2 Experiment to show transverse nature of light waves

Take a tourmaline crystal T_1 (Fig. 12.3). cut parallel to the crystallographic axis and send a beam of ordinary light to pass normally through it. Now, take another similar crystal T_2 of tourmaline to place it parallel to the first. It is seen that light received by

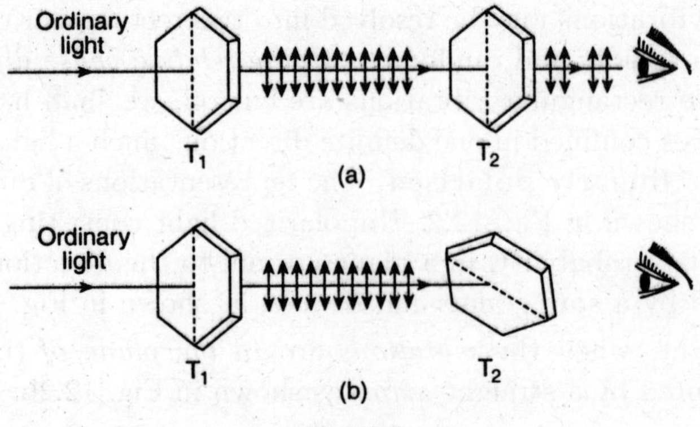

Fig. 12.3

T_2 is *almost completely transmitted by it.* In this position of the crystal-pair (parallel position) the intensity is thus *maximum.*

Now, rotate T_2 about an axis parallel to incident light; the transmitted beam gradually *decreases* in intensity and becomes *zero* when the *axis of T_1 and T_2 are at right angles* to each other. *No light is transmitted through the second crystal T_2.* On further rotation of T_2, light again starts transmission and the intensity becomes maximum when both T_1 and T_2 are again *parallel.*

This simple experiment demonstrates the *important* fact that on its passage through the first crystal, light acquires a *property* so as to prevent its passage through the second at a definite orientation. When light passes through T_1, it absorbs vibrations in all directions except those || to the crystallographic axis. The unabsorbed vibrations are thus only transmitted and the transmitted light from T_1 has vibrations in *one direction* only (i.e., **polarised**). The experiment also demonstrates that *light waves are transverse* in character.

12.3 Plane of vibration and plane of polarisation

The plane in which the vibrations of polarised light are confined is known as the **plane of vibration**—the plane $ABCD$ that *contains the directions of vibration and also the direction of propagation* as shown in Fig. 12.4. This plane contains the electric vector $\vec{E}$ and the propagation vector $\vec{k}$.

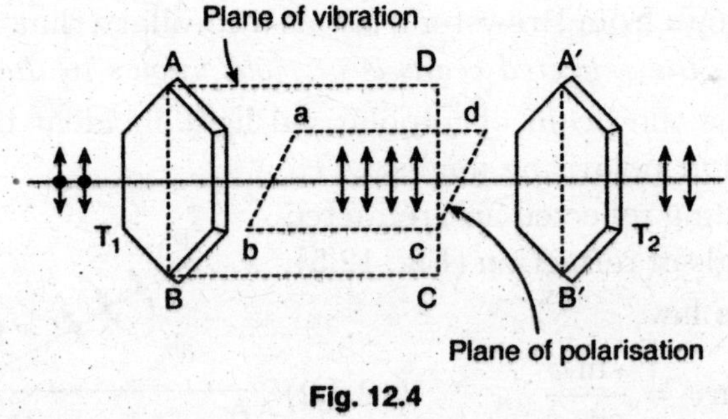

Fig. 12.4

The plane perpendicular to the plane of vibration and hence containing no vibration is known as the **plane of polarisation**. This plane *abcd* passes *through the direction of propagation and is at right angles to the plane of vibration.*

• We shall now discuss some of the methods of producing plane polarised light. They usually fall into the following *three* broad classes.

1. By simple **reflection**
2. By **double refraction**
3. By **dichroism** or **selective absorption**.

Any method of producing polarised light consists essentially in dividing the unpolarised light into two components polarised at right angles to each other and then eliminating one of them so that the component that is left is a plane polarised beam.

The device that divides the unpolarised light into two components and eliminates one of the two is called **a polariser**.

12.4 Polarisation by reflection : Brewster's law

Malus observed that when unpolarised light is incident on the surface of a transparent material, both the reflected and the refracted beams are partially plane polarised.

For a particular angle of incidence, known as **the polarising angle**, *the reflected beam is completely plane polarised, the plane of vibration being at right angles to the plane of incidence.*

Brewster's law—Brewster found that there exists a simple relation between the polarising angle ϕ and the refractive index μ of the reflector relative to the surrounding medium. This relation is given by

$$\boxed{\mu = \tan\phi} \tag{12.4.1}$$

and is known as **Brewster's law.**

The polarising angle of glass relative to air is 57°, for taking $\mu_{\text{glass}} = 1.5$, Brewster's law gives $\phi = \tan^{-1}(1.5) \sim 57°$.

Corollary—It follows from Brewster's law as a corollary that *when light is incident at the polarising angle, the reflected beam is at right angles to the refracted beam.*

Proof. Let PO be the beam of unpolarised light incident on the surface AB of a transparent material at polarising angle ϕ, OR and OT the corresponding reflected and refracted beam. Let r be the angle of refraction (Fig. 12.5).

$\therefore$ From Brewster's law,

$$\mu = \tan\phi = \frac{\sin\phi}{\cos\phi} \tag{12.4.2}$$

From Snell's law, again, we have

$$\mu = \frac{\sin\phi}{\sin r} \tag{12.4.3}$$

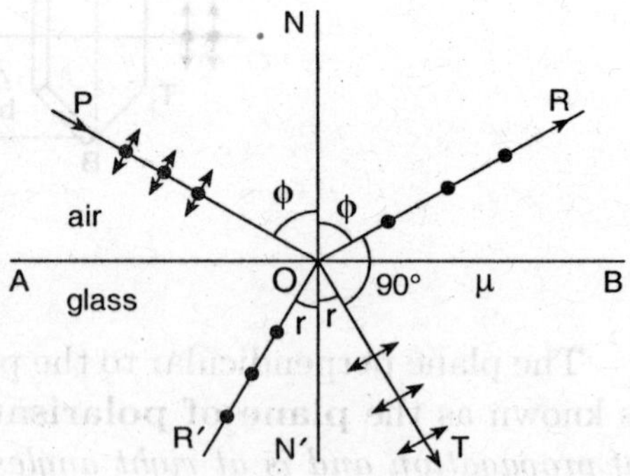

Fig. 12.5

From (12.4.2) and (12.4.3) therefore

$$\frac{\sin\phi}{\cos\phi} = \frac{\sin\phi}{\sin r}$$

$$\Rightarrow \quad \cos\phi = \sin r = \cos(90° - r)$$

$$\therefore \quad \phi = 90° - r$$

$$\Rightarrow \quad \boxed{r + \phi = 90°} \tag{12.4.4}$$

This proves that *the reflected and the refracted beams are at right angles* (Fig. 12.5) *to each other.*

• The converse can also be readily proved, that is, assuming that when light is incident at polarising angle ϕ on the surface of a transparent material, the reflected beam is at right angles to the reflected one, *Brewster's law* can be obtained. This is left as an exercise to the reader.

• Brewster's law is obeyed even when light is **reflected at the surface of a rarer medium.** If light is incident on glass plate at polarising angle r along TO, the refracted ray will be along OP making ϕ with the normal and the reflected ray along OR'.

$\therefore$ Index of medium relative to air, μ, is given by : $\sin r/\sin\phi = 1/\mu$.

$$\therefore \tan r = \frac{\sin r}{\cos r} = \frac{\sin r}{\sin\phi} = \frac{1}{\mu}$$

The ray OR' will be plane polarised in the plane of incidence.

12.5 Production and detection of polarised light by reflection

A simple instrument that can both *produce* and *detect* plane polarised light by reflection is **Biot's polariscope.**

It consists essentially of two polished glass plates P_1, P_2 (Fig. 12.6) blackened on their back and mounted at the two ends of a cylindrical tube. P_1 and P_2 can be rotated about the axis of the tube and also about a diameter of the cross-section of the tube—the amount of rotation can be read from an attached circular scale.

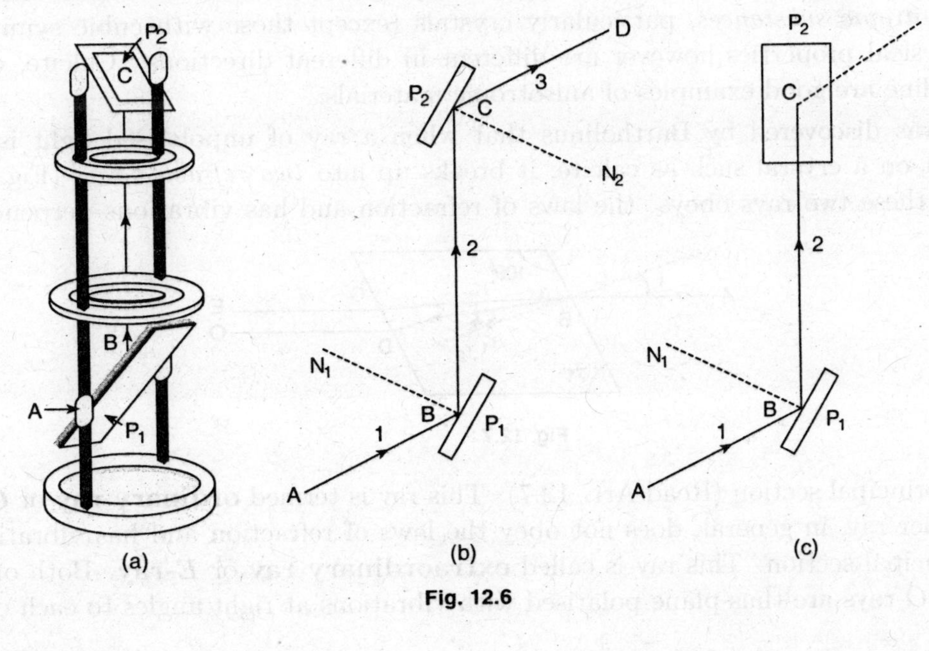

Fig. 12.6

When the planes of P_1 and P_2 are parallel, the light is reflected from P_1 and P_2 as shown in Fig. 12.6b. If the incident ray 1 is incident at *polarising angle* (57°), the reflected ray 2 is *completely plane polarised* with vibrations normal to the plane of incidence of P_1. But P_2 is parallel to P_1. So their planes of incidence are also parallel and the ray 2 incident on P_2 has its vibrations perpendicular to the plane of incidence of P_2. And since only the vibrations perpendicular to the plane of incidence are reflected, the intensity of the ray 3 reflected from P_2 will be *maximum.*

When however the plane P_2 is rotated about the ray 2 to have an orientation normal to P_1, the vibrations in ray 2 become $\parallel$ to the plane of incidence of P_2. For reason already stated in the previous paragraph, *no light* is reflected from P_2 in this orientation (Fig. 12.6c). So, by rotating P_2, one would find that the intensity of ray 3 varies with *zero as minimum.* This only implies that the ray 2 is *completely plane polarised.*

In case the light is incident on P_1 at an angle *other than the polarising angle,* the ray 2 will only be *partially polarised.* So, in this situation, the intensity of ray 3, on rotating P_2, will very between a *maximum* and a *non-zero minimum.*

Thus to produce plane polarised light, the inclination of P_1 relative to the incident ray 1 is to be adjusted so that on rotating P_2, the intensity of ray 3 varies and is *zero* when $P_2 \perp P_1$.

- Plate P_1 that makes the ray 2 polarised is called the **polariser** and the plate P_2, which demonstrates that the ray 2 is polarised, is called the **analyser** or detector.

12.6 Double refraction

In an *isotropic refracting medium,* the physical properties are the same in all directions. In *anisotropic substances,* particularly crystals (except those with cubic symmetry), the physical properties however are different in different directions. Calcite, quartz, tourmaline are good examples of anisotropic materials.

It was discovered by Bartholinus that when a ray of unpolarised light is made incident on a crystal such as calcite, it breaks up into *two refracted rays* (Fig. 12.7). One of these two rays obeys the laws of refraction and has vibrations perpendicular

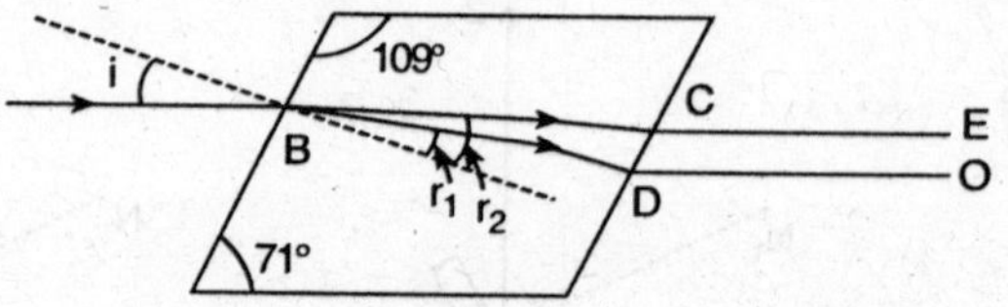

Fig. 12.7

to the principal section (Read Art. 12.7). This ray is termed **ordinary ray** or ***O*-ray.** The other ray, in general, does not obey the laws of refraction and has vibrations in the principal section. This ray is called **extraordinary ray** or ***E*-ray**. Both of these *E*- and *O*-rays are thus plane polarised with vibrations at right angles to each other.

This optical phenomenon of splitting of a single incident ray into two refracted rays is called **double refraction** or **birefringence** and the crystals that exhibit the phenomenon are known as *doubly refracting crystals* or *birefringent.*

Double *refraction can be easily* demonstrated by the following simple experiment. On a white sheet of paper, put a dot mark with ink and place a calcite crystal over it. Look through the top face to see two images side by side. Rotate the crystal slowly along the direction shown (Fig. 12.8) and observe that one of the images remains stationary while the other rotates in the direction of rotation of the crystal. The stationary image is the *ordinary image*, the rotating one the *extraordinary image.*

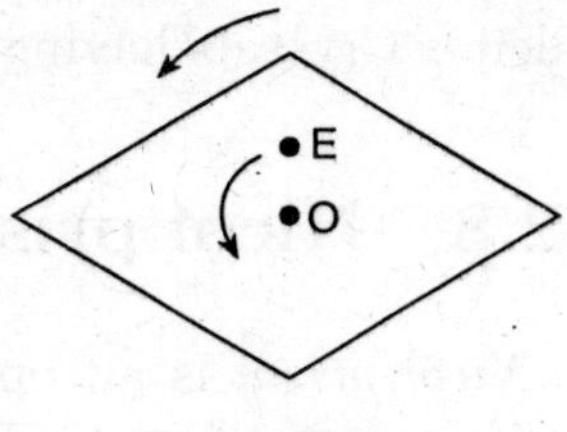

Fig. 12.8

12.7 Geometry of calcite : Optic axis and principal section

Calcite is a *colourless, transparent* crystal and is chemically hydrated *calcium carbonate*, $CaCO_3$. It is also known as *iceland spar.*

Like other crystals, calcite is also bounded by plane faces inclined to one another by fixed angles. Crystallogaphically, it belongs to rhombohedral class of hexagonal system. The *six faces* of the rhomb are *parallelograms*, each with angles 101°55′ and 78°5′ (Fig. 12.9). There are two diagonally opposite corners *A* and C_1 where the *three obtuse angles of* 101°55′ *meet.* These are called the *blunt corners.* At *six other* corners, there are one obtuse and two acute angles.

Optic axis—*A line passing through any one of the blunt corners (A or C_1) and equally inclined to the three faces that meet at the corners gives the direction of the* **optic axis** *of the crystal.*

Optic axis of a crystal is *a direction and not any particular line* so that any line parallel to the one described above is also the optic axis. Thus an optic axis can be drawn through every point in the crystal.

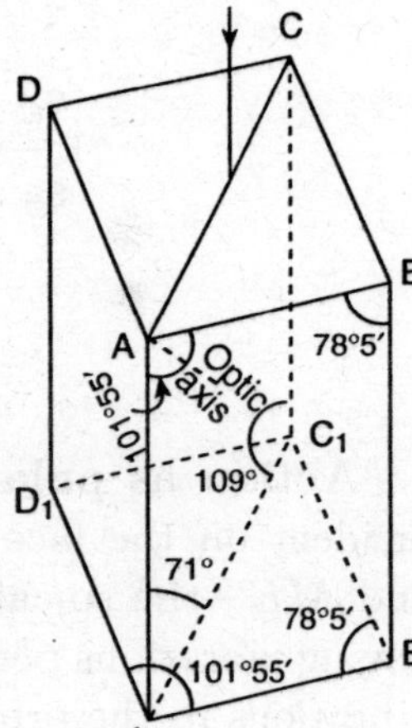

Fig. 12.9

The **characteristics of the optic axis** are :

1. *Optic axis is a direction and not any particular line.*
2. *The crystal is symmetrical about the optic axis.*
3. *The velocities of O-ray and E-ray along the optic axis are equal.*

Principal section (plane)—Principal section of a crystal is the *section or plane that passes through the optic axis and perpendicular to the two opposite refracting faces.*

Since there are *infinite* number of lines ∥ to direction of optic axis, there are also an *infinite* number of *principal sections.* One such is ACC_1A_1 (Fig. 12.9). A principal section cuts the crystal surfaces in a parallelogram with angles 71° and 109°.

• Crystals having **one optic axis** are called **uniaxial crystals** e.g., quartz and calcite. Crystals having **two optic axes** are called **biaxial crystals**, e.g., mica.

12.8 Nicol prism

A *Nicol prism* is an optical device, made out of a calcite crystal whose property of birefringence is used in Nicol prism for both production and analysis of polarised light.

Construction—A calcite crystal $ABCD$ is shaped, from a natural crystal (Fig. 12.10), having a *length to breadth* ratio as 3 : 1. Next, the end-faces AB and CD are properly ground to make the angles in the principal section 68° and 112° in place of 71° and 109° respectively. The crystal is then cut into two halves along the plane $A'D'$, *perpendicular to both the principal section and the end-faces.* The two surfaces are polished *optically flat* and then cemented together by canada balsam—a transparent liquid with index $\mu = 1.55$ for sodium light. The crystal is finally enclosed in a tube blackened inside and the construction of Nicol prism is complete.

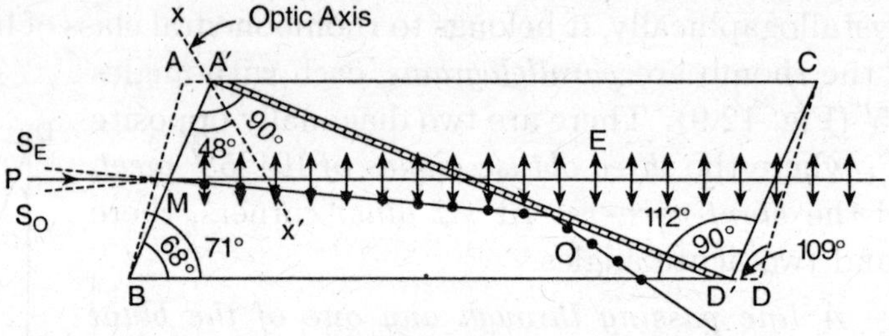

Fig. 12.10

Action as polariser—Let a ray PM of unpolarised light, almost ∥ to BD', be incident on the face $A'B$. Inside the crystal, it splits into two refracted rays, MO and ME—the so-called *O*-ray and *E*-ray. Both the rays are *palne polarised*—*O*-ray having vibrations perpendicular to the principal section of the crystal and *E*-ray having vibrations in the principal section.

Now, the index μ_c of canada balsam lies between μ_e and μ_0, i.e., $\mu_e < \mu_c < \mu_0$. So, when the *O*-ray reaches the balsam layer, it passes from a *denser* to a *rarer* medium. Since the length of the crystal is large, the *O*-ray is usually incident at the calcite-balsam surface at an angle greater than its *critical* (69°) and thus gets *totally reflected* to be finally absorbed by the tube enclosing the crystal.

The *E*-ray however is transmitted by the calcite-balsam surface to emerge from the Nicol as plane polarised light with vibrations parallel to the principal section.

Nicol as analyser—When an unpolarised ray is incident on a Nicol P (Fig. 12.11a), the emergent ray, as we have just seen, is plane polarised with vibrations in the principal section of P. If this emergent ray falls on a second Nicol A whose principal section is parallel to that of P, the vibrations will be in the principal section of A and the ray, behaving as E-ray in it, will be transmitted *completely.* The intensity of the emergent light will be a *maximum.*

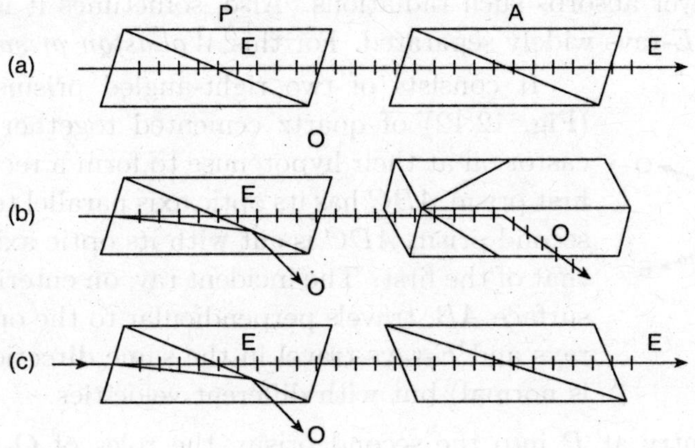

Fig. 12.11

If now A is rotated to make its principal section perpendicular to that of P (Fig. 12.11b), the vibrations of the incident polarised light will be perpendicular to its principal section and the ray will behave as O-ray inside A. So it gets lost by *total reflection* at calcite-balsam surface. *No light* will emerge from A and the Nicols P and A are said to be **crossed.**

If A is further rotated till its principal section is again parallel to that of P (Fig. 12.11c), the intensity of the emergent light will again be a *maximum.* So, for analysing plane polarised light the procedure is as follows : if the given light on viewing through *a rotating Nicol shows intensity variation with zero minimum*, the given light is *plane polarised.*

• Nicol P is called the **polariser** and the Nicol A is called the **analyser**.

Uses—The Nicol prism finds applications both as a *polariser* and as an *analyser.*

• **Limitations**—A Nicol prism works only when the incident beam is either *slightly convergent or slightly divergent* and *cannot* be used in highly convergent or divergent beams. Let us explain why.

If the incident ray makes an angle *much smaller* than $\angle PMB$ (Fig. 12.10) with the face $A'B$, the O-ray strikes the calcite-balsam surface at an angle smaller than the critical (69°). O-ray will thus be transmitted and the emergent light will *not* be plane polarised.

Conversely, if the incident ray makes an angle *much larger* then $\angle PMB$, the E-ray will become more and more parallel to the optic axis and its index increases to become higher than that of balsam. E-ray will also undergo *total reflection* from calcite-balsam surface and there will be *no emergent ray.* So to get plane polarised light, the incident beam should *not* be *too divergent.*

• With chosen dimensions, the semi-vertical angle $S_E PS_O$ of the incident light-cone must not be > 14° and it is to make the ray PM lie within 14° that the end faces are grounded.

Wollaston prism—Nicol prism cannot be used with ultra-violet light since the canada balsam layer absorbs such radiations. Also, sometimes it is desirable to have both O-rays and E-rays widely separated. For this *Wollaston prism* is used.

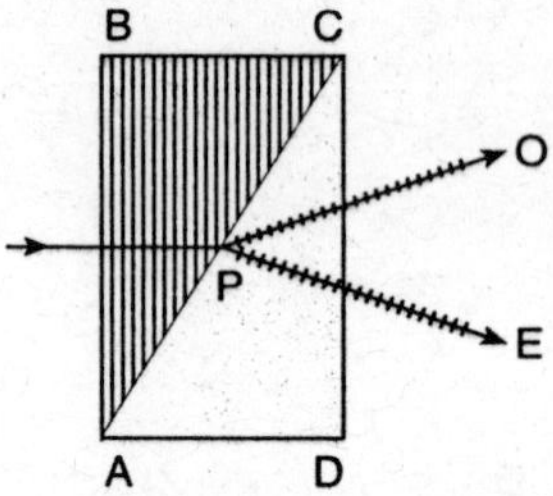

Fig. 12.12

It consists of two right-angled prisms ABC and ADC (Fig. 12.12) of quartz cemented together with glycerine or castor oil at their hypotenuse to form a rectangular slab. The first prism ABC has its optic axis parallel to the face AB. The second prism ADC is cut with its optic axis perpendicular to that of the first. The incident ray, on entering normally to the surface AB, travels perpendicular to the optic axis so that O-rays and E-rays travel in the same direction (since incidence is normal) but with different velocities.

Just at the entry at P into the second prism, the roles of O-ray and E-ray are interchanged—the O-ray becomes E-ray and conversely. So both the rays in prism ACD are deviated resulting in greater and greater separation between them.

12.9 Malus law

Malus obtained a relation that shows how the intensity of light transmitted by the analyser varies with the angle made by its plane of transmission with that of the polariser. This relation is called **Malus law**.

Statement—Malus law states that *if a completely plane polarised light is incident on an analyser, the intensity of the emergent light varies as the square of the cosine of the angle between the planes of transmission of the polariser and the analyser.*

Proof. Let a be the amplitude of light transmitted by polariser and θ the angle between the planes of transmission of the polariser and the analyser (Fig. 12.13). This plane polarised light is incident on the analyser.

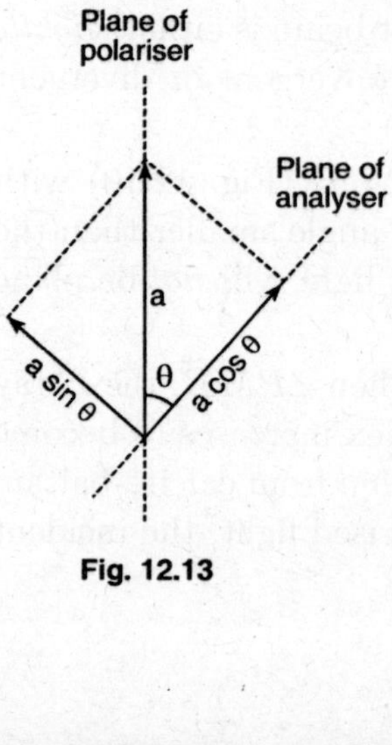

Fig. 12.13

Now, a may be resolved into two components : $a\cos\theta$ and $a\sin\theta$, respectively *along* and *perpendicular* to the plane of transmission of the analyser. Of these, the component $a\sin\theta$ is eliminated and only the parallel component $a\cos\theta$ is freely transmitted by the analyser.

∴ Intensity of light emerging from the analyser is

$$I_\theta = a^2\cos^2\theta = I\cos^2\theta \qquad (12.9.1)$$

where $I = a^2$ is the intensity of polarised light incident on analyser. The relation (12.9.1) is the **Malas law**.

Special cases—If the polariser and the analyser are *parallel* to each other, then $\theta = 0°$ or $180°$. So, $I_\theta = I$.

If however the polariser and the analyser are *perpendicular* to each other, then $\theta = 90° \Rightarrow I_\theta = 0$.

Experimental verification—A beam of unpolarised light is passed through a pair of polaroids—a polariser and an analyser. The emergent beam is next passed into a photocell that measures the light intensity. With light beam as axis, the analyser is rotated and the *orientation* corresponding to *maximum* intensity of emergent beam is marked as $\theta = 0$. Next the intensity I_θ for various values of θ, relative to the above orientation, are measured *photo-electrically*. A plot of I_θ against $\cos^2\theta$ is made. The graph thus obtained is found to be a *straight line* which verifies Malus law.

Importance—The importance of Malus law lies in explaining a number of phenomena. We discuss only two of them here.

1. Even an ideal polariser can at best have a 50% transmission—Let I_0 be the intensity of unpolarised light incident on a polarising sheet and θ the angle between the polarising direction of sheet and that of vibration of light vector of incident light. So, by Malus law, the intensity of transmitted polarised light is

$$I = I_0 \cos^2\theta$$

The incident light being unpolarised, its light vector has random orientations of vibration in a plane at right angles to the direction of propagation. So to compute I, the average value of $\cos^2\theta$ in the above equation is to be taken

$$\therefore \quad I = I_0\,\overline{\cos^2\theta} = \tfrac{1}{2}I_0 \qquad (12.9.2)$$

The polarising sheet thus reduces the intensity of unpolarised light by 50%, implying that *even an ideal polariser can at best have a transmission of* 50%.

2. Unpolarised light, *if incident on two crossed polarising sheets, is* **not transmitted**. *On insertion of a third sheet however between them, light may be transmitted.* The phenomenon can be easily explained by Malus law.

Let A and B be two crossed polarising sheets. Then the intensity of the plane polarised light transmitted by A is

$$I_A = \tfrac{1}{2}I_0 \quad \text{(by \textbf{1} above)}$$

where I_0 is the intensity of unpolarised light incident on A,

When the third sheet C is inserted between A and B (but *not parallel* to either), the intensity of light transmitted by C is

$$I_C = I_A\cos^2\theta = \tfrac{1}{2}I_0\cos^2\theta$$

where θ is the angle between A and C.

Since A and B are *crossed*, the angle between C and B is $(90° - \theta)$. So the intensity of light transmitted by B will be

$$\begin{aligned} I_B &= I_C \cos^2(90° - \theta) \\ &= I_C \sin^2\theta \\ &= \tfrac{1}{2} I_0 \cos^2\theta \sin^2\theta \end{aligned} \tag{12.9.3}$$

which obviously is **non-zero.**

Special cases—From above, we see that when $\theta = 0°$, $I_B = 0$, as expected; if $\theta = 90°$, $I_B = 0$.

Similarly, if $\theta = 45°$, we have from (12.9.3)

$$I_B = \frac{1}{2} I_0 \left(\frac{1}{2}\right)\left(\frac{1}{2}\right) = \tfrac{1}{8} I_0$$

12.10 Dichroism : Polaroids

Dichroism—Some doubly refracting crystals and minerals exhibit the property of absorbing one of the doubly refracting rays strongly and allowing the other to pass through with little loss. This phenomenon of *selective absorption* of light rays is called **dichroism** and the crystals that show this property are termed **dichroic crystals**. Tourmaline is a good example of such a crystal and absorbs ordinary or O-ray much more strongly then the extraordinary or E-ray. Thus if an unpolarised light is passed through a dichroic crystal like tourmaline of proper thickness, one of the components (O-ray) is totally absorbed and the other (E-ray) transmitted in appreciable amount (Fig. 12.14). The emergent beam will thus be *linearly polarised.* A thin ($\sim$ 1 mm) crystal of tourmaline with its faces cut *parallel to optic axis* may be used as a polariser. However, its use as a polariser is *limited* as the polarised light is coloured owing to unequal absorption of light of various wavelengths.

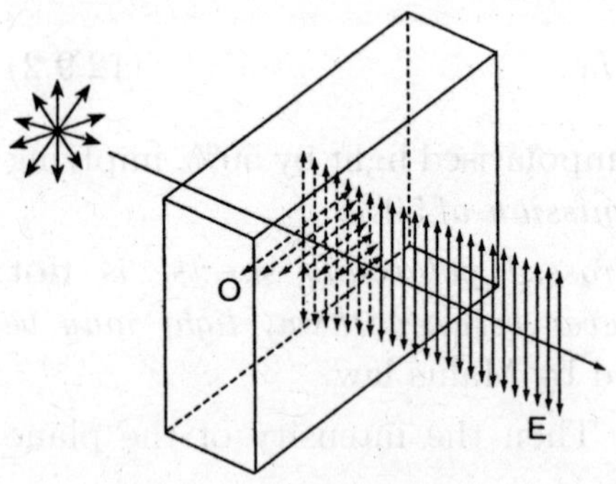

Fig. 12.14

Polaroid—A *polaroid* is a thin transparent film that can produce plane polarised light on transmission through it. The film is a thin sheet of *nitro-cellulose* packed with dichroic ultra-microscopic crystals (*herapathite*) with their *optic axis all parallel* so that they act together as a *single crystal* of large dimensions. Such crystals are *highly dichroic* and completely absorbs one of the beams within a thickness of 0.13 mm. It has largely replaced Nicol prism for producing and analysing polarised light in the lab.

Long-sized polaroids are now manufactured from polyvinyl alcohol film by *heating* and *stretching* it 3 to 8 times its original length. Stretching orients the complex molecules with long axis along the direction of stress. This makes them doubly refracting. The film is next impregnated with iodine when it becomes *dichroic*. These are called ***H*-polaroids**. These are colourless and transmits more light than herapathite polaroid.

Land and Rogers discovered another kind of polaroids called ***K*-polaroids**. When the stretched polyvinyl alcohol film is heated with a catelyst like HCl acid (instead of iodine impregnation), it slightly darkens and becomes strongly dichroic and very stable.

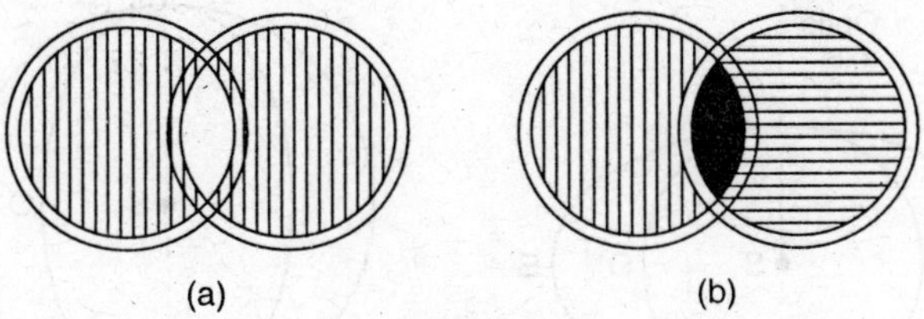

Fig. 12.15

When two pieces of polaroids are *parallel*, the light transmitted by the first is also transmitted by the second. If *crossed*, there is a perfect extinction of light (Fig. 12.15).

Applications—Polaroids are cheaper than Nicols and are used in the lab. for the production and analysis of plane-polarised light.

They are also widely used in *sun-glasses* to cut off glare of light from horizontal surfaces. *K*-polaroids are extensively used in *head-lights* and *wind screens* of cars to cut-off the dazzling light of an approaching car from opposite direction. They are also employed in *windows of trains and planes* to control intensity of light entering into them and *in viewing stereoscopic motion pictures.*

12.11 Huygens' theory of double refraction

According to Huygens' wave theory, each point on a wavefront becomes the source of new disturbance and send secondary wavelets. The envelope of all the secondary wavelets at any instant gives the new wavefront which is the locus of particles vibrating in the same phase. *Huygens' theory in its original form cannot explain double refraction in uniaxial crystals.* He therefore extended his theory of secondary wavelets to explain this phenomenon. We give below his extended theory in a *summary fashion*, emphasising only on its **salient features**.

1. When a wavefront is incident on a doubly refracting crystal, every point on the crystal becomes the origin of *two wavefronts*, *O* and *E*—one the *ordinary* and the other *extraordinary*—to account for the two types of rays.

2. Ordinary wavefront corresponds to ordinary rays which obey the laws of refraction and have the *same velocity in all directions.* The *ordinary wavefront is* thus **spherical**.

3. Extraordinary wavefront corresponds to extraordinary rays which do not obey the laws of refraction and have *different velocities in different directions.* The extraordinary *wavefront* is thus an *ellipsoid of revolution*, with the optic axis as the axis of revolution.

4. The sphere and the ellipsoid touch each other at points that lie on the optic axis of the crystal since the *velocity of the ordinary and the extraordinary ray is the same along the optic axis.* So the uniaxial crystals *do not* show any *double refraction* along the optic axis.

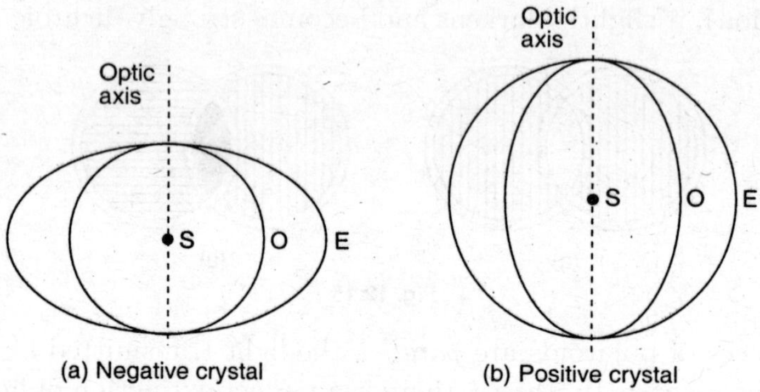

(a) **Fig. 12.16** (b)

5. In crystals e.g., calcite and tourmaline, called **negative crystals**, the ellipsoid lies outside the sphere, S being the source (Fig. 12.16a) implying that in such crystals the *extraordinary wavefront travels faster* than the ordinary wavefront, except along the optic axis.

In some other crystals, e.g., quartz and ice, called the **positive crystals**, the sphere is outside the ellipsoid (Fig. 12.16b) implying that in such crystals the *ordinary wavefront travels faster* than the extraordinary wavefront, except however along the optic axis.

12.12 Huygens' construction of wave surfaces (uniaxial crystal)

For construction of wave surfaces on the plane of incidence in uniaxial crystals we discuss *three* distinctly different cases.

Case 1. Optic axis inclined to upper face but lying in the plane of incidence—Let a plane wavefront AB be incident obliquely on the upper face of a *uniaxial negative crystal*, say calcite, cut such that the optic axis lies in the plane of incidence along AX, but inclined to the upper face at some angle (Fig. 12.17).

The point A of the wavefront AB, where it first strikes the crystal surface becomes the centre of both ordinary and extraordinary wavelets O and E. By the time the disturbance from B reaches C, the ordinary wavefront travels a distance AD and the extraordinary wavefront travels over AF.

$$\therefore \quad \frac{BC}{c} = \frac{AD}{v_0} = \frac{AF}{v_e}$$

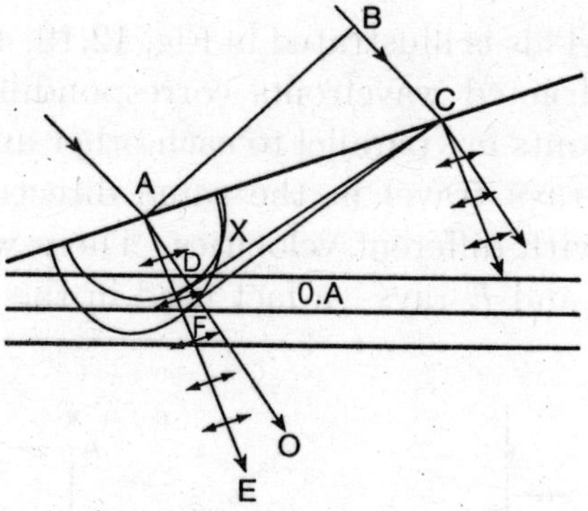

Fig. 12.17

where c = light velocity in air, v_0 = velocity of O-ray in the crystal and v_e = velocity of extraordinary light in the crystal.

$$\therefore \quad AD = BC(v_0/c) = BC/\mu_0;$$

$$AF = BC(v_e/c) = BC/\mu_e$$

μ's being the indices for the two types of rays.

With A as centre and BC/μ_0 as radius, draw a circle cutting the optic axis at X. The circle gives the position of O-wave surface. In negative crystals, $\mu_e < \mu_0$, and $(\mu_e)_{\max} = \mu_0$ along the optic axis. AX is thus the semi-minor axis of ellipse. The semi-major axis is given by $BC/(\mu_e)_{\min}$. If an ellipse is drawn with given minor and major axes touching the circle at X, it gives the position of E-wave surface.

Tangents CD and CF drawn from C on ordinary and extraordinary wave surfaces represent the ordinary and extraordinary refracted wavefronts. As they advance in the crystal with different velocities, *double refraction* occurs.

Case 2. Optic axis parallel to upper face but lying on the plane of incidence—Here two different situations arise : (i) *where the incidence is oblique* and (ii) *when the incidence is normal.*

Oblique incidence—As the optic axis is parallel to the crystal-face and lies in the plane of incidence, we can follow similar method of construction as in **Case 1.**, and get the positions of O-wave surface and the E-wave surface (Fig. 12.18). The circle and the ellipse will touch at X along AX, the optic axis, CD and CF are the ordinary and extraordinary refracted wavefronts corresponding to the incident wavefront AB. Plainly, the O-ray and the E-ray travel with different velocities in different directions giving rise to *double refraction.*

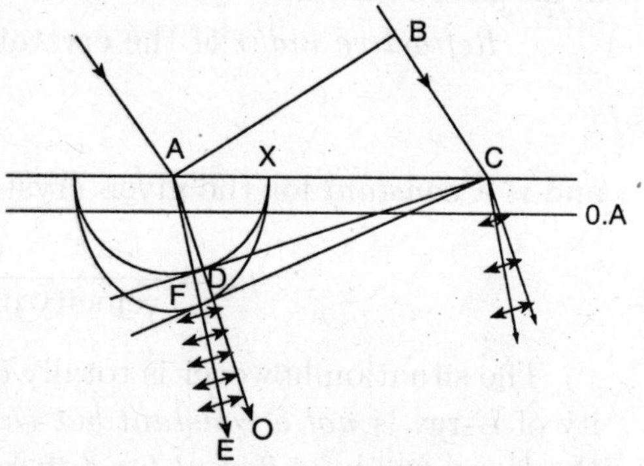

Fig. 12.18

Normal incidence—This is illustrated in Fig. 12.19. CD and FG are the ordinary and the extraordinary refracted wavefronts corresponding to the incident wavefront AB. The refracted wavefronts are parallel to each other and also to the crystal surface. Although the O- and E-rays travel in the same direction yet there will be *double refraction* as they travel with different velocities. There will thus be a path difference between the emergent O- and E-rays—a fact used in the construction of $\lambda/4$ and $\lambda/2$ plates.

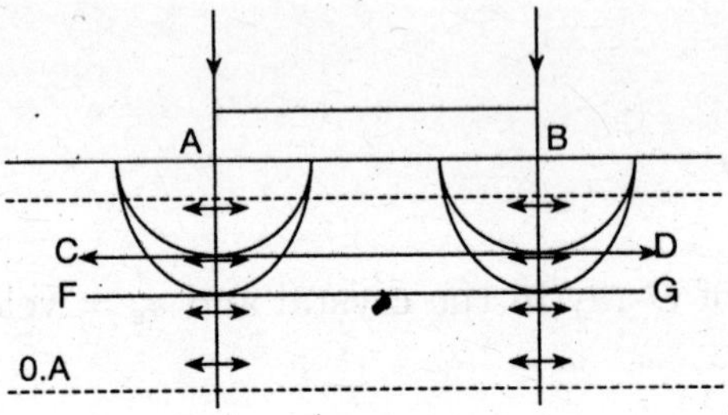

Fig. 12.19

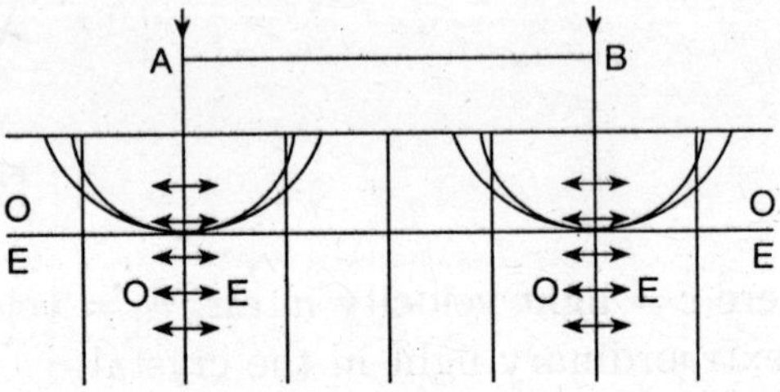

Fig. 12.20

Case 3. Normal incidence, optic axis perpendicular to the upper face—In this case, AB is the incident wavefront. Since the incidence is normal, there is *no double refraction* as O-ray and E-ray travel with the same velocity along the optic axis, the direction of travel. This is illustrated in Fig. 12.20. The two rays (O and E) coincide in the crystal—OO or EE.

12.13 Refractive indices of a crystal

The refractive index μ of a substance is the ratio of the velocity of light c in vacuum to the wave velocity of light v in that substance.

$$\therefore \quad \mu = c/v$$

In a crystal, the wave velocity v_0 of the ordinary ray is a constant being the same in all directions and for all orientations of the optic axis.

$\therefore$ *Refractive index* of the crystal *for O-ray* is thus given by

$$\mu_0 = c/v_0$$

and is a *constant* for the given crystal.

$$\therefore \quad \mu_0 = \frac{\text{velocity of light in vacuo}}{\text{velocity of } O\text{-ray in crystal in any direction}}$$

The situation however is totally different for the extraordinary ray. The wave velocity of E-ray is *not a constant but varies with directions.* So the index of the crystal for the E-ray will be *different for different directions.* If v_e be the wave velocity of E-ray in the direction *perpendicular to the optic axis,* then the refractive index μ_e of the crystal for E-ray along this direction is

$$\therefore \quad \mu_e = \frac{\text{velocity of light in vacuo}}{\text{velocity of } E\text{-ray in a direction} \perp \text{optic axis}}$$

$$\therefore \quad \mu_e = c/v_e$$

and is known as the **principal refractive index** for E-rays.

Since both O-rays and E-rays move with the same velocity along the optic axis, both will have the same refractive index in this direction.

- For a **negative crystal** (e.g., calcite), the velocity of E-ray perpendicular to the optic axis is *maximum* $\Rightarrow \mu_e < \mu_0$.
- For a **positive crystal** (e.g, quartz), the velocity of E-ray perpendicular to the optic axis is *minimum* $\Rightarrow \mu_e > \mu_0$.

Refractive index for E-rays in any direction—To define the index for E-rays in any direction in a crystal we have to distinguish between *ray-velocity* and *wave-velocity.* Fig. 12.21 shows that the direction OC of E-ray and the normal ON to the E-wavefront TN do not coincide except along and perpendicular to the optic axis. *The rate of increase of OC* (the line joining the origin O of a secondary wave and the point of contact C of the wavefront with that secondary wave) is called *ray-velocity* of the wave. The rate of increase of ON (normal to the wavefront from origin O of secondary wave) gives the *wave-velocity* of the wave. The two always differ but become equal only along and perpendicular to optic axis.

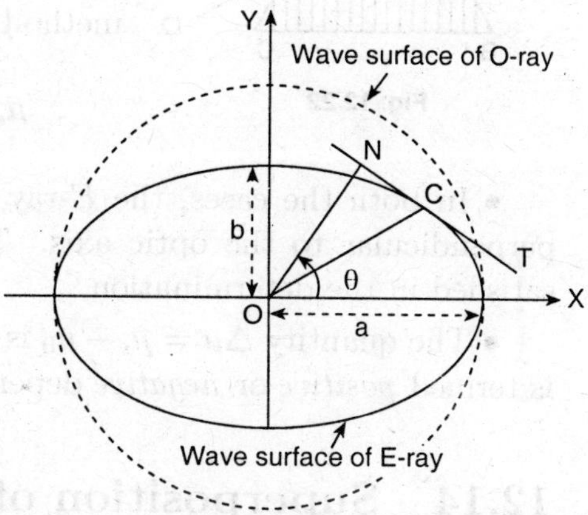

Fig. 12.21

$\therefore$ For any direction making θ with optic axis, the refractive index μ_θ for E-ray is given by

$$\mu_\theta = \frac{\text{velocity of light in vacuum, } c}{\text{wave velocity for } E\text{-ray along this direction}}$$

Now, from coordinate geometry, we have

$$ON^2 = p^2 = a^2 \cos^2\theta + b^2 \sin^2\theta \tag{12.13.1}$$

where θ = angle between the normal ON and the major axis of the ellipse.

But, $p = v_\theta t$, $a = v_0 t$, $b = v_e t$ so that from (12.13.1), we obtain

$$v_\theta^2 = v_0^2 \cos^2\theta + v_e^2 \sin^2\theta \tag{12.13.2}$$

Dividing both sides of (12.13.2) by c^2, we obtain

$$\boxed{\frac{1}{\mu_\theta^2} = \frac{\cos^2\theta}{\mu_0^2} + \frac{\sin^2\theta}{\mu_e^2}} \tag{12.13.3}$$

The importance of this relation lies in the fact that the index μ_θ for E-rays in any direction can be readily calculated from a knowledge of μ_0 and μ_e, the principal index of E-ray.

Determination of μ_0 and μ_e—The uniaxial crystal prism ABC (Fig. 12.22), *cut with optic axis perpendicular to the edge*, is properly mounted on the prism table of the spectrometer such that the double images (O-image and E-image) of the slit are clearly seen by refraction through the prism. Each image is adjusted in succession by rotating the prism to its position of minimum deviation for O-ray and E-ray respectively and are recorded.

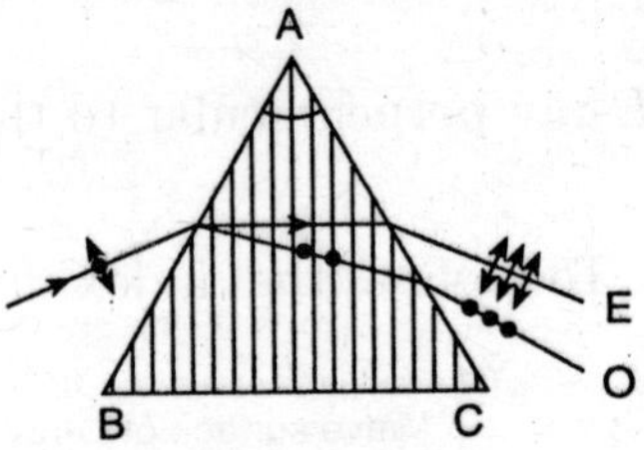

Fig. 12.22

The angle of prism A is also measured by the usual method. Then μ_e and μ_0 are computed from the formula

$$\mu_e = \frac{\sin(A + D_{me})/2}{\sin A/2}; \quad \mu_0 = \frac{\sin(A + D_{mo})/2}{\sin A/2}$$

• In both the cases, the E-ray moves through the prism parallel to the base and perpendicular to the optic axis. The very condition used in the definition of μ_e is satisfied in the determination.

• The quantity $\Delta\mu = \mu_e - \mu_0$ is often called the *birefringence* and a uniaxial crystal is termed *positive* or *negative* depending on the sign of $\Delta\mu$.

12.14 Superposition of two plane polarised waves with vibrations at right angles

Let a plane polarised monochromatic light be incident *normally* on a *calcite crystal, cut with its optic axis parallel to its faces* and A be the amplitude of vibration that makes an angle θ with optic axis so that the vibration is along PA (Fig. 12.23).

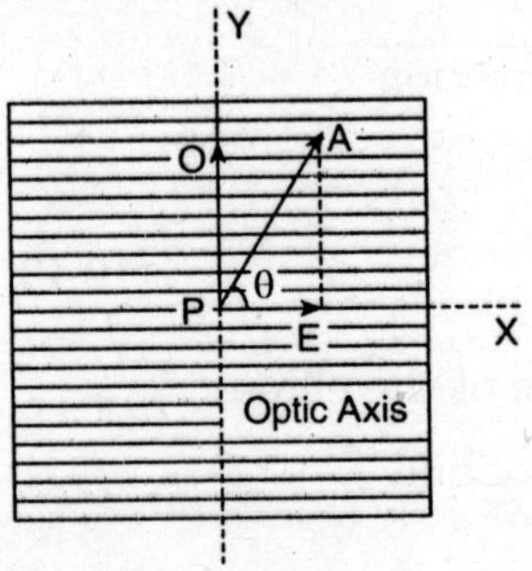

Fig. 12.23

On entering the crystal at P, the light breaks up into (i) O-wave of amplitude $b = A\sin\theta$ with vibrations normal to optic axis and (ii) the E-wave of amplitude $a = A\cos\theta$ with vibrations along the optic axis. The O- and E-waves tavel in the crystal in the *same direction* but with *different velocities.* On emergence, therefore, a path difference $t(\mu_0 - \mu_e)$ would occur between them, where t is the thickness of the crystal and μ_0, μ_e are the indices of O-ray and E-ray for the crystal. Calcite is a *negative* crystal and $\mu_0 > \mu_e$.

$$\therefore \quad \text{Phase difference, } \delta = \frac{2\pi}{\lambda} t(\mu_0 - \mu_e)$$

So, the equations of E- and O-waves may be written as

$$x = A\cos\theta\sin(\omega t + \delta) = a\sin(\omega t + \delta) \tag{12.14.1}$$

$$y = A\sin\theta\sin\omega t = b\sin\omega t \tag{12.14.2}$$

where $\omega =$ the frequency of vibration; $a = A\cos\theta$ and $b = A\sin\theta$.

From (12.14.1), we obtain

$$\frac{x}{a} = \sin(\omega t + \delta) = \sin\omega t\cos\delta + \cos\omega t\sin\delta$$

$$= \frac{y}{b}\cos\delta + \sqrt{1 - \frac{y^2}{b^2}}\sin\delta, \quad \text{using (12.14.2)}$$

$$\Rightarrow \quad \left(\frac{x}{a} - \frac{y}{b}\cos\delta\right)^2 = \left(1 - \frac{y^2}{b^2}\right)\sin^2\delta$$

$$\therefore \quad \boxed{\frac{x^2}{a^2} + \frac{y^2}{b^2} - \frac{2xy}{ab}\cos\delta = \sin^2\delta} \tag{12.14.3}$$

which, in general, is the equation of an **ellipse** confined in a rectangle of sides $2a$ and $2b$. The nature of the resultant emerging light however depends on the value of the phase difference δ.

Special cases

1. When $\delta = 0, 2\pi, 4\pi, \cdots$, i.e., $\delta = 2n\pi$, $n = 0, 1, 2\cdots$, we have $\sin\delta = 0$ and $\cos\delta = 1$. So, from (12.14.3), we obtain

$$\left(\frac{x}{a} - \frac{y}{b}\right)^2 = 0 \Rightarrow \frac{x}{a} - \frac{y}{b} = 0$$

$$\therefore \quad \boxed{y = \frac{b}{a}x} \tag{12.14.4}$$

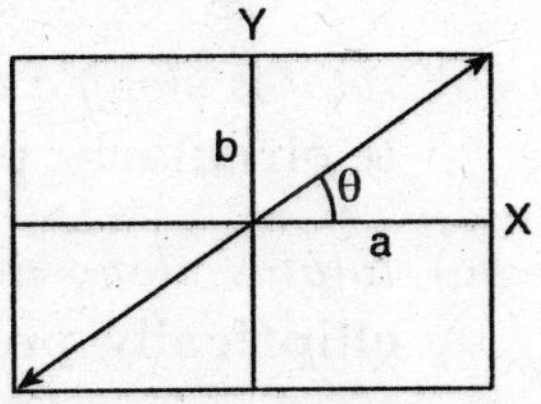

Fig. 12.24

This is an equation of a **straight line** *through the origin with a positive slope* b/a. So the resultant emergent light will be **plane polarised** *with vibrations in the same plane as the original incident light* (Fig. 12.24).

2. When $\delta = \pi, 3\pi, 5\pi, \cdots$; i.e., $\delta = (2n+1)\pi$, $n = 0, 1, 2, \cdots$, we have $\sin\delta = 0$, $\cos\delta = 1$. So, from (12.14.3),

$$\left(\frac{x}{a} + \frac{y}{b}\right)^2 = 0 \Rightarrow \frac{x}{a} + \frac{y}{b} = 0$$

$$\therefore \quad \boxed{y = -\frac{b}{a}x} \tag{12.14.5}$$

This again is the equation of a **straight line** *through the origin with a negative slope* $-b/a$. So, the resultant emergent light is **plane polarised**, the direction of vibration making an angle $-\theta$ with x-axis (Fig. 12.25).

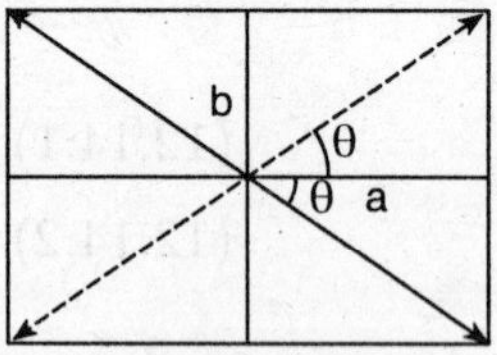

Fig. 12.25

3. When $\delta = \pi/2, 3\pi/2, 5\pi/2, \cdots$ i.e., $\delta = (2n+1)\frac{\pi}{2}$, $n = 0, 1, 2, \cdots$, $\sin\delta = 1$ and $\cos\delta = 0$. So from (12.14.3), we shall obtain

$$\boxed{\frac{x^2}{a^2} + \frac{y^2}{b^2} = 1} \qquad (12.14.6)$$

This is the equation of a **symmetrical ellipse** *with major and minor axis coinciding with the coordinate axes.* The emergent light will thus be **elliptically polarised** (Fig. 12.26), the axes of the ellipse being along and perpendicular to the optic axis.

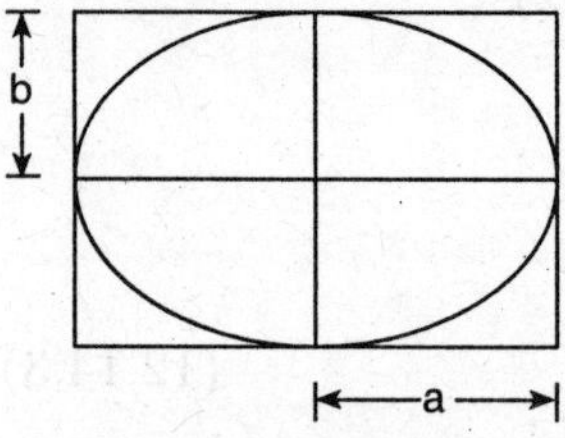

Fig. 12.26

4. When $\delta = \pi/2$, or $3\pi/2$ *and* $\theta = 45°$ i.e., incident light making an angle 45° with the optic axis, then, we have

$a = A\cos 45°$; $b = A\sin 45° \;\Rightarrow\; a = b$. So from (12.14.6),

$$\boxed{x^2 + y^2 = a^2} \qquad (12.14.7)$$

This is the equation of a **circle** and the emergent light will be **circularly polarised**. This is the basis on which a *quarter wave plate* works.

Summarising : In polarised light, *if the light vector*

(a) *vibrates along a straight line, it is a* **plane** *or* **linearly polarised light**,

(b) *rotates along a circle, i.e., does not change its magnitude while in rotation, it will be* **circularly polarised**,

(c) *rotates along an ellipse, i.e., changes its magnitude during rotation, it will be* **elliptically polarised**.

• **An important note**—The elliptically polarised light and the circularly polarised light may again be both **left-handed** or *anticlockwise* and **right-handed** or *clockwise, according as the rotating light vector rotates anticlockwisely* or *clockwisely.*

For instance, when $\delta = \pi/2, 5\pi/2, 9\pi/2, \cdots$, the equation of motion of E- and O-waves become

$$\left.\begin{aligned} x &= a\cos\omega t \\ y &= b\sin\omega t \end{aligned}\right\} \qquad (12.14.8)$$

using (12.14.1) and (12.14.2).

And the resulting elliptic vibration ($z = 0$ is the plane of observation) becomes

$$\frac{x^2}{a^2} + \frac{y^2}{b^2} = 1$$

But the equations (12.14.8) show that this ellipse is traced by the tip of the light vector *rotating anticlockwise* (Fig. 12.27a), relative to an observer towards whom the light travels and the light is said to the **left-handed elliptically polarised** (LEP). Note that $\vec{r} = \hat{i}x + \hat{j}y$ rotates *counterclockwise* with increasing time.

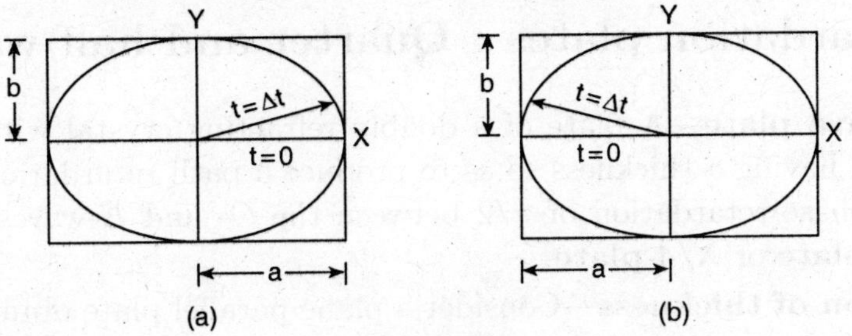

Fig. 12.27

If however $\delta = 3\pi/2, 7\pi/2, 11\pi/2, \cdots$ the equations for E- and O-waves are, from (12.14.1) and (12.14.2)

$$x = -a\cos\omega t$$
$$y = b\sin\omega t \qquad (12.14.9)$$

and lead again to an elliptically polarised light, since here also

$$\frac{x^2}{a^2} + \frac{y^2}{b^2} = 1$$

But the ellipse is *described clockwise* (Fig. 12.27b) relative to an observer towards whom light travels and the light is **right-handed elliptically polarised** (REP). Here $\vec{r} = \hat{i}x + \hat{j}y$ rotates *clockwise* with increasing time.

Note that half of the allowed δ-values for elliptic polarisation, as given in **special case 3**, correspond to LEP and the other half to REP.

In a similar fashion arise the *left-handed* and *right-handed* **circularly polarised** *light*. For instance, over and above the two conditions just cited (i.e., $\delta = \pi/2$ and $\delta = 3\pi/2$), if $\theta = 45°$, we shall get in the first case : $a = b$ and the equations for E- and O-waves are

$$x = a\cos\omega t, \quad y = a\sin\omega t \qquad (12.14.10)$$

and the resultant vibration becomes : $x^2 + y^2 = a^2$, representing a **circularly polarised** wave. But the *circle is described anticlockwise* relative to an observer to whom the light travels. So the light is **left-handed circularly polarised** (LCP).

In the second case, $a = b$ and the equations for E- and O-waves are

$$x = -a\cos\omega t; \quad y = a\sin\omega t \qquad (12.14.11)$$

which again leads to a *circle* : $x^2 + y^2 = a^2$, but the *circle is* now *described clockwise* relative to an observer to whom the light travels and the light is **right-handed circularly polarised** (RCP).

- We shall later see *how to convert a left handed elliptically or circularly polarised light into a right-handed one and conversely.*

In this context, we shall have to use a class of optical device, called **retarders** or **retardation plates** such as quarter wave plate, half wave plate etc. which will introduce path retardations between the E- and O-waves.

12.15 Retardation plates : Quarter and half wave plates

A. Quarter wave plate—A plate of a doubly refracting crystal with faces parallel to optic axis and having a thickness so as to produce a path retardation of $\lambda/4$ or, for that matter, a phase retardation of $\pi/2$ between the O- and E-waves is known as a **quarter wave plate** or **λ/4-plate.**

Computation of thickness—Consider a plane parallel plate obtained by cutting a doubly refracting crystal with faces parallel to optic axis. Let a beam of light of wavelength λ be incident normally on it (Fig. 12.28). The beam breaks up into O- and E-waves, both travelling along the same path perpendicular to the faces, but with different velocities.

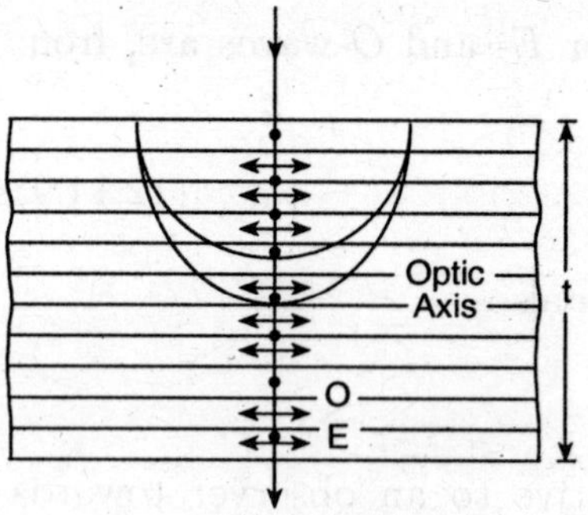

Fig. 12.28

For **negative crystals** (e.g., calcite), E-waves move faster than O-waves so that $\mu_O > \mu_e$, where μ_O is the principal index of the crystal for O-wave, μ_e that of the E-waves. If t be the thickness of the plate, the path difference between the two waves, on emergence, is

$$\Delta = (\mu_O - \mu_e)t$$

For the plate to become a quarter wave plate, we must have $\Delta = \lambda/4$,

$$\therefore \quad (\mu_O - \mu_e)t = \frac{\lambda}{4}$$

$$\Rightarrow \quad \boxed{t = \frac{\lambda}{4(\mu_O - \mu_e)}} \tag{12.15.1}$$

For **positive crystals** (e.g., quartz), $\mu_e > \mu_O$, so that

$$\boxed{\boldsymbol{t = \frac{\lambda}{4(\mu_e - \mu_O)}}} \tag{12.15.2}$$

So, a plate of thickness given by either (12.15.1) or (12.15.2) will serve as a *quarter-wave plate* for a given wavelength λ.

• Obviously, the thicknesss is *λ-specific*, i.e., with *change in λ, the thickness* of the quarter wave plate will *require a corresponding change.* Note that the above t-value is the **minimum thickness** required of a $\lambda/4$ plate.

Applications—Quarter wave plates are used for the *production of circularly and elliptically polarised light.* Further, in conjunction with a Nicol, it becomes a versatile analyser and can *analyse all kinds of polarised light,* as we shall see later.

B. Half-wave plate—A doubly refracting crystal plate having a thickness so as to produce a path difference $\lambda/2$ or, for that matter, a phase difference of π between the O- and the E-waves is called a **half-wave plate** or *$\lambda/2$-plate.*

Computation of thickness—Let t be the thickness of a half-wave plate. Then by definition, we have for a **negative** *crystal* (i.e., calcite)

$$(\mu_0 - \mu_e)t = \frac{\lambda}{2}$$

$$\Rightarrow \quad \boxed{t = \frac{\lambda}{2(\mu_0 - \mu_e)}} \tag{12.15.3}$$

For a **positive** *crystal* (e.g., quartz), the expression obviously becomes

$$\boxed{t = \frac{\lambda}{2(\mu_e - \mu_0)}} \tag{12.15.4}$$

• Note that the above expressions for t gives the **minimum thickness** of $\lambda/2$ plate.

If a *linearly polarised* light passes through a $\lambda/2$ plate, the emergent light is *also linearly polarised.* But the direction of its vibration is inclined at 2θ with the incident light, θ being the angle between the incident vibration and the principal section of the plate. A $\lambda/2$-plate is therefore used as *half-shade device* in polarimeters to divide the field of view into two halves presenting side by side.

• Both $\lambda/4$- and $\lambda/2$-plates are often made of quartz by cutting it parallel to optic axis. Quartz is a positive crystal with no cleavage planes. So it is to be cut and the faces polished to make them optically flat.

12.16 Babinet's compensator

A $\lambda/4$-plate or a $\lambda/2$-plate produces only a fixed path-retardation between the E- and O-waves for a given λ and hence is useful for light of a particular wavelength. Different plates are to be used for different wavelenths.

To remove this drawback, Babinet designed a **compensator** by which any desired path-difference can be introduced. This important optical device is useful for the production and analysis of elliptically polarised light (EP).

Construction—It consists essentially of two wedges A and B of quartz crystal, having equal small acute angles. They are placed with their hypotenuse-planes in contact so as to form a rectangular block (Fig 12.29a). Of the wedges, the *left* one is cut with its *optic axis perpendicular* to the refracting edge, while the *right* one

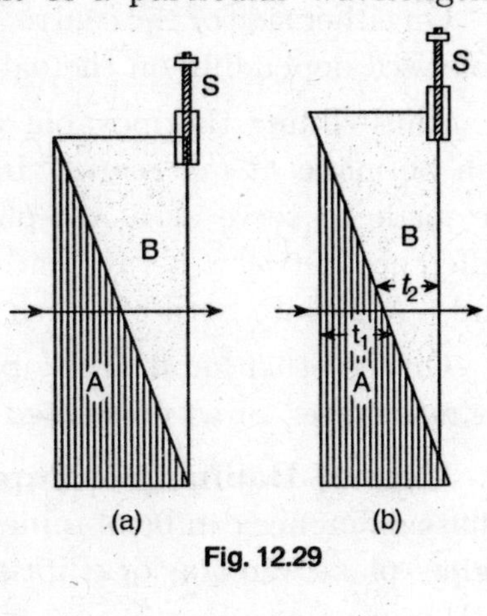

Fig. 12.29

with its *optic axis parallel* to it. Thus the O-ray in one will behave as E-ray in the other and vice versa. One of the wedges (A) is fixed in position and the other (B) slidable in its own plane by a micrometer screw S (Fig. 12.29b).

Theory—When plane polarised light falls normally on the first wedge A, with its plane of vibration inclined at θ with optic axis, it breaks up into O-ray and E-ray. The E-ray, parallel to the optic axis, moves slower than O-ray, perpendicular to optic axis. The angle of the wedge is so small that the separation between E- and O-rays is ignored. On entry into the second wedge B, the O-ray becomes the E-ray and conversely. So, they interchange velocities in transit from one wedge to the other and cancel each other's effect.

Let t_1, t_2 be the thicknesses of the wedges traversed by a ray and μ_e, μ_0 the indices of quartz for E- and O-ray respectively.

$\therefore$ Path-differences introduced between the components by the first and the second wedge respectively are :

$$\Delta_1 = t_1(\mu_e - \mu_0); \quad \Delta_2 = t_2(\mu_0 - \mu_e)$$

$\therefore$ Resultant path-difference, $\Delta = \Delta_1 + \Delta_2 = (\mu_e - \mu_0)(t_1 - t_2)$.

and phase difference, $$\delta = \frac{2\pi}{\lambda}(t_1 - t_2)(\mu_e - \mu_0).$$

At the **centre** of compensator, $t_1 = t_2 \Rightarrow \Delta = \delta = 0$, i.e., the emergent light is *plane polarised* in the the original plane. If the compensator be placed between two crossed Nicols, a dark band (central dark band) is obtained in this position. With crossed Nicols fixed in positions, if the micrometer screw is rotated, the phase difference go on changing and again a dark band appears when $\delta = 2\pi$. If the shift for this in micrometer be $2b$, then for a shift x, a phase difference

$$\delta = \frac{2\pi}{2b}x = \frac{\pi x}{b}$$

will be introduced.

On either side of the centre, Δ gradually increases and the emergent light is variously polarised depending on the values of $(t_1 - t_2)$.

Thus sliding the movable wedge, relative to the fixed one, any value of $(t_1 - t_2)$ can be made *at the centre*. In this way and at that position, the compensator may be made to serve as a $\lambda/4$-plate, $\lambda/2$-plate etc. by introducing the required phase difference between E-rays and O-rays. This is its *positive advantage* over a $\lambda/4$-plate and a $\lambda/2$-plate both of which are λ-specific.

On the other hand, if any phase difference exists between E-rays and O-rays, it can be neutralised or compensated by it. Hence it is called a *compensator*.

Uses of Babinet's compensator—Since any desired value of path-difference or phase difference can be obtained at the centre of the compensator, all types of polarised light—plane, circular or elliptic—can be obtained with its help. It can also be used for a complete analysis of elliptically polarised light by proper calibration of the device to which we shall refer elaborately when the occasion arises. A complete analysis entails finding of (i) phase difference between O and E components, (ii) the positions of the major and minor axes, and (iii) the ratio of the axes.

12.17 Production of circularly and elliptically polarised light

A. Circularly polarised light : A *circularly polarised* light can be produced by superposing two coherent linear vibrations, of equal amplitudes, at right angles to each other with a phase difference of $\pi/2$ or a path difference of $\lambda/4$. This is done with the help of a *quarter wave plate.* For instance,

$$x = a\cos(\omega t + \pi/2)$$

$$\text{and} \quad y = a\cos\omega t$$

give, on superposition, a *circular* vibration : $x^2 + y^2 = a^2$.

The **experimental set up** is shown in Fig. 12.30. A parallel beam of monochromatic light is allowed to fall on Nicol N_1 so that the emergent light is plane polarised. N_2 is a second Nicol *at cross* with N_1 so that *no light is transmitted* by N_2.

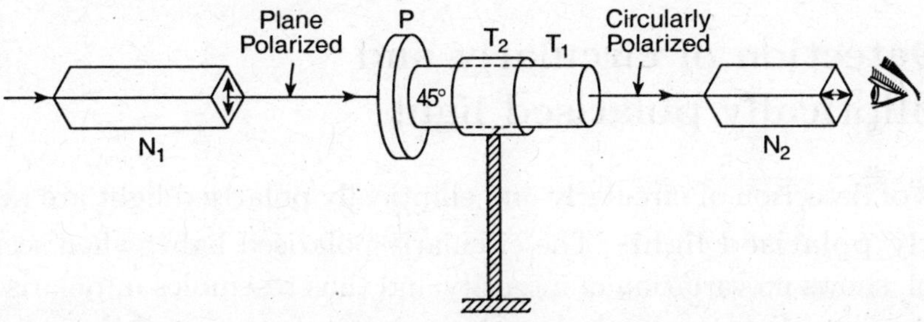

Fig. 12.30

Now, a $\lambda/4$-plate P, mounted on a tube T_1, the circumference of which is graduated in degrees, is inserted between N_1, N_2 and held *normal* to the incident beam. Tube T_1 can rotate in an outer fixed tube T_2 mounted on a stand so that the $\lambda/4$ plate can rotate about a horizontal axis through any desired angle. With $\lambda/4$-plate inserted, some light may appear in the field of view of N_2. The $\lambda/4$-plate is next rotated till the field of view of N_2 is again *dark.* This ensures that *vibrations of light incident on $\lambda/4$-plate are parallel to its optic axis* and both E- and O-waves travel in the same direction but with different velocities to be refused transmission through N_2 *at cross* with N_1. The $\lambda/4$-plate is now rotated through 45° with its optic axis so that the amplitudes of E- and O-waves are equal and also a *phase difference* of $\pi/2$ is introduced between the two waves. The resulting beam emerging out of $\lambda/4$-plate will therefore be **circularly polarised** (CP).

B. Elliptically polarised light—An elliptically polarised light can be produced by superposing two coherent linear vibrations of *unequal amplitudes*, at right angles to each other, with a phase difference of $\pi/2$. The help of a $\lambda/4$-plate is used here also.

For instance, the linear vibrations

$$x = a\cos(\omega t + \pi/2)$$
$$y = b\cos\omega t$$

yield an elliptic vibration : $x^2/a^2 + y^2/b^2 = 1.$

The **experimental set-up** (Fig. 12.30) is the same as in the production of circularly polarised light. A parallel monochromatic light beam is allowed to fall on Nicol N_1. Next, the emerging plane polarised light falls on a second Nicol N_2, *at cross* with N_1. So the field of view of N_2 is *dark* . The $\lambda/4$ plate P is now inserted between N_1, N_2 and held normal to the incident beam. The field of view of N_2 may now show some light. The $\lambda/4$ plate is now rotated till the field view of N_2 is dark again. In this position, vibrations of incident light on $\lambda/4$-plate are $\parallel$ optic axis of the plate and so perpendicular to N_2.

The $\lambda/4$-plate is again rotated so that vibrations of incident light on it makes any angle *other than* 45°. *This ensures unequal amplitudes of E- and O-rays and the light from the $\lambda/4$-plate is* **elliptically polarised.**

12.18 Detection of circularly and elliptically polarised light

The methods of detection of circularly and elliptically polarised light are rather simple.

Circularly polarised light—The circularly polarised light, when seen through a rotating Nicol, shows no variation in intensity and thus resembles unpolarised light. So, to detect if the given light is circularly polarised, it is first passed through a $\lambda/4$-plate with its *optic axis in any position.*

The light that will emerge out of the $\lambda/4$-plate will become *plane polarised.* This plane polarised light is next examined by a Nicol prism. If the Nicol, on rotation, shows a variation in intensity with *zero minimum*, then the original light under test is **circularly polarised.**

Elliptically polarised light—To detect if the light emerging out of $\lambda/4$ plate is elliptically polarised or not, it is first examined with a rotating Nicol which is adjusted to the position of *maximum intensity.*

Next, another $\lambda/4$-plate is interposed between the first $\lambda/4$-plate and the analysising Nicol such that *its optic axis is parallel to the principal section* of the Nicol, adjusted for *maximum intensity.* If the light is *elliptically polarised*, it will now become *plane polarised* and on rotating a Nicol, there will be a variation in intensity between *maximum* and *zero minimum.*

12.19 Conversion of one type of polarised light into another

We discuss below the various ways by which one type of polarised light may be converted into another type.

1. Elliptically polarised light → plane polarised light → circularly polarised light—Insert a $\lambda/4$-plate with optic axis parallel to the principal section of an analysing Nicol adjusted for maximum intensity. Then the optic axis of the $\lambda/4$-plate is also $\|$ major axis of the elliptic vibration. In this situation, the $\lambda/4$-plate introduces a further phase difference between the two vibrations at right angles to each other. The light emerging from $\lambda/4$-plate becomes **plane polarised.**

Now, **to convert the plane polarised light into circularly polarised light**, alow it to fall normally on another $\lambda/4$-plate such that the direction of vibration of plane polarised light makes 45° with the direction of the optic axis of $\lambda/4$-plate. The $\lambda/4$-plate breaks the incident wave into E- and O-waves of equal amplitude and introduces a phase difference of $\pi/2$ between them. On emergence from the plate, the O- and E-waves combine to give **circularly polarised light.**

2. A left-handed elliptically polarised light to a right-handed one (and vice versa)—For this, a phase difference of π must be introduced. So it is allowed to be incident normally on a half-wave plate such that its vibrations make an angle θ **excluding** $0°, 45°, 90°$ with the direction of the optic axis. The emergent light will become **right-handed elliptically polarised**, as the phase difference between the component waves (E and O) is $\delta = \pi/2 + \pi = 3\pi/2$.

3. A left-handed circularly polarised light into a right-handed one (and vice versa)—A phase difference of π must be introduced. So it is allowed to pass normally through a half-wave plate such that its vibrations make an angle of 45° with the optic axis. The emergent light from the $\lambda/2$-plate will be **right-handed circularly polarised light**, as the phase difference between the component waves is $\delta = \pi/2 + \pi = 3\pi/2$.

4. A circularly polarised light into a plane polarised light—If a circularly polarised light is passed through a $\lambda/4$-plate such that its vibrations make an angle of 45° with its optic axis, the light energing from the plate will be **plane polarised.** This is because the phase difference between the components (E and O) becomes $\delta = \pi/2 + \pi/2 = \pi$.

12.20 Applications of Babinet's compensator

1. Production of elliptic and circularly polarised light—The prism B (Fig. 12.29) is shifted by $x = b/2$ from its position of coincidence with prism A; then the phase difference introduced is $\pi/2$ (a shift $2b$ produces a phase difference of 2π). If the direction of vibration of the incident plane polarised light (obtained from a polariser) makes an angle $\alpha(\neq 45°)$ with the optic axis of A, then the emergent light will consist of *two rectangular* vibrations of *unequal amplitude* and *differening in phase* by $\pi/2$. The result is **elliptically polarised** light.

If however $\alpha = 45°$, the two rectangular vibrations will have *equal amplitude* and the emergent light will be **circularly polarised**.

2. Analysis of elliptically polarised light—This entails (a) *determination of phase difference between the two vibrations* giving rise to elliptic light and (b) the *determination of the positions and ratio of the axes* of elliptic light.

Action for (a) : First, the compensator is inserted between two *crossed* Nicols. Using white light and rotating the micrometer screw, the central dark band is made to concide with the cross-wire of observing eyepiece. Polariser is then removed and the compensator is allowed to receive the elliptic light under test. The central dark band gets shifted and is brought back to its original position by shifting the screw through x.

$$\therefore \quad \text{Required phase difference,} \quad \phi = \frac{\pi}{b}x,$$

$2b$ being the shift corresponding to a phase difference by 2π.

Action for (b) : First, the compensator is inserted between two *crossed* Nicols. Using white light and rotating the micrometer screw, the central dark band is made to coincide with the cross-wire of observing eyepiece. The micrometer screw is then turned through $b/2$ to introduce a phase difference of $\pi/2$ at the centre.

The central dark band is no more on the cross-wire. Polariser is now removed and compensator illuminated by elliptically polarised light under test. The compensator is *rotated in its own plane* till the central dark band is in the cross-wiere. Analyser is also rotated to make the bands distinct. *The axes of the elliptically polarised light are now parallel to the optic axis of the wedges of the compensator* (Fig. 12.31).

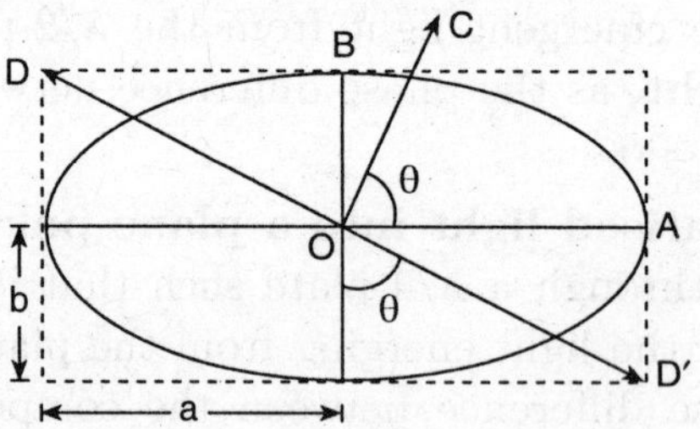

Fig. 12.31

The situation, as shown in Fig. 12.31, indicates that OA and OB are the optic axes of the wedges, OC the direction of principal section of analyser and DD' the direction of light emerging from compensator at the cross-wire. Plainly, the *tangent of the angle θ made by the principal section of analyser with the optic axis of the wedge gives the ratio of the axes* :

$$\tan\theta = \frac{a}{b}$$

To determine θ, the analyser is rotated until the bands disappear to give uniform illumination. The angle of rotation is θ.

12.21 Scheme for a complete analysis of polarised light

According to a theorem due to Stokes, only **seven** possible states of polarisation of a beam of light can occur. These are :

1. **unpolarised or natural light**
2. **plane or linearly polarised light**
3. **elliptically polarised light**
4. **circularly polarised light**

and the rest **three** are **mixtures**—each of unpolarised and any of the other three.

On passing the light through an analyser, e.g., a Nicol, that can rotate about the direction of light as axis, one can detect *elliptically* or *plane polarised* light since the intensity of the transmitted light varies with the orientation of the analyser. But if thee is *no variation with orientation*, then the original light may be *either unpolarised* or *circularly polarised* or *a mixture of the two.*

If however the Nicol is used *in conjunction with a quarter-wave plate*, one can make a **complete, systematic** *qualitative analysis of light.* The procedure is as follows.

Test I. Using a rotating Nicol—Insert the Nicol and rotate it. There are *three* possibilities.

(a) The *intensity* of the transmitted light is *independent* of the *orientation* of the analyser. This indicates that the light is *either unpolarised* or *circularly polarised* or *their mixture.* For, unpolarised light has vibrations of equal amplitudes in all directions and the rotating Nicol cannot show any variation of intensity. And circularly polarised light is the resultant of two mutually perpendicular vibrations of equal emplitudes and rotating Nicol transmits equal intensity in all positions due to symmetry of vibrations about the direction of propagation.

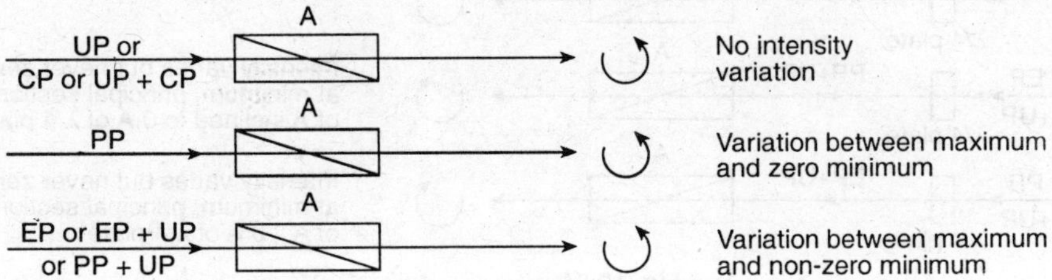

Fig. 12.32

(b) The intensity of transmitted light varies and *no light* is transmitted *for one* particular *orientation* of the analyser. This indicates that light is *completely plane polarised.* For, plane polarised light has vibrations only in one direction and when the direction is perpendicular to the principal section of Nicol, Nicol stops transmission.

(c) The *intensity* of transmitted light *varies* with the orientation of the analyser but *never falls to zero.* This indicates that the light is *either elliptically polarised* or *a mixture of elliptically polarised and unpolarised light* or *a mixture of plane polarised and unpolarised light.*

Fig. 12.32 represents the situations pictorially.

Test II. To investigate **I(b)**, the light is first passed normally through a $\lambda/4$-plate followed by an analyser which is rotated. Again, there are the *same three possibilities.*

(a) If the *intensity* is *independent of orientation* of the analyser, it indicates that the given light is *unpolarised.*

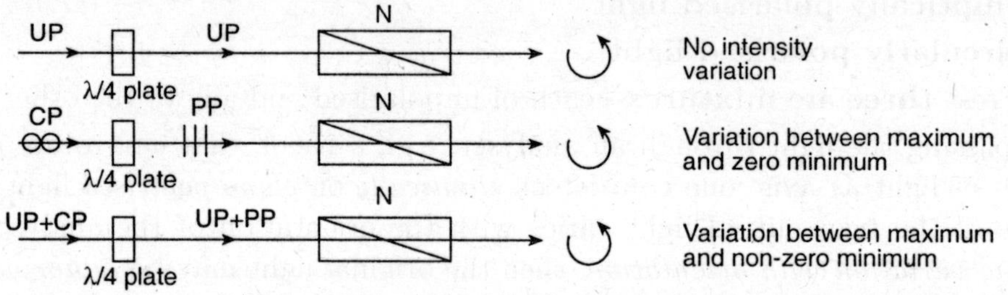

Fig. 12.33

(b) If there is *no transmission* of light for *one orientation of analyser*, then the original beam is *circularly polarised.*

(c) If the *intensity varies*, but *never falls to zero,* then it is a *mixture of circularly polarised and unpolarised light.*

Fig. 12.33 pictorially represents the situations.

Test III. To investigate **I(c)** further, again a $\lambda/4$-plate and an analyser are used, but now the *$\lambda/4$-plate is placed with its optic axis parallel to the principal section of the analyser adjusted for maximum intensity* and examined by rotating the Nicol. The *possibilities* are (Fig. 12.34) :

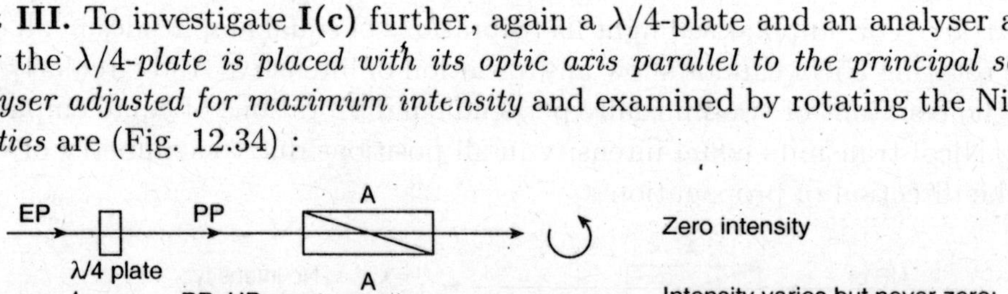

Fig. 12.34

(a) If the *light is extinguished for one setting of the analyser*, it indicates that the given light is *elliptically polarised.*

(b) If the *light-intensity varies but never falls to zero* and the position of *maximum* intensity *occurs at the same setting* of Nicol, it indicates that the light is a *mixture of plane polarised and unpolarised light.*

(c) If however the *maximum intensity occurs at a different setting* of the Nicol, the given light is *a mixture of elliptically polarised and unpolarised light.*

• It would be helpful to understand the above conclusions, if the reader **remembers** the following :

In the passage through a $\lambda/4$-plate, (i) circularly polarised light becomes plane polarised and (ii) plane polarised light may become, depending on the orientation of the $\lambda/4$-plate, elliptically, circularly or plane polarised light. Further, an elliptically polarised light may emerge from a $\lambda/4$-plate as plane polarised light or elliptically polarised light. Unpolarised light, however, remains unpolarised.

12.22 Rotatory polarisation (Optical activity)

Some crystals possess the property that when plane polarised light passes through them along the optic axis, they rotate the plane of polarisation about the direction of propagation through a certain angle. This phenomenon is termed *rotatory polarisation* or **optical activity**. The angle of rotation depends on (i) the thickness of the crystal, (ii) its nature and (iii) the wavelength of the light used. The substances that rotate the plane of polarisation are called **optically active**.

There are *two* types of optically active substances. Those which rotate the plane of polarisation *clockwise* while looking *against* the direction of light are called **dextro-rotatory** or **right-handed** while those that rotate the plane of polarisation *anticlockwise* on looking *against* the light are called **laevo-rotatory** or **left-handed**.

The optical activity of crystala is due to structural dissymetry. The atoms or molecules consituting the optically active crystals are arranged on a spiral which may be left-handed or right-handed. Some crystals have both the varieties, one being the mirror image of the other. Example is quartz.

Some substances e.g., cane sugar, in solution show the optical activity. In such cases, the activity is associated with the structural asymmetry of the molecules themselves. There may be some preferential direction of rotation associated with a molecule. While the molecules in a solution are oriented at random, their rotations do not cancel out since the sense of rotation is the same for molecules having the same type of dissymmetry. There are others where optical activity is associated with both type of dissymmetry.

Experimental detection (demonstration)—The set-up is shown in Fig. 12.35. If plane polarised light emerging from Nicol N_1 is examined through another Nicol

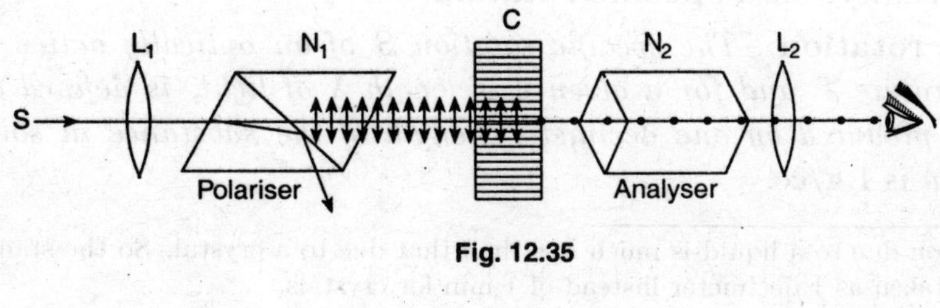

Fig. 12.35

N_2, *at cross* with N_1, the field of view is totally *dark*. Now a quartz plate C, cut with optic axis perpendicular to its face (so that there is no double refraction), be inserted between N_1 and N_2 such that light is incident normally on C, then some light passes through N_2. To cut off the light completely again, N_2 is to be rotated through a certain angle (whose value depends on the thickness of C). This proves that light emerging from quartz still remain plane polarised but its plane of polarisation has been rotated by the plate through a certain angle. Quartz is thus *optically active*.

12.23 Biot's studies : Specific rotation

Biot conducted a series of experiments on optical rotation using different substances and came out with the following laws.

1. The *angle of rotation* θ of the plane of polarisation, for a given wavelength λ, is *directly proportional to the length* l of the active substance *traversed* and is not affected by turning the substance around. So the rotation will be annulled if the ray is reflected back through it.

2. For *solutions* and *vapours*, the *angle of rotation* θ for a given path length is *proportional to the concentration* of the solution and vapour.

3. The *combined rotation* produced by a *number of active substances* having different thicknesses is the *algebraic sum of the individual rotations*—clockwise and anticlockwise rotations are taken with opposite signs.

4. The *angle of rotation* θ is (approximately) *proportional inversely* to the *square of the wavelength* λ. For quartz,

$$\theta = A + B/\lambda^2 \tag{12.23.1}$$

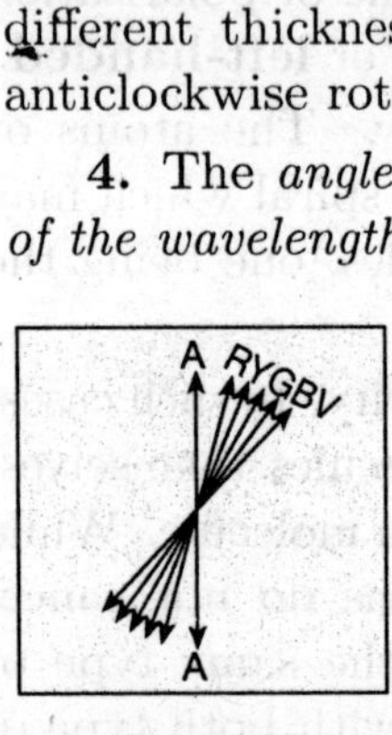

Fig. 12.36

where A and B are constants.

This implies that if plane polarised white light with vibrations along AA (Fig. 12.36) be incident normally on a quartz plate, the different colours will be rotated by different angles and the field of view will appear coloured. This phenomenon is called **rotatory dispersion.**

5. *The angle of rotation also depends on the temperature* of the active substance.

In the light of the above characteristics of the phenomenon, we now define an *important* quantity called **specific rotation.**

Specific rotation—*The specific rotation S of an optically active substance at a given temperature T and for a given wavelength λ of light, is defined as the rotation (in degrees) produced by one decimeter*[1] *length of the substance in solution when its concentration is* 1 g/cc.

[1]The rotation due to a liquid is much less than that due to a crystal. So the standard path length for a liquid is taken as 1 decimeter instead of 1 mm for crystals.

If θ be the rotation in degrees produced by l decimeter length of a solution and c be the concentration in g/cc, then

$$[S]_\lambda^T = \frac{\theta}{l \times c} = \frac{\text{rotation in degrees}}{\text{length in decimetre} \times \text{concentration in g/cc}}$$

$$\Rightarrow \boxed{\theta = S \times l \times c} \tag{12.23.2}$$

- c = simple density for pure liquids. If l be expressed in centimeter, then however the relation (12.23.2) takes the form

$$\theta = s \times (l/10) \times c.$$

- For optically active crystals, *specific rotation is defined as the rotation produced by the crystal plate of thickness* 1 mm.
- The *product of the specific rotation and the molecular weight* of the active substance is known as **molecular rotation.**

$\therefore$ **Molecular rotation = specific rotation × molecular weight**

12.23.1 An important theorem

Statement—*A linear vibration may be considered to be equivalent to two oppositely directed circular vibrations.*

Proof. Let $x = 0$ and $y = a\cos\omega t$ be a linear vibration along the y-axis in the xy-plane, a being the amplitude and $\omega/2\pi$ the frequency of vibration.

The above equations may be re-written as

$$x = \tfrac{1}{2}a\sin\omega t - \tfrac{1}{2}a\sin\omega t = x_1 + x_2$$

$$y = \tfrac{1}{2}a\cos\omega t + \tfrac{1}{2}a\cos\omega t = y_1 + y_2$$

and can be taken as the resultant of the following two vibrations (the principle of superposition) :

$$x_1 = \tfrac{1}{2}a\sin\omega t; \quad y_1 = \tfrac{1}{2}a\cos\omega t \tag{12.23.3}$$

$$x_2 = -\tfrac{1}{2}a\sin\omega t; \quad y_2 = \tfrac{1}{2}a\cos\omega t \tag{12.23.4}$$

Now, equations (12.23.3) are the components of a *clockwise circular vibration* ($\because\ x_1^2 + y_1^2 = (a/2)^2$) in two perpendicular directions. Equations (12.23.4) similarly denote the components of *an anticlockwise circular vibration* along the same two directions. Since these are derived from a linear vibration, *the linear vibration is equivalent to two oppositely directed circular vibrations of the same frequency.*

12.24 Fresnel's theory of optical rotation

This is based on the principle (just proved) that a linear vibration is equivalent to the resultant of two oppositely directed (anticlowise and clockwise) circular vibrations of the same frequency.

Assumptions—In an attempt to explain the optical rotation, Fresnel made the following assumptions.

1. The plane polarised light on entering a substance gets split into *two circularly polarised* waves—one clockwise and the other anticlockwise.

2. The above two waves move with the same speed in an optically non-active substance, while in active substances they travel with different speeds—the clockwise moves faster in the dextro-rotatory substance; in the laevo-rotatory substance, the anticlockwise one moves faster. This results in a *phase difference* being developed between them as they move in the substance.

3. On emergence, the *two circular components recombine* to produce the smae *plane polarised* light with the difference that the plane of polarisation is rotated relative to the incident one by an angle depending on the phase difference.

Explanation—Let the plane polarised light fall normally on a quartz plate cut perpendicular to the optic axis, the first face being in xy-plane. Let the vibration be

$$y = a \cos \omega t$$

which, on just entering the crystal, splits into the following two equal and opposite circular motions (Fig. 12.37a) : OL is the circularly polarised vector rotating anticlockwise (left-handed) and OR is the circularly polarised vector rotating clockwise (right-handed), OA being their resultant vector.

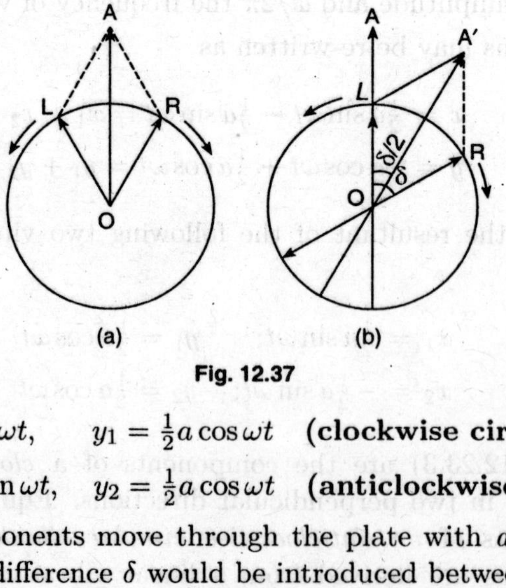

Fig. 12.37

$$x_1 = \tfrac{1}{2} a \sin \omega t, \quad y_1 = \tfrac{1}{2} a \cos \omega t \quad \textbf{(clockwise circular)}$$

$$x_2 = -\tfrac{1}{2} a \sin \omega t, \quad y_2 = \tfrac{1}{2} a \cos \omega t \quad \textbf{(anticlockwise circular)}$$

These circular components move through the plate with *different* speeds so that, on emergence, a phase difference δ would be introduced between them. Assuming the clockwise component moves faster (Fig. 12.37b), the emergent circular components are :

$$\left.\begin{aligned} x_1 &= \tfrac{1}{2}a\sin(\omega t+\delta) \\ y_1 &= \tfrac{1}{2}a\cos(\omega t+\delta) \end{aligned}\right\} \qquad \left.\begin{aligned} x_2 &= -\tfrac{1}{2}a\sin\omega t \\ y_2 &= \tfrac{1}{2}a\cos\omega t \end{aligned}\right\}$$

On emergence, OR is the right-handed and OL the left-handed circularly polarised vector with resultant OA'.

∴ Resultant displacements along the axes are :

$$x = x_1 + x_2 = \frac{1}{2}a\,[\sin(\omega t+\delta) - \sin\omega t] = a\cos\left(\omega t + \frac{\delta}{2}\right)\sin\frac{\delta}{2} \qquad (12.24.1)$$

$$y = y_1 + y_2 = \frac{1}{2}a\,[\cos(\omega t+\delta) + \cos\omega t] = a\cos\left(\omega t + \frac{\delta}{2}\right)\cos\frac{\delta}{2} \qquad (12.24.2)$$

Dividing (12.24.1) by (12.24.2), we obtain

$$\boxed{\frac{x}{y} = \tan\frac{\delta}{2}} \qquad (12.24.3)$$

which represents a *straight line inclined at $\delta/2$ to the y-axis* (Fig. 12.37b). So the emergent light from the plate is plane polarised with vibrations inclined at $\delta/2$ to the y-axis, i.e., relative to the incident vibrations.

Now, plainly, we have

$$\delta = \frac{2\pi}{\lambda} \times \text{path diff.} = \frac{2\pi}{\lambda}(\mu_A - \mu_C)d \qquad (12.24.4)$$

where μ_A, μ_C are the indices of quartz along the optic axis for anticlockwise and clockwise circular components respectively and d the thickness of the crystal.

∴ **Rotation of the plane of polarisation,**

$$\boxed{\theta = \frac{1}{2}\delta = \frac{\pi}{\lambda}(\mu_A - \mu_C)d} \qquad (12.24.5)$$

12.25 Experimental demonstration of Fresnel's idea

To demonstrate experimentally his idea, Fresnel built a rectangular block $ACFE$ of *alternate* prisms of *right-handed* and *left-handed* quartz, all having their *optic axis perpendicular to the end faces AC* and EF (Fig. 12.38).

A plane polarised ray incident normally on the face AC will break up, if Fresnel is correct, into two oppositely directed circular components—L (left-handed) and R

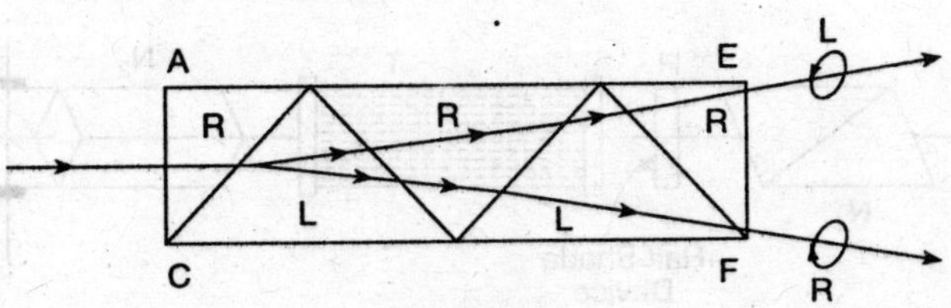

Fig. 12.38

(right-handed)—which travel with *unequal velocities* but in the *same direction.* The R-component moves faster in the right-handed (R) prism and slower in the left-handed (L) one, and conversely for the L-component. So, when these components pass from R-prism to L-prism, the material of L prism acts as a denser medium for R-component and as a rarer medium for L-component. In passing through L-prism, the R-component bends towards the base (downward) and L-component upward. In passing through R-prism, the L-component bends towards the base (upward) and R-component downward. Thus, the separation between L and R-components goes on increasing as they pass through the various prisms of the block. A detectable separation is thus produced between the two circularly polarised waves.

On analysis by a $\lambda/4$ plate and a Nicol prism, they were found to be circularly polarised in opposite sense, confirming Fresnel's concept.

• On passing through the $\lambda/4$ plate, both the waves become plane polarised with vibrations perpendicular to each other. Hence, on passing now through the rotating Nicol, they are alternately cut-off with every 90°-rotation of the Nicol showing that the emerging waves are circularly polarised in opposite sense.

12.26 Polarimeters

A *polarimeter* is an optical instrument used for accurately measuring the optical rotation of the plane of polarisation of active substances. When a polarimeter is calibrated directly to measure the percentage of sugar in the solution, it is called a **saccharimeter**.

Polarimeter may be of different kinds, but in all of them, the *basic polarimeter is identical.* It is only to know exactly the correct position for complete darkness, different attachments are used in different polarimeters. We shall describe here

1. **Laurent half-shade polarimeter**
2. **Bi-quartz polarimeter**, and
3. **Lippich polarimeter.**

12.26.1 Laurent half-shade polarimeter

The essential optical parts are illustrated in Fig. 12.39a Monochromatic light from a source (e.g., Na-vapour lamp), after passing through a narrow slit S, is rendered

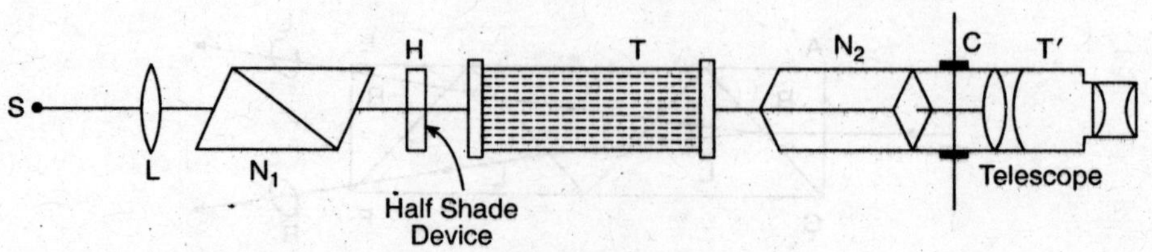

Fig. 12.39a

parallel by a converging lens L and allowed to fall on a polarising Nicol N_1. Light emerging from N_1 is plane polarised and passes through the half-shade device H (called *Laurent's plate*) and thence through the glass tube containing the active solution onto the analysing Nicol N_2. The emerging light is viewed through the telescope T. N_2 can be rotated about the light as axis (axis of tube) and its rotation in degrees measured by a vernier on the graduated circular scale C.

Working—Assuming H to be absent, the position of N_2 is adjusted till the field of view is totally dark, with tube T empty. In this position, N_2 is *at cross* with N_1. The position of N_2 is read. Next, T is filled with experimental solution and placed in position. This rotates the plane of polarisation of light from N_1 through some angle θ so that some light is now transmitted by N_2 and the field appears illuminated. N_2 is now rotated till the field is again dark and its reading noted. This happens when N_2 is rotated through θ in the direction of optical rotation. So the difference of the two readings of N_2 gives the *rotation* of the plane of polarisation.

Action of half-shade—The difficulty in the above procedure is that the eye *cannot judge* the position of complete darkness accurately and the field appears dark over a wide range when the analyser N_2 is rotated.

To *remove this difficulty,* the half-shade device is inserted after the polarising Nicol N_1. It consists of two *semicricular plates* ACB and ADB (Fig. 12.39b) cemented along the dia AB. Plate ACB is of *quartz, cut parallel to optic axis,* while the plate ADB is of *glass.* The thickness of ACB is such that it introduces a phase difference of π between the O and E vibrations, i.e., it is a *half-wave plate.* The thickness of ADB is such that it absorbs the *same* amount of light as does the quartz plate.

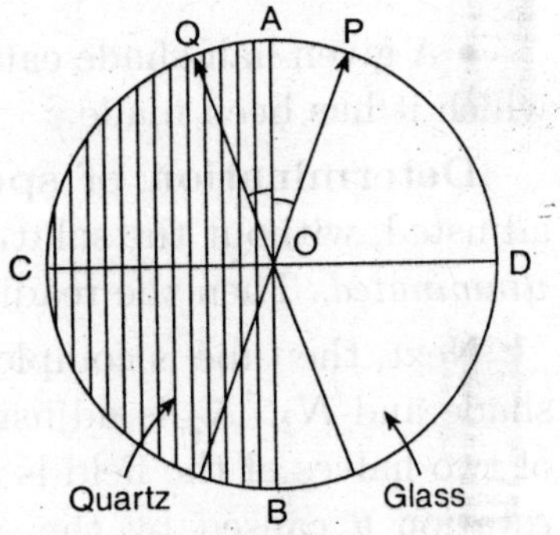

Fig. 12.39b

The plane polarised light from N_1 falls normally on half-shade device with vibrations along OP, say (Fig. 12.39b). Through the glass-half, it emerges as such with vibrations still along OP. In the quartz-half, the light gets split into two—E-component parallel to optic axis OA and the O-component perpendicular to optic axis, i.e., along OD. In quartz, O-ray travels faster and gains, on emergence, a phase of π over E-ray. So, O-component vibrates along OC (instead of OD) and E-component still having vibrations along OA. The resultant vibrations of light emerging from quartz is thus along OQ such that $\angle AOQ = \angle AOP$.

If now the analyser N_2 has its *principal section* $N_2ON_2 \parallel COD$ (Fig. 12.40a), both halves of the field appear *slightly* but *equally* illuminated, since OE and OF of the emergent vibrations OP and OQ along the principal section of N_2 are *small* but *equal.* If N_2 is slightly rotated *clockwise* (Fig. 12.40b), OE decreases appreciably and OF increases. So the *right half of the field* becomes *darker*, the *left half brighter.* Converse is the case for slight *anticlockwise* rotation (Fig. 12.40c). The eye con judge more accurately the position of *equal illumination* of two halves than the position of complete darkness. So, the use of half-shade enables one to set N_2 more accurately.

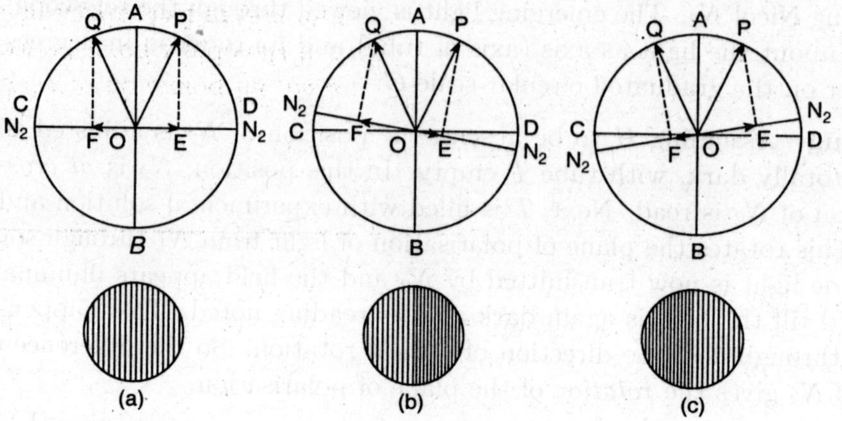

Fig. 12.40

• A given half-shade can be used for a particular wavelength only (λ-specificity) for which it has been made.

Determination of specific rotation of sugar solution—The analyser N_2 is adjusted, without the solution, for the two halves of the field to be slightly but *equally illuminated.* Then the reading on the circular scale is taken.

Next, the tube is completely filled with sugar solution and placed between the half-shade and N_2. N_2 is adjusted again and the reading for *small* but *equal illumination* of two halves of the field is noted. The difference of the two readings gives the optical rotation θ caused by the given l decimeter long sugar solution. Then the specific rotation

$$(S)_T^\lambda = \frac{\theta}{l \times c}$$

where c is the concentration of the solution in g/c.c. and is assumed known.

Determination of strength of sugar solution—If m g of sugar is dissolved in V c.c. of an optically inactive solvent (e.g., water), then $c = m/V$. So, from the above equation,

$$S = \frac{\theta}{l \times (m/V)} \Rightarrow c = \frac{m}{V} = \frac{\theta}{S \times l}$$

where S = specific rotation of sugar solution for the same wavelength and temperature.

• The value of $(S)_T^\lambda$ may be obtained from the table of constants.

12.26.2 Bi-quartz polarimeter

It consists of a *narrow slit, a condensing lens, a polarising Nicol, bi-quartz plate, a tube* containing the optically active solution, an *analysing Nicol* and a *telescope* fitted with a *circular scale* graduated in degrees and a *vernier* just as in a half-shade polarimeter.

The only difference is that the half-shade device is replaced here by a *bi-quartz plate* and the Na-*light by white light.* The basic polarimeter is thus identical with that of half-shade. The arrangement is also the same. We shall therefore only explain the action of the bi-quartz plate.

Action of bi-quartz plate—It consists of *two similar semicircular plates of quartz*—one *right-handed* (R) and the other *left-handed* (L), both *cut perpendicular to the optic axis* and cemented together along the diameter PP' and hence the name. The thickness of each plate is about 3.75 mm so that each rotates the plane of polarisation of yellow light through 90°, one clockwisely and the other anticlockwisely. When white light, made linearly polarised by the polarising Nicol, passes through the bi-quartz normally, it propagates along the *optic axis* and rotatory dispersion occurs in each plate. If the vibrations of incident light be along PP', then after the passage through bi-quartz, the vibrations of yellow light (Y) are along a direction nromal to PP' and those of red (R) and blue light (B) are along different directions as shown in Fig. 12.41. If the principal section of the analysing Nicol be parallel to PP', the yellow light will get quenched in both halves of the field, but the red and blue will be present in equal proportion in each half. The two halves will then appear *equally illuminated* with a reddish-violet tint, called *tint of the passage.* The tint is so sensitive that a slight rotation of Nicol from this position will make one half blue and the other half red.

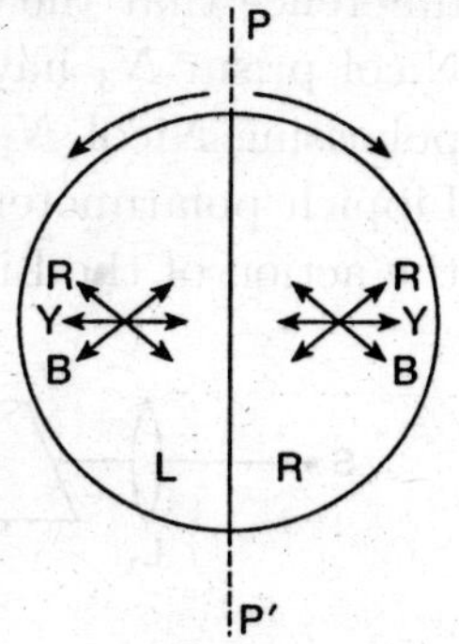

Fig. 12.41

Determination of specific rotation—The position of the sensitive 'tint of passage' is obtained twice by rotating the analyser : First, without the active solution and secondly, with it. The difference of the two readings of the analyser gives the optical rotation θ caused by the active solution. Once θ is known, the specific rotation S is obtained from the usual formula

$$S = \frac{\theta}{l \times c}$$

where l is the length of the solution in *decimeter* and c the concetration in g/cc.

12.26.3 Relative merits of half-shade and bi-quartz

The **half-shade** device is *fairly sensitive* and gives, when adjusted for equal illumination of two halves, *accurate observation* since slight rotation of the analyser changes the intensity of the two halves abruptly. It is also *suitable for colour-blind* persons.

It however suffers from the *disadvantage* that it is suitable only for *one particular wavelength* for which the path-difference between E and O-waves is $\lambda/2$ on transmission through it. It is usually constructed for yellow light of sodium.

The **bi-quartz**, on the other hand, is *more sensitive* than the half-shade device as slight rotation of analyser changes the colour of two halves. Further, it *works well with a white light source.* It is, in fact, *considered as a highy sensitive device for measuring optical rotation.*

It suffers however from two *disadvantages* : (i) *colour-blind persons cannot work* with bi-quartz. Even *different persons* with normal vision will get *different readings* as the colour-sensitivity of eyes is person-specific, (ii) usually and especially when the active substance is coloured, it is *not possible to obtain accurate setting* of the analyser for which the tint in both the halves of the field is the same; this is *due to rotatory dispersion* caused by the active substance itself.

12.26.4 Lippich polarimeter

Constructionwise, the Lippich polarimeter consists essentially of the same optical parts as in any basic polarimeter, such as half-shade or bi-quartz polarimeter, with the only difference that the half-shade plate/biquartz plate has been replaced here by a small Nicol prism N_3 having its principal plane inclined at a small angle α to that of the polarising Nicol N_1 and covers *only half of the field of view.* The optical parts of a Lippich polarimeter have been illustrated in Fig. 12.42. We shall only describe here the action of the Lippich device.

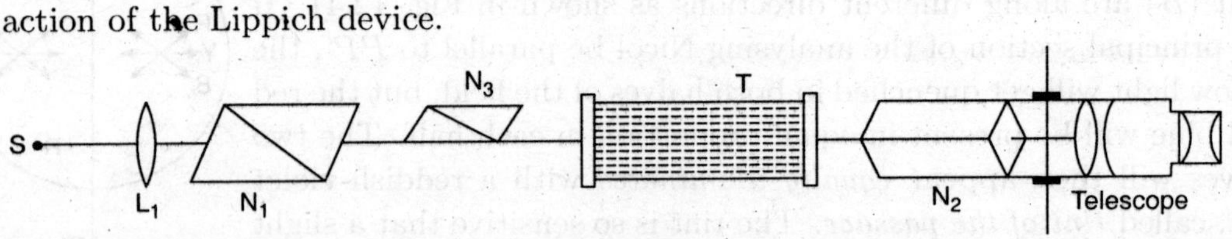

Fig. 12.42a

Lippich device—It consists, as already stated, of a small Nicol N_3 with its principal plane inclined at a small angle α with that of the polarising Nicol N_1. It covers *only half of the field of view* which is thus divided into two halves separated from each other by a sharp line which is the image of the edge of the Nicol N_3.

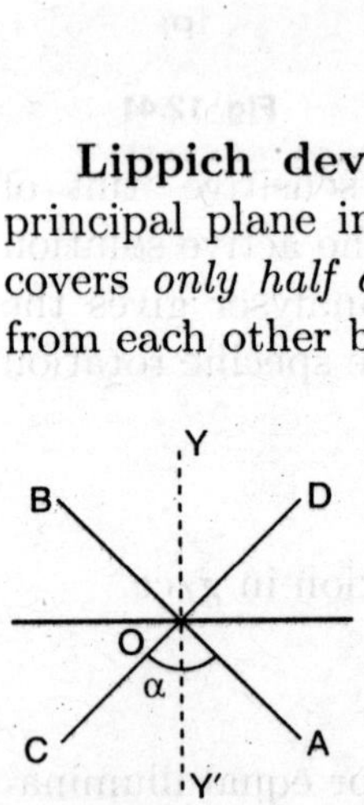

Fig. 12.42b

Let the plane of vibration of N_1 be along AB and that of N_3 along CD (Fig. 12.42b), the angle between the planes being α. When analyser N_2 is rotated such that its plane of vibration is along AB, the left half will be brighter than the right one. If however its plane of vibration is along CD, the right half will be brighter. When the plane of vibration of N_2 is along YY', the bisector of $\angle AOC$, the two halves are *equally illuminated*—the position at which it should be set. As the angle α is small, the setting is very sensitive.

Three-field system—The improved form of Lippich polarimeter has a *three* field system as shown in Fig. 12.43.

The two-field system has the problem that if the eye is off the axis then, even for equal brightness position, one side appears brighter than the other. So two Nicols, N_3 and N_4, instead of N_3 alone, are used just behnd N_1. The planes of vibration of N_3, N_4 are parallel and make a small angle with that of N_1. For a given position of analyser N_2, the field has *three* parts—light passing through N_1 and N_2 illuminates the central portion, the two other portions which are equally bright, being illuminated by the light passing through N_1, N_2 and one of N_3 and N_4.

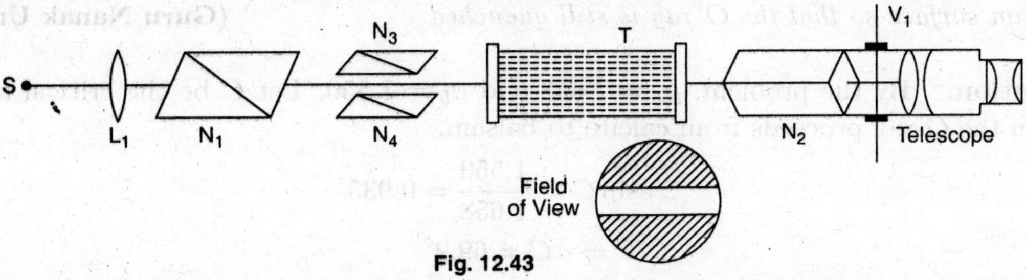

Fig. 12.43

• Under Art 12.26 the three main methods of *increasing the sensitivity of a pair of crossed Nicols* in polarimetry are discussed—(i) *half-shade device*, (ii) *bi-quartz device* and (iii) *Lippich device*.

12.27 Illustrated Examples

▶ **Example 1.** *When light is incident at an angle* 60° *to the normal, the reflected ray is plane polarised. What is the* (i) *refractive index of the transparent reflecting medium,* (ii) *angle of refraction corresponding to the angle of incidence* 60° *and* (iii) *the angle between the reflected and the refracted components?* **(Punjab Univ.)**

Solution. Here $i = 60°$. Since the reflected ray is plane polarised, i is the *polarising angle* ; so from Brewster's law :

(i) $\mu = \tan i = \tan 60° = \mathbf{1.73}$

(ii) Also, Brewster's law gives that if light is incident at polarising angle, then the reflected and refracted rays are at right angles.

$$\therefore \quad i + r = 90°$$

$$\therefore \text{ angle of refraction } r = 90° - i$$
$$= 90° - 60° = \mathbf{30°}$$

(iii) Again, since $i + r = 90°$, the angle between the reflected and the refracted components is **90°**.

▶ **Example 2.** *What is the refractive index of glass if the light of wavelength* 546 nm *is plane polarised when reflected at an angle of* 60° *?* **(Calcutta Hons.)**

Solution. From Brewster's law, the reuired refractive index μ is given by

$$\mu = \tan i_p = \tan 60° = \sqrt{\mathbf{3}}$$

▶ **Example 3.** *The refractive indices of calcite and canade balsam are* 1.658 *and* 1.550 *respectively for O-ray. Calculate the possible minimum inclination with canada balsam surface so that the O-ray is still quenched.* **(Guru Nanak Univ.)**

Solution. By the problem, $\mu_c = 1.658$ and $\mu_b = 1.550$. Let C be the critical angle when the O-ray proceeds from calcite to balsam.

$$\therefore \sin C = \frac{1.550}{1.658} = 0.935$$

$$\Rightarrow C = 69.2°$$

$\therefore$ O-ray when incident on balsam at an angle $< 69.2°$ will not be quenched.

$\therefore$ Required minimum possible inclination for quenching is

$$= 90° - 69.2 = \mathbf{20.8°}$$

▶ **Example 4.** *An unpolarised plane light wave of intensity* 10 mW/cm^2 *passes through two Nicols with the principal sections at* 30° *to each other. Calculate the intensity of the transmitted wave.* **(Calcutta Hons.)**

Solution. Let I_0 be the intensity of incident wave and I_1 that of the polarised light produced by the first Nicol. Since the incident unpolarised wave can be resolved into two orthogonal linear vibrations

$$I_1 = \tfrac{1}{2} I_0 = 5 \text{ mW/cm}^2$$

The intensity of light transmitted through 2nd Nicol is

$$I_2 = I_1 \cos^2\theta = 5 \times \cos^2 30°$$

$$= \mathbf{3.75 \text{ mW/cm}^2}$$

▶ **Example 5.** *Two polaroids are adjusted so as to obtain maximum intensity. Through what angle should one polaroid be rotated to reduce the intensity to* (i) *half and* (ii) *one-fourth.* **(Himachal Univ.)**

Solution. From Malus law, we have $I_1 = I_0 \cos^2\theta \Rightarrow I/I_0 = \cos^2\theta$.

Now, (i) when $I_1/I_0 = \frac{1}{2}$, $\cos^2\theta = \frac{1}{2} \Rightarrow \cos\theta = \pm 1/\sqrt{2} \Rightarrow \theta = \mathbf{\pm 45°, \pm 135°}$

(ii) When $I_1/I_0 = 1/4$, $\cos^2\theta = 1/4 \Rightarrow \cos\theta = \pm\frac{1}{2} \Rightarrow \theta = \mathbf{\pm 60°, \pm 120°}$

Required angles of rotation are $\mathbf{\pm 45°, \pm 135°}$ and $\mathbf{\pm 60°, \pm 120°}$.

▶ **Example 6.** *Two Nicols are crossed to each other. One of them is then rotated through* 60°*. Calculate the percentage of incident unpolarised light that will pass through the optical system.*

Solution. Initially, the Nicols are at 90° to each other. On rotating one by 60°, the angle between the principal planes of Nicols = 90° − 60° = 30°.

Now, only O-component of incident light is transmitted through the first Nicol. So, the intensity of transmitted light

$$I = \tfrac{1}{2}I_0, \quad \text{where } I_0 = \text{incident intensity}$$

$\therefore$ From Malus law, $I_1 = I\cos^2 30° \Rightarrow I_1/I = \cos^2 30 = 3/4$

$$\Rightarrow \quad \frac{I_1}{\frac{1}{2}I_0} = \frac{3}{4} \quad \Rightarrow \quad \frac{I_1}{I_0} = \frac{1}{2}\times\frac{3}{4} = \frac{3}{8}$$

$$\therefore \quad \text{Percentage transmission} = \tfrac{3}{8}\times 100 = \mathbf{37.5\%}$$

▶ **Example 7.** *The critical angle of light in a certain substance is 45°. What is its polarising angle?*

Solution. The incident light moves in denser medium. If μ be the refractive index of the rarer (second) medium and c the critical angle, then

$$\mu = \sin c = \sin 45° = 1/\sqrt{2} = 0.707$$

If now ϕ_p be the polarising angle, then from Brewster's law, we obtain

$$\tan\phi_p = \mu \quad \Rightarrow \quad \phi_p = \tan^{-1}\mu = \tan^{-1}(0.707) = \mathbf{35.26°}$$

▶ **Example 8.** *A thin polaroid is interposed between a pair of two crossed polaroids and made to rotate at a rate ω about their common axis. If an unpolarised light of intensity I_0 falls on the first polaroid, then show that the transmitted intensity is*

$$I = \frac{I_0}{16}(1 - \cos 4\omega t)$$

Solution. On transmission through the first polaroid, the intensity is $\frac{1}{2}I_0$ where $I_0 =$ intensity of unpolarised light. At any instant t, the angle between the axes of the first and middle polaroid is $\theta = \omega t$.

$\therefore$ By Malus law, the intensity of light on transmission through middle polaroid is

$$I_1 = \frac{I_0}{2}\cos^2\theta = \frac{I_0}{2}\cos^2\omega t$$

The angle between the axes of the middle and last polaroids $= 90° - \theta$. So the intensity of transmitted light is :

$$I = I_1\cos^2(90° - \theta) = \frac{I_0}{2}\cos^2\omega t\sin^2\omega t$$

$$= \frac{I_0}{8}\sin^2 2\omega t = \mathbf{\frac{I_0}{16}(1 - \cos 4\omega t)}$$

▶ **Example 9.** *Calculate the thickness of $\lambda/4$-plate for light of λ = 5893 A, given μ_0 = 1.544 and μ_e = 1.553.* **(Calcutta Hons; Himachal Univ.)**

Solution. Here $\lambda = 5893 \times 10^{-10}$ m, $\mu_O = 1.544$ and $\mu_e = 1.553$. Since $\mu_e > \mu_O$, the $\lambda/4$-plate is of a *uniaxial positive crystal.*

$$\therefore \text{ Thickness, } t = \frac{\lambda}{4(\mu_e - \mu_O)} = \frac{5893 \times 10^{-10}}{4(1.553 - 1.544)} = \frac{5893 \times 10^{-10}}{4 \times 0.009}$$

$$= \mathbf{0.1637 \times 10^{-4} \ m}$$

► **Example 10.** *Calculate the thickness of a half-wave plate for sodium light, given $\mu_O = 1.54$ and the ratio of velocity of O-component and E-component is 1.007. Is the crystal positive or negative?* **(Poona Univ.)**

Solution. Here $\lambda = 5893 \times 10^{-10}$ m; $\mu_O = 1.54$. Also, by the problem, $v_O/v_e = 1.007 \Rightarrow \mu_e/\mu_O = 1.007$ where v's are velocities

$$\therefore \mu_e = 1.007\mu_O = 1.007 \times 1.54 = 1.551$$

Since $\mu_e > \mu_O$, the crystal is **positive**.

Thickness of $\lambda/2$-plate is given by

$$t = \frac{\lambda}{2(\mu_e - \mu_O)} = \frac{5893 \times 10^{-10}}{2(1.551 - 1.540)}$$

$$= \mathbf{0.2679 \times 10^{-4} \ m}$$

► **Example 11.** *A beam of linearly polarised light is changed into a circularly polarised light by passing it through a slice of crystal* 0.003 cm *thick. Find the difference in the indices of two rays in the crystal assuming it to be the minimum thickness that will produce the effect* ($\lambda = 6 \times 10^{-7}$ m). **(Indian Adm. Service)**

Solution. Here $\lambda = 6 \times 10^{7}$ m; $t = 0.003$ cm $= 3 \times 10^{-5}$ m. Now, since a linearly polarised light changes into a circularly polarised one, the crystal plate has a thickness equal to that of a quarter wave plate.

$$\therefore \quad t = \text{minimum plate thickness} = \frac{\lambda}{4(\mu_e \sim \mu_O)}$$

$$\therefore \quad \mu_e \sim \mu_O = \frac{\lambda}{4t} = \frac{6 \times 10^{-7}}{4 \times 3 \times 10^{-5}} = \mathbf{0.005}$$

► **Example 12.** *A sheet of cellophane is a half-wave plate for light of* $\lambda = 4 \times 10^{-5}$ cm. *Assuming that there is negligible variation in indices of refraction with* λ, *how would the sheet behave with respect to the wavelength* 8×10^{-5} cm?

Solution. If t be the thickness of the half-wave plate,

$$t = \frac{\lambda}{2(\mu_e \sim \mu_O)} = \frac{4 \times 10^{-5}}{2(\mu_e \sim \mu_O)} = \frac{8 \times 10^{-5}}{4(\mu_e - \mu_O)}$$

which corresponds to thickness formula for a quarter wave plate of $\lambda = 8 \times 10^{-5}$ cm. So, for $\lambda = 8 \times 10^{-5}$ cm, the sheet would behave as a **quarter-wave plate**.

▶ **Example 13.** *A beam of plane polarised light of $\lambda = 6000$ A is incident on a thin quartz-plate cut with faces parallel to the optic axis. Calculate the minimum thickness of the plate which introduces a phase difference of 60° between the O-rays and the E-rays. Given $\mu_e = 1.553$ and $\mu_O = 1.544$.*

Solution. A phase difference of 360° corresponds to a path-difference of λ.

$\therefore$ Phase difference of 60° $\equiv$ a path difference $(\lambda/360) \times 60$ or $\lambda/6$.

$$\therefore \ t(\mu_e - \mu_O) = \lambda/6 \ \Rightarrow \ t = \frac{\lambda}{6(\mu_e - \mu_O)} = \frac{6000 \times 10^{-10}}{6 \times (1.553 - 1.544)}$$

$$= \mathbf{0.111 \times 10^{-5}\ m}$$

▶ **Example 14.** *Write an expression for a linearly polarised wave of angular frequency ω moving along $+z$ direction with plane of vibration making 60° to the xz-plane.*

Solution. Let E_0 be the amplitude of the wave. So, its x and y-components are

$$\left.\begin{aligned} E_{0x} &= E_0 \cos 60° = \frac{1}{2} E_0 \\ E_{0y} &= E_0 \sin 60° = \frac{\sqrt{3}}{2} E_0 \end{aligned}\right\}$$

$\therefore$ The required expression for the linearly polarised wave may by written as :

$$\vec{E}(z,t) = \hat{\imath}\frac{E_0}{2}\cos(kz - \omega t) + \hat{\jmath}\frac{\sqrt{3}}{2}E_0\cos(kz - \omega t)$$

▶ **Example 15.** *Find the state of polarisation when the x and y components of the electric field are*

(i) $E_x = E_0 \sin(\omega t + kz)$; $E_y = E_0 \cos(\omega t + kz)$

(ii) $E_x = E_0 \cos(\omega t + kz)$; $E_y = (E_0/\sqrt{2}) \cos(\omega t + kz + \pi)$. **(Calcutta Hons.)**

Solution. (i) We have $E_x^2 + E_y^2 = E_0^2$. This represents a circular virbation. At any position, say $z = 0$, the vector

$$\vec{E} = \hat{\imath}E_x + \hat{\jmath}E_y = \hat{\imath}E_0 \sin \omega t + \hat{\jmath}E_0 \cos \omega t$$

rotates *clockwisely* and the light is **circularly polarised.**

(ii) Here $E_y = -(E_0/\sqrt{2})\cos(\omega t + kz) = (-1/\sqrt{2})E_x$. This represents a **linearly polarised** light making an angle $\tan^{-1}(-1/\sqrt{2})$ with the x-axis.

► **Example 16.** *Two linearly polarised light waves are in phase but have different amplitudes. They are represented by*

$$\vec{E}_1(z,t) = \hat{\imath}A_1\cos(kz-\omega t) + \hat{\jmath}B_1\cos(kz-\omega t)$$
$$\vec{E}_2(z,t) = \hat{\imath}A_2\cos(kz-\omega t) + \hat{\jmath}B_2\cos(kz-\omega t)$$

Show that $\vec{E}_1 + \vec{E}_2 = \vec{E}$ is also linearly polarised. Find its polarisation direction. **(Calcutta Hons.)**

Solution. $\vec{E}(z,t) = \vec{E}_1(z,t) + \vec{E}_2(z,t)$

$$= \left[\hat{\imath}(A_1+A_2) + \hat{\jmath}(B_1+B_2)\right]\cos(kz-\omega t)$$

The amplitude $\hat{\imath}(A_1+A_2)+\hat{\jmath}(B_1+B_2)$ being a constant, the resultant wave is also linearly polarised.

Let $\vec{E} = \hat{\imath}E_x + \hat{\jmath}E_y$, then from above it follows that

$$E_x = (A_1+A_2)\cos(kz-\omega t)$$
$$E_y = (B_1+B_2)\cos(kz-\omega t)$$

$$\therefore \quad \frac{E_y}{E_x} = \frac{B_1+B_2}{A_1+A_2} \quad\Rightarrow\quad E_y = \frac{B_1+B_2}{A_1+A_2}E_x$$

So, it is a **linear polarised vibration** in the xy-plane making an angle $\tan^{-1}[(B_1+B_2)/(A_1+A_2)]$ with x-axis.

► **Example 17.** *The adjoining diagram shows the section of an extraordinary wave surface in a positive uniaxial crystal with optic axis along OX. Show that the ray refractive index of the medium at an angle θ to the optic axis is given by*

$$\mu_\theta = \sqrt{\mu_0^2\cos^2\theta + \mu_e^2\sin^2\theta}$$

where μ_0 and μ_e are respectively the refractive index for the O-ray and the principal refractive index of E-ray. **(Calcutta Hons.)**

Solution. The equation of the wave surface (Fig. 12.44) is given by

$$\frac{x^2}{a^2} + \frac{y^2}{b^2} = 1 \tag{12.27.1}$$

But, $x = r\cos\theta$, $y = r\sin\theta$, $b = v_e t$, v_e being the wave velocity for E-ray along OY (perpendicular to optic axis) and $r = OP = v_\theta t$, where v_θ is the ray velocity of E-ray along OP. Again, since the wave surface for O-ray is spherical, touching the wave surfaoe for E-ray along optic axis OX, $a = v_0 t$ where v_0 is the wave velocity of O-ray, being the same in every direction.

$\therefore$ Eq. (12.27.1) may be written as :

$$\frac{v_\theta^2 t^2 \cos^2\theta}{v_0^2 t^2} + \frac{v_\theta^2 t^2 \sin^2\theta}{v_e^2 t^2} = 1$$

or, $$\frac{\cos^2\theta}{v_0^2} + \frac{\sin^2\theta}{v_e^2} = \frac{1}{v_\theta^2}$$

or, $$\frac{c^2}{v_0^2}\cos^2\theta + \frac{c^2}{v_e^2}\sin^2\theta = \frac{c^2}{v_\theta^2}$$

or, $$\mu_0^2\cos^2\theta + \mu_e^2\sin^2\theta = \mu_\theta^2$$

$$\therefore \quad \mu_\theta = \sqrt{\mu_0^2\cos^2\theta + \mu_e^2\sin^2\theta}$$

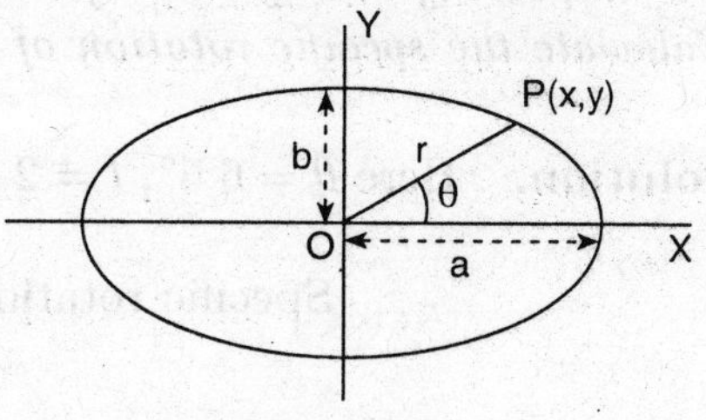

Fig. 12.44

where $\mu_\theta = c/v_\theta$, the ray refractive index for E-ray at an angle θ to the optic axis.

▶ **Example 18.** *The rotation in the plane of polarisation* ($\lambda = 5893\,\text{A}$) *in a certain substance is* $10°$ per cm. *Calculate the difference between the refractive indices for right-handed and left-hangded circularly polarised light in the substance.* **(Mysore Univ.)**

Solution. Let t cm be the thickness of the substance and μ_R and μ_L the indices for right and left circularly polarised light in it. Then the rotation θ in the plane of polarisation is given by

$$\theta = \frac{\pi}{\lambda}(\mu_R \sim \mu_L)t \;\Rightarrow\; \mu_R \sim \mu_L = \frac{\theta}{t}\cdot\frac{\lambda}{\pi}$$

Here $\theta/t = (10 \times 2\pi/360)$ rad/cm and $\lambda = 5893\,\text{A} = 5893 \times 10^{-8}$ cm.

$$\therefore \quad \mu_R \sim \mu_L = \frac{10 \times 2\pi}{360} \times \frac{5893 \times 10^{-8}}{3.142}$$

$$= \mathbf{4.09 \times 10^{-9}}$$

▶ **Example 19.** *A solution of camphor in alcohol in a tube* 25 cm *long is found to affect the rotation of the plane of vibration of light passing through it by* $36°$. *What must be the density of camphor in* g/cc *of the solution? Specific rotation of camphor is* $+56°$. **(Guru Nanak Univ.)**

Solution. Here, rotation $\theta = 36°$, specific rotation, $S = +56°$, l = length of solution = 25 cm = 2.5 decimeter.

Let c be the concentration in g/cc. Then, we have

$$c = \frac{\theta}{S \cdot l} = \frac{36}{56 \times 2.5} = 0.257\,\text{g/cc}$$

$\therefore$ Required density of camphor = **0.257 g/cc.**

► **Example 20.** *The plane of polarisation of plane polarised light is rotated through* 6.5° *in passing through a length of* 2.0 *decimeter of sugar solution of* 5% *concentration. Calculate the specific rotation of the sugar solution.*

Solution. Here $\theta = 6.5°$, $l = 2.0$ dm, $c = 5\% = 0.05$ g/cc.

$$\therefore \text{ Specific rotation, } S = \frac{\theta}{l \cdot c} = \frac{6.5°}{2.0\,\text{dm} \times 0.05\,\text{g/cc}}$$

$$= \mathbf{65° \ (dm)^{-1} \ (g/cc)^{-1}}$$

► **Example 21.** *A certain length of* 5% *solution causes an optical rotation of* 20°. *How much length of* 10% *solution of the same substance will cause rotation of* 35°*?*

(Kanpur Univ.)

Solution. Let S be the specific rotation of the solution, l_1 the length of 5% solution causing 20° rotation and l_2 the length of 10% solution causing 35° rotation.

$$\therefore \quad S = \frac{\theta_1}{l_1 \cdot c_1} = \frac{\theta_2}{l_2 \cdot c_2} \quad \Rightarrow \quad \frac{20°}{l_1 \times 5\%} = \frac{35°}{l_2 \times 10\%}$$

$$\therefore \quad l_2 = \frac{5 \times 35}{10 \times 20} l_1 = \mathbf{\frac{7}{8} l_1}$$

► **Example 22.** *A* 20 cm *length of a certain solution causes right-handed rotation of* 38° *and* 30 cm *of another solution causes left-handed rotation of* 24°. *What optical rotation will be caused by* 30 cm *length of a mixture of the above solutions in the ratio* 1 : 2 *by volume?*

(Rajasthan Univ.)

Solution. Optical rotation, $\theta \propto l$, length of substance traversed for a given λ.

$$\text{Rotation by 20 cm of sol. 1} = +38° \text{ (right-handed)}$$

$$\text{Rotation by 30 cm of sol. 2} = -24° \text{ (left-handed)}$$

$$\Rightarrow \quad \text{Rotation by 40 cm of sol. 2} = -24° \times \frac{40}{30} = -32°$$

Since rotation of a mixture of active substances is the algebraic sum of individual rotations, we write as :

rotation by (20 cm of 1 + 40 cm of 2) = +38° − 32° = 6° i.e., mixture in the ratio 1 : 2.

$\therefore$ rotation by 30 cm of the mixture = **+3° (right-handed)**

► **Example 23.** *The optical rotation produced by a material is* 30°/mm *at* $\lambda = 5000$ A *and* 50°/mm *at* $\lambda = 4000$ A. *In the visible region, rotation is given by* $\theta = a + b/\lambda^2$. *Find 'a' and 'b' and the least thickness of the material which when interposed between two crossed Nicols will produce maximum transmission for* $\lambda = 5000$ A.

Solution. We have $\theta = a + b/\lambda^2$. So, by the problem, we can write

$$30°/\text{mm} = a + b/(5000\,\text{A})^2; \quad 50°/\text{mm} = a + b/(4000\,\text{A})^2 \tag{12.27.2}$$

Subtracting, $20°/\text{mm} = \dfrac{b}{10^6}\left(\dfrac{1}{16} - \dfrac{1}{25}\right)\text{A}^{-2} = \dfrac{9b}{4 \times 10^8}\text{A}^{-2}$

On simplification, $\mathbf{b = 8.9 \times 10^8}$ **deg. mm^{-1} A^2**

Substituting for b in any of the set of equations (12.27.2)

$$\mathbf{a = -5.6\ \text{deg. mm}^{-1}}$$

The maximum transmission occurs when the interposed material rotates the plane of polarisation of light of $\lambda = 5000$ A by 90°.

1 mm thickness rotates by 30°. So required **thickness** = (90/30) mm = **3 mm.**

Review Questions

1. Distinguish between polarised and unpolarised light. Explain the term : plane of polarisation and plane of vibration. **(Kurukshetra and Himachal Univ.)**
2. What is dichroism? How does it produce plane polarised light? Use a dichroic cyrstal to analyse plane polarised light. **(Punjab Univ.)**
3. Explain the method of producing plane polarised light by reflection. **(Benares and Guru Nanak Univ.)**
4. State and derive Brewster's law. What does the law become when the rays of light travel from a denser to a rarer medium? **(Himachal Univ.)**
5. Show, by Brewster's law, that light incident on a transparent substance at polarising angle gives reflected and refracted rays at right angles to each other. **(Calcutta and Burdwan Univ.)**
6. What is birefringence? How does it explain double refraction? Give some applications of birefringence. **(Guru Nanak and Punjab Univ.)**
7. What do you understand by 'double refraction'? What are ordinary and extraordinary rays? How will you show that these are plane polarised?
8. Define 'optic axis' and the 'principal section' of a crystal. What is an uniaxial crystal? What do you mean by 'positive' and 'negative' crystals'? Give one example for each. What is the difference between the two?

9. Describe the construction and action of a Nicol prism. Explain how a Nicol prism is used to produce and analyse plane polarised light. **(Calcutta and Cochin Univ.)**

10. State and explain Malus law. How will you prove the law experimentally. **(Guru Nanak Univ.)**

11. Give Huygens' theory of double refraction. Draw diagrams and discuss double refraction through uniaxial crystals due to plane wave when the optic axis is (a) in clined to upper face but lying in the plane of incidence, (b) parallel to the upper face but lying in the plane of incidence, (c) parallel to the upper face but perpendicular to the plane of incidence and (d) perpendicular to the upper face. **(Poona and Guru Nanak Univ.)**

12. Explain the method of finding the refractive indices of the ordinary and the extraordinary rays. **(Punjab Univ.)**

13. What are polaroids? What do you mean by *H*-polaroid and *K*-polaroid. Give some of their applications. **(Kurukshetra and Himachal Univ.)**

14. What is meant by an elliptically and a circularly polarised light? Explain them with necessary theory. **(Himachal and Punjab Univ.)**

15. Give the construction and the theory of (a) a quarter wave plate and (b) a half-wave plate. **(Burdwan and Cochin Univ.)**

16. How are elliptically and circularly polarised light produced experimentally? **(Himachal Univ.)**

17. Explain how the state of polarisation of a given beam of light can be determined with the help of a Nicol prism and a quarter wave plate.

18. How would you analyse plane polarised, circularly polarised and elliptically polarised light? **(Punjab Univ.)**

19. How can you convert (a) elliptically polarised light into circularly polarised light and (a) plane polarised light into circularly polarised light?

20. How would you distinguish between elliptically polarised light and a mixture of plane polarised and unpolarised light? **(Benares Hindu Univ.)**

21. How is circularly polarised light distinguished from the unpolarised light? **(Benares Hindu and G.N. Univ.)**

22. Explain how would you produce and detect beams of circularly and elliptically polarised light. **(Kurukshetra Univ.)**

23. Describe the construction and explain the working of a Babinet compensator. What is its advantage over a $\lambda/4$ or $\lambda/2$ plate?

24. Describe how you would completely analyse an elliptically polarised light by a Babinet compensator.

25. How can you convert (a) a left-handed elliptically polarised light into a right-handed one and (b) a left-handed circularly polarised light into a right-handed one? **(Punjab and Punjabi Univ.)**

26. If μ_0 is refractive index for O-rays, μ_e the principal refractive index for E-rays, then show that the refractive index μ_θ for E-rays in a direction θ with optic axis is given by

$$\frac{1}{\mu_\theta^2} = \frac{\cos^2\theta}{\mu_0^2} + \frac{\sin^2\theta}{\mu_e^2}$$

27. Explain clearly (a) an optically active substance and (b) specific rotation. **(Calcutta and Vidyasagar Univ.)**

28. Show that a rectilinear vibration is equivalent to two opposite circular vibrations.

29. What is rotatory polarisation? Give Fresnel's theory of optical rotation. Discuss the dependence of rotation on λ and the experimental evidence in support of the theory. **(Agra, Vikram and Meerut Univ.)**

30. Describe the different factors on which the optical rotation, as studied by Biot, depend. Hence explain the phenomenon of 'rotatory dispersion'.

31. Describe the construction and working of a Laurent half-shade polarimeter, explaining the action of half-shade device. How would you use it to determine the specifc rotation of sugar solution. **(Calcutta, Allahabad and Lucknow Univ.)**

32. Explain the working of a bi-quartz polarimeter. Discuss the relative merits of a bi-quartz and a half-shade polarimeter. **(Indore, Calcutta and Saugar Univ.)**

Short Answer and Objective Questions

1. 'Polarisation requires that vibrations are transverse'—Comment on this statement.

2. Light waves can be polarised but sound waves cannot. Why? **(Himachal Univ.)**

3. Explain the terms : (a) polarisation of light, (b) plane of polarisation and (c) plane of vibration.

4. State and prove Brewster's law. State and explain Malus law.

5. Explain the terms : Optic axis, principal plane, uniaxial crystal, positive and negative crystals.

6. What is birefringence? Explain double refraction and ordinary and extraordinary rays.

7. What is dichroism? What are polaroids? Give some of their applications. What is meant by H-polaroid?

8. Write a short note on Wollaston prism. **(Punjab Univ.)**

9. What do you conclude about the nature of light from polarisation? **(Calcutta and Guru Nanak Univ.)**

10. What is the greatest importance of polarisation? **(Guru Nanak Univ.)**

11. What is the difference between positive and negative crystals? **(Kurukshetra Univ.)**

12. Explain what do you understand by the principal refractive index. **(Punjab Univ.)**

13. A black dot is marked on a white piece of paper. It is then viewed through a calcite crystal from the top :
    (a) How many images are expected to be seen and why?
    (b) Is it possible to see one image only? Explain your answer. **(Mysore Univ.)**

14. A Nicol prism cannot be used for highly convergent and highly divergent beam. Explain why.

15. What do you mean by elliptically and circularly polarised light?

16. Give the theory and construction of a quarter wave plate.

17. How can you convert plane polarised light into a circularly polarised light?

18. Explain, with a neat diagram, the working of a Babinet's compensator.

19. What is the nature of the emergent light when circularly polarised light is passed through (a) a quarter-wave plate and (b) a half-wave plate. **(Benares Hindu Univ.)**

20. How will you detect plane polarised light from partially polarised light? **(Cochin Univ.)**

21. Indicate the state of the emerging light in the following three cases :
    (a) a beam of plane polarised light is passed through a quarter-wave plate.
    (b) a beam of elliptically polarised light is passed through a quarter-wave plate.
    (c) a beam of plane polarised light is passed through a $\lambda/4$-plate such that its vibrations make an angle of 45° with the optic axis of the plate on incidence.

22. A rotating Nicol fails to distinguish between circularly polarised light and unpolarised light'.—Explain this and discuss how to distinguish between the two. **(Meerut Univ.)**

23. 'A rotating Nicol fails to distinguish between elliptically polarised and partially polarised light'. Explain this and discuss how to distinguish between the two. **(Agra Univ.)**

24. If you are given a $\lambda/4$-plate, a $\lambda/2$-plate and a simple glass plate how will you proceed to distinguish them from each other?

25. What is optical activity? Explain specific rotation and give its unit.

26. Name the factors on which the optical rotation depends. What are meant by dextro-rotation and laevo-rotation? Name a substance which can show both types of rotation.

27. What do you mean by 'rotatory dispersion'? Can it occur with monochromatic plane polarised light?

28. What is the principle on which Fresnel's explanation of optical rotation is based? Write down the expression for rotation obtained from the theory.

29. Find for what thickness no light will be transmitted when a quartz piece (cut perpendicular to optic axis) is interposed between a pair of parallel Nicols? Why is it necessary to cut the plate perpendicular to optic axis?

30. What are the disadvantages of a polarimeter using bi-quartz arrangement?

31. If θ be the polarising angle, the refractive index μ of the material is given by

    (a) $\sin\theta$, (b) $\tan\theta$, (c) $\cos\theta$, (d) $\cos^{-1}\theta$. **(Himachal Univ.)**

32. The refractive index of water is 1.33. The angle of polarisation of light reflected from the surface of a pond is

    (a) 53.1°, (b) 36.9°, (c) $\cos^{-1} 1.33$, (d) None of these.

33. The sum of the angle of refraction and of polarisation for a medium is

    (a) > 90°, (b) < 90°, (c) = 90°, (d) None of these.

34. Plane polarised light can be produced by

    (a) Simple reflection, (b) Nicol's prism, (c) Pile of plates, (d) All of these.

35. Nicol prism is based on

    (a) refraction, (b) scattering, (c) double refraction, (d) none of these.

36. The E-ray travels with the same velocity as O-ray
   (a) along the optic axis, (b) perpendicular to optic axis
   (c) along the principal plane, (d) none of these.

37. What will be the state of polarisation of the emerging light when a beam of circularly polarised light is passed through a $\lambda/4$-plate?
   (a) plane polarised, (b) elliptically polarised,
   (c) circularly polarised, (d) unpolarised.

38. In elliptically polarised light, the amplitude of vibration changes
   (a) in direction only, (b) in magnitude only,
   (c) both in magnitude and direction, (d) none of these.

39. When a $\lambda/2$-plate of quartz is placed between two crossed polaroids such that $\lambda/2$-plate is oriented at 45° to the axis of polaroids, the combination will transmit
   (a) no light, (b) $\frac{1}{4}$ of the intensity of incident light,
   (c) half the intensity of incident light, (d) none of these.

40. Calcite crystal is
   (a) dextro-rotatory, (b) laevo-rotatory,
   (c) both of these, (d) none of these.

41. The rotation of the plane of polarisation does *not* depend on
   (a) wavelength of light, (b) temperature,
   (c) thickness of the material, (d) none of these.

42. Of the following statements, which one is *not* correct?
   (a) A quartz crystal may be right-handed
   (b) A quartz crystal may be left-handed
   (c) A quartz crystal may be both right-handed and left-handed
   (d) None of the above is true.

Problems

1. The refractive index of glass is 1.5. Calculate the Brewster's angle for it and also the angle of refraction. **(Benares Hindu Univ.)** [56.3°; 33.7°]

2. Determine the Brewster's angle for glass of refractive index 1.5 immersed in water of refractive index 1.33. **(Punjab Univ.)** [48.5°]

3. The critical angle for refraction for a transparent substance to air is 40°. Calculate the polarising angle of the substance. [57.3°]

4. The polarising angle of a medium is 60°. What is the critical angle for it. [25°16′]

5. If the angle between a polariser and an analyser is 60°, what will be the intensity of light transmitted having I_0 as the original intensity of the incident light? **(Himachal Univ.)** [$0.25I_0$]

6. Two Nicol prisms are so arranged that the amount of light transmitted through them is maximum. What will be the percentage reduction in intensity of incident light when the analyser is rotated through (a) 30°, (b) 45°, (c) 60°, (d) 90°? [25%, 50%, 75%, 100%]

7. An analysing Nicol examines two adjacent plane polarised beams of light A and B with their plane of polarisation at right angles to each other. In one position of the analyser, the beam B shows zero intensity. A rotation by 30° from this position shows the beams to be matched (i.e., equal). What is the intensity ratio of the beams? [1 : 3]

8. Explain how you can rotate the direction of polarisation of a linearly polarised light by 90°, using two polaroid filters. Find the final intensity in terms of incident intensity I_0. [Polaroids to be kept with transmission axes at 45°; $I_0/4$]

9. A plane polarised light ($\lambda = 5893$ A) is incident on the quartz plate cut with optic axis parallel to the face. If the ratio of transmitted intensities of E and O-rays is 3 : 1, find the angle made by light vibration with the optic axis. [30°]

10. Calculate the thickness of a half-wave plate for light of wavelength $\lambda = 5000$ A; $\mu_0 = 1.544$ and $\mu_e = 1.533$. **(Calcutta and Punjab Univ.)** [0.227×10^{-4} m]

11. Calculate the thickness of a quartz plate for C-line of wavelength 6563 A. Given $\mu_0 = 1.54181$. Is the crystal positive? [3.63×10^{-5} m; Yes]

12. Determine the wavelength of light used when a $\lambda/4$-plate of thickness 1.2×10^{-5} cm is used for detection. Given $\mu_0 = 1.65$ and $\mu_e = 1.64$. Is the crystal plate positive? [4800 A; No]

13. A plane polarised beam of light is incident on a quartz plate cut with faces parallel to the axis. Find the least thickness for which O-ray and E-ray will combine to form plane polarised light. Given, $\mu_0 = 1.5442$ and $\mu_e = 1.5533$ and $\lambda = 5000$ A. [2.748×10^{-5} m]

14. The faces of a quartz plate are parallel to the optic axis of the crystal. What is the thinnest possible plate that would make the ordinary and the extraordinary rays

of $\lambda = 5890$ A, a half-wave apart on their exit? What multiple of this thickness would give the same result? Given, $\mu_e = 1.553$ and $\mu_0 = 1.544$.

[3.27×10^{-5} m; n-multiples, $n = 1, 2, \cdots$]

15. A plane polarised light of $\lambda = 600$ nm changes to a circularly polarised light on passing through a quartz crystal cut parallel to optic axis. Calculate the minimum thickness to produce such effect. Given, $\mu_e - \mu_0 = 0.005$. [0.003 mm]

16. Elliptically polarised light is incident normally on a $\lambda/4$-plate. Find the state of polarisation of the emergent light if the major axis of the ellipse makes an angle (a) 0°, (b) 90°, (c) 30° with the principal plane of the quarter wave plate.

[(a) and (b) plane polarised; (c) elliptic]

17. Describe the state of polarisation of the wave represented by

$$\vec{E}(z,t) = \hat{\imath}E_0\cos(kz - \omega t) - \hat{\jmath}E_0\sin(kz - \omega t)$$ [left circularly polarised]

18. A left circularly polarised light beam ($\lambda = 580$ nm) is incident on a doubly refracting crystal of thickness 0.0058 mm with the optic axis cut parallel to the surface. What will be the state of polarisation of the emergent beam? Given, $\mu_0 - \mu_e = 0.15$. **(Calcutta Hons.)** [right circularly polarised]

19. A right circularly polarised beam ($\lambda = 525$ nm) is incident normally on a doubly refracting crystal with optic axis parallel to the surface. The thickness of the crystal is 0.003 mm. Given $\mu_0 - \mu_e = 0.175$. Find the state of polarisation of the emergent beam. **(Calcutta Hons.)** [right circularly polarised]

20. Give the equation for a left-handed elliptically polarised light propagating along $+z$ direction and with its major axis at 135° with x-axis.

[$\vec{E}(z,t) = \hat{\imath}E_0\cos(kz - \omega t) + \hat{\jmath}E_0\sin(kz - \omega t + 3\pi/4)$]

21. A plane light wave is propagating along $+z$ axis. The wavefront at $z = 0$ is the xy-plane. At an instant t, the equations of the wave disturbance on xy-plane are given by

$$x = A_x\cos\omega t; \quad y = A_y\cos(\omega t + \phi)$$

ϕ being an ordinary function of time for unpolarised wave. With the above notation, write a set of equations of wave disturbances at $z = 0$ for (a) plane polarised wave and (b) a circularly polarised wave. **(Calcutta Hons.)**

[(a) $x = A_x\cos\omega t$; $y = 0$, (b) $x = A\cos\omega t$; $y = A\sin\omega t$]

22. A 22 cm long tube containing 88 cc of sugar solution produces an optical rotation of 9.9° when placed in a polarimeter. If the amount of sugar in the solution is 6 g, find the specific rotation. [66°]

Chapter 13

Coherence

13.1 Concept of coherence

If a wave appears to be a pure sine wave for an infinitely extended time or an infinitely extended space, it is called *a perfectly coherent wave.* The *characteristic* of such a wave is that a definite relationship between the phases of the wave at a certain time and at a given time later, or at a certain space-point and at a given space-point later exists. A *perfectly coherent wave* however is an *idealisation,* for a real light source always emits light in short pulses. Light waves that are pure sine waves only in a limited space or for a limited time are only *partially coherent waves.*

From above, it is apparent that there are two different criteria as regards coherence—the criterion of *time* or **temporal coherence** and the criterion of *space* or **spatial coherence.** We shall discuss them hereunder at a greater length.

13.1.1 Temporal coherence

The oscillating electric field E of a perfectly coherent light wave has a *constant* amplitude at any point in space, but its phase will vary with time linearly. As a function of t, the field E will appear as illustrated in Fig. 13.1. Obviously, it is an *ideal* sinusoidal function of time and does *not* agree with real cases.

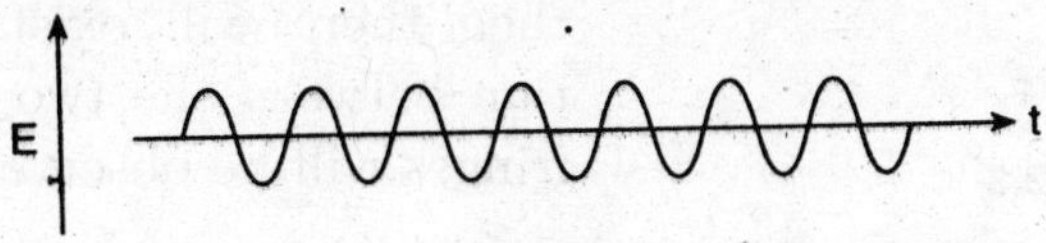

Fig. 13.1

No real source can ever produce an ideal sinusoidal field for all time, since when an excited atom makes a transition back to its initial state, it emits a light pulse of extremely short duration $\sim 10^{-10}$s. The field therefore is sinusoidal for an interval of $\sim 10^{-10}$s only. It undergoes abrupt phase changes thereafter for, different atoms of the source emit radiation in a random fashion. Fig. 13.2 illustrates the field due to a real source of light.

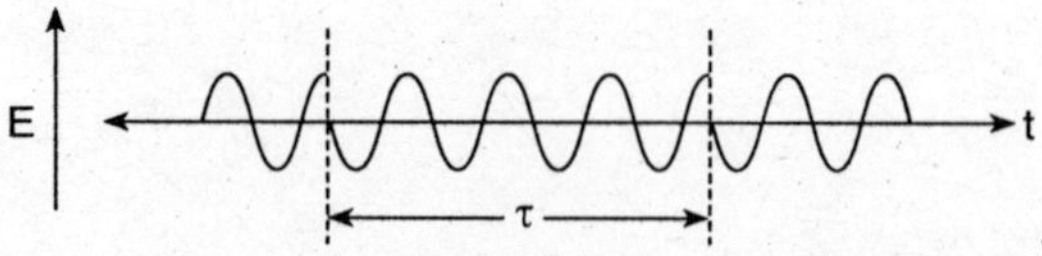

Fig. 13.2

The average time-interval for which the field of light vector remains sinusoidal, and therefore a definite phase relationship exists, is termed **temporal coherence** *of the light beam* or **coherence time**, symbolised by τ_c (or Δt).

The distance over which the field is sinusoidal is termed the **coherent length** L_c *of the light beam* and is given by

$$\boxed{L_c = \tau_c \times c} \tag{13.1.1}$$

where c is the speed of light in vacuum.

Illustration—We illustrate the coherence time or coherent length with fringes produced in a Michelson's interferometer (Fig. 13.3) about which we already read in Chapter : *Interference of light.*

The two beams 1 and 2 can produce stationary interference pattern only when there exists a definite phase relationship between them. With M_2', the image of the mirror M_2, produced by the semi-silvered plate P, the Michelson's arrangement is equivalent to an air-film sandwiched between two parallel reflecting surfaces—M_2' and the mirror surface M_1. If d be the separation between M_1 and M_2, $2d$ will be the path difference between the two interfering beams. Now, if

$$2d \ll L_c$$

then there will exist a definite phase relation between the two beams and interference fringes will be observed.

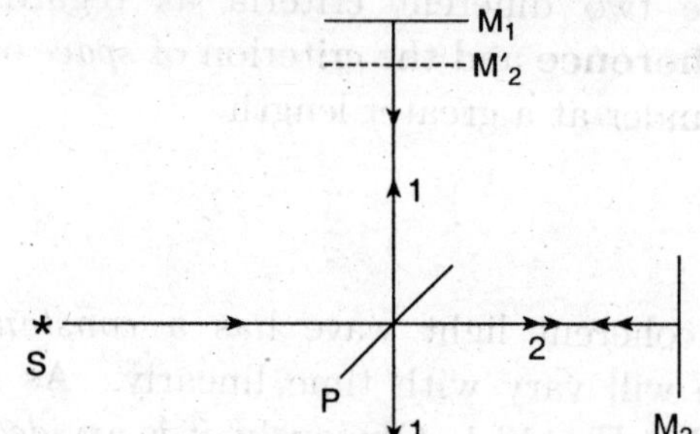

Fig. 13.3

Conversely, if the path-difference between the two interfering beams be much much greater than the coherent length, i.e.,

$$2d \gg L_c$$

the fringes will *not* be observed as there will be *no* definite phase relationship between the two beams.

So, starting with equal path lengths, if by moving one of the mirrors, the separation d is gradually increased, the fringes become poorer and poorer in contrast and eventually at a certain stage completely disappear. The path difference corresponding to the disappearance gives an *estimate of coherent length.*

• While for yellow light of sodium, the coherent length is about 3 cm, it is 30 cm for krypton red light and few kilometers for laser light.

13.1.2 Spatial coherence

The **spatial coherence** *refers to the phase relationship between the radiation fields at different space points.*

Let the light waves be emitted from the source S (Fig. 13.4) and A, B be two space points on a line joining them with S. The phase relationship between the points A and B depends on (i) the distance AB and (ii) the temporal coherence of the beam.

If $AB \ll L_c$, the coherent length, there will be a definite phase relationship between A and B. Consequently, there will be a *high degree of coherence* between the points A and B.

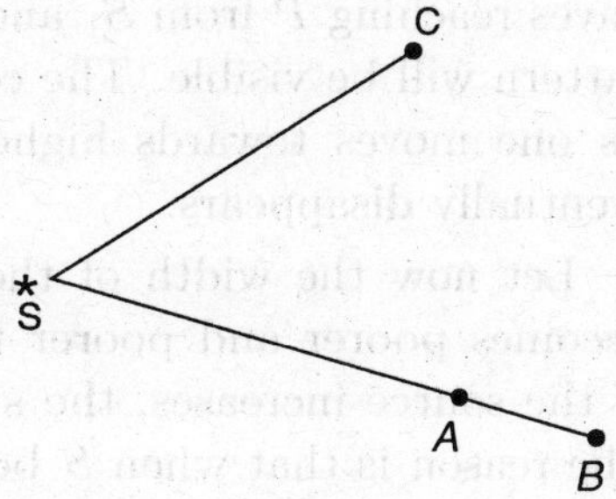

Fig. 13.4

Conversely, if the distance $AB \gg L_c$, the coherent length, there will be *no* coherence between the points A and B.

We now consider the points A and C which are *equidistant* from the source S, but unlike the previous case, *not* on the same line joining with S. If S is a true *point source*, light waves will reach A and C exactly in the same phase. In other words, the two space points will be in *perfect spatial coherence.* In case the source is an extended one, the two points A and C will no longer be in coherence as illustrated by Young's double slit experiment.

Illustrations—In reference to Fig. 13.5, light from a *narrow* slit S falls on two slits S_1, S_2, situated *symmetrically* with respect to S and equidistant from it. Since the light beams from S_1 and S_2 are from the *same* original beam, a *constant* phase difference will be maintained at all points on the screen and a stationary interference pattern will be observed.

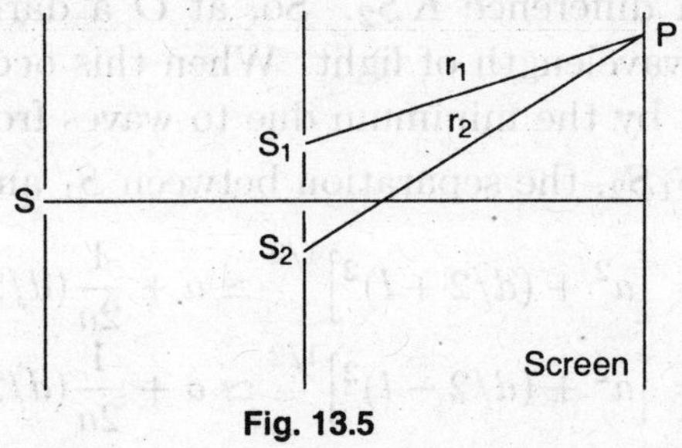

Fig. 13.5

Consider a point P on the screen and let $S_1P = r_1$ and $S_2P = r_2$. The interference pattern is seen around P at time t is due to the superposition of waves from S_1 and S_2

at times $t - r_1/c$ and $t - r_2/c$ respectively. If

$$\frac{r_2 - r_1}{c} \ll \tau_c,$$

the waves arriving at P from S_1 and S_2 will have a definite phase relationship and interference of good contrast is observed.

If, however,

$$\frac{r_2 - r_1}{c} \gg \tau_c,$$

waves reaching P from S_1 and S_2 will have no fixed phase relation and no interference pattern will be visible. The central fringe for which $r_1 = r_2$ will have a good contrast. As one moves towards higher order, the pattern becomes poorer and poorer, and eventually disappears.

Let now the width of the slit S be gradually increased. Then also the pattern becomes poorer and poorer in contrast and eventually vanishes, that is, as the size of the source increases, the spatial coherence on the screen changes into incoherence. The reason is that when S becomes wide, the slits S_1 and S_2 receive light waves from *different*, and hence *independent*, parts of the source and do not remain coherent with each other.

Mathematical analysis—Let the double-slit be illuminated by two independent point sources S and S', separated by l (Fig. 13.6). To find the least value of l at which the pattern on the screen would disappear.

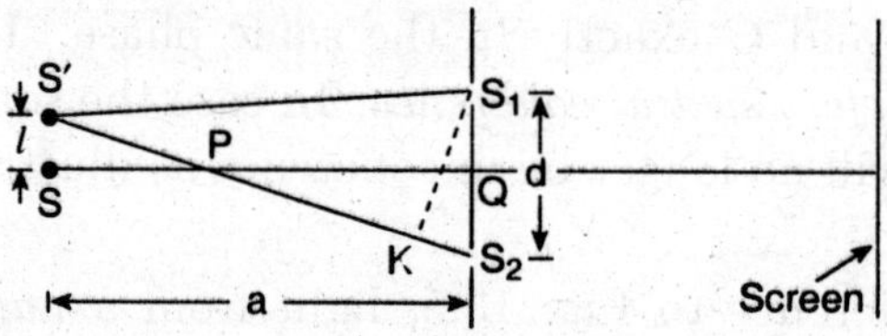

Fig. 13.6

There would be a bright fringe at O on the screen due to interference of waves from S reaching O *via* S_1 and S_2, having zero path difference. Waves from S' reaching O *via* S_1 and S_2, have path difference KS_2. So, at O a dark fringe is obtained when $KS_2 = \lambda/2$, λ being the wavelength of light. When this occurs, the maximum due to waves from S is destroyed by the minimum due to waves from S'.

Let $a = SQ$ and $d = S_1S_2$, the separation between S_1 and S_2. Then,

$$S'S_2 = \left[a^2 + (d/2 + l)^2\right]^{1/2} \simeq a + \frac{1}{2a}(d/2 + l)^2$$

$$S'S_1 = \left[a^2 + (d/2 - l)^2\right]^{1/2} \simeq a + \frac{1}{2a}(d/2 - l)^2,$$

assuming $a \gg d$ and also l.

$$\therefore\ KS_2 = S'S_2 - S'S_1 = ld/a.$$

So, for the fringes to disappear, we must have

$$KS_2 = \lambda/2 \simeq ld/a$$

$$\Rightarrow \boxed{l = \frac{\lambda a}{2d}}$$

It thus follows that if the double-slit is illuminated with an extended source with linear dimension exceeding $\lambda a/2d$, no interference pattern appears on the screen.

If we have an extended source of linear dimension $\lambda a/d$, then for every point on the source, there is an *independent* point at a distance $\lambda a/2d$ and the above analysis shows that there would be *no interference pattern.* For an extended source, fringes of good contrast will be observed only if

$$l \ll \frac{\lambda a}{d} \tag{13.1.2}$$

If θ be the angle subtended by the extended source SS' at Q, then $\theta = l/a$ and (13.1.2) becomes

$$d \ll \lambda/\theta \tag{13.1.3}$$

But, if on the other hand, we have

$$d \sim \lambda/\theta$$

then the fringes will be of very poor contrast or disappear altogether.

The distance $\lambda/\theta = l_w$ gives the *lateral distance* over which the beam may be assumed to be *spatially coherent* and is known as **lateral coherence width**.

- For a circular source, l_w is given by

$$l_w = \frac{1.22\lambda}{\theta}$$

- The analysis points out that by using non-coherent sources like the sun, the electric bulb etc., an almost spatially coherent beam can be produced by passing the light through a small hole of size $\ll l_w$.

13.2 Line width : Purity of spectral line and temporal coherence

A spectral line is said to be *perfectly pure* if it consists of a *single frequency.* This is only an *ideal* condition and is true only theoretically. Every spectral line has a finite width implying that it corresponds to a continuous distribution of wavelength in an interval, however narrow, between λ and $\lambda + d\lambda$. The *width* of a spectral line is a measure of the *purity* of line; smaller the width, purer is the line.

In Michelson's interferometer, the fringe pattern disappears, as already mentioned, when the path difference between the interfering beams exceeds the coherence length.

Thus, the concept of temporal coherence is directly related to the width or purity of the spectral line.

Relation with temporal coherençe—We adopt the criterion that when the path-difference between the interfering beams equals the coherent length L_c, the Michelson's rings due to two very close wavelengths λ_1 and λ_2 are completely out of step at the centre, i.e., a *bright* ring of λ_1 coincides with a *dark* ring of λ_2.

$$\therefore \quad L_c = n\lambda_1 = \left(n + \frac{1}{2}\right)\lambda_2 \tag{13.2.1}$$

We eliminate n from equation (13.2.1) by substituting $n = L_c/\lambda_1$ to get

$$\therefore \quad L_c = \left(\frac{L_c}{\lambda_1} + \frac{1}{2}\right)\lambda_2 \quad \Rightarrow \quad \frac{L_c}{\lambda_2} - \frac{L_c}{\lambda_1} = \frac{1}{2}$$

$$\therefore \quad L_c \frac{\lambda_1 - \lambda_2}{\lambda_1\lambda_2} = \frac{1}{2}$$

$$\Rightarrow \quad L_c = \frac{\lambda_1\lambda_2}{2(\lambda_1 - \lambda_2)} \tag{13.2.2}$$

If instead of discrete wavelengths λ_1 and λ_2, the beam has a continuous distribution of wavelengths from λ_1 to λ_2, then the fringe pattern disappears if

$$L_c = \frac{\lambda^2}{\Delta\lambda} \quad \Rightarrow \quad \Delta\lambda = \frac{\lambda^2}{L_c} \tag{13.2.3}$$

Thus, if the interference fringes become indistinct when the path-difference exceeds the coherent length L_c, we may conclude that *the spectral line of mean wavelength* λ has a *width or spread*

$$\Delta\lambda \simeq \frac{\lambda^2}{L_c} = \frac{\lambda^2}{c\tau_c}$$

Thus, the temporal coherence τ_c of the beam is directly related to the spectral width or spread $\Delta\lambda$ of the mean wavelength λ.

Since $\nu = c/\lambda \;\Rightarrow\; |\Delta\nu| = (c/\lambda^2)\Delta\lambda$, we get from above

$$\Delta\nu \simeq \frac{c}{L_c} \simeq \frac{1}{\tau_c} \tag{13.2.4}$$

where τ_c is the *coherence time*.

So, *narrower the frequency-spread, longer is the coherence time.* A perfectly sharp monochromatic (purest) spectral line ($\Delta\nu = 0$) would correspond to an infinite interval of time, $\tau_c = \infty$.

- The quantity $\Delta\nu/\nu$ represents the *monochromaticity* (or spectral purity) of the source.

13.3 Visibility of fringes and degree of coherence

The *visibility* V of interference fringes is defined as follows :

$$V = \frac{I_{max} - I_{min}}{I_{max} + I_{min}}$$

where I_{max} and I_{min} are respectively the maximum and minimum intensity.

The visibility V is a *measure of the degree of coherence* of the interfering light waves.

When two wave trains of equal intensities *overlap* in the entire finite lengths, the minimum intensity $I_{min} = 0$. So, the visibility $V = 1$ or 100% and the waves are said to be in *perfect coherence* with highest contrast between the maxima and the minima.

If however the two wave trains do *not overlap* at all $I_{max} = I_{min}$ and so $V = 0$. The waves have *no coherence* and fringes are not visible.

The intermediate situation is the *partial overlapping* of the two wave trains as illustrated in Fig. 13.7.

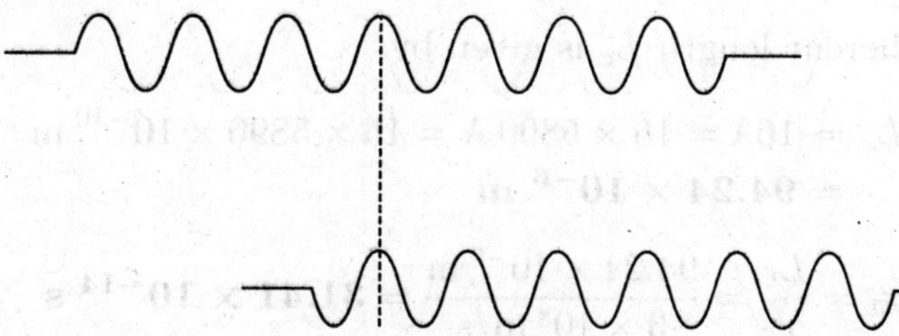

Fig. 13.7

In this case, interference is possible but the contrast between the fringes is far less. The degree of contrast depends on the degree of overlapping.

Relation between visibility and coherence—Let two points on a screen be illuminated by two bundles of light to produce *equal* intensities I_0 and assume that each bundle has *two* parts a and b. Of these, the parts a are completely *coherent* and parts b completely *incoherent*. Let the parts a produce an intensity

$$I_a = \beta I_0$$

where β is known as the *degree of coherence*.

$\therefore$ Parts b will produce an intensity, $I_b = (1 - \beta)I_0$.

Obviously, interference occurs due to parts a only and give fringes whose maxima is 4 times the individual contributions (see Note at the end).

$$\therefore \text{ Maximum intensity} = 4I_a; \quad \text{Minimum intensity} = 0$$

Due to the b parts, a uniform intensity $2I_b$ will be superposed on the above interference pattern. In the pattern, therefore,

$$\begin{aligned} I_{max} &= 4I_a + 2I_b = 4\beta I_o + 2(1-\beta)I_0 \\ &= 2(1+\beta)I_0 \end{aligned}$$

$$I_{min} = 0 + 2I_0 = 2(1-\beta)I_0$$

$$\therefore \quad V = \frac{2(1+\beta)I_0 - 2(1-\beta)I_0}{2(1+\beta)I_0 + 2(1-\beta)I_0} = \frac{4\beta I_0}{4I_0} = \beta$$

The visibility or the degree of contrast of the fringes produced by two waves of equal intensities is equal to the degree of coherence between the waves.

Note. $I_{max} = (a_1 + a_2)^2$, where a = amplitude. If $a_1 = a_2 = a$, say then we have $I_{max} = 4a^2 = 4 \times I_{indiv}$.

13.4 Illustrated Examples

► **Example 1.** *Light of wavelength* $\lambda = 5890\,\text{A}$ *has wave trains* 16λ. *Calculate its coherent length and coherence time.*

Solution. The coherent length L_c is given by

$$L_c = 16\lambda = 16 \times 5890\,\text{A} = 16 \times 5890 \times 10^{-10}\,\text{m}$$
$$= \mathbf{94.24 \times 10^{-6}\ m}$$

$$\text{Coherent time, } \tau_c = \frac{L_c}{c} = \frac{94.24 \times 10^{-6}\,\text{m}}{3 \times 10^8\,\text{m/s}} = \mathbf{31.41 \times 10^{-14}\ s}$$

► **Example 2.** *The Doppler width of a source at* $\lambda = 6058\,\text{A}$ *is* $0.0055\,\text{A}$. *Find out the coherent length.* **(Calcutta Hons.)**

Solution. Here frequency $\nu = c/\lambda \Rightarrow |\Delta\nu| = \frac{c}{\lambda^2}\Delta\lambda$

$$\therefore \quad \text{Coherence time, } \tau_c = \frac{1}{\Delta\nu}$$

$$\Rightarrow \text{coherent length, } L_c = c \times \tau_c = c\frac{1}{\Delta\nu} = \frac{c \cdot \lambda^2}{c \cdot \Delta\lambda} = \frac{\lambda^2}{\Delta\lambda}$$

$$\therefore \quad L_c = \frac{(6058 \times 10^{-10})^2}{0.0055 \times 10^{-10}} = \mathbf{0.67\ m}$$

► **Example 3.** *A quasi-monochromatic light source gives radiations of wavelength (mean)* $\lambda = 660\,\text{nm}$ *and has a band width* $\Delta\nu = 10^9$ Hz. *Find* (i) *coherence time,* (ii) *coherent length and* (iii) *frequency-stability.*

Solution. $$\text{Coherence time, } \tau_c = \frac{1}{\Delta\nu} = \frac{1}{10^9} = \mathbf{10^{-9}\ s.}$$

$$\text{Coherent length, } L_c = c \times \tau_c = 3 \times 10^8 \times 10^{-9} = \mathbf{0.3\ m}$$

$$\text{Frequency, } \nu = \frac{c}{\lambda} = \frac{3 \times 10^8}{660 \times 10^{-9}} = 4.5 \times 10^{14}\,\text{Hz}$$

$$\therefore \quad \text{Frequency-stability} = \frac{\Delta\nu}{\nu} = \frac{10^9}{4.5 \times 10^{14}} = \mathbf{2.2 \times 10^{-6}}$$

▶ **Example 4.** *For sodium light of wavelength* $\lambda = 5890$ A *and coherence time* $\tau_c = 10^{-10}$s*; find the monochormaticity of the source.* **(Calcutta Hons.)**

Solution. Frequency, $\nu = \dfrac{c}{\lambda} \Rightarrow |\Delta\nu| = \dfrac{c}{\lambda^2}\Delta\lambda$

Now, monochromaticity,

$$|\Delta\lambda| = \frac{\lambda^2\Delta\nu}{c} = \frac{\lambda^2}{c \cdot \tau_c} \quad \left(\because\ \tau_c = \frac{1}{\Delta\nu}\right)$$

$$= \frac{(5890 \times 10^{-10})^2}{3 \times 10^8 \times 10^{-10}}\,\text{m} = \mathbf{0.1156\ A}$$

▶ **Example 5.** *A* He–Ne *laser has a coherent length of* 10 m. *What is the coherence time?* **(Calcutta Hons.)**

Solution. Coherence time, $\tau_c = L_c/c$, where L_c is the coherent length and c the velocity of light in vacuo.

$$\therefore \quad \tau_c = \frac{10\,\text{m}}{3 \times 10^8\,\text{m/s}} = \mathbf{3.33 \times 10^{-8}\ s}$$

▶ **Example 6.** *Red light of* Cd *of mean wavelength* 6438 A *and line width* 0.015 A *is used in a Michelson interferometer for the initial setting of zero path-difference. Then one of the mirrors is slightly moved till the fringes first disappear. What is the distance by which the mirror is moved?*

Solution. Here $\lambda = 6438\,\text{A} = 6438 \times 10^{-10}$ m and $\Delta\lambda = 0.015\,\text{A} = 0.015 \times 10^{-10}$ m

$$\therefore \quad \text{Coherent length, } L_c = \frac{\lambda^2}{\Delta\lambda} = \frac{(6438 \times 10^{-10})^2}{0.015 \times 10^{-10}} = 0.30\,\text{m}$$

When the (optical) path difference $2d$ exceeds L_c, the fringes will (first) disappear and this occurs when the mirror moves over d.

$$\therefore \quad \text{Required shift in mirror} = \tfrac{1}{2} \times 0.30 = \mathbf{0.15\ m}$$

▶ **Example 7.** *One of the most ideal line of* Cd-red *has wavelength* 6438 A *and a coherent length* $L_c = 20$ cm. *Calculate* (i) *the line width, and* (ii) *the coherence time.*

Solution. Here $\lambda = 6438\,\text{A} = 6.438 \times 10^{-7}$ m, $L_c = 20$ cm $= 0.2$ m and $c = 3 \times 10^8$ m/s.

$$\therefore \quad \text{Line width, } d\lambda = \frac{\lambda^2}{L_c} = \frac{(6.438 \times 10^{-7})^2}{0.20} = 0.02 \times 10^{-10}\ \text{m} = \mathbf{0.02\ A}$$

$$\text{Frequency-spread,} \quad \Delta\nu = \frac{c}{L_c} = \frac{3 \times 10^8}{0.2} = 1.5 \times 10^9\ \text{Hz}$$

$$\therefore \quad \text{Coherence time, } \tau_c = \frac{1}{\Delta\nu} = \frac{1}{1.5 \times 10^9} = \mathbf{0.67 \times 10^{-9}\ s}$$

► **Example 8.** *Two light beams having intensities in the ratio 1 : 9 produce interference fringes of intensity 0.3. What information do you get about the degree of coherence?* **(Allahabad Univ.)**

Solution. Visibility of fringes, $V = \dfrac{I_{max} - I_{min}}{I_{max} + I_{min}}$.

If the intensity ratio of the two coherent beams be α, then we have

$$V = \frac{2\sqrt{\alpha}}{1+\alpha}$$

Proof: With a's as amplitudes, $I_{max} = (a_1 + a_2)^2$; $I_{min} = (a_1 - a_2)^2$

$\Rightarrow I_{max} - I_{min} = 4a_1a_2$ and $I_{max} + I_{min} = 2(a_1^2 + a_2^2)$. Also, $\alpha = a_1^2/a_2^2$.

$$\therefore \quad V = \frac{4a_1a_2}{2(a_1^2 + a_2^2)} = \frac{2a_1a_2}{a_2^2(1 + a_1^2/a_2^2)} = \frac{2(a_1/a_2)}{1 + a_1^2/a_2^2} = \frac{2\sqrt{\alpha}}{1+\alpha}$$

$$\text{Here, } \alpha = 1/9 \Rightarrow V = \frac{2\sqrt{(1/9)}}{1 + (1/9)} = 0.6$$

But the given value of visibility is **0.3** i.e., half the value of two coherent beams. So, the degree of coherence is only **50%**.

► **Example 9.** *Assuming $\Delta\nu = 1/\Delta t$ where $\Delta\nu$ is the frequency band and Δt the coherence time, derive the expression for the coherent length Δl of a wave in terms of line width $\Delta\lambda$.*

Solution. Wavelength, $\lambda = c/\nu$, with usual meanings of the symbols. Differentiating,

$$|\Delta\lambda| = \frac{c\Delta\nu}{\nu^2} = \frac{\lambda^2 \Delta\nu}{c}$$

$$\therefore \quad \Delta\lambda = \frac{\lambda^2}{c \cdot \Delta t} = \frac{\lambda^2}{\Delta l} \quad (\because \ \Delta\nu = 1/\Delta t)$$

where $\Delta l = c \cdot \Delta t =$ the coherent length of the wave.

$$\therefore \quad \text{Coherent length,} \quad \boxed{\Delta l = \frac{\lambda^2}{\Delta\lambda}}$$

Review Questions

1. Explain the concept of coherence. Discuss temporal coherence and spatial coherence, illustrating them with the help of suitable experiments.
(Kanpur Univ.)

2. Explain how the width and purity of a spectral line is related with the concept of temporal coherence. Show that the frequency-spread of a spectral line is of the order of inverse of coherence time.

3. How does the visibility of fringes depend on the degree of coherence?
(Allahabad Univ.)

4. Distinguish between the spatial coherence and temporal coherence. What are coherence length and coherence time? **(Delhi Univ.)**

5. Explain what you understand by line width and the frequency spread.
(Himachal Univ.)

6. Write a note on temporal coherence. Explain spatial coherence due to a point source and an extended source of light.

7. Explain spatial coherence with the help of Young's double slit experiment. Find a relation between the linear dimension and spatial coherence of a source for a given separation between the slits.

Short Answer and Objective Questions

1. Define the following terms : coherence time and coherent length.
2. Establish how the frequency-spread is related with coherence time.
3. Explain spatial coherence due to a point source.
4. What exactly is the difference between temporal coherence and spatial coherence?
5. What is monochromaticity? Is there any relationship between natural width of a spectral line and coherence time?
6. Is strictly monochromatic light realizable practically? Explain your answer.
7. Can two independent laser beams produce an interference pattern?
(Allahabad Univ.)

8. The frequency-spread of a line, $\Delta\nu$ is given by

(a) $\Delta\nu = c/\tau_c$, (b) $\Delta\nu = \tau_c/c$, (c) $\Delta\nu = \tau_c$, (d) $\Delta\nu = 1/\tau_c$.

9. The coherent length L_c is given by

(a) $L_c = c/\tau_c$, (b) $L_c = \tau_c/c$, (c) $L_c = c\tau_c$, (d) None.

10. The coherent length for the red line of Cadmium of $\lambda = 6.4 \times 10^{-7}$ m is 30 cm. The coherence time is

(a) 10^{-9} s, (b) 10^{-7} s, (c) 10^{-8} s, (d) None.

Problems

1. One of the Krypton orange lines has a wavelength $\lambda = 6058$ A. If its coherent length is 20 cm, find the line width and the coherence time. [1.5×10^9 Hz; 0.6×10^{-9}s]

2. The coherence length for the red line of Cd of wavelength 6.438×10^{-7} m is 30 cm. Calculate the number of oscillations corresponding to the coherence length. [4.66×10^{-5}]

3. In Young's double slit experiment, the pinhole is of diameter 1 mm. Calculate the distance between S_1 and S_2 (Fig. 13.5) to have distinct interference fringes if $SS_1 = SS_2 = 100$ cm and $\lambda = 5 \times 10^{-7}$ m? [Small compared to 6.1×10^{-4} m]

4. A pinhole of diameter 1 mm is placed at a distance of 1 m from two pinholes. Find the lateral coherence width of the pin hole which is illuminated with light of $\lambda = 5000$ A. [0.06 cm]

5. If a perfectly monochromatic light beam of $\lambda = 623.8$ nm be chopped into 0.1 ns pulses by a shutter, calculate the resultant line width, band width and coherence length. [0.13 A, 10^{10} Hz; 3 cm]

6. Calculate the frequency band width for white light having the range 4×10^{14} Hz to 7.5×10^{14} Hz. Also, find the associated (a) coherent time and (b) coherence length. [3.5×10^{14} Hz; 2.86×10^{-15}s; 8.58×10^{-7} m]

7. The coherence length of sodium D-line ($\lambda = 5890$ A) is 2.5 cm. Calculate the coherence time and the spectral width of the line. [0.83×10^{-10}s; $\Delta\lambda = 0.138$ A]

8. Light of wavelength $\lambda = 660$ nm has wave trains 20λ. Calculate its (a) coherence length and (b) coherence time. [13.2 μm; 4.4×10^{-14}s]

9. The orange Krypton line of wavelength 605.8 nm has a coherence length of ~ 20 cm. Calculate (a) the line width and (b) the frequency stability. [~ 0.01A; $\sim 1.5 \times 10^{-6}$]

Chapter 14

Laser and Holography

14.1 Introduction

Laser is one of the most fascinating development in physics of the recent past. It is an acronym for Light Amplification by Stimulated Emission of Radiation (LASER) and is a device for producing a strong, highly monochromatic, directional and coherent beam of light of high power density. Its working depends on the phenomenon of what is called '*stimulated emission*' of radiation, the theory of which was worked out as early as 1917 by Einstein. He was considering the equilibrium between matter and electromagnetic radiation in a black body chamber at a constant temperature where energy-exchange occurs due to absorption and spontaneous emission of radiation by atoms. He observed that the absorption and emission processes alone cannot explain the equilibrium and predicted that there must be an additional process, now termed *'stimulated emission'*.

The prediction attracted little attention until in 1954 two Russian physicists N. Basov and A.M. Prokhorov, and the American physicist C.H. Townes discovered almost simultaneously and independently the phenomenon of Microwave Amplifiction by Stimulated Emission of Radiation (MASER), using ammonia, NH_3. In 1958, Townes and A.L. Schawlow showed that the 'maser' principle can also be extended to the field of optical or visible radiation. Finally, in 1960, a device called *'optical maser'* or *'laser'* was first built by T.H. Maiman using ruby as the active medium, and since then it opened up a completely new and exciting field of development in optics. Although a source of light, it differs widely from traditional light sources. It is not used for illumination purposes unlike other usual light sources. Lasers closely resemble radio and microwave transmitters and generate a highly coherent and directional light beam of extreme monochromaticity.

14.2 Absorption and emission of radiation by matter

Absorption of radiation—An atom has a number of possible quantised energy states characterised by its principal quantum number n ($= 1, 2, 3, \cdots,$). It remains in the ground state with ninimum energy E_1 in absence of external influences. On being subjected to some action, say irradiation by photons of right frequency ν, it transits to a higher energy state E_2, absorbing energy $h\nu$ of the radiation. This is called *absorption* or *excitation* (Fig. 14.1) for which the right frequency ν is given by

$$\nu = \frac{E_2 - E_1}{h}$$

• Absorption is necessarily a *stimulated* or *induced process*, the absorbed photon being the stimulating photon.

Spontaneous emission of radiation—Consider now an atom initially in the excited state E_2 (Fig. 14.2). An atom stays in the excited state usually for a short period $\sim 10^{-8}$s and returns, *of its own*, to the initial state E_1, emitting a photon of frequency ν. The process is termed *spontaneous emission* or *de-excitation* and the photon energy is given by

$$h\nu = E_2 - E_1$$

If there is an assembly of atoms, the radiation (photons) emitted by each atom, due to spontaneous transition, has a *random direction* (no directivity) and a *random phase.* The emitted light is therefore *non-coherent* in nature.

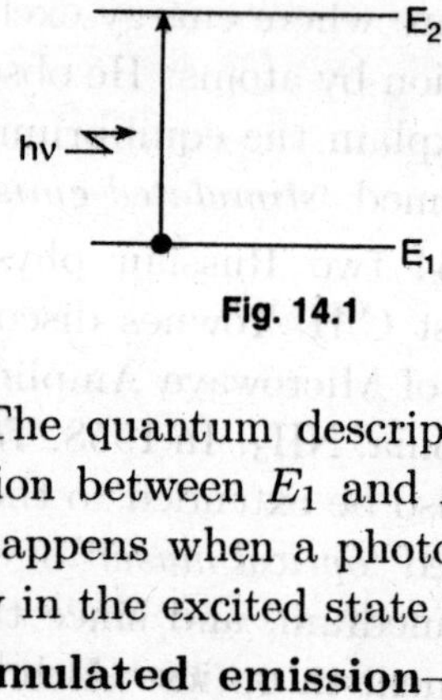

Fig. 14.1

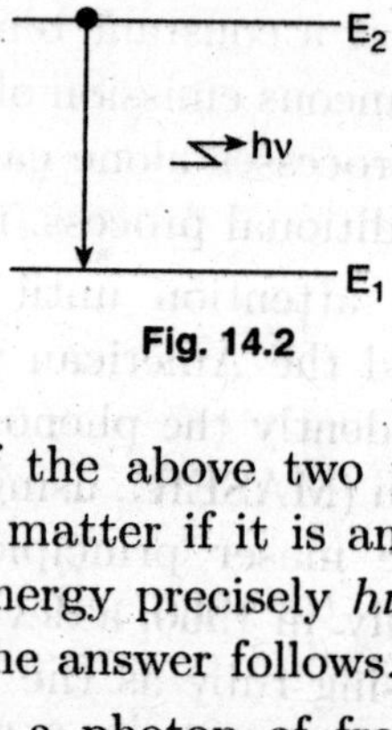

Fig. 14.2

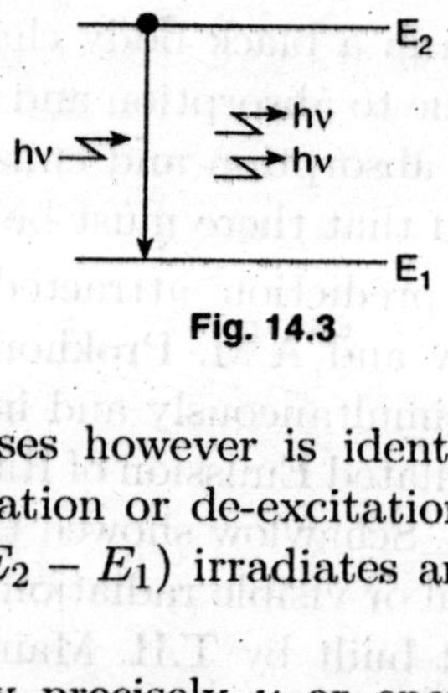

Fig. 14.3

• The quantum description of the above two processes however is identical—a transition between E_1 and E_2, no matter if it is an excitation or de-excitation. But what happens when a photon of energy precisely $h\nu$ $(= E_2 - E_1)$ irradiates an atom already in the excited state E_2? The answer follows.

Stimulated emission—When a photon of frequency precisely ν or energy $h\nu$ $(= E_2 - E_1)$ irradiates an atom, already in the excited E_2, it cannot excite the atom which is already excited. Instead, it produces the equivalent effect : it de-excites the atom. So, *under the influence of the electromagnetic field of a photon of frequency ν incident on it, it makes a transition to the lower energy state E_1*, emitting an additional photon of the same frequency ν (Fig. 14.3). So, now there are two photons, one original and the other emitted. This type of transition is called *stimulated* (or *induced*) *emission* of radiation in contrast to spontaneous emission.

• While in spontaneous emission photons are emitted in random directions, in stimulated emission the photon always leaves the atom *in the direction of the incident stimulating photon.* More remarkable is that the incident and stimulated photons are *coherent* and add completely to *amplify* the incident beam. If these two photons are now incident on two other atoms in the state E_2, it will result in stimulated emission of two more photons and so on *in chains.* So, if a large number of excited atoms is involved, the stimulated emission generates *an intense beam of high coherence and directivity.*

14.3 Einstein's theory : A and B coefficients

Transition between various energy states is essentially a *statistical process* and one cannot predict which particular atom at a given instant will transit from one energy state to another. But if a very large number of atoms are involved, we can calculate, applying the *probability theory,* the rate of relative transitions between two energy states fairly accurately. This is what Einstein exactly did, assuming atomic system to be in equilibrium with e. m. radiation.

Calculation—Let an assembly of atoms be in *thermal equilibrium* at a temperature T with radiation of frequency ν and energy density $u(\nu)$. Let N_1 and N_2 be the number of atoms per unit volume at any instant in state 1 and 2 respectively. The probable rate P_{12} of *absorption transition* $1 \to 2$ depends on the states 1 and 2 and is proportional to the energy density of radiation $u(\nu)$.

$$\therefore \quad P_{12} = B_{12}u(\nu),$$

the proportionality constant B_{12} is called the *Einstein's coefficient of absorption of radiation.*

$\therefore$ Number of atoms in state 1 that absorb a photon and rise to state 2 per unit time is given by

$$N_1P_{12} = N_1B_{12}u(\nu)$$

The probable rate of *spontaneous transition* $2 \to 1$ depends on the states 1 and 2 and on A_{21}, the probability of spontaneous transition $2 \to 1$.

$\therefore$ Number of atoms in state 2 that drop to state 1 by spontaneous transition per unit time is plainly

$$N_2A_{21}$$

being *independent of* the energy density $u(\nu)$ of radiation.

There may again be a downward *stimulated transition* from $2 \to 1$ due to the electromagnetic radiation field. The probability of such emission transition is proportional to the energy density $u(\nu)$ of stimulation, apart from the dependence on states 1 and 2, and may be written as $B_{21}u(\nu)$, where B_{21} is *the probability of stimulated emission.*

$\therefore$ No. of atoms undergoing stimulated transition $2 \to 1$ per unit time is

$$N_2B_{21}u(\nu)$$

For *equilibrium,* the absorption and emission per unit time must occur equally

$$\therefore \quad N_2A_{21} + N_2B_{21}u(\nu) = N_1B_{12}u(\nu)$$

$$\therefore\ u(\nu) = \frac{N_2 A_{21}}{N_1 B_{12} - N_2 B_{21}} = \frac{A_{21}}{B_{21}} \cdot \frac{1}{\frac{N_1}{N_2}\left(\frac{B_{12}}{B_{21}}\right) - 1}$$

$$= \frac{A_{21}}{B_{21}} \cdot \frac{1}{(B_{12}/B_{21}) e^{h\nu/kT} - 1} \tag{14.3.1}$$

$$(\because\ N_1/N_2 = e^{h\nu/kT})$$

This relation for the energy density of radiation of frequency ν must be in accord with *Planck's radiation formula*

$$u(\nu) = \frac{8\pi h\nu^3}{c^3} \cdot \frac{1}{e^{h\nu/kT} - 1} \tag{14.3.2}$$

Comparing (14.3.1) and (14.3.2), we obtain

$$B_{12} = B_{21} \quad \text{and} \quad A_{21}/B_{21} = 8\pi h\nu^3/c^3$$

and, in general, $\quad B_{mn} = B_{nm}$ and $A_{nm}/B_{nm} = 8\pi h\nu^3/c^3$

Thus, *the probabilities of stimulated absorption and stimulated emission are equal.* Further, the second equality implies that *the probabilities of spontaneous emission increases as* ν^3, i.,e., *rapidly with energy difference* between the involved states.

The A's and B's are respectively known as **Einstein's A and B coefficients.** They cannot be determined by classical electromagnetic theory. However, B can be calculated quantum mechanically using Dirac's theory and thence the value of A can be obtained.

• The *reciprocal of* A_{nm} is a measure of the *life-time* of the upper state against the spontaneous decay to the lower one.

14.4 Population inversion : Different methods

Normally, due to thermal equilibrium, the higher energy states are much less populated since by Boltzmann law $N_2/N_1 = e^{-(E_2-E_1)/kT}$ i.e., $N_2 < N_1$. So, even if $B_{mn} = B_{nm}$, the stimulated emission is normally unlikely to occur. It can be initiated and sustained, only *if by some means a large number of atoms are made available in appropriate excited energy states i.e.,* $N_n > N_m$.

The situation in which the number of atoms in the higher state exceeds that in the lower state is known as **population inversion**. And *the process of achieving population inversion is called* **pumping**.

There are many different methods or types of pumping. The principle is to *selectively transfer one form of energy to a specific energy state of the atom.* There are broadly *four* different methods.

1. *Optical pumping* or excitation by photons,
2. *Excitation with* the help of *electrons,*
3. *Inelastic collision* between atoms, and
4. Excitation by *chemical method.*

Optical pumping—Where high energy light source e.g., Xenon flash lamp supplies sufficient photon–flux in a given spectral range to excite the atoms. The atoms, by selective absorption of the radiation, are raised from the lower state to the appropriate excited state. The excited atoms then make non-radiative transition (energy transferred to lattice thermal motion) to a metastable energy state and remain in this metastable state for a relatively longer time. So N_2 is now more than N_1, i.e., we have *population inversion.* Optical pumping is used in ruby laser.

Excitation by electrons—Where electrons are released from gaseous atoms by high voltage electric discharge through the gas. The intense electric field accelerates these electrons to high speeds and they collide with neutral atoms, a fraction of which make thereby a transition to the excited state. This method is used in some gas lasers.

Inelastic collision between atoms—If an electric discharge passes through a gas having two kinds of atoms X and Y, then some X's are raised to excited state X^* and some Y's to Y^*. Let one of them, say X^* be in a *metastable state*—a state with a mean life-time longer than normal mean life ($\sim 10^{-8}$s) of an excited atom. If excitation energies of X^* and Y^* are *nearly* equal, then some Y-atoms in the ground state may, by inelastic collision with X^*, transit to the excited state :

$$X^* + Y \rightarrow Y^* + X$$

and the number of Y^*-atoms continually increases because of metastability of X^*. This method of pumping is used in Helium-Neon gas laser.

Chemical excitation—Often an atom or a molecule is a product of a chemical reaction and is produced in an excited state, e.g., HF. The HF-molecule is produced in an excited state when hydrogen and fluorine combine. The number of such excited molecules (or atoms) may be considerably more than those in the normal state and gives rise to population inversion.

• The situation of population inversion ($N_n > N_m$) indicates that the Kelvin temperature T in the equation

$$\frac{N_n}{N_m} = e^{-(E_n - E_m)/kT}$$

must be made *negative.* For this, a state of population inversion is sometimes referred to as a *negative Kelvin temperature* (!).

For details on negative Kelvin temperature, consult author's 'Thermal Physics'.

14.5 Basic laser system

Whatever be the active substance, solid dielectric, gas, semiconductor etc., a basic laser system consists of the following components (Fig. 14.4).

1. The *active* or *laser medium, A*, which is either a solid, liquid, gas or semiconductor junction, capable of sustaining stimulated emission.
2. An *external source of excitation, E* that supplies pumping energy to cause population inversion, and

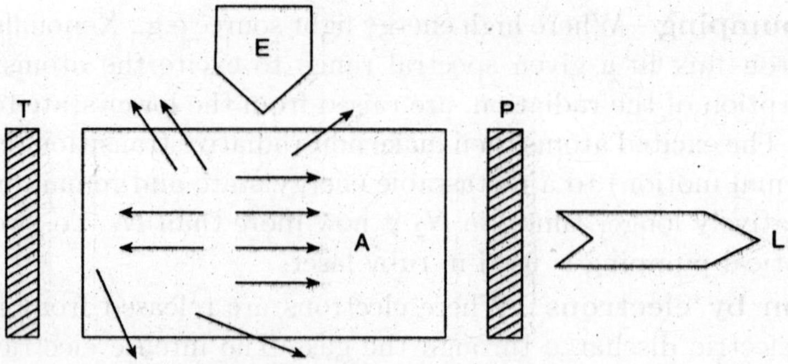

Fig. 14.4

3. An *optical resonator*—To sustain laser oscillation, a part of the output must be fed back into the system. Such a positive feed back is brought about by placing the active medium between a pair of plane parallel mirrors, facing each other, one at each end of the medium; one of them is a total reflector T and the other a partial one P that allows a part of the generated laser beam L to pass out. The arrangement is known as an *optical resonator*.

Action—To cause population inversion, the medium A is fed with pump energy E which is then released from excited atoms by stimulated emission. A photon released by one excited atom stimulates another it encounters in its path to release a second photon; the two coherent photons add completely to a beam of twice the intensity. As the beam courses through the medium, its amplitude rapidly increases by more and more stimulated emissions. The reflector T reverses the beam allowing it another passage (in fact, multiple reversions may occur) through the excited medium for further amplification. On reaching P, a part L of the beam escapes as a *laser beam*.

Stimulated photons that are emitted inclined to the axis are lost through the sides of the system so that the final emergent beam is always along the axis.

14.6 Optical resonator : *Q*-value

An optical resonator used in a laser can have a number of different configurations. Of them, the simplest one is already described in Art. 14.5 and is now seldom used, for the mirrors require critically precise alignment which is a problem.

Among the other alternative configurations (Fig. 14.5a), mention may be made of (i) *confocal configuration* where two identical concave mirrors face each other being separated by a distance equal to their radius; (ii) *concentric* or *spherical configuration* where also two concave mirrors of the *same radius* face each other and are separated by a distance equal to *twice their radius*; (iii) *hemispherical cavity* where a concave spherical mirror is placed at one end and a plane mirror is placed at its *centre of curvature*; and (iv) a *long-radius cavity* having two concave mirrors of *long* but *equal radius* of curvature.

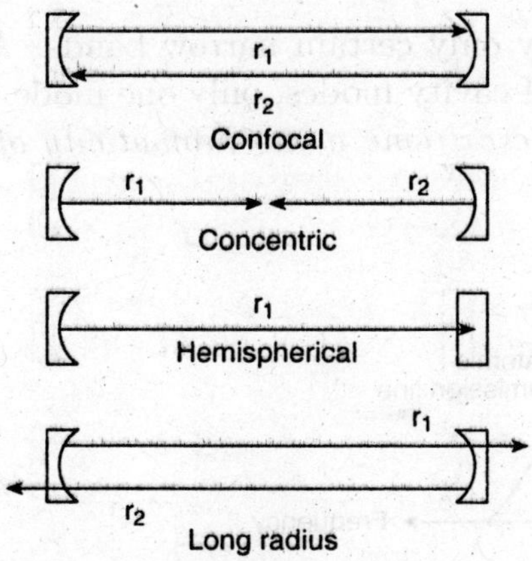

Fig. 14.5a

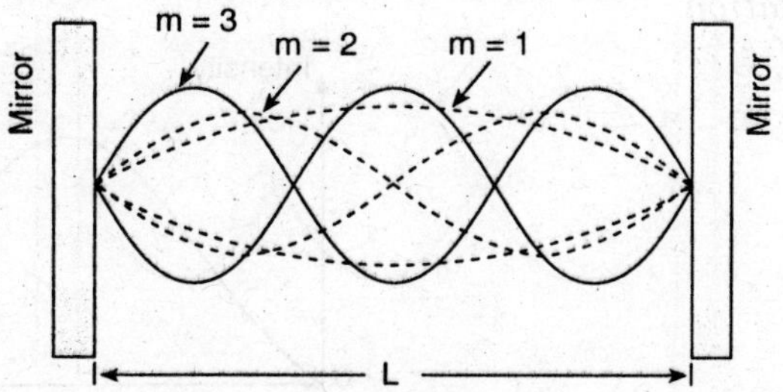

Fig. 14.5b

• In confocal configuration, the alignment problem is much less; in hemispherical cavity it is easier to align but the power output is comparatively low; the long-radius cavity is the most common among the commercial lasers.

Standing waves—The radiation field inside the resonator gives rise to a stationary wave pattern with a *node at each mirror.* When the cavity resonates, the distance L between the mirrors has clearly an *integral multiple* of half wavelengths (Fig. 14.5b).

$$\therefore \quad L = m\frac{\lambda}{2}, \quad m = 1, 2, 3, \cdots$$

where m corresponds to the *axial mode number.*

Expressed in terms of frequency,

$$\boxed{\nu_m = \frac{(c/n)}{\lambda} = \frac{mc}{2Ln}} \tag{14.6.1}$$

where c is the speed of light in vacuum and n the index of the active medium.

So, an infinite number of longitudinal cavity modes are possible, each with a distinct frequency ν_m. The *frequency-separation* $\Delta\nu$ between two consecutive modes is given by

$$\Delta\nu = \nu_{m+1} - \nu_m = \frac{c}{2Ln}[(m+1) - m]$$

$$\text{or,} \quad \boxed{\Delta\nu = \frac{c}{2Ln}} \tag{14.6.2}$$

The important point to be noted is that the resonant modes of the cavity are *narrower in frequency than the width of a single spontaneous emission line* that may engulf several resonant cavity modes. The broad emission line and the narrow cavity modes are illustrated in Fig. 14.6. From the comparatively broad range of frequencies of an emission line, the cavity selects and amplify only certain narrow bands. And it is possible to have, by adjustment of separation of cavity modes, only one mode in the band-width of an emission line. *This explains the extreme monochromaticity of laser radiation.*

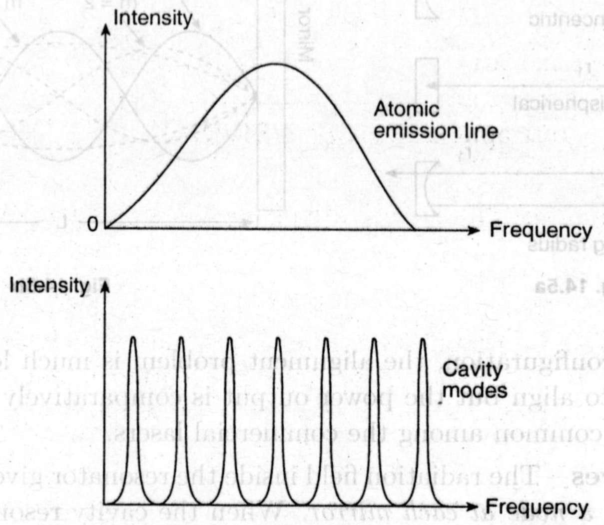

Fig. 14.6

• Laser cavities may be both *stable* and *unstable.* In *stable resonator*, the beam is close to the axis and bounces back and forth between the mirrors without much loss. In an *unstable resonator*, however, the beam moves away from the axis at each reflection and leaves the resonator altogether in a short time.

• There are a few **transverse modes** as well, in addition to the longitudinal modes discussed. The fields are nearly normal to the axis of the cavity, and are thus called *transverse electric and magnetic* or ***TEM*** *mode.* The most widely used mode is the lowest order ***TEM*** *mode.*

***Q*-value or *Q*-factor**—The resonant mode of a resonator is not a sharp line but has a finite, though small, frequency-spread. To this spread are linked the various losses involved with the modes in the cavity. The usual losses are mainly due to *absorption, scattering, diffraction* etc.

A measure of these losses is usually expressed in terms of what is called **quality factor** Q or the ***Q*-value** of the cavity. *Q is defined as*

$$Q = 2\pi \times \frac{\text{maximum energy stored per cycle}}{\text{energy dissipated per cyle}},$$

in the mode under reference.

An alternative expression for the quality factor Q is given by

$$Q = \frac{\omega}{\Delta\omega}$$

where $\Delta\omega$ is the linewidth, so that a *high Q ⇒ low loss and narrow width* of a mode.

14.7 Threshold condition for laser action

A laser consists of an active medium placed between two mirrors forming an optical resonator. The radiation that bounces back and forth between the mirrors is amplified by the active medium. The radiation also suffers losses. For sustained laser action, the losses are to be adequately compensated by induced transitions in the active medium. The important losses are due to absorption by active medium, transmission, scattering and diffraction by mirrors, scattering by optical inhomogeneities in the medium etc. For the compensation referred to, a *minimum population inversion density* in required. This is called *threshold population inversion*. The losses may be included in a single parameter t_p, the life time of a photon in the cavity. Its reciprocal, $1/t_p$ is then *a measure of the rate of loss of energy.*

In an assembly of atoms with two energy states E_1 and E_2, let there be N_1 and N_2 atoms per unit volume in states E_1 and E_2 respectively.

The transition probability for stimulated emission is $B_{21}u(\nu)$ and that for stimulated absorption is $B_{12}u(\nu)$, where $u(\nu)$ is the energy density of radiation.

Let a laser beam of energy density $u(\nu)$ propagate through a laser medium (Fig. 14.7) along $+x$ direction and consider a thickness dx and a cross-section of area S.

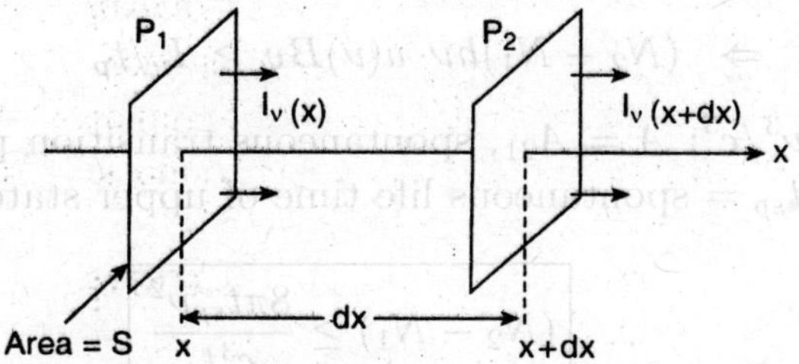

Fig. 14.7

∴ Energy loss per sec due to stimulated absorption in the volume $S\,dx$ is

$$B_{12}u(\nu)\cdot N_1(S\,dx)h\nu$$

Energy gain per sec due to stimulated emission in the same volume is

$$B_{21}u(\nu)\cdot N_2(S\,dx)h\nu$$

∴ Rate of increase of energy in the volume $S\,dx$ is given by

$$[B_{21}N_2 - B_{12}N_1]u(\nu)h\nu\cdot S\,dx = (N_2 - N_1)Bu(\nu)h\nu\cdot S\,dx$$

$$(\text{using } B_{12} = B_{21} = B)$$

Now, net energy leaving the volume $S\,dx$ per sec is

$$[I_\nu(x+dx) - I_\nu(x)]\,S = (dI_\nu/dx)\cdot dx\,S$$

where $I_v(x)$ is the intensity at x and $I_v(x+dx)$ is that at $x+dx$.

This must equal the rate of increase of energy due to stimulated transition.

$$\therefore \quad \frac{dI_\nu}{dx} = (N_2 - N_1)h\nu \cdot u(\nu)B$$

The rate of increase of intensity is given by

$$\left(\frac{dI_\nu}{dt}\right)_{\text{gain}} = \frac{dI_\nu}{dx}\frac{dx}{dt} = \frac{dI_\nu}{dx}v = (N_2 - N_1)h\nu \cdot u(\nu)Bv \qquad (14.7.1)$$

where v is the velocity of light in the medium.

For, $(dI_\nu/dt)_{\text{gain}} > 0$, $N_2 > N_1$ $\Rightarrow$ that for the growth of laser intensity we must have *population inversion.*

The total rate of loss of intensity is

$$\left(\frac{dI_\nu}{dt}\right)_{\text{loss}} = \frac{I_\nu}{t_p} \qquad (14.7.2)$$

So, the *condition of sustained laser action dictates*

$$\left(\frac{dI_\nu}{dt}\right)_{\text{gain}} \geq \left(\frac{dI_\nu}{dt}\right)_{\text{loss}}$$

$$\Rightarrow \quad (N_2 - N_1)h\nu \cdot u(\nu)Bv \geq I_\nu/t_p \qquad (14.7.3)$$

Now, $A/B = 8\pi h\nu^3/c^3$; $A = A_{21}$, spontaneous transition probability from $2 \to 1$. Also, $1/A_{21} = 1/A = t_{sp}$ = spontaneous life time of upper state and $I_\nu = u(\nu)$.

$$\therefore \quad \boxed{(N_2 - N_1) \geq \frac{8\pi t_{sp}\nu^2}{c^3 t_p}} \qquad (14.7.4)$$

which gives the ***threshold condition*** for laser action.

• In deriving (14.7.4), it is assumed that the atom interacts with radiation of a particular frequency. But, in fact, the emission lines have finite width due to Doppler effect, natural breadth, pressure broadening etc. and the atom can react with radiations over a frequency range. The strength of interaction is a function of frequency $g(\nu)$, called *shape function.* So, with this effect, (14.7.4) becomes

$$(N_2 - N_1) \geq \frac{8\pi t_{sp}\nu_0^2}{c^3 g(\nu_0)t_p}, \qquad (14.7.5)$$

ν_0 being the value of the *central frequency.*

• For threshold population inversion, therefore, (i) t_p should be as *long* as possible i.e., losses to be minimum, (ii) t_{sp} of the upper level should be *very short,* (iii) $\Delta\nu$, the atomic linewidth should be kept at *minimum* by keeping the temperature low, as $\Delta\nu \propto 1/g(\nu_0)$ and (iv) the lasing material should be so selected as to keep N_1 *small.*

14.8 Typical lasers

Various types of laser are now being used. They fall mainly into *four* categories : *solid dielectric, gas, semiconductor* and *tunable lasers.* They emit red, blue, blue-green or invisible radiation ranging from microwave to ultraviolet. Some produce continuous waves (cw), others are flashed or pulsed. The pulse rate is very short, being millisecond (10^{-3}s), nanosecond (10^{-9}s) or picosecond (10^{-12}s). The pulse may be normal or Q-switched, i.e., release of energy in a single giant controllable pulse. The power ranges from 25–400 mJ for a duration of 10 s–5 min. The highest power laser can produce short bursts of energy at rates $> 10^{13}$W by mode locking and oscillator-amplifier techniques.

Solid lasers—Solid dielectric lasers include ruby at 6940 A, Nd : YAG (Yttrium-Aluminium Garnet), i.,e., YAG with Nd as impurity at $\lambda = 1.064\,\mu$m and glass : Nd, i.e., glass with Nd as impurity at $\lambda = 1.064\,\mu$m.

Gas lasers—Gas-lasers include He-Ne at 6328 A and argon ion and krypton ion emitting in blue-green and red part. Another important gas laser is CO_2-laser developed by C.K.N. Patel.

Semiconductor lasers—Such lasers are made of a single crystal of a suitable impurity semiconductor such as GaAs (gallium arsenide) emitting near infrared (8300–8500A) radiation.

Tunable lasers—They are of two kinds : (i) *dye lasers* (pumped by argon and nitrogen lasers) and (ii) *parametric oscillators* (where a non-linear crystal, e.g., Li NbO_2 (lithium niobate) is pumped by Nd : YAG lasers). The dye lasers use active fluorescent material (organic dye), e.g., fluorescin or rhodamine 6G or B in a solvent, emitting radiation ranging from 500–1000 A.

14.9 Pulsed ruby laser

Maiman built the first ever successful solid state laser using a ruby crystal.

Construction—Ruby is Al_2O_3 doped with Cr^{+3} ions (0.04%) which are involved in lasing. The crystals are grown in special furnaces and then shaped into *cylindrical* rod AB (Fig. 14.8) about 1 cm in dia and 5 cm in length. The end faces A and B are polished *optically flat* and *parallel.* The ends are also *silvered*—the end B *partially*

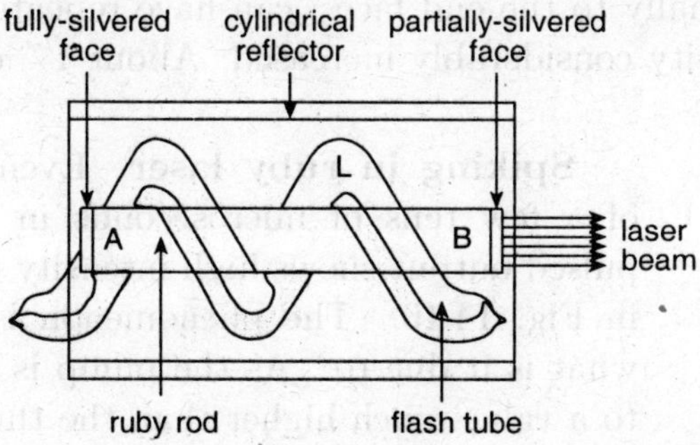

Fig. 14.8

and the end A *totally.* The purpose is to totally reflect the laser radiation, produced in the crystal, from the face A. The radiation thus proceeds backward to B which, being semi-silvered, allows a small fraction (1%) of the intense radiation to emerge out as *laser beam.* The major part is reflected back into the crystal. The curved surface is also coated to make it reflecting.

A helical Xenon flash lamp L whose output is rich in yellow-green light and which is to be flashed for optical pumping surrounds, along its axis, the ruby rod. To keep the rod cool, a cylindrical glass tube through which liquid N_2 circulates (not shown) also surrounds the rod

Working—When L is switched on, a flash from it excites a large number of Cr-ions from the ground state E_1 to the wide-band excited state E_3 by absorption of light of $\lambda = 5500$ A (Fig. 14.9), Cr-ions have an intermediate metastable state E_2 and soon they decay to the state E_2 by *radiationless transition* and remain there relatively longer (several millisec) to transit finally to the ground state by spontaneous emission of $\lambda = 6943$ A. The flash lamp thus increases the population in the metastable state E_2 *via* the wide band energy state E_3.

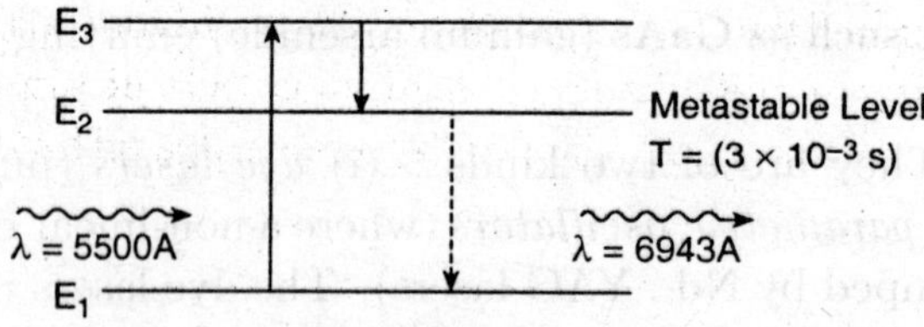

Fig. 14.9

When Cr-ions in E_2 make spontaneous transition to ground state, the emitted photons at $\lambda = 6943$Å interact with excited ions in state E_2 and induce transition to produce lasing.

The role of reflectors is to increase the energy density of radiation of frequency ν_{12} considerably and to increase the probability of stimulated emission. Light of $\lambda = 6943$ A is repeatedly reflected from the two silvered faces and in doing so travels over a considerable distance increasing the probability of these photons to interact with other Cr-ions. More and more of them are raised to the metastable state E_2. Since the beam, incident normally to the end faces, can have repeated reflections, the axial beam only has its intensity considerably increased. About 1% of this emerges as laser beam.

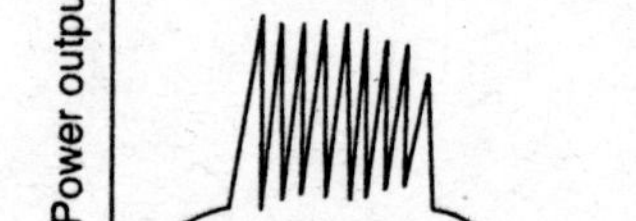

Fig. 14.10

Spiking in ruby laser—Even in the short duration of a few tens of microseconds in which ruby lases, the pulsed output shows high intensity emissions as illustrated in Fig. 14.10. The phenomenon is called **spiking**. But what is it due to? As the pump is suddenly switched 'on' to a value much higher than the threshold, the population inversion builds up and overshoots the threshold value. As a result the photon number builds up readily to a value much higher than the steady state value. Thus, the rate of depletion of upper level through stimulated emission is also much higher than the pump rate. So the inversion is below threshold and lasing ceases. This stops the emission for few μs during which flash lamp again pumps up atoms in the ground state to the upper level and laser oscillation re-starts. And the process gets repeated.

• Ruby laser is a *pulsed laser*—it emits laser radiation in pulses.

• The metastable state E_2 is in fact a *doublet*, of separation 14A, giving rise to two lines $\lambda_1 = 6943$ A and $\lambda_2 = 6929$ A in the laser output. But the spectral line 6943 A predominates.

14.10 Continuous He-Ne laser

Ali Javan and co-workers developed the first gas laser—Helium-Neon laser—in 1961. It is one of the most popular *continuous wave lasers.*

Construction—It is in effect a *discharge tube* (Fig. 14.11) about 1 m long, 4–6 mm in dia, filled with a *mixture of pure* He–Ne *gas* in the ratio 5 : 1 at a total pressure of 1 torr. The ratio 5 : 1 is rather critical. The gain of the system varies directly with tube length and the intensity varies with diameter. The relative intensity of the laser radiation depends largely on it.

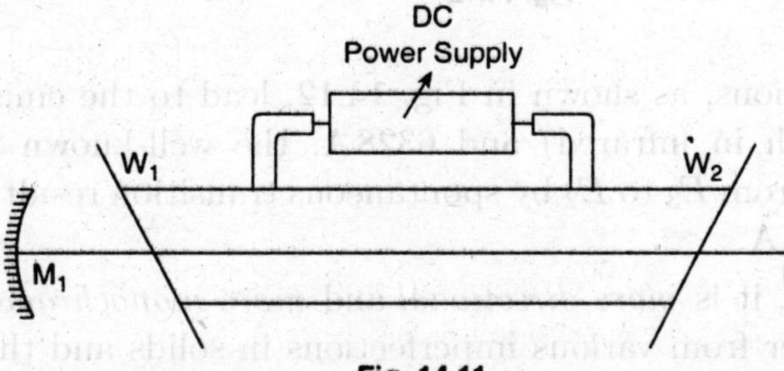

Fig. 14.11

The other parts are a high voltage D.C. power supply, two spherical mirrors M_1, M_2 (99.5% reflectivity), each of radius of curvature of the cavity length and mounted on a rotatable support to bring them into alignment. Two optically flat windows W_1 and W_2, called *Brewster's windows*, are cemented to the ends of the tube at Brewster's angle $\theta = \tan^{-1} n$, where n is the index of the material of the window, to provide a plane polarised beam.

This laser operates continuously at low power (5–10 W), the output lies between 1 mW to 50 mW and gives waves at 6328 A and 11520 A.

Principle—Due to discharge in the gas-mixture, some He-atoms are raised by *electron collision* to the metastable state F_2 and F_3 (Fig. 14.12) and the number gradually builds up since no radiative transition to lower state is possible. Now, some of the excited states of Ne are *lower than but close to* the above metastable state of He. So, when He-atoms in states F_2, F_3 collide with Ne-atoms in ground state E_1, energy-exchange occurs. This results in excitation of Ne-atoms to E_4 and E_6 levels and de-excitation of He-atoms to ground state level. States F_2, F_3 being unstable, the process of energy transfer has a high probability. The discharge in the tube thus continuously populates the Ne-levels E_4 and E_6. So, population inversion occurs between E_4 (or E_6) and the lower-lying E_5 and E_3 levels.

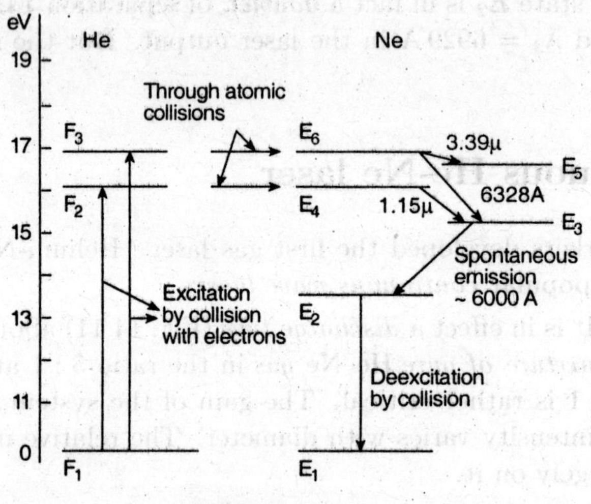

Fig. 14.12

The various transitions, as shown in Fig. 14.12, lead to the emission of radiations 3.39 μm, 1.15 μm (both in infrared) and 6328 A, the well-known red light of laser. Excited Ne-atoms fall from E_3 to E_2 by spontaneous transition resulting in the emission of photons of $\lambda \sim 6000$ A.

• Being a gas laser, it is *more directional* and *more monochromatic* than the solid state lasers which suffer from various imperfections in solids and the heating effect of the flash lamp used.

14.11 Semiconductor laser

A semiconductor laser consists of a heavily doped *p-n* junction, usually made from GaAs. While the conduction band in *n*-region has free electrons, the valence band in *p*-region has holes. Now, a large number of holes is equivalent to a large number of empty states into which electrons may fall. A population inversion thus exists, and lasing may occur when the junction is forward-biased, electrons from the *n*-region and holes from the *p*-region flow into the junction and recombine to produce radiation.

If the junctional geometry satisfies some dimensional requirements, the photons may stimulate further release of photons leading to laser radiation.

To obtain population inversion both p and n sides are heavily doped. The shape of the diode is *cubic* with junction in a horizontal plane through the centre (Fig. 14.13).

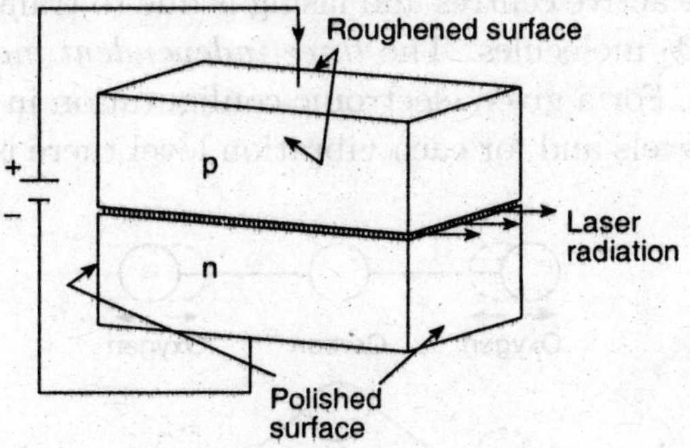

Fig. 14.13

To the top and bottom surfaces, there are contact points for the passage of ohmic current. The two opposite faces, perpendicular to the plane of junction, are polished and the remaining two are deliberately kept rough so as to eliminate lasing in directions other than the main one.

• This laser is also known as *injection laser* as lasing is made by charge carriers injected into the *p-n* junction.

• Semiconductor lasers have the *advantage* of small size and hence *portability*, and *high efficiency*. It is a *tunable laser* as the operating frequency can be changed by exploiting external parameters to change the band gap. The intensity can also be modulated by modulating the forward current.

14.12 Carbon di-oxide laser

This is a four-level laser system, operating at 10.6 μm in far infra-red region and is one of the most useful and efficient lasers.

Construction and action—A typical CO_2-laser is illustrated in Fig. 14.14. It is essentially a discharge tube of length about 25 cm and having a bore of cross-section

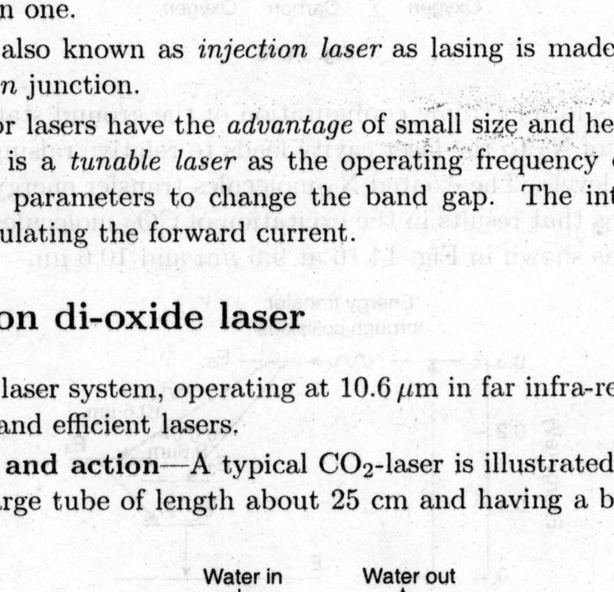

Fig. 14.14

~ 1.5 mm^2. The tube is filled with a gas mixture consisting of CO_2, molecular nitrogen and helium in the ratio 1 : 4 : 5. Water vapour is also present as an additive. CO_2 molecules constitute the active centres and lasing is due to transitions between different vibrational states of CO_2 molecules. The *three independent modes* of vibration of CO_2 are shown in Fig. 14.15. For a given electronic configuration in a molecule, there exists a number of vibration levels and for each vibration level there is a number of rotational levels.

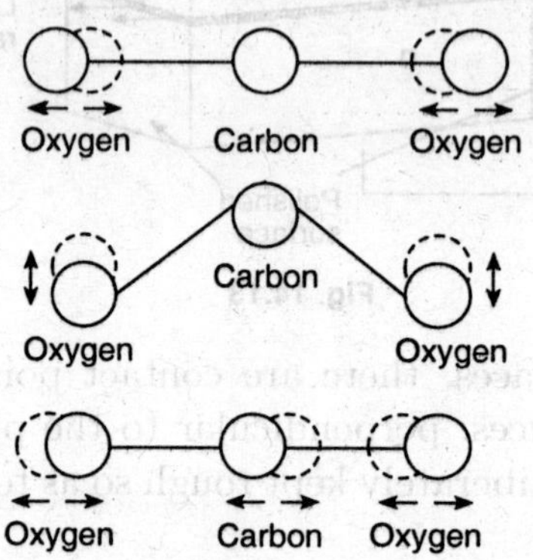

Fig. 14.15

The energy levels for electronic configuration of the ground state are illustrated in Fig. 14.16. The use of N_2 to the laser cavity leads to relative raising of CO_2-molecules to the desired laser levels. The excited N_2-molecules transfer energy to CO_2 molecules by resonant collisions that results in the excitation of CO_2 molecules. Important lasing transitions occurs, as shown in Fig. 14.16 at 9.6 μm and 10.6 μm.

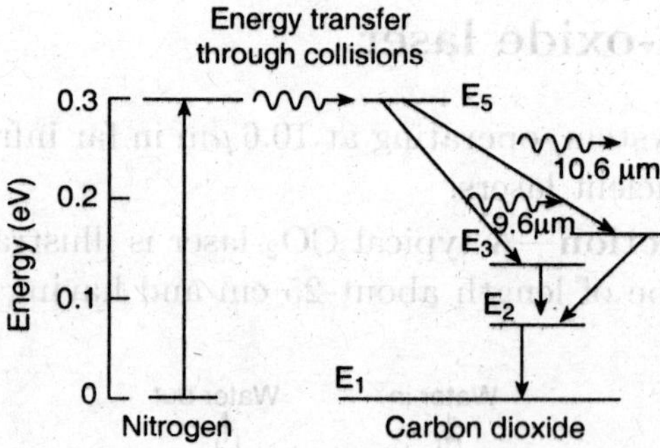

Fig. 14.16

14.13 Nd : YAG laser

This is one of the popular four-level laser systems where the trivalent neodymium Nd^{3+} ion is doped in Ytrium-Aluminium-Garnet ($Y_3Al_5O_{12}$), an optically isotropic crystal, commonly known as YAG. The active centres are the Nd^{3+} ions and except housing the Nd^{3+} ions, the crystal atoms have no role in the lasing action.

Construction—A typical Nd : YAG laser is illustrated in Fig. 14.17. It consists essentially of an elliptically cylindrical reflector, about 10 cm in length and 12 mm in dia, in which the laser rod resides along one of its focal lines. Along the other focal line, the flash tube is placed. Both ends of the laser rod are appropriately polished and silvered so as to serve as an optical resonator. The light that leaves one of the foci of the ellipse passes through the other focus on reflection from the silvered surface of the reflector. The whole of the radiation of the flash tube is thus focussed on the laser rod.

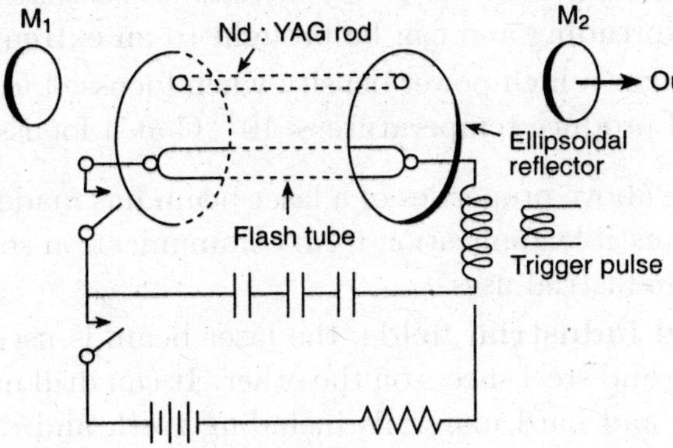

Fig. 14.17

Action—It is essentially a four-level system with terminal laser level E_2 much above the ground level (Fig. 14.18). The optical pumping is made with light having wavelength ranging from 500–800 nm from a krypton arc lamp when the ground state Nd^{3+} ions are excited to the multiple energy level at E_4. The metastable level E_3 is the upper laser level, while E_2 is the lower laser level. As the excited Nd^{3+} ions make quick downward transitions from the upper energy bands, E_3 gets rapidly populated. E_2 being much above the ground level, it cannot be populated by Nd^{3+} ions through thermal transition from the ground level. Thus population inversion occurs readily between E_3 level and E_2 level. The laser emission is in infra-red region having $\lambda \simeq$ 1.06 μm. Being a 4-level laser, the population inversion can be sustained in the face of continuous lasing.

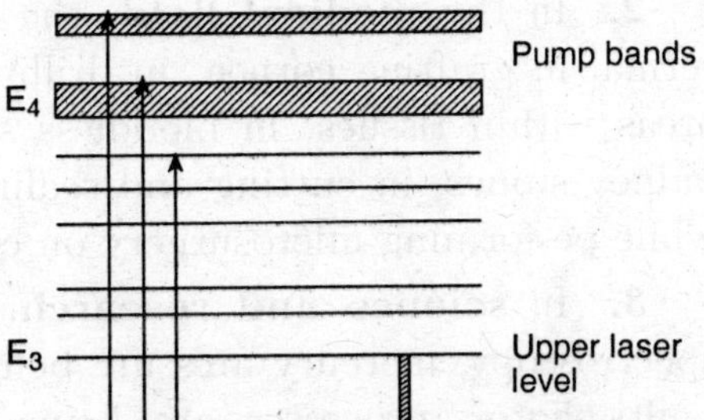

Fig. 14.18

14.14 Properties and uses of a laser beam

A laser beam possesses a number of characteristic properties *not found* in light beam from other sources and these properties impart special importance to laser.

Properties—A laser beam is endowed with a number of important and unique characteristics. It is

1. *completely spatially coherent*, with the waves all exactly in phase, and hence an interference pattern may be obtained even with beams from separate lasers.
2. *almost perfectly monochromatic*, i.e., *highly temporally coherent*, with ultra narrow band width.
3. *almost perfectly parallel*,, i.e., *extremely directional* so that it can travel a long distance without spreading and can be brought to an extremely sharp focus.
4. *highly intense* and has a high power density when focussed; can vaporise even the hardest metal and produce temperature $\sim 10^4\,°C$ at a focussed point.

Applications—The above properties of a laser beam has made it a versatile device having an almost inexhaustible application from communication science to biology and medicine, let alone the industrial uses.

1. In **technical and industrial fields**, the laser beam is used for cutting fabrics for clothing in one hand, and steel sheets on the other. It can drill minute holes in paper clips, single hair strand and hard materials including teeth and diamond. Extremely fine wires in cables are drawn through diamond hole. Laser welding melts and joins metallic rods. Laser heat-treatment can harden surfaces of engine crankshafts and the cylinder walls. Laser beam is used to vaporise unwanted materials during manufacture of electronic circuits on semiconductor chips.

2. In the **medical field**, the laser beam is used for '*spot welding*' of detached retina, in grafting cornea, in drilling holes in bones, in destroying specific cancerous areas within tissues, in bloodless surgery like vocal cord operation, stomach ulcers, kidney stones, in cutting and sealing small blood vessels during brain operation and while performing microsurgery on cells and chromosomes.

3. In **science and research**, laser has become a very useful tool. In Raman spectroscopy, mercury arcs are being fast replaced by He-Ne lasers. In spectroscopy, multi-photon processes are being studied in determining the symmetry of wave functions involved in various transitions, with laser. Experiments with Michelson's interferometer using laser show that fringes are obtained even with path-difference of 9 m (against 3 m with ordinary light), confirming the temporal coherence of laser. Laser is also being used in ionisation and dissociation studies of gases, in kinetic studies of fast reactions, in spectrochemical analysis of solids in conjunction with mass and emission spectrometer and in investigation of energy transfer by exciting a specific vibration level and observing its decay in polyatomic molecules.

4. **Other uses** include : 'inertial confinement' of plasma and determination of its temperature; detection and the destruction of enemy missiles; to aim at the enemy at night; laser-rifles, laser pistols and laser bombs are also being made; detection of nuclear explosions and earthquakes, vaporisation of solid fuel of rockets and study of the surfaces of distant planets and satellites; accurate measurent of very long distances; the distance between the earth and the moon has been measured to an accuracy of 15 cm.

Three-dimensional motion picture production by *laser holography* (Art. 14.15) is a fascinating new field. Laser has also revolutionised *communication technology* in conjunction with *optical fibres.* Control of the motion of moving aircrafts, rockets or missiles to their destination is also a reality with laser. Non-linear optical properties of different materials can also be investigated with the help of laser beams.

14.15 Holography : Introduction

Holography is an unusual method of producing an image by what is called *wave front reconstruction.* In conventional photography, the camera consisting of a lens-system, focuses the light reflected from an illuminated three-dimensional object on a photographic plate (film) where a two-dimensional image of the object is formed. Each point of the object is related to a conjugate point in the image. On developing the plate (or film), the image can be seen on illumination by an ordinary incoherent light. The drawback of this conventional photography is that the focussing takes place only in a particular plane and all other planes are out of focus. The details of the field nearer and further than the focussed plane thus get lost. The photographic plate records only the *intensity distribution* while the *phase distribution* at the plane of the photographic plate is totally lost. Hence the 3D-character of the object i.e., the complete field of view, as originally observed, cannot be obtained.

Dennis Gabor in 1947 invented a radically different type of photography, called **holography**, requiring highly coherent light. It however attracted little attention until the highly coherent light of laser became available in 1960. The word '*holography*' originates from the Greek word '*holos*' meaning the '*whole*' and '*graphy*' meaning the '*writing*'. Holography thus means '**complete recording**'.

In holography, the image of the object to be photographed is not recorded, but the light waves—their amplitude and phase distributions, as reflected from the object, are recorded. The photographic record is called a **hologram** that bears no semblance with the object. Nevertheless, it contains in a kind of 'optical code' all the information about the object. When illuminated by a coherent light source, the hologram produces a three-dimensional image of the original object. The formation of the image from the hologram is called **reconstruction process**.

• Gabor's idea was first successfully translated into holography by Leith and Upatnicks fifteen years later in 1962, soon after the availability of highly coherent laser beam.

14.16 Basic principle of holography

The light waves reflected from an object are characterised by their amplitudes and phases. When plane monochromatic light waves fall on a point object, as illustrated in Fig. 14.19, the reflected wavefronts are all spherical with centre at the point object.

If, however, the object is a three-dimensional solid, it consists of a collection of a large number of points and each point reflects its own set of spherical waves (Fig. 14.20). As a result, the wave pattern reflected from the whole object is highly complex. It can be recorded on a photographic plate as such, for the plate will record the intensity i.e., amplitude squared, but not the phase.

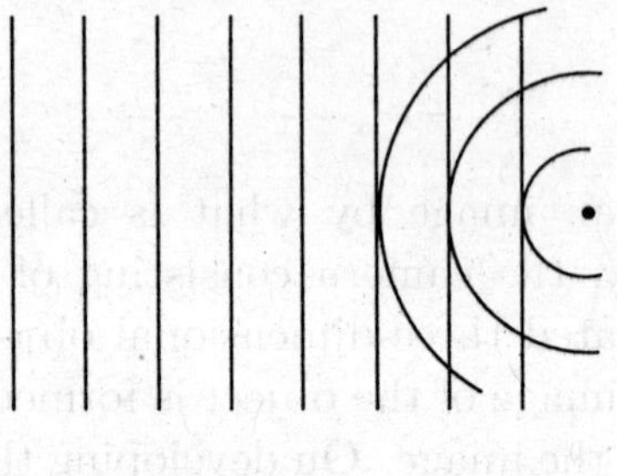

Fig. 14.19

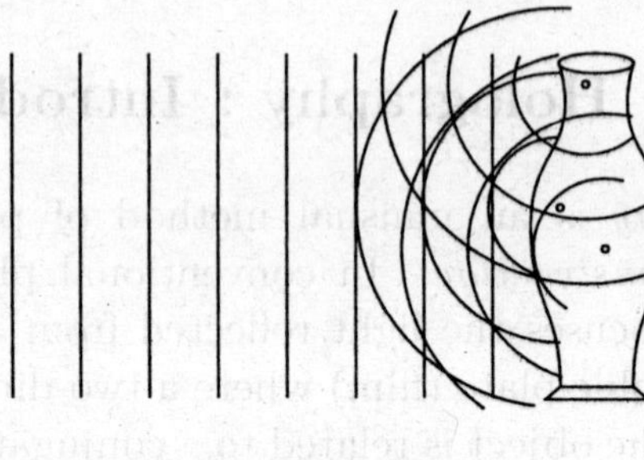

Fig. 19.20

The difficulty was overcome by Gabor by mixing wave pattern with a known monochromatic coherent wave called the '*reference wave*'. The reflected waves from the object and the reference wave interfere so as to modify the intensity at all points strictly in accordance with the phase relationship between the waves—the original wave and the reference wave. The *phase variations* in the wave pattern reflected from the object are converted into *intensity variations.*

- In Gabor's original experiments, the reference beam and the object beam were *coaxial.* Leith and Upatnicks however made an advancement by using the reference beam at an *offset angle* and thus made possible the recording of holograms of three-dimensional objects.

14.17 Recording of a hologram

The experimental set up of an *off-axis arrangement* for recording a hologram is illustrated in Fig. 14.21 below.

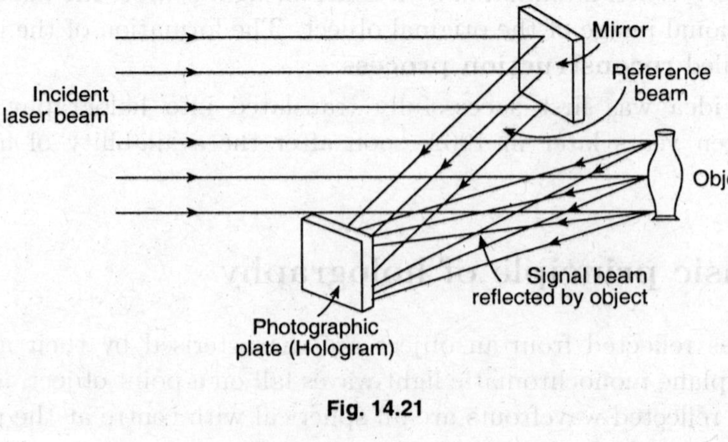

Fig. 14.21

Holography being, in principle, an *interference-based technique*, the object is illuminated by a highly monochromatic and coherent beam of laser obtained by dividing a broad laser beam into two beams—**reference beam** and **object beam**—by a beam splitter. Each point on the object scatters the object beam to the entire photographic plate P. The other part of the incident laser beam is allowed to fall upon a plane mirror M that reflects it towards the same photographic plate. This is the *reference beam.* Thus the plate P is exposed *simultaneously* to the reflected beam and the reference beam. Since both the beams are from the same laser wavefront, they interfere on the plate where we get a complicated pattern of lines and swirls, the 'developed negative' of which is called **a hologram.**

When seen under magnification, the hologram appears not as a distinct image of the object but to consist of highly irregular fringes bearing no apparent relationship with the object. However, it contains information regarding the object wave, both in respect of *the amplitude and the phase*, in the *fringe-contrast* and the *fringe-spacing.* And the image of the object is to be *reconstructed* finally from the *hologram.* Holography thus entails two processes : (1) production of the hologram and (2) the subsequent reconstruction of image from hologram.

14.18 Reconstruction of image from hologram

Whenever needed, the object may be viewed by *reconstruction of the image* from the hologram. This is essentially a reverse process of making a hologram and is illustrated in Fig. 14.22.

In the **reconstruction process**, the hologram is illuminated by a collimated laser beam identical to the reference beam used in forming the hologram. The beam undergoes diffraction at the hologram and one of the diffracted beams emerging from the hologram, when projected back, appears to diverge from an apparent object located in the position of the original object (removed long since). This beam forms a *virtual image* observed by looking through the hologram, as if through a window. The other

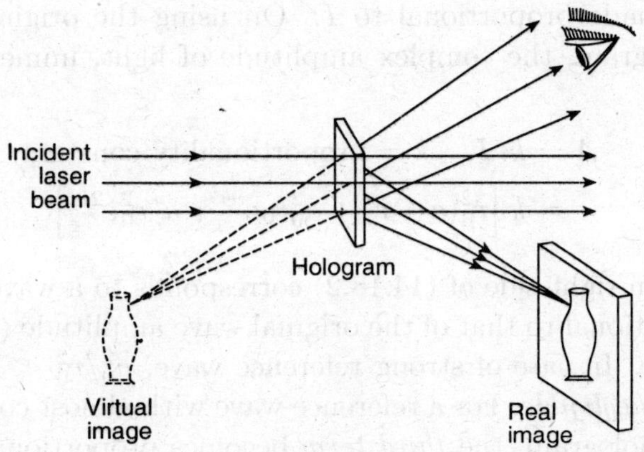

Fig. 14.22

beam produces a *real image* that can be recorded in a photographic plate, without the aid of lenses, just by placing a light-sensitive medium at that position. By moving the positions of the eye, the different perspective of the object can be seen in the real image. The virtual image also possesses all the characteristics like parallax etc.

- The non-diffracted direct beam (i.e., zero order) however gives no information, for it has only amplitude-variation, but nothing about the phase-variation.

- This type of hologram is called a *transmission hologram* as the image is seen by looking through it.

- The advantage of the *off-axis configuration* is that the two images are separated.

14.18.1 Theory of holography : A short mathematical note

The general theory of holography is much involved and would not be described here. We present below a very simplified account and that too briefly.

Let $a = a_0 e^{i\theta}$ be the complex amplitude of light wave reflected i.e., scattered by the object and striking the hologram, and $r = r_0 e^{i\phi}$ be that of the reference wave. Now, since the scattered wave and the reference wave are phase correlated, the complex amplitude of their resultant is

$$A = a + r$$

$\therefore$ Intensity on the hologram plane is given by

$$\begin{aligned} I &= AA^* = (a+r)(a^*+r^*) \\ &= aa^* + rr^* + ar^* + a^*r \\ &= a_0^2 + r_0^2 + 2a_0r_0\cos(\phi - \theta) \end{aligned} \tag{14.18.1}$$

The resultant intensity is thus the sum of the two intensities and the interference term has the information regarding the relative phase.

By an appropriate process of development, the transmittance of amplitude of the hologram can be made proportional to I. On using the original reference beam to illuminate the hologram, the complex amplitude of light, immediately on emergence from hologram, is

$$\begin{aligned} A &= krI, \quad k = \text{proportionality constant} \\ &= k\left[r_0^2(a+r) + a_0^2 r_0 e^{i\phi} + a^* r_0^2 e^{2i\phi}\right] \end{aligned} \tag{14.18.2}$$

The *first term* on right side of (14.18.2) corresponds to a wavefront whose complex amplitude is proportional to that of the original wave amplitude $(a+r)$. This *gives rise to* 3D-*virtual image.* In case of strong reference wave, $a_0^2/r_0^2 \ll 1$ and this makes the *second term* rather *negligible.* For a reference wave with almost constant amplitude and phase all over the hologram, the *third term* becomes proportional to the conjugate of the object wave amplitude. This *gives rise to the real image which can be photographed.*

14.19 Features of a hologram

A hologram has a number of uniquely *exciting* and *distinguishing* features of its own. These are enumerated below.

1. During recording, the plate receives light from the side of the object facing the plate and also from the adjoining sides. So if the observer, while viewing the hologram, moves the head sideways, the original object is completely seen in 3D-*form*. For instance, if some scene is holographed, observer feels a parallax effect between near and far objects in the scene, and even if an object is hidden just behind another, one can see the hidden object through the hologram by looking around the obstructing object. All the visual properties of the original scene is therefore preserved in the hologram.

2. During recording, each point on the hologram receives light from all parts of the object and thus stores information about the entire object. If, therefore, a hologram breaks into pieces, each one would reconstruct the entire image, only the resolution would be less.

3. Reconstruction always gives a positive. If a hologram is copied by contact printing, the opaque parts in the hologram become transparent in the copy and conversely. If now the image is reconstructed from the copy, it will still be a positive and completely indistinguishable from the image reconstructed from the original hologram of the object.

4. Waves travelling into a *thick* photographic emulsion may interfere with those reflected from successive layers. A 3D-standing wave pattern will thus be formed. Such *volume holograms*, (see also Art 14.20), if viewed in white light, give reconstruction in full colour.

14.20 Some advances in holography

Some *advances* have of late been made in holography and these include

1. Reflection holography
2. Volume holography
3. White light reflection holography
4. Rainbow holography, etc.

Reflection holography—In its recording process, the object beam and the reference beam reach the recording medium from opposite directions (Fig. 14.23) and combine within the medium to form interference fringes which are recorded in thick emulsions. In reconstruction, the illuminating beam is scattered back by the hologram towards the viewer. It is also called, and justifiably, **thick emulsion holography**.

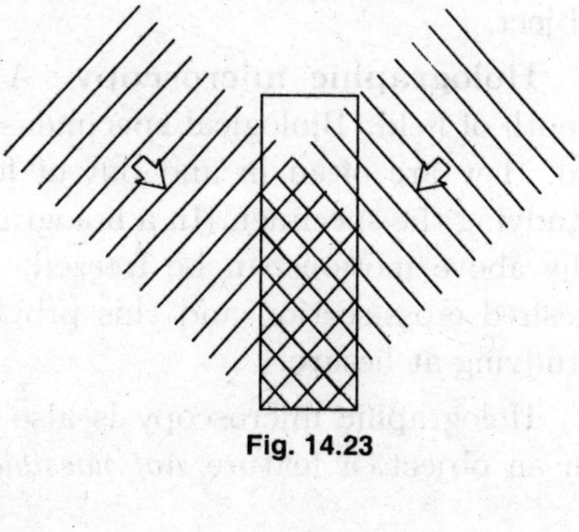

Fig. 14.23

Volume holography—Developed by Denisyuk, the process is similar to Lippman's colour photography. Here the object beam is reflected from the object and it propagates back to overlap the incoming reference beam forming stationary wave pattern. The fringes are recorded by photo-emulsion throughout the thickness to form a **volume hologram**. An interdependence of λ and θ exists *via* Bragg's law : $2d \sin\theta = m\lambda$. So, by changing θ or λ, a number of holograms can be stored in the medium. Different and mutually incoherent laser beam may be used to produce different component holograms to impart a *multicolour effect* in the image.

White light reflection holography—In this method, developed by Stroke and Labeyrie, the hologram is produced with coherent light, but in reconstruction an ordinary white light beam with wavefront similar to original coherent beam is used. With coherent sources of different wavelengths, a number of holograms are stored in a single film. On reconstruction by white light, the image seen in reflection is multicoloured.

Rainbow holography—This type of colour photography, developed in 1969 by Stephen Benton, is also called *white light transmission hologram.* When viewed with a white light point source, a bright colour image can be observed, if viewed in correct plane. In case it is missed, different shades of colour can still be seen.

14.21 Applications of holography

Holography has not only revolutionised photography, but has also influenced interferometry, microscopy, acoustics and many others, resulting in various useful applications in medicine and technology.

Holographic interferometry—Using a special technique, called **double exposure**, holography is utilised in testing stresses, strains and surface deformations in objects. For this, the original (unstressed) object and the object after deformation are recorded in two successive holograms in the same photographic plate with the same original reference beam. When the images are recostructed, the two object waves interfere and produce interference fringes which are characteristic of the strain suffered by the object. A quantitative study of the fringe pattern gives the distortions in the object.

Holographic microscopy—A conventional high power microscope has a small depth of field. Biological specimens are usually suspended in a fluid and move about. So, they are often in and out of focus of the micrscope, causing much difficulty in studying the specimen. In a hologram taken through the specimen and the microscope, the above motion can be freezed. The reconstructed 3D-image can be seen at any desired cross-section and this provides a high depth of field and an opportunity for studying at liesure.

Holographic microscopy is also used to study time-varying phenomena occurring in an object, a feature *not possible* with ordinary microscopy. If an object and its hologram, made sometime earlier, are illuminated by the same laser beam as used in making the hologram, the wave from the object and the reconstructed wave from the hologram will interfere. Any temporal change in the object that occurred in the mean time will produce fringes indicating changes that could not be detected by microscopic inspection.

Acoustic holography—It is not essential that the waves used in making the hologram and for reconstruction should have an identical wavelength. Only the apparent depth of the object increases in case shorter wavelength is used in recostruction.

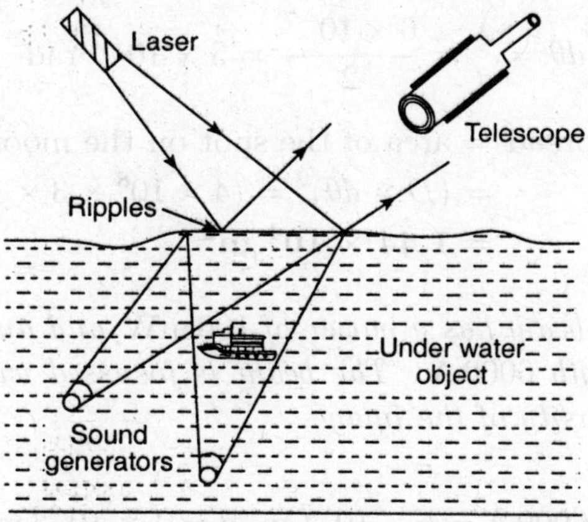

Fig. 14.24

Similarly, the original and the reconstructing wave must not necessarily have the same physical nature, e.g., ultrasonics (*hf*-sound waves) can be used to construct a hologram and the acoustic hologram can be reconstructed into a pictorial image with a laser beam. In a certain technique, two submerged coherent acoustic generators give rise to the reference and the signal, scattered by an object. (Fig. 14.24). These two generate ripples on a calm surface of water. The ripple pattern is the hologram which may be photographed and reconstructed when required.

Acoustic holography, in fact, has been used to get a 3D-image of human internal organs. A coherent ultrasonic beam is split into two parts—one serving as reference beam, the other being scattered by the internal organ under study. The scattered beam serves as the object beam. The two beams interfere to form a hologram. From the reconstructed 3D-image, one can make an in-depth study of the concerned organ.

Other applications—Among other applications of holography, mention may be made of holographic cinema, spatial filtration and character recognition, long distance holography with microwaves etc.

Holography is also useful to provide a high capacity system for the image storage and their re-examination.

Holographic optical elements (HOE) are diffractive devices which are holograms of typical optical components such as lenses or mirrors. HOEs perform precisely the same type of operation as the optical elements but through diffraction. HOEs are used in bar code scanners, office copy machines etc.

14.22 Illustrated Examples

► **Example 1.** *A laser beam of* $\lambda = 5000$ A *on earth is focussed by a lens (or mirror) of diameter* 2 m *onto a crater on the moon. The distance of the moon is* 4×10^8 m. *How big is the spot on the moon? Neglect the effect of the earth's atmosphere.* **(IAS)**

Solution. Given, $d = 2$ m, $D = 4 \times 10^8$ m and $\lambda = 6000$ A $= 6 \times 10^{-7}$ m. Therefore, the angular spread

$$d\theta = \frac{\lambda}{d} = \frac{6 \times 10^{-7}}{2} = 3 \times 10^{-7} \text{ rad}$$

$$\begin{aligned} \text{Areal spread} &= \text{area of the spot on the moon} \\ &= (D \times d\theta)^2 = (4 \times 10^8 \times 3 \times 10^{-7})^2 \\ &= \mathbf{1.44 \times 10^4\ m^2} \end{aligned}$$

► **Example 2.** *A laser beam has a power of* 100 mW *and an aperture of* 4×10^{-3} m. *It emits light of wavelength* 6000 A. *The beam is focussed with a lens of focal length* 0.1 m. *Calculate the intensity of the image.*

Solution. Given : $\lambda = 6000$ A $= 6 \times 10^{-7}$ m, $d = 4 \times 10^{-3}$ m and $f = 0.1$ m.

$$\therefore \text{ Angular spread, } d\theta = \frac{\lambda}{d} = \frac{6 \times 10^{-7}}{4 \times 10^{-3}} = 1.5 \times 10^{-4} \text{ rad}$$

$$\therefore \text{ Areal spread} = (d\theta \times f)^2 = (1.5 \times 10^{-4} \times 0.1)^2 = 2.25 \times 10^{-10} \text{ m}^2$$

$$\therefore \text{ Intensity of image} = \frac{\text{power}}{\text{area}} = \frac{100 \times 10^{-3} \text{ W}}{2.25 \times 10^{-10} \text{ m}^2} = \mathbf{4.44 \times 10^8\ W/m^2}$$

► **Example 3.** *A laser beam of wavelength* $\lambda = 6000$ A, *power* 10 mW *and angular spread* 1.5×10^{-4} rad *is focussed by a lens of focal length* 10 cm. *Find* (i) *radius,* (ii) *power density of the image and* (iii) *the coherence width.* **(Calcutta Hons.)**

Solution. The radius of the image, $r = f\theta$. Here $f = 10$ cm, $\theta = 1.5 \times 10^{-4}$ rad.

$$\therefore \quad r = 10 \times 1.5 \times 10^{-4} = \mathbf{1.5 \times 10^{-3}\ cm}$$

Area of the image $= \pi r^2$ so that the power density of the image is

$$\frac{\text{power}}{\text{area}} = \frac{10 \times 10^{-3} \text{ W}}{\pi \times (1.5 \times 10^{-3})^2 \text{ cm}^2} \simeq \mathbf{1.4\ kW/cm^2}$$

$$\text{Coherence length} \quad l_w = \lambda/\theta = \frac{6000 \times 10^{-10} \text{ m}}{1.5 \times 10^{-4} \text{ rad}} = \mathbf{4\ mm}$$

► **Example 4.** *Find the ratio of the rates of stimulated and spontaneous emissions at $T = 10^3$K for visible radiation of frequency 5×10^{14} Hz and microwave frequency 10^9 Hz. Comment on the result.*

Solution. The ratio of spontaneous to stimulated emission rates is

$$R = e^{h\nu/kT} - 1$$

(i) For visible light, $$\frac{h\nu}{kT} = \frac{6.62 \times 10^{-34} \times 5 \times 10^{14}}{1.38 \times 10^{-23} \times 10^3} \simeq 24$$

$$\therefore \quad (R)_{\text{visible}} = e^{24} - 1 \gg \mathbf{1}$$

(ii) For microwave, $$\frac{h\nu}{kT} = \frac{6.62 \times 10^{-34} \times 10^9}{1.38 \times 10^{-23} \times 10^3} \simeq 4.8 \times 10^{-6}$$

$$\therefore \quad (R)_{\mu\text{-wave}} = e^{4.8 \times 10^{-6}} - 1 \simeq \mathbf{4.8 \times 10^{-6}}$$

The result shows that the probability of stimulated emission is negligible compared to that of spontaneous emission in the visible region. In microwave region however the reverse is the case. This explains why *maser* is a forerunner of *laser*.

► **Example 5.** *In a* He-Ne *laser transition from* $3S \to 2P$ *level gives a laser emission of* $\lambda = 632.8$ nm. *If the* $2P$ *level has energy* 15.2×10^{-19} J, *calculate the pumping energy required, assuming no loss.* **(Calcutta Hons.)**

Solution. Pumping energy = energy of $2P$ level + photon energy

$$\begin{aligned}\therefore \quad (E)_{\text{pump}} &= 15.2 \times 10^{-19}\,\text{J} + h(c/\lambda) \\ &= 15.2 \times 10^{-19}\,\text{J} + \frac{6.62 \times 10^{-34} \times 3 \times 10^8}{632.8 \times 10^{-9}}\,\text{J} \\ &= \mathbf{11.46\ eV} \quad (1\,\text{eV} = 1.6 \times 10^{-19}\,\text{J})\end{aligned}$$

► **Example 6.** *A relative population (Boltzmann ratio) of 1/e is representative of the ratio of populations in two energy states at room temperature (= 27°C). Determine the wavelengths of the radiation emitted at that temperature.*

Solution. With usual meanings of the symbols, we have

$$N_2/N_1 = e^{-(E_2 - E_1)/kT}. \text{ But, here, } N_2/N_1 = e^{-1}$$

$$\therefore \quad \frac{E_2 - E_1}{kT} = 1 \;\Rightarrow\; E_2 - E_1 = kT = h\nu = hc/\lambda$$

$$\therefore \quad \lambda = \frac{ch}{kT} = \frac{(3 \times 10^8)(6.62 \times 10^{-34})}{(1.38 \times 10^{-23})(273 + 27)} = \mathbf{4.8 \times 10^{-5}\ m}$$

Review Questions

1. Explain clearly the terms 'absorption', 'spontaneous emission' and 'stimulated emission'. **(Rejasthan Univ.)**

2. What is stimulated absorption? Explain the difference between spontaneous and stimulated emissions. **(Punjab Univ.)**

3. What do you mean by Einstein's A, B coefficients? Show that the ratio

$$\frac{A_{nm}}{B_{nm}} = \frac{8\pi h\nu^3}{c^3},$$

the symbols used having their usual significance.

4. What are the important features of stimulated emission? Discuss the essential requirements for producing laser action. **(Allahabad Univ.)**

5. Describe with suitable diagrams the principle, construction and working of a ruby laser. What is 'spiking'? **(Calcutta Hons.)**

6. Discuss, with suitable disgrams, the construction and working of a He-Ne gas laser. **(Bangalore Univ.; Calcutta Hons.)**

7. What is population inversion? Discuss the different methods of achieving the population inversion. How is population inversion achieved in He-Ne gas laser?

8. Describe briefly a CO_2 laser and a Nd:YAG laser and explain their principle of action.

9. How do you interpret high monochromaticity and high directionality of laser radiation?

10. What is an optical resonator? Discuss the role played by it in a laser system.

11. Derive the threshold condition for the laser action.

12. What are the characteristic properties of a laser beam? Describe its important applications.

13. Explain the basic principle of holography. What is the essential difference between a conventional photograph and a hologram?

14. Discuss how a hologram is prepared and viewed. Describe the important features of a hologram. What is the advantage of off-axis configuration over the coaxial configuration?

15. Explain the wavefront reconstruction of the image from a hologram. Discuss holographic microscopy and interferometry.

Short Answer and Objective Questions

1. Explain what is meant by stimulated emission of radiations.
2. Explain what are population inversion and pumping.
3. What do you understand by 'a metastable state'? Give an example.
4. Why do you need population inversion in a laser?
5. Both 'laser' and 'maser' are acronyms. Write them in full.
6. The coefficient of stimulated emission and absorption are equal. Nevertheless, emission and absorption are observed—Explain. **(Calcutta Hons.)**
7. What do you mean by the statement–'$1/A_{21}$ is a measure of the life-time of upper state against the spontaneous decay to the lower one'.
8. Describe briefly a semiconductor laser. **(Kurukshetra Univ.)**
9. What are the various uses of a laser? **(Punjabi Univ.)**
10. Indicate the different configurations of an optical resonator used in a laser.
11. What is the meaning of 'holography'? Why is it so called? Who was the progenitor of the idea?
12. What are virtual and real images in relation to the reconstruction of the image from a hologram?
13. Light amplification by stimulated emission of radiation is known as

    (a) Light amplifier, (b) maser, (c) laser, (d) none.
14. Metastable state has a mean life-time of more than

    (a) 10^{-3}s, (b) 10^{-5}s, (c) 10^{-4}s, (d) 10^{-2}s.
15. The ratio of Einstein's A, B coefficients i.e., A_{21}/B_{21} say, is proportional to

    (a) ν, (b) ν^2, (c) ν^3, (d) ν^4.
16. Holography was originally an idea of

    (a) Gabor, (b) Snitzer, (c) Leith, (d) Maiman.
17. Holography is equivalent to n-successive transformations. The value of n is

    (a) 1, (b) 2, (c) 3, (d) none.

Problems

1. Calculate the ratio of stimulated to spontaneous emission rates for the wavelength $\lambda = 5893$ A at 200°C.

2. Show that in the $2P \rightarrow 1S$ transition in H-atom, the value of the coefficient $B_{21} \simeq 4.1 \times 10^{20} \text{m}^3\text{J}^{-1}\text{s}^{-2}$.

3. A leaser beam has a power density of 100 mW. It has an aperture 5×10^{-3} m and emits a light of $\lambda = 6943$ A. The beam is focussed with a lens of focal length 0.1 m. Calculate the area and the intensity of the image. [2.89×10^{-10} m^2; 3.46×10^8 W/m^2]

4. A laser beam has a wavelength of $\lambda = 7.2 \times 10^{-7}$ m and an aperture 5×10^{-3} m. The laser beam is sent to the moon, the distance of which from the earth is 4×10^8 m. Calculate (a) the angular spread and the areal spread when the beam reaches the moon. [1.757×10^{-4} rad; 49.39×10^8 m^2]

5. A laser beam has a power of 50 mW and an aperture of 5×10^{-3} m. It emits radiation of wavelength 7200 A. The beam is focussed with a lens of focal length 0.1 m. Calculate the intensity of the image. [2.411×10^3 W/m^2]

6. Let a diffraction limited laser beam of diameter $D = 2$ cm and the wavelength $\lambda = 6000$ A, be aimed at a target 1 km away. Calculate the beam-diameter when it strikes the target. [3.66 cm]

7. A laser beam of dia 0.5 cm and $\lambda = 5000$ A, is directed to the moon, 4×10^5 km away from the earth. Calculate the linear spread of the beam on the surface of the moon. [40 km]

8. The energy-gap between two levels corresponds to wavelength $\lambda = 5000$ A. Find the ratio of populations of the two states in thermal equilibrium at room temperature. [$N_2/N_1 \simeq e^{-96}$]

9. The relative population of two energy states at room temperature, $T = 300$ K is $1/e$. Determine the wavelength of radiation emitted due to transition between the states. [48 μm]

10. An assembly of atoms has two energy states E_1 and E_2. It is in equilibrium with a radiation field of frequency $\nu = (E_2 - E_1)/h$. If the source of radiation is suddenly cut off, show that the population of upper state decreases exponentially with a time constant $1/A$, where A is the spontaneous transition probability.

Chapter 15

Fibre Optics

15.1 Introduction

To design and improve communication system for sending messages over long distances have been the constant endeavour of specialists in the field. In the recent past (1970), transparent dielectric fibres have been developed to carry optical signals efficiently from one place to another. These are known as *optical fibres.* Prior to the introduction of fibre optics, electromagnetic waves down to microwave region were used for the purpose of communication. But since the amount of information carried by e. m. waves is proportional to its frequency, only about 100 pulses per sec at best could be transmitted by a wave of frequency 100. The introduction of optical fibre brought about a revolution in the field. Now, by using laser light through fibre, million pulses per second can be transmitted. About 2000 telephone signals, appropriately mixed with light waves, may now be simultaneoulsy transmitted through a typical optical fibre. The importance of fibre optics can thus be hardly overestimated. With the development of powerful coherent laser radiation, a new vista has been opened up; the higher frequency and more information carrying capacity are the two main reasons for introducing laser in communication.

But light waves cannot travel far in the open atmosphere as the energy is progressively attenuated by dissipation. This would distort the signal. So some kind of guiding channel is necessary just as electric current is guided in a conducting path through a metallic wire. It is the optical fibre that provides the necessary *wave guide* for light.

The diameter of optical fibres is usually large compared to the wavelength of optical signals. The wave nature of light may thus be neglected to study the properties of optical fibres and the known laws of geometrical optics may be applied. In fibres of thin variety, the diameter of the fibre becomes of the order of the wavelength and the wave nature of light cannot be neglected. In such cases, light propagates as an e. m. wave along the fibre, similar to *microwave propagation along a wave guide.*

Fibre optics thus *may be defined as a technology related to transportation of optical energy* (light energy) *in guided media, especially the transparent dielectric fibres of glass, plastic etc.*

15.2 Basic principle of fibre optics

As *optical fibre* is a very thin cylindrical fibre of glass or any transparent dielectric medium such as plastic, having a radius ~ a micrometer (10^{-6} m). A bundle of such thin fibres forms what is called a *light pipe*, consisting of many thousands of very long fine-quality glass or quartz fibres (Fig. 15.1). The fibres are coated with a layer of material of lower refractive index.

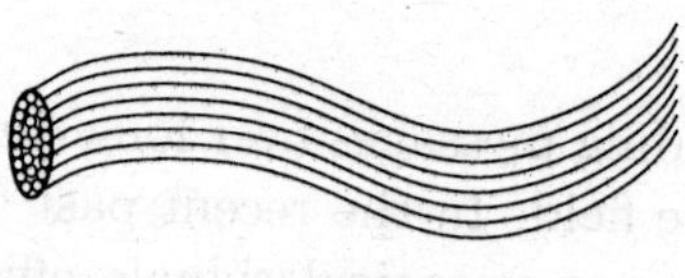

Fig. 15.1

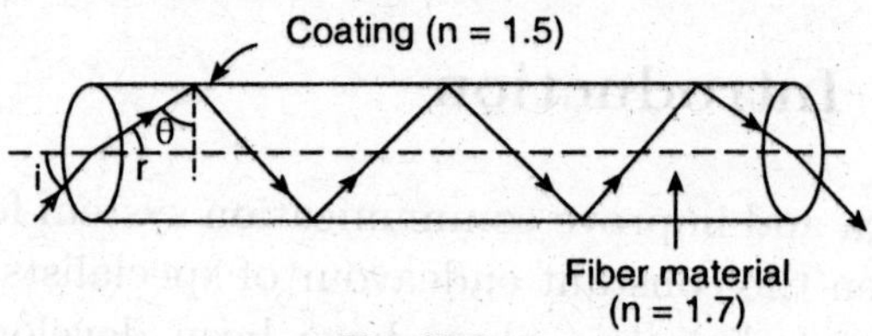

Fig. 15.2

Total internal reflection is the basic principle of optical fibres. When light is incident at one end of the fibre wall, it makes a nearly glancing incidence because of the small radius of the fibre. Let i be the angle of incidence of the ray with the axis (Fig. 15.2) and r the angle of refraction. If θ be the angle at which the ray is incident on the fibre boundary, then $\theta = 90° - r$. Let n_1 be the refractive index of the fibre. If $\theta > \theta_c$, the critical angle, where $\theta_c = \sin^{-1}(1/n_1)$, then the ray will undergo total internal reflection. The ray is thus transmitted along the fibre by repeated internal reflections (Fig. 15.3a) until its emergence by the other end. Even if the fibre is bent in a complicated shape (Fig. 15.3b), the transmission occurs without any appreciable loss.

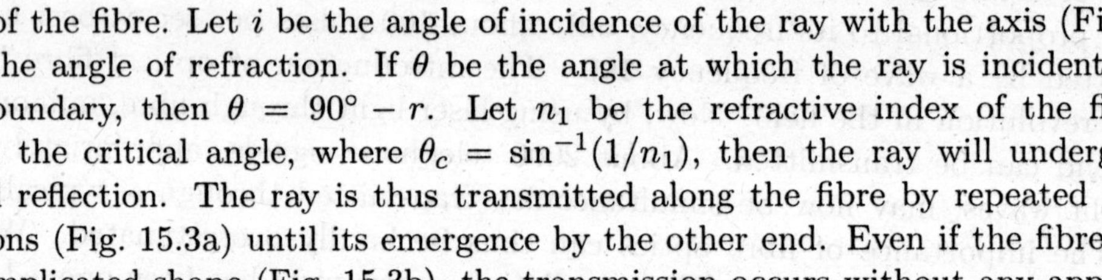

Fig. 15.3

Thus the light ray is guided through the fibre from one end to the other without any energy being lost due to refraction.

- It is important to note that there must be very little *absorption* of light during transit over a long distance inside the fibre. This is achieved by *purification* and *special preparation* of the material used.

- Bundles of optical fibres are widely used for *optical signal transmission* (and also for *receiving electrical signals* on conversion back into light).

15.3 Structure and classification

The fibres used in optical communication are wave guides made of transparent dielectric and they function to guide infrared and visible light over long distances.

Structure—Structurally, an optical fibre consists of an inner cylinder usually made of glass and is known as the *core*. It is the core that carries the light. The dia of core varies from about 5 μm to 100 μm.

The core is surrounded by another coaxial cylindrical shell of refractive index lower than that of the core. This is called the *cladding* whose diameter is usually about 125 μm. The function of cladding is to retain the light within the core through total internal reflections.

To provide greater strength and protection, a layer of soft plastic coating of dia about 250 μm surrounds the cladding. This coating is the *primary coating*. A second layer of coating, called *secondary coating*, is often used outside the primary.

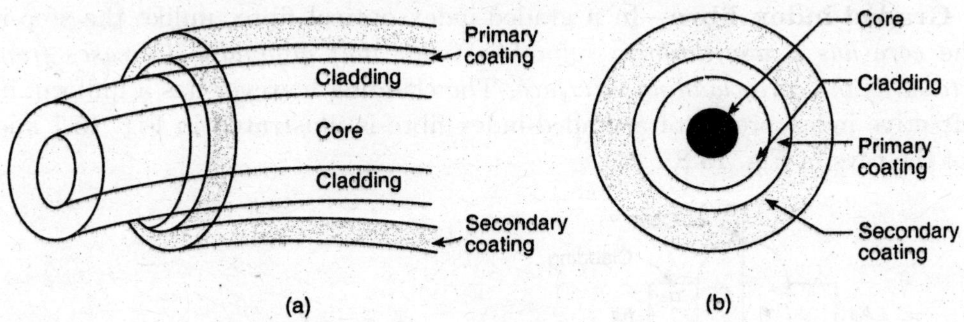

Fig. 15.4

The core, the cladding and the coatings of an optical fibre are illustrated diagrammatically in Fig. 15.4a along with a cross-sectional view in Fig. 15.4b.

Classification—There are mainly *two* types of optical fibres, based on refractive index profile of the core, namely (i) *step-index fibre* and (ii) *graded-index fibre*.

1. Step-index fibre—In step-index fibre, the refractive index n_1 of the core is uniform throughout the core. Similarly, the refractive index n_2 of the cladding is also uniform. Necessarily, $n_1 > n_2$. Let a and b be the radius of the core and cladding respectively so that $b > a$. The index profile and the radii of the core and the cladding of a step-index fibre is illustrated in Fig. 15.5.

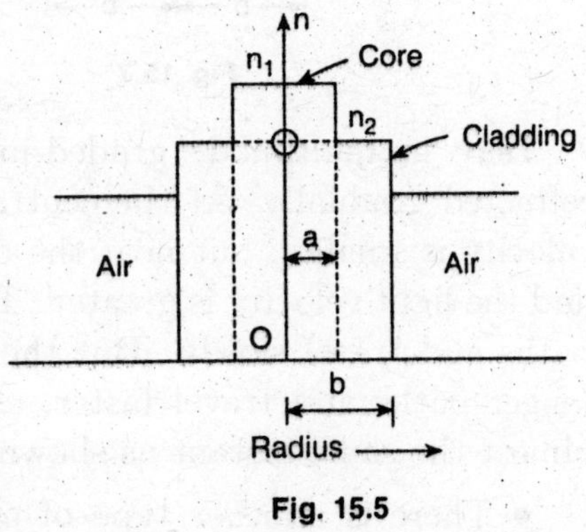

Fig. 15.5

Depending on the launch angle of the incident light into the fibre, there may be a large number of different zig-zag paths of the rays or modes by which energy travels down the core. In Fig. 15.6, we have restricted to two rays only, entering at different angles of incidence with the fibre-axis and emerging out of the fibre at different times. This is then called a *step-index multimode* fibre. The rays travelling along different paths take different times to reach the other end of the fibre leading to an widening of input pulse as it traverses the fibre. This is called *intermodal dispersion* that *distorts* the signal to be transmitted.

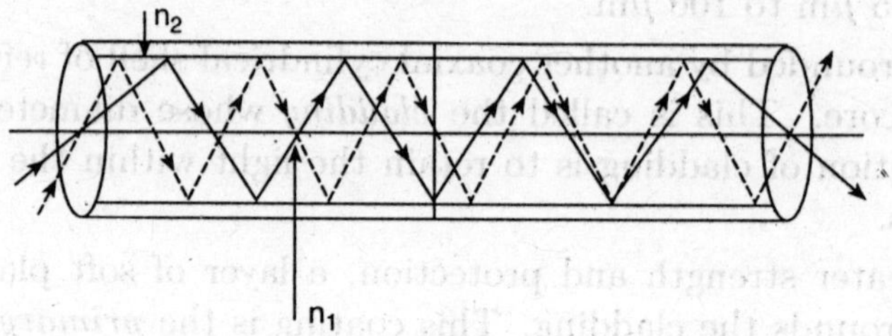

Fig. 15.6

2. Graded-index fibre—In a graded-index optical fibre, unlike the step-index one, the *core has a non-uniform refractive index that gradually decreases from the centre towards the core-cladding interface.* The cladding however has a uniform index. The refractive index profile of a graded-index fibre is illustrated in Fig. 15.7 and the paths of the rays in Fig. 15.8.

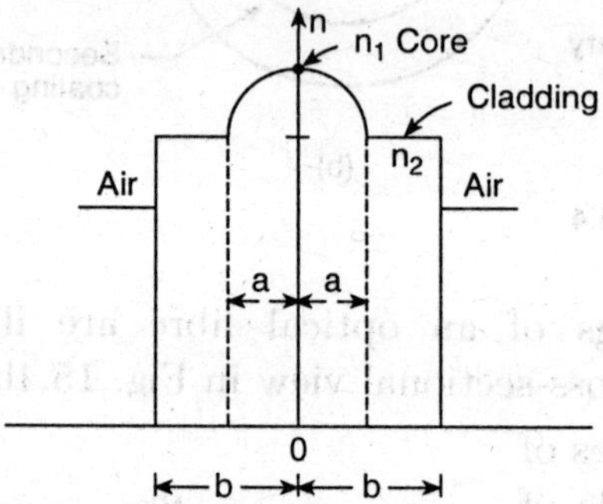

Fig. 15.7

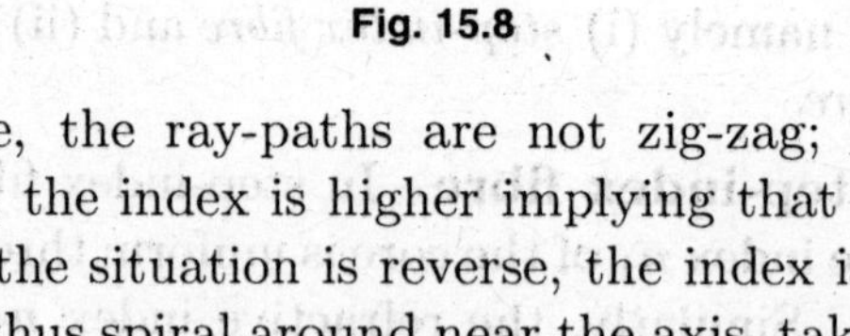

Fig. 15.8

Here in multimode graded-index fibre, the ray-paths are not zig-zag; they are refracted gradually. At the central region, the index is higher implying that the light velocity is smaller; but near the cladding the situation is reverse, the index is smaller and the light velocity is greater. The rays thus spiral around near the axis, take shorter paths and travel slowly. But the rays spiralling around the vicinity of cladding take longer paths and travel faster. As a result, all the rays tend to recross the axis at almost the same instant as shown in Fig. 15.8.

- There is another type of optical fibres, called *monomode optical fibres* having *narrow cores* ($\sim 10\ \mu$m) which *allow only one mode to propagate*, as shown in Fig. 15.9. Here the rays travel only parallel to the axis and the problem of intermodal dispersion is practically absent.

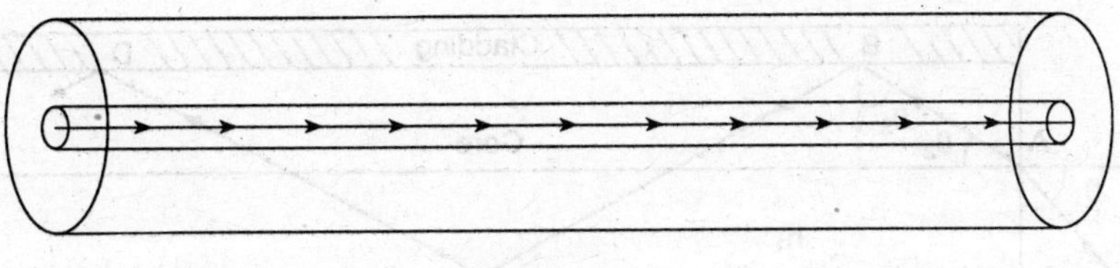

Fig. 15.9

• Of the different types of optical fibres, the multimode step-index fibre is the oldest. Its *advantages* are that it is least expensive, rugged and can be infused easily with light. But its *serious drawback* is its intermodal dispersion and is unsuitable for long distance communication.

The multimode graded-index fibre is medium-priced and intermodal dispersion is not serious. It is thus widely used in medium distance communication.

The monomode fibres are relatively expensive and require laser sources. There is no intermodal dispersion. They are most suitable for long distance communication.

• If many uncladded fibres are packed to form a bundle, then light may leak through from one fibre to the other. This is termed *cross-talk.* In modern fibres, therefore, cladding is used. *Cross-talk* may also occur if a fibre is bent too sharply for then the angle of incidence at core-cladding interface may be less than the critical angle and a part of light may escape.

• If fibres are properly aligned in a bundle, it is called a *coherent bundle*; if not, the bundle is an *incoherent one.*

• The **reason for using glass** for optical fibres are : (i) It has a range of accessible temperatures where its viscosity is variable; it does not solidify at a fixed freezing point but gradually becomes harder and harder and in the transition region can be easily drawn into a thin fibre. (ii) Pure silica is characterised by extremely low 'loss' e.g., in commercial silica fibres 96% of power gets transmitted through 1 km of optical fibre. (iii) Its intrinsic strength is as high as 2×10^{16} lb/inch^2. Further, silica is very cheap and abundantly available.

15.4 Acceptance angle and numerical aperture

We consider a step-index optical fibre. Let n_1, n_2 be the indices of the core and cladding respectively and n_0 be that of the outside medium.

Let a ray be incident on the entrance aperture of the fibre at an angle θ_1 with its axis and θ_2 be the corresponding angle of refraction (Fig. 15.10).

$$\therefore \quad n_0 \sin\theta_1 = n_1 \sin\theta_2 \tag{15.4.1}$$

To retain the ray inside the core, the angle of incidence θ_3 at the core-cladding interface must *not* be less than the critical angle θ_c.

$$\text{But,} \quad \theta_3 = 90^\circ - \theta_2 \tag{15.4.2}$$

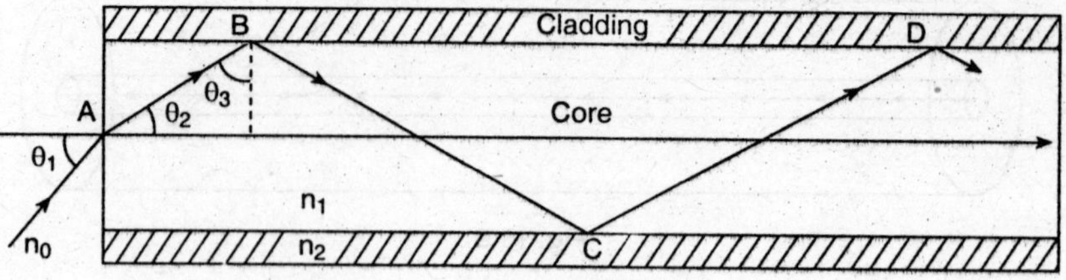

Fig. 15.10

Now, if θ_1 is increased, θ_2 also increases and so θ_3 decreases. There is thus a *maximum value of* θ_1 for which θ_3 is *not* less than θ_c and internal reflection at core-cladding interface occurs. This angle is called the *acceptance angle*, θ_A or θ_{max}.

The **acceptance angle** is thus *defined as the maximum value of launch angle of incident light into the fibre for which the ray is totally internally reflected at core-cladding interface and gets transmitted without loss.*

- The acceptance angle is a measure of the light gathering power of the fibre.
- A cone of light of semi-angle θ_A is called the **acceptance cone**. *Maximum acceptance angle* is $2 \times \theta_A$.

Determination of θ_A—We have from (15.4.1) and (15.4.2),

$$n_0 \sin\theta_1 = n_1 \sin\theta_2 = n_1 \sin(90 - \theta_3) = n_1 \cos\theta_3$$

$$\Rightarrow \quad n_0 \sin\theta_A = n_1 \cos\theta_c \tag{15.4.3}$$

for $\theta_1 = (\theta_1)_{max} = \theta_A$ and $\theta_3 = \theta_c$, the critical angle.

$$\therefore \quad n_0 \sin\theta_A = n_1\sqrt{1 - \sin^2\theta_c} = n_1\sqrt{1 - n_2^2/n_1^2} = \sqrt{n_1^2 - n_2^2}. \tag{15.4.4}$$

since, $\sin\theta_c = n_2/n_1$.

The quantity $n_0 \sin\theta_A$ or $\sin\theta_A$ (when the outside medium is *air*) is defined as the **numerical aperture** (N.A) of the fibre. *It is a measure of the light gathering power of the fibre.* If N.A is large it becomes easier to launch power into the fibre. Since for all practical purposes $(n_1^2 - n_2^2) > 1$, we can define the *numerical aperture* as

$$\text{N.A} = \sqrt{n_1^2 - n_2^2} \tag{15.4.5}$$

15.4.1 Fractional index change

Another important parameter, in this context, is the *fractional refractive index change.*

It is *defined* as the fractional difference Δ between the indices of the core and the cladding of the fibre.

$$\therefore \quad \boxed{\Delta = \frac{n_1 - n_2}{n_1}} \tag{15.4.6}$$

The parameter Δ is always positive since n_1 is always greater than n_2 to satisfy one of the conditions for total internal reflection. For guiding the light rays effectively through a fibre $\Delta \ll 1$ and typically $\Delta \sim 0.01$.

Relation between Δ and N.A—We have, from (15.4.4),

$$\text{NA} = \sqrt{n_1^2 - n_2^2}$$

Now, we write

$$\begin{aligned} n_1^2 - n_2^2 &= (n_1 + n_2)(n_1 - n_2) \\ &= \frac{1}{2}(n_1 + n_2)\frac{(n_1 - n_2)}{n_1} 2n_1 \\ &\simeq (n_1)\Delta 2n_1 \qquad \left[\because \tfrac{1}{2}(n_1 + n_2) \simeq n_1\right] \\ &= 2n_1^2\Delta \end{aligned}$$

$$\therefore \quad \boxed{NA = \sqrt{2n_1^2\Delta} = n_1\sqrt{2\Delta}} \qquad (15.4.7)$$

15.5 Intermodal dispersion in a step-index fibre

Fig. 15.10 shows a step-index fibre with core of uniform index n_1 and the cladding having an uniform index n_2 ($n_1 > n_2$). Depending on the launch angle, a large number of different zig-zag paths occurs, along which light energy travels down the fibre. The axial ray takes the shortes time, t_{min} and

$$t_{\text{min}} = \frac{\text{axial length}}{\text{velocity of light in fibre}} = \frac{L}{c/n_1} \qquad (15.5.1)$$

If a ray refracts into the fibre making θ_2 with the axis, then the path l traversed by it will be

$$l = \frac{L}{\cos\theta_2}$$

The path length will thus be longest when the ray is incident at the core-cladding interface at critical angle, i.e., $\theta_3 = \theta_c$. But $\theta_2 = 90° - \theta_3$. So the maximum time of travel is given by

$$\begin{aligned} t_{\text{max}} &= \frac{l}{(c/n_1)} = \frac{Ln_1}{c\cos\theta_2} = \frac{Ln_1}{c\cos(90° - \theta_c)} = \frac{Ln_1}{c\sin\theta_c} \\ &= \frac{Ln_1}{c\,(n_2/n_1)} = \frac{Ln_1^2}{cn_2} \quad (\because \sin\theta_c = n_2/n_1) \end{aligned} \qquad (15.5.2)$$

∴ The *maximum time interval* between the rays at the output end will be

$$\Delta t = t_{\text{max}} - t_{\text{min}} = \frac{Ln_1^2}{cn_2} - \frac{L}{(c/n_1)} = \frac{Ln_1}{c}\left(\frac{n_1}{n_2} - 1\right) \qquad (15.5.3)$$

The speed of the rays inside the fibre is c/n_1. So it will *spread in space* by a length

$$\Delta L = \frac{c}{n_1} \times \Delta t = L\left(\frac{n_1}{n_2} - 1\right) \qquad (15.5.4)$$

using equation (15.5.3).

Informations in fibre optics are usually coded digitally and are sent in series of pulses through the fibre. The difference in transit time has a deleterious effect in the form of broadening of the shape of the pulses. To read the transmitted signal, the interval between the consecutive pulses must be large compared to the spread-out width Δt. The intermodal dispersion thus limits the rate of signal transmission in fibres.

15.6 Ray path in a graded-index (GRIN) fibre

A graded-index fibre is characterised by its refractive index decreasing smoothly from the axis radially outward. The index distribution may, as shown in Fig. 15.11, be represented by the parobolic equations :

$$\begin{aligned} n^2(r) &= n_1^2[1-(r/a)^2] \quad \text{Core region,} \quad |r| < r_0, \\ n^2 &= n_1^2[1-(r_0/a)^2] \quad \text{Cladding region,} \quad |r| > r_0 \end{aligned} \tag{15.6.1}$$

Because of the continuous change in refractive index, a ray inside the fibre gets bent continuously towards the axis as shown in Fig. 15.11.

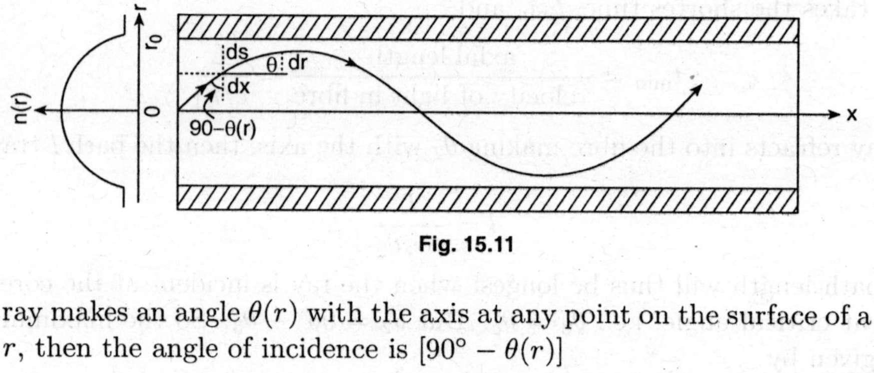

Fig. 15.11

If the ray makes an angle $\theta(r)$ with the axis at any point on the surface of a cylinder, of radius r, then the angle of incidence is $[90° - \theta(r)]$

$$\therefore \quad n(r)\sin[90° - \theta(r)] = \text{constant, by Snell's law}$$

$$\text{or,} \quad n(r)\cos\theta(r) = \text{an } invariant \text{ of the ray path} = \beta, \text{ say.} \tag{15.6.2}$$

If ds be an infinitesimal arc length along the curve, then

$$(ds)^2 = (dr)^2 + (dx)^2 \Rightarrow (ds/dx)^2 = (dr/dx)^2 + 1 \tag{15.6.3}$$

From (15.6.1) and (15.6.2),

$$\frac{dx}{ds} = \cos\theta = \frac{\beta}{n(r)}$$

So, from (15.6.3),

$$\left(\frac{dr}{dx}\right)^2 = \frac{n^2(r)}{\beta^2} - 1$$

Differentiating the above equation with respect to x, we obtain

$$2\frac{dr}{dx}\cdot\frac{d^2r}{dx^2} = \frac{1}{\beta^2}\frac{dn^2}{dr}\cdot\frac{dr}{dx}$$

$$\Rightarrow \quad \frac{d^2r}{dx^2} = \frac{1}{2\beta^2}\frac{dn^2}{dr}$$

Using the value of $n^2(r)$ for the *core region*, we obtain the differential equation

$$\frac{d^2r}{dx^2} = -\left(\frac{n_1}{a\beta}\right)^2 r, \tag{15.6.4}$$

having the solution of the following form :

$$r(x) = A\sin(kx+\phi), \quad k = n_1/a\beta \tag{15.6.5}$$

• Let the ray be incident at $(x=0, r=0)$ making an angle θ_1 with x-axis. Then, we may take $\phi = 0$, $\beta = n_1\cos\theta_1$ and $k = 1/a\cos\theta_1$. To find A of (15.6.5), we note

$$\left.\frac{dr}{dx}\right|_{x=0} = \tan\theta_1 = Ak \quad \Rightarrow \quad A = \frac{1}{k}\tan\theta_1 = a\sin\theta_1$$

The ray path in the meridian plane is thus sinusoidal and may be written as

$$r(x) = A\sin kx = a\sin\theta_1\sin\left(\frac{x}{a\cos\theta_1}\right) \tag{15.6.6}$$

Angle of acceptance—Let the refractive index at the axis be n_1 and that of the cladding be n_2. Applying Snell's law at O (Fig. 15.12), we obtain

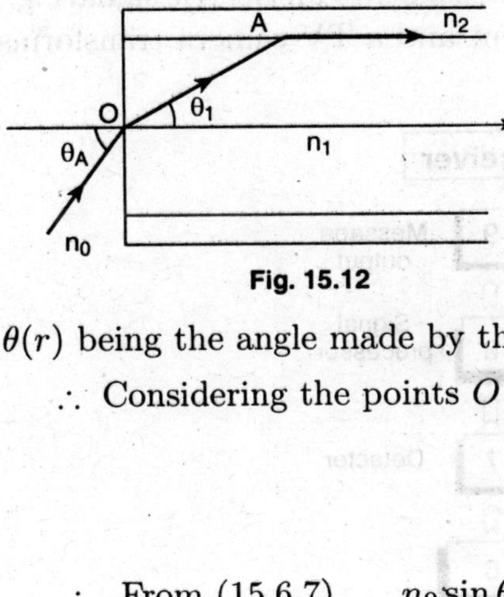

Fig. 15.12

$$n_0\sin\theta_A = n_1\sin\theta_1 \tag{15.6.7}$$

where n_0 is the index of the environment in which the fibre is kept and θ_A the *angle of acceptance.*

Also, from Snell's law, we have along the ray path,

$$n(r)\cos\theta(r) = \text{invariant}$$

$\theta(r)$ being the angle made by the ray with the axis of the fibre.

∴ Considering the points O and A, we have

$$n_1\cos\theta_1 = n_2\cos 0°$$

$$\therefore \quad \cos\theta_1 = n_2/n_1 \tag{15.6.8}$$

∴ From (15.6.7) $\quad n_0\sin\theta_A = n_1\sqrt{1-\cos^2\theta_1}$

$$= n_1\sqrt{1-n_2^2/n_1^2}, \quad \text{using (15.6.8)}$$

$$= \sqrt{n_1^2 - n_2^2} \tag{15.6.9}$$

$$\Rightarrow \sin\theta_A = \sqrt{n_1^2-n_2^2}, \quad \text{if } n_0 = 1 \text{ i.e., the fibre is in air.}$$

• It must be borne in mind that the above equation (15.6.9) applies only to meridional rays incident at axis. For other rays, however, the value of θ_A may not agree with what is given by (15.6.9).

15.7 Fibre optics communication system

A fibre optics communication system consists basically of the following units :

1. A transmitter that transforms an electric signal (information signal) to be transmitted into an optical system.

2. A receiver whose function is to convert the optical signal back into the original electric form, and

3. An information channel (fibre transmission line) that guides the optical signal from the transmitter to the receiver.

These *three basic constituents* forming the fibre optics communication system is illustrated by a block diagram in Fig. 15.13.

Fig. 15.13

Each of the receiver and the transmitter again has a number of different parts, each with a definite function. These parts are shown in block diagram in Fig. 15.14.

1. Message origin—It converts a non-electrical message into an electric signal e.g., a microphone converts sound waves into electric current and a TV-camera transforms images into electric current.

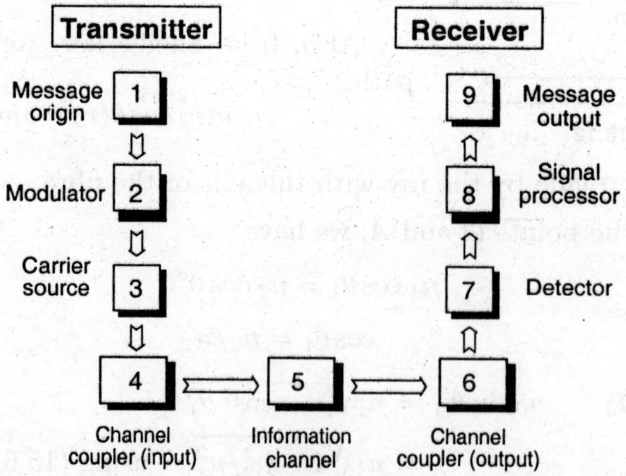

Fig. 15.14

2. Modulator—This converts electrical messages into proper modulation format and impresses the signal onto the carrier waves. The format may be *analog* or *digital*. In fibre optics, information is contained in the variation of optical power so that *intensity modulation* occurs.

3. Carrier source—It generates carrier waves. While in radio-communication, these are provided by an oscillator, in fibre optic system a light emitting diode (LED) or a laser diode (LD) is used.

4. Channel coupler (input)—It feeds power into the information channel. While the antenna is the channel coupler in a TV or a radio broadcasting, in fibre system the coupler transfers the modulated light beam from the source into the optical fibre.

5. Information channel—It provides a path between the transmitter and the receiver. In fibre optic system, a glass (or plastic) fibre is the channel. The channel may be guided or unguided. For instance, the atmosphere is an unguided channel for radio waves and fibre is a guided channel. Other guided channels are two-wire lines, coaxial cable etc.

6. Channel coupler (output)—This routes the information collected from information channel into the receiver. In radio communication, it is the antenna of the receiver. In fibre system, it is a simple butt connection, directing the emergent light from fibre into light detector.

7. Detector—The function of the detector is to separate the information from the carrier waves. In fibre system, a photo-detector is used and its output current contains the information.

8. Signal processor—This includes amplification, and filtering of the undesired frequencies in analog transmission, and in digital system there is an additional decision circuit.

9. Message output—In order to hear the message, the electric signal is transformed into sound waves as in radio and for viewing the message, as in TV, the cathode ray tube is used.

15.8 Advantages of optical fibre communication

The *advantages* of optical fibre communication over the conventional one are too many and some of them are as under.

1. It possesses far *more information carrying capacity* in comparison to the conventional radio and microwave carriers. Losses are extremely low and the *clarity of the signals* transmitted is *much better.*

2. Because of the *smaller size* and *light weight* of the system, it is highly suitable in *aeronautical and space* applications.

3. The *cost of cables* per unit length is much *lower* than that of the metallic counterpart, as fibre is made of glass whose *raw material* silica is both *cheap* and available in *plenty.*

4. Optical fibres, be it of glass or plastic, are insulators and current cannot flow through them either due to transmitted signal or external radiation. Also, the wave in the fibre is trapped and light cannot couple into the fibre from the side. There is practically no leakage, no cross-talk and the fibre is *well protected from interference and coupling with signals in other fibres or communication channels.*

5. Fibres neither pick up nor propagate extraneous e. m. pulses. They do not also radiate energy within them so that no intruder can detect the signal being transmitted. The *security* and *privacy* are the two vital features of this communication system.

6. While a wire communication link may easily short-circuit the high voltage lines by falling across them, there is no such problem with fibres. Fibre glass can *withstand temperatures up to* 800°C without being affected. Also, *corrosion* from either water or chemicals is *much less severe* for glass than for copper. Therefore, additional equipment for protection against grounding or voltage problem is not needed.

• The progress in the field, both scientific and technological, has indeed been phenomenal. We are now in the fifth generation of optical fibre communication system. Of the recent developments, mention may be made of optical amplifiers and wavelength division multiplexing (WDM) leading to a communication system virtually with 'zero' loss and 'infinite' bandwidth. The proliferation of 'Internet' is making an ever-increasing demand on communication links and the optical fibre system is successfully responding to it.

15.9 Applications of fibre optics

The most important application of fibre optics is to *transmit communication signals* through light pipes and its advantages over the conventional system have just been described at length. Other applications are given below in brief.

1. The fibre optic system is widely used in *defence services* as high privacy can be maintained. The *military applications* include communication, command and control links on ships and aircrafts, data links for satellite earth stations etc. Digital data generated by computers are most suited for transmission by the fibre system. It is also used in signalling.

2. Its applications in *science* and *industry* are numerous. Fibre-optic sensors measure temperature, pressure, liquid levels, rotary and linear positions etc. It is also used in cable television, space vehicles, ships and submarine cables. Optical fibres are also widely used in security and alarm systems, electronic instrumentation systems, industrial automation and process control.

Fibre-optic sensors have many advantages compared to conventional ones, viz. lower cost, smaller size, more accuracy, greater flexibility and reliability. They are also immune to external e.m. interference and can be used in explosive and hazardous environments.

3. The other important application is in the medical field for *diagnostic purpose*—the **endoscopy**—meaning *to see within.* To view areas difficult to reach, e.g., inside of a human body, the endoscopy is used.

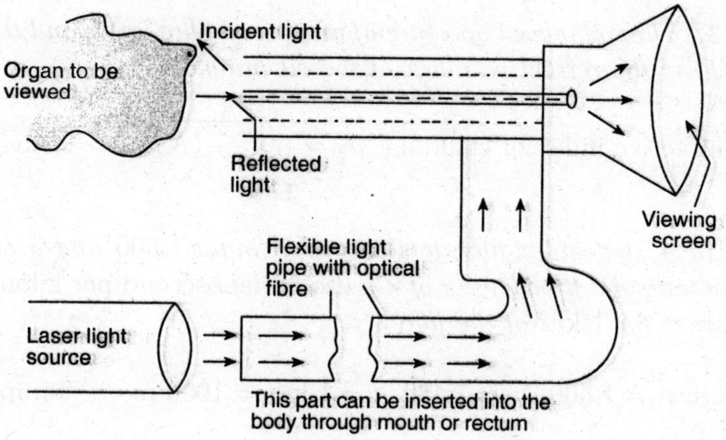

Fig. 15.15

An endoscope consists of a bunch of optical fibres (fibrescope), i.e., light pipes carry laser beam to the inside of the body. For example, a patient's stomach can be viewed by inserting one end of a light pipe into the stomach through the mouth. Light is sent through one set of fibres in the pipe to illuminate the inside of the stomach. The light reflected from inside travels back through another set of fibres in the pipe. The viewer gets the image at the other end on the screen. The process is illustrated in Fig. 15.15.

15.10 Illustrated Examples

▶ **Example 1.** *Find the numerical aperture (NA) of a step-index fibre when the refractive index of the core is* 1.51 *and that of the material used for cladding is* 1.47.

Solution. Here, by the problem, $n_1 = 1.51$, and $n_2 = 1.47$.

∴ Numerical aperture is given by

$$\text{NA} = \sqrt{n_1^2 - n_2^2} = \sqrt{1.51^2 - 1.47^2}$$
$$= \sqrt{(1.51 + 1.47)(1.51 - 1.47)}$$
$$= \sqrt{2.98 \times 0.04} = \mathbf{0.3452}$$

▶ **Example 2.** *Calculate the angle of acceptance of the fibre referred to in* Ex. 1 *when it is placed in air.*

Solution. Let θ_A be the required angle of acceptance. Then we have,

$$n_0 \sin\theta_A = \text{N.A}$$

Here, N.A = 0.3452 (Ex. 1) and $n_0 = 1$ (air). So, here, we obtain

$$\sin\theta_A = 0.3452 \;\Rightarrow\; \theta_A = \sin^{-1}(0.3452) = \mathbf{20.19°}$$

► **Example 3.** *The numerical aperture of an optical fibre is* 0.5 *and the core refractive index is* 1.54. *Find the refractive index of the cladding.*

Solution. Refractive index of cladding, $n_2 = \sqrt{n_1^2 - (NA)^2} = \sqrt{1.54^2 - 0.5^2}$

$$= \mathbf{1.46}.$$

► **Example 4.** *A step-index fibre has a core of index* 1.500 *and a cladding of index* 1.489. *Find the temporal broadening of a pulse in* nanosecond per kilometre. *Also, find the spread in space for* 1 km *of the fibre.*

Solution. Here $n_1 = 1.500$, $n_2 = 1.489$, $L = 1$ km $= 1000$ m. ∴ Temporal broadening Δt is given by

$$\Delta t = \frac{Ln_1}{c}\left(\frac{n_1}{n_2} - 1\right) = \frac{1000 \times 1.500}{3 \times 10^8}\left(\frac{1.500}{1.489} - 1\right) \text{ s/km}$$

$$= \mathbf{36.9 \text{ ns/km}}$$

The spread in space, ΔL is given by

$$\Delta L = \Delta t \cdot \frac{c}{n_1} = L\left(\frac{n_1}{n_2} - 1\right)$$

$$= 1000\left(\frac{1.500}{1.489} - 1\right) = \mathbf{7.4 \text{ m}}$$

► **Example 5.** *In a graded-index fibre, the radial distribution of index is given by*

$$n(r) = 1.52 - 2r^2$$

with r in mm *up to a radius* $r_0 = 0.2$ mm. *Calculate the acceptance angle of the fibre.*

Solution. We have here, $n_1 = 1.52$ and $n_2 = 1.52 - 2 \times (0.2)^2 = 1.52 - 0.08 = 1.44$

$$\therefore \quad \text{Acceptance angle, } \theta_A = \sin^{-1}\sqrt{n_1^2 - n_2^2}$$

$$= \sin^{-1}\sqrt{(1.52)^2 - (1.44)^2}$$

$$= \sin^{-1}\sqrt{2.96 \times 0.08} = \mathbf{29.12°}$$

► **Example 6.** *An optical fibre is of diameter* 100 μm. *If the core is of refractive index* 1.5, *find the axial distance travelled by a ray incident at* 30° *between two successive internal reflections.*

Solution. Fig. 15.16 shows a section of the fibre along with the relevant angles; θ = angle of incidence, θ_1 = angle of refraction. D = diameter of the fibre and x = axial distance travelled by the ray between two successive internal reflections.

From the figure, we have, $\quad D/x = \tan\theta_1 \Rightarrow x = D/\tan\theta_1$

But from Snell's law, $\quad \sin\theta/\sin\theta_1 = n_1 \Rightarrow \sin\theta_1 = \sin\theta/n_1$

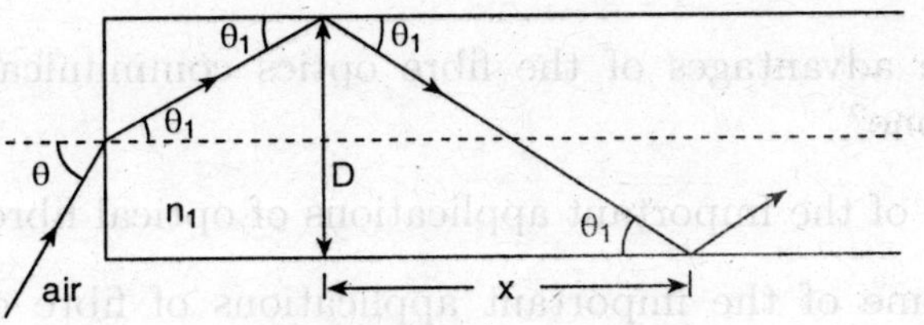

Fig. 15.16

$$\therefore \quad x = \frac{D\cos\theta_1}{\sin\theta_1} = \frac{D\sqrt{1-\sin^2\theta_1}}{\sin\theta_1}$$

$$= \frac{D\sqrt{1-\sin^2\theta/n_1^2}}{\sin\theta/n_1} = \frac{D\sqrt{n_1^2-\sin^2\theta}}{\sin\theta}$$

Substituting the given values,

$$x = \frac{100\times10^{-6}\sqrt{1.5^2-1/4}}{1/2} = \mathbf{2.82\times10^{-4}\ m}$$

▶ **Example 7.** *For a typical step-index (multimode) fibre with $n_1 = 1.45$ and $\Delta = 0.01$, calculate the angle of acceptance.*

Solution. $\sin\theta_A = NA = n_1\sqrt{2\Delta} = 1.45\times\sqrt{2\times0.01}$

$$= 1.45\times\sqrt{0.02} = 0.205 = \sin 12^\circ \quad \therefore \quad \theta_A = \mathbf{12^\circ}.$$

Review Questions

1. What is an optical fibre? Explain the working principle of an optical fibre as a wave-guide for light.
2. Describe the structure of a typical optical fibre giving the necessary diagrams.
3. Describe, with necessary diagram, the step-index and graded-index optical fibre. In what respect do they differ?
4. What are meant by acceptance angle and numerical aperture of a fibre. Find their expressions and significance for a step-index fibre.
5. What is a graded-index fibre? Discuss its advantages over a step-index one.
6. What are the different classes of fibres? Discuss their relative merits and demerits.
7. What is intermodal dispersion? Derive its expression for a step-index fibre. How does it affect the rate of signal propagation through a fibre?
8. Show that the intermodal dispersion in a step-index fibre limits the rate of signal transmission in fibres.
9. Describe schematically, using block diagram, the basic elements of a fibre optics communication system.

10. What are the advantages of the fibre optics communication system over the conventional one?

11. Mention some of the important applications of optical fibres.

12. Enumerate some of the important applications of fibre optics communication system.

Short Answer and Objective Questions

1. What is an optical fibre? Describe a typical optical fibre with diagram.

2. What do you mean by 'core' and 'cladding'? Whose refractive index is more and why? Give an idea of the dia of core and cladding.

3. What is the essential difference between 'step-index' and 'graded-index' type of optical fibres? Draw their index profiles.

4. Indicate what are meant by 'mono-mode' and 'multi-mode' optical fibre?

5. Show, by a block diagram (no description), the essential parts of a fibre optics communciation system.

6. Enumerate some of the advantages of fibre optics communication system.

7. What is 'endoscopy'? Explain the use of fibre optics in endoscopy for diagnostic purposes.

8. Give the expression for the numerical aperture (NA) of an optical fibre and show that it is related to fractional index change Δ, by the equation

$$NA = n_1\sqrt{2\Delta}$$

9. "Usually optical guides are analysed using the laws of geometrical optics, whereas wave guides in the microwave region are analysed by the wave theory of light".— Discuss why it is so.

10. Optical fibres are largely made of glass. Suggest some reasons for this.

11. Fibre optics is based on

(a) refraction, (b) polarisation, (c) dispersion, (d) total internal reflection.

12. If n_1 and n_2 be the indices of the core and the cladding respectively, then

(a) $n_2 > n_1$, (b) $n_2 \geq n_1$, (c) $n_1 \geq n_2$, (d) $n_1 > n_2$.

13. The numerical aperture (NA) of a step-index fibre is given by

(a) $\sqrt{n_2^2 - n_1^2}$, (b) $n_1^2 - n_2^2$, (c) $\sqrt{n_1^2 - n_2^2}/n$, (d) none of these.

14. A measure of the light gathering power of an optical fibre is the
    (a) numerical aperture, (b) fractional index change,
    (c) acceptance angle, (d) none of these.

Problems

1. Determine the numerical aperture of a step-index fibre when the refractive index of the core, $n_1 = 1.5$ and that of cladding, $n_2 = 1.48$. [0.244]

2. Determine the maximum angle for entrance of light in a step-index fibre of indices as indicated in Q. 1 above, if it is placed in air. [14.13°]

3. A step-index optical fibre has refractive indices of the core and cladding as 1.44 and 1.40 respectively and is surrounded by air. Calculate the numerical aperture. **(GATE)** [0.337]

4. Calculate the maximum angle for entrance of light into the fibre as described in Ex. 3 above. [19.69°]

5. An optical fibre is having core index of refraction of 1.45. What will be the critical angle of the fibre? Find also its numerical aperture. [9.46°; 0.242]

6. In a step-index fibre the refractive index of the core and the cladding are 1.4746 and 1.4600 respectively. What is the intermodal delay for an impulse passing through it if the length of the fibre is 10 km? [491.5 ns]

7. An optical fibre has a diameter of 50 μm. Its core is of index 1.6. Calculate the number of total internal reflections that a ray incident at 30° will suffer in moving through 1 m of the fibre. [6580]

8. A step-index fibre has a core of refractive index 1.5 and a cladding of index 1.4. The fibre is used in an environment of index 1.33. Calculate the numerical aperture and the acceptance angle. [0.538; 23.9°]

9. The refractive index of the core of an optical fibre is 1.54. Find the refractive index of the cladding for an acceptance angle of 25°. [1.481]

10. An optical fibre cable has an acceptance angle of 30° and a core index of refraction of 1.4. Find the refractive index of the cladding. [1.308]

11. A single clad optical fibre has an index of 1.62 for the core and 1.52 for the clad. Find its numerical aperture. When immersed in air, what is its acceptance angle? What is the maximum acceptance angle? [0.55; 33°22′; 66°44′]